PRENTICE HALL

Biology
The Living World

Peter Alexander, Ph.D.
Professor of Biology and Computer Science
St. Peter's College
Jersey City, New Jersey

Mary Jean Bahret
Biology Teacher
Newburgh Free Academy
Newburgh, New York

Judith Chaves
Biologist
Formerly with University of Vermont
Burlington, Vermont

Gary Courts
Biology Teacher
Miamisburg High School
Miamisburg, Ohio

Naomi Skolky D'Alessio
Biology Teacher and Chairperson
of Science Department
University School of Nova University
Fort Lauderdale, Florida

PRENTICE HALL
A Division of Simon & Schuster
Englewood Cliffs, New Jersey

Biology: The Living World Program

Biology Textbook with Teacher's Edition
Biology Laboratory Manual with Teacher's Edition
Biology Teacher's Resource Book
Biology Color Overhead Transparencies
Biology Computer Test Bank

CONTENT REVIEWERS

Rod Casto, Ph.D.
Graduate Research Associate
Department of Physiology
College of Medicine
University of Florida
Gainesville, Florida

George W. Cox, Ph.D.
Professor of Biology
San Diego State University
San Diego, California

James S. Findley, Ph.D.
Professor of Biology and Director of
 Museum of Southwestern Biology
University of New Mexico
Albuquerque, New Mexico

Malcolm J. Fraser, Jr., Ph.D.
Assistant Professor of Biology
Department of Biology
University of Notre Dame
Notre Dame, Indiana

Dennis E. Freer, Ph.D.
Director of Clinical Chemistry
Edward W. Sparrow Hospital
Lansing, Michigan

Porter M. Kier, Ph.D., Sc.D.
Curator of Paleobiology
U.S. National Museum of
 Natural History
Washington, DC

C. Sam Levings, Ph.D.
Professor of Biology
Department of Genetics
North Carolina State University
Raleigh, North Carolina

Cynthia A. Needham, Ph.D.
Director of Microbiology,
 University Hospital
Assistant Professor of Microbiology
Boston University School of Medicine
Boston, Massachusetts

Philip S. Perlman, Ph.D.
Associate Professor of Genetics
The Ohio State University
Columbus, Ohio

Mickey C. Smith, Ph.D.
Professor of Health Care Administration
School of Pharmacy
University of Mississippi
University, Mississippi

ABOUT THE COVER

*Front: These scarlet macaws are among the largest members of the parrot family.
(R. C. Simpson/Tom Stack & Associates)*

Back: This day lily flower remains open for but a single day. (Imagery)

Photo credits begin on p. 854.

© 1989, 1986 by Prentice-Hall, Inc., Englewood Cliffs, New Jersey 07632 (formerly published as Silver Burdett Biology). All rights reserved. No part of this book may be reproduced in any form or by any means without permission in writing from the publisher. Printed in the United States of America.

ISBN 0-13-077603-3

10 9 8 7 6 5

Prentice-Hall of Australia, Pty. Ltd., Sydney
Prentice-Hall Canada Inc., Toronto
Prentice-Hall Hispanoamericana, S.A., Mexico
Prentice-Hall of India Private Ltd., New Delhi
Prentice-Hall International (UK) Limited, London
Prentice-Hall of Japan, Inc., Tokyo
Prentice-Hall of Southeast Asia Pte. Ltd., Singapore
Editora Prentice-Hall do Brasil Ltda., Rio de Janeiro

CONTRIBUTING WRITERS

Thomas M. Baird, Program Manager, Energy Management Center, Port Richey, FL

Mary Deisher Cooke, Biology Teacher and Science Chairperson, Clover Hill High School, Midlothian, VA

Francis A. Hughes, Biology Teacher, East Senior High School, Columbus, IN

David E. LaHart, Senior Instructor, Florida Solar Energy Center at Cape Canaveral, Branch of University of Central Florida, Orlando, FL

Michael Levine, Ph.D., Assistant Professor of Psychology, Kenyon College, Gambier, OH

J. Michael McCormick, Ph.D., Professor of Biology, Montclair State College, Montclair, NJ

Philip S. Perlman, Ph.D., Associate Professor of Genetics, The Ohio State University, Columbus, OH

Carl M. Raab, Assistant Principal and Science Supervisor, Fort Hamilton High School, Brooklyn, NY

FIELD TEST TEACHERS AND REVIEWERS

Peter L. Barsness, Biology Teacher and Science Department Chairperson, Stevens Point Area Senior High School, Stevens Point, WI • **Nancy Blaszko,** Biology Teacher, Immaculate Heart of Mary High School, Westchester, IL • **Tiny Burgess,** Dwight Morrow High School, Englewood, NJ • **Leonard Blessing,** Biology Teacher, Millburn High School, Millburn, NJ • **Richard Booher,** Biology Teacher, Jonathan Dayton Regional High School, Springfield, NJ • **Harry Bullock,** Biology Teacher, Mendham High School, Mendham, NJ • **Jean Carpenter,** Science Teacher, Blevins Jr. High School, Fort Collins, CO • **Nancy Cox,** Biology Teacher, Spring High School, Spring, TX • **David O. Funderburke,** Ph.D., Administrative Coordinator for Life Science, Fernbank Science Center, Atlanta, GA • **Don Gabriel,** Biology Teacher, Musak High School, Monroe, CT • **Reeves Geraghty,** Biology Teacher, Mendham High School, Mendham, NJ • **Caren Gough,** Biology Teacher, Half Hollow Hills High School, Dix Hills, NY • **H.M. Hummel,** Ph.D., Biology Teacher, Cooper High School, Minneapolis, MN • **Jack Kelley,** Biology Teacher, El Camino High School, Sacramento, CA • **Earl F. King,** Biology Teacher, Dover-Sherborn High School, Dover, MA • **Joseph G. Krajkovich,** Ph.D., Supervisor of Science, Edison Township Board of Education, Edison, NJ • **Daniel W. McGary,** Program Coordinator for Science, School District of Lancaster, Lancaster, PA • **Danny L. McKenzie,** Ph.D., Assistant Professor of Biology, Southwest Texas State University, San Marcos, TX • **Stephan Michlovitz,** Science Supervisor, Franklin High School, Somerset, NJ • **Kathleen Nee,** Biology Teacher, James Caldwell High School, Caldwell, NJ • **Mary Rose Neff,** Biology Teacher, Summit High School, Summit, NJ • **Sr. Pauline O'Dwyer,** Science Teacher, Holy Angels School, Colma, CA • **Roger Orwick,** Biology Teacher and Department Chairperson, Frostproof High School, Frostproof, FL • **James A. Robertson,** Biology Teacher and Department Chairperson, Salem High School, Salem, VA • **Angelina Romano,** Biology Teacher, Sayreville War Memorial High School, Parlin, NJ • **Dorothy Sippo,** Biology Teacher, Dwight Morrow High School, Englewood, NJ • **Virginia Sullivan,** Biology Teacher, Dwight Morrow High School, Englewood, NJ • **David V. Thomas,** Biology Teacher, Flowing Wells High School, Tucson, AZ • **Robert K. Webb,** Science Teacher and Department Chairperson, Blevins Jr. High School, Fort Collins, CO • **Virginia Wheeler,** Biology Teacher and Department Head, Forest Park Senior High School, Forest Park, GA • **Michele Whitehead,** Biology Teacher, Immaculate Heart of Mary School, Westchester, IL • **William Woodside,** Biology Teacher, Summit High School, Summit, NJ • **Jean B. Worsley,** Biology Teacher and Science Department Chairperson, Harding High School, Charlotte, NC

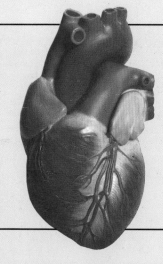

Prentice Hall expresses special thanks to Mr. Marcus Sommer of Somso Biological Works, and Dr. Kenneth Perkins of Carolina Biological Supply Company for making possible the use of the unique Somso models.

The models are designed and manufactured by Somso Biological Works, Coburg, West Germany, and are available through Carolina Biological Supply Company, 2700 York Road, Burlington, North Carolina 27215. Catalogues are available upon request.

Contents

UNIT I Introducing Biology *1*

1. **Studying Life** *2*
   THE STUDY OF SCIENCE • SCIENTIFIC MEASUREMENT • THE TOOLS OF A BIOLOGIST

2. **Understanding Life** *24*
   EARLY OBSERVATIONS OF THE NATURAL WORLD • CHARACTERISTICS OF LIVING THINGS

3. **Basic Chemistry** *36*
   THE NATURE OF MATTER • INTERACTIONS OF MATTER • COMPOUNDS IN LIVING THINGS • ENERGY AND MATTER

4. **Cell Biology** *58*
   HISTORY OF CELL THEORY • CELL STRUCTURE AND FUNCTION • CELLULAR TRANSPORT

5. **Cell Chemistry** *78*
   CONTROLLING CELLULAR ACTIVITIES • ENERGY FOR CELLS • THE PROCESS OF PHOTOSYNTHESIS

6. **Cell Reproduction** *98*
   FORMATION OF BODY CELLS • FORMATION OF SEX CELLS • THE GENETIC CODE • PROTEIN SYNTHESIS

UNIT II Continuity of Life *121*

7. **Basic Genetics** *122*
   GENETICS AND GREGOR MENDEL • OTHER DISCOVERIES IN GENETICS • CHROMOSOME THEORY OF HEREDITY

8. **Applied Genetics** *146*
   CHANGES IN HEREDITARY MATERIAL • GENETIC DISORDERS IN HUMANS • APPLICATIONS OF GENETICS • GENE EXPRESSION

9. **Evolution** *168*
   EVIDENCE FOR EVOLUTION • THEORIES OF EVOLUTION • MECHANISMS OF EVOLUTION

10. **History of Life** *190*
ORIGIN OF LIFE • PATTERNS OF EVOLUTION • HUMAN ANCESTRY

11. **Classification** *212*
HISTORY OF CLASSIFICATION • CLASSIFYING ORGANISMS • IDENTIFYING LIVING THINGS

UNIT III Microbiology *231*

12. **Viruses and Monerans** *232*
CHARACTERISTICS OF VIRUSES • BACTERIA AND OTHER MONERANS

13. **The Protists** *254*
INTRODUCTION TO PROTISTS • ALGAL AND FUNGAL PROTISTS • PROTOZOANS: ANIMALLIKE PROTISTS

14. **The Fungi** *274*
CHARACTERISTICS OF FUNGI • CLASSIFICATION AND REPRODUCTIVE PATTERNS • ACTIVITIES OF FUNGI

15. **Microbial Disease** *292*
THE NATURE OF DISEASE • AGENTS OF DISEASE • DEFENSES AGAINST DISEASE

UNIT IV The Plants *315*

16. **Algae to Ferns** *316*
THE SIMPLEST PLANTS • NONVASCULAR LAND PLANTS • VASCULAR PLANTS

17. **The Seed Plants** *336*
CLASSIFYING SEED PLANTS • ROOTS AND STEMS • LEAVES: STRUCTURE AND FUNCTION

18. **Transport in Plants** *356*
WATER BALANCE • MOVEMENT OF FLUIDS WITHIN PLANTS • PLANT RESPONSES TO WATER

19. **Plant Reproduction** *374*
ASEXUAL REPRODUCTION • SEXUAL REPRODUCTION • FROM SEED TO PLANT

20. **Growth and Behavior** *394*
FACTORS AFFECTING PLANT GROWTH • GROWTH AND DEVELOPMENT • PLANT BEHAVIOR

UNIT V Invertebrates *415*

21. Sponges to Worms *416*
SIMPLE INVERTEBRATES • FLATWORMS AND ROUNDWORMS • SEGMENTED WORMS

22. Mollusks and Echinoderms *438*
THE MOLLUSKS • THE ECHINODERMS

23. The Arthropods *458*
WHAT ARE ARTHROPODS? • KINDS OF ARTHROPODS • ARTHROPOD ADAPTATIONS

24. Comparing Invertebrates *480*
BODY PLANS OF INVERTEBRATES • SYSTEMS FOR BODY MAINTENANCE • CONTROL SYSTEMS • REPRODUCTION AND DEVELOPMENT

UNIT VI Vertebrates *501*

25. Fish and Amphibians *502*
THE PHYLUM CHORDATA • THE CLASSES OF FISH • THE AMPHIBIANS

26. Reptiles and Birds *522*
THE REPTILES • THE BIRDS • BIRD BEHAVIOR

27. The Mammals *542*
ABOUT MAMMALS • CLASSIFICATION OF MAMMALS • BEHAVIOR OF MAMMALS

28. Comparing Vertebrates *562*
BODY PLANS OF VERTEBRATES • SYSTEMS FOR BODY MAINTENANCE • CONTROL SYSTEMS • REPRODUCTION AND DEVELOPMENT

UNIT VII Human Biology *583*

29. Body Structure *584*
THE BODY FRAMEWORK • THE MUSCULAR SYSTEM • THE INTEGUMENTARY SYSTEM

30. Nutrition and Digestion *604*
NUTRITION AND DIET • THE DIGESTIVE SYSTEM • DIGESTION AND ABSORPTION OF FOOD • PROBLEMS OF NUTRITION

31. Circulation *624*
THE FUNCTIONS OF BLOOD • BLOOD AND LYMPH FLOW • CONTROL OF CIRCULATION • PROBLEMS OF CIRCULATION

32. **Respiration and Excretion** *644*
THE RESPIRATORY SYSTEM • RESPIRATORY PROBLEMS •
THE EXCRETORY SYSTEM

33. **The Nervous System** *662*
STRUCTURE OF THE NERVOUS SYSTEM • CONDUCTION OF IMPULSES • THE SENSES

34. **The Endocrine System** *684*
ENDOCRINE GLANDS AND HORMONES • ENDOCRINE DISORDERS •
ENDOCRINE CONTROL MECHANISMS

35. **Reproduction and Growth** *704*
REPRODUCTIVE SYSTEMS • DEVELOPMENT BEFORE BIRTH •
DEVELOPMENT AFTER BIRTH

UNIT VIII Ecology *725*

36. **The Environment** *726*
ORGANISMS AND THE ENVIRONMENT • ABIOTIC FACTORS
IN THE ENVIRONMENT • BIOTIC FACTORS IN THE ENVIRONMENT

37. **Ecological Changes** *744*
CHANGES IN POPULATIONS • THE BEHAVIOR OF POPULATIONS •
CHANGES IN COMMUNITIES

38. **Biomes** *762*
THE EARTH'S BIOMES • TERRESTRIAL BIOMES • AQUATIC BIOMES

39. **Your Environment** *784*
MATERIAL RESOURCES • POLLUTION IN THE ENVIRONMENT • ENERGY RESOURCES

APPENDIX A. **Safety in the Laboratory** *803*
APPENDIX B. **Biological Word Parts** *804*
APPENDIX C. **Five-Kingdom System for Classifying Organisms** *805*
GLOSSARY *809*
INDEX *833*

To the Student

You are probably opening this book for the first time at the beginning of the school year in September. However, in many parts of the United States, September marks the beginning of the end of "nature's year." In New England, for example, the maple leaves will soon show their autumn colors.

In the central part of the United States, ducks, geese, and other birds are flying south to escape the cold winter months. Farmers are harvesting crops they tended over summer's warm days.

In California, you can observe flocks of monarch butterflies flying toward Mexico. You might be surprised to learn that their delicate butterfly wings carry them thousands of kilometers to a warm climate. Everywhere in the natural world, plants, animals, and other living things prepare for change as the earth moves around the sun.

Life on Earth is ever-changing. Sometimes the changes are easily observed. At other times the changes are hidden from the casual, and even the serious, observer.

As you read this book, you will learn about the nature of science and some of the major theories in biology. You will find that there are many unanswered questions in biology. Scientists have proposed answers to some of these questions. You will be able to consider for yourself which answers seem most reasonable.

As with most textbooks, this book is organized into units and chapters. There is a unit on plants, a unit on animals with backbones, and several other units. Each chapter within a unit has an opening photograph to help you start thinking about some of the ideas that will be discussed in the chapter.

The many illustrations and photographs are closely related to the idea development that occurs in the text. Use these photographs and illustrations to help you understand the ideas being presented.

At the end of each chapter is a Chapter Review. It includes a summary of the main ideas covered in the chapter. The Chapter Review also includes a variety of questions. Some questions will test your understanding of biological terms. Others will test your understanding of the concepts covered in the chapter. Additional questions will ask you to apply your knowledge to problems that relate to the basic concepts of the chapter.

The text includes a variety of "extras." Each chapter has a laboratory investigation and each unit has a skills page. Special features vary from chapter to chapter. Some are biographies of scientists. Others discuss issues in science, discoveries, computer use in biology, and careers in the biological sciences.

Today, there are developments in biology that may affect your future. Understanding these developments can help you recognize their possible effects on your life. Thus, studying biology can help you deal with the challenges and opportunities of a changing world.

UNIT I

Introducing Biology

The Earth teems with life. You can find living things in the depths of the ocean and in the outer limits of the atmosphere, on the earth's surface and just below it. Each kind of living thing, like this sloth, is unique. However, living things share similarities of structure and perform similar functions. In this unit, you will learn about the structures and functions common to all living things.

Chapters

1 *Studying Life*
2 *Understanding Life*
3 *Basic Chemistry*
4 *Cell Biology*
5 *Cell Chemistry*
6 *Cell Reproduction*

1. Studying Life

This albatross has returned to the island on which it was born, walking on land for the first time in two years! Here it will raise its young and then again take to the air. The albatross will return to its island home about the same time every year to raise its young.

The albatross is a remarkable living thing. However, it shares many characteristics with all life forms. In this chapter, you will explore some of the characteristics shared by all living things.

CHAPTER OBJECTIVES

After completing this chapter, you will be able to
- **Describe** how the scientific method is used by scientists to answer questions about the natural world.
- **Discuss** the steps of the scientific method.
- **Describe** the International System of Measurement.
- **Identify** tools used by scientists in their work.
- **Explain** why safety in the laboratory is important.

The Study of Science

1•1 WHAT IS SCIENCE?

Science is a method of obtaining knowledge about nature. The earth, space, nonliving things, and living things are all part of nature. Science involves the examination of nature to understand and describe aspects of it. Science seeks to answer certain kinds of questions about nature. Examples of such questions are "Why do monarch butterflies migrate?" and "Can a person live a long time with an artificial heart?" The answers to the questions usually lead to many more questions. For example, as scientists learn about migration, some may ask, "How do migrating butterflies find their way?"

Nature is studied in all branches of science—in biology, chemistry, physics, and so on. And in all branches of science, nature is studied in the same manner. Scientists use the scientific method in attempting to understand and explain nature. The **scientific method** is a means of gathering information and testing ideas. It is the way a scientist tries to find answers to questions about nature. Although the procedure can vary, the scientific method consists of these steps: making observations, forming hypotheses, testing explanations, and drawing conclusions. It is the use of the scientific method that separates science from other fields of study.

1•2 SCIENTIFIC OBSERVATIONS

Observation, usually the first step in the scientific method, is the examination of something in nature. Observations may be made directly with the senses. For example, a scientist listening to crickets might observe that they chirp faster on some nights than on others. Or a scientist watching bats might observe that they hunt tiny insects at night. Observations may also be made by the use of instruments that extend normal sensory perception. An important instrument in biology is a microscope. A microscope magnifies objects and allows a person to observe objects that are not visible to the unaided eye.

SECTION OBJECTIVES

- Explain the steps involved in applying the scientific method.
- Identify the differences between a scientific theory and a scientific law.
- Describe areas of biological study.

FIGURE 1•1 These monarch butterflies are resting before continuing on their winter migration. Science seeks to answer questions about butterfly migration.

4 STUDYING LIFE

FIGURE 1·2 This hungry bat has captured a cricket. Note the thin bones and membranes that make up the bat's wings.

To be useful, a scientist's observations must be accurate, or free from error. Suppose a scientist who is counting the different kinds of insects that land on a flower mistakes a moth for a bumblebee. The scientist's observations will be in error.

A scientist also must be careful that his or her existing opinions and emotions do not influence the observations. An existing idea or opinion that influences an observation is a **bias.** A scientist's observations should be free of bias. For example, suppose a scientist with a fear of bats interprets bat behavior as always aggressive towards people. The scientist's "observation" about bat behavior would be biased because it is influenced by a personal opinion about bats.

Information obtained from observations and from other sources is called **data.** Observations made by a scientist must be recorded in writing, on film, on tape, or by another means. Such recorded observations make up a scientist's data.

1·3 HYPOTHESES

An observation, or a series of observations, often leads a scientist to ask one or more additional questions. If a scientist first observes bats hunting insects in the dark, the scientist might then ask, "How do bats detect the tiny insects on which they feed in the dark?" To answer that question, the scientist may first read about bats and their behavior. If the scientist cannot find the answer in the writings of other scientists, then further observations about bats must be carefully made. Look at Figure 1·3. Observe the bat's head and wings closely. Make any observations you think might help you determine how a bat feeds at night.

After a question has been stated, the scientist answers the question by stating a **hypothesis** (hī-POTH-uh-sihs). A hypothesis is a possible answer to a question about nature based on the observations, reading, and knowledge of a scientist.

What hypothesis can you form about how bats hunt in the dark? One hypothesis is that bats use their vision when hunting at night. The next step in the scientific method is to test the hypothesis.

1·4 EXPERIMENTATION

The scientific testing of a hypothesis is called **experimentation.** A scientist must design an experiment to test the hypothesis that is proposed.

An experiment generally includes two groups, or setups, from which observations can be made. One setup is called the control group. The other setup is called the experimental group. The experimental group differs from the control group in only one factor or condition. The only condition that is different is the one that is being tested by the experiment. The condition that distinguishes the experimental group from the control group is called the **variable factor.**

Remember the hypothesis that bats use their vision to help them hunt insects at night? What experiment must be devised to test this

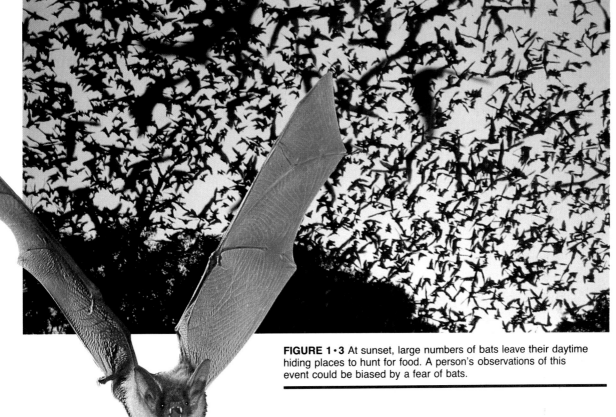

FIGURE 1·3 At sunset, large numbers of bats leave their daytime hiding places to hunt for food. A person's observations of this event could be biased by a fear of bats.

hypothesis? One experimental design to test this hypothesis would involve collecting 300 bats of the same kind. The bats would be banded for identification, as shown in Figure 1·4. Then they would be divided into two groups of 150 bats. The bats in one group would have their eyes covered to prevent them from seeing. These bats would be the experimental group. The bats in the other group—the control group—would not have their eyes covered. What would the variable factor be in this experiment?

All the bats would be released at the same time in the same area. The next day, all the banded bats that returned would be collected. Their stomach contents would be analyzed to determine what they had eaten. The number of bats that returned and the stomach contents would be recorded as data, and the data would be analyzed. Suppose the data showed that 146 bats from the experimental group returned and 136 from the control returned. Suppose also that the analysis of their stomach contents showed there was no difference in the foods eaten by the control group and the experimental group. The scientist would use these data to draw conclusions about the hypothesis.

FIGURE 1·4 Bats huddle together to sleep. Some of these bats are banded with metal rings for identification.

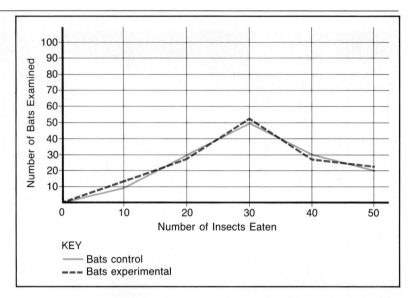

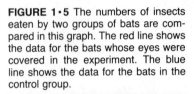

FIGURE 1·5 The numbers of insects eaten by two groups of bats are compared in this graph. The red line shows the data for the bats whose eyes were covered in the experiment. The blue line shows the data for the bats in the control group.

While an experiment is in progress, accurate observations must be recorded about both the experimental group and the control group. The recorded observations are the data provided by the experiment. Enough data must be obtained from the experiment to ensure confidence that what is taking place is not a one-time occurrence. For example, in trying to determine whether bats use their eyesight to hunt insects at night, a scientist must work with groups of bats. The scientist cannot examine the feeding habits of one bat and obtain useful data. A large sample size ensures that what is being observed is what normally occurs under the conditions that exist in the experiment.

After the data have been recorded, they must be organized and analyzed. Often data are presented in a table or a graph. Both methods allow the data to be understood quickly and easily. Look at Figure 1·5. Was there a significant difference in the amounts of insects eaten by bats with eyes covered and bats with eyes uncovered? Until recently, scientists had to spend many hours arranging and analyzing data. Today, computers reduce the time spent in the process of organizing data.

1·5 CONCLUSIONS

The data obtained from an experiment are studied to determine whether or not they support the original hypothesis. If the data support the hypothesis, the scientist has obtained evidence that the hypothesis is valid. If the data do not support the hypothesis, the conclusion is that the hypothesis is incorrect.

In the experiment with the bats, both the experimental group and the control group had eaten similar amounts and kinds of food. So a logical conclusion would be that vision does not play a major role in the ability of bats to hunt at night. In this case, the data do not support the hypothesis.

At this point, a scientist might formulate another hypothesis. For example, a scientist might state that hearing plays a role in the ability

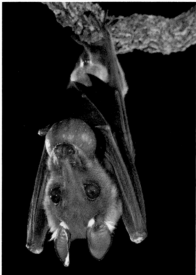

FIGURE 1·6 Bats and dolphins both rely on their hearing to locate food. These dolphins are pursuing a school of fish. This bat has already found its meal—a piece of fruit.

of bats to hunt at night. To find out, the scientist would cap the ears of the bats in the experimental group to prevent those bats from hearing. Suppose the results showed that many of the bats in the experimental group did not return, and those that did return had not eaten. Only the control bats had eaten. Do the data from this experiment support the scientist's hypothesis that hearing plays a role in the ability of bats to hunt at night?

1·6 THEORIES AND LAWS

If a hypothesis is repeatedly supported by further experimentation and verified by other scientists, it may become a **theory.** A theory is a hypothesis, or explanation of some part of nature, that has repeatedly been supported by evidence. The germ theory of disease, for example, states that certain diseases are caused by tiny organisms. It also states that a disease can spread from person to person when these organisms move from person to person. A theory usually serves as a basis for further experimentation. For example, the germ theory of disease led to experiments that resulted in the development of vaccines and other methods to prevent disease.

The popular meaning of the word theory, which implies an unsupported idea, differs from the scientific meaning. In science, a theory is an explanation in which there is a high degree of confidence. Scientific theories are supported by data.

Scientific theories can change. In some cases, theories are replaced by new theories. In other cases, new data cause theories to be modified. The atomic theory—the theory that explains the structure of atoms—has been modified several times. Scientific theories are always subject to further testing and modification.

In addition to theories, there are also **scientific laws,** or principles. A law describes an aspect of nature. But unlike a theory, a law does not explain an aspect of nature. For example, Allen's law states that a particular animal's body parts are shorter in cold climates than

STUDYING LIFE 7

8 STUDYING LIFE

FIGURE 1·7 Allen's law is illustrated by these two animals. The arctic hare lives in snowy polar regions (top), and the blacktailed jack rabbit lives in hot, dry deserts (bottom). Note the difference in the size of the animals' ears.

in warm climates. Thus, as shown in Figure 1·7, a rabbit living in a polar region has shorter ears than a rabbit living in a desert. A rabbit's ears contain many tiny blood vessels. As heat passes from the tiny blood vessels in the rabbit's ears into the surrounding air, the rabbit's body temperature decreases. Thus a rabbit's ears provide a means by which the rabbit is kept cool. So long ears are an advantage if a rabbit lives in a desert and a distinct disadvantage if a rabbit lives in a polar region.

1·7 BRANCHES OF BIOLOGICAL STUDY

The word **biology** contains the parts *bio-* and *-logy*. The prefix *bio-* means "life." The suffix *-logy* means "study of." By combining the meanings, biology can be defined as "the study of life."

Because the natural world is so complex, no one biologist can study all aspects of the natural world in depth. So biologists limit their field of study. Some biologists study plants; others study humans. Sometimes the field that a biologist studies seems very narrow. For example, a scientist may spend an entire lifetime studying a small group of earthworms.

You are probably familiar with several different fields of biological study. Other fields, however, may be new to you. **Microbiologists** study microorganisms, which are organisms too small to be seen with the unaided eye. Some microorganisms can cause disease in humans. Other microorganisms are beneficial. For example, some bacteria are used to tan leather; others are used to flavor certain kinds of food.

Botanists study plants. Within their field, botanists often specialize in only one kind of plant.

Zoologists study animals. Because zoology is such a wide field of study, many zoologists confine their field of expertise to one animal group. For example, **ornithologists** study birds. **Entomologists** study

FIGURE 1·8 Zoologists have attached a radio transmitter in a collar around this moose's neck. The transmitter gives off a signal that permits scientists to track the moose as it wanders in search of food. In time, the zoologists will have a good idea of this moose's travels and of how much land a moose needs in order to survive.

Discovery

Sharkskin and Yacht Racing

Sometimes the solution to a problem is difficult to discover. At other times a solution might be very obvious. And sometimes the solution comes from an unexpected source. You might be interested to learn how observing a shark helped to win a yacht race.

In 1984, the United States entry in the America's Cup races lost to its Australian challenger. For the first time the America's Cup Trophy passed out of American hands. After the loss, American yacht designers worked to develop an improved yacht that would recapture the trophy. Yacht designers thought that the hull of a racing yacht should be very smooth to let the boat slip effortlessly through the water. But one scientist had a different idea. He decided to study the shark for help in designing the new yacht.

The scientist had observed that a shark is one of the fastest fish in the sea even though its skin is very rough. In fact, for many years sharkskin was used in place of sandpaper to smooth wood surfaces before they were varnished.

The scientist wondered how the shark could move so quickly in the water with its rough skin. And he also wondered if, likewise, a rough finish on the bottom of a racing yacht would improve its speed.

A new paint was developed, one with a rough finish when it dried. This new paint was applied to the bottom of the American yacht that was to be entered in the America's Cup races in 1987. The paint did improve this yacht's speed because it cut down on the small currents of water that adhered to the hull, thus slowing it down. And it proved to be one of the important reasons that the yacht eventually won the America's Cup and returned it to the United States. So you might say that this famous event in 1987 was won with the help of a shark and the insight of an observant scientist.

insects. **Geneticists** study the field of biology that deals with how characteristics are passed from an organism to its offspring. They study all organisms in the natural world.

Knowledge of life on Earth is ever increasing. But there is another field of biology that studies far-ranging life forms. **Exobiologists** study life in space. Presently, exobiologists are searching for life forms away from Earth. In the future, as people venture far from their home planet, exobiology may become an important and interesting area of biological research.

The word **technology** (tehk-NAHL-uh-jee) is often in the news today. Technology can be defined as the use of scientific knowledge to improve the quality of human life. Technology is sometimes called

STUDYING LIFE

10 STUDYING LIFE

WORD PART	MEANING
bio-	life
entomo-	insect
exo-	outside
gen-	beginning, origin
micro-	small
-ology	study of
ornitho-	bird
zoo-, -zoa	animal

FIGURE 1·9 Knowing word parts and their meanings helps you understand scientific terms. A more complete list of word parts is found in Appendix B.

applied science. A bioengineer is a technologist. A bioengineer applies a scientific knowledge of engineering to solve biological problems.

There are many branches of biology. To learn their names would be a monumental task. But a knowledge of some common biological prefixes and suffixes can help you decipher even the most confusing words. Look at the chart in Figure 1·9. This chart shows some of the many commonly used prefixes and suffixes. Also see Appendix B for a more detailed list.

REVIEW QUESTIONS

1. What is science? What is biology?
2. What are the steps in the scientific method?
3. How do the popular and scientific meanings of theory differ?
4. How does a scientific law differ from a scientific theory?
5. Suppose you observed an ant carrying a grain of sugar to an anthill. Soon after, you saw hundreds of ants moving to the spot where the first ant had found the sugar. What hypothesis would you suggest to explain how the first ant communicated the location of the sugar to the other ants? How might you test your hypothesis?

Scientific Measurement

SECTION OBJECTIVES

- Identify the basic units of the International System of Measurement.
- Distinguish between weight and mass.

1·8 INTERNATIONAL SYSTEM OF MEASUREMENT

All scientific observations and experiments produce data. Frequently, collecting data involves making measurements. A scientist studying the chirping of crickets might measure the air temperature. A scientist studying sleeping behavior in humans might measure the volume of air inhaled. What kinds of measurements are the scientists making in Figure 1·10 ?

Scientists are extremely careful when making measurements and recording observations so that the data they collect will be accurate. Scientists also want other scientists to be able to understand their data. To help ensure that data are accurate and can be understood, all scientists use the same system of measurement. The system of measurement used in science is the **International System of Measurement.** The abbreviation for this system is SI. SI comes from the French name for the system, *Le Système International d'Unités.* SI is the modern version of what you may call the metric system. In the next few sections, you will learn about the SI units commonly used in biology. Figure 1·11 shows some of the prefixes used in SI.

FIGURE 1·10 Scientists must make careful measurements when collecting data. All scientists use the same system of measurement—The International System of Measurement (SI).

1·9 LENGTH

In the International System of Measurement, the **meter** (m) is the standard unit of length. A newborn child is about 0.5 m long. The average adult male is about 1.8 m tall; the average adult female is about 1.6 m tall.

All other units of length in SI are based on the meter. Each of these units is named by adding a prefix to the word meter. Prefixes in SI indicate the size of the unit compared with the standard for the measurement. Look at Figure 1·11 on this page. According to the table,

FIGURE 1·11 Prefixes in SI indicate the size of the unit compared with the standard for the measurement.

PREFIX	SYMBOL	MEANING	MULTIPLIER
mega-	M	one million	1 000 000
kilo-	k	one thousand	1 000
centi-	c	one hundredth	0.01
milli-	m	one thousandth	0.001
micro-	μ	one millionth	0.000 001
nano-	n	one billionth	0.000 000 001

STUDYING LIFE 11

the prefix *kilo-* means one thousand. So a kilometer (km) is 1000 meters. A millimeter (mm) is one thousandth of a meter. A millimeter is usually the smallest division on a metric ruler. The length of an eyelash would be measured in millimeters. The micrometer (μm), one millionth of a meter, and the nanometer (nm), one billionth of a meter, are used to measure extremely small objects. For example, viruses might be measured in nanometers.

1·10 WEIGHT AND MASS

Figure 1·12 shows an astronaut on Earth and on the moon. You may be surprised to learn that the astronaut does not weigh the same in both locations. **Weight** is a measure of the gravitational force one object has on another object. On Earth, the weight of an object is determined by the gravitational pull of Earth on that object. Because the gravitational pull of the moon is less than that of Earth, the weight of an object on the moon will be less than its weight on Earth. An astronaut, therefore, weighs less on the moon than on Earth. Weight is measured on a scale. The unit of weight in SI is the newton (N).

The **mass** of an object refers to the quantity, or amount, of matter in the object. Mass is measured on a balance against one or more

FIGURE 1·12 Although this astronaut weighs less on the moon than on the Earth, his mass is unchanged. Weight changes when the force of gravity changes. Mass changes only if matter is lost or gained.

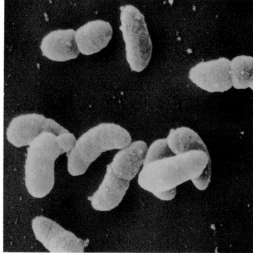

FIGURE 1·13 While the length of this mouse might be measured in centimeters, the length of these bacterial cells would probably be measured in micrometers.

objects of known mass. Thus an astronaut's mass, as measured on a balance, would be the same on Earth and on the moon. The unit of mass in SI is the **kilogram** (kg). A gram (g) is one thousandth of a kilogram. A milligram (mg) is one thousandth of a gram.

The mass of an organism such as a horse would be measured in kilograms. A mouse's mass would be measured in grams. Drugs, such as aspirin, and vitamins are commonly measured in milligrams.

1·11 VOLUME

Volume is the amount of space an object takes up. Volume measurements are based on linear measurements. A cubic centimeter (cm^3) is a cube that is 1 cm long, 1 cm wide, and 1 cm high. The formula for volume is volume = length × width × height. So a cube 1 cm on each side has a volume of 1 cm^3.

The **liter** (L) is the standard unit of volume in the International System of Measurement. One liter equals 1000 milliliters (mL). Another unit of volume is the cubic centimeter (cm^3, or cc). For all liquids, one milliliter equals one cubic centimeter, or 1 mL equals 1 cc. Liquid medicines are often prescribed in mL's or cc's.

For liquid water, an important relationship exists between volume and mass. A volume of 1 milliliter, or 1 cubic centimeter, of water has a mass of one gram. One liter of water has a mass of 1000 grams.

Figure 1·14 shows a 100-mL graduated cylinder. A graduated cylinder is used to measure liquid volumes. Look at the close-up of the water in the graduated cylinder. You will observe that the level of the water is not flat. There is a slight dip in the middle. The dip is called the meniscus. When you determine the volume of a liquid in a graduated cylinder, you must read the amount that corresponds to the bottom of the meniscus. What is the volume of water in this cylinder?

FIGURE 1·14 To determine the amount of liquid in a graduated cylinder, read the amount that corresponds to the meniscus, the slight dip at the top of the liquid's surface.

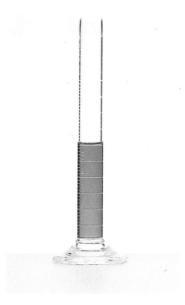

STUDYING LIFE   13

1·12 TIME AND TEMPERATURE

The **second** (s) is the basic unit of time in SI. The prefix *milli-* is often used with the word second. A millisecond (ms) is one thousandth of a second. The blink of an eye lasts about 300 ms (0.3 s).

Temperature is measured in **degrees Celsius** (SEHL-see-uhs). Look at Figure 1·15. You can see that the freezing point of water is 0° C and the boiling point of water is 100° C. Normal body temperature is 37° C, and a comfortable room temperature is 21° C.

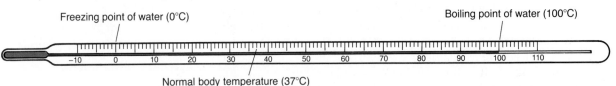

FIGURE 1·15 A Celsius thermometer is used to measure temperature. In SI, temperature is measured in degrees Celsius.

REVIEW QUESTIONS

6. Why is the International System of Measurement used in science?
7. If you ran 10 km, how many meters would you have run?
8. How many grams are there in a mass of 50 kg?
9. An average adult has about 6 L of blood. If such an adult donates 500 mL of blood, how many milliliters of blood will he or she have left?
10. What is the difference between mass and weight?

The Tools of a Biologist

SECTION OBJECTIVES

- Describe some historical advances in the development of microscopes.
- Distinguish between the light microscope and the electron microscope.
- State some rules that promote safety in the laboratory.
- Describe ways in which computers are used by biologists.

1·13 EARLY MICROSCOPES

You know now that the success of the scientific method depends on the quality of the observations that are made. Perhaps you have seen photographs of a zoologist sitting in a rain forest making observations of gorillas that feed and sleep only a few meters away. Most biologists who work in the field take detailed notes. These notes may later form the basis of a scientific paper or article. But what kinds of observations can be made of living things that are too small to be seen with the unaided eye?

For these kinds of observations, biologists must rely on special tools. During your course of study, you will become familiar with some of the tools of the biologist. These tools, when properly used, will provide a glimpse of a part of the natural world you may not have known existed.

FIGURE 1·16 This biologist is studying gorillas in their natural environment. The success of her research depends on the quality of the observations she makes.

One of the most important tools of biologists is the **microscope**. A microscope makes small things appear larger. Microscopes use light that passes through a series of lenses to magnify objects that are being examined. Lenses are specially ground to bend the rays of light that pass through them. It is the system of lenses in a microscope that actually magnifies the image of the object being viewed. But the microscopes used today are far different from early microscopes.

Look at Figure 1·17. This microscope was made more than 300 years ago by Anton van Leeuwenhoek in Holland. It looks rather like a small Ping-Pong paddle. Anton van Leeuwenhoek was an amateur scientist. He was very curious about the world around him, especially the world of the very small. In order to examine this world, he built simple, single lens microscopes.

He placed an object to be viewed in one of his microscopes. A viewer held the paddle to the eye and viewed the specimen with the aid of light that passed through a small round opening in the paddle and through the small lens that van Leewenhoek ground. Since his microscopes could not be reused, van Leeuwenhoek made a new microscope for each different object he viewed. A collection of his original microscopes, as well as drawings of his observations, are preserved in London. Today, scientists are amazed at the quality of van Leeuwenhoek's work. As you can see in the illustration, van Leeuwenhoek made drawings that are very accurate. He was even able to observe microscopic objects that would remain unseen by others until the invention of better microscopes many years after his death.

FIGURE 1·17 Anton van Leeuwenhoek built many simple microscopes like this one (right). These microscopes were able to magnify objects about 200 times. He made many drawings of his observations. Today, scientists are able to identify many of the organisms in his drawings (left).

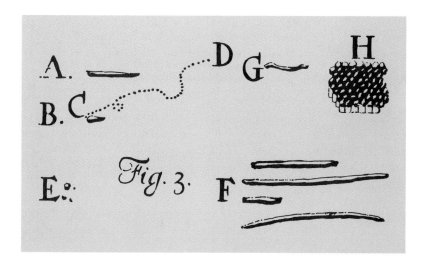

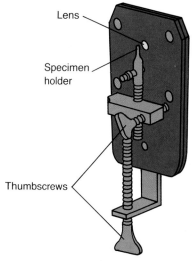

16 STUDYING LIFE

1·14 THE COMPOUND LIGHT MICROSCOPE

Modern microscopes are unchanged in principle from the microscopes used by van Leeuwenhoek in the 1680s, although they are much improved in design and easier to use. Light is still passed through the object being viewed, and the image of the object is enlarged by a series of lenses. Unlike the microscopes made by van Leeuwenhoek, however, today's microscopes use disposable glass slides on which a specimen is mounted. They also offer greater powers of magnification.

Figure 1·18 shows a typical compound light microscope. Locate these parts of the microscope: eyepiece, objectives, coarse adjustment, stage, diaphragm, fine adjustment, arm, and base. What is the function of the diaphragm? What is the function of the objectives?

There are several objectives, or lenses, on a compound microscope. Magnification is determined by multiplying the magnifying

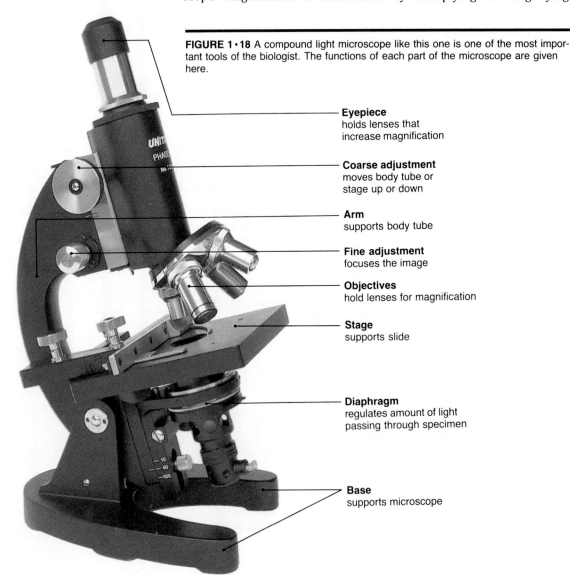

FIGURE 1·18 A compound light microscope like this one is one of the most important tools of the biologist. The functions of each part of the microscope are given here.

Eyepiece
holds lenses that increase magnification

Coarse adjustment
moves body tube or stage up or down

Arm
supports body tube

Fine adjustment
focuses the image

Objectives
hold lenses for magnification

Stage
supports slide

Diaphragm
regulates amount of light passing through specimen

Base
supports microscope

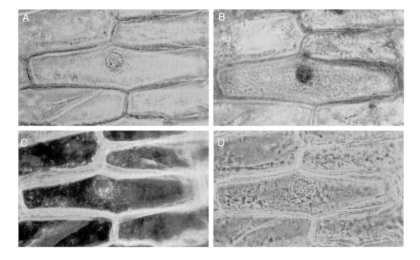

FIGURE 1·19 The details of the unstained cell are difficult to see (A). An iodine stain dyes certain structures yellowish-brown (B). Dark-field illumination gives greater contrast between structures within the cell and the background (C). Although many structures in the cell are transparent or colorless, they all change the light that passes through them in different ways. Phase-contrast illumination converts these changes in light into differences in color and contrast (D).

power of the eyepiece lens by the magnifying power of the objective being used. Magnifications often are indicated by a number that expresses the degree of magnification followed by an X, as in 200X.

Magnifying power is an important feature of a microscope. However, the ability to distinguish details of specimens is also important. The extent to which a microscope can distinguish two objects that are close together is called resolution, or resolving power. High resolution is needed to see details of very small cell parts. The best light microscopes cannot resolve objects that are closer together than 0.2 μm. Objects closer than 0.2 μm are seen as one object. Some microscopes have a special type of high magnification lens, called an oil-immersion lens, which increases resolution. When you use an oil-immersion lens, a drop of special oil is placed directly on a cover slip over the specimen to be viewed. The objective is then lowered into the oil. The oil changes the angle at which light enters the objective and thus increases the resolving power of the microscope.

In many cases, the details of an object may be difficult to see because the object is colorless or transparent. Different microscopic techniques are used to make such objects visible. Figure 1·19 shows onion cells prepared and viewed in different ways to make various parts visible. One technique involves the use of stains or dyes to color certain cell parts. Notice how the onion cells look in Figure 1·19B, which shows cells stained with iodine. Another technique, called dark-field illumination, allows the observation of certain objects that cannot be seen easily by other methods. Figure 1·19C shows this technique. The phase-contrast microscope makes it possible to see different parts in cells without using stains. A phase-contrast microscope changes the angle at which light waves pass through a specimen. In Figure 1·19D you can see how this technique makes some parts of the specimen appear lighter or darker than others. Scientists use all of these techniques in their study of cells.

FIGURE 1·20 The top photograph shows a fruit fly as it might appear through a magnifying glass. Note its large red eyes. The texture of the fruit fly's eyes and the details of its head become clear when magnified 60 times (center). A scanning electron microscope produced this detailed, three-dimensional image of one of the fruit fly's eyes (bottom).

1·15 THE ELECTRON MICROSCOPE

Compound light microscopes are important tools for biologists. However, the invention of the electron microscope in the early 1930s has made much higher magnification with good resolution possible. An electron microscope is one that uses a beam of electrons instead of a beam of light to magnify objects.

There are two main types of electron microscope—the transmission electron microscope and the scanning electron microscope. The transmission electron microscope (TEM) produces an image by passing a beam of electrons through a specimen. Magnifications up to 300 000 times and resolutions of 0.2 μm are possible. However, the TEM has one major disadvantage. Because a specimen must be sliced extremely thin and placed in a vacuum, only dead cells can be observed.

The scanning electron microscope (SEM) forms an image by scanning the surface of a specimen with a beam of electrons. Surfaces of objects viewed with the SEM can be seen in detail. Notice in Figure 1·20 that objects viewed with an SEM have a three-dimensional appearance. Magnification is not nearly as great with the SEM as it is with the TEM. As with the TEM, the specimen to be viewed must usually be killed. However, it has recently become possible to photograph living specimens with the SEM.

1·16 COMPUTERS

Computers are used in many different fields of endeavor today. You probably have operated a computer yourself. What is a computer and how does it work?

A computer is a device that can store, find, and process data. The first computer, developed in the 1940s, filled an entire room. Despite its size, it could carry out only simple calculations. Today, computers that fit in the palm of the hand can carry out many different kinds of complex calculations.

The quantity of data that computers can process and the speed and accuracy with which computers work make them invaluable tools. Computers have become a vital resource in fields such as biochemistry, agriculture, and medicine.

In medicine, computers are often used to help diagnose disease. The CAT scan uses computerized X-rays to make a series of images of the body. Physicians can view the internal structure of a human body with a CAT scan, without ever having to perform an operation.

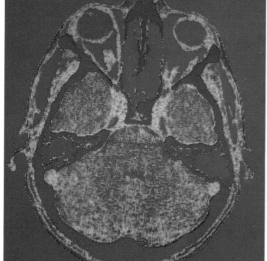
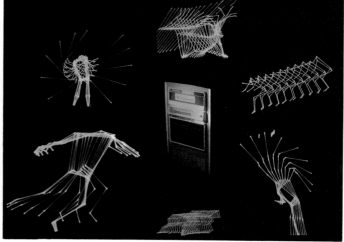

FIGURE 1·21 Modern technology aids biologists in many ways. This CAT scan picture (left) shows the inside of a person's head. The eyes appear as greenish circles at the top of this cross-sectional view. In the montage photograph (right), the movements of athletes are shown broken down and analyzed by a computer.

FIGURE 1·22 Common safety symbols.

An apron should be worn to avoid damaging clothing with chemicals that may stain.

Wear safety goggles when working with any substances that may injure your eyes.

Handle sharp instruments with care. Always cut away from you.

Use caution when using chemicals in this investigation.

Exercise caution when using electrical appliances and outlets.

Use caution to avoid burning the skin or clothing.

Computers are also used to analyze movements of the human body. Look at the right photograph in Figure 1·21. It shows a computer's analysis of different athletes in motion. A special camera is used to photograph an athlete. Then, a computer analyzes the photographs. The computer produces images that show the components of the athlete's movement. The athlete's position at any one point can be studied.

1·17 SAFETY IN THE SCIENCE LABORATORY

To better understand the facts and concepts you will learn this year, you will work in the laboratory. Scientists know that it is very important to follow safety procedures when working in the laboratory. So they take as many precautions as possible to protect themselves and their fellow workers.

The work that you will do in the laboratory involves investigations that have been done many times before. When done properly, these investigations are perfectly safe. But if not done properly, the investigations can lead to accidents. How can you avoid accidents? The answer is simple.

Each investigation in this text is marked with appropriate safety symbols. The symbols alert you to the need for special safety precautions. Familiarize yourself with these safety symbols, shown in Figure 1·22. Then follow the instructions given for that symbol whenever you see it. More detailed instructions are given in Appendix A at the back of your textbook.

Before you work in the laboratory for the first time, make sure you read and understand the safety rules supplied by your teacher. If you do not understand a safety rule, ask your teacher to explain the rule to you. Never try an experiment on your own without first asking your teacher's permission. When you are not sure how to proceed with an investigation, ask your teacher. As you work in the laboratory, you may want to suggest additional rules to make the laboratory a safe place for you and your classmates.

A general rule to follow when working in the science laboratory is: ALWAYS FOLLOW YOUR TEACHER'S DIRECTIONS AND THE DIRECTIONS IN YOUR TEXTBOOK EXACTLY AS STATED.

REVIEW QUESTIONS

11. Why are microscopes used to study cells?
12. How did the work of Anton van Leeuwenhoek add to scientific knowledge about the structure of organisms?
13. Describe the basic differences between the compound light microscope and the electron microscope.
14. What are three ways in which computers are used by biologists?
15. What are two important safety rules you should follow while working in the laboratory?
16. How is scientific knowledge in part based on the kinds of tools that are available for scientists to use?

Investigation

How Is the Scientific Method Used to Solve a Problem?

Goals
After completing this activity, you will be able to
- Describe the steps used in the scientific method.
- Explain the meaning of the terms *control group* and *experimental group*.

Materials (for groups of 3)
6 petri dishes
glass-marking pencil
10 water-soaked sunflower seeds
10 dry sunflower seeds
10 water-soaked lima beans
10 dry lima beans
10 water-soaked orange seeds
10 dry orange seeds
paper toweling   scissors
tap water   lightproof box

Procedure

A. With a glass-marking pencil, label the lid of each dish as follows: Seed A, Seed A Control; Seed B, Seed B Control; Seed C, Seed C Control; Label the dishes with your name and date.

B. Cut the paper toweling into circles the size of the bottom of a petri dish. Place two layers of toweling in each dish.

C. To the dishes marked *Seed A, Seed B,* and *Seed C,* add enough tap water to completely moisten the paper toweling. Do not add any water to the dishes marked *Control.*

D. Place the 10 water-soaked sunflower seeds in a circle in the dish marked *Seed A* and the 10 dry sunflower seeds in a circle in the dish marked *Seed A Control.* Place the water-soaked lima beans in the dish marked *Seed B* and the dry lima beans in the dish marked *Seed B Control.* Place the water-soaked orange seeds in the dish marked *Seed C* and the dry orange seeds in the dish marked *Seed C Control.*

E. Place all the petri dishes in a lightproof box in a warm place.

F. Predict what will happen in each of the petri dishes. Write your prediction in the form of a hypothesis.

G. Check each dish every day for 5 days to note any changes. In the dishes with moistened paper toweling, add water if the toweling becomes dry.

H. Copy the chart below. Each day, record your observations of any changes. At the end of 5 days, draw a sketch of the seeds in each dish.

I. Analyze the data collected. Decide whether the data support your hypothesis.

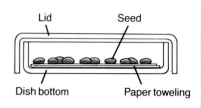

Questions and Conclusions

1. After 5 days, in which dish or dishes did you see the greatest amount of sprouting? In which did you see the least amount?
2. What did you observe in the dishes marked *Control?* What is a control?
3. What is the purpose of a control group?
4. In this activity, what was the *experimental group?*
5. In the dishes in which seeds sprouted, did all 10 seeds in a given dish sprout the same amount? Give a reason for any differences you saw.
6. Compare your results with those of the rest of the class. If the results were not the same, give reasons for the differences.
7. What were the scientific problems being studied in this activity? State the steps used to solve these problems.

	DAY 1	DAY 2	DAY 3	DAY 4	DAY 5
SEED A					
SEED B					
SEED C					
SEED A Control					
SEED B Control					
SEED C Control					

Chapter Review

SUMMARY

1·1 Science is a method of obtaining knowledge about nature.

1·2 The scientific method, a way of obtaining knowledge about nature, depends upon accurate, unbiased observations.

1·3 A hypothesis is an explanation based on observations of some part of nature.

1·4 A hypothesis must be tested by experiments with controls.

1·5 An experiment may provide evidence to support a hypothesis, or it may provide evidence that a hypothesis is incorrect.

1·6 A hypothesis supported by much evidence can become a theory. A theory explains an aspect of nature. Scientific theories can change if new evidence is found. A law describes an aspect of nature. It does not explain an aspect of nature.

1·7 Biology is the study of living things. However, this area of study is so broad that many biologists limit their field of study.

1·8 All scientific observations and experiments produce data. The International System of Measurement (SI) is used by scientists throughout the world to ensure accuracy in measurements and understanding in the exchange of scientific data.

1·9 The SI unit of length is the meter.

1·10 Weight is the measure of gravitational force on an object. Mass is the measure of the amount of matter in an object. The kilogram is the SI unit of mass.

1·11 Volume is the amount of space an object takes up. The liter is the standard unit of volume in the International System of Measurement.

1·12 The second is the SI unit of time. Temperature is measured in degrees Celsius.

1·13 Biologists rely on the microscope to observe microscopic life.

1·14 The compound light microscope is an important tool used by biologists. Various techniques are used to improve the quality of observations made with the light microscope.

1·15 An electron microscope produces a magnified image by passing a beam of electrons through a specimen or by scanning the surface of a specimen with electrons.

1·16 Computers are important because they can rapidly store, find, and process data.

1·17 Safety rules promote good laboratory work habits and make laboratory work safe.

BIOLOGICAL TERMS

bias
biology
botanist
data
degrees Celsius
entomologist
experimentation
exobiology
geneticist
hypothesis
International System of Measurement
kilogram
liter
mass
meter
microscope
microbiologist
observations
ornithologist
science
scientific law
scientific method
second
technology
theory
variable factor
weight
zoologist

USING BIOLOGICAL TERMS

Choose the term that does not belong and explain why.
1. meter, liter, mass, kilogram
2. experimentation, bias, hypothesis, data
3. kilogram, meter, liter, data
4. botany, zoology, microbiology, meteorology
5. liter, milliliter, cubic centimeter, gram

Supply the term from the list that best fits the definition.

6. A hypothesis supported by evidence.
7. A statement describing an aspect of nature.
8. The unit of time in the SI system.
9. The scientific way to test a hypothesis.
10. Sometimes referred to as applied science.

UNDERSTANDING CONCEPTS

1. How can the scientific method be used to answer questions about nature? (1·1)
2. Briefly explain an approach that could be used to make reliable observations. (1·2)
3. Why would a hypothesis made by a scientist be considered an educated guess? (1·3)
4. Why is it important to have one variable factor in an experimental group? (1·4)
5. What happens if the data collected do not support a hypothesis? (1·5)
6. What is the difference between a theory and a law? (1·6)
7. Why is biology divided into small fields of study? (1·7)
8. Why is it important for scientists to use a standard system of measurement? (1·8)
9. What is the difference between a meter, a kilometer, and a centimeter? (1·9)
10. Would a person weigh less on the moon than on Earth? Would a person have more mass on the moon or on Earth? (1·10)
11. What important relationship exists between volume and mass for liquid water? (1·11)
12. In what SI units are temperatures measured? (1·12)
13. Anton van Leeuwenhoek was not a trained scientist, yet he made important contributions to scientific knowledge. How did he contribute to scientific knowledge? (1·13)
14. How does the resolution of a microscope affect the observations made by a microbiologist? (1·14)
15. How has the electron microscope changed the way biologists view the natural world? (1·15)
16. Why is a computer an important tool for biology research today? (1·16)
17. What are two good safety rules to follow when you work in a laboratory? (1·17)

APPLYING CONCEPTS

1. Some people claim that astrology is a science. How would you convince them that it is not?
2. A student places a jar of fresh, clear pond water on a sunny windowsill. A week later the student notices that the water has turned green! What hypothesis might you suggest to explain this change of color? How could you test this hypothesis in your home? What is the control in your experiment?
3. For practice in using SI measurements, try estimating the sizes of various objects. For example, estimate the length and width of this book to the nearest centimeter. Use a metric ruler to make accurate measurements. Compare these measurements to your estimates. Try this procedure with other objects.
4. Why has knowledge of microscopic life increased so rapidly in the past fifty years?
5. A very large rocket is needed to send astronauts to the moon. However, a small rocket can bring astronauts back from the moon's surface. Why is this so?

EXTENDING CONCEPTS

1. Obtain a small, young geranium plant or houseplant that has a single stem or that can be trimmed to have a single stem. Transplant the plant into a square plastic flower pot. Test the plant's response to its environment by changing the position of the pot that holds the plant every few weeks. For example, from the upright position, place the pot on one side, followed by the opposite side, and then back to the upright position. Continue to repeat this procedure or your own variation, making sure that the plant shows a definite response before you change the pot's position. Can you predict what the plant will look like after several months?
2. What is your height, vertical reach, and arm span in meters? Locate a part of your hand that is one centimeter in size. What part of your hand is 0.5 millimeter thick?
3. There is an old folk notion that fresh cucumber slices repel cockroaches. Design an experiment to test this hypothesis. It might help you to know that most cockroaches are active at night.

READINGS

Janovy, John, Jr. *On Becoming a Biologist.* Harper & Row, 1985.

Johnson, Sylvia A. *Bats.* Lerner, 1985.

2. Understanding Life

17,000 years ago, ancient peoples painted these animals on the walls of the Lascaux caves in France. These drawings, alive with color, reveal the artists' desire to understand the living world around them. Today, biologists use the methods of science to explain many of the secrets of nature that have long mystified people. In this chapter, you will explore the nature of life and examine some of the structures and functions of living things.

CHAPTER OBJECTIVES

After completing this chapter, you will be able to

- **Discuss** spontaneous generation.
- **Compare** biogenesis and abiogenesis.
- **Describe** the experiments that proved that life only comes from living things.
- **Identify** the characteristics of living things.

Early Observations of the Natural World

2·1 SPONTANEOUS GENERATION

Many people in ancient civilizations studied nature and proposed hypotheses for what they observed. However, these hypotheses were seldom, if ever, tested. So reliable answers to questions about the natural world were not obtained until the scientific method came into widespread use.

One of the most fascinating questions about nature had to do with the origin of living things. From what did life originate? A popular belief that existed for more than 1600 years was that life could arise spontaneously. **Spontaneous** (spahn TAY nee uhs) **generation** is a hypothesis that states that living things can originate from nonliving matter. Another name for spontaneous generation is **abiogenesis.** The Greek teacher Aristotle (383–322 B.C.) believed in spontaneous generation and based his beliefs in part on his observations.

Aristotle had observed a pond during a long drought. The water in the pond had dried up, leaving only mud on the bottom. No fish were

SECTION OBJECTIVES

- Describe some of the classical experiments whose evidence disproved the hypothesis of spontaneous generation.
- Explain how the scientific method was used to disprove the hypothesis of spontaneous generation.

FIGURE 2·1 This woodcut, made in 1725, shows sheep arising from a tree (left). This lamb tree is a product of early travellers' garbled or exaggerated descriptions of cotton plants (right).

present. When the drought was over and the water in the pond had been replenished, Aristotle observed the presence of fish. On the basis of these observations, Aristotle concluded that the fish had been produced from the nonliving mud on the pond's bottom. Aristotle reasoned that the fish could not have been produced by other fish because all the previously existing fish must have died when the pond dried up.

Aristotle also believed that flies were produced by the rotting flesh of dead animals. He thought that other types of insects originated from wood, dried leaves, and even horsehair. Do you think it was logical for Aristotle to arrive at this belief in abiogenesis on the basis of the observations he made? Would abiogenesis have been widely accepted if it had been tested by the scientific method?

2•2 BIOGENESIS AND THE WORK OF FRANCESCO REDI

Until the mid-1600s, the hypothesis of spontaneous generation was accepted by most people. But the Italian physician and scientist Francesco Redi (1626–1697) was not convinced that flies came from rotting flesh. Redi had observed that flies often landed on rotting meat and soon after tiny, wormlike organisms called maggots appeared on the meat. These maggots ate the rotting flesh.

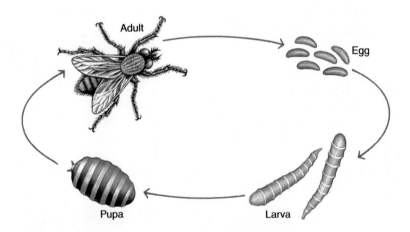

FIGURE 2•2 The life cycle of a fly. The larvae of the fly were the maggots that Redi observed in some of the jars.

Eventually the maggots stopped eating and moving. They appeared to change into small oval structures, which Redi collected and placed in covered glass jars. Later, flies emerged from the oval structures. The new flies resembled the flies that Redi had previously observed on the rotting meat. Redi hypothesized that the new flies developed from the maggots that had been seen on the rotting meat. In other words, the new flies were the offspring of the original flies—not of the rotting flesh.

Redi devised an experiment to test his hypothesis and determine whether maggots would develop if no flies came into contact with rotting meat. He placed pieces of meat into eight separate jars. He left four of the jars open to the air. He sealed the other four jars. Redi observed that flies moved into and out of the four open jars. After a short period of time, the meat in the four open jars contained tiny maggots. No maggots were observed on the meat in the four sealed jars. Redi concluded that maggots appeared on rotting meat only if flies laid their eggs on the meat.

Redi's opponents, who were supporters of spontaneous generation, came to a different conclusion about the results of Redi's experiment. They contended that maggots did not develop in the sealed jars because there was no air entering these jars. The lack of air, they argued, had prevented spontaneous generation from occurring.

To disprove their contention and prove his hypothesis, Redi designed another series of experiments. He covered some of the jars with netting, a material with holes too small for adult flies to pass through, but large enough to permit air to pass through freely. See Figure 2·3. No maggots appeared in the jars that were covered with netting.

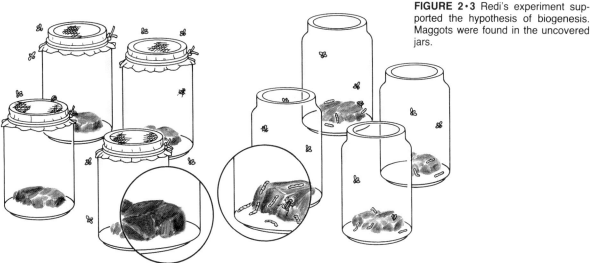

FIGURE 2·3 Redi's experiment supported the hypothesis of biogenesis. Maggots were found in the uncovered jars.

Redi's experiments supported the hypothesis of biogenesis (bī oh JEHN uh sihs). **Biogenesis** is a hypothesis that living things come only from other living things. Redi's experiments provided evidence against the hypothesis of spontaneous generation.

The supporters of spontaneous generation accepted the hypothesis that flies came from flies. However, they still believed that microorganisms (mī kroh OR guh nihz uhmz), small organisms visible only through a microscope, were produced by spontaneous generation.

2·3 NEEDHAM, SPALLANZANI, AND PASTEUR

Among the supporters of the hypothesis of spontaneous generation was John Needham (1713–1781), an English scientist. Needham performed numerous experiments in which he boiled broths of meat and vegetables. He then let the broths stand in loosely corked flasks. Needham correctly believed that boiling would kill all the microorganisms that had been present. After a few days Needham observed that the broths contained microorganisms. Needham concluded that the microorganisms must have developed from the broth. Needham's findings supported the hypothesis of spontaneous generation of microorganisms. Needham did not realize that microorganisms were able to enter because he had not sealed the flasks completely.

Lazzaro Spallanzani (1729–1799) was an Italian scientist who repeated Needham's experiments. Spallanzani took special care in boiling the mixtures and in filling the flasks. Then he sealed all of the flasks. He noticed that no microorganisms grew in the sealed flasks. Spallanzani then broke the seals on some of the flasks. Microorganisms began to grow in these flasks.

He offered this experiment as evidence that spontaneous generation does not occur. Still the supporters of spontaneous generation pointed out that air had been excluded from the sealed flasks. They claimed that air was essential for spontaneous generation to take place. The biogenesists, however, believed that air was the source of contamination and had to be excluded.

It was not until 1864 that Louis Pasteur (1822–1895), a French scientist, ended the controversy. Pasteur had shown that microorganisms were present on the dust particles in the air. He decided to test the hypothesis of spontaneous generation. Pasteur began by placing broth in several flasks. He then heated the necks of some of the flasks and drew out the glass into swan-necked shapes, as shown in Figure 2·5. The remaining flasks had straight necks. Pasteur then boiled the

FIGURE 2·4 Spallanzani's experiment showed that microorganisms do not grow in boiled, sealed flasks. His experiment provided additional evidence to show that spontaneous generation does not occur.

FIGURE 2·5 Pasteur's experiment disproved the idea of spontaneous generation. Microorganisms were not found in the swan-necked flasks.

broth in all the flasks, allowing steam to escape from the necks of the flasks. The escaping steam showed that the swan-necked flasks were open to the air. The flasks with straight necks were exposed to the air and sealed. Microorganisms grew only in the straight-necked flasks.

The swan-necked shapes allowed air to enter, but dust particles were trapped at the low points of the necks. Since no microorganisms developed in those flasks, Pasteur concluded that the growth of microorganisms was directly dependent on contamination by microorganisms on the dust particles in the air. Pasteur's work supported the hypothesis of biogenesis.

REVIEW QUESTIONS

1. What is spontaneous generation?
2. What observations led to the hypothesis of spontaneous generation?
3. Why did microorganisms develop in Needham's experiments?
4. How did Pasteur demonstrate that spontaneous generation did not occur in the broths?
5. A friend tells you that some plants arise from the soil by spontaneous generation. Design an experiment to test the hypothesis of spontaneous generation in plants.

Characteristics of Living Things

2•4 LIFE IS CELLULAR

It is not always easy to determine if an object is living or nonliving. However, biologists recognize certain characteristics that all living things possess. It is the appearance of these characteristics in an object that qualifies the object as living.

Biologists have found that all living things have the same basic structure on a microscopic level. All living things are made of cells. The **cell** is the unit of structure and function of all living things. Some organisms are made up of but a single cell. An amoeba is an example of a one-celled organism. One-celled organisms are also referred to as **unicellular** organisms. Other organisms, such as oak trees and baboons, are composed of many cells. Organisms composed of many cells are referred to as **multicellular** organisms.

2•5 LIVING THINGS USE ENERGY

The sun is the ultimate source of energy for life on the earth. Plants are able to use the sun's energy directly. Other organisms, however, cannot use the energy of sunlight directly. They must obtain the sun's energy in an indirect way. For example, animals survive by using

SECTION OBJECTIVES

- Identify the cell as a unit of structure and function of living things.
- Define life in terms of the characteristics of living things.

FIGURE 2·6 Diatoms are unicellular, or single-celled, organisms (left). A mandrill baboon is a multicellular, or many-celled, organism (right).

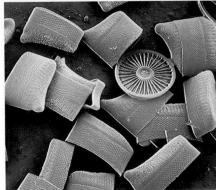

FIGURE 2·7 Growth is an increase in living mass. These tiny orchid seeds (top) will eventually grow into flowering orchid plants like this one (bottom).

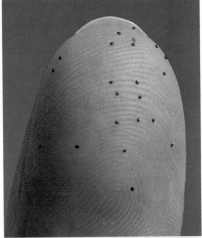

the energy trapped in plants. A lamb eats grass and thereby obtains the energy trapped in the grass. A wolf gets its energy by eating the lamb. But it, too, is ultimately dependent upon the energy trapped in the grass. As you can see, organisms continue to live only as long as they are able to get and use energy.

2·6 LIVING THINGS GROW

The orchid seed in Figure 2·7, finer than a speck of dust, will eventually grow into a flowering plant. An infant penguin will eventually grow into an adult penguin. All living things grow. **Growth** is an increase in living mass. Unicellular organisms grow only by increasing the mass of their single cell. Multicellular organisms grow by increasing the mass of each cell as well as by increasing the total number of cells in their bodies.

Most living things do not grow at a uniform rate throughout their entire life. Growth occurs in stages. The growth of unicellular organisms depends to a great extent on the amount of food materials they take in. For example, the more food an amoeba takes in, the faster it grows. Multicellular organisms also need food for growth. The growth of multicellular organisms occurs in stages that end in the death of the organism.

An organism has a definite **life span,** or length of life. Life span varies from one kind of organism to another. A mayfly lives as an adult for only a single day. Certain trees can live for hundreds, and even thousands, of years.

2·7 LIVING THINGS METABOLIZE

Living things need energy to grow, to replace cells that are damaged or worn out, and to reproduce. Chemical activities supply the energy and the materials that are needed by living things. The sum of all the different chemical activities carried out by a living thing is called **metabolism** (muh TAB uh lihz uhm). Metabolism includes the activities of breaking down food, obtaining energy, and using raw materials to build new living matter.

FIGURE 2·8 All organisms have a definite life span, or length of life. The life span of an adult mayfly is only one day (left). The life span of a bristlecone pine is much longer—about 5500 years (right).

Digestion includes the metabolic processes that are involved in breaking down food into simple substances. **Respiration** involves the processes by which most cells obtain energy as a result of combining food with oxygen. **Synthesis** (SIHN thuh sihs) includes the processes by which living things combine simple substances to form more complex substances. These three processes—digestion, respiration, and synthesis—make up metabolism.

Organisms regulate their metabolic processes, maintaining conditions that are suitable for life. The balancing of metabolic processes to maintain the conditions needed for life is called **homeostasis** (hoh mee oh STAY sihs). Homeostasis involves maintaining a flow of needed substances, producing energy, and removing wastes.

FIGURE 2·9 This giraffe is using its long tongue to feed on leaves (top). Materials from digested leaves will be synthesized into living materials inside the giraffe. This harbor seal (bottom) must regulate its body processes in order to stay warm and alive in its icy environment.

2·8 LIVING THINGS REPRODUCE

Reproduction is the process by which new individuals are produced. Because all living things have a definite life span, reproduction will not ensure an individual's survival. But reproduction is necessary for the survival of a group of living things. If the members of a group of living things failed to reproduce, the **extinction** (ehk STIHNK shuhn), or elimination, of that group would occur.

FIGURE 2·10 Many unicellular organisms reproduce asexually. Here, a single paramecium is dividing in half to produce two new individuals (left). Most plants and animals reproduce sexually. This baby penguin has two parents (right).

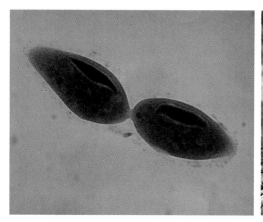

32 UNDERSTANDING LIFE

There are two types of reproduction. In **asexual reproduction,** a single parent gives rise to one or more individuals. Asexual reproduction is carried out by unicellular and some multicellular organisms. **Sexual reproduction** involves the joining of two specialized sex cells. These sex cells may be produced by different parents or the same parent. The joining of these two sex cells leads to the development of a new individual. Most plants and animals reproduce sexually.

2·9 LIVING THINGS RESPOND TO THEIR ENVIRONMENT

The **environment** plays an important role in the survival of living things. The environment includes all living and nonliving factors that may affect an organism. Look at the two photographs in Figure 2·11. One photograph shows zebras grazing on an African plain. The zebra's environment is made up of all the rocks, plants, soil, water, air, and other components of its surroundings.

The zebra needs many things from its environment in order to survive. However, its survival is by no means assured. This is because other things are also part of a zebra's environment, and these things are not always beneficial. For example, microorganisms may affect the zebra's health. Lions may pose a threat to a zebra's life, as shown in the other photograph. In order for the lions to survive in their environment, some zebras and other animals must perish.

The ability of an organism to react to changes in its environment is a characteristic of life. A change that can cause a reaction is called a **stimulus** (STIHM yuh luhs). A reaction of an organism to a stimulus is called a **response.** You can see from Figure 2·11 that a zebra's ability to respond quickly and correctly to a hungry lion may help the zebra survive. The fact that not all zebras will respond quickly and correctly helps guarantee the lion's survival as well.

FIGURE 2·11 In order to survive, zebras must respond quickly and correctly to their environment (top). A zebra that cannot respond well will not survive (bottom).

2·10 LIVING THINGS MOVE

All living things are capable of movement. Birds and bees can fly; a dog comes when you call its name. And although you may not have realized it, plants move too. A plant is responding to stimuli when its leaves and stem grow toward the sun and its roots move down through the earth. The sensitive plant in Figure 2·12 can even move when its leaves are touched. Movement is a characteristic of all living things.

FIGURE 2·12 Plants usually move slowly. A sensitive plant responds to stimuli more quickly than most plants. Within a second of being touched, the plant's leaves fold together.

REVIEW QUESTIONS

6. What are the characteristics of living things?
7. Describe how living things grow?
8. Why is the ability to reproduce considered a characteristic of living things?
9. How are homeostasis and metabolism related?
10. A snowball will increase its mass if rolled on snow. How does this differ from growth in living things?

Investigation

How Is a Microscope Used to Gather Information?

Goals

After completing this activity, you will be able to
- **Explain** how the parts of a microscope work.
- **Describe** how a microscope is used to examine very small objects.

Materials (for groups of 2)

microscope slide
cover slip pond water
forceps dropper newspaper
lens paper paper towels

Procedure

A. Place your microscope firmly on your work table. If your light source is attached to your microscope, plug it in. If you have no built in light source, use the mirror to focus light through the microscope stage.

   CAUTION: Do not focus the mirror with sunlight as the source of light. Use an electric light source.

B. You will examine a small piece of a newspaper page. Tear a small piece of newspaper from the large sheet. You should be able to find the letter "e" on the sample. Place the newspaper sample on the slide so that the "e" is in the center of the slide. Use a dropper to put one drop of water on the piece of newspaper. Now cover the drop with a cover slip. You should try to make sure that there are no air bubbles under the cover slip.

C. Place the slide on the microscope stage so that the letter "e" can be read correctly as you look down on it. Make sure that there is no water on the bottom of the slide when you place it on the microscope stage.

D. With your low-power lens in place, carefully use the coarse adjustment to move the lens as close to the slide as you can. You should move the low-power lens while looking at the microscope from the side. The low-power lens should not touch the slide.

E. Now look through the eyepiece. Try to keep both eyes open while you view an object. Focus the low-power lens by moving it away from the surface of your slide. Use the fine adjustment to focus until the letter "e" can be seen clearly.

F. Move the slide to the right and then to the left.

G. After you have made a sketch of what you observed, remove the slide. Wash the slide and the cover slip carefully. Dry the slide. Place a drop of pond water on the slide and put a cover slip over the pond water. Now observe the pond water using the techniques you practiced with the letter "e."

H. After you have completed your observations, remove the slide from the microscope and clean it. Clean the lenses of the microscope.

   CAUTION: Use only the special lens paper your teacher will supply. Make sure that the light source is disconnected and the low-power lens is in position before you replace the microscope in its storage cabinet.

Questions and Conclusions

1. Describe the appearance and position of the letter "e" under the microscope.

2. What happened to the letter "e" when you moved the slide to the left? To the right?

3. What did you observe in the sample of pond water? Draw several of the microorganisms you observed.

4. Suppose you wanted to follow an organism in the pond water that was moving to the right. Which way would you move the slide? Why?

5. How would you go about finding the names of some of the pond water organisms you observed? How would accurate observations and clear drawings help you identify these organisms?

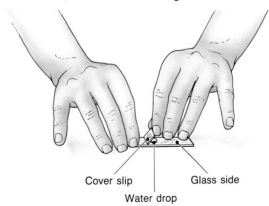

Cover slip Glass side
Water drop

UNDERSTANDING LIFE

Chapter Review

SUMMARY

2·1 Aristotle believed that life came from non-life. He believed in abiogenesis, or spontaneous generation.

2·2 Francesco Redi's controlled experiments provided evidence against the hypothesis of spontaneous generation.

2·3 Pasteur's and Spallanzani's scientific studies of microorganisms firmly established the hypothesis of biogenesis as the explanation of how living things are produced.

2·4 All living things are made of cells. The cell is the unit of structure and function of all living things.

2·5 Living things are characterized by their ability to use energy. The sun is the ultimate source of energy for life on Earth.

2·6 Growth involves changing raw materials into new living material.

2·7 Metabolic activities such as digestion, respiration, and synthesis supply an organism with energy and materials needed for life.

2·8 Reproduction involves the production of offspring in living things.

2·9 The ability to respond to changes in the environment is a characteristic of living things.

2·10 All living things, including plants, have the ability to move.

BIOLOGICAL TERMS

abiogenesis
asexual reproduction
biogenesis
cell
digestion
environment
extinction
growth
homeostasis
life span
metabolism
multicellular
reproduction
respiration
response
sexual reproduction
spontaneous generation
stimulus
synthesis
unicellular

USING BIOLOGICAL TERMS

From the list of biological terms, select one that fits each definition below.

1. All the living and nonliving factors that affect an organism.
2. The balanced regulation of life processes.
3. The ability to increase in size.
4. The hypothesis that life comes only from existing life.
5. The hypothesis that life comes from nonliving things.
6. The process of breaking down food into simpler substances.
7. The method by which cells obtain energy by combining food with oxygen.
8. The length of time an organism usually lives.
9. The basic unit of life.
10. The process by which living things combine simpler substances into more complex substances.

UNDERSTANDING CONCEPTS

1. What are some of the observations of Aristotle that lead him to believe in abiogenesis? Based on the observations he made, was Aristotle's conclusion reasonable? (2•1)
2. What was the variable factor in Redi's experiment in which he placed meat in jars and covered some of the jars with netting? (2•2)
3. Describe Pasteur's experiment that was designed to test the claim of abiogenesists that air was necessary for the development of living things from nonliving things. (2•3)
4. What is the basic unit of which all living things are made? (2•4)
5. Explain why the sun is considered the ultimate source of energy for life on the earth. (2•5)
6. What two methods of increasing living mass are part of the growth process? (2•6)
7. How is the growth of a nonliving object, such as an icicle, different from the growth of a living thing? (2•6)
8. How are the life activities of respiration and synthesis related to the process of metabolism? (2•7)
9. How are the characteristics of living things related to homeostasis? (2•7)
10. How does asexual reproduction differ from sexual reproduction? (2•8)
11. The appearance of a shark is a stimulus to a school of small fish. How might the small fish respond? (2•9)
12. How do plants show the ability to move? (2•10)

APPLYING CONCEPTS

1. Suppose you observed some drawings of animals on a cave wall. What kinds of information might you learn from these drawings?
2. Describe two ways technology has contributed to the knowledge of biogenesis.
3. A farmer threw some old shirts into a corner of a barn in which corn was stored. Later, the farmer observed some baby mice nestled in the folds of the shirts. The farmer thought that the shirts had given rise to the mice. Propose a logical explanation for the farmer's observations. How would you test your hypothesis?
4. In one experiment Pasteur used flasks that had swan-shaped necks. He tipped some of the flasks so that broth ran into the low points of the necks and then ran back into the flasks. Pasteur later observed microorganisms in these flasks. Explain this observation.
5. Choose an animal that is commonly seen in your environment. Remember that insects are animals. Observe the animal and record behaviors that indicate it is exhibiting the characteristics of life you studied in this chapter.

EXTENDING CONCEPTS

1. Pasteur studied wine spoilage for wine growers in France. You might like to investigate this topic. How did Pasteur explain why wine sometimes spoiled?
2. Viruses are considered to be on the borderline of living and nonliving things. Use library materials to find out why viruses are difficult to classify. Keep in mind the characteristics of living things you learned about in this chapter.
3. The sensitive plant, *Mimosa pudica,* moves quickly. In this way it is unlike most other plants. You might like to grow this plant from seeds. After you have grown it and demonstrated its ability to move, find out how and why the plant moves so quickly when it is touched.

READINGS

Hay, John. *The Immortal Wilderness.* Norton, 1987.

Horowitz, Norman H. *To Utopia and Back: The Search for Life in the Solar System.* Freeman, 1986.

Johnson, Cathy. *The Local Wilderness: Observing Neighborhood Nature Through an Artist's Eye.* Prentice Hall, 1987.

Smith, John Maynard. *The Problems of Biology.* Oxford University Press, 1986.

Wallace, Bruce, and George M. Simmons Jr. *Biology for Living.* Johns Hopkins University Press, 1987.

3. Basic Chemistry

Few animals are as busy as a bee. A great deal of activity takes place in and around a beehive. The bees take nectar from flowers and carry it to their hive. In the bee's body the nectar undergoes a series of changes that result in the formation of honey. In this chapter, you will learn about the kinds of chemical substances and changes that are characteristic of living things.

CHAPTER OBJECTIVES

After completing this chapter, you will be able to

- **Describe** the current model of the structure of the atom.
- **Explain** chemical properties and reactions using the current model of the atom.
- **Identify** the structure, formation, and function of some important organic compounds
- **Distinguish** between physical and chemical changes in matter.
- **Relate** mass to energy.

The Nature of Matter

3•1 ATOMS

Matter is anything that has mass and takes up space. All matter is made up of the same basic unit—the atom. An **atom** is the smallest particle of matter that can exist and still have the properties of a particular kind of matter. The term *atom* is a Greek word that means "indivisible." The Greek thinker Democritus (460–370 B.C.) believed that matter could be subdivided only so many times. He called the smallest portion into which matter could be divided the atom because he considered it to be indivisible.

Early scientists believed that the atom was a single solid mass. Today, scientists know that the atom can be divided into smaller units. The atom is composed of many subatomic particles. The most important of these particles are protons, neutrons, and electrons. Each subatomic particle is distinguishable by its mass, its charge, and its location in the atom. A *proton* is a positively charged subatomic particle found in the *nucleus,* or central part, of an atom. A *neutron* is an uncharged subatomic particle also found in the nucleus of an atom. Each proton and neutron has a mass of one *atomic mass unit* (amu). An *electron* is a negatively charged subatomic particle with a much smaller mass than that of a proton or a neutron. An electron is in constant motion. An electron moves rapidly around the nucleus of the atom.

Over the past 100 years, scientists have suggested many hypotheses to explain the arrangement of the particles in the atom. These hypotheses have been used to develop models that illustrate the placement of each subatomic particle. A *model* is a representation of how something, such as an atom, looks or behaves. Many different models have been suggested for the atom. For example, at one time scientists believed that the electrons moved around the nucleus in orbits, much as the planets move in orbits around the sun.

SECTION OBJECTIVES

- Describe the current model of the atom.
- Distinguish between elements, ions, and compounds.
- Calculate the atomic number and mass number of an atom given the number of protons and neutrons.

38 BASIC CHEMISTRY

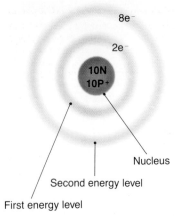

P+ – proton   N–neutron   e⁻ – electron

FIGURE 3·1 This is a model of a neon atom. It shows the number of electrons that surround the nucleus and the number of neutrons and protons in the nucleus.

The currently accepted model of the atom shows electrons moving around the nucleus in specific energy levels. An energy level is a region around the nucleus in which an electron moves. The exact location of an electron in an energy level is not known. However, electrons are more likely to be in some places than in others. Therefore, energy levels are described as electron clouds. An *electron cloud* is the space around the nucleus in which an electron is *likely* to be located. Because exact locations of electrons are not known, electrons in the figures in this text are indicated by the symbol e^- rather than by dots.

A model of an atom shows not only the number of electrons but also the number of protons and neutrons in the nucleus. Figure 3·2 shows the hydrogen, helium, and lithium atoms. Look at the hydrogen atom. How many protons and how many neutrons are in the hydrogen nucleus? How many protons and how many neutrons are in the lithium nucleus?

Compare the number of protons with the number of electrons in each of the atoms shown. You will observe that in each case the number of protons equals the number of electrons. Since the protons are positively charged, and the electrons are negatively charged, the net charge on each atom is zero.

FIGURE 3·2 Atoms of hydrogen, helium, and lithium differ in the number of subatomic particles they contain.

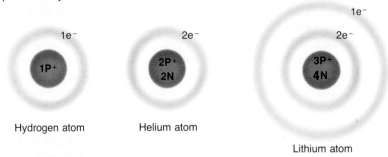

3·2 ELEMENTS

An **element** is matter that contains only one type of atom. There are 92 naturally occurring elements. About 14 others have been made by scientists in the laboratory. Figure 3·3 lists 10 of the most common elements found in the human body. Note that each element is represented by a symbol. A *symbol* is a shorthand method of representing an element. In most cases the symbol comes from the first letter of the English name of the element. For example, the symbol for nitrogen is N. The symbol for phosphorus is P. In some cases the symbol comes from two letters in the English name. Examples are Cl, for chlorine, and Mg, for magnesium. In other cases the symbol comes from the name for the element in another language. The symbol for gold, Au, comes from the Latin word for gold, *aurum*.

Look at the third column in Figure 3·3. You will notice that the first four elements listed—hydrogen, carbon, nitrogen, and oxygen—make up the greatest percentage (96.3 percent) of the human body. Biologically important substances almost always contain one or more of these elements.

Two terms are often used to describe an element. One of these terms is the atomic number. The **atomic number** of an element is the number of protons in the nucleus of the element. Each element has a unique atomic number. The other term used to describe an element is mass number. The **mass number** is the number of protons and neutrons. Look at Figure 3·2. Helium has two protons and two neutrons. So the atomic number of helium is two. Its mass number is four. What is the atomic number of lithium? What is its mass number? As you can see in the following statements, atomic number and mass number provide much information about an atom.

- **The atomic number is equal to the number of protons.**
- **Since all atoms are electrically neutral, the number of protons is equal to the number of electrons.**
- **The mass number is equal to the number of protons plus the number of neutrons.**
  or
  The number of neutrons is equal to the mass number minus the atomic number.

Figure 3·4 shows some of the elements found in the human body. Apply the above statements to determine the number of protons, electrons, and neutrons in an atom of carbon. You will find that there are six protons, six electrons, and six neutrons. Now determine the number of protons, electrons, and neutrons in phosphorus.

FIGURE 3·3 Some of the elements in the human body, their symbols, and their percents by mass are shown in this table. Oxygen makes up the greatest percentage of the human body.

ELEMENT	SYMBOL	MASS (%)
hydrogen	H	9.5
carbon	C	18.5
nitrogen	N	3.3
oxygen	O	65.0
fluorine	F	trace
sodium	Na	0.2
magnesium	Mg	0.1
silicon	Si	trace
phosphorus	P	1.0
sulfur	S	0.3

FIGURE 3·4 These are examples of some of the elements found in the human body. In the body these elements are combined with other elements.

Magnesium
Atomic number 12
Mass number 24

Carbon
Atomic number 6
Mass number 12

Phosphorus
Atomic number 15
Mass number 31

Sulfur
Atomic number 16
Mass number 32

Many elements exist in several forms. The forms exist as atoms with the same number of protons but different numbers of neutrons. Because the numbers of neutrons are different, the mass numbers of such atoms also differ. Atoms of the same element that have different mass numbers are called **isotopes** (ī suh tohps).

iso- (equal)
topos (place)

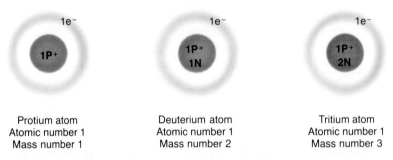

FIGURE 3·5 Protium, deuterium, and tritium are three isotopes of hydrogen. These isotopes differ from each other in the number of neutrons in the nucleus.

Figure 3·5 shows models of the three isotopes of hydrogen. Each isotope has one electron and one proton. The forms differ in the number of neutrons and therefore in the mass number. The most common isotope, protium (PROH tee uhm), has a mass number of 1 because it has one proton and no neutrons. The isotope called deuterium (doo TIHR ee uhm) has a mass number of 2. It has one proton and one neutron. Tritium (TRIHT ee uhm) has a mass number of 3. How many neutrons does it have?

The isotopes of some elements are not stable. That is, the nuclei of such isotopes change spontaneously into different nuclei by a process called radioactive decay. For example, some uranium atoms change into atoms of lead. Some radioactive isotopes have practical applications. They are used to treat cancer, develop new drugs, and trace chemical reactions in cells and in organisms.

3·3 IONS

If the number of positive charges (protons) equals the number of negative charges (electrons), the atom is neutral. Atoms, however, can gain or lose electrons. An atom that has gained one or more electrons has a negative charge. An atom that has lost one or more electrons has a positive charge. An **ion** is an atom that has gained or lost one or more electrons and so has a negative or positive charge.

How do ions form? Energy levels of atoms have maximum numbers of electrons they can hold. The first energy level, which is nearest the nucleus, can hold a maximum of two electrons. The second energy level, which is farther from the nucleus, can hold a maximum of eight electrons. The number of electrons in an atom's outer energy level influences whether an atom will gain or lose electrons.

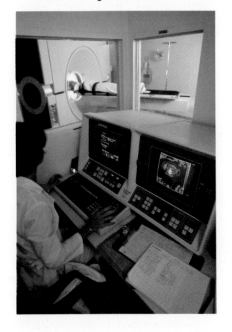

FIGURE 3·6 Radioactive materials have many uses in medicine. This CAT scan (computer axial tomography) machine can make thousands of images of the human body. The images can be studied for signs of disease.

An element such as chlorine readily forms a negative ion. Figure 3·7 shows a model of the neutral chlorine atom. Notice that there are seven electrons in this atom's outer energy level. The chlorine atom is electrically neutral. Under certain conditions the chlorine atom gains one electron and becomes negatively charged. The chlorine atom thus becomes a negatively charged ion. This ion is called the chlor*ide* ion. The shorthand method for indicating the negative chloride ion is Cl^-. A model of the chloride ion is also shown in Figure 3·7. How does the chloride ion differ from the chlorine atom?

FIGURE 3·7 Compare the chlorine atom with the chloride ion. The chloride ion has gained an electron. It is a negative ion.

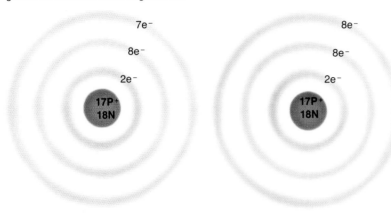

Neutral chlorine atom (Cl)   Negatively charged chloride ion (Cl^-)

A sodium atom is neutral. However, the element sodium readily forms positive ions. Notice in Figure 3·8 that the neutral sodium atom has one electron in its outer energy level. Under certain conditions, the atom loses this one electron. Then the outer energy level has eight electrons, the maximum number of electrons it can hold.

When a sodium atom loses one electron, the atom has a net positive charge. It is then called a sodium ion. The shorthand method for indicating the positive sodium ion is Na^+. Figure 3·8 compares the neutral sodium atom with the positively charged sodium ion.

3·4 COMPOUNDS

When certain atoms give up electrons and others gain electrons, positive and negative ions are formed. Positive and negative ions tend to attract each other. The sodium ion and the chloride ion are oppositely charged and attract each other. The sodium atom gives up the one electron in its outer energy level, and the chlorine atom picks up this electron. A sodium ion and a chloride ion are formed. These two ions attract each other and bond, or join, to form a new substance.

The new substance is called a compound. A **compound** is matter composed of two or more elements chemically bonded. The formation

FIGURE 3·8 Compare the sodium atom with the sodium ion. The sodium ion has lost an electron. It is a positive ion.

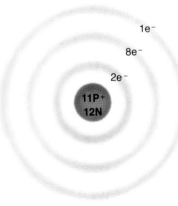

Neutral sodium atom (Na)

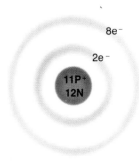

Positively charged sodium ion (Na^+)

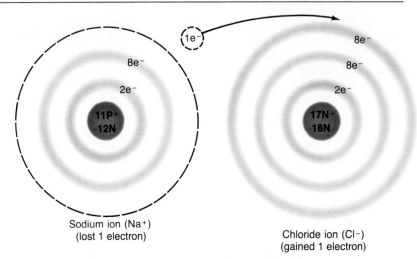

Sodium ion (Na+)
(lost 1 electron)

Chloride ion (Cl−)
(gained 1 electron)

FIGURE 3·9 An ionic compound forms when sodium and chlorine combine chemically. An electron is transferred when this bond is formed.

of the compound sodium chloride is shown in Figure 3·9. When the sodium ion and the chloride ion bond, an ionic compound is formed. An **ionic compound** is a compound formed when two or more ions chemically bond. The bond formed is called an ionic bond. An **ionic bond** is formed when atoms transfer electrons.

3·5 MOLECULES

Compounds can also be formed when atoms share electrons. A chemical bond formed when electrons are shared is called a **covalent** (koh VAY luhnt) **bond.** A compound formed when electrons are shared is a **covalent compound.** Two or more atoms joined by one or more covalent bonds form a molecule. A **molecule** consists of two or more atoms covalently bonded.

Some elements exist in nature as molecules. These molecules contain two or more of the same kind of atoms covalently bonded. Hydrogen gas is an example of an element that forms a covalently bonded molecule. It is a diatomic molecule, which means that it contains two atoms of the same kind. Figure 3·10 shows a model of a hydrogen molecule. Notice that each atom has one electron located in the first energy level. The two atoms share the electrons. The shared electrons travel around the nuclei of both atoms.

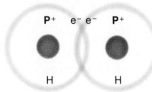

FIGURE 3·10 This molecule of hydrogen is formed when two atoms of hydrogen share an electron. This is a covalent bond.

REVIEW QUESTIONS

1. Describe the kinds of information that may be obtained from the current model of the atom.
2. Define the term *isotope*. Explain how protium differs from tritium.
3. Describe how the chlorine atom differs from the chloride ion.
4. Distinguish between an ionic bond and a covalent bond.
5. Bromine has 35 protons and 44 neutrons. What is its mass number?

Interactions of Matter

3·6 CHEMICAL FORMULAS

Scientists use chemical formulas to represent compounds and molecules. A *chemical formula* is a combination of chemical symbols and numbers that indicate the number and kinds of units (atoms or ions) in a molecule or a compound. For example, H_2 is the formula for a molecule of hydrogen gas. The subscript 2 means that there are two atoms of hydrogen in this molecule. The formula H_2O means that there are two hydrogen atoms and one oxygen atom in a molecule of water. The formula $C_6H_{12}O_6$ represents a molecule of glucose, a sugar. How many of each kind of atom are present in this molecule?

Any chemical formula may be preceded by a number called a coefficient. A *coefficient* indicates the number of molecules. In the formula $3H_2O$, the coefficient 3 means three molecules of water.

Scientists have developed the *structural formula,* which shows atoms and the bonds that hold the atoms together. Some structural formulas are shown below. The line between the H's represents a single bond, the pair of shared electrons that forms the covalent bond between the atoms of hydrogen. Oxygen gas, O_2, has six electrons in the outer energy level. The two lines between the O's represent a double bond, or two pairs of shared electrons. Nitrogen gas, N_2, has five electrons in the outer energy level. The three lines between the N's represent a triple bond, or three pairs of shared electrons.

$$H-H \qquad O=O \qquad N\equiv N$$
Hydrogen molecule   Oxygen molecule   Nitrogen molecule

3·7 CHEMICAL REACTIONS AND EQUATIONS

Most elements have unfilled outer energy levels. Such atoms can lose, gain, or share electrons to achieve full energy levels. There is a tendency for electrons to be transferred or shared to achieve stability. A stable atom is one that has a filled outer energy level.

A **chemical reaction** is a change in which one or more new substances form. A chemical reaction occurs when electrons are transferred between atoms or become shared by atoms. For example, electrons may be transferred between the elements magnesium and chlorine. When magnesium and chlorine react, magnesium chloride forms. Magnesium, which has two electrons in its outer energy level, transfers the two electrons and becomes an ion with a +2 charge. Chlorine, which has seven electrons in its outer energy level, will take on one electron. Two chlorine atoms are needed to accept both electrons from magnesium. The formula for magnesium chloride is $MgCl_2$. This formula means that one magnesium ion bonds with two chloride ions to form magnesium chloride.

SECTION OBJECTIVES

- Describe some basic interactions of matter.
- Identify kinds of mixtures.
- Explain the pH scale and identify the pH of some common substances.

FIGURE 3·11 The burning of magnesium is an example of a chemical reaction. The light produced by burning magnesium is so bright that early photographs were taken with flashbulbs filled with shredded magnesium.

BASIC CHEMISTRY

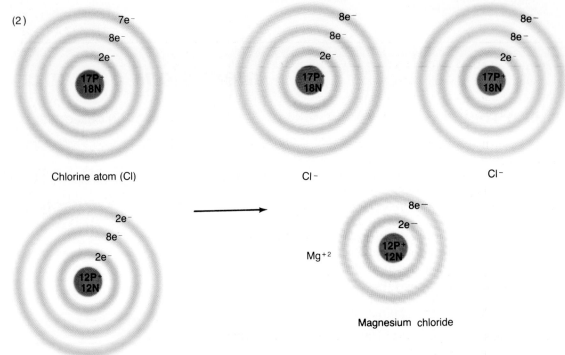

FIGURE 3·12 The formation of magnesium chloride is an example of a chemical reaction. The bonds formed in this reaction are ionic bonds.

Scientists use a shorthand method to write chemical reactions. A **chemical equation** is the symbolic language used to describe a chemical reaction. The chemical reaction to show the formation of magnesium chloride is below.

$$\text{Mg} + \text{Cl}_2 \longrightarrow \text{MgCl}_2$$
(magnesium)   (chlorine) yields (magnesium chloride)

An important part of understanding biology is knowing how to read chemical equations. Chemical equations will be used in this book to describe many processes that occur in living things.

3·8 MIXTURES

Sometimes substances are mixed together but are not chemically combined. A **mixture** is two or more distinct substances that are mixed together but not chemically combined. A salad is an example of a mixture. Each substance in a mixture retains its own properties. Another characteristic of a mixture is that the amount of each substance within the mixture can vary. In a salad, for example, the amount of carrots or the amount of lettuce can vary. Most of the matter on earth is in the form of mixtures. Air, ocean water, and soil are all mixtures.

Mixtures can be described as being either homogeneous (hoh-muh JEE nee uhs) or heterogeneous (heht uhr uh JEE nee uhs). A *homogeneous mixture* is one in which the substances are evenly distributed throughout the mixture. A *heterogeneous mixture* is one in which the substances are not evenly distributed.

FIGURE 3·13 This is an example of a heterogeneous mixture. You can see that the quartz and gold in this rock are not evenly distributed.

44   BASIC CHEMISTRY

A **solution** is an example of a homogeneous mixture. In a solution one substance is dissolved in another. The substance that is dissolved is the **solute** (SAHL yoot). The substance in which the solute is dissolved is the **solvent** (SAHL vuhnt). Particles in a solution do not settle out.

Water is one of the most common solvents. It is the solvent in which most chemical reactions occur in living things. Some covalent compounds, such as sugar, can dissolve in water. In the human body, sugar molecules are dissolved in body fluids, such as blood.

Many ionic compounds also dissolve in water. When an ionic compound dissolves in water, the individual ions of the ionic compound separate from each other. The separation of the positive and negative ions of an ionic compound that is dissolved in water is called *dissociation* (dih soh see AY shuhn). Sodium chloride (table salt) that is dissolved in water dissociates into sodium ions and chloride ions.

$$NaCl \longrightarrow Na^+ + Cl^-$$

Most ionic compounds in the blood and in the cells are dissociated and are present in the form of ions.

A **suspension** is a heterogeneous mixture that contains particles distributed within a liquid, gas, or solid. The particles in a suspension are too large to remain evenly distributed and will settle out unless the suspension is stirred constantly. Soil mixed in water forms a suspension. After a few hours most of the particles of soil will settle out.

Another kind of mixture is called a colloid (KAHL oid). A **colloid** is a mixture composed of particles dispersed in a medium. The particles are intermediate in size between those in a solution and those in a suspension. They do not settle out. Gelatin in water forms a colloid. A colloid such as gelatin can exist in one of two states: sol or gel. In the sol state, the colloid is a liquid. In the gel state, the colloid forms a jellylike material.

3•9 ACIDS AND BASES

Most chemical reactions in living things take place in water in the cell. The nature of water affects these chemical reactions. Water exists mainly as molecules. However, a small number of water molecules dissociate, or separate, into two ions: a positive hydrogen ion (H^+) and a negative hydroxide ion (OH^-). The dissociation of water may be expressed as follows:

$$H_2O \longrightarrow H^+ + OH^-$$

Water dissociates into hydrogen ions and hydroxide ions. In pure water, the number of hydrogen ions and the number of hydroxide ions are equal. A solution in which the number of hydrogen ions equals the number of hydroxide ions is called a *neutral solution.*

Other compounds also dissociate. For example, hydrogen chloride (HCl) dissociates into hydrogen ions (H^+) and chloride ions (Cl^-). A chemical that produces hydrogen ions in a water solution is

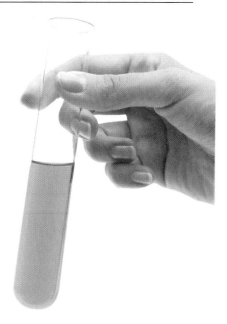

FIGURE 3•14 A solution is a homogeneous mixture in which one substance is dissolved in another.

kolla (glue)
-oid (like)

46 BASIC CHEMISTRY

an **acid**. When HCl gas is added to water, hydrochloric acid (HCl) is formed. Acids can be found in the body. For example, hydrochloric acid is released in the stomach.

Sodium hydroxide (NaOH) dissociates into sodium ions (Na^+) and hydroxide ions (OH^-). A water solution of sodium hydroxide has more hydroxide ions than hydrogen ions. A chemical that produces hydroxide ions in a water solution is a **base**. Therefore, sodium hydroxide is a base.

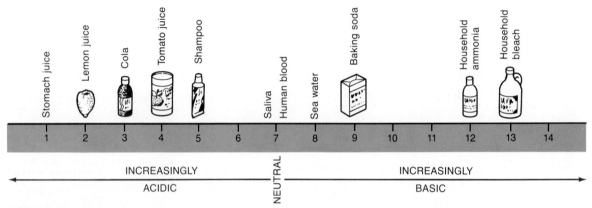

FIGURE 3·15 This pH scale shows the pH of several common substances. As you can see, the juice produced in your stomach is a strong acid.

Scientists have devised a scale for measuring how acidic or basic various solutions are. This scale, called the pH scale, indicates the concentration of hydrogen ions in a solution. The pH scale ranges from 0 to 14. Any solution with a pH of 7 is neutral; the number of hydrogen ions equals the number of hydroxide ions. A solution with a pH below 7 is acidic; the number of hydrogen ions is greater than the number of hydroxide ions. A solution with a pH above 7 is basic; the number of hydroxide ions is greater than the number of hydrogen ions. The lower the pH number, the more acidic the solution. The higher the pH number, the more basic the solution.

Figure 3·15 shows the pH scale and some common substances. What is the pH of lemon juice? Is it an acid or a base? What is the pH of baking soda? Is it an acid or a base?

REVIEW QUESTIONS

6. Define the term *chemical formula*. Explain what is meant by $C_6H_{12}O_6$.

7. Explain what is meant by $Mg + Cl_2 \longrightarrow MgCl_2$. What is a statement like this one called?

8. Give one example of each of the following: (a) a solution, (b) a colloid, (c) a suspension.

9. Define *acid;* define *base*. Give an example of each.

10. Indicate which of the following are likely to be bases and which acid: (a) KOH, (b) HNO_3, (c) NH_4OH, (d) HBr.

Investigation

How Can the Relative Strength of Two Bases Be Determined?

Goals
After completing this activity, you will be able to
- Describe one way in which the relative strength of two bases can be determined.
- Describe how an indicator can be used to determine when a base has been neutralized.

Materials (for groups of 3)
vinegar   red litmus paper
blue litmus paper
two 100-mL beakers   dropper
250-mL beaker   base A   base B
indicator   three 100-mL graduates
goggles   glass-marking pencil

Procedure

A. Wear safety goggles.

B. Place 50 mL of base A in a 100-mL beaker. With a glass-marking pencil, label the beaker *Base A*. Test the base with litmus paper. (Blue litmus paper turns red when placed in an acidic solution. Red litmus paper turns blue when placed in a basic solution.) On a separate sheet of paper, copy the chart shown below and record the color of the litmus paper.

CAUTION: If chemicals spill on your skin or clothing, flush the affected area with running water.

C. Add 15 mL of the indicator to the beaker. (An indicator is a substance that shows the presence of another substance by changing color.) Record the color of the solution.

D. Test the solution in the beaker with the red litmus paper and the blue litmus paper. Record the color of the litmus paper.

CHANGES IN LITMUS PAPER	RED	BLUE
Base A		
Base A + indicator		
Base B		
Base B + indicator		
Vinegar		

COLOR OF SOLUTION	
Base A only	
Base A + indicator	
Base B only	
Base B + indicator	

E. Test the vinegar with the red litmus paper and the blue litmus paper. Record the color of the paper.

F. The strength of a base can be expressed as the amount of acid needed to neutralize it. You will add vinegar to the base until it is neutralized. Swirl the liquid in the beaker. With the dropper add 1 drop of vinegar at a time until the solution becomes colorless. In a chart on a separate sheet of paper, record the number of drops of vinegar needed to turn the solution colorless and keep it colorless.

DROPS OF VINEGAR TO NEUTRALIZE BASE	
Base A	
Base B	

G. Test the solution in the beaker with the litmus paper and record the color of the strips of paper.

H. Repeat steps A–G for base B.

I. Follow your teacher's instructions for the proper disposal of all materials.

Questions and Conclusions

1. Compare the color of the red and blue litmus paper when it is placed in base A, base B, and vinegar.
2. How many drops of vinegar were needed to neutralize base A? Base B?
3. According to your results, which base (A or B) was stronger? Explain the reason for your answer.
4. How could an indicator be used to find the amount of base needed to neutralize an acid?
5. People with indigestion often have excess stomach acid. How is this condition often treated?

BASIC CHEMISTRY

Compounds in Living Things

SECTION OBJECTIVES

- Distinguish between inorganic and organic compounds.
- Identify the structure, formation, and function of carbohydrates, lipids, proteins, and nucleic acids.

3·10 INORGANIC AND ORGANIC COMPOUNDS

Compounds can be divided into two groups—those that contain carbon and those that do not. Traditionally, compounds that contain carbon are called **organic compounds**. Compounds that lack carbon are generally called *inorganic compounds*. However, there are some compounds that contain carbon and are referred to as inorganic compounds. Carbon dioxide (CO_2) is an example of an inorganic compound that contains carbon.

People often think of the term *organic* as referring to living things and the term *inorganic* as referring to nonliving things. However, many inorganic compounds are important to life. Water and oxygen are two examples.

Water (H_2O) is an inorganic compound that is essential to life on the earth. In fact water is the most abundant compound in living organisms. The bodies of living organisms are 65 to 95 percent water. In the cells, water is the solvent in which chemical reactions take place. In photosynthesis, a food-manufacturing process that occurs in green plants, water combines with carbon dioxide to form sugar. Within the bodies of many organisms, water serves as a transport medium, distributing materials from one place to another in the body. Water also dilutes many of the harmful wastes produced as a result of metabolism. Because water changes temperature less rapidly than most other substances, it helps to regulate body temperature. When scientists investigate other planets for the possibility of life, one of the materials they look for is water.

Oxygen (O_2) is an inorganic molecule that is necessary for life in many living things. For most organisms, oxygen is needed in the cellular process in which foods are broken down for the release of energy.

The original source of carbon in living organisms is carbon dioxide. Using carbon dioxide and water as raw materials, plants make carbon compounds, such as sugars. Animals eat plants that contain these carbon compounds and, in turn, incorporate carbon into their own body tissues. Meat-eating animals obtain the carbon by eating animals that have eaten plants.

3·11 BONDS IN ORGANIC COMPOUNDS

Carbon has four electrons in its outer energy level. Because of this arrangement of electrons, carbon forms compounds by sharing its four outer electrons. Carbon forms four covalent bonds when it bonds with other elements. The most common elements with which carbon bonds are hydrogen, oxygen, and nitrogen. Figure 3·17 illustrates the number of covalent bonds each of these atoms forms. Carbon atoms also readily bond with other carbon atoms.

FIGURE 3·16 Carbon exists in several different forms. Graphite in a pencil is pure carbon, as is diamond. The value of these different forms of carbon varies greatly.

BASIC CHEMISTRY

Carbon
4 covalent bonds

Hydrogen
1 covalent bond

Oxygen
2 covalent bonds

Nitrogen
3 covalent bonds

One of the simplest carbon compounds is methane (CH_4). The structural formula of a molecule of methane is shown below. How many covalent bonds are in a methane molecule?

FIGURE 3·17 Covalent bonds formed by carbon, hydrogen, oxygen, and nitrogen are shown in this diagram. The number of covalent bonds an atom of an element can form depends upon the number of electrons in the outer energy level of the atom.

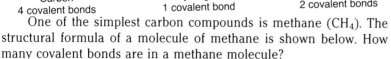

Methane (CH_4)

Butane (C_4H_{10})

Isobutane (C_4H_{10})

Carbon atoms can easily bond to each other, forming chains of carbon atoms. Look at the butane and isobutane structural formulas. Both butane and isobutane have the same formula, C_4H_{10}, but they have different structures. Butane and isobutane are different compounds, each with its own distinct characteristics. Compounds such as these are called isomers (ī suh muhrz). **Isomers** are compounds with the same molecular formula but different molecular structures. Isomers may have different chemical properties.

3·12 CARBOHYDRATES

Organic compounds in living organisms contain the elements carbon, hydrogen, oxygen, and, in some cases, nitrogen. The compounds form four groups: carbohydrates, lipids, proteins, and nucleic acids.

Carbohydrates (kahr boh HĪ drayts) are organic compounds that are made of carbon, hydrogen, and oxygen. Carbohydrates are the body's main sources of energy. Sugars and starches are carbohydrates. The simplest carbohydrates are the **monosaccharides** (mahn uh SAK uh rīdz), or single sugars. The most common monosaccharides have the formula $C_6H_{12}O_6$ and are isomers. Notice that the ratio of hydrogen atoms to oxygen atoms is 2:1. This 2:1 ratio is characteristic of carbohydrates.

Glucose is a monosaccharide with the formula $C_6H_{12}O_6$. Produced by green plants during photosynthesis, it supplies the energy for a cell's metabolic activities. In the human body, glucose is the sugar found dissolved in the blood. The average adult has about 5 g of glucose in the bloodstream. This is enough of a supply of glucose to provide the energy the body needs for about 15 minutes. The body keeps receiving a steady supply of glucose from food.

FIGURE 3·18 Carbohydrates are organic compounds made of carbon, hydrogen, and oxygen. All of the foods shown here contain carbohydrates.

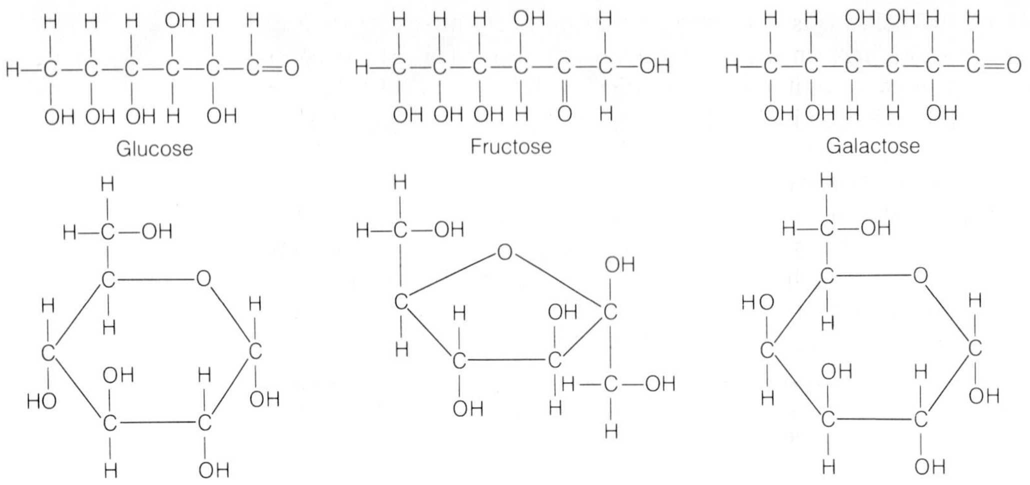

FIGURE 3·19 Each structural formula represents a molecule of a simple sugar. Notice that all of the simple sugars have the same formula, $C_6H_{12}O_6$.

Fructose and galactose are two other monosaccharides. These sugars are found in fruits, honey, and vegetables. Because they both have the formula $C_6H_{12}O_6$, they are isomers of glucose. Look at Figure 3·19. How do glucose, fructose, and galactose molecules differ?

The disaccharides (dī SAK uh rīdz) make up another group of carbohydrates. **Disaccharides** are double sugars produced when two monosaccharides chemically combine. One common disaccharide is *lactose,* which is composed of the sugars glucose and galactose. Lactose, called milk sugar, is found in the milk of mammals. Two other disaccharides are *sucrose* and *maltose.* Sucrose, commonly called table sugar, is composed of glucose and fructose. Maltose is composed of two glucose molecules.

di- (two)
saccharum (sugar)

The most common disaccharides have the formula $C_{12}H_{22}O_{11}$. They are produced as a result of the following chemical reaction.

$$\underset{\begin{pmatrix}\text{mono-}\\\text{saccharide}\end{pmatrix}}{C_6H_{12}O_6} + \underset{\begin{pmatrix}\text{mono-}\\\text{saccharide}\end{pmatrix}}{C_6H_{12}O_6} \longrightarrow \underset{\begin{pmatrix}\text{di-}\\\text{saccharide}\end{pmatrix}}{C_{12}H_{22}O_{11}} + \underset{(\text{water})}{H_2O}$$

poly- (many)

A third group of carbohydrates is composed of the polysaccharides (pahl ee SAK uh rīdz). **Polysaccharides** are large molecules formed when many monosaccharides bond together. The most common polysaccharide is *starch,* which is made up of many glucose molecules bonded together. Although starch is made up of glucose molecules, it is not sweet and does not have the characteristics of sugar. Starch is produced by plants. Plants store starch as an energy source. The energy that a sprouting plant needs for growth comes from starch stored in the seed.

Another common polysaccharide is *glycogen* (GLĪ kuh juhn), which is similar to starch but is found in certain animal cells. Like starch, glycogen is made up of glucose molecules. When the sugar level in animal cells is low, glycogen molecules are broken down into glucose molecules. In the human body, glycogen may be stored in muscles, but most of it is stored in the liver. When the body's glucose level is low, glycogen is changed into glucose in the liver and released into the bloodstream.

Cellulose (SEHL yuh lohs) also is a polysaccharide. Like starch and glycogen, cellulose is formed of many glucose molecules linked together. The cellulose molecule is much larger and more complex than either the starch or glycogen molecule. Cellulose is a tough, stringy substance that forms the cell wall that surrounds a plant cell. The cellulose in plant cells gives plants their rigid structure.

FIGURE 3·20 When glucose and fructose are joined together, sucrose, or table sugar, is formed. Cellulose is formed of a long chain of glucose molecules.

3·13 LIPIDS

Lipids are organic compounds that store energy. They include fats and oils. Like carbohydrates, lipids are composed of carbon, hydrogen, and oxygen. However, in lipids the ratio of hydrogen to oxygen atoms is greater than 2:1. You can recognize lipids because they are greasy, oily, or waxy. The body uses lipids as sources of energy. Substances such as fats and cooking oils are lipids. Many lipids in cells are large molecules that form when three fatty acid molecules and one glycerol molecule combine.

A *fatty acid molecule* is composed of a long chain of carbon and hydrogen atoms chemically bonded to a carboxyl group. A *carboxyl group* consists of a carbon atom linked to an oxygen atom and a hydroxyl (−OH) group. It is represented as −COOH.

The structural formula of a fatty acid is shown in Figure 3·21. To the right is the chain of carbon and hydrogen atoms, identified as a hydrocarbon chain. You will notice a series of dots in the middle of the chain. These dots indicate that the number of carbons in the hydrocarbon chain varies. Find the carboxyl group in the fatty acid molecule. It is the carboxyl group that makes the molecule acidic.

Glycerol (GLIHS uh rohl) is the other kind of molecule found in a lipid molecule. Glycerol is a three-carbon molecule that has a hydroxyl (−OH) group attached to each carbon atom.

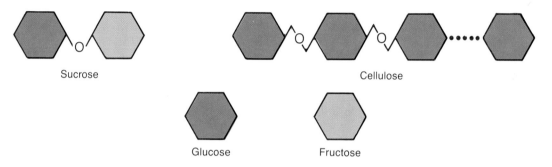

FIGURE 3·21 This illustration represents the structural formula of a fatty acid. The number of carbon atoms, which can vary, determines the identity of the fatty acid.

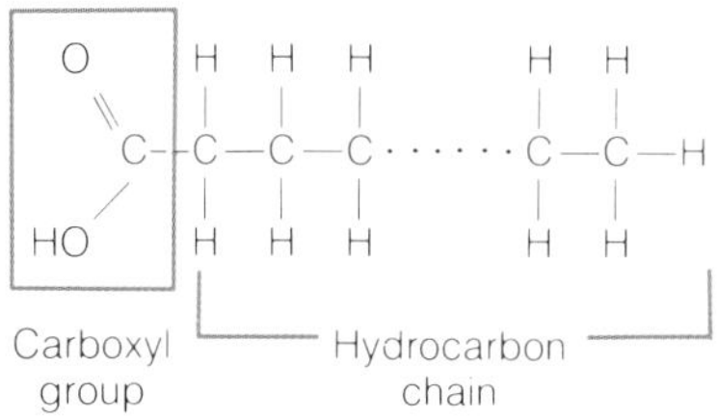

Both fats and oils are lipids. Fats are usually of animal origin. Fats have single bonds between the carbon atoms of the hydrocarbon chain. Lipids with single bonds between carbon atoms are saturated lipids. Oils are usually of plant origin. Oils have double bonds between some carbon atoms of the hydrocarbon chain. Lipids with double bonds between carbon atoms are unsaturated lipids.

3·14 PROTEINS

A major group of organic compounds is proteins. **Proteins** are large, complex organic compounds that contain the elements carbon, hydrogen, oxygen, and nitrogen. Some proteins also contain sulfur. Proteins are very abundant organic compounds in cells. Some proteins form part of a cell's structure. Other proteins regulate and control a cell's functions.

Proteins contain hundreds of repeating units known as amino (uh-MEE noh) acids. **Amino acids** are the building blocks of proteins. About 20 different amino acids are used to make proteins, but all have the same core structure. This core structure is illustrated in Figure 3·22. Each amino acid has an amino group ($-NH_2$), a carboxyl group ($-COOH$), and a hydrogen atom bonded to a central carbon atom. In every amino acid there is also another atom or group of atoms bonded to the central carbon. This side group, which is represented by the letter R, is what makes each amino acid different.

FIGURE 3·22 The core structure of an amino acid consists of an amino group, a carboxyl group, a hydrogen atom bonded to a central carbon atom, and a side group. A chain of amino acids makes up a protein molecule.

Profiles

Lloyd N. Ferguson (1918–   )

Dr. Lloyd Noell Ferguson is a man with two primary interests in his life: chemistry and sharing his knowledge of chemistry with others through teaching. Throughout his career, Dr. Ferguson has pursued excellence in both of these areas.

Lloyd Ferguson was born in Oakland, California, in 1918. He graduated from the University of California at Berkeley in 1940 with a bachelor's degree in chemistry. He received a doctorate in chemistry in 1943 from the same institution.

Dr. Ferguson began his teaching career at Howard University in Washington, D.C., where he helped to establish the first Ph.D. program in the university's history. He traveled to Kenya, where he helped to establish the chemistry department at the University of Nairobi. From 1965 to the present, Dr. Ferguson has been a professor of organic chemistry at California State University in Los Angeles.

As a teacher, Dr. Ferguson has received numerous awards in acknowledgment of his excellence and service in the field. Among these awards are an excellence in teaching award from the Chemical Manufacturers Association in 1974, the National Organization of Black Chemists and Chemical Engineers Outstanding Teaching Award in 1979, and the Outstanding Professor Award from the California State University System.

Dr. Ferguson has authored many articles for professional journals as well as six college textbooks. Additionally, Dr. Ferguson is deeply involved in research—into the use of chemotherapy as a treatment for certain types of cancer and into the chemical processes involved in the sense of taste.

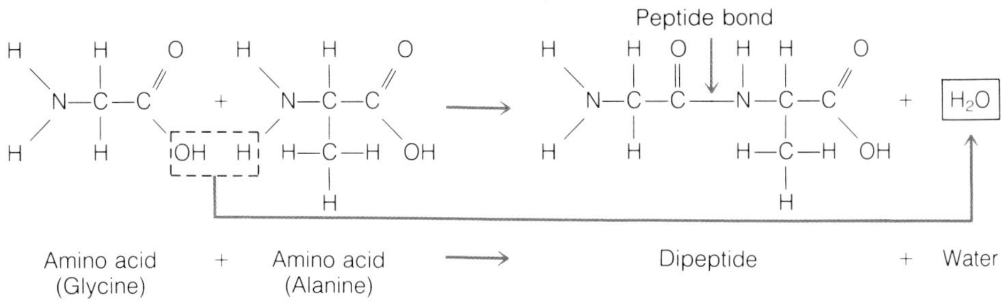

FIGURE 3·23 A dipeptide molecule forms when the carboxyl group of one amino acid bonds with the amino group of another amino acid.

Amino acids bond by the process shown in Figure 3·23. The carboxyl group of one amino acid bonds with the amino group of another amino acid to form a dipeptide. A *dipeptide* (dī PEHP tīd) is an organic compound formed when the carboxyl group of one amino acid bonds with the amino group of another amino acid. The covalent bond between the amino acids is called a *peptide bond.* Many amino acids may bond together to form a large molecule called a *polypeptide* (pahl ee PEHP tīd). A polypeptide usually has between 6 and 40 amino acids. A protein molecule is longer and may consist of hundreds of amino acids bonded together.

Proteins exist in a great variety of forms. Yet all proteins are built of combinations of 20 different amino acids. How can such a small number of amino acids give rise to so many different types of proteins? Proteins differ because of the differences in number and arrangement of amino acids. A small change in the order of amino acids can produce a major change in the structure and function of a protein.

3·15 NUCLEIC ACIDS

Nucleic (noo KLEE ihk) acids are another group of organic compounds. **Nucleic acids** are the hereditary materials of cells. Nucleic acids are among the largest and most complex of the organic compounds. They contain the elements carbon, hydrogen, oxygen, nitrogen, and phosphorus. There are two types of nucleic acids: deoxyribonucleic acid, or DNA, and ribonucleic acid, or RNA. These two nucleic acids store genetic information and control the production of proteins in the cells.

REVIEW QUESTIONS

11. Name three inorganic compounds that are essential to life.
12. Explain why butane and isobutane are isomers.
13. Define the term *carbohydrate* and give two examples.
14. What polysaccharide is found in seeds? How do plants use it?
15. The composition of an organic compound is found to be 41 percent carbon, 7 percent hydrogen, 33 percent oxygen, and 19 percent nitrogen. Is this organic compound most likely a carbohydrate, a lipid, or a protein?

Energy and Matter

SECTION OBJECTIVES

- State the Law of Conservation of Mass—Energy.
- Distinguish between potential and kinetic energy.

FIGURE 3·24 Melting ice is an example of a physical change. The burning marshmallow is an example of a chemical change.

3·16 CHANGES IN THE STATE OF MATTER

Matter is anything that has mass and occupies space. In living organisms, matter undergoes physical or chemical changes. A *physical change* is one in which the state, color, size, or texture of a substance changes but the kind of substance remains the same. For example, the evaporation of water is a physical change. When the water boils it changes from a liquid to a gas, but it is still water. In the same way, liquid water turns to solid ice when it freezes. How is water changing in Figure 3·24?

A *chemical change* is one in which matter changes from one substance to another. For example, the production of starch from molecules of glucose is an example of a chemical change. The number and arrangement of atoms in molecules of starch differ from those in molecules of glucose. Chemical changes involve the rearrangement of atoms. Chemical bonds are broken and new ones are formed. A totally new substance with different properties is produced during a chemical change. Figure 3·24 (bottom) also shows a chemical change. Why is the toasting of a marshmallow considered to be an example of chemical change?

3·17 CONSERVATION OF MASS-ENERGY

A chemical reaction can be expressed in the form of a chemical equation. The substances that react with each other are the **reactants.** The substances that are produced as a result of the reaction are the **products.** The total number of each kind of atom of the reactants on the left side of the equation must equal the total number of atoms of the products on the right side.

During chemical reactions, the amount of matter remains the same. The total quantity of the reactants equals the total quantity of the products. This phenomenon is called the *Law of Conservation of Mass:* During a chemical reaction, mass is neither created nor destroyed. When writing a chemical equation, the Law of Conservation of Mass is followed.

In chemical reactions, energy also is neither created nor destroyed. This phenomenon is summed up in the *Law of Conservation of Energy:* The total quantity of energy in a chemical reaction is always the same; energy is neither created nor destroyed.

In the early 1900s, Albert Einstein showed that mass can be changed to energy in nuclear reactions. According to Einstein's work, mass and energy are equivalent. Scientists had to develop a new law that combines the two conservation laws previously defined. This new law is the *Law of Conservation of Mass-Energy:* Mass and energy always are conserved, and their sum cannot be increased or decreased.

FIGURE 3·25 Potential energy is high when the jumper is at the top of the bar. Kinetic energy is high as the jumper begins to fall towards the earth.

3·18 TYPES OF ENERGY

All changes in matter involve energy. Energy is the capacity to do work or to cause change. Energy is classified as either potential or kinetic. *Potential energy* is energy of position. *Kinetic energy* is energy of motion. Look at Figure 3·25, which shows a high jumper just as he crosses the bar and after he has begun to fall back to the ground. The energy the high jumper needed to cross the bar was stored as potential energy. At the moment the high jumper hovered over the bar, he had high potential energy because of his position. As he began to fall, his potential energy was converted to kinetic energy.

Energy exists in many forms. Heat, radiant energy, chemical energy, electrical energy, and nuclear energy are some examples. Energy can be changed from one form to another. As a marshmallow burns, chemical energy is changed to heat energy and light energy. Stored chemical energy is an example of potential energy.

Energy conversions are continuously occurring inside the cells. For example, chemical energy is changed to mechanical energy when a muscle cell contracts. During photosynthesis, the sun's radiant energy is converted to chemical energy and stored in the glucose molecules. When glucose is broken down, much of the stored chemical energy is released as heat. The energy changes that take place in cells are necessary for an organism to carry out all of its life processes.

REVIEW QUESTIONS

16. Distinguish between a physical change and a chemical change.
17. Name the components of a chemical reaction.
18. State the Law of Conservation of Mass-Energy.
19. Distinguish between potential energy and kinetic energy.
20. Suppose 20 kg of oxygen is used to burn 20 kg of wood. Energy is given off as heat and light. If all the ashes and gases are collected, how would this mass compare with that of the original substances?

BASIC CHEMISTRY

Chapter Review

SUMMARY

3·1 The atom is the basic unit of matter. It is composed of protons, neutrons, and electrons.

3·2 There are 92 naturally occurring elements. An element can be described by its atomic number and by its mass number.

3·3 Atoms that gain or lose electrons become ions.

3·4 A compound is a substance formed by the chemical combination of two or more elements.

3·5 A molecule is a unit made up of two or more atoms covalently bonded.

3·6 A chemical formula indicates the number and kinds of atoms or ions in a compound.

3·7 A chemical reaction is a change in which new substances form. Chemical reactions involve the transfer or sharing of electrons.

3·8 A mixture is a combination of two or more substances that are not chemically combined.

3·9 The pH of a solution is a measure indicating whether the solution is acidic, basic, or neutral.

3·10 Organic compounds are compounds that contain carbon. Inorganic compounds do not contain carbon.

3·11 Carbon forms four covalent bonds. Carbon atoms can bond to one another, forming chains of various lengths.

3·12 Carbohydrates are organic compounds containing carbon, hydrogen, and oxygen and that supply energy for an organism.

3·13 Lipids are organic molecules that are a source of energy. The basic units of many lipids are fatty acids and glycerol.

3·14 Proteins are large molecules that are important structural and regulatory substances of cells. The basic unit of proteins is the amino acid.

3·15 Nucleic acids are complex molecules that determine the hereditary information of cells.

3·16 Matter can undergo physical changes and chemical changes.

3·17 The Law of Conservation of Mass-Energy states that during a chemical change, mass and energy can be neither created nor destroyed.

3·18 Energy can be transformed from one form to another.

BIOLOGICAL TERMS

acid
amino acids
atom
atomic number
base
carbohydrates
chemical equation
chemical reaction
colloid
compound
covalent bond
covalent compound
disaccharides
element
ion
ionic bond
ionic compound
isomers
isotopes
lipids
mass number
mixture
molecule
monosaccharides
nucleic acids
organic compounds
polysaccharides
products
proteins
reactants
solute
solution
solvent
suspension

USING BIOLOGICAL TERMS

1. Name three groups of carbohydrates.
2. If a/an _____ gains or loses an electron, it becomes a/an_____. Supply the two terms.
3. What is an acid?
4. How is a solute different from a solvent?
5. How is a product different from a reactant?
6. Identify the four major groups of organic compounds found in organisms.
7. What is an element?
8. What are isotopes?
9. What do the atomic number and the mass number tell about an atom?
10. What are proteins?

UNDERSTANDING CONCEPTS

1. Describe the current model of the atom, including the arrangement of the three major types of subatomic particles. (3·1)
2. Find the element with the symbol Mg in Figure 3·3. The atomic number of this element is 12. Its mass number is 24. Write the name of the element and the number of protons, neutrons, and electrons it has. (3·2)
3. Explain how the elements potassium (K) and fluorine (F) would form ions. (3·3)
4. How does an element differ from a compound? (3·4)
5. How does an ionic bond differ from a covalent bond? (3·5)
6. The chemical formula for a particular compound is CO_2. Using Figure 3·3, determine the kinds of elements in this compound. Also determine how many atoms of each element make up a molecule of the compound. (3·6)
7. Study the following chemical equation. Write a word description of the equation. (3·7)
$$2Na + Cl_2 \rightarrow 2NaCl$$
8. For a solution of salt in water, identify the solute and the solvent. (3·8)
9. Substance A has a pH of 4. Substance B has a pH of 11. Which substance is a base? Which substance produces a high level of H^+ in solution? (3·9)
10. Which of the following compounds is organic? Explain your answer. (3·10)
    a. HNO_3 b. C_2H_5OH c. H_2SO_4
11. What kinds of chemical bonds does carbon form in compounds? (3·11)
12. What is the function of carbohydrates in living things? (3·12)
13. What kinds of molecules make up lipids? (3·13)
14. What functions do proteins serve in living things? (3·14)
15. What are the functions of nucleic acids in cells? (3·15)
16. Distinguish between a chemical change and a physical change. (3·16)
17. How do the laws of conservation of mass and conservation of energy apply to chemical reactions? (3·17)
18. Why is energy important to living things? (3·18)

APPLYING CONCEPTS

1. Earlier models of the atom have been revised and/or replaced by more accurate models. The electron cloud model, which is the most current model, may be replaced. How is this situation an example of the way theories may change in science?
2. Sugar and sand are added to water. How can you separate these three substances?
3. Calcium (Ca) combines with chlorine (Cl) to form calcium chloride, an ionic compound. Calcium has two electrons in its outer energy level. Chlorine has seven electrons in its outer energy level. Determine the formula for calcium chloride.
4. A small amount of sucrose ($C_{12}H_{22}O_{11}$) is heated in a test tube. A vapor is given off that forms a clear liquid when cooled. A black residue remains in the tube. What do you think is this residue?

EXTENDING CONCEPTS

1. Using jelly beans and toothpicks, make a structural model of a molecule of glucose.
2. Prepare a food chart. List various foods and show whether each food is high in carbohydrates, lipids, or proteins.
3. Prepare a report on the medical use of radioactive isotopes as tracers and in treating certain kinds of disorders.
4. Prepare a report, including illustrations, on how the model of the atom has changed during the twentieth century.

READINGS

Baker, J.J., and G.E. Allen. *Matter, Energy and Life: An Introduction to Chemical Concepts,* 4th ed. Addison-Wesley, 1981.

Rosenfeld, Albert. "The Great Protein Hunt." *Science 81,* January/February 1981, p. 64.

4. Cell Biology

Although most cells are microscopic, some can be easily seen with the unaided eye. The largest cell in the world is the yolk of an ostrich egg. It can be as large as 17 cm by 13.5 cm, larger than an orange. The paramecium you see here is microscopic. However, size is not the only way in which cells differ. Cells differ also in structure and function. Yet in many ways, cells are very much alike. In this chapter, you will learn more about these building blocks of life.

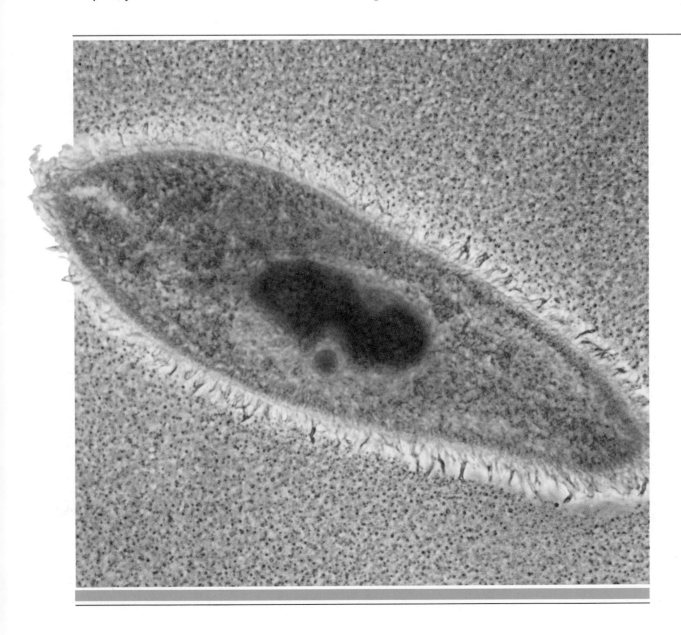

CHAPTER OBJECTIVES

After completing this chapter, you will be able to

- **Discuss** the parts of the cell theory.
- **Compare** prokaryotic and eukaryotic cells.
- **List** the major cell parts and give their functions.
- **Describe** some of the processes that occur within cells.
- **Explain** how molecules enter and leave cells.

History of the Cell Theory

4·1 THE DISCOVERY OF CELLS

The first microscopes were built around 1600. The Italian scientist Galileo observed insects with a microscope that he made. Galileo's microscope was a compound microscope. A *compound microscope* contains two lenses, one mounted at each end of a hollow tube. Two Dutch spectacle makers, Jans and Zacharias Janssen, are also credited with having developed early compound microscopes.

However, it was the English scientist Robert Hooke who improved the design of the compound microscope. Hooke observed many objects with his microscope, including thin slices of cork. What Hooke saw reminded him of small rooms, or cells, such as those in a monastery. In 1665, in a publication called *Micrographica,* Hooke used the word *cells* to describe the "boxes" he had observed in the cork. However, Hooke had not observed living cells. Instead, he had seen the walls of cells that had once been alive. Even so, Hooke is credited with being the first person to observe and identify cells. Hooke's drawing of a slice of cork is shown in Figure 4·1.

4·2 THE CELL THEORY

By the nineteenth century, microscopes had been greatly improved. Scientists were able to study structures that had not previously been observed in cells. In 1833, the Scottish botanist Robert Brown discovered that a central structure was present in the cells of orchid leaves. This structure now is called the *nucleus* (NOO klee uhs). A few years later the term **protoplasm** (PROH tuh plaz uhm) was used to refer to the living material within cells. In 1838, the German botanist Matthew Schleiden, as a result of observations of plant tissue, proposed the hypothesis that all plants are made up of cells. The next year the German zoologist Theodor Schwann, after observations of animal tissue, extended this hypothesis by proposing that all animals are also made up of cells. Schwann also proposed that the life processes of

SECTION OBJECTIVES

- Explain the main points of the cell theory.
- Distinguish between prokaryotic and eukaryotic cells.

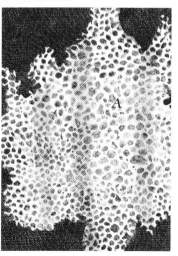

FIGURE 4·1 These cork cells were drawn by Robert Hooke. Actually, these are the cell walls that remained after the cork cells died.

FIGURE 4·2 According to the cell theory, all organisms are made up of cells. Each individual in this group of *Vorticella* is composed of a single cuplike cell (left). This sea otter has tangled itself in seaweed so that it will not be washed out to sea as it naps (right). Both the otter and the seaweed are made up of many cells.

organisms take place within cells. In 1858, Rudolf Virchow provided evidence that cells reproduce to form new cells.

The results of many investigations, including the work of Schleiden, Schwann, and Virchow, led to the development of the cell theory. The **cell theory** can be summarized by these statements.

- **All organisms are made up of one or more cells.**
- **Cells are the basic units of structure and function of all organisms.**
- **New cells come from existing cells by cell reproduction.**

Some theories are based primarily on the hypothesis and investigations of one individual. However, the cell theory was based on the discoveries of many biologists. Today, the cell theory is recognized as one of the major theories in biology. It has served as a basis for biologists seeking new knowledge about cells and their properties.

4·3 KINDS OF CELLS

Most cells contain structures that carry out specific functions. These specialized structures are called **organelles.** Today, cells are classified into two groups according to whether or not they contain specialized membrane-bound organelles. A *membrane* is a structure that encloses a cell or part of a cell. What differences can you observe between the two types of cells in Figure 4·3? Prokaryotic (proh kar ee AHT ik) cells are simple cells that do not contain membrane-bound organelles. They are small cells, averaging 1 micrometer (μm) in diameter. Bacteria are classified as **prokaryotes** (proh KAR ee ohts), or organisms that are made of prokaryotic cells. Prokaryotes are the oldest known forms of life.

pro- (before)
karyo (nucleus)

eu- (true)

Cells that contain membrane-bound organelles are called eukaryotic (yoo kar ee AHT ik) cells. Organisms that consist of eukaryotic cells are called **eukaryotes** (yoo KAR ee ohts). Eukaryotic cells are larger than prokaryotic cells, averaging 20 μm in diameter. Plants, fungi, and animals are eukaryotes.

Both eukaryotes and prokaryotes contain nucleic acid. The information for controlling a cell's activities is contained in the nucleic acid. The nucleic acid in a eukaryote is enclosed within the membrane-bound structure called the nucleus. The **nucleus** is the organelle that controls a cell's activities. Notice in Figure 4·3A that the prokaryote does not contain a nucleus. The cell's nucleic acid is not contained within a nuclear membrane.

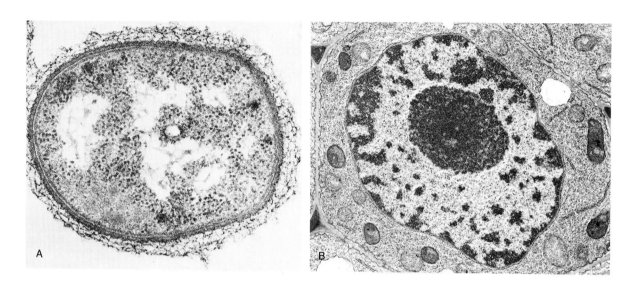

FIGURE 4·3 This prokaryotic cell (left) is magnified 60 000 times. This eukaryotic cell (right) is magnified 9945 times.

The jellylike material found within prokaryotic and eukaryotic cells is called **cytoplasm** (sī tuh plaz uhm). A number of organelles can be seen in the cytoplasm of the eukaryotic cell. Most of the activities that take place in organelles in eukaryotic cells are associated with the cell membrane in prokaryotic cells.

REVIEW QUESTIONS

1. Why is a microscope needed for cell study?
2. How did the discoveries of Hooke add to scientific knowledge about the structure of organisms?
3. Which three scientists are most closely identified with the development of the cell theory?
4. Summarize the main points of the cell theory in your own words.
5. Scientists have recently found some organisms composed of cytoplasm and organelles, including many nuclei. The cytoplasm is not divided into small cells, each with a nucleus. Modify the cell theory to include such organisms.

Cell Structure and Function

SECTION OBJECTIVES

- Identify the major cell parts.
- Describe the functions of the major cell parts.

4·4 THE NUCLEUS

Eukaryotic cells are more complex than prokaryotic cells. The membrane-bound organelles in a eukaryotic cell allow for division of labor within the cell. *Division of labor* is the distribution of life activities among the organelles of the cell. Each organelle is specialized to carry out a particular activity.

Figure 4·4 shows a generalized animal cell. Not every animal cell has all of the organelles shown here. The nucleus is usually the most conspicuous organelle. The nucleus is surrounded by a two-layered membrane called the **nuclear membrane.** Notice the *pores,* or openings, in the nuclear membrane. Molecules pass between the nucleus and the cytoplasm through these pores.

Within the nucleus is an irregularly shaped body called the **nucleolus** (noo KLEE uh luhs). The nucleolus makes and stores *RNA,* a type of nucleic acid. RNA is used in making proteins in the cell. Also within the nucleus is a material called chromatin. **Chromatin** (KROH-muh tihn) is made of protein and the nucleic acid called *DNA.* During cell division, the chromatin material takes the form of threadlike structures known as **chromosomes** (KROH muh sohmz).

4·5 CYTOPLASMIC ORGANELLES

- *Mitochondria* Within the cytoplasm are structures called mitochondria (mi tuh KAHN dree uh). **Mitochondria** (sing., *mitochondrion*) carry out the chemical reactions that release energy for cellular activities.

FIGURE 4·4 This is a typical animal cell. Notice that the nucleus is the most conspicuous structure shown.

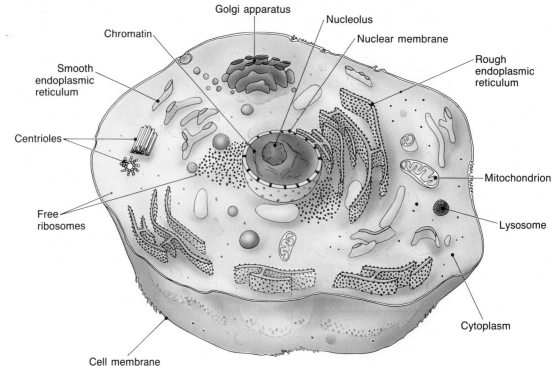

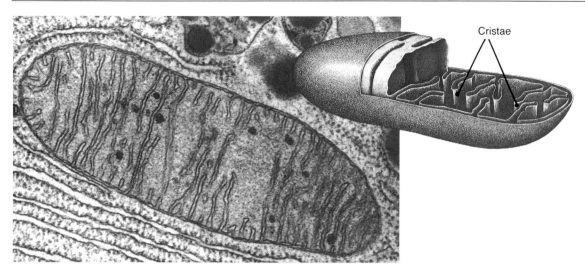

FIGURE 4·5 A mitochondrion is magnified 50 000 times in this electron micrograph. The inset drawing shows the three dimensional shape of the mitochondrion.

Notice in Figure 4·5 the two separate membranes that make up a mitochondrion. The outer membrane is smooth, but the inner membrane folds to form fingerlike projections called *cristae* (KRIHS tee). Some of the chemical reactions that release energy from food take place on the surfaces of the cristae. Because they produce energy, the mitochondria often are called the "powerhouses" of the cell.

Mitochondria are most numerous in cells that are continuously working, such as the cells that make up heart muscle. Thousands of mitochondria might be present in each of these cells. In what other kinds of cells would you expect to find many mitochondria?

● *Endoplasmic Reticulum* The **endoplasmic reticulum** (ehn doh-PLAZ mihk reh TIHK yuh luhm) is a system of membranes that runs throughout the cytoplasm. Endoplasmic reticulum (ER) is shown in Figure 4·6. The endoplasmic reticulum extends from the nuclear membrane to the cell membrane. The membranes of the ER provide passageways for the movement of materials throughout the cell.

● *Ribosomes* Some membranes of the ER have a bumpy appearance. This type of ER is called *rough ER*. The rough appearance is due to the tiny structures called ribosomes (RĪ buh sohmz). **Ribosomes** are the organelles on which proteins are made. Proteins made on ribosomes of the rough ER can be transferred throughout the cell. Some proteins may be transferred to the cell membrane and released from the cell. Ribosomes also occur free in the cytoplasm. Proteins manufactured on the free ribosomes are released directly into the cytoplasm. Membranes of the ER that lack ribosomes are referred to as *smooth ER*. Certain kinds of lipids are made by the smooth ER.

● *Golgi Apparatus* Figure 4·7 shows a Golgi apparatus (GAHL-jee ap uh RAY tuhs), an organelle similar in appearance to the ER.

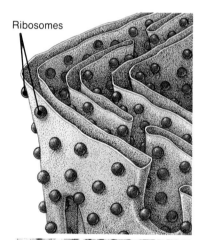

FIGURE 4·6 An endoplasmic reticulum (bottom) has been magnified 45 000X in this electron micrograph. The drawing (top) shows the three dimensional structure of an endoplasmic reticulum.

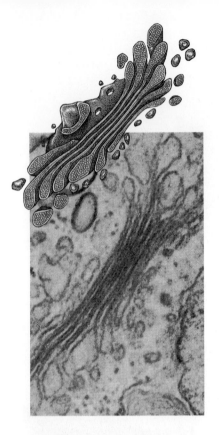

FIGURE 4·7 A Golgi apparatus is magnified 55 000 times in this electron micrograph. The inset drawing shows the structure of the Golgi apparatus.

FIGURE 4·8 This electron micrograph shows a cross-section of the tail of a moth sperm cell. The small circles in the light-colored part of tail are microtubules. Microtubules are pipelike structures that are involved in the movement of some cells.

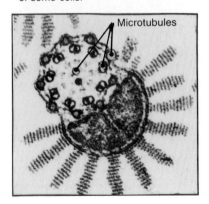

This organelle resembles a stack of flattened sacs. The sacs are formed of membranes. The **Golgi apparatus** is an organelle that prepares materials for secretion from the cell. *Secretion* is a process in which substances produced in cells are released to the surrounding environment through the cell membrane. Proteins and lipids synthesized in the endoplasmic reticulum may be carried to the Golgi apparatus. The Golgi apparatus concentrates the protein or lipid molecules by removing water. The product is then wrapped in a membrane derived from the Golgi apparatus. The concentrated product then moves to the cell membrane where it is released.

● *Vacuoles* Organelles called **vacuoles** (VAK yoo ohlz) are fluid-filled structures that may contain a variety of substances. In animal cells, vacuoles are usually small. Several of them may be scattered throughout the cytoplasm. These vacuoles usually serve as temporary storage sites for materials. In unicellular organisms, vacuoles have a variety of specialized functions. Food may be digested within certain vacuoles. Other vacuoles, called *contractile vacuoles,* operate like pumps, removing excess water or waste materials from the cell.

● *Lysosomes* Also within the cytoplasm are organelles called lysosomes (LĪ suh sohmz). **Lysosomes** contain digestive enzymes. Digestive enzymes aid in the breakdown of large molecules, such as starches, lipids, and proteins. One function of lysosomes is to digest foreign particles, such as bacteria, that enter the cell. The digestive enzymes destroy the bacteria and release the products of digestion into the cytoplasm. Another function of lysosomes is the destruction of worn-out cell parts. The products formed can then be reused by the cell. Normally, lysosomes digest complex molecules without damaging the cell. But sometimes the membrane around the lysosome ruptures, causing the cell to digest itself.

● *Cytoskeleton* The **cytoskeleton** is made up of structures that maintain the cell's shape and control certain kinds of cell movement. Microfilaments and microtubules are the two organelles that make up the cytoskeleton.

Microfilaments are very thin threadlike fibers found within the cell. Microfilaments are made of protein. They are often found in sheets or bundles below the cell membrane. A major function of microfilaments is the movement of materials within the cell by producing a flowing motion of the cytoplasm. Microfilaments also enable some unicellular organisms, like the ameoba, to move about by a flowing motion.

Organelles called **microtubules** are pipelike structures composed of protein. The arrangement of the microtubules in a cell helps to give a cell its shape. Microtubules are involved in the ability of certain cells to move. Many unicellular organisms move by means of hairlike structures called **cilia** (SIHL ee uh). Others move by using a taillike structure called the **flagellum** (fluh JEHL uhm). Microtubules make up the basic structures of flagella and cilia. Microtubules extend from the cell into a flagellum or into the cilia.

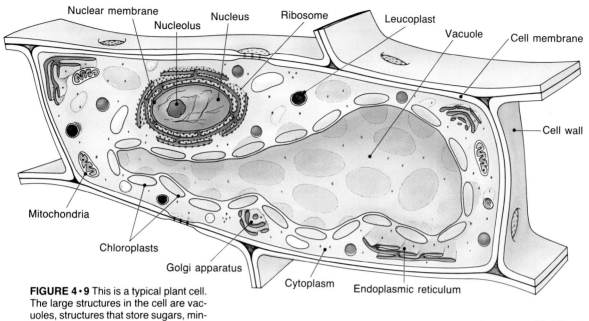

FIGURE 4·9 This is a typical plant cell. The large structures in the cell are vacuoles, structures that store sugars, minerals, and protein dissolved in water.

4·6 ORGANELLES OF PLANT CELLS

The organelles discussed thus far are found in both animal and plant cells. Other organelles are found only in plant cells or are more conspicuous in plant cells.

In a plant cell, one large vacuole takes up most of the cell space. This central vacuole pushes the cytoplasm up against the cell membrane. The large vacuole stores a variety of substances, often in water solution. These substances include sugars, minerals, and proteins.

● *Plastids* Within the cytoplasm of plant cells are organelles called plastids (PLAS tihdz). Some **plastids** function as chemical factories and others function as storehouses for food and colored pigments. In the cells of green plants, the most common plastid is the chloroplast (KLOR-uh plast). The **chloroplast** is the organelle in which food is made. The structure of a chloroplast is shown in Figure 4·10. Notice the layers of membranes that look like stacks of coins. Each stack of membranes is called a *granum* (pl., *grana*). A gellike substance called *stroma* surrounds the grana. The green pigment *chlorophyll* (KLOR uh fihl) is concentrated in the grana. Chlorophyll traps the energy of sunlight, which is used to make food.

Other plastids called **leucoplasts** (LOO kuh plasts) are storage plastids. They may contain proteins, lipids, or starch. When fruits ripen or when leaves change color in the fall, chloroplasts and leucoplasts may be converted to another kind of plastid called a chromoplast (KROHM uh plast). **Chromoplasts** are plastids that contain red, yellow, or orange pigments.

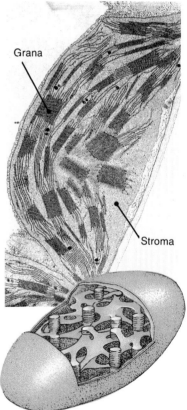

FIGURE 4·10 A chloroplast is magnified 17 100 times in this electron micrograph. Notice the stacked membranes in the grana.

CELL BIOLOGY 65

● **Cell Wall** Each plant cell has a structure outside the cell membrane called the cell wall. The **cell wall** gives shape and rigidity to a plant cell. The cell wall is made up mostly of cellulose, a complex carbohydrate. Some cell walls also contain pectin or lignin, materials that give strength to the walls. The cell wall allows air, water, and dissolved materials to pass through. Openings in the cell wall allow the cell membrane to come into contact with the membranes of neighboring cells. Materials can be transferred from one cell to another through these openings. Cell walls also are found in prokaryotic cells and in fungi. But the walls of prokaryotic cells are different in composition from those of fungi and other eukaryotic cells.

4•7 COMPARISON OF CELLS

Recall that prokaryotic cells are primitive cells that lack membrane-bound organelles. Bacteria are classified as prokaryotes. Bacteria have no membrane surrounding their nucleus. They have nuclear material floating in the cytoplasm. Eukaryotic cells are more advanced and contain membrane-bound organelles. Their nuclear material is enclosed within a nuclear membrane. Plants, fungi, and animals consist of eukaryotic cells. A prokaryotic cell is usually surrounded by a cell wall, whereas only some kinds of eukaryotic cells are surrounded by cell walls. Prokaryotes are generally small cells; eukaryotes are generally large.

Most biologists think that the line between prokaryotic cells and eukaryotic cells is the major dividing line in the biological world. The most important difference between prokaryotes and eukaryotes is the presence of the nuclear membrane in eukaryotes. Figure 4•11 lists the major cell structures in prokaryotes and eukaryotes. What type of

FIGURE 4•11 This table summarizes the functions of major cell structures. As you can see, many more structures are found in eukaryotic cells.

STRUCTURE	FOUND IN PROKARYOTES	FOUND IN EUKARYOTES	FEATURE/FUNCTION
Cytoplasm	X	X	Gellike material within the cell
Nucleus		X	Control center of cell
Nucleolus		X	Synthesizes and stores RNA
Mitochondrion		X	Releases energy for cell activity
Endoplasmic reticulum		X	Provides passageways from nucleus to outside of cell
Ribosome	X	X	Synthesizes proteins
Golgi apparatus		X	Prepares materials for secretion; manufactures complex organic molecules
Lysosome		X	Stores digestive enzymes
Microfilament		X	Aids cellular movement
Microtubule		X	Aids cellular movement; gives structure and shape
Plastid		X	Synthesizes chemicals; stores food and pigments
Vacuole		X	Fluid-filled storage compartment
Cell wall	X	X	Gives rigidity to cells

eukaryotic cell structure is found only in plant cells? Between which cells is the difference greater—plant cells and animal cells or prokaryotic and eukaryotic cells?

REVIEW QUESTIONS

6. How do organelles allow for division of labor within cells?
7. What is the function of the nucleus in eukaryotic cells?
8. What is the difference between smooth ER and rough ER?
9. How are the structures of mitochondria and chloroplasts similar?
10. Suppose you wish to view slides of cells that have a great many chloroplasts. What specimens would you choose? Give reasons for your choice(s).

Cellular Transport

4·8 THE CELL MEMBRANE

Many metabolic activities take place within the cell. Sugars are broken down for energy, proteins are made from simpler materials, and various wastes are produced. The cell needs a constant supply of materials to carry out its life processes. It must get rid of waste products before they build up and harm the cell. By exchanging materials with its environment, and thereby obtaining nutrients and disposing of wastes, the cell can continue to function.

The **cell membrane** helps regulate the passage of materials between the cell and its environment. The cell membrane may prevent some substances, such as large proteins and lipids, from entering the cell. The membrane permits the passage of simple sugars, oxygen, water, and carbon dioxide. For this reason the cell membrane is said to be *selectively permeable.*

A model of the structure of the cell membrane is shown in Figure 4·12. The cell membrane is about 7.5 to 10 nanometers (nm) thick. The membrane is made up almost entirely of protein and lipid molecules. You can see that the molecules that make up the lipid layers have two regions—the "head end" and the "tail end." These lipid molecules are arranged in two layers in the cell membrane. Notice that the tails in each layer point to the tails in the other layer. This arrangement causes the cell membrane to appear as if it is made up of three layers when photographed under high magnification.

Various proteins are embedded in the lipid layers. Some of these proteins make contact with only the internal or external environment of the cell. Others penetrate through the membrane. The proteins may also change their positions in the membrane. The different types of

SECTION OBJECTIVES

- Describe the structure of the cell membrane.
- Explain the processes by which substances enter and leave cells.

proteins associated with the membrane play important roles in the movement of materials across the membrane. In Figure 4•12 notice the channel that is located within one of the protein molecules. Such channels are pores that allow certain substances to pass through the membrane.

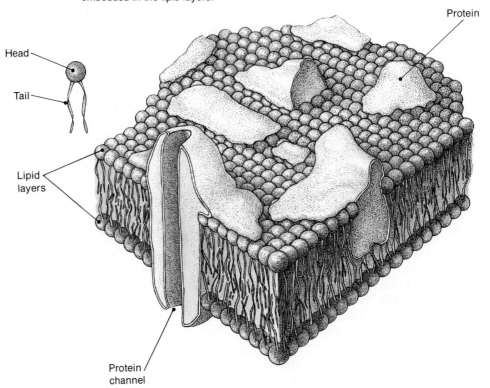

FIGURE 4•12 Cell membranes are made up of two layers of lipids. Proteins are embedded in the lipid layers.

4•9 TYPES OF CELLULAR TRANSPORT

The constant movement of substances in both directions across the cell membrane is called cellular transport. **Cellular transport** is the process by which needed materials enter the cell while waste materials and cell secretions leave. Cellular transport may be either active or passive. **Active transport** requires the use of energy by the cell to move materials across the cell membrane. Active transport will be explained in more detail in section 4•13.

Passive transport does not require the use of energy by the cell to move materials across the cell membrane. Passive transport depends on the kinetic energy of the particles in matter. Atoms, ions, and molecules of all substances are in constant motion. In solids, the particles vibrate in one place. However, the particles of liquids and gases move around more freely and randomly. Such particles travel in straight lines until they collide with other particles. The collisions cause the particles to change direction. One type of passive transport is called diffusion.

4·10 DIFFUSION

The random movement of molecules in liquids and gases is responsible for a process called diffusion (dih FYOO zhuhn). **Diffusion** is the movement of atoms, molecules, or ions from a region where they are more concentrated to a region where they are less concentrated. You are probably familiar with many cases of diffusion. For example, suppose someone spills some perfume on a desk near the front of your classroom. Within a short time students near the spill will smell the perfume. Soon the scent will be noticeable all over the room.

Diffusion of a solid in a liquid is shown in Figure 4·13. A sugar cube has been dropped into a cup of water. Molecules of sugar are concentrated in the cube. As the sugar dissolves, the sugar molecules collide with each other and with the water molecules. What happens to the sugar molecules in the water?

Diffusion continues until the sugar molecules are distributed evenly throughout the water. Once this occurs, the concentration of sugar molecules remains constant. The molecules will continue to move, but the concentration will not change. This state of constant concentration maintained by the motion of molecules is known as *dynamic equilibrium*. Once equilibrium has been reached, further movement has no net effect on the even distribution of sugar molecules throughout the water.

A diffusing substance moves down a concentration gradient. A gradient may be thought of as degree of steepness, as in the gradient of a hill. A **concentration gradient** is a measure of the difference in concentration of a substance in two regions. The rate of diffusion is related to the size of the concentration gradient. For example, the greater the concentration gradient, the greater the rate of diffusion from the region of greater concentration to the region of lesser concentration. Saying that a substance moves from a region of high concentration to a region of low concentration is the same as saying that it moves down a concentration gradient.

Some substances, including oxygen and carbon dioxide, pass through pores in the cell membrane by simple diffusion. Usually oxygen molecules are more highly concentrated outside the cell, and carbon dioxide molecules are more highly concentrated inside the cell. Oxygen diffuses into the cell while carbon dioxide diffuses outward.

The cell constantly uses up oxygen and produces carbon dioxide. Therefore, the concentration of oxygen in the cell is lower than the concentration outside the cell. The concentration of carbon dioxide in the cell is always higher than outside the cell. As a result, oxygen continues to diffuse into the cell, and carbon dioxide continues to diffuse out.

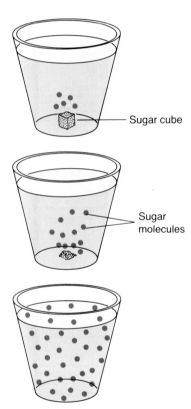

FIGURE 4·13 Over time, the molecules of sugar in the cube diffuse in the water. Eventually, the molecules will be evenly distributed throughout the water.

4·11 OSMOSIS

Water can pass through the cell membrane. The movement of water across a selectively permeable membrane is called **osmosis** (ahz MOH sihs). In osmosis, water moves from a region of high concentration to a region of lower concentration. Thus osmosis is a special kind of passive transport, the diffusion of water.

How can the concentration of water in two regions be compared? The concentration of water is determined by the amount of material dissolved in the water. If very little material is dissolved in the water, the concentration of water can be considered high. For example, if a solution contains 1 g of salt in 1000 g of water, the concentration of water is high. If a solution contains 100 g of salt in 1000 g of water, the concentration of water is lower than that in the previous solution.

Figure 4·14 shows a demonstration of osmosis. A glass funnel has been partly filled with a solution of sugar water. A selectively permeable membrane has been fastened over the top of the funnel. The membrane allows water but not sugar to pass through. The funnel has been inverted in a beaker of water. At this time, where is the water more concentrated—in the beaker or in the funnel? The total number of water molecules in each milliliter of liquid is greater in the beaker than in the funnel. Water is more concentrated in the beaker. Therefore, a concentration gradient exists for the water.

Water begins moving by osmosis into the funnel. Even though water moves through the membrane, the larger sugar molecules cannot pass through in the opposite direction. As osmosis continues, the sugar water rises in the funnel tube, and the solution becomes more dilute. After a while, the concentration of water in the funnel is almost equal to the concentration of water in the beaker.

In living organisms, water enters and leaves cells by osmosis. Figure 4·15 shows red blood cells and plant cells in an isotonic (ī-suh TAHN ihk) solution. An **isotonic solution** is one in which the concentration of dissolved substances inside the cell is *equal* to the concentration outside the cell. In the bloodstream, the fluid around a red blood cell contains as much dissolved material as does the blood cell. Because the concentration is the same on both sides of the cell membrane, water moves inward and outward at the same rate. Under isotonic conditions, red blood cells and plant cells maintain their shapes.

In Figure 4·15, the two kinds of cells have been placed in a hypotonic (hī puh TAHN ihk) solution. A **hypotonic solution** is one in which the concentration of dissolved materials in the water outside the cell is *lower* than the concentration inside the cell. Thus the concentration of water is higher outside the cell than inside. A red blood cell placed in distilled water is in a hypotonic solution because distilled water contains no dissolved materials. Because of the greater concentration of water outside the cell membrane, water moves into the cell. This has caused the blood cells to swell. If water continues to move into the cells, the cells will burst.

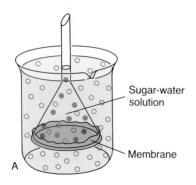

FIGURE 4·14 There are fewer water molecules in the funnel at the beginning of this experiment. In time the water molecules move through the membrane and into the funnel. Notice that the level of water in the funnel rises.

hypo- (under)

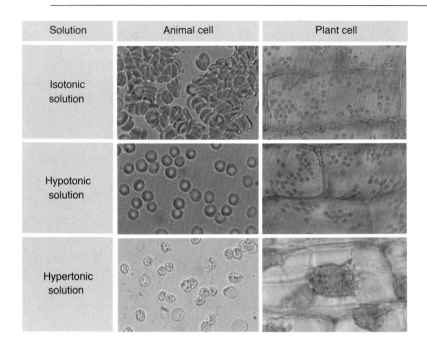

FIGURE 4·15 Water moves into plant and animal cells in a hypotonic solution. Eventually, red blood cells will burst. The cell walls of plant cells help these cells keep their shape as water moves in. In a hypertonic solution, water moves out of plant and animal cells.

Look at the plant cell in the hypotonic solution. Mature plant cells are generally in a hypotonic environment because of the materials dissolved within the cell. These dissolved substances decrease the internal water concentration, so water moves into the cell by osmosis. The intake of water causes the cell contents to push against the cell wall. However, the cell does not burst because the cell wall is strong enough to prevent the cell from being pushed out further. The outward pressing of water against the cell wall is **turgor** (TER guhr) **pressure.** Turgor helps to give firmness and stiffness to stems and leaves.

Figure 4·15 also shows cells in a hypertonic (hī puhr TAHN ihk) solution. A **hypertonic solution** is one in which the concentration of dissolved substances is *greater* in the water outside the cell than in the water inside the cell. A saltwater solution is hypertonic to the cells shown. In a hypertonic solution, water moves out of the cell by osmosis. As a result, a red blood cell shrivels. In a plant cell, the cell's contents shrink away from the cell wall. The shrinking of a plant cells contents due to loss of water is known as **plasmolysis** (plaz MAHL uh sihs). Plasmolysis results in wilting of stems and leaves.

hyper- (over)

4·12 FACILITATED DIFFUSION

Another type of passive transport across a cell membrane is accomplished with the aid of carrier molecules. **Carrier molecules** in the cell membrane permit specific molecules on one side of the membrane to pass through to the other side. Investigations have revealed that these carriers are proteins. The exact way in which these protein molecules function, however, is not fully known. The diffusion of materials across the cell membrane by the carrier molecules is called **facilitated diffusion.** The process of diffusion is facilitated, or made easier, by the carriers. Facilitated diffusion involves the movement of substances down a concentration gradient. In facilitated diffusion, however, the substances move more rapidly than by simple diffusion. For example, glucose moves into red blood cells by facilitated diffusion. It diffuses hundreds of times more rapidly than other sugars that have similar properties and only slightly different chemical structures. Apparently only particular kinds of molecules can be transported by carriers.

4•13 ACTIVE TRANSPORT

Transport of some materials into and out of cells takes place *against* a concentration gradient. In such cases cell energy must be used to move substances from regions of lower concentration to regions of higher concentration. The process of active transport uses cell energy to move atoms, molecules, and ions against a concentration gradient. In a resting human, 30 to 40 percent of all energy expended is needed for active transport of materials into cells.

The exact way in which active transport works is not known. Carrier molecules in the cell membrane may be involved. A model used to explain active transport is shown in Figure 4•16. A carrier molecule in the cell membrane has an active site into which only certain molecules can fit. After a molecule enters the carrier molecule, energy is released by the cell, causing the carrier molecule's shape to change. The carrier molecule is thought to deliver the carried molecule into the cell. After the carried molecule is released, the carrier molecule returns to its original shape and is ready to continue the process.

There is evidence that glucose, amino acids, and certain ions sometimes are moved into cells by active transport. Wastes leave some cells in this manner. Active transport is also the process by which some mineral ions in soil water enter plant roots. There is greater concentration of minerals within the cells of the root than in the water in soil. Energy must be used by the root cells to enable minerals to move into the root cells against the concentration gradient. You will study how materials move within plants in Chapter 18.

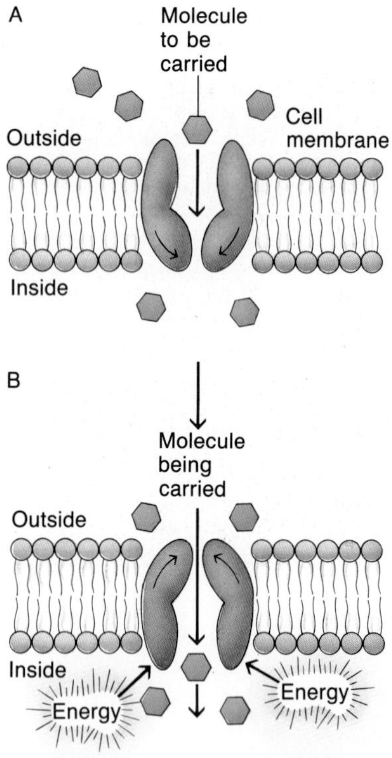

4•14 ENDOCYTOSIS AND EXOCYTOSIS

The transport processes discussed so far involve the passage of small molecules across the cell membrane. Cells also have ways of getting large molecules, groups of molecules, and even whole cells across the cell membrane.

Endocytosis (ehn doh sī TOH sihs) is the process by which cells take in large materials that cannot pass through the cell membrane. There are two types of endocytosis. A cell takes in small particles or liquid droplets in the process of **pinocytosis** (pih noh sī TOH sihs). Figure 4•17 shows a cell taking in material by pinocytosis. Follow the diagrams as the process is described.

FIGURE 4•16 Carrier molecules in a cell membrane change shape to permit other molecules to enter the cell. The carried molecule is "delivered" into the cell.

endo- (inside)
cyto- (cell)

pino- (to drink)

Discovery

Cell Techniques

For many years, cell biologists have been interested in details of the structure of the nucleus. Their research was limited, however, by the equipment and techniques that were available. As a result, the structure of the nucleus was oversimplified for years.

When the nucleus was first investigated in some detail, it was studied with the light microscope. The nucleus at that time was thought basically to be a bag of chromosomes. After the electron microscope was developed, the focus of the research shifted from the chromosomes to the pores of the nuclear membrane. In all the research, specific parts of the nucleus were studied independently of other aspects of the nucleus.

Recently, researchers have found that chromosomes seem to have specific contact points with the nuclear membrane. The nucleus, therefore, has been under detailed study to see if there is a structural reason for these contact points.

To find out if the nucleus contains some internal structure other than chromosomes, scientists have developed methods to isolate the nucleus for study. *Cell fractionation* is a procedure by which cell parts are separated. Cells are ground up in a blender and then the liquid is poured into tubes. The tubes are spun at different speeds in an instrument called a *centrifuge*. Cell parts that are larger and denser move to the bottom of the centrifuge tubes. Nuclei can easily be separated from the rest of the cell parts because they are larger and denser than other cell parts. Then the nuclei can be broken down further.

When the nucleus was broken open, researchers found a fiberlike protein material in the nucleoplasm, or liquid portion of the nucleus. This protein component of the nucleus has been named the *nuclear matrix*. The function of the nuclear matrix has not yet been determined. However, the nuclear matrix is thought to have a role in the precise position and movement of chromosomes within the nucleus.

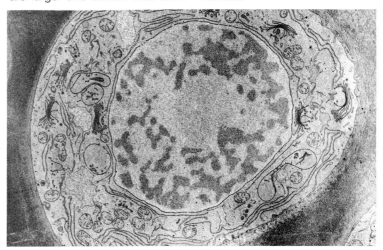

1. A particle to be taken into the cell becomes bound to the cell membrane. The cell membrane may become indented and form a thin channel.
2. The particle may then drop to the bottom of the channel.
3. The lower part of the channel containing the particle may break away from the rest of the cell membrane, forming a pouch called a *vesicle* (VEHS uh kuhl). The vesicle becomes a separate structure inside the cell. The particle in the vesicle can then be digested by the cell.

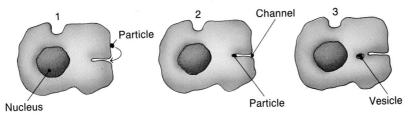

FIGURE 4·17 Small particles are taken into a cell during pinocytosis. A vesicle is formed when a particle reaches the bottom of a channel.

-phage (to eat)

The second type of endocytosis is called phagocytosis (fag uh sī-TOH sihs). **Phagocytosis** is the process by which large, solid materials are taken into a cell. Phagocytosis has been observed in certain unicellular organisms and in animal cells such as white blood cells. An amoeba surrounding a green alga is shown in Figure 4·18.

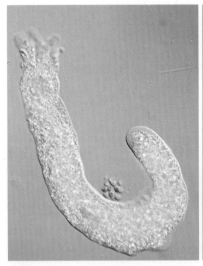
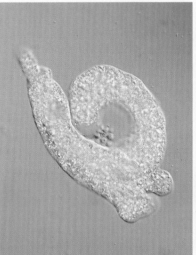
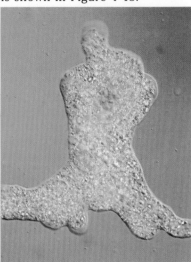

FIGURE 4·18 Large solid materials are taken into a cell by phagocytosis. The amoeba in this illustration is taking in a smaller organism by this process.

1. The amoeba extends its cell membrane, forming pseudopods that move around the green alga.
2. The amoeba encloses the alga in a pocket.
3. The pocket breaks away from the membrane and becomes a large vesicle that moves into the cytoplasm. The vesicles formed in phagocytosis are much larger than those formed in pinocytosis.

Exocytosis (ehks uh sī TOH sihs) is the reverse of endocytosis. **Exocytosis** is the release of large molecules or groups of molecules from the cell. The materials released may be wastes. Or they may be useful secretions delivered to the cell membrane by the Golgi apparatus. A vesicle that contains secretions moves to the cell membrane. There the vesicle fuses with the membrane. The membrane breaks down at that place, and the contents of the vesicle are released.

REVIEW QUESTIONS

11. What kinds of molecules make up the cell membrane?
12. Define the process of diffusion and give an example of the process.
13. Explain why a red blood cell shrivels when placed in a salt solution but swells when placed in distilled water.
14. In what ways are facilitated diffusion and active transport alike? In what ways are they different?
15. Ringer's solution is a special solution used to prepare slides of cells of animals. When placed in this solution, the cells retain their normal shape—that is, they appear as they do in life. In relation to the animal cell, is Ringer's solution isotonic, hypertonic, or hypotonic? Why?

Investigation

What Is the Effect of a Salt Solution on Green Plant Cells?

Goals
After completing this activity, you will be able to
- Describe the effect of a salt solution on green plant cells.
- Identify and label the vacuole, chloroplast, cytoplasm, and cell wall of a green plant cell.
- Describe the effects of plasmolysis on green plant cells.

Materials (for groups of 2)
forceps microscope
cutting of *Elodea* dropper
1 slide 1 cover slip
10% salt solution paper towel
aquarium water

Procedure

A. Place a drop of aquarium water in the center of a slide. Remove a leaf from the tip of a piece of *Elodea*. Using forceps, place the leaf in the drop of water.

B. Add the cover slip. Using the low power of the microscope, look for a thin area where the cells can be seen clearly. View the cells under high power. On a separate sheet of paper, draw a few *Elodea* cells. Label the vacuole, chloroplast, cytoplasm, and cell wall. Switch back to low power. Remove the slide from the microscope.

C. Using a dropper, add a drop of 10% salt solution to the right edge of the cover slip. Place a small piece of paper towel at the left edge of the cover slip to draw the salt solution under the cover slip. Repeat this step with a second drop of salt solution.

D. Replace the slide on the stage of your microscope. Wait 3

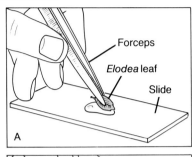
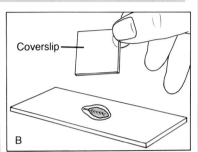

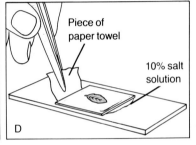

minutes. Observe the *Elodea* cells under low power. Switch to high power. On a separate sheet of paper, make another drawing of the cells of *Elodea*. Label the chloroplast, cell wall, cell membrane, and cytoplasm. Save your slide for further observation to see if there have been other changes.

E. Follow your teacher's instructions for the proper disposal of all materials.

Questions and Conclusions

1. Describe the cells of *Elodea* when they were observed in the drop of water. Describe how the *Elodea* cells changed when you added the salt solution.

2. Did water move into or out of the cell? Give reasons for the direction of movement.

3. Could you see the cell membrane before you added the salt solution? Why or why not?

4. Explain which term—*osmosis* or *diffusion*—better describes the process you observed in this activity.

5. (a) Review the terms *hypotonic solution, isotonic solution,* and *hypertonic solution* in the text. Which of these terms describes the salt solution used in this activity? Explain your answer.
 (b) Aquarium water is an isotonic solution for *Elodea* cells. Predict what would happen if you replaced the salt solution with aquarium water in the slide you saved in step D. If you have time, test your prediction.

6. Review the meaning of the word *plasmolysis* in your text. Explain whether this term applies to any of the changes that you have observed.

7. How might the use of salt for snow removal affect plants that grow next to sidewalks and roadsides?

Chapter Review

SUMMARY

4·1 The invention of the microscope led to the discovery of cells.

4·2 The cell theory states that cells are the basic unit of structure and function of all living things.

4·3 Eukaryotic cells have organelles surrounded by separate membranes. Prokaryotic cells do not have such organelles.

4·4 The cell nucleus contains nucleic acids necessary for cell division and protein synthesis.

4·5 The life activities of eukaryotic cells are divided among a number of cytoplasmic organelles.

4·6 Certain cell parts, such as cell walls and plastids, are found in plant cells but not in animal cells.

4·7 The differences between prokaryotic and eukaryotic cells are more numerous than those between plant and animal cells.

4·8 The cell membrane is the boundary between the cell and its environment. The cell membrane helps to regulate the transport of materials into and out of the cell.

4·9 Cellular transport, the process by which materials enter and leave the cell, may be either active or passive.

4·10 Some substances pass through a cell membrane by diffusion.

4·11 In osmosis, water moves through a selectively permeable membrane from a region of higher concentration to one of lower concentration.

4·12 Facilitated diffusion is a process by which materials cross a cell membrane with the aid of carrier molecules.

4·13 Active transport is a process by which materials move across a cell membrane against the concentration gradient. It requires the use of carrier molecules and the use of energy by the cell.

4·14 Endocytosis and exocytosis are processes by which cells take in and release large materials that cannot pass through the cell membrane.

BIOLOGICAL TERMS

active transport
carrier molecules
cell membrane
cells
cell theory
cellular transport
cell wall
chloroplast
chromatin
chromoplasts
chromosomes
cilia
concentration gradient
cytoplasm
cytoskeleton
diffusion
endocytosis
endoplasmic reticulum
eukaryotes
exocytosis
facilitated diffusion
flagellum
Golgi apparatus
hypertonic solution
hypotonic solution
isotonic solution
leucoplasts
lysosomes
microfilaments
microtubules
mitochondria
nuclear membrane
nucleolus
nucleus
organelles
osmosis
passive transport
phagocytosis
pinocytosis
plasmolysis
plastids
prokaryotes
protoplasm
ribosomes
turgor pressure
vacuoles

USING BIOLOGICAL TERMS

1. How is a cell membrane different from a cell wall?
2. What are mitochondria?
3. What is active transport?
4. Distinguish between a nucleus and a nucleolus.
5. What is a chromoplast?
6. The term *prokaryote* means "before a nucleus." Why might this name have been chosen?
7. Define *turgor pressure*.
8. Distinguish between the processes of endocytosis and exocytosis.

9. Several terms contain the word part *cyt-* (-*cyt*-). List the terms and, for each, explain why the word part is appropriate.
10. What is a flagellum?

UNDERSTANDING CONCEPTS

1. Although Rober Hooke "named" cells, he did not actually observe a cell. What did Robert Hooke observe? (4·1)
2. Experiments that provided evidence *against* the hypothesis of spontaneous generation provided support *for* the cell theory. Explain what is meant by this statement. (4·2)
3. Distinguish between prokaryotic cells and eukaryotic cells. (4·3)
4. What structures are found within the nucleus? (4·4)
5. What are the functions of the mitochondria and the ribosomes? (4·5)
6. Distinguish between the functions of microfilaments and microtubules. (4·5)
7. Identify the functions of the plastids found in plant cells. (4·6)
8. Look at Figure 4·11. Compare prokaryotes and eukaryotes with regard to the number of cell structures involved in the life processes of each type of cell. (4·7)
9. Explain why the cell membrane is said to be selectively permeable. (4·8)
10. Distinguish between the two types of cellular transport. (4·9)
11. Explain how the process of diffusion is illustrated by a sugar cube dissolving in water. (4·10)
12. Explain why red blood cells burst when placed in a hypotonic solution, whereas plant cells do not. (4·11)
13. Give a reason why facilitated diffusion is a form of passive transport. (4·12)
14. What is meant by the statement, "Active transport takes place against a concentration gradient?" (4·13)
15. Distinguish between the processes of pinocytosis and phagocytosis. (4·14)

APPLYING CONCEPTS

1. Salt is sometimes sprinkled in the cracks in a sidewalk to prevent the growth of grass. Based on what you have learned in this chapter, explain why salt will kill grass.
2. Osmosis can take place in dead cells and tissues, but active transport can occur only in living cells. Explain why this is so.
3. Although cells were first seen in the seventeenth century, most of the current knowledge of cells has been accumulated within the past 50 years. Explain why knowledge of cells has increased so rapidly during this short time.

EXTENDING CONCEPTS

1. Several careers involve the study of cells. Persons in these careers may be referred to as cytologists, microbiologists, bacteriologists, and microscopists. Interview a person having one of the titles named. Find out what this person does in a typical day. Find out what education is needed to enter the field.
2. Build a three-dimensional model of an animal cell or a plant cell. Materials you might use to construct the cell parts include colored clay, beads, pipe cleaners, plastic foam, and yarn.
3. Using at least three references, prepare a report on *one* of these scientists: Robert Hooke, M. J. Schleiden, T. Schwann, or Rudolf Virchow. Find out when and where the scientist's discoveries were made, how news of the discoveries was reported, and what the significance of each finding was.

SUGGESTED READINGS

Allen, Robert Day. "The Microtubule as an Intracellular Engine." *Scientific American,* February 1987, p. 42.

Dautry-Varsat, Alice, and Harvey F. Lodish. "How Receptors Bring Proteins and Particles into Cells." *Scientific American,* May 1984, p. 52.

Nachmias, Vivianne T. *Microfilaments,* Carolina Biology Reader. Carolina Biological Supply Co., 1984.

5. Cell Chemistry

Special light-emitting organs are scattered over this deep-sea fish's body. Chemical reactions inside these organs produce light. Most organisms, of course, cannot use chemical energy to produce their own light. But all living things require energy to carry on the processes of life. You use energy to run, walk, and play sports. You use energy even when you sleep. The chemical processes going on within your body even provide the energy you need to read this book.

CHAPTER OBJECTIVES

After completing this chapter, you will be able to
- **Explain** the basic reactions that enable cells to grow and use energy.
- **Compare** models of enzyme function.
- **Discuss** the process of respiration.
- **Describe** the different ways ATP is produced from glucose.
- **Identify** the steps of photosynthesis.
- **Relate** respiration and photosynthesis.

Controlling Cellular Activities

5·1 BASIC CELLULAR REACTIONS

All cells, whether they exist alone or as part of a larger organism, carry out certain basic life functions. These functions include taking in nutrients, growing, reproducing and getting rid of wastes. Cells obtain energy from food to carry out these basic functions.

Sources of food vary for different kinds of living things. Living things that make their own food are called **autotrophs** (AW tuh trohfs). Most autotrophs use the energy of sunlight for food making. Green plants, algae, and some bacteria are autotrophs. These organisms may have specialized organelles where food making takes place.

Other living things cannot make their own food. Living things that must take in food from other sources are known as **heterotrophs** (HEHT uhr uh trohfs). Animals and fungi are examples of heterotrophs. They depend on autotrophs or other heterotrophs for food. Once food is either made or taken in by a living thing, much of it is broken down for energy in the cells.

The processes that take place in cells are physical and chemical processes. In a cell there is constant activity at the molecular level. Some molecules are being put together and others are being taken apart every second. The total of all the reactions taking place in a cell is called *metabolism*. Those reactions in which simpler substances join to form more complex substances are called **anabolic** (an uh BAHL ihk) **reactions.** Reactions in which cells build protein molecules, for example, are anabolic reactions.

Another type of reaction is a catabolic (kat uh BAHL ihk) reaction. **Catabolic reactions** are reactions in which complex substances are broken down into simpler substances. Proteins, polysaccharides, and other large molecules are broken down into simpler molecules, such as amino acids and monosaccharides, by catabolic reactions. Thus metabolism consists of anabolic and catabolic reactions.

SECTION OBJECTIVES

- Describe the processes that make up a cell's metabolism.
- Discuss the role of enzymes in cellular reactions, including some models of enzyme function.

auto- (self)
trophos (feeder)
hetero- (other)

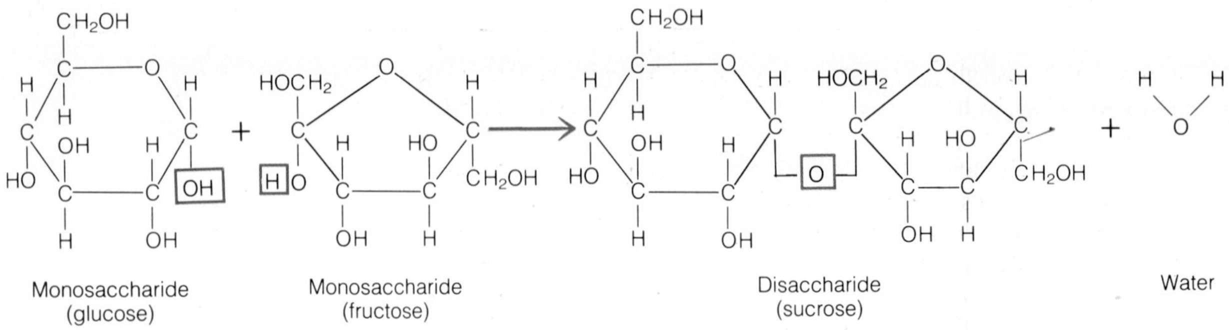

[Chemical structure diagram showing glucose + fructose → sucrose + water]

Monosaccharide (glucose) + Monosaccharide (fructose) → Disaccharide (sucrose) + Water

FIGURE 5·1 Sucrose or table sugar is made when a molecule of glucose combines with a molecule of fructose. This is an anabolic reaction.

Figure 5·1 shows an example of an anabolic reaction that takes place in cells. Notice that a molecule of glucose and a molecule of fructose have been linked. A larger molecule, sucrose, has thus been formed. The glucose and fructose are joined by bonding to an oxygen atom. What other product has been formed in the reaction? Where did this product come from?

Notice that in the reaction an −OH group has been removed from one molecule. A hydrogen atom has been removed from the other molecule. Together these form a water molecule. An anabolic reaction that involves the removal of water is called **dehydration synthesis.** *Dehydration* means the removal of water; *synthesis* means the joining of parts to form a whole. Thus dehydration synthesis forms a new substance (in this case a sucrose molecule) by joining its parts (fructose and glucose), removing water in the process.

Polysaccharides are long chains of monosaccharides built by dehydration synthesis. Glycogen (found in animal cells) and starch (found in plant cells) are polysaccharides made up of glucose units. Each time a monosaccharide, glucose, is added to a chain of glucose units, a water molecule is removed.

Proteins are long chains of amino acids, also built by dehydration synthesis. Water molecules are formed as the reaction occurs.

Figure 5·2 shows a catabolic reaction. This is a reaction in which complex substances are broken down. A catabolic reaction that involves the addition of water is called **hydrolysis** (hī DRAHL uh sihs). Figure 5·2 shows the hydrolysis of the disaccharide maltose. By the addition of water, the disaccharide is broken into two glucose mole-

hydro- (water)
lysis (to loosen)

FIGURE 5·2 Maltose can be broken down into two molecules of glucose by the addition of a molecule of water. This is an example of a catabolic reaction.

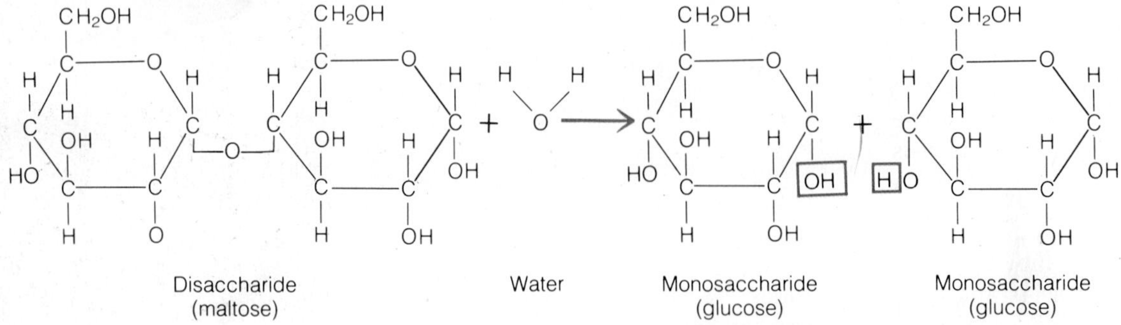

[Chemical structure diagram showing maltose + water → glucose + glucose]

Disaccharide (maltose) + Water → Monosaccharide (glucose) + Monosaccharide (glucose)

80

cules. Cells can use glucose as an energy source. The digestion of most foods takes place by hydrolysis. For example, starch, a polysaccharide, is broken down into glucose by hydrolysis.

In hydrolysis large molecules found in living cells are broken down. Carbohydrates, lipids, and proteins are all broken into smaller, more usable molecules by hydrolysis. Dehydration synthesis reverses the process and builds large carbohydrate, lipid, and protein molecules out of simple substances. Hydrolysis and dehydration synthesis are the major catabolic and anabolic reactions that take place in living cells.

5·2 CONTROL OF CELL REACTIONS

A chemical reaction that needs or uses energy is known as an **endergonic** (ehn duhr GAHN ihk) **reaction**. In plants, light energy is used to produce food. Thus the production of food in plants is an endergonic reaction.

Some reactions give off energy. A chemical reaction that releases energy is called an **exergonic** (ehk suhr GAHN ihk) **reaction**. Often this energy is given off in the form of heat. In living things, exergonic reactions provide the energy to carry out the cell's activities.

Many chemical reactions, whether they are endergonic or exergonic, need the addition of energy to start. The burning of wood is an exergonic reaction. Once lighted, wood gives off energy as it burns. However, wood does not start burning by itself. Heat must be supplied to the wood to start the burning. The energy that is needed to start a chemical reaction is called **activation energy**. Once started, an exergonic reaction continues to give off energy. In many exergonic reactions, energy is given off quickly. The amount of activation energy needed to start an exergonic reaction is usually much less than the energy given off by the reaction.

Perhaps you have burned sugar in a saucepan. In this chemical reaction the sugar is broken down. However, the reaction cannot begin unless an outside energy source is added. Heat must be added to the sugar. Then the sugar is broken down and its energy released.

Exergonic reactions take place in cells. For example, cells break down sugar and energy is released. In cells, however, there is not enough heat to start reactions. High heat would kill cells. Also, the quick release of energy, as occurs in burning, does not take place in cells. How do cells start and control reactions?

The rates of chemical reactions within the cells are controlled by substances called catalysts. A **catalyst** (KAT uh lihst) is a substance that alters the rate of a chemical reaction without itself being changed or used up. The catalysts in cells are special proteins called **enzymes** (EHN zymes). Enzymes make reactions possible by reducing the amount of activation energy needed. Enzymes also control the rate at which reactions take place so that the energy produced can be given off slowly. Thus enzymes allow reactions to take place at temperatures low enough to prevent harm to cells.

FIGURE 5·3 Exergonic reactions give off heat. The burning of wood is an example of an exergonic reaction.

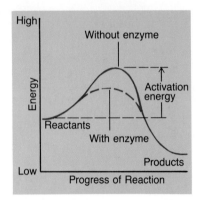

FIGURE 5·4 Enzymes lower the amount of energy necessary to start a chemical reaction. Enzymes are catalysts because they control the rate of a chemical reaction without being changed or used up.

FIGURE 5·5 Many enzymes in your body are named after the substrates on which they act.

SUBSTRATE	ENZYME
Sucrose	Sucrase
Lactose	Lactase
Maltose	Maltase
Urea	Urease
Ribonucleic acid	Ribonuclease

Figure 5·4 illustrates how an enzyme controls a chemical reaction. The graph shows the energy change in an exergonic reaction. Notice that the products have less energy than the reactants. The hill in the graph shows the amount of activation energy needed for the reaction to take place. When no enzyme is present, the activation energy is high (solid line). Notice how much less activation energy is needed when an enzyme is present (dashed line).

The thousands of reactions that occur in living things are controlled by thousands of enzymes. Different reactions are controlled by different enzymes. The substance on which an enzyme acts is called the **substrate.** A substrate is changed into one or more new products through a reaction.

An enzyme may be named after the substrate on which it acts. The suffix -*ase* is added to part of the name of the substrate. Sometimes an enzyme is named for the kind of general compound on which it acts. What, do you think, is the substrate for a protease?

Enzymes are not used up in reactions and can be used over again. A single enzyme can catalyze from 100 to 30 000 000 reactions in a minute! However, a particular enzyme acts on one specific substrate. So a particular enzyme can control only one kind of reaction.

In some enzyme-controlled reactions, small molecules called *coenzymes* join with the enzymes to control the reactions. Unlike enzymes, coenzymes are not proteins. But like enzymes, they are not changed during the reactions. Some vitamins, such as B_1, B_2, B_6, and K, are coenzymes. A reaction that needs a coenzyme will not take place if the coenzyme is not present.

5·3 ENZYME MODELS

The shape and structure of an enzyme determine the reaction that an enzyme can catalyze. As shown in Figure 5·6, part A, in a chemical reaction an enzyme unites for a time with the specific substrate on which it can act. An enzyme differs in shape and structure from other enzymes. The **active site,** a special region of the enzyme, can join with a specific substrate or substrates. The substrate has a molecular shape that fits the active site of the enzyme. The enzyme joins with the substrate to form an **enzyme-substrate complex,** or **E-S.** At the active site the enzyme and substrate fit together. The idea that enzyme action is based on a fit between enzyme and substrate brought about this **lock-and-key model.** The enzyme and the substrate fit together like a lock and its key. When the E-S is formed, the activation energy is lowered. This lower activation energy allows the reaction to take place more quickly than it would without the enzyme being present.

After the reaction has taken place, the products are released. The active site is then free to join with other molecules of the substrate. The enzyme is free to act as a catalyst for another reaction.

There is a second model of enzyme activity that differs somewhat from the lock-and-key model. In the **induced-fit model,** the sub-

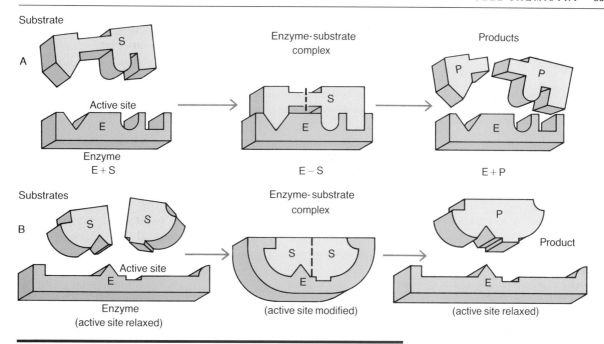

FIGURE 5·6 Scientists have proposed several models that explain how enzymes work. The lock-and-key model (A) and the induced-fit model (B) are shown here.

strates alter the structure of the enzyme's active site. Figure 5·6, part B shows that when the substrates are close to the active site, they cause changes in the site's structure that allow the active site to fit the substrates. This change is similar to the way that you change the shape of an empty glove by "inducing" it to fit your hand. As in the lock-and-key model, the active site is such that it fits only a specific substrate (or specific substrates).

5·4 FACTORS AFFECTING ENZYME ACTIVITY

Enzymes are proteins and protein structure determines function. Thus the factors that affect protein structure also affect enzyme function, or activity. Temperature and pH affect protein structure and thus enzyme function. Substrate concentration also affects the activity of enzymes.

Most chemical reactions take place faster at higher temperatures. Figure 5·7 shows how the rate of an enzyme-controlled reaction changes with temperature. What happens to the rate of the reaction as the temperature gets higher? Can you explain what is happening?

Proteins are heat sensitive. High temperatures break certain bonds within protein molecules. This causes the proteins to change shape. This change in the shape of a protein molecule is known as *denaturation* (dee nay chuh RAY shuhn). Denaturation changes the shape of the active site of an enzyme, causing the enzyme to no longer fit its substrate. Therefore the rate of the reaction declines above a certain temperature.

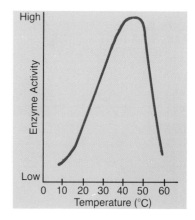

FIGURE 5·7 The rate of an enzyme controlled reaction rises as the temperature increases. At a certain point, however, the rate drops sharply. As you can see, the optimum temperature of enzyme activity is near the normal human body temperature of 37°C.

84 CELL CHEMISTRY

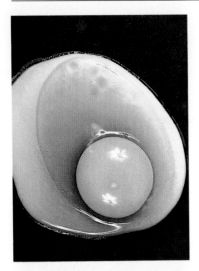

FIGURE 5·8 Egg whites go from a liquid to a solid as they are cooked. Cooked egg white is an example of a denatured protein.

Many scientists think that the serious effects of very high fever in humans are related to the denaturation of enzymes. Important cellular reactions may take place too slowly during high fever. Denaturation of protein usually is not reversible.

The activity of an enzyme is also affected by pH. Different enzymes work best at low pH (acid), neutral pH, or high pH (basic). The effect of pH that is too high or too low is similar to the effect of high temperature. As a result of denaturation, the enzymes no longer control the reactions. Figure 5·9 shows the effect of pH on different enzymes. Which graph shows the activity of pepsin, an enzyme that works in the acidic environment of the stomach?

Enzyme activity also can be affected by the concentration of substrate. As the concentration of substrate increases, enzyme activity increases *up to a certain point*. But beyond that point, which differs for each enzyme, enzyme activity stops increasing and remains constant. There is a limit to the amount of enzyme that is available to join with substrate molecules. The substrate molecules cannot join with an enzyme until an active site on an enzyme is free.

An enzyme's ability to function can be interfered with by different kinds of chemicals. For example, a compound similar in shape to the substrate may bind to an enzyme's active site. As long as the compound remains bound, the enzyme cannot function. A chemical that interferes with enzyme function is called an *inhibitor*.

FIGURE 5·9 The activity of enzymes can be affected by pH. Enzyme A would not work well in an acidic environment.

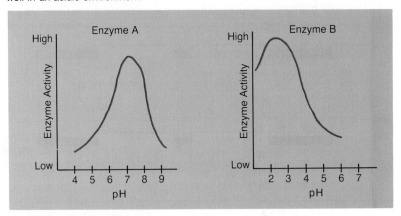

REVIEW QUESTIONS

1. Distinguish between anabolic and catabolic reactions.
2. Compare dehydration synthesis and hydrolysis.
3. Why are enzymes needed to control chemical reactions within cells?
4. Describe the lock-and-key model of enzyme function. How does it help to explain denaturation of an enzyme?
5. The pH of the small intestine is basic. What is likely to happen to the enzyme pepsin that enters the small intestine from the stomach?

Investigation

What Are the Actions of Certain Enzymes?

Goals
After completing this activity, you will be able to
- Demonstrate how the action of an enzyme is specific for a certain substrate.
- Identify the substrate in an enzyme-substrate reaction.
- Use certain nutrient tests to show the presence of starch and protein.

Materials (for groups of 3 or 4)
3 test tubes, each with 2 mL solidified gelatin
paper towel   3 empty test tubes
test-tube rack
6 mL starch solution
glass-marking pencil
Lugol's iodine solution
meat tenderizer   2 stirring rods
saliva   37°C water bath
100-mL beaker   thermometer
10-mL distilled water   hot plate
10-mL graduate

Procedure

A. On a separate sheet of paper, set up a data chart. In the chart list (a) test tubes 1–6, (b) the substrate—gelatin or starch, (c) the substance added, and (d) the changes. Record your observations as you do the steps of the activity.

B. In the beaker collect enough saliva from several students to equal 5 mL. Add 5 mL distilled water to the saliva.

C. From your teacher obtain 3 test tubes, each with 2 mL solidified gelatin. With a glass-marking pencil number these tubes 1, 2, and 3. As shown in Figure A add 2 mL saliva to tube 2. To tube 3 add enough meat tenderizer to cover the surface of the gelatin.

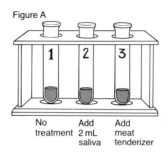

Figure A

D. Three minutes after adding the saliva and meat tenderizer to the tubes with gelatin, gently poke the surface with a stirring rod. Wipe the rod with a paper towel after each tube is tested. At 1-minute intervals repeat the poking until you see a change in the gelatin surface in any tube.

E. In your data chart, record any changes you observe in the gelatin.

F. With a glass-marking pencil, number three clean test tubes 4, 5, and 6. Use the graduated cylinder to pour 2 mL of starch solution into each tube.

G. Treat each tube of starch as shown in Figure B. Use a paper towel to wipe the stirring rod after the starch is mixed in each tube. Allow the tubes to warm in a 37°C water bath. See Figure C for the proper way to warm the tubes. After 10 minutes, add two drops of iodine to tubes 5 and 6 as shown in Figure B.

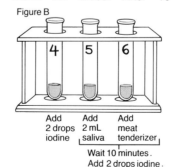

Figure B

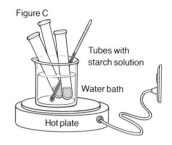

Figure C

H. In your data chart, record any changes you observed in the tubes of starch after treatment with iodine, saliva, and meat tenderizer.

I. Follow your teacher's instructions for the proper disposal of all materials.

Questions and Conclusions
1. The color observed in test tube 4 indicates the presence of starch. Of test tubes 5 and 6, which tube shows no evidence of starch?
2. In which tube of gelatin is the surface of the gelatin different from the untreated gelatin in tube 1?
3. According to your observations in this activity, which enzyme source has an enzyme that reacts with starch? With protein? Give evidence to support your answers.
4. Identify the substrates used in this activity.
5. How does this activity show that enzymes are specific for certain substrates?
6. How does this activity show the effects of enzymes on substrates?

CELL CHEMISTRY   85

Energy for Cells

SECTION OBJECTIVES

- Explain why ATP is important to living systems.
- Compare different methods of producing ATP from glucose.

5·5 ADENOSINE TRIPHOSPHATE

The main source of energy for living things is *glucose,* a six-carbon sugar. Chemical energy is stored in glucose and in other organic molecules that can be changed to glucose. Cells can use the energy to do work such as pulling (muscle cells), carrying impulses (nerve cells), moving nutrients (plant root cells), and making protein and other compounds needed in the cell.

Energy is give off when glucose is broken down in cells. The energy is given off in a series of enzyme-controlled steps. Most of the energy given off is stored in another chemical compound. The compound that stores the energy in cells is **adenosine triphosphate** (uh DEHN uh seen trī FAHS fayt), or **ATP.**

The complex ATP molecule is shown in Figure 5·10. Adenosine has two parts—adenine, a base also found in DNA and RNA, and ribose, a five-carbon sugar also found in RNA. Each of the three phosphate groups is made of one phosphorus atom joined to four oxygen atoms. Hydrogens are bonded to some of the oxygen atoms. Note the wavy lines between the phosphate groups. These wavy lines represent *high-energy bonds.* The chemical energy stored in compounds can be thought of as being stored in the bonds. When an enzyme breaks the end phosphate group away from ATP, a great amount of energy is given off for use by the cell. The molecule that remains when ATP loses a phosphate group is **adenosine diphosphate** (uh DEHN uh-seen dī FAHS fayt), or **ADP.**

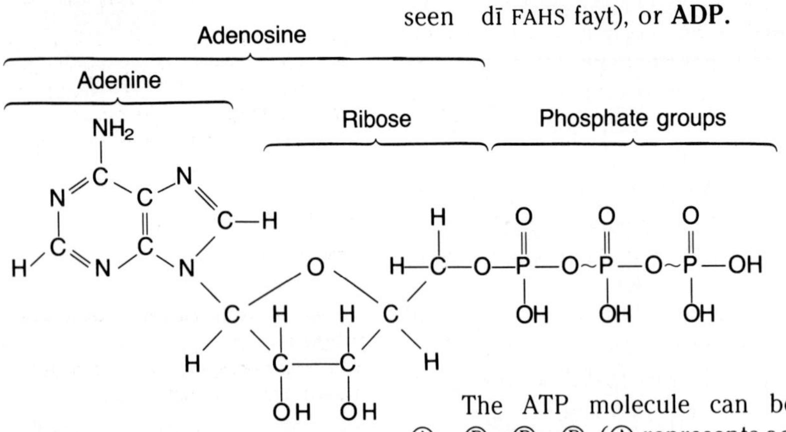

FIGURE 5·10 Energy is stored in the body as ATP. The structural formula of the ATP molecule is shown here. Three phosphate groups are attached to an ATP molecule.

The ATP molecule can be shown in shorthand form as Ⓐ—Ⓟ~Ⓟ~Ⓟ. (Ⓐ represents adenosine; Ⓟ represents phosphate.) The reaction in which ATP forms ADP and Ⓟ provides usable energy for the cell and can be written as follows.

$$Ⓐ—Ⓟ~Ⓟ~Ⓟ \longrightarrow Ⓐ—Ⓟ~Ⓟ + Ⓟ + \text{energy}$$

If the chemical reaction shown is reversed, ADP and Ⓟ can be joined to form ATP. What happens to the energy that enters this reverse reaction? Where does the energy come from?

A cell has a constant need for energy. Thus a cell must be able to produce ATP continuously from ADP and ⓟ, which are present in the cell. The energy needed to form ATP from these materials comes from food, usually glucose. The ATP formed is a more usable source of energy than the food was. ATP breaks down and gives off energy much more readily than food does. Thus ATP is a valuable, easily used carrier of chemical energy for cells.

5·6 CELLULAR RESPIRATION

Some of the energy released when glucose is broken down in a living cell is used to make ATP. In most cells this process requires oxygen. The chemical breakdown of glucose that uses oxygen or some other inorganic substance is called **cellular respiration**. Cellular respiration that needs oxygen is called **aerobic** (air OH bihk) **respiration**.

The breakdown of glucose in aerobic respiration involves a series of reactions. However, the overall reaction can be represented by the following equation.

$$C_6H_{12}O_6 + 6O_2 \xrightarrow{\text{enzymes}} 6CO_2 + 6H_2O + ATP$$
(glucose) (oxygen) (carbon dioxide) (water) (energy)

Note that the glucose is broken down into carbon dioxide and water in aerobic respiration.

• *Glycolysis* Aerobic respiration takes place in two stages. In the first stage glucose is split into two molecules of pyruvic (pī ROO vihk) acid. Pyruvic acid is a three-carbon compound. Two ATP molecules are used in the reaction, but four ATP molecules are made. In addition, hydrogens, along with electrons, are moved to a coenzyme called nicotinamide adenine dinucleotide (NAD^+), forming NADH. The production of ATP by changing glucose to pyruvic acid is called **glycolysis** (glī KAHL uh sihs). Glycolysis occurs in the cytoplasm of the cell.

Glycolysis is *anaerobic* (an air OH bihk), or does not require oxygen. However, the pyruvic acid made by glycolysis is used in the second stage of aerobic respiration. The second stage is aerobic; it does require oxygen.

Glycolysis gives off only about 10 percent of the energy available from glucose. This energy is stored in the form of ATP and NADH. The remaining energy from glucose is given off by the breakdown of each molecule of pyruvic acid into water and carbon dioxide.

Figure 5·12 shows the first step in the breakdown of pyruvic acid, the changing of the three-carbon pyruvic acid to a two-carbon compound. This two-carbon compound is acetic (uh SEE tihk) acid. It is bound to a coenzyme called coenzyme A, or coA for short. The name for the entire compound is acetyl-coA (uh SEE tuhl koh AY). Notice that a molecule of carbon dioxide is formed when acetyl-coA is formed. Also, the hydrogen that is removed from the pyruvic acid, along with electrons, becomes bound to NAD^+ and forms NADH.

FIGURE 5·11 This gymnast needs to have a good source of energy to complete her routine. ATP is the source of energy required during exercise.

an- (without)
aeros (air)

FIGURE 5·12 Before the citric acid cycle can begin, pyruvic acid must be broken down into acetic acid.

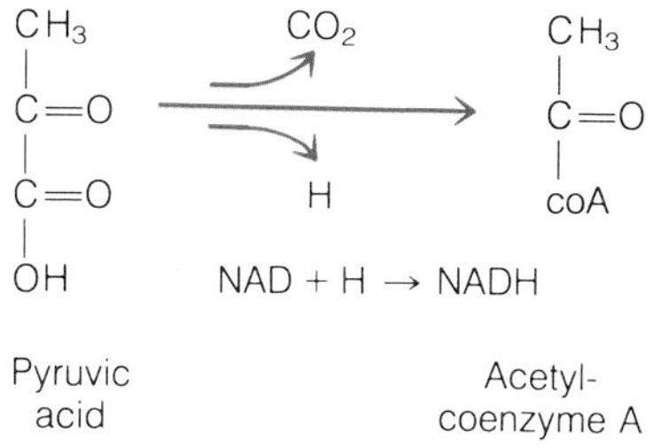

Pyruvic acid → Acetyl-coenzyme A
(with CO_2 released and NAD + H → NADH)

- **Citric Acid Cycle** The acetyl-coA enters a series of reactions called the **citric acid cycle,** which completes the breakdown of glucose. Follow the steps in Figure 5·13 as you read about the cycle.

1. Acetyl-coA joins with a four-carbon compound (oxaloacetic acid) to form a six-carbon compound (citric acid).

2–4. In these reactions citric acid is changed back to oxaloacetic acid. Note that at some points CO_2 is given off, NADH (or a similar hydrogen carrier, coenzyme $FADH_2$) is generated, and ATP is produced. The cycle continues.

You will recall that one glucose molecule breaks down to two pyruvic acid molecules during glycolysis. The glucose molecule is completely broken down after both molecules of pyruvic acid enter the reactions of the citric acid cycle.

The citric acid cycle can break down substances other than acetyl-coA. Some substances formed from the breakdown of lipids and proteins can enter the citric acid cycle at different points. Energy is obtained as such substances are further broken down in the cycle.

FIGURE 5·13 The citric acid cycle breaks down pyruvic acid, a product of glycolosis. As pyruvic acid is broken down, energy, carbon dioxide, and electrons for the electron transport chain are produced.

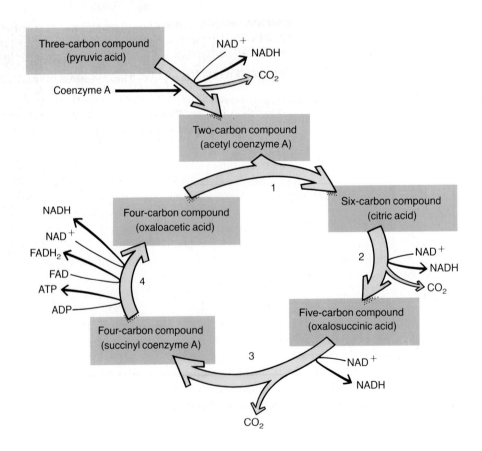

- *Electron Transport Chain* The CO_2 formed in the citric acid cycle is a waste product that is given off. A molecule of ATP is produced during each cycle. However, most of the energy from the glucose is now being carried by molecules of NADH and $FADH_2$. As these molecules undergo a series of changes, electrons are given off and the energy in their chemical bonds is released. These electrons undergo a series of transfers on the cristae in the mitochondria between compounds that are electron carriers. This series of electron carriers is called the **electron transport chain.** One of these carriers is a coenzyme. The rest are iron-containing compounds called *cytochromes*. Each carrier is lower in energy than the carrier before it. The energy that is given off as the electrons are transferred from one carrier to another is used to form ATP.

Figure 5•14 shows the electron transport chain. ATP is formed as electrons high in energy are passed down the chain and become lower in energy. The electrons and hydrogen, first carried by NADH and $FADH_2$, are finally passed to oxygen. Oxygen is thus the final electron acceptor in the chain. Using Figure 5•14, determine what product forms from oxygen receiving these electrons and hydrogen.

The electron transport chain results in the production of 32 molecules of ATP per glucose molecule broken down. There is a net gain of two ATP's from glycolysis and two ATP's from the citric acid cycle. Thus there is a net gain of 36 ATP's when one glucose molecule is broken down into carbon dioxide and water.

It is clear from the above figures that most of the ATP, or usable energy, formed during cellular respiration is formed during the aerobic, or oxygen-requiring stage, which takes place in the mitochondria. In the electron transport chain, oxygen is the final electron acceptor. Thus living things that break down glucose by means of aerobic respiration are dependent on the intake of oxygen for energy.

Perhaps you have taken part in an aerobic exercise class. If exercise is truly aerobic, the energy needs for the exercise are met by aerobic respiration. Since energy production and energy use are in balance, aerobic exercise can be performed without muscle fatigue.

Not all forms of respiration need oxygen, however. Certain bacteria break down food by a means of respiration that is anaerobic. Respiration in which the final electron acceptor in the electron transport chain is an inorganic substance other than oxygen, such as sulfate (SO_4^{-2}) or nitrate (NO_3^-), is called **anaerobic respiration.** Anaerobic respiration yields less ATP than does aerobic respiration.

5•7 FERMENTATION

Respiration is only one way of breaking down glucose. Fermentation (fer mehn TAY shuhn) is another anaerobic way of breaking down glucose and producing usable energy. Recall that in respiration the final electron acceptor is an *inorganic* substance, usually oxygen. **Fermentation** is the breakdown of glucose and the release of energy in which *organic* substances are the final electron acceptors.

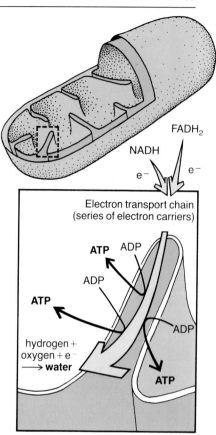

FIGURE 5•14 ATP is produced in mitochondria (top). Electrons from NADH and $FADH_2$ provide the energy for ATP production (bottom).

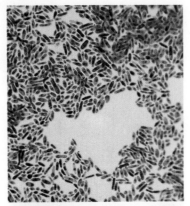

FIGURE 5·15 The fermentation of corn by yeast (top) in industrial plants produces ethanol. Ethanol can be added to gasoline and used to power automobiles and other vehicles.

Some living things, such as certain kinds of bacteria, get energy only by fermentation. They do not need oxygen. In fact, certain kinds of bacteria cannot live in the presence of oxygen. Some cells, such as human muscle cells, can produce energy for a short time by fermentation. However, fermentation in such cases is an "emergency" way to make energy when oxygen is in short supply.

The first part of fermentation is glycolysis. As in respiration, two molecules of pyruvic acid are formed. This gives a net gain of two molecules of ATP. Then, depending on the living thing, the pyruvic acid is changed either to alcohol and CO_2 or to lactic acid. It is in this second part that fermentation is different from respiration.

Fermentation that forms alcohol and CO_2 as products is called **alcoholic fermentation.** In alcoholic fermentation, pyruvic acid is changed to ethyl alcohol and carbon dioxide. The reaction can be shown as follows.

$$C_6H_{12}O_6 \rightarrow 2C_2H_5OH + 2CO_2 + 2ATP$$
$$\text{(glucose)} \quad \begin{pmatrix}\text{ethyl}\\\text{alcohol}\end{pmatrix} \quad \begin{pmatrix}\text{carbon}\\\text{dioxide}\end{pmatrix} \quad \text{(energy)}$$

Yeast cells carry on alcoholic fermentation. Carbon dioxide given off by fermentation in the yeast cells causes bread dough to rise. The air spaces in bread are caused by bubbles of CO_2. Then the dough is baked.

Recall that there is a net gain of only two ATP's in the fermentation of one molecule of glucose. Most of glucose's potential energy has not been moved to ATP but is, instead, in the alcohol. This potential energy makes ethyl alcohol a valuable commercial fuel.

Many kinds of cells can change pyruvic acid to lactic acid. Fermentation that forms lactic acid is called **lactic acid fermentation.** The reaction for lactic acid fermentation is shown as follows.

$$C_6H_{12}O_6 \rightarrow 2CH_3CHOHCOOH + 2ATP$$
$$\text{(glucose)} \quad \text{(lactic acid)} \quad \text{(energy)}$$

Lactic acid fermentation, like alcoholic fermentation, is anaerobic. The net gain in energy is also two ATP's for every glucose molecule broken down. The remaining energy is stored in the lactic acid molecule. Lactic acid fermentation is important in the making of many dairy products, such as most cheeses, buttermilk, and yogurt.

Lactic acid fermentation is the sole means of glucose breakdown in some living things. It also can take place within the cells of humans and other living things in which glucose is normally broken down by means of aerobic respiration. Under most conditions, glycolysis is followed by the aerobic breakdown of pyruvic acid. But when not enough oxygen is present, as in the muscle cells of a sprinting runner, lactic acid fermentation takes place. The buildup of lactic acid in a muscle produces muscle fatigue and the burning sensation felt by those who exercise strenuously. Recovery from the fatigue involves energy production by aerobic respiration. Lactic acid eventually is carried by the blood to the liver, where it is changed to glucose.

As you can see in Figure 5·16, fermentation produces a net gain of two molecules of ATP from one molecule of glucose. In contrast, cellular respiration produces a net gain of 36 molecules of ATP from one molecule of glucose. Obviously, the glucose breakdown in aerobic respiration is more complete and gives more usable energy (ATP) than the glucose breakdown in fermentation. However, fermentation is still important to aerobic organisms. Muscle cells would be unable to contract if there were no way to make energy when oxygen is not present.

REVIEW QUESTIONS

6. What molecule is the primary source of energy for cells? Where does it come from?
7. ATP is sometimes called the "energy currency" of the cell. How can you best explain this analogy?
8. Name two differences between cellular respiration and fermentation.
9. How do the two types of fermentation differ?
10. Predict what would happen if a cell lacked the enzyme that controls the conversion of pyruvic acid to acetyl-coA.

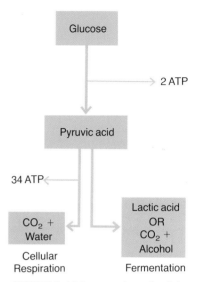

FIGURE 5·16 A comparison of cellular respiration and fermentation is shown here. Cellular respiration produces more energy than fermentation.

The Process of Photosynthesis

5·8 CONDITIONS NEEDED FOR PHOTOSYNTHESIS

Most autotrophs make their own food by using light energy. The light energy is changed to chemical energy stored in glucose. The process of **photosynthesis** (*photo,* light; *synthesis,* putting together) allows these autotrophs to make their own food. Living things depend directly or indirectly on photosynthesis for food. Some heterotrophs, such as cows and deer, eat plants. Other heterotrophs, such as wolves and coyotes, eat plant eaters. So the energy captured by autotrophs is passed throughout all living things.

Photosynthesis is a complex process. However, the overall reaction can be summarized as follows.

$$6CO_2 + 6H_2O + \text{light energy} \xrightarrow[\text{chlorophyll}]{\text{enzymes}} C_6H_{12}O_6 + 6O_2$$
(carbon dioxide) (water)   (glucose)   (oxygen)

Is photosynthesis an exergonic reaction or an endergonic reaction?

The reaction summarizes the conditions needed to carry out photosynthesis. Sunlight is the energy source. **Chlorophyll,** a green pigment in the cells of photosynthetic autotrophs, captures light energy. Carbon dioxide and water are the raw materials. Enzymes and coenzymes control the formation of glucose from these raw materials.

SECTION OBJECTIVES

• Describe how autotrophs make food.

• Discuss the cyclic relationship between the processes of respiration and photosynthesis.

5·9 LIGHT AND PIGMENTS

Light energy is needed for photosynthesis to occur. Light is a form of *radiant energy*. Radiant energy is energy that travels as waves. Light, radio waves, infrared waves, ultraviolet waves, and X rays are all forms of radiant energy. But only light waves are used by autotrophs to make food.

When light strikes a material, some of the light energy is absorbed and changed to other forms of energy. An example of such a change is the rise in temperature of a car seat when sunlight shines on it for a time. In this case the light energy changes to heat energy. When sunlight strikes molecules of chlorophyll in a cell, some of the light energy is absorbed by the chlorophyll. This energy eventually is changed into chemical energy in molecules of glucose.

When a beam of light passes through a prism, it is broken up into colors. The colors make up the *visible spectrum*. The colors of the visible spectrum are red, orange, yellow, green, blue, indigo, and violet. The colors of the spectrum that the pigment chlorophyll absorbs best are blue and red. Chlorophyll appears green because it reflects much of the green light that falls on it, as shown in Figure 5·17. Thus green light is not as important in photosynthesis as are some of the other colors of light.

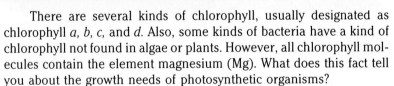

FIGURE 5·17 Leaves appear green because green light is reflected from leaves to your eyes. The other colors are absorbed by the leaf. Red and blue light are used by the plant as energy for photosynthesis.

There are several kinds of chlorophyll, usually designated as chlorophyll *a, b, c,* and *d.* Also, some kinds of bacteria have a kind of chlorophyll not found in algae or plants. However, all chlorophyll molecules contain the element magnesium (Mg). What does this fact tell you about the growth needs of photosynthetic organisms?

Pigments called carotenoids (kuh RAHT uh noidz) are also found in autotrophs. Carotenoids may be orange, yellow, or red in color. These pigments usually are masked by the green color of chlorophyll. However, they are seen in leaves in the autumn, when the amount of chlorophyll in the leaves is reduced. The carotenoids also absorb light, but they are less important than chlorophyll in this process.

5·10 LIGHT REACTIONS

Photosynthesis takes place in two major parts. The chemical reactions of step one depend on light energy. The reactions that depend on light are called **light reactions.** The chemical reactions of step two do not depend directly on light. The reactions that do not depend directly on light are called the **dark reactions.** The light reactions take place in the grana of the chloroplasts. The dark reactions take place in the stroma. The light reactions are described below.

1. Light energy is absorbed by chlorophyll and other pigment molecules in the grana of the chloroplast.
2. This increases the energy of certain electrons in the pigment molecules. Electrons with the increased energy are said to be "excited," or "activated," electrons. They are said to be at a higher energy level. As the electrons of the pigments return to their "unexcited" state, or to a lower energy level, they give off energy.
3. Electrons return to a lower energy level by being passed down an electron transport chain, much like in cellular respiration. In the electron transport chain energy is released from electrons and ATP is formed. In other words, the energy of the electrons becomes energy in a usable form (ATP) in the chloroplast. The ATP produced during the light reactions is used in the dark reactions.

ATP production is not the only result of the light reactions. In the light reactions, water is split into oxygen and hydrogen ions. The oxygen is given off. Sometimes you can see small bubbles on leaf surfaces of water plants. Those bubbles contain oxygen that has been given off in the light reactions of photosynthesis. The hydrogens formed when water is split in the light reactions join with an electron carrier to form NADPH. NADPH is used in the dark reactions.

Figure 5·18 shows the light and dark reactions of photosynthesis. Water is used in the light reactions and oxygen, ATP, and hydrogens are formed. The oxygen is given off. The ATP and hydrogens formed in light reactions are used in dark reactions.

FIGURE 5·18 Plants give off oxygen during the light reactions. During the dark reactions, plants make food in the form of glucose.

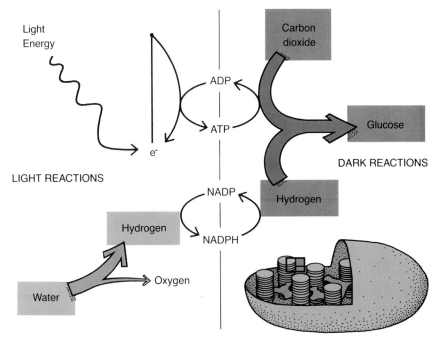

94 CELL CHEMISTRY

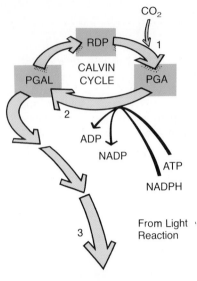

FIGURE 5·19 The Calvin cycle is part of the dark reactions. As you can see, glucose and starch are produced.

5·11 DARK REACTIONS

In the dark reactions of photosynthesis, which take place in the stroma of the chloroplasts, carbon dioxide is used and glucose is formed. The dark reactions can take place in the light, but light is not needed. The dark reactions include a series of reactions called the *Calvin cycle*. Each step in the dark reactions is controlled by an enzyme. Follow the steps of the dark reactions in Figure 5·19.

1. Carbon dioxide joins with a compound called RDP (ribulose diphosphate), producing another compound, PGA (phosphoglyceric acid).
2. Using the ATP and hydrogens (carried by NADPH) formed during the light reactions, molecules of a compound called PGAL (phosphoglyceraldehyde) are formed from PGA.
3. Glucose is formed from PGAL in another series of reactions.

5·12 FACTORS AFFECTING PHOTOSYNTHESIS

The rate at which photosynthesis takes place in plants is not always the same. The graph in Figure 5·20A shows how the rate of photosynthesis changes as the light intensity increases. Increasing the light increases the amount of light energy present for photosynthesis. But what happens when the light intensity rises beyond 9000 lumens? The rate of photosynthesis does not increase. This is because the enzyme-controlled reactions, which are not speeded up directly by increasing light intensity, are taking place at maximum rates.

Figure 5·20B shows how the rate of photosynthesis changes as the temperature increases and the light intensity is high. In this case the rate increases to a certain point and then drops off. At about what temperature does the rate of photosynthesis decrease dramatically? What accounts for this rapid decrease? In this case, heat denatures enzymes and this leads to a decrease in the rate of photosynthesis.

FIGURE 5·20 Light intensity and temperature affect the rate of photosynthesis. The rate of photosynthesis levels off at about 9000 lumens, and drops sharply at a temperature of 35°C.

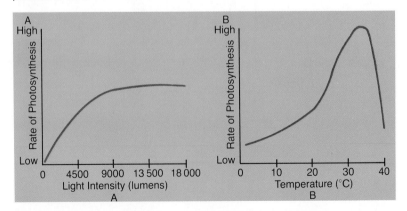

Artificial Photosynthesis
Discovery

Many scientists are studying photosynthesis to learn new ways to use solar energy. They want to understand how the chloroplast does its job so well. Scientists can then apply that knowledge to build devices that use solar energy to make fuel such as gasoline.

Some scientists are isolating chlorophyll from plants and combining it with other chemicals to carry out a process they call artificial photosynthesis. In its use of chlorophyll, this process resembles photosynthesis in plants.

Other scientists are using proteins from chloroplasts to make fuel. They isolate the chloroplast proteins that convert light energy into chemical energy and bond them to plastic sheets. When these sheets are exposed to light and chemicals, fuels are produced.

Finally, other scientists are trying to change plants so that they can make fuel that we can use. This can be done by treating plants with chemicals that change the photosynthetic activity of the plants.

5·13 RESPIRATION VS. PHOTOSYNTHESIS

Photosynthesis provides energy for living things by changing light energy to chemical energy. Glucose can be used by the plant that formed it or by an animal that eats the plant. Respiration provides a way of getting the energy from glucose.

During respiration, oxygen is taken from the air. The oxygen is used by the cell, and carbon dioxide is formed. Respiration provides the energy for both plant and animal cells to carry on their life activities. If there were no way of replacing the oxygen or removing the carbon dioxide from the atmosphere, most life would end. The process of photosynthesis removes carbon dioxide from the atmosphere and forms oxygen.

FIGURE 5·21 This figure compares photosynthesis and respiration. As you can see, respiration is the reverse of photosynthesis.

	PHOTOSYNTHESIS	RESPIRATION
Reactants	CO_2, H_2O	$C_6H_{12}O_6$, O_2
Products	$C_6H_{12}O_6$, O_2	CO_2, H_2O
Equation	$6CO_2 + 6H_2O + E \rightarrow C_6H_{12}O_6 + 6O_2$	$C_6H_{12}O_6 + 6O_2 \rightarrow 6CO_2 + 6H_2O + E$

REVIEW QUESTIONS

11. What is the overall reaction for photosynthesis?
12. What is the function of the pigments involved in photosynthesis?
13. How are the products of the light reactions of photosynthesis used in the dark reactions?
14. How are respiration and photosynthesis part of one cycle?
15. Predict how a decreasing amount of water would effect on the rate of photosynthesis.

Chapter Review

SUMMARY

5·1 Cells break down organic molecules by hydrolysis and build up other organic molecules by dehydration synthesis.

5·2 Many reactions in living systems require activation energy. Organic catalysts called enzymes lower the amounts of activation energy needed for cellular reactions.

5·3 Every reaction in a cell is controlled by an enzyme. Enzymes are proteins and can be used over again. The active site of an enzyme fits the structure of a specific substrate.

5·4 Enzyme activity is affected by pH, temperature, and substrate concentration.

5·5 Cells convert the chemical energy in glucose to the more readily available chemical energy stored in ATP.

5·6 In cellular respiration, glucose breaks down to produce ATP. Aerobic respiration uses oxygen and produces ATP, water, and carbon dioxide.

5·7 Fermentation, an anaerobic process, converts glucose to either alcohol and CO_2 or lactic acid and produces small amounts of ATP.

5·8 Carbon dioxide and water are the raw materials needed by autotrophs to carry out the process of photosynthesis.

5·9 The pigment chlorophyll is located in chloroplasts. Chlorophyll can absorb sunlight for the process of photosynthesis.

5·10 In the light reactions of photosynthesis, the energy of sunlight is converted to chemical energy as water is split and oxygen released. NADPH and ATP are produced.

5·11 The dark reactions of photosynthesis do not require light. Products of the light reactions, ATP and NADPH, are used in the dark reactions. Glucose is formed in the dark reactions.

5·12 The rate of photosynthesis is influenced by the amounts of light and heat.

5·13 In photosynthesis, energy is used to combine carbon dioxide and water into glucose, which is used by plants and animals. In respiration, glucose is broken down into carbon dioxide and water, releasing energy.

BIOLOGICAL TERMS

activation energy
active site
adenosine diphosphate
adenosine triphosphate
aerobic respiration
alcoholic fermentation
anabolic reactions
anaerobic respiration
autotrophs
catabolic reactions
catalyst
cellular respiration
chlorophyll
citric acid cycle
dark reactions
dehydration synthesis
electron transport chain
endergonic reaction
enzymes
enzyme-substrate complex
exergonic reaction
fermentation
glycolysis
heterotrophs
hydrolysis
induced-fit model
lactic acid fermentation
light reactions
lock-and-key model
photosynthesis
substrate

USING BIOLOGICAL TERMS

1. Distinguish between aerobic respiration and anaerobic respiration.
2. Distinguish between an anabolic reaction and a catabolic reaction.
3. Distinguish between adenosine diphosphate and adenosine triphosphate.
4. Distinguish between an autotroph and a heterotroph.
5. Distinguish between an exergonic reaction and an endergonic reaction.

Choose the term that does not belong and explain why it does not belong.

6. enzyme, active site, autotroph, substrate
7. chlorophyll, light reaction, dark reaction, heterotroph

8. glycolysis, catalyst, fermentation, aerobic respiration
9. lactic acid fermentation, anaerobic respiration, photosynthesis, alcoholic fermentation
10. glycolysis, dark reactions, electron transport chain, citric acid cycle

UNDERSTANDING CONCEPTS

1. How is dehydration synthesis the opposite of hydrolysis? (5·1)
2. Give an example of an endergonic reaction and an exergonic reaction. (5·2)
3. Give an example of using activation energy to start a reaction. (5·2)
4. How do enzymes affect the chemical reactions occurring in cells? (5·2)
5. How is the lock-and-key model used to explain the way enzymes work? How does this model differ from the induced-fit model of enzyme activity? (5·3)
6. In terms of enzyme function, why is a high fever dangerous to an organism? (5·4)
7. How is ATP useful to a cell? (5·5)
8. Use the following components to write the chemical reaction for aerobic respiration: $C_6H_{12}O_6$, CO_2, H_2O, O_2, enzymes, energy. (5·6)
9. Athletes in competitive sports may experience extreme muscle fatigue. Explain this fatigue in terms of lactic acid. (5·7)
10. How are fermentation and aerobic respiration alike? How are they different? (5·7)
11. Use the following components to write the chemical equation for photosynthesis: $C_6H_{12}O_6$, light, O_2, H_2O, CO_2, enzymes, chlorophyll. (5·8)
12. Describe the role of chlorophyll in the processes of photosynthesis. (5·9)
13. What is the source of the oxygen that is released during the light reactions of photosynthesis? (5·10)
14. What happens to the carbon dioxide (CO_2) involved in photosynthesis? (5·11)
15. How does an increase in light intensity affect the rate of photosynthesis? (5·12)
16. Compare the chemical equations for respiration and photosynthesis. (5·13)

APPLYING CONCEPTS

1. Predict what would happen to the earth's atmosphere if green plants were to disappear. How would animal life be affected?
2. Describe what would happen in an enzyme-controlled reaction if conditions changed as follows: (a) temperature is decreased; (b) temperature is increased; (c) amount of enzyme is increased; (d) strong acid is added.
3. All food that is eaten by humans contains energy captured during photosynthesis. Explain how this is so.
4. In wine making, there is an increase in temperature in the vats in which fermentation of grape juice occurs. Explain why this temperature increase occurs.
5. Scientists discover that *substance X* has a molecular shape similar to that of pyruvic acid. How might substance X interfere with respiration?

EXTENDING CONCEPTS

1. Design an experiment that would determine which color of light in the visible spectrum is best utilized by green plants in the process of photosynthesis.
2. Cyanide (CN^-) is a poison. Find out how cyanide affects cells.
3. Using reference materials, prepare a report on how certain types of algae and bacteria can live in hot springs at temperatures most organisms cannot tolerate.
4. Write a report on an enzyme deficiency disease in humans, such as alcaptonuria or phenylketonuria. Explain the nature of the disease in terms of enzyme function.

READINGS

Dickerson, R.E. "Cytochrome c and the Evolution of Energy Metabolism." *Scientific American,* March 1980, p. 136.

Steinberg, S. "Genes Shed Light on Photosynthesis." *Science News,* August 1983, p. 102.

Youvan, Douglas, and Barry L. Marrs. "Molecular Mechanisms of Photosynthesis." *Scientific American,* June 1987, p. 42.

6. Cell Reproduction

This photograph shows a small portion of a computer generated image of a special molecule. This molecule is called DNA. It makes up part of the nuclear material in a cell. DNA is special because locked within its chemicals are the "directions" that tell a cell what proteins to make, how large to grow, and when to divide. In fact, DNA controls all the activities of cells. In this chapter you will learn more about this molecule and its role in providing the instructions that make your body, you.

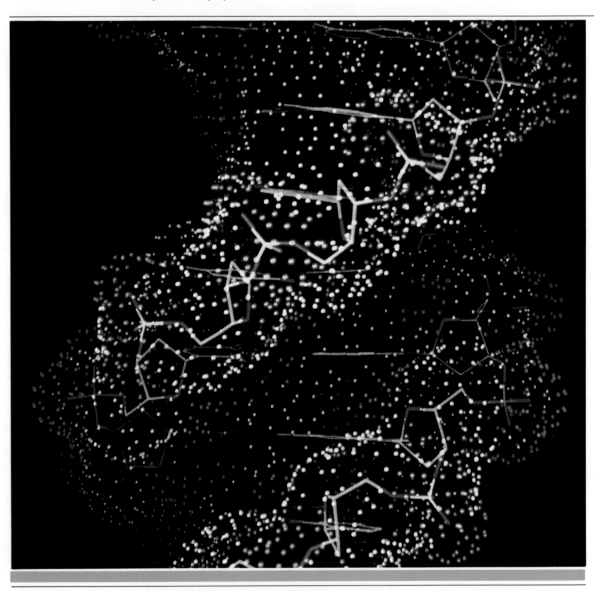

CHAPTER OBJECTIVES

After completing this chapter, you will be able to

- **Identify** the steps of mitosis and meiosis.
- **Compare** mitosis and meiosis.
- **Contrast** DNA and RNA.
- **Discuss** the double-helix model of DNA.
- **Explain** how DNA makes an exact copy of itself.
- **Describe** how nucleic acids control protein synthesis.

Formation of Body Cells

6·1 WHEN CELLS DIVIDE

When you get cut, the wound heals. When a starfish loses an "arm," it grows a new one. A single-celled organism living in a pond can produce so many offspring that the pond changes color in just a few days. What is similar about all these events? In all these events cells reproduce; cells produce new cells.

The cell theory states that living things are made of cells and that all cells come from preexisting cells. New cells are produced by the process of cell division, the dividing of a cell to form two new cells. When a cell divides, both the nucleus and the cytoplasm divide. The cell that divides is called the *parent cell*. The cells that are formed are called *daughter cells*. Cell division provides new cells for growth, the healing of wounds, and the replacement of injured parts of the body.

The nucleus of a cell controls all cell activities. It also controls the development of all cell characteristics. The information to control these activities and characteristics is carried by *chromosomes*. During cell division the chromosomes are passed to the new cells that are formed. Each kind of living thing is made of cells that contain a certain number of chromosomes. For example, the body cells of an onion plant contain 16 chromosomes. Figure 6·1 shows the number of chromosomes for several organisms.

Notice that the numbers for all the organisms listed are even numbers. Chromosomes in the body cells of many organisms occur in pairs. Such pairs of like chromosomes are called **homologous** (hoh MAHL uh guhs) **chromosomes.** In cat cells there are 19 pairs of homologous chromosomes. How many pairs of homologous chromosomes are there in human cells? In a cell that contains chromosomes in homologous pairs, the total number of chromosomes in the cell is called the **diploid number.** The numbers in Figure 6·1 are diploid numbers. A diploid number is shown by the symbol *2n.*

SECTION OBJECTIVES

- Describe the steps of mitosis.
- Discuss the importance of mitosis.

FIGURE 6·1 As you can see in this table, different organisms have different numbers of chromosomes in their body cells.

ORGANISM	DIPLOID NUMBER
Cat	38
Cow	60
Dog	78
Fruit fly	8
Goldfish	94
Grasshopper	14
House fly	12
Human	46
Onion	16
Plum	48
Rice	12
Sunflower	34

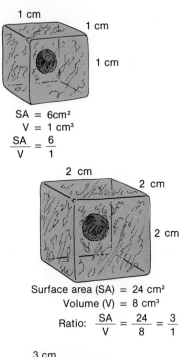

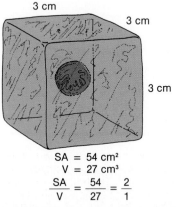

FIGURE 6·2 As a cell grows, its volume increases at a faster rate than its surface area. As you can see, the ratio of surface area to volume changes.

inter- (between)

A daughter cell that has the same number of chromosomes the parent cell had can carry out the same activities of the parent cell and has the same characteristics as well. In the type of cell division called **mitosis** (mī TOH sihs), each daughter cell receives the same number of chromosomes the parent cell had.

The chromosomes in the cell are duplicated before a cell undergoes mitosis. You will learn how chromosomes are duplicated in the second half of this chapter. When mitosis begins, two complete sets of chromosomes are present. The events of mitosis separate the two sets of chromosomes. Each daughter cell ends up with one full set of chromosomes. Although the chromosomes are duplicated before mitosis, they cannot be seen until mitosis begins.

6·2 WHY CELLS DIVIDE

Several factors are involved in triggering cell division, and scientists are still studying this process. Cell size is an important factor in causing a cell to divide. A cell will not continue to grow indefinitely. When a cell reaches a certain size, it will divide.

Figure 6·2 can help you understand why size is a factor in triggering mitosis. Both the volume and the surface area are given for each cube-shaped cell in the drawing. Think of the drawings as showing a growing cell. As you can see, both surface area and volume increase as the cell gets larger. However, how does the ratio of surface area to volume change as the cell gets larger?

As a cell grows, the cell's surface area does not increase as fast as its volume. Amino acids, oxygen, and other materials needed by the cell must pass into the cell through the cell membrane. Carbon dioxide and other wastes produced by the cell must pass out of the cell through the cell membrane. As a cell grows, a point is reached at which there is too little cell-membrane surface area to allow the needed amounts of materials to move in and out. For the cell, the amount of materials that must cross the cell membrane is too great for the surface area. When a cell divides, the surface area increases with respect to the volume; the ratio of surface area to volume increases.

6·3 INTERPHASE

Mitosis is described in terms of the movement of chromosomes. The chromosomes can be seen through a microscope. Scientists divide mitosis into phases to make it easier to study. Remember, however, that mitosis is a continuous series of events.

Interphase is the period in which cells carry out many activities other than mitosis. Interphase is not part of the period of mitosis. Cells often grow in size during interphase. They carry out such life activities as synthesis and the movement of materials into and out of the cells. Many kinds of materials, including enzymes and other types of proteins, are formed within the cells during interphase. Some of the materials formed during this period are stored for use in the next mitosis.

As you can see in Figure 6·3, the only visible structure within the nucleus of a cell in interphase is the nucleolus. The chromatin is spread throughout the nucleus like fine threads. In animal cells, structures called *centrioles* (SEHN tree ohlz) can be seen outside the nucleus. Most plant cells do not have centrioles.

During interphase a very important event takes place. Chromosomes are duplicated. The DNA within the chromosomes *replicates,* or duplicates (Section 6·12). This replication of DNA results in the doubling of the number of chromosomes. At this point, the cell has two identical sets of chromosomes. The stage is set for mitosis to begin.

6·4 THE STAGES OF MITOSIS

Mitosis involves four stages. As you read about each stage, compare the drawings of mitosis with the photographs of a cell undergoing mitosis.

1. **Prophase** is the first stage of mitosis. During prophase, the chromosomes can be seen through a microscope. At the beginning of prophase, the chromosomal material, or chromatin, gradually appears as shortened, distinct rods. This shortening of the chromatin is one of the first observable signs that mitosis has begun. In animal cells the centrioles begin to move to opposite sides of the cell.

Each chromosome is made of two distinct strands that are called **chromatids** (KROH muh tihdz). Each pair of chromatids is held together by a **centromere** (SEHN truh mihr). Locate the chromatids and centromeres in prophase of Figure 6·4.

As the chromosomes become visible, other events take place within the cell. The nuclear membrane and the nucleolus gradually disintegrate. A new structure, the spindle, appears. The **spindle** is a three-dimensional structure shaped somewhat like a football. It consists of microtubules that extend across the cell. The fibers of the spindle appear to guide the movements of the chromosomes during mitosis. In animal cells, the centrioles appear to control the formation of the spindle. Recall that most plant cells do not have centrioles, although they have a spindle. Animal cells also have another structure that plant and other cells do not have. This structure is the aster. The aster is made of microtubules that radiate out from the centrioles.

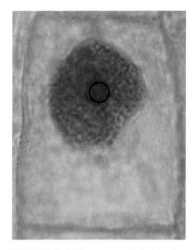

FIGURE 6·3 The area circled in the center of the nucleus is the nucleolus. This onion root-tip cell has been magnified about 1800 times.

pro- (before)

FIGURE 6·4 During prophase, the chromosomes become visible. Each chromosome consists of 2 chromatids joined together at the centromere (right). This whitefish cell has been magnified about 780 times (left).

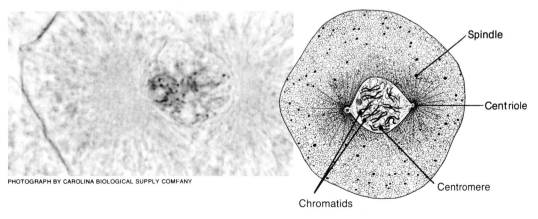

meta- (between)

2. Metaphase is the second stage of mitosis, during which the paired chromatids move to the center, or *equator,* of the cell. The chromatids are arranged in a row at right angles to the fibers in the spindle. The centromere of each pair of chromatids is attached to a spindle fiber. The spindle fibers may cause the paired chromatids to line up at the equator of the cell. During metaphase, the chromatids are thick and often are coiled around each other.

FIGURE 6·5 During metaphase, the chromosomes line up along the equator of the cell.

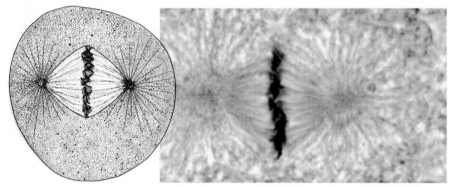

ana- (up)

3. Anaphase is the third stage of mitosis, during which the paired chromatids separate. At the start of anaphase, the centromere of each pair of chromatids divides. The paired chromatids separate into individual chromosomes. The separated chromosomes move toward opposite poles, or ends, of the spindle. Each chromosome moves "centromere first," as if being pulled by the attached spindle fiber. All chromosomes move to the poles of the cell at nearly the same time. If mitosis proceeds correctly, an equal number of chromosomes moves to each pole of the cell.

FIGURE 6·6 During anaphase, the centromeres divide and the paired chromatids separate into individual chromosomes, which are then pulled to opposite ends of the cell.

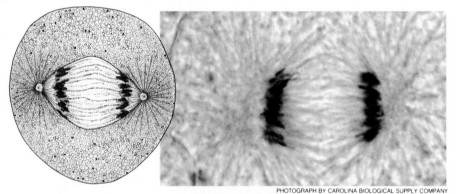

telos (end)

4. Telophase is the last stage of mitosis, during which the chromosomes again become threadlike. They lengthen and become as indistinct as they were at the beginning of prophase. The spindle breaks apart, a nucleolus reappears, and a nuclear membrane forms around each mass of chromatin.

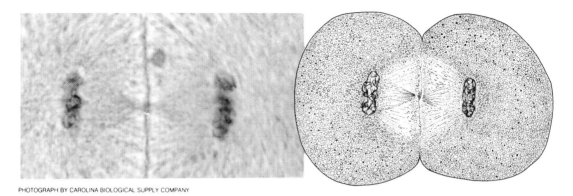

In animal cells during telophase, the cytoplasm pinches together along the equator until two daughter cells are formed. In plant cells, a delicate membrane called the *cell plate* starts to form in the middle of the spindle. The cell plate appears to form from the center outward. A new cell wall later forms on either side of the cell plate. Two new daughter cells are then formed.

The division of the cytoplasm is called *cytokinesis* (sī toh kih NEE-sihs). It appears to be a separate process from division of the nucleus. Mitosis is sometimes considered to be only the division of the nucleus. In a few types of cells, mitosis, or nuclear division, may take place without cytoplasmic division. In this case, cells with several nuclei result.

In both plants and animals mitosis results in the formation of two daughter cells that are identical to each other and to the parent cell from which they came. The new daughter cells enter interphase.

All of the phases of mitosis are shown in Figure 6·8. The phases of mitosis are not completed in the same time. Anaphase takes the least amount of time. Prophase and telophase are the longest of the actively dividing stages. The length of time a cell is in interphase varies greatly. In general, interphase is much longer than mitosis.

The length of time needed for a cell to divide completely depends upon the kind of cell and the environmental conditions under which it exists. Mitosis in the root tips of young plants occurs almost continuously. The cells of some insects may complete mitosis in as little as 10 minutes. Most cells complete mitosis in one to two hours. Mammalian cells grown in a laboratory may require 24 hours to complete one division.

6·5 SIGNIFICANCE OF MITOSIS

Mitosis is an important type of cell division because it forms daughter cells that have the same number of chromosomes the parent cell had. Since the parent cell's chromosome number is kept the same in the daughter cells, these cells receive a set of DNA that is a duplicate of that found in the parent cell. Thus the daughter cells are able to carry out all of the activities that the parent cell carried out.

FIGURE 6·7 During telophase, the chromosomes again become threadlike and the cytoplasm divides.

FIGURE 6·8 The stages of mitosis are summarized in this diagram.

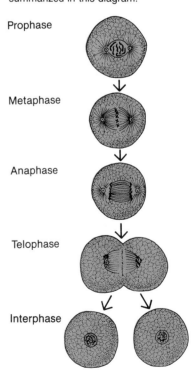

The daughter cells produced by mitosis are *identical* to each other and to the parent cell from which they formed. The duplicate sets of chromosomes in the daughter cells contain the same information. Mitosis is a way of increasing the number of cells without changing the traits of the cells.

Asexual reproduction is the formation of new cells or organisms from only one parent cell. In unicellular organisms, mitosis is a means of producing many offspring that are identical. Recall the single-celled pond organism mentioned earlier. How was this organism able to change the color of a pond in just a few days?

REVIEW QUESTIONS

1. What is meant by the term *diploid number of chromosomes*?
2. Describe what occurs in each of the stages of mitosis.
3. Distinguish between mitosis in plant cells and in animal cells.
4. What is the significance of mitosis?
5. Suppose a certain kind of cell undergoes mitosis when the ratio of surface area to volume is 2:1. The cell is 3 μm long, 2 μm wide, and 2 μm high. Find the ratio of surface area to volume. Is the cell ready to undergo mitosis? (Hint: To find cell surface area, you must add the surface areas of all the sides. Also, recall that the formula for volume is l × w × h.)

Formation of Sex Cells

SECTION OBJECTIVES

- Explain the steps and importance of meiosis.
- Compare how meiosis occurs in males and females.

6·6 WHAT IS MEIOSIS?

Sexual reproduction usually requires two parents. Each parent contributes one of the two cells that will fuse to form the offspring's first cell. The specialized cells that join during sexual reproduction are called **gametes** (GAM eets), or sex cells. In a female, the gametes are *ova*, or *eggs*. In a male, the gametes are *sperm*. The union of an egg and a sperm is called fertilization. The cell that is formed from the union of an egg and a sperm is called a *zygote* (ZĪ goht).

Recall that *body cells* in most organisms contain the diploid number of chromosomes. If each gamete contained the diploid number, the fusion of two gametes during fertilization would result in a zygote with twice the diploid number. Each time offspring were produced, they would contain twice the number of chromosomes the parents had. This doubling does not occur. The number of chromosomes in the body cells of a certain kind of organism is maintained generation after generation. Gametes do not have the diploid number of chromosomes and must be formed by a kind of cell division that is different from mitosis.

Meiosis (mī OH sihs) is cell division in which the chromosome number is reduced by one half and gametes are formed. Meiosis involves division of a cell that begins with the diploid number of chromosomes. Although the cell undergoes two successive divisions, the chromosomes are duplicated only once. The two divisions result in four daughter cells. Each cell contains one half the number of chromosomes of the parent cell. Half the diploid number of chromosomes is the **monoploid number.** The monoploid number is also called the *haploid number.* While diploid is represented by *2n,* monoploid is represented by *n.* When two gametes with the *n* number of chromosomes join, the zygote formed will have *2n* chromosomes.

Human body cells contain 46 chromosomes. Forty-six is the *2n* number. Human gametes formed by meiosis each contain the monoploid number of chromosomes—23. When a sperm joins with an egg, a zygote with 46 chromosomes *(2n)* results.

Interphase before meiosis is similar to interphase before mitosis. Proteins and other compounds are made, energy is stored and used, and materials are exchanged with the environment. The materials needed to carry out meiosis are stored in the cell during interphase. Before the cell begins the first division of meiosis, the DNA in the chromosomes of the cell's nucleus replicates. The cell then contains two complete sets of chromosomes and is ready to begin meiosis.

6·7 THE STAGES OF MEIOSIS

Meiosis consists of two successive divisions. Each of these divisions is divided into phases similar to those of mitosis. The first division is called *meiosis I,* and the second is called *meiosis II.* Each stage of the first division of meiosis is followed by the Roman numeral I. Each phase of the second division is followed by the numeral II.

1. In *prophase I,* the first prophase of meiosis, the chromatin shortens and thickens. Each chromosome is visible as two chromatids joined by a centromere. The nuclear membrane and nucleolus break apart. The spindle forms between opposite poles of the cell.

An event occurs during prophase I of meiosis that does not occur during mitosis. As the chromosomes become visible, pairs of homologous chromosomes, called **homologs,** line up. Homologs carry the same kinds of genetic information and in the same order. The homologs in each pair become closely entwined. You can illustrate this process by pressing the palms of your hands together. With each finger representing a chromosome, the matching fingers represent a pair of homologs. The pairing of homologs in prophase I is called **synapsis** (sih NAP sihs). Each chromosome is made up of two chromatids. The four chromatids of a pair of homologs are called a *tetrad.*

When homologs pair during synapsis, their chromatids often twist around one another. Sometimes the chromatids break and exchange pieces. The exchange of pieces of chromatid material between homologs during meiosis is called **crossing over.**

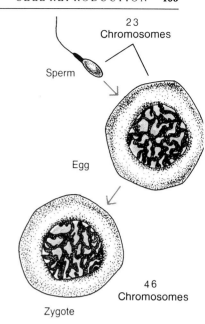

FIGURE 6·9 The fertilization of an egg with 23 chromosomes by a sperm with 23 chromosomes results in a zygote with 46 chromosomes. The sperm and the egg cells are not drawn to scale and are shown here much larger than life.

FIGURE 6·10 Crossing over occurs during meiosis. Parts of one chromosome are joined with another chromosome during crossing over.

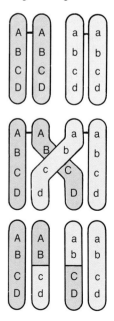

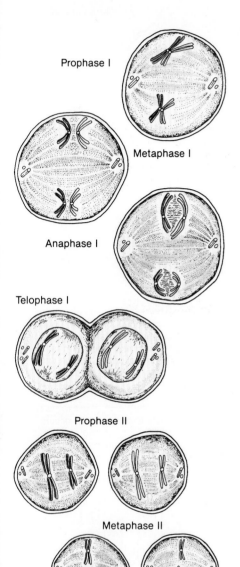

2. During *metaphase I* the tetrads line up along the equator of the cell. Homologs are paired along the equator. The tetrads line up at right angles to the spindle fibers. Each chromosome is attached to a spindle fiber.

3. During *anaphase I* the homologous pairs of chromosomes separate. One chromosome of each pair moves to one pole of the cell. The other chromosome of each pair moves to the opposite pole. Each chromosome is still made of two chromatids joined at a centromere. The chromatids do not separate at this time, as they do in mitosis.

4. During *telophase I* the cytoplasm divides, forming two cells. Each of the cells contains one member of each pair of homologous chromosomes. The chromosome number has been reduced to monoploid, one half the diploid number. Remember that each chromosome still is made of two chromatids joined by a centromere. A nuclear membrane forms around the chromosomes in each new cell.

Following telophase I the first cell division of meiosis is complete. The two cells enter a phase called *interkinesis* (ihn tuhr kih NEE sihs). Interkinesis is similar to interphase, but the chromosomes do not replicate. The second cell division of meiosis usually occurs quickly. This division is similar to mitosis in that the chromatids separate. The following phases of the second cell division of meiosis occur in both of the cells formed by the first cell division.

5. During *prophase II*, the second cell division of meiosis begins. The nuclear membrane and nucleolus break apart. The chromosomes shorten and become visible. Each chromosome is made of two chromatids joined by a centromere.

6. During *metaphase II* the chromatids, still attached by centromeres, move to the equator of the cell.

7. During *anaphase II* the chromatids separate. One chromatid from each chromosome moves to one pole of the cell. The other chromatid moves to the opposite pole.

8. During *telophase II* the cytoplasm divides, forming two cells, each with the monoploid number of chromosomes. A nuclear membrane forms around the chromosomes in each daughter cell.

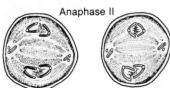

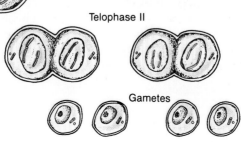

FIGURE 6·11 The phases of meiosis are summarized in this diagram. Since the chromosomes segregate randomly, the resulting gametes might contain different chromosomes.

6·8 MALE AND FEMALE GAMETES

The formation of gametes by meiosis is called **gametogenesis** (gam uh tuh JEHN uh sihs). In males, gametogenesis results in the formation of sperm cells. The formation of sperm cells is known as **spermatogenesis** (sper mat uh JEHN uh sihs). The production of sperm takes place in the testes, the male reproductive organs. Figure 6·12 shows the major steps in the formation of sperm. A specialized diploid cell in the testes, the *primary spermatocyte* (sper MAT uh sīt), undergoes meiosis. After telophase I, two *secondary spermatocytes* are formed. Each secondary spermatocyte undergoes the second meiotic division, resulting in the production of four *spermatids*. The spermatids mature into monoploid sperm cells.

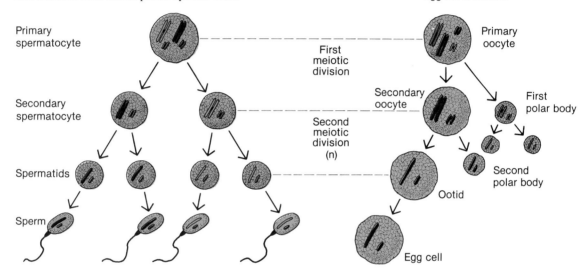

FIGURE 6·12 Spermatogenesis (left) and oogenesis (right) result in the formation of gametes. Note that four sperm cells are produced, but only one egg cell is formed.

In females, egg cells are formed in the female reproductive organs, the ovaries. The formation of gametes in females is called **oogenesis** (oh uh JEHN uh sihs), shown in Figure 6·12. In the ovary, a specialized diploid cell called the *primary oocyte* (OH uh sīt) undergoes meiosis. At the end of the first meiotic division, two unequal-sized cells have been formed—a large cell called the *secondary oocyte* and a very small cell called the *first polar body*. During the second meiotic division, the secondary oocyte divides to form two monoploid cells—a large *ootid* (OH uh tihd) and a small cell called the *second polar body*. The first polar body may divide to form two more polar bodies. The ootid gets almost all of the cytoplasm from the original primary oocyte. The polar bodies eventually disintegrate. The ootid develops into the monoploid egg cell, or the *ovum*.

You can see that both oogenesis and spermatogenesis result in monoploid gametes. However, there are some differences between the two forms of gametogenesis. Spermatogenesis forms four equal-sized sperm. Oogenesis forms one large egg cell as well as three polar bodies, which disintegrate. Only the egg cell is a functional gamete.

6·9 COMPARING MEIOSIS AND MITOSIS

Meiosis and mitosis have a number of features in common. The movement of the chromosomes through the various phases follows a generally similar pattern. Both processes result in the production of new cells. However, there are some very important differences between meiosis and mitosis. Some of these differences are summarized in Figure 6·13.

FIGURE 6·13 This table shows some differences between mitosis and meiosis.

MITOSIS	MEIOSIS
Occurs in most types of eukaryotic cells.	Occurs in formation of gametes in eukaryotic cells.
No pairing of homologs occurs.	Homologous chromosomes are paired in synapsis, and crossing over may occur at this time.
Chromosome number is maintained.	Chromosome number is reduced from diploid to monoploid.
One division.	Two divisions.
Two daughter cells produced.	Four daughter cells produced.
Daughter cells are identical to each other and to parent cell.	Daughter cells contain varying combinations of chromosomes and are not identical to parent cell.

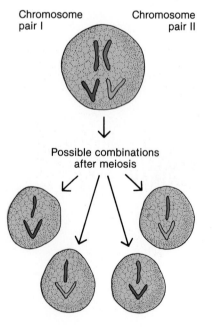

FIGURE 6·14 Possible combinations of chromosomes that could result if a cell with two pairs of chromosomes underwent meiosis.

Mitosis is most often associated with growth and asexual reproduction. It forms identical new cells. These new cells may increase the size of some tissue. New cells formed by mitosis may also replace cells lost by wear or injury. The new cells formed by mitosis are almost always the same as those they replace.

Meiosis, on the other hand, is associated with sexual reproduction. Gametes are produced through two meiotic divisions, which reduce the chromosome number from diploid to monoploid.

The separation of homologous pairs of chromosomes occurs at random. For example, suppose one homolog of a pair moves to the left pole. Either homolog of any other pair could also move to that pole. Figure 6·14 illustrates the possible combinations of chromosomes that would result if a cell having only two pairs of chromosomes underwent meiosis. In a human, with 23 pairs of chromosomes, there are over 8 million possible combinations of chromosomes for every cell formed by meiosis! The random separation of homologs adds to the great variation in characteristics found in organisms that reproduce sexually. Crossing over, or the exchange of chromatin segments, further adds to the variability of the daughter cells formed by meiosis.

REVIEW QUESTIONS

6. What is the effect of meiosis and fertilization on the number of chromosomes?
7. Explain what may occur between chromosomes during synapsis.
8. Describe gametogenesis in male and female humans.
9. How does meiosis differ from mitosis?
10. You are observing a living cell dividing and see that homologous chromosomes are paired along the equator of the cell. Are you observing mitosis or meiosis? Which stage of cell division are you observing?

The Genetic Code

6•10 THE ROLE OF DNA

Why is it so important that chromosomes be passed on from parent cell to daughter cells? Chromosomes are formed of **genes,** the segments of DNA that are the units of inheritance. Genes control the development of traits, such as hair color, blood type, skin color, and eye color. Genes are commonly said to be "on" chromosomes. Genes, however, are actually parts of chromosomes. The genes are linked on the chromosomes much like beads on a string.

Genes are segments of DNA, and thus it is the chemical **DNA** that controls the development of traits and cellular activities. Experiments with bacteria that cause pneumonia have provided evidence that DNA is the material of heredity. Some of these bacteria cause a severe form of pneumonia. These disease-causing bacteria have outer coats. Other, closely related bacteria lack the outer coats and are harmless. The presence or absence of the coat and also the ability to cause disease are inherited characteristics. The ability to cause disease can be transferred from the disease-causing bacteria to the harmless ones by transferring DNA. The harmless bacteria then develop coats and the ability to cause disease. The offspring of these changed bacteria also have coats and can cause disease. This experiment and others have shown that DNA is the genetic material.

5•11 THE STRUCTURE OF DNA

The knowledge that DNA is the hereditary material led many scientists to study the structure of DNA. In 1953 an American biologist, James Watson, and a British biophysicist, Francis Crick, proposed a model for the structure of DNA.

DNA is a very large molecule, but it is built from only a few kinds of simple chemicals. A DNA molecule is composed of units called **nucleotides** (NOO klee uh tīdz). Identify the components of a nucleotide in Figure 6•15. Each nucleotide contains a *phosphate group,* a

SECTION OBJECTIVES

- Describe the structure of a molecule of DNA.
- Explain how DNA replicates.

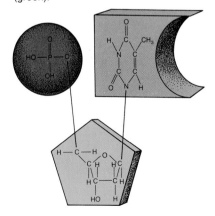

FIGURE 6•15 A nucleotide consists of a phosphate group (red), deoxyribose (orange), and a nitrogenous base (green).

110 CELL REPRODUCTION

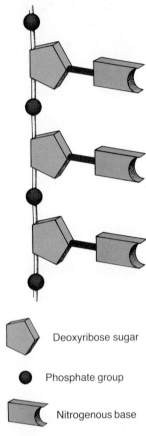

Deoxyribose sugar

Phosphate group

Nitrogenous base

FIGURE 6·16 The linkage of nucleotides in DNA occurs when bonds form between the phosphate group of one nucleotide and the sugar of the next nucleotide.

five-carbon sugar called *deoxyribose* (dee ahk suh RĪ bohs), and a base that contains nitrogen called a *nitrogenous* (nī TRAHJ uh nuhs) *base*. Nucleotides are joined by bonds between the phosphate group of one nucleotide and the sugar of the next nucleotide. The phosphate of the second nucleotide is bonded to the sugar of the third nucleotide, and so on. As a result, a long chain of nucleotides is formed—each nucleotide bonded from phosphate to sugar. Figure 6·16 shows how nucleotides are linked.

The nitrogenous bases extend from the sugars of the phosphate-sugar chain. Four bases occur in DNA. These bases are **adenine** (AD uh neen), **cytosine** (SĪ tuh seen), **guanine** (GWAH neen), and **thymine** (THĪ meen).

A DNA molecule is composed of two chains of nucleotides joined by weak hydrogen bonds between the nitrogenous bases. The chains of nucleotides spiral around a common center. The spiral shape of the molecule is a *double helix*. To picture the shape of a double helix, imagine a flexible ladder that can be twisted. The sides of the ladder represent the bonded sugars and phosphates of the nucleotides. The rungs of the ladder represent the bonded pairs of bases. Figure 6·17 shows the "twisted flexible ladder" structure of a DNA molecule.

Recall that the two chains of nucleotides in a DNA molecule are held together by bonds between the bases. Only specific pairs of bases bond together. Adenine (A) always bonds to thymine (T). Cytosine (C) always bonds to guanine (G). Because only specific pairs of bases will bond, the sequence of bases in one chain of nucleotides determines the sequence of bases in the other chain. The two chains of nucleotides are complements of each other. The pairing of bases in the complementary chains of nucleotides is the basis of DNA replication.

6·12 DNA REPLICATION

Chromosomes are duplicated during interphase, before cell division begins. The process in which a DNA molecule makes copies of itself is called **DNA replication.**

Many free nucleotides, each made of a phosphate, a deoxyribose sugar, and a nitrogenous base are found in the nucleus of a cell. These free nucleotides are the building blocks from which new DNA is built. Follow the steps in the replication of DNA as shown in Figure 6·17.

1. The double helix untwists. Now the DNA molecule resembles a ladder. Recall that the DNA molecule consists of two chains of nucleotides. The rungs of the DNA "ladder" are paired bases, while the uprights are the sugars and phosphates of the nucleotides. After the DNA untwists, bonds between the bases of the DNA molecule are broken. The two nucleotide chains begin to separate, starting at one end and moving to the other, like the opening of a zipper.

2. Each chain, or half of the DNA molecule, serves as a pattern, or template, for the formation of a new chain of a DNA molecule.

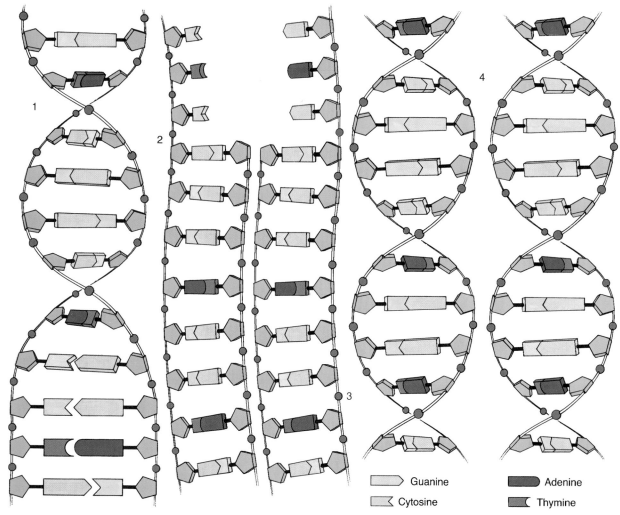

FIGURE 6·17 A DNA molecule makes copies of itself in a process called replication. This process results in two new identical DNA molecules.

Bases in the free nucleotides in the nucleus join with the correct bases on the two chains of nucleotides. Remember that adenine always bonds to thymine, and cytosine always bonds to guanine. These base-pairing rules insure that the newly formed copies of DNA are accurate copies of the original.

3. Bonds form between the phosphates and sugars of the nucleotides that have paired with the DNA chains. The result of replication is that two identical copies of the original molecule of DNA are formed.
4. The two new molecules of DNA become twisted and again take the form of a double helix.

REVIEW QUESTIONS

11. What evidence indicates that DNA is the genetic material?
12. Name the three parts of a nucleotide.
13. Describe the double helix model of DNA.
14. Explain how DNA replicates.
15. A section of DNA has the following sequence of bases: G, A, C, C, T, A, G, T, T, A, A, G. Determine the complementary sequence.

Protein Synthesis

SECTION OBJECTIVES

- Explain how DNA determines the formation of RNA.
- Describe how the order of bases in RNA determines how proteins are synthesized.

Discovery

RNA as an Enzyme

In biological systems, chemical reactions usually involve proteins called enzymes. Enzymes interact with specific chemicals and greatly increase the rate of a chemical reaction. Usually an enzyme is not changed during a reaction.

In 1982, Dr. Thomas Cech discovered that RNA can be an enzyme. He found that a type of RNA molecule taken from the nucleus of a protozoan causes a change in the length and structure of that RNA even when no proteins are present. Dr. Cech called this type of enzyme a *ribozyme* (for RNA enzyme). Unlike most enzymes, this ribozyme can act on itself. This action produces smaller RNA molecules that no longer function as enzymes.

In 1983, Dr. Sidney Altman and Dr. Norman Pace reported a second type of ribozyme. They studied a small RNA molecule from bacteria. This RNA molecule causes the shortening of a different RNA molecule. Like most enzymes this ribozyme acts repeatedly and does not change.

6·13 TRANSCRIPTION

Enzymes are proteins that control all of the chemical reactions of living things. Other types of proteins are used as building materials. Cells are made partly of protein. What controls the making of all these different proteins? The information for making all proteins is stored in the DNA molecules of the chromosomes. The sequence of bases in DNA molecules is a chemical code for the sequence of amino acids in proteins. A segment of DNA that codes for a particular protein is called a gene.

The synthesis of thousands of different proteins may seem too complex to be controlled by the sequence of bases in a DNA molecule. But think about all the objects and ideas that can be expressed in the English language by the changing combinations of the 26 letters in the alphabet. The four nitrogenous bases of DNA—adenine (A), guanine (G), thymine (T), and cytosine (C)—make up the DNA "alphabet." A molecule of DNA can be made of thousands of nucleotides each with one of these bases. How can these four bases be arranged to form a code?

The genetic code is made up of three-letter "words." The four bases join into three-letter "words"—AGC, CGT, and so on—and result in 64 different "words," or groups. Sixty-four combinations are more than enough to code for the 20 different amino acids. Such sequences of three nucleotide bases are called *triplets*. Each DNA triplet codes for an amino acid. Most amino acids are coded by more than one triplet. The arrangement of bases on a DNA molecule codes for the sequence of amino acids that make up a particular protein.

RNA is a nucleic acid made of a single chain of nucleotides. The sugar in RNA is *ribose* (RĪ bohs), which differs slightly from the deoxyribose of DNA. RNA and DNA differ in the kinds of bases in the nucleotides. In place of the base thymine, RNA contains the base **uracil** (YUR uh suhl). Uracil (U) bonds only with adenine. Three types of RNA are found in cells.

Messenger RNA, or **mRNA,** carries the instructions to make a particular protein from the DNA in the nucleus to the ribosomes. Molecules of mRNA are assembled according to the code contained in the DNA. Transfer RNA, or **tRNA** carries amino acids to the ribosomes. The tRNA is found in the cytoplasm of cells. Ribosomal RNA, or **rRNA,** is one of the chemicals of which ribosomes are composed.

The DNA in the nucleus of a cell contains instructions to make thousands of different proteins. When a certain protein is needed, mRNA is formed from the information in the DNA. The process of producing mRNA from the instructions in DNA is called **transcription** (tran SKRIHP shuhn). Study Figure 6·18 as you go through these steps.

1. The portion of DNA that contains the code for the needed protein untwists and separates. This process is similar to the beginning of DNA replication. The result is that the bases are exposed.
2. Free RNA nucleotides found in the nucleus pair with the exposed DNA bases. Uracil rather than thymine pairs with adenine. Triplets on the DNA strand result in the formation of complementary triplets on the mRNA molecule. A sequence of three nucleotides on an mRNA molecule coding for an amino acid is called a **codon** (KOH dahn).
3. The mRNA molecule is completed by the formation of bonds between the RNA nucleotides. The mRNA molecule separates from the DNA molecule. The completed mRNA molecule, which carries the code to make a single type of protein, leaves the nucleus. It passes through the nuclear membrane and moves to the ribosomes in the cytoplasm.

FIGURE 6·18 The transcription of DNA results in the formation of a molecule of mRNA. The mRNA carries the instructions to make a protein from the nucleus to the ribosomes.

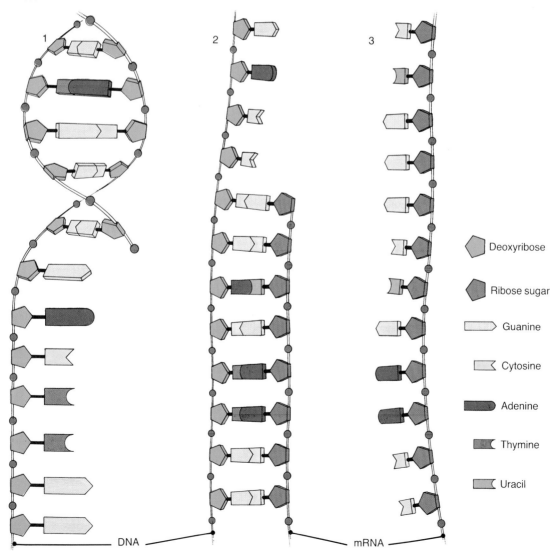

6•14 TRANSLATION

The assembly of a protein molecule according to the code in an mRNA molecule is known as **translation.** The process is similar to changing a message from a foreign language into English. The assembly of proteins is called translation because it involves a change from the nucleic acid "language" (sequence of bases) to the protein "language" (sequence of amino acids). When translation is complete, the information stored in DNA is expressed as a finished protein.

In the cytoplasm, mRNA moves to the ribosomes. Before a protein molecule can be synthesized, amino acids must be brought to the ribosomes. The needed amino acids are scattered throughout the cytoplasm. The correct amino acids are found and brought to the mRNA by transfer RNA (tRNA).

Strands of tRNA are shorter than those of mRNA and have a cloverleaf shape (Figure 6•19). Notice that on one loop of a tRNA molecule is a set of three bases called an **anticodon.** An amino acid is carried at the opposite end of the tRNA. There is a specific tRNA molecule for each possible codon on an mRNA molecule. The bases of tRNA anticodons are complementary to the bases of the mRNA codons.

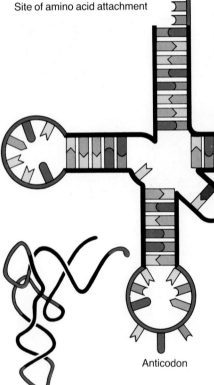

FIGURE 6•19 This shows a model of a tRNA molecule. Strands of tRNA bring amino acids to the ribosome where they are assembled into proteins.

Site of amino acid attachment

Anticodon

The amino acid glutamine is coded by the DNA triplets GTT and GTC. The mRNA codons that correspond to these DNA triplets are CAA and CAG. From this you can see that tRNA molecules with the anticodons GUU and GUC transport glutamine to the ribosomes.

Figure 6•20 shows the steps in the translation of the mRNA message. Follow the translation of the message as you read the steps.

1. One end of an mRNA molecule attaches to a ribosome.
2. Transfer RNA molecules in the cytoplasm pick up the amino acids. They are coded for by the anticodons they carry. With certain amino acids attached, tRNA molecules move to the point at which mRNA is attached to the ribosome.
3. A tRNA molecule with the right anticodon links to the complementary codon on the mRNA. The amino acid attached to the tRNA molecule is held in position.
4. As the mRNA moves along the ribosome, the next codon contacts the ribosome. The next tRNA moves into position with its amino acid. Adjacent amino acids are linked by a *peptide bond*.
5. The first tRNA molecule is released. The next codon comes into place, and the next amino acid is positioned.

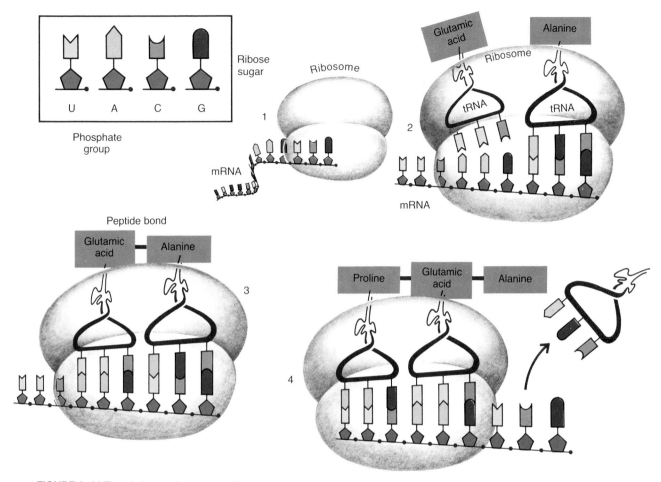

FIGURE 6·20 Translation produces a specific protein when amino acids are joined in a particular order.

Steps 3 through 5 are repeated until the entire message is translated. In this way a chain of amino acids is formed. A protein molecule is built from one or more chains of amino acids.

In summary, DNA codes messenger RNA; messenger RNA carries the information needed for synthesis of a certain protein to the ribosomes, where the protein is made. This process can be shown in the following equation.

$$\text{DNA} \xrightarrow{\text{transcription}} \text{mRNA} \xrightarrow{\text{translation}} \text{protein}$$

REVIEW QUESTIONS

16. List some functions of proteins.
17. How is mRNA produced?
18. How do tRNA molecules differ from one another?
19. Describe the process of translation.
20. Suppose one of the bases in an mRNA codon were changed. What effect might this change have on translation?

Investigation

How Do Genes Direct the Synthesis of Proteins?

Goals

After completing this activity, you will be able to

- Determine the order of bases in a DNA molecule needed to code for a segment of a protein molecule.
- Determine the codons needed for the assembly of a segment of a protein molecule.

Materials

2 sheets of yellow paper
2 sheets of green paper
1 large sheet of white paper
tape   scissors

Procedure

A. Fold the two sheets of yellow paper into 16 boxes as shown. Label the top two rows of boxes on each yellow sheet with G and C to represent the bases guanine and cytosine. Label the bottom two rows with the letters A and T to represent the bases adenine and thymine. Recall that the arrangement of bases (G, C, A, T) in a DNA molecule codes for the sequence of amino acids that makes up a particular protein. Cut each box out on the folds. Then cut each box into the shapes indicated.

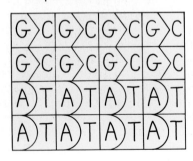

B. Fold the two sheets of green paper into 16 boxes as shown. Label the two top rows of each green sheet with G and C and the two bottom rows with A and U. These letters represent guanine, cytosine, adenine, and uracil, the bases in RNA. Cut each box out on the folds. Then cut each box into the shapes shown.

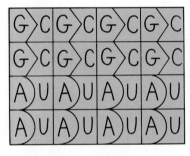

C. Proteins are made up of combinations of 20 different amino acids. The bases in DNA are arranged in groups of threes (triplets), such as GTA, TAC, and so forth. Examine the table, which lists 10 amino acids and some of the DNA triplets that code for them. (There may be several triplets that code for a certain amino acid; only one triplet is given here.)

TABLE 1. The Genetic Code

DNA TRIPLETS	AMINO ACID
CGA	Alanine (Ala)
TTG	Asparagine (Asn)
GTC	Glutamine (Gln)
CCA	Glycine (Gly)
AAT	Leucine (Leu)
TTC	Lysine (Lys)
AAA	Phenylalanine (Phe)
AGA	Serine (Ser)
TGA	Threonine (Thr)
CAA	Valine (Val)

D. Below is a segment of a protein molecule. The amino acids making up this segment are shown. Use the genetic code table to assemble the DNA triplets that could have coded for this protein segment. Use the shapes cut from the yellow paper to assemble in the proper order the bases of the DNA molecule that could have coded the protein segment. Tape these shapes to the sheet of white paper.

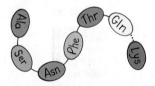

E. Determine the codons from the base sequence in your DNA molecule. Remember that a codon is a sequence of three nucleotides on an mRNA molecule that codes for an amino acid. Use the shapes cut from green paper to assemble the codons. Tape these shapes to the piece of white paper.

Questions and Conclusions

1. State the names and order of bases in the DNA molecule that could have coded for the protein segment.
2. List the codons required to form the protein segment.
3. Which step in this activity represents transcription?
4. What would happen to the protein if the first DNA triplet were CCA instead of CGA?
5. Determine the anticodon base sequence for the tRNA's that would complement the mRNA molecule.

Careers in Biology

Biochemist

Biochemists are scientists who study the chemical processes of living things, such as growth and reproduction. For example, some biochemists study the effects of different chemicals on the human body. Other biochemists develop methods to help doctors diagnose and treat disease.

Many biochemists work for chemical manufacturers. Others work for colleges, agriculture agencies, and small research firms.

A biochemist must have a bachelor's degree, with a major in biochemistry or chemistry. However, most biochemists also have at least a master's degree, with a specialty in an area like genetics. For more information write to the American Society of Biological Chemists, 9650 Rockville Pike, Bethesda, MD 20014.

Cytotechnologist

Cytology—the study of the structure and function of cells—has important medical applications. A change in the size, shape, or color of cell structures might indicate a disease such as cancer. A person who is trained to recognize these cell abnormalities is called a cytotechnologist. A cytotechnologist prepares and stains slides, and uses microscopes and other sophisticated equipment to identify disease conditions of cells.

Many cytotechnologists work in hospitals, university medical laboratories, government health agencies, or private research centers. A cytotechnologist in a hospital, for example, performs laboratory tests on a tissue sample from a patient and then helps a doctor interpret the results.

A career as a cytotechnologist requires at least two years of college, with courses in biology, and the successful completion of a one-year training and certification program administered by a hospital. For more information write to the American Society for Medical Technology, 330 Meadowfern Drive, Houston, TX 77067.

Biology Teacher

Biology teachers help high school students discover the complex world of living things. They introduce students to the great variety of plants and animals and the ways in which these organisms function and interact. Biology teachers also help students learn about scientific methods and the value of research.

To become a biology teacher, a person must have a bachelor's degree, with courses in biology and education. Usually, experience as a student teacher is also required to obtain a state teaching license. For more information write to the National Association of Biology Teachers, 11250 Roger Bacon Drive, Reston, VA 22090.

How To Find Out About Careers

School and community libraries have a wealth of information about many kinds of jobs and careers. Two very helpful sources are published by the United States Government. They are the *Occupational Outlook Handbook* and the *Chronicle of Occupational Briefs*. These publications provide the following information about specific jobs: a general description of duties; education and training requirements; personal qualifications necessary; working conditions; and related jobs. Both publications are updated regularly.

In some areas, large corporations and businesses in the community provide career information days for interested high school students. Your school counseling or placement office may have information on these special career days.

Chapter Review

SUMMARY

6·1 Cell division, or mitosis, produces new cells for growth and for the replacement of worn-out tissue. During mitosis the chromosomes are passed to the new cells that are formed.

6·2 After reaching a certain size, a cell may divide.

6·3 During interphase, many activities occur and the DNA within a cell replicates. The cell contains two identical sets of chromosomes at the end of interphase.

6·4 Mitosis involves a series of steps that results in the formation of two daughter cells, each with a complete set of chromosomes.

6·5 As a result of mitosis, the daughter cells are able to carry out all of the activities of the parent cell.

6·6 Meiosis is cell division in which the chromosome number is reduced by one half.

6·7 Meiosis includes two divisions which result in four daughter cells containing the monoploid number of chromosomes.

6·8 Gametogenesis in human males results in four sperm cells, each with the monoploid number of chromosomes. Gametogenesis in human females results in a large egg cell and three tiny polar bodies, each with the monoploid number of chromosomes.

6·9 During meiosis, exchange of genetic material may occur between chromosomes. This exchange does not occur during mitosis. The daughter cells of meiosis contain varying combinations of chromosomes; the daughter cells of mitosis are identical.

6·10 DNA is the genetic material.

6·11 DNA is composed of two spiraling linked chains of nucleotides.

6·12 When DNA replicates, the two molecules formed are identical copies of the original.

6·13 Messenger RNA is produced by DNA. mRNA carries the information needed to make a single protein. Each three-base sequence, or codon, codes for a specific amino acid.

6·14 On the ribosomes, proteins are synthesized as transfer RNA delivers amino acids into position. The anticodons on the tRNA match the codons on the mRNA. The sequence of codons in the mRNA determines the sequence of amino acids in the protein being assembled.

BIOLOGICAL TERMS

adenine
anaphase
anticodon
centromere
chromatids
codon
crossing over
cytosine
diploid number (2n)
DNA
DNA replication
gametes
gametogenesis
gene
guanine
homologous chromosomes
homologs
interphase
meiosis
metaphase
mitosis
monoploid number (n)
mRNA
nucleotides
oogenesis
prophase
RNA
rRNA
spermatogenesis
spindle
synapsis
telophase
thymine
transcription
translation
tRNA
uracil

USING BIOLOGICAL TERMS

Identify the sentences in which the italicized words are used correctly. Correct any incorrect sentences.

1. The type of cell division that reduces the number of chromosomes from the diploid number to the monoploid number is *mitosis*.
2. A *centromere* is made up of a sugar, a phosphate group, and a nitrogenous base.
3. The stage of mitosis in which chromosomes shorten and thicken and the nuclear membrane disappears is called *anaphase*.
4. *Synapsis* is the pairing of homologous chromosomes during meiosis.
5. *RNA* is composed of double strands of nucleotides.

118 CELL REPRODUCTION

UNDERSTANDING CONCEPTS

1. How do daughter cells formed by mitosis compare with each other and with the original parent cell? (6·1)
2. The body cells of many organisms contain the diploid number of chromosomes. What does this statement mean? (6·1)
3. As a cell grows, how does the ratio of surface area to volume change? Why is this change significant for the cell? (6·2)
4. What event occurs during interphase that is a precondition for mitosis? (6·3)
5. Name each phase of mitosis and briefly describe what occurs in each stage. (6·4)
6. Describe the structure and function of the spindle. (6·4)
7. How do plant cells differ from animal cells in the process of division of the cytoplasm? (6·4)
8. Compare the length of time required for each phase of mitosis. (6·4)
9. What is the significance of the fact that mitosis maintains genetic continuity from one cell generation to the next? (6·5)
10. What is the end result of mitosis in unicellular organisms? (6·5)
11. Gametes are produced by meiosis. How do gametes differ from body cells? (6·6)
12. What is the chromosome number of a human body cell?—of a human gamete? (6·6)
13. Describe what happens to homologous chromosomes during meiosis I. (6·7)
14. Describe what happens during synapsis of prophase I. (6·7)
15. Describe what occurs during meiosis II. (6·7)
16. Compare spermatogenesis and oogenesis. (6·8)
17. How do meiosis and mitosis differ? (6·9)
18. What evidence led to the idea that DNA is the hereditary material? (6·10)
19. Describe the structural model of DNA. (6·11)
20. Which bases in the DNA molecule will form bonds? (6·11)
21. Explain how DNA replication results in the formation of DNA molecules that are identical to the original. (6·12)
22. What is meant by the term *triplet* in reference to the DNA molecule? (6·13)
23. What is the function of a gene? (6·13)
24. Discuss the production of mRNA through transcription. (6·13)
25. Compare the structure of RNA with that of DNA. (6·13)
26. Explain how mRNA and tRNA are involved in the translation of the genetic code. (6·14)

APPLYING CONCEPTS

1. When formed, most molecules of DNA are not "totally new" either structurally or informationally. Explain this statement.
2. By what basic process does a zygote develop into an organism made up of, perhaps, millions of cells?
3. Is meiosis likely to occur in organisms that only reproduce asexually? Why or why not?
4. Knowledge of the amino acid sequence of a protein does not mean that the exact sequence of nucleotides on the corresponding segment of DNA is known. Why not?
5. Suppose an error occurred during one of the steps of meiosis. How might such an error affect the offspring that would eventually develop?

EXTENDING CONCEPTS

1. Build a model of the DNA molecule. Using additional material, demonstrate DNA replication.
2. James Watson, Francis Crick, Maurice Wilkins, and Rosalind Franklin all made significant contributions to the understanding of the nature of the genetic material. Report how one of these scientists contributed to our knowledge of heredity.
3. Cancer is uncontrolled cell division. Do a report on a recent hypothesis as to how cancer cells arise.

READINGS

Asimov, Isaac. *How Did We Find Out About DNA?* Walker, 1985.

"Biological Information Storage: How to Fold Proteins." *Science News,* April 14, 1984, p. 231.

Darnell, James. "The Processing of RNA." *Scientific American,* October 1983, p. 90.

Unit I Skills

INTERPRETING ELECTRON MICROGRAPHS

Much of our knowledge of the structure and parts of cells comes from electron micrographs, or photographs taken with an electron microscope. Although these pictures are two-dimensional, they can help scientists understand cell structure in three dimensions.

To prepare an object for examination under the transmission electron microscope (TEM), the specimen must first be stained. Then the specimen is embedded in a special type of plastic so that it can be sectioned. A device called a microtome is used to slice the specimen into very thin sections. Now the specimen can be examined under the TEM. In the microscope a beam of electrons forms an image of the specimen on either a fluorescent screen or a photographic plate.

Researchers can learn the size, shape, and location of various cell structures by reading electron micrographs. However, these photographs must be analyzed carefully. Examining just one picture will not provide complete information about the cell. For example, some cell structures may not be present in a particular slice or section. Also, most cell structures are so large that one section may show only a small part of that structure. Imagine trying to determine the shape or size of a sausage if all you could see was one thin slice of it.

To further complicate matters, some cell parts are irregular in shape. The structures may be branched, cup-shaped, twisted, and so on. A single slice of a structure that is branched usually will not reveal the branch point. In fact, in some cases the section will contain a part of each branch, but the branches will not appear to be connected. On this page you can see electron micrographs of two different sections of microtubules. Examine the pictures and answer the questions below on a separate sheet of paper; do not write in this book.

1. What shapes do you see in the pictures?
2. Why do the microtubules look different in each picture?
3. If the microtubules were cut on the diagonal (an angle less than 90°), what shape might you see?

An important way of learning the actual shape and size of cell structures is to examine photographs of many sections from a single cell. Your teacher will demonstrate an experiment with apples to help you understand how this technique works. You will see apples sliced in two different ways. Spread out the pieces of each apple on a table in the order in which they were cut. Look at the location and shape of each apple's internal parts. Then answer the questions below on a separate sheet of paper.

4. What conclusions can you draw about the shape and distribution of the seeds within the apple?
5. Look at one crosswise section and one lengthwise section of apple. Draw what you see in the center of each slice. How are the two pictures different?

Scientists study sections of cells in a manner similar to that in which you studied apples. When a series of sections of a cell are kept in order and analyzed that way, they are called serial sections. Some cell structures are large enough to appear in a number of adjacent sections. Such a cell structure can be then traced through the set of photographs to discover its actual three-dimensional form.

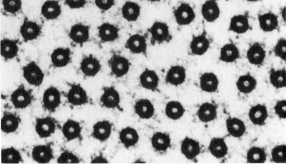

UNIT II

Continuity of Life

No one has to tell you that baby swans are similar to their parents. Why is it that all living things always produce other living things that are like themselves? In this unit you will find out how the messages of heredity are passed from one generation to another.

Chapters

7 *Basic Genetics*
8 *Applied Genetics*
9 *Evolution*
10 *History of Life*
11 *Classification*

7. Basic Genetics

Living things always come from other living things like themselves. The similarities that exist between parents and offspring are not just coincidental; they result from the hereditary material that parents pass down. Since no two living things are exactly alike, parents must be able to pass on different kinds of traits. In this chapter, you will explore how parents pass on traits to their offspring.

CHAPTER OBJECTIVES

After completing this chapter, you will be able to

- **Explain** how the work of Gregor Mendel contributed to an understanding of how traits are inherited.
- **Relate** the laws of probability to genetics.
- **Discuss** how multiple alleles and polygenic inheritance affect the expression of certain traits.
- **Relate** chromosomes to genes.
- **Describe** the importance of sex-linked traits.
- **Explain** how crossing over affects gene linkages.

Genetics and Gregor Mendel

7•1 WHAT IS GENETICS?

For hundreds of years, people have noticed that certain physical features of plants, animals, and humans were the same in both parent and offspring. Early humans noticed many things that were the same as well as different between members of one generation and the next. Yet, an understanding of the biological principles explaining similarities and differences between generations has come about only recently. The first real explanation appeared just a little over 100 years ago. However, most of the present knowledge of how living things pass on certain features to their young was learned since 1900.

Among living things, researchers have found thousands of traits that are *inherited,* or passed on from parents to their young. A feature that a living thing can pass on to its young is called a **trait** (trayt). In humans, you have surely noticed that eye color, hair color, facial features, and body build can be inherited by children from their parents. The passing of traits from parents to their young is called **heredity** (huh REHD uh tee). The branch of biology dealing with the study of heredity is **genetics** (juh NEHT ihks). Many of the great scientific discoveries of this century have been in the field of genetics. Genetics is still one of the most active areas of scientific research.

7•2 MENDEL'S EARLY EXPERIMENTS

The foundation for the modern study of genetics was laid by an Austrian monk, Gregor Mendel (1822–1884). Mendel lived in a monastery in what is now the city of Brno, Czechoslovakia. He attended the University of Vienna for two years, where he studied biology and mathematics. While there, he became interested in improving plants by crossing organisms that were different in one or more traits. This interest was to lead him to discover the basic principles of how living things inherit traits.

SECTION OBJECTIVES

- Describe Mendel's garden pea experiments that led to the principle of dominance and the principle of segregation.
- Apply principles of probability to solve genetics problems.

Mendel kept a small garden in the monastery where he performed experimental crosses with garden peas. Garden peas were a good choice because they have a number of contrasting traits that can easily be distinguished. In Figure 7·1 you can see seven pairs of contrasting traits that Mendel chose for his studies. As you can see, he chose three pairs of contrasting traits associated with pea seeds, two with pea pods, and two with pea stems.

The structure of the garden pea flower was also ideal for Mendel's breeding experiments. You can see the structure of a pea flower in Figure 7·2. Pea plants reproduce sexually. Find the stamens (male reproductive structures) and the pistil (female reproductive structure) within the flower. Pollen formed at the tip of the stamens contains the male gametes, or sex cells. Female gametes are formed within the pistil. Before fertilization can take place, pollen must be moved from a stamen to a pistil by a process called *pollination* (pahl uh NAY shuhn). Most flowering plants are cross-pollinated by the wind or insects. **Cross-pollination** is a process by which pollen formed by a flower of one plant is moved to the pistil in a flower of another plant of the same kind. Notice in Figure 7·2 that the stamens and pistil in a pea flower are contained within closed petals. This condition usually prevents cross-pollination. Thus, in garden peas, pollen is normally transferred from the stamen to the pistil of the same flower. The transfer of pollen within the same flower or between flowers of the same plant is called **self-pollination**. The fact that pea flowers usually self-pollinated was very important in Mendel's crossing experiments. Although pea flowers could be artificially cross-pollinated, the closed petals prevented pollen of other pea flowers from interfering with the experimental results.

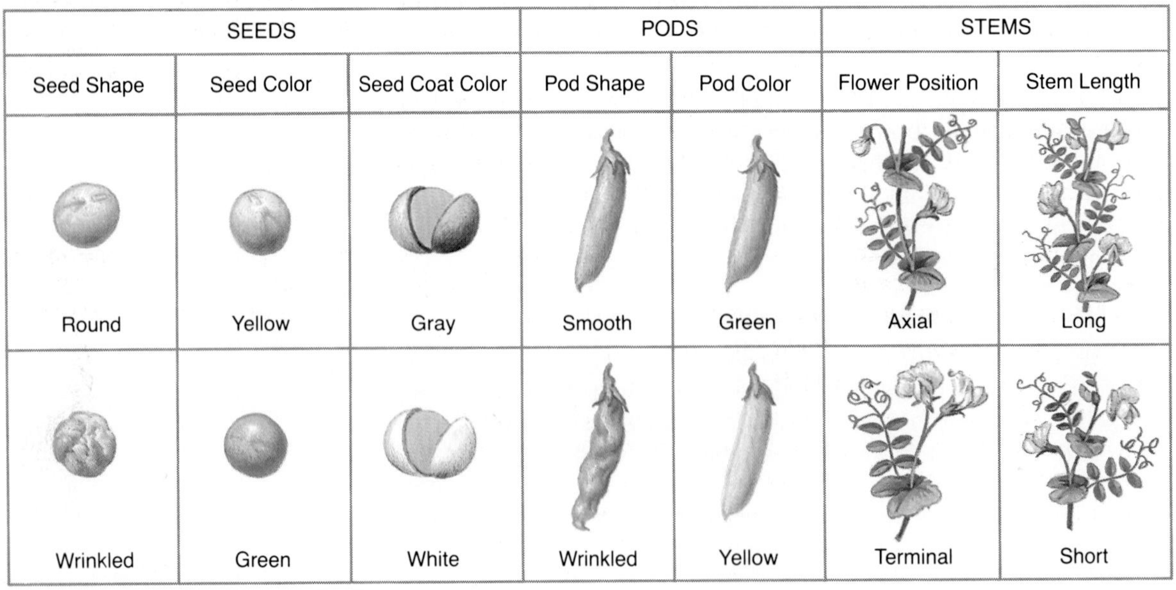

FIGURE 7·1 These are the seven garden pea traits that were studied by Mendel. Notice that the traits occur as contrasting pairs.

SEEDS			PODS		STEMS	
Seed Shape	Seed Color	Seed Coat Color	Pod Shape	Pod Color	Flower Position	Stem Length
Round	Yellow	Gray	Smooth	Green	Axial	Long
Wrinkled	Green	White	Wrinkled	Yellow	Terminal	Short

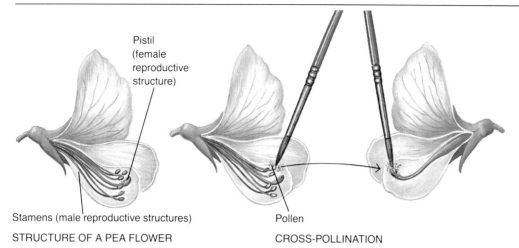

FIGURE 7·2 Parts of the pea flower have been removed to make it easier to see reproductive structures. A paint brush is used to transfer pollen from stamen to pistil.

Mendel began his experiments by developing a number of types, or lines, of plants that were *pure* for each of the seven pairs of traits you examined in Figure 7·1. Remember, Mendel chose these traits because each has contrasting forms. It was easy for him to distinguish between the contrasting forms of the traits. For example, seed color was either yellow or green, pod shape was either smooth or wrinkled, and so on. A **pure line** is a group of living things that produce young having only one form of a trait in each generation. For example, one variety of peas that was pure for seed color produced only yellow-seeded peas generation after generation. Another pure variety produced only green-seeded peas in each generation. By letting the peas self-pollinate for several generations, Mendel produced seven pairs of pure lines. The self-pollinating nature of peas was a critical feature that allowed Mendel to establish these pure lines.

After establishing pure lines, Mendel then made hundreds of crosses by transferring the pollen from the stamens of plants having one trait to the pistils of plants having the contrasting trait. An example of one of Mendel's crosses is shown in Figure 7·3. In this experiment he crossed plants that produced round seeds with those that produced wrinkled seeds. The pure-line plants that he used in making these crosses are considered the **parental generation.** The parental generation (P_1) is the group of organisms used to make the first cross in a breeding experiment. After new seeds developed, Mendel examined them to see how they looked. Would they all be round? Would they all be wrinkled? Would some be round and some be wrinkled? If you look at Figure 7·3, you will see the results. Notice that all the seeds produced from the cross were round. Mendel also found that it made no difference whether the pollen from a round-seeded plant had been placed on the pistil of a wrinkled-seeded plant, or the pollen from a wrinkled-seeded plant had been used to pollinate a round-seeded plant. Only round-seeded peas were among the offspring.

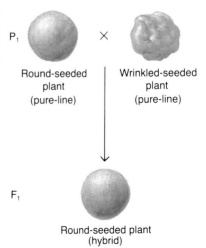

FIGURE 7·3 Mendel's cross between a pure round-seeded plant and a pure wrinkled-seeded plant produced a plant that had round seeds.

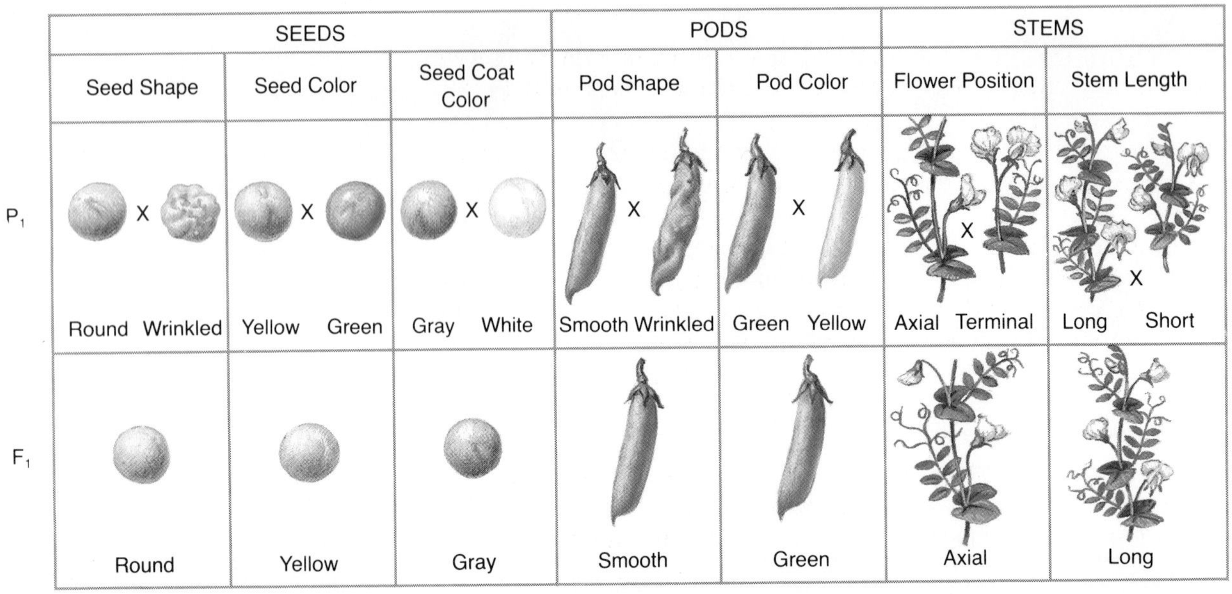

	SEEDS			PODS		STEMS	
	Seed Shape	Seed Color	Seed Coat Color	Pod Shape	Pod Color	Flower Position	Stem Length
P₁	Round × Wrinkled	Yellow × Green	Gray × White	Smooth × Wrinkled	Green × Yellow	Axial × Terminal	Long × Short
F₁	Round	Yellow	Gray	Smooth	Green	Axial	Long

FIGURE 7·4 Mendel's seven parental crosses are shown in the top row. The results of these crosses are shown in the bottom row.

filius (son)
filia (daughter)
hybrida (mongrel)

The round-seeded peas produced by Mendel's parental cross were first-generation organisms. The offspring from a parental cross are called the **first filial** (FIHL ee uhl) **generation** (F_1). All of these F_1 round-seeded plants were hybrids. A **hybrid** (HĪ brihd) is the offspring of two parents that differ in one or more inherited traits. When Mendel crossed a pure-line round-seeded plant with a pure-line wrinkled-seeded plant, he performed a cross known as a monohybrid (mahn uh-HĪ brihd) cross. A **monohybrid cross** is a cross that involves one pair of contrasting traits.

The outcome of all Mendel's parental crosses is shown in Figure 7·4. Notice that in every case, all F_1 individuals resembled only one parent. Mendel was not ready to draw any conclusions until he had followed these traits through another generation.

In his second set of experiments, Mendel allowed the F_1 generation to self-pollinate. The offspring from the self-pollination of the F_1 generation are known as the **second filial generation** (F_2). Mendel found that some F_2 pea seeds were round, and some seeds were wrinkled. The results of the self-pollination of the hybrid round-seeded plants in the F_1 generation are shown in Figure 7·5. What happened to the trait that was "lost" in the F_1 generation?

FIGURE 7·5 When a hybrid round-seeded plant had self-pollinated, ¾ of its offspring produced round seeds and ¼ produced wrinkled seeds. The trait that was "lost" in the F_1 generation reappeared.

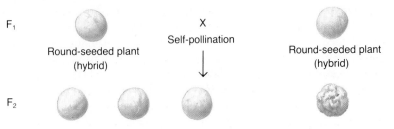

Figure 7·6 shows the actual results of three of Mendel's seven monohybrid crosses. Similar results were found in each of the other four monohybrid crosses. In the column "F_2 Generation" you can see that the trait that did not appear in the F_1 generation showed up in the F_2 generation in each monohybrid cross. In the column "Actual F_2 Ratios," the numbers of plants obtained for each contrasting trait are expressed as mathematical ratios. A ratio is a mathematical expression that shows a relationship between two or more numbers. In the cross between hybrid round-seeded peas, the ratio of round-seeded peas to wrinkled-seeded peas can be expressed as follows.

$$\frac{5474 \text{ round-seeded peas}}{1850 \text{ wrinkled-seeded peas}} = \frac{2.96}{1}$$

If 2.96 is rounded off to 3, the ratio may be expressed as 3:1. In each F_2 generation, about ¾ of the offspring were like the F_1 parents. The other ¼ showed the form of the trait that had been absent before.

FIGURE 7·6 This table shows the results of three of Mendel's seven monohybrid crosses. As you can see, Mendel worked with large numbers of F_2 individuals.

F_1	SELF-POLLINATION	F_2 GENERATION	ACTUAL F_2 RATIOS	APPROXIMATE F_2 RATIOS
Hybrid round seeds	× Hybrid round seeds	5474 round seeds 1850 wrinkled seeds	2.96:1	3:1
Hybrid yellow seeds	× Hybrid yellow seeds	6022 yellow seeds 2001 green seeds	3.01:1	3:1
Hybrid gray seed coats	× Hybrid gray seed coats	705 gray seed coats 224 white seed coats	3.15:1	3:1

7·3 MENDEL'S RESULTS EXPLAINED

Mendel developed a set of hypotheses to explain his results. His insight into the processes controlling heredity was amazing because each of his hypotheses has since been shown to be correct. First, Mendel suggested that every hereditary trait was controlled by two separate factors, one from each parent. Mendel's **factors** were units of heredity that the pea plants passed on to future generations. Mendel called these hereditary factors *elemente*, but today they are called genes. You have learned that a gene is a segment of DNA that acts as a unit of inheritance. Although Mendel did not use the term *gene*, scientists today use it to refer to Mendel's factors. Scientists now know that genes are units of hereditary material found in chromosomes. Chromosomes and the genes on them are passed from parents to offspring by way of gametes.

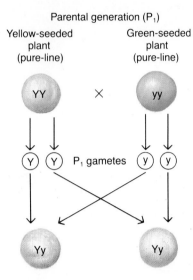

FIGURE 7·7 Like other traits, the color of pea seeds is controlled by a pair of genes. Each parent contributes one gene to the offspring.

Mendel established the practice of using letters to represent the paired genes controlling inherited traits. As an example of this practice, consider the parental cross of yellow-seeded and green-seeded peas. The gene for yellow seeds may be represented by the capital letter *Y*. The contrasting gene for the green seed trait is then represented by *y*, the lower case of the same letter. If genes occur in pairs, as Mendel believed, then each pea plant must have two genes determining its seed color. However, the genes in a pair may be alike (*YY* or *yy*) or they may be different (*Yy*). A living thing in which the two genes for a given trait are alike is **homozygous** (hoh muh ZI guhs). A pea plant having the gene pairs *YY* or *yy* is homozygous, or pure, for seed color. An organism in which the two genes for a given trait are different is **heterozygous** (heht uhr uh ZĪ guhs). A pea plant that is heterozygous, or hybrid, for seed color is shown by the letters *Yy*.

The results of a cross that led Mendel to a second major hypothesis are shown in Figure 7·7. Notice that the pure lines of yellow-seeded (*YY*) and green-seeded (*yy*) plants are homozygous. Now look at the P_1 gametes. Mendel reasoned that only one gene from each parent passed into a gamete. Thus each gamete from a yellow-seeded parent contained only one *Y* gene, while each gamete produced by a green-seeded parent contained only one *y* gene. When these gametes combined as a result of fertilization, you can see that only one combination of genes was possible in the F_1 generation, *Yy*.

● **Dominance** Mendel concluded that one gene determined the expression of a trait in the hybrids of the F_1 generation. This prevented the contrasting form of that trait from showing up, or being expressed. In a hybrid organism, a gene that prevents the expression of another gene is said to be **dominant** (DAHM uh nuhnt). The gene that is not expressed is called **recessive** (rih SEHS ihv). In the cross shown in Figure 7·7, the gene for yellow seeds is dominant over the recessive gene for green seeds. Thus, all the seeds produced in the F_1 generation were yellow. Even though recessive genes were not expressed in the F_1 generation, they were still present, and they remained unchanged. Look at Figure 7·8. You can see that about ¼ of the F_2 generation had green seeds. Even though the genes for green seeds were hidden in the F_1 generation, they were passed on to the F_2 generation, where they were expressed—that is, the trait became visible. Mendel's hypothesis that one gene could cause one form of a trait to appear and could stop the contrasting form from appearing is now known as the principle of dominance. The **principle of dominance** can be stated in this way: In a hybrid organism, one gene determines the expression of a particular trait and prevents the expression of the contrasting form of that trait.

● **Segregation** Mendel developed a third hypothesis to explain why the traits that disappeared in the F_1 generation reappeared in the F_2 generation. He reasoned that in any cross, each parent pea plant gave only one gene to each gamete that was formed. In other words, the genes were separated, or segregated, from each other during gamete formation. They recombined when fertilization took place. Study Figure 7·8. Notice that the yellow-seeded plants produced two kinds of male gametes and two kinds of female gametes. When these gametes joined during fertilization, male gametes carrying the green seed trait (*y*) fertilized female gametes carrying the green seed trait (*y*) about one out of four times. As a result of this, the green seed trait that had disappeared in the F_1 generation reappeared in the F_2 generation. What ratio of dominant to recessive traits did Mendel find in the F_2 generation?

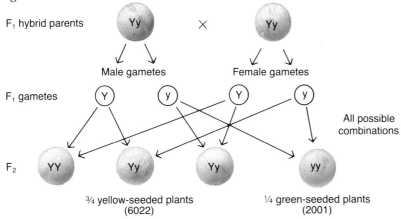

FIGURE 7·8 The self-pollination of hybrid yellow-seeded plants produced both yellow-seeded plants and green-seeded plants.

The hypothesis that a gamete receives only one member of a pair of genes is now known as Mendel's **principle of segregation**. The principle of segregation (sehg ruh GAY shuhn) can be stated in this way: When gametes form, the genes that control a particular trait are separated into different gametes.

● *Genotype and Phenotype* Biologists often need to make a distinction between the physical appearance of an organism and its genetic makeup. **Genotype** (JEEN uh tīp) is the genetic makeup of an organism. For example, *RR* represents the genotype of a pea plant with round seeds if the plant is homozygous for round seeds. **Phenotype** (FEEN uh tīp) is the outward appearance of an organism. In Mendel's pea plants, the phenotype for either genotype *RR* or genotype *Rr* is round seeds. Suppose *A* represents the gene for axial flowers, which is dominant, and *a* represents the gene for terminal flowers, which is recessive. What is the genotype for a pure-line axial-flowered plant? What phenotype is produced by the genotype *Aa*? Why will the genotype *aa* always result in the terminal flower phenotype?

7·4 PROBABILITY

Mendel was not the first person to produce hybrids. However, he was the first to produce and classify thousands of hybrids and apply mathematical analysis to his data. Mendel used probability in his reasoning. **Probability** is the study of the operation of the laws of chance. *Chance* refers to the likelihood of a certain event happening, such as having "heads" come up when a coin is tossed.

Probability is usually expressed by means of fractions or percentages, as in this expression.

$$\text{probability} = \frac{\text{number of chances for an event}}{\text{number of possible events}}$$

When a coin is flipped, one of two possible events can occur. It can come up either heads or tails. The probability of obtaining heads can be expressed by the fraction ½. There is one chance out of two possibilities that heads will come up.

Two important principles of probability are useful in studying genetics. The first principle is called the rule of independent events. The **rule of independent events** is as follows: Previous events do not affect the probability of later occurrences of the same event. Suppose you toss a nickel, and it comes up heads. What is the probability of getting heads if you toss the nickel again? The first time the nickel is tossed, the probability of getting heads is ½. On a second toss or any toss after that, the probability of tossing heads is still ½. Each toss of a nickel is independent of any other toss of the same nickel.

The second principle of probability useful in genetics is the product rule. The **product rule** may be stated like this: The probability of independent events occurring together is equal to the product of the probabilities of these events occurring separately. If you toss a penny and a nickel together, what is the chance of both coins coming up heads? You can find the answer by multiplying the separate probabilities for each coin coming up heads. The probability of the penny landing heads up is ½, and the probability of the nickel showing heads is ½. The probability for heads on both coins is the product of their separate probabilities.

probability of penny coming up heads × probability of nickel coming up heads = probability of both coming up heads
½   ½   ¼

FIGURE 7·9 The possible combinations resulting from tossing a penny and a nickel. Notice that every combination has a 1/4 chance of occurring.

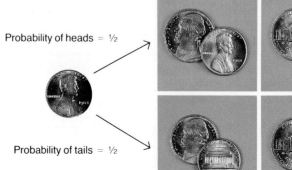

Investigation

How Are the Laws of Probability Applied to Genetics?

Goals
After completing this activity, you will be able to
- Demonstrate the laws of probability by tossing coins.
- Use the laws of probability to predict the outcome of certain events.
- Apply the laws of probability to genetics problems.

Materials (for groups of 2)
2 coins   masking tape   scissors

Procedure

A. On a separate sheet of paper, copy the chart shown below.

B. Work with a partner. You will toss two coins at the same time. Your partner will tally the results—heads/heads, heads/tails, or tails/tails. First toss both coins 10 times and record your results. Then toss both coins 50 times and record the results. Finally, toss both coins 100 times and record the results in Chart 1.

C. Determine the ratio of heads/heads to heads/tails to tails/tails for each series of tosses (10 tosses, 50 tosses, and 100 tosses).

D. Cut off 4 pieces of tape just big enough to fit on the coins. Stick the tape on both sides of the coins. On one side write *T*, and on the other side write *t*. Consider that these coins represent traits in pea plants. *T* stands for *tall* pea plants; *t* stands for *short* pea plants. Tossing two coins together represents the crossing of hybrid pea plants as in Mendel's F_1 cross.

E. Work with a partner. Toss the two labeled coins together 10 times, 50 times, and 100 times. On a separate sheet of paper, copy the chart below and record your results.

F. One student in the class will record the results of 100 tosses by 10 pairs of students. Record this information on your chart following *Results of 1000 tosses*.

G. From Mendel's experiments with pea plants, you know that tallness is dominant over shortness. Thus, you should know the phenotype of the following: TT, Tt, and tt. From the data in Chart 2, find the ratio of TT to Tt to tt for each group of tosses.

H. Follow your teacher's instructions for the proper cleanup of all materials.

Questions and Conclusions

1. When you tossed two coins 100 times in step B, which combination occurred most often—heads/heads, heads/tails, or tails/tails? What was the ratio of heads/heads to heads/tails to tails/tails?

2. When you tossed the coins that represented the traits for tallness and shortness, what was the ratio of TT to Tt to tt for 100 tosses? For 1000 tosses?

3. What are the phenotypes of the following pea plants: TT, Tt, and tt? From the results in Chart 2, what is the ratio of tall plants to short plants?

4. From the recorded results of 1000 tosses, what is the ratio of tall plants to short plants?

5. In one of Mendel's many experiments with peas, he had the following results: 787 tall plants, 277 short plants (total: 1064 plants). How does his ratio of tall plants to short plants compare with the results found in your class?

6. How do the results of a small sample (100 tosses) compare with the results of a large sample (1000 tosses)?

CHART 2. Results of Tossing Labeled Coins

	TT	Tt	tt
10 TOSSES			
50 TOSSES			
100 TOSSES			
1000 TOSSES			

CHART 1. Results of Coin Tossing

	HEADS/HEADS	HEADS/TAILS	TAILS/TAILS
10 TOSSES			
50 TOSSES			
100 TOSSES			

7·5 PROBABILITY AND MENDEL'S RESULTS

Rules of probability can be used to help predict the results of simple genetic crosses. One method for figuring probabilities is to construct a box called a Punnett square. A **Punnett** (PUHN iht) **square** is a chart showing the possible combinations of genes among the offspring of a cross. It is named for the geneticist R. C. Punnett, who first suggested its use.

Study Figure 7·10 to understand how a Punnett square is used. The outcome of a cross involving pure-line red-flowered pea plants and pure-line white-flowered pea plants is shown like this.

$$RR \quad \times \quad rr$$
Pure-line red-flowered pea plants   Pure-line white-flowered pea plants

1. Letters representing the gametes produced by one parent are written above the columns on the top of the chart. The top of the chart shows that a gamete has a ½ chance of receiving one of the R genes from the red-flowered parent and a ½ chance of receiving the other R gene.

2. Letters representing the gametes produced by the other parent are written next to the rows on the left side of the chart. Notice that the gametes produced by the white-flowered parent have a ½ chance of receiving either of the two r genes.

3. The inside squares are filled in to show the genotypes that can result when the gametes combine in fertilization. In each square is written the letters for the gametes from the top and the side of the chart. The combinations of letters in the boxes show all the possible genotypes of offspring resulting from the cross. You can see that the genotypes of all F_1 individuals are the same. All the F_1 plants, or 100 percent, have the heterozygous genotype Rr.

4. The inside squares also help to show the ratio of phenotypes that is obtained when a cross is made. Since the Rr genotype produces only red-flowered pea plants, 100 percent of the F_1 plants should show the dominant phenotype. Did Mendel find only the dominant phenotype among the F_1 offspring in each of his crosses?

FIGURE 7·10 This figure shows how to set up a Punnett square to illustrate a cross between a pure red-flowered plant and a pure white-flowered plant. What is the phenotype of the offspring?

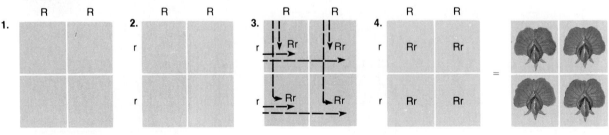

R = dominant gene for red-flower trait
r = recessive gene for white-flower trait

Genotype = 100% Rr
Phenotype = 100% red-flowered

The Punnett square in Figure 7·11 shows the probabilities of obtaining certain phenotypes and genotypes in the F_2 generation. Notice that the proportion of the resulting genotypes are ¼ *RR*, ½ *Rr*, ¼ *rr*. Both *RR* and *Rr* genotypes produce red-flowered plants. Only the *rr* genotype produces white-flowered plants. Therefore, the ratio of phenotypes in the F_2 generation should be three red-flowered plants to one white-flowered plant. Was this result observed by Mendel?

Mendel reasoned that all red-flowered plants in the F_1 generation had the genotype *Rr*. He also assumed that the genotype of all white-flowered plants was *rr*. So he predicted that a cross of F_1 plants (*Rr*) with white-flowered plants (*rr*) should produce about an equal number of red-flowered and white-flowered offspring. This is what he found when he performed his experiment, as shown in Figure 7·12.

Mendel's cross with the F_1 hybrid and the homozygous recessive plant was a test cross. A cross between a living thing showing a dominant trait but of uncertain genotype and a living thing that is homozygous recessive for a trait is called a **test cross.** Today plant and animal breeders use a test cross to find whether a living thing showing a dominant phenotype is homozygous or heterozygous for the trait. In guinea pigs, the black coat trait is dominant over the white coat trait. However, a black guinea pig can be either homozygous (*BB*) or heterozygous (*Bb*) for that trait. How can you find the difference?

A Punnett square is a useful tool for finding the probable results of genetic crosses. However, the expected results may be quite different from the actual results. The results of a genetic cross may match the probable expected results only when there is a large number of crosses or many offspring. The Punnett square method can be used to solve genetic problems. Study the following sample problems.

● **Sample Problem 1** In mice, black hair is dominant over white hair. A white-haired male is mated with a heterozygous black-haired female. What are the probable genotypic and phenotypic ratios that could result from this mating?

Solution

a. Write the genotypes of the animals being mated.

    bb × *Bb*
  Genotype   Genotype of
   of male   female

b. Use these genotypes to complete a Punnett square.

		Male gametes	
		b	b
Female	B	Bb	Bb
gametes	b	bb	bb

c. Summarize the genotypic and phenotypic ratios.

Genotypic ratio = ½ *Bb* : ½ *bb*
Phenotypic ratio = ½ black-haired : ½ white-haired

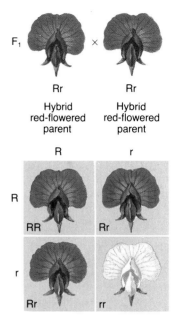

Genotypic ratio = ¼ RR : ½ Rr : ¼ rr
Phenotypic ratio = ¾ red-flowered :
                   ¼ white-flowered

FIGURE 7·11 Mendel's cross between hybrid red-flowered plants.

FIGURE 7·12 Mendel's test cross which showed that the red-flowered plant had genotype *Rr*.

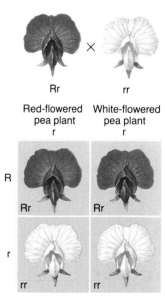

Genotypic ratio = ½ Rr : ½ rr
Phenotypic ratio = ½ red-flowered :
                   ½ white-flowered

● **Sample Problem 2** In guinea pigs, rough coat (*R*) is dominant over smooth coat (*r*). Two rough-coated guinea pigs are mated, and they produce a litter of four rough-coated and two smooth-coated guinea pigs. What are the genotypes of the parents?

Solution: Work backwards to find the genotypes of the parents.

a. Write a complete or partial genotype for the parents and offspring. Since the rough-coat trait is dominant, each parent and each of the 4 rough-coated offspring must have at least one gene for the rough-coat trait.

Parents: *R?*
4 rough-coated offspring: *R?*

Since the smooth-coat trait is recessive, the 2 smooth-coated offspring must be homozygous recessive (*rr*). If either had just one *R* gene in its genotype, it would show the rough-coat phenotype.

2 smooth-coated offspring: *rr*

b. Work backwards to determine the genotypes of the parents.

R?   *R?*
(Rough-coated parent) × (Rough-coated parent)
R?   *rr*
(4 rough-coated offspring) + (2 smooth-coated offspring)

In the 2 smooth-coated offspring, one gene for smooth-coat comes from each parent. Each parent must be heterozygous (*Rr*).

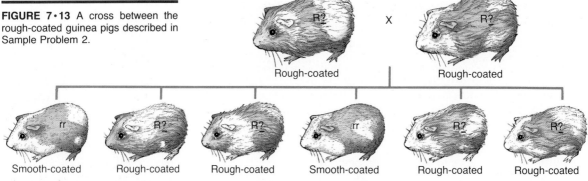

FIGURE 7·13 A cross between the rough-coated guinea pigs described in Sample Problem 2.

REVIEW QUESTIONS

1. State Mendel's principle of dominance and principle of segregation.
2. What is the probability of tossing a quarter and getting four tails in a row?
3. In cats, tabby (black and gray striped) is dominant over black. How could the owner of a tabby cat determine the genotype of the cat?
4. In dogs, the wirehair trait (*W*) is dominant over the smooth-hair trait (*w*). A female dog heterozygous for wirehair is mated with a smooth-haired male. What genotypes and phenotypes can be expected?
5. In fruit flies, gray body color (*G*) is dominant over black body color (*g*). A gray male is mated with a black female; they produce 197 gray offspring. What are the probable genotypes of the parents and the offspring?

Other Discoveries in Genetics

7·6 MENDEL'S CROSSES WITH TWO FACTORS

Mendel also traced the inheritance of two traits at a time. He started by producing plants that were pure for two traits. For instance, by self-pollination he developed plants that were homozygous dominant for both round seed shape and yellow seed color (*RRYY*). He also developed plants that were homozygous recessive for wrinkled seed shape and green seed color (*rryy*).

Mendel then crossed plants that were pure-breeding for both seed shape and color (*RRYY* × *rryy*). Since round and yellow seed traits were dominant over wrinkled and green seed traits, all F_1 plants had round, yellow seeds. The Punnett square in Figure 7·14 shows that these results were to be expected. Since the F_1 plants were hybrid for two traits (*RrYy*), they were called *dihybrids*.

Next, Mendel let the dihybrids self-pollinate to produce an F_2 generation. A cross involving two different sets of traits is called a **dihybrid** (dī HĪ brihd) **cross.** The ratio of these phenotypes is about $9:3:3:1$, as shown in Figure 7·15. The phenotypic ratio of $9:3:3:1$ could result only if each of the F_1 plants produced the four kinds of gametes shown in Figure 7·15. Also the gametes must have been produced in roughly equal proportions. Apparently the genes for the two different traits were separated independently of each other. *R* was separated from *r*, and *Y* was separated from *y*. Each *R* or *r* then had an equal chance of getting together with a *Y* or *y* to produce one of the *RY*, *Ry*, *rY*, or *ry* gametes. Study the 16 possible combinations of gametes shown in Figure 7·15. Notice that these combinations confirm Mendel's dihybrid ratio.

SECTION OBJECTIVES

- Describe Mendel's experiments that led to the principle of independent assortment.
- Discuss some variations of Mendel's principles of heredity.

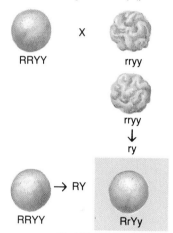

FIGURE 7·14 When Mendel crossed a pea plant with round yellow seeds with a plant with wrinkled green seeds, the offspring had round yellow seeds. The offspring were hybrid for two traits.

FIGURE 7·15 The F_1 pea plants were hybrid for both seed color and seed shape. When these dihybrid plants self-pollinated, the F_2 offspring had four different seed phenotypes.

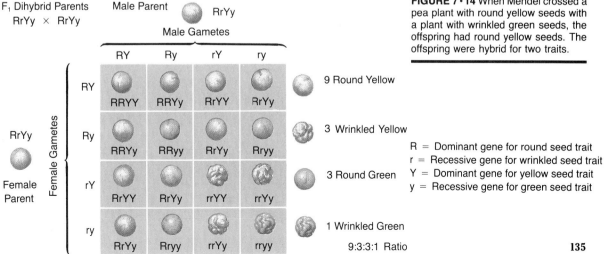

R = Dominant gene for round seed trait
r = Recessive gene for wrinkled seed trait
Y = Dominant gene for yellow seed trait
y = Recessive gene for green seed trait

9:3:3:1 Ratio

If you tossed some coins in the air, each coin would land head or tail independent of the others. The results for one coin do not affect the other coins. This is similar to the distribution of genes in a dihybrid cross. As you can see from Figure 7·15, the genes in the gamete sort out independent of each other. This independent sorting out of different genes during gamete formation is known as the **principle of independent assortment.** This principle states: During gamete formation, the genes for one trait are separated and distributed to the gametes independently of the genes for other traits.

7·7 MULTIPLE ALLELES

Each of the traits studied by Mendel was controlled by one of two possible forms of a gene. For example, there were two forms of the gene controlling seed color in peas. Pea seeds were colored green or yellow, depending on the particular combination of genes in their genotype. The different forms of the gene for a particular trait are called **alleles** (uh LEELS). Alleles of a particular trait are at the same location on homologous chromosomes.

allelon (of each other)

There are many cases in which there are more than two alleles for a particular trait in a population. Traits for which there are three or more alleles are said to be controlled by **multiple alleles.** Multiple alleles exist when there are more than two alleles for a particular trait. An example of multiple alleles in humans occurs in blood types. Although an individual can have only two alleles for blood type, there are three alleles for blood type among humans.

The three blood type alleles are usually designated as I^A, I^B, and i. Both I^A and I^B are dominant over i. I^A and I^B show no dominance over each other. As shown in Figure 7·16, six genotypes and four phenotypes can result from these alleles. A person having either genotype I^AI^A or I^Ai has type A blood. A person with type B blood has a genotype of either I^BI^B or I^Bi. A person with type AB blood has both dominant alleles, I^AI^B, in the genotype. A person with type O has both recessive genes, ii. A knowledge of human blood types has made safe blood transfusions possible.

FIGURE 7·16 This table shows the genotypes for human blood types. Human blood types are determined by multiple alleles.

GENOTYPE	BLOOD TYPE
I^AI^A or I^Ai	A
I^BI^B or I^Bi	B
I^AI^B	AB
ii	O

7·8 POLYGENIC INHERITANCE

The garden pea traits studied by Mendel could be clearly seen. Ripe pea seeds were either yellow or green, the pods were either smooth or wrinkled. The stems were clearly long or short. Mendel may not have been aware that characteristics in many living things may vary between two extremes because they are under the control of a number of genes. Characteristics such as height, weight, and skin color in humans show *variations* from one individual to another. These variable characteristics are controlled by several pairs of independent genes. In **polygenic inheritance** (pahl ee JEHN ihk ihn HER uh tuhns) a single trait is determined by the interaction of two or more pairs of genes. The genes for such traits are found at different locations on a chromosome or on different chromosomes.

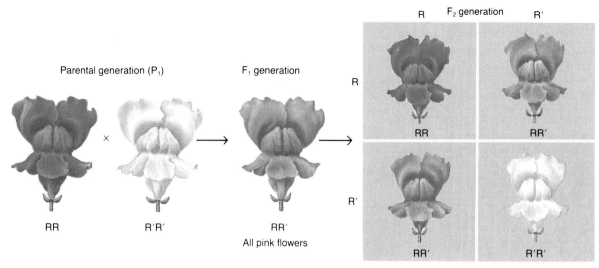

FIGURE 7·17 Flower color is not always the result of the expression of dominant or recessive alleles. Unlike pea plants, red flower color is not dominant over white in snapdragons.

7·9 INCOMPLETE DOMINANCE

In Mendel's hybrid pea plants, one allele was clearly dominant over the other. Today scientists know of many situations in which two alleles affect the phenotypes of hybrid organisms. A condition in which a hybrid has a phenotype intermediate between the contrasting traits of its parents is called **incomplete dominance**. Incomplete dominance takes place in snapdragon flowers. Red flower color is not dominant over white, and white is not dominant over red. A cross between a pure red snapdragon and a pure white snapdragon results in F_1 offspring with flowers of an intermediate color—pink. The F_2 generation will have red, pink, and white flowers in the ratio of 1:2:1 as seen in Figure 7·17.

REVIEW QUESTIONS

6. State Mendel's principle of independent assortment.
7. In summer squash, white fruit color (*W*) is dominant over yellow fruit color (*w*). Disc-shaped fruit (*D*) is dominant over spherical fruit (*d*). A plant with homozygous white spherical fruit is crossed with a plant with yellow spherical fruit. What are the possible genotypes and phenotypes of the offspring fruit?
8. If a person with type O blood marries a person with type AB blood, what types of blood can their children have?
9. Define *polygenic inheritance* and then list several examples of polygenic inheritance in humans.
10. In radishes, a cross between plants that produce round radishes (*R*) and long radishes (R^1) results in oval radishes. What would be the genotypic and phenotypic ratios in a cross between a long radish plant and an oval radish plant?

Chromosome Theory of Heredity

SECTION OBJECTIVES

- Describe the experiments that demonstrated the relationship of chromosomes to genes.
- Explain the importance of sex-linked traits.
- Discuss how crossing over leads to variations in the genetic material.

7·10 CHROMOSOMES AND GENES

Mendel published a paper on his work on inheritance in 1866. However, other scientists did not recognize the importance of his work. It was not until 1900 that his paper was rediscovered by three European scientists working separately. This was 16 years after Mendel's death. Each of these researchers gave Mendel full credit for the brilliant work he had done.

Early in this century, Walter S. Sutton, an American graduate student at Columbia University, read Mendel's paper. Sutton was studying the process of meiosis in grasshopper sperm cells. He noticed similarities between the behavior of the chromosomes and Mendel's "factors." These similarities are listed in Figure 7·18. What conclusion might be drawn from Sutton's observations?

FIGURE 7·18 Sutton observed that there were important similarities between behavior of chromosomes and Mendel's "factors." This table summarizes his observations.

CHARACTERISTICS OF CHROMOSOMES	CHARACTERISTICS OF MENDEL'S FACTORS
Chromosomes occur in pairs.	Mendel's factors occur in pairs.
Chromosomes segregate during meiosis.	Mendel's factors segregate during gamete formation.
Pairs of chromosomes assort independently of other chromosome pairs.	Mendel's factors assort independently.

At that time, the function of chromosomes was unknown. Sutton studied the similarities between Mendel's factors and the movement of chromosomes in meiosis. He hypothesized that the chromosomes were the carriers of the factors, or genes, described by Mendel. Sutton was not able to prove that genes really were carried by chromosomes. Other scientists discovered the proof a few years later. However, Sutton's work led to the formulation of the chromosome theory of heredity early in this century. The **chromosome theory of heredity** states that chromosomes are the carriers of genes.

7·11 SEX-LINKED GENES

In 1906, Thomas Hunt Morgan began a series of genetic studies using the fruit fly, *Drosophila melanogaster*. The fruit fly was a good choice for genetic experiments because the females laid hundreds of eggs at a time, and only 10 days were needed to produce a new generation. Also, hundreds of the small organisms could be grown in small bottles or vials on an inexpensive food medium.

Another feature that made *Drosophila* useful for Morgan's genetic studies was their small number of chromosomes. Fruit flies have only four pairs of chromosomes. In examining the cells of the fruit flies, Morgan noticed a difference in the chromosomes of the male and female. In Figure 7·19 you can see the differences that Morgan observed. Notice that the female has four pairs of matching chromosomes, but the male has only three matching pairs. One of the chromosomes making up the fourth pair in the male is rod-shaped, but the other is hooked. The pairs of chromosomes that are different in the two sexes are the **sex chromosomes.** The sex chromosomes determine the sex in most animals. The matching sex chromosomes in the female are labeled *X*. In the male, the rod-shaped chromosome is labeled *X*, but the hooked one is labeled *Y*. Chromosomes other than the sex chromosomes are known as **autosomes** (AW tuh sohms).

Sex determination in *Drosophila* and many other organisms is a matter of chance. Figure 7·20 shows that male gametes contain either an X or a Y chromosome, but a female gamete contains only an X chromosome. At fertilization, on average, half of the offspring receive two X chromosomes, producing females. The other half receive an X and a Y, and this combination produces a male. Which parent actually determines the sex of the offspring?

Shortly after the beginning of his studies, Morgan found that Mendel's rules could not be applied to the inheritance of some traits in fruit flies. One of the most striking features of fruit flies was their red eyes. One day Morgan noticed a single white-eyed male in one of his bottles. This male was mated with a red-eyed female. All of the F_1 flies had red eyes, which indicated that the white-eyed trait was recessive. Morgan then allowed hybrid flies to mate among themselves to produce an F_2 generation. Much to his surprise, all of the white-eyed flies were males! Why were there no white-eyed females?

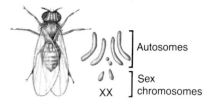

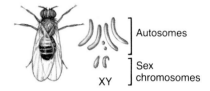

FIGURE 7·19 Female *Drosophila* have two identical sex chromosomes, while the males' sex chromosomes are different. Both sexes have four pairs of chromosomes.

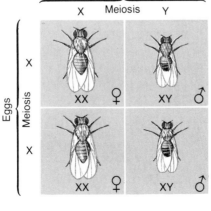

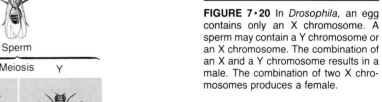

FIGURE 7·20 In *Drosophila*, an egg contains only an X chromosome. A sperm may contain a Y chromosome or an X chromosome. The combination of an X and a Y chromosome results in a male. The combination of two X chromosomes produces a female.

Morgan's explanation was that the gene for eye color was located on the X chromosome, but the Y chromosome carried no eye color gene. Study the P_1 cross in Figure 7·21.

1. Morgan hypothesized that the red-eyed female had a dominant gene for red eyes (R) on each X chromosome. The genotype of the red-eyed female is written $X^R X^R$.
2. Morgan hypothesized that the male had a recessive gene (r) for white eyes on the X chromosome, but no gene for eye color on the Y chromosome. The genotype of the male is written $X^r Y$.
3. All F_1 offspring, male and female, had red eyes. The females were heterozygous for red eyes.

The reasoning Morgan applied to his F_1 cross is shown in Figure 7·21. A cross of the F_1 hybrids should have resulted in an F_2 generation made up of about 25 percent white-eyed males. Morgan had found about this percentage in his experiment.

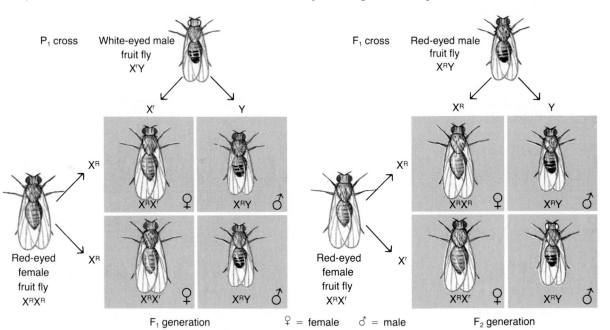

FIGURE 7·21 Morgan discovered that the inheritance of eye color in fruit flies did not follow Mendel's rules. This figure shows Morgan's crosses involving the sex-linked gene for eye color in *Drosophila*.

To test his hypothesis that the gene for eye color was located on the X chromosome, Morgan did another experiment to see if white-eyed females could be produced. He predicted that a cross of white-eyed males ($X^r Y$) with F_1 females ($X^R X^r$) should produce some white-eyed females. When he crossed a white-eyed male with one of the F_1 females he obtained:

	Female	*Male*
Red-eyed	129	132
White-eyed	88	86

The presence of white-eyed females among the offspring provided further evidence that the genes for eye color in *Drosophila* are located on the X chromosomes. Remember, Sutton had earlier hypothesized that genes are located on chromosomes. Morgan's experiment provided evidence that Sutton was correct.

The inheritance of eye color in *Drosophila* is a sex-linked characteristic. A **sex-linked characteristic** is a trait determined by a gene or genes carried on the X chromosome but not on the Y chromosome. Several traits inherited by humans are known to be sex-linked. An example is red-green colorblindness. A person with this condition cannot distinguish the colors of red and green. Red-green color vision is controlled by genes located on the X chromosome. However, there are no genes for color vision on the Y chromosome. The gene for normal color vision is dominant over the gene for colorblindness.

If C is used to represent the gene for normal color vision, and c is used to represent the gene for colorblindness, the following genotypes can be identified.

$X^C X^C$ = female, normal color vision
$X^C X^c$ = female, normal color vision but carrier of colorblindness
$X^c X^c$ = female, colorblind
$X^C Y$ = male, normal color vision
$X^c Y$ = male, colorblind

Figure 7·22 shows how red-green colorblindness may be passed by a mother to her sons, even though the father has normal color vision. Note that all daughters have normal color vision, but the probability that a son is colorblind is 1/2. Colorblindness appears more often among males than females.

7·12 LINKAGE AND CROSSING OVER

There are only four pairs of chromosomes in *Drosophila,* but there are many inherited traits. Thus, a number of genes must be located on each chromosome. The occurrence of a number of genes on the same chromosome is called **linkage.** Genes which are linked on the same chromosome are inherited together, and they make up a *linkage group.* Mendel established the principle that gene pairs assort independently. However, he did not encounter linkage in any of his experiments. The pea plant has seven pairs of chromosomes, and each of the traits studied by Mendel was located on a different chromosome. Because they are on the same chromosome, linked genes are usually passed on together to the next generation. However, under certain conditions they do not stay together. Sometimes parts of homologous chromosomes break off and switch places as seen in Figure 7·23. Recall that this process is called crossing over. Notice that crossing over results in a recombination of linked genes.

Crossing over takes place during meiosis. During interphase, each chromosome forms two chromatids which are joined together. The homologous chromosomes then come together during synapsis to form tetrads. These tetrads are units of four chromatids lying side by side. During synapsis, the chromatids may become twisted around each other and break apart, resulting in an exchange of segments.

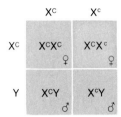

FIGURE 7·22 Although both parents may have normal color vision, a son may be colorblind if his mother is a carrier of this sex-linked trait.

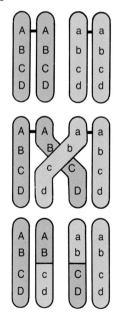

FIGURE 7·23 As you can see, when crossing over occurs in meiosis, parts of homologous chromosomes are exchanged.

While conducting some crosses with linked genes in *Drosophila*, Morgan and his associates had some surprising results. These results led to the discovery of crossing over. Most fruit flies have gray bodies and long wings. The genes controlling these traits are dominant, (*G*) and (*L*), and they are located on the same chromosome. Morgan mated flies having these dominant genes (*GGLL*) with fruit flies that all had recessive genes for black bodies and short wings (*ggll*). The F_1 generation was made up of offspring all having gray bodies and long wings. Figure 7·24 shows that this outcome was to be expected in the F_1.

The F_1 flies were then crossed. Figure 7·24 shows that if the genes for body color and wing length were linked on the same chromosome, then two types of offspring could be expected. About 75 percent of the F_2 offspring should have been gray with long wings, and the other 25 percent should have been black with short wings. However, a few of the F_2 flies were gray with short wings, and some were black

Profiles

Barbara McClintock (1902–   )

In October 1983, Barbara McClintock became the third woman to receive an unshared Nobel Prize. At the age of 81, Dr. McClintock received the Nobel Prize for Physiology and Medicine for her more than 40 years of research on mobile genetic elements, or "jumping genes," in Indian corn (maize).

Barbara McClintock was born in Hartford, Connecticut, in 1902. Although her mother thought college was no place for a woman, Barbara McClintock entered Cornell University at the age of 17. She chose to study botany and received a Ph.D. in plant genetics from Cornell in 1927. It was during her course of study at Cornell that Barbara McClintock began her research into and lifelong involvement with Indian corn.

Dr. McClintock's reputation as a plant geneticist grew during the 1920s and 1930s as a result of her research on the genetic structure of maize. After leaving Cornell, Dr. McClintock continued her research at the California Institute of Technology and at the University of Missouri. In 1942 she accepted a position with the Carnegie Institute's Cold Spring Harbor Laboratory on Long Island, where she still works.

During the 1940s and 1950s, Dr. McClintock's research on Indian corn led her to discover that genes on a chromosome can move about, causing variations and possibly mutations in an offspring. These findings were met with doubt and disbelief by fellow geneticists, who thought that genes held specific, immovable positions on a chromosome. One must remember that Dr. McClintock's early research was done a mere 20 years after the rediscovery of Mendel's papers on heredity. She began her research before DNA had been identified as the chemical basis of the gene.

With knowledge of the structure and function of DNA, and the development of the electron microscope, Dr. McClintock's research on "jumping genes" slowly came to be accepted by her colleagues. In 1981 she began receiving the major awards and recognition she so richly deserved.

Today, Barbara McClintock continues her research in her laboratory at Cold Spring Harbor, seemingly unaffected by her recent recognition and fame.

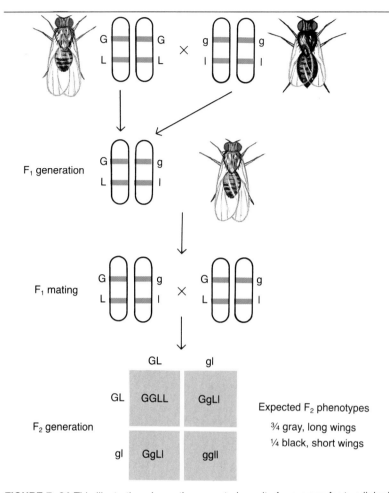

FIGURE 7·24 This illustration shows the expected results for a cross for two linked traits that Morgan was studying. However, his actual results were different because crossing over occurred.

with long wings. The only way in which these unexpected phenotypes can be explained is to assume that crossing over had occurred. As a result, in a few gametes the gene for gray body (*G*) was linked to the gene for short wings (*l*). In a few gametes the gene for black body (*g*) was linked to the gene for long wings (*L*). We now know that the process of crossing over can result in new gene combinations. These new gene combinations produce a greater variety of offspring.

REVIEW QUESTIONS

11. What is the chromosome theory of heredity?
12. Describe how sex is determined in *Drosophila*.
13. What does the term *sex-linked characteristic* mean?
14. What is the significance of the process of crossing over?
15. Under what circumstances can the son of a colorblind father also be colorblind?

Chapter Review

SUMMARY

7·1 Genetics is the branch of biology dealing with the study of heredity; heredity is the passing of traits from parents to their offspring.

7·2 Mendel's classic genetic experiments involved the interpretation of the crossing of garden pea plants having seven pairs of contrasting traits.

7·3 The principle of dominance states that in hybrids the dominant form of a particular gene determines the expression of a trait and prevents the expression of the recessive form of the gene. The principle of segregation states that during gamete formation, the genes that control a particular trait are separated into different gametes.

7·4 Rules of probability can be used to predict the outcome of genetic crosses.

7·5 A Punnett square, which is a chart showing the possible combinations of genes among the offspring of a cross, can be used to predict the outcome of genetic crosses.

7·6 As a result of his studies of inheritance in dihybrids, Mendel developed the principle of independent assortment which states that, during gamete formation, the genes for one trait are separated and distributed to the gametes independently of the genes for other traits.

7·7 Traits that have three or more alleles are said to be controlled by multiple alleles.

7·8 Polygenic inheritance is a type of inheritance in which a single trait is determined by the interaction of two or more pairs of genes.

7·9 Incomplete dominance is a condition in which a hybrid has a phenotype intermediate between the contrasting phenotypes of its parents.

7·10 Chromosomes have been identified as the carriers of genes.

7·11 Some inherited traits in fruit flies, humans, and other organisms are controlled by genes on the X chromosome.

7·12 In crossing over, chromosomes may exchange parts, resulting in recombination of linked genes.

BIOLOGICAL TERMS

alleles
autosomes
chromosome theory of heredity
cross-pollination
dihybrid cross
dominant factors
first filial generation
genetics
genotype
heredity
heterozygous
homozygous
hybrid
incomplete dominance
linkage
monohybrid cross
multiple alleles
parental generation
phenotype
polygenic inheritance
principle of dominance
principle of independent assortment
principle of segregation
probability
product rule
Punnett square
pure line
recessive
rule of independent events
second filial generation
self-pollination
sex chromosomes
sex-linked characteristic
test cross
trait

USING BIOLOGICAL TERMS

1. Distinguish between the terms *phenotype* and *genotype*.
2. Define the term *dihybrid cross*.
3. Using a dictionary, find out the meaning of the word parts *hetero-* and *homo-*. Explain the significance of these word parts in the terms *heterozygous* and *homozygous*.
4. Distinguish between the first filial generation and the second filial generation.
5. What are alleles?
6. Distinguish between the terms *genetics* and *heredity*.
7. From the list of key terms, supply the term the geneticist would use in place of the following: *characteristic, mating to determine parentage, the way an organism looks,* and *the genetic makeup of an organism*.
8. Define the chromosome theory of heredity.

9. Distinguish between the terms *dominant* and *recessive*.
10. Name the rule or law that states that previous events do not affect the probability of later occurrences of that same event.

UNDERSTANDING CONCEPTS

1. Identify several traits that are inherited. (7·1)
2. Explain why garden pea plants were well suited for Mendel's genetic experiments. (7·2)
3. Explain how the principle of dominance applies to the results of a cross of homozygous yellow-seeded pea plants and homozygous green-seeded plants. You may use a Punnett square in your answer. (7·3)
4. How can you determine the probability of getting double tails when you toss two pennies? (7·4)
5. Set up a complete Punnett square showing the cross of a pea plant heterozygous for gray-seed coat and a pea plant homozygous for white-seed coat. Explain the meaning of the combination of letters within the boxes and outside the boxes of the Punnett square. (7·5)
6. In pea plants, long stems are dominant over short stems. Green pods are dominant over yellow pods. A heterozygous long-stemmed, green-podded plant is crossed with a homozygous short-stemmed, yellow-podded plant. Use your own system of letters and a Punnett square to show the possible genotypes and phenotypes of such a cross. (7·6)
7. Define the term *multiple alleles* and give an example of the occurrence of multiple alleles in humans. (7·7)
8. Kernel color in wheat is controlled by several pairs of genes. Identify the type of inheritance involved. (7·8)
9. Give an example of incomplete dominance and explain how incomplete dominance differs from dominance. (7·9)
10. What evidence led Sutton to hypothesize that genes are located on chromosomes? (7·10)
11. Suppose a woman who has normal vision but is a carrier of colorblindness marries a colorblind man. Develop a Punnett square to show the possible genotypes and phenotypes of their offspring. (7·11)
12. Explain through words and diagrams how the process of crossing over occurs. How does crossing over affect linkage? (7·12)

APPLYING CONCEPTS

1. A woman with normal vision has two daughters that are colorblind. Give the possible genotypes of this woman, her husband, and the two daughters. Use C for normal vision and c for colorblindness.
2. In horses, black (B) is dominant over chestnut (b). The trotting gait (T) is dominant over the pacing gait (t). If a horse heterozygous for both traits is mated to a chestnut pacer, what is the probability of their producing a colt that is a chestnut pacer?
3. A certain couple has three daughters. The mother is expecting a fourth child. What are the chances that the child will be a boy? Give an explanation for your answer.
4. Is it possible for a woman with type AB blood and a man with type B blood to have children with type A blood? Use a Punnett square to explain your answer.

EXTENDING CONCEPTS

1. Interview a person working in the field of genetics. Be sure to find out the kinds of things the person does in a typical day, the educational background required for the work, and the reasons he or she entered the field. You may want to ask the person if you may tape-record the interview and photograph him or her at work.
2. From a biological supply house, obtain F_2 corn kernels or tobacco seeds carrying recessive genes for the albino trait. Plant the kernels and then examine any differences in the seedlings when they germinate. Interpret your findings in terms of what you now know about inheritance.

SUGGESTED READINGS

Keller, E.F. *A Feeling for the Organism: The Life and Work of Barbara McClintock.* W.H. Freeman and Co., 1983.

Zegura, Stephen L. "Blood Test." *Natural History,* July 1987, p. 8.

8. Applied Genetics

When you think of a skunk, you think of a small animal that is black and has a broad white stripe on its back. This color pattern is inherited. These animals are albino skunks. How are they different from other skunks? Why did they not inherit normal coloration? In this chapter you will learn how the normal expression of a gene is changed.

CHAPTER OBJECTIVES

After completing this chapter, you will be able to

- **Explain** how gene mutations and chromosomal alterations determine how traits are expressed.
- **Explain** how genetic disorders are transmitted, detected, and treated.
- **Give examples** of some human genetic disorders.
- **Describe** how people have used genetic principles to modify inheritance.
- **Discuss** how different factors affect gene expression.

Changes in Hereditary Material

8·1 NATURE OF MUTATIONS

Mitosis increases the number of somatic (soh MAT ihk), or body, cells. Meiosis leads to the formation of gametes, or sex cells. In both cases the new cells should receive the correct number of chromosomes and genes. However, changes may occur in chromosomes and genes. Such changes in hereditary material are called **mutations** (myoo TAY shuhnz). Mutations are passed to new cells formed during cell division. Some mutations produce no visible effects. Others produce drastic effects in an organism and sometimes affect an organism's offspring as well.

Mutations may involve the structure or the number of chromosomes or the chemical nature of the genes. A change in the structure or number of chromosomes is called a **chromosomal alteration.** It usually causes changes in the phenotype. A change in the chemical nature of the DNA is called a **gene mutation.** A gene mutation may or may not be visible.

Mutations occur randomly. The causes of most naturally occurring spontaneous mutations remain unknown. However, substances and forms of energy that increase the frequency of mutation have been identified. Something that can cause mutations is called a **mutagen** (MYOO tuh juhn). Various forms of radiation, such as X rays and ultraviolet rays, are mutagens. Certain chemicals, such as asbestos, benzene, and formaldehyde, have been identified as mutagens.

Somatic mutations occur in the body cells of an organism. Such mutations are passed on only to cells that come from the original mutant cell. They are never passed on to offspring. **Germ mutations** occur in the reproductive cells of an organism. Such mutations can be passed on to offspring. Many mutations that produce noticeable effects are harmful and interfere with an organism's ability to function. The effects of some mutations are severe enough to cause death.

SECTION OBJECTIVES

- Describe a gene mutation.
- Explain what a chromosome mutation is and identify several kinds of chromosomal alterations.

Sometimes mutations are helpful to an organism. In such cases a mutation makes the organism better able to survive in a certain environment. Can you think of an example of a mutation that would be helpful to a human?

8·2 ALTERATIONS IN CHROMOSOMES

As you have read, mutations affecting the chromosomes are called *chromosomal alterations.* There are two kinds of chromosomal alterations: changes in the normal number of chromosomes and changes in the structure of a chromosome itself.

Sometimes during meiosis, paired chromosomes do not separate. The failure of one or more pairs of chromosomes to separate during meiosis is called **nondisjunction** (nahn dihs JUHNGK shuhn). Nondisjunction may take place in autosomes or in sex chromosomes. If nondisjunction takes place, the gametes formed will have either too many or too few chromosomes. If such gametes take part in fertilization, the offspring will not have the correct number of chromosomes in their cells. Therefore nondisjunction can result in abnormalities in the offspring.

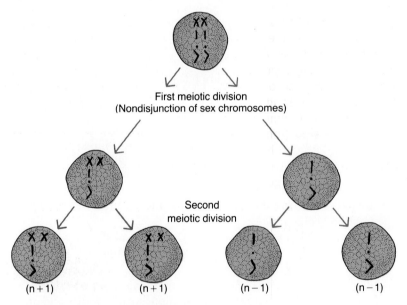

FIGURE 8·1 When nondisjunction occurs during meiosis, the resulting sex cells will have the wrong numbers of chromosomes.

Figure 8·1 shows nondisjunction occurring during meiosis. How many chromosomes are there in the primary sex cell? During which meiotic division does nondisjunction take place? Find the gametes that have an extra chromosome ($n + 1$) and those that lack a chromosome ($n - 1$). If any of these gametes are involved in fertilization, the resulting zygote would have either one more or one less than the expected number of chromosomes. What would these chromosome numbers be? The effects of nondisjunction in humans will be discussed later in the chapter.

Sometimes when cells fail to divide normally the resulting cells contain more than one set of chromosomes. The condition in which cells have extra sets of chromosomes is called **polyploidy** (PAHL ee-ploi dee). Polyploidy can occur during meiosis when whole sets of homologous chromosomes do not separate. Gametes are normally monoploid (n); however, when whole sets of chromosomes do not separate, the resulting gametes are diploid ($2n$). Fertilization of such gametes produces zygotes with extra sets of chromosomes. The zygotes may be $3n$, $4n$, or other multiples of n. Sometimes polyploidy can occur during mitosis. In this situation chromosomes separate, but the cell fails to divide. The result is a cell with double the number of chromosomes ($4n$). Figure 8·2 shows a polyploid plant.

Sometimes during meiosis changes occur in the structure of single chromosomes. A section of a chromosome is sometimes moved to a nonhomologous chromosome, as shown in Figure 8·3A. Although all the genes are present in the resulting gamete, some are misplaced. Such genes may not be able to function in the new position. A chromosomal alteration involving the transfer of a chromosome segment from one chromosome to another nonhomologous chromosome is called a **translocation** (trans loh KAY shuhn).

Duplications and deletions are other types of chromosomal alterations (Figure 8·3B). A **duplication** is an alteration in which a segment of a chromosome is present more than once on a chromosome. Duplications occur as a result of the exchange of unequal chromosome sections during crossing over. An alteration caused by a loss of a segment of a chromosome is called a **deletion** (dih LEE shuhn). Deletions may occur during crossing over as well as when chromosomes break and small fragments remain unattached.

An **inversion** (ihn VER zhuhn), shown in Figure 8·3C, occurs when a chromosome breaks and reattaches so that the sequence of genes is reversed. Inversions interfere with normal interactions that must occur between genes.

FIGURE 8·2 A polyploid geranium (bottom) has more sets of chromosomes than a normal geranium (top). The flowers of the polyploid plant are larger and have more petals.

FIGURE 8·3 Translocation (A), duplication and deletion (B), and inversion (C) result in changes to the structure of a chromosome.

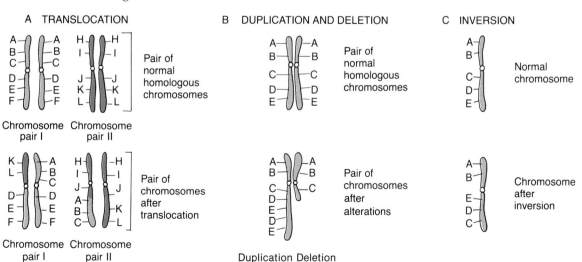

8·3 MUTATIONS IN GENES

Chromosomal alterations are not the only kind of changes that occur in genetic material. Mutations that involve changes in the chemical nature of genes also occur. Recall that a gene is a segment of a DNA molecule. Genetic information is determined by the order of the nitrogenous bases in a gene. This code controls protein synthesis. A change in the base sequence may change the structure of the protein that is synthesized. Gene mutations may result from the addition or deletion of nitrogenous bases and from the substitution of one base for another.

Gene mutations, also called *point mutations*, can take place during the replication of DNA. Figure 8·4 shows how one base may be substituted for another. Find the mutation in the newly formed series of nucleotides. The mutation will be passed on to succeeding cells each time cell division takes place. Also, this error in the genetic code may lead to a change in the protein being synthesized.

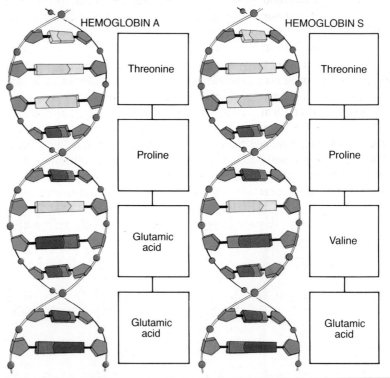

FIGURE 8·4 A point mutation can cause a change in the protein that is being synthesized.

REVIEW QUESTIONS

1. Distinguish between a chromosomal alteration and a gene mutation.
2. Identify some possible mutagens.
3. How does nondisjunction lead to changes in genetic information?
4. Describe the types of chromosomal alterations that involve changes in the structure of chromosomes.
5. Why are somatic mutations not passed on to future generations?

Genetic Disorders in Humans

8·4 DISORDERS RESULTING FROM NONDISJUNCTION

Some genetic disorders result from nondisjunction. Nondisjunction in humans can result in gametes with one more ($n + 1$) or one less ($n - 1$) chromosome than normal. Instead of 23 chromosomes, these abnormal gametes may have 24 or 22 chromosomes. When fertilization occurs between a normal gamete and one having an extra chromosome, the resulting zygote has 47 chromosomes ($2n + 1$). The condition in which there are three homologs of a particular chromosome instead of a normal pair is called **trisomy** (TRĪ soh mee). When fertilization occurs between a normal gamete and a gamete that lacks a chromosome, the resulting zygote has 45 chromosomes ($2n - 1$). The condition in which there is only one chromosome of a homologous pair is called **monosomy** (MAHN uh soh mee). In both trisomy and monosomy, the abnormal chromosome number will be maintained with each mitotic cell division.

Notice that chromosomes in Figure 8·5 are arranged in pairs. Such a picture of chromosomes arranged as homologs is called a **karyotype** (KAR ee oh tīp). Also notice that there is an extra chromosome number 21 in part B. Compare the karyotype in part B with that in part A. The abnormal condition resulting from the extra chromosome is called *trisomy 21,* or **Down's syndrome**. Persons with this genetic disorder have several traits in common: short in height; round, full face; enlarged and creased tongue; and noticeable eye folds. These persons are also mentally retarded. The risk of having a child with Down's syndrome increases with the age of the mother. Down's syn-

SECTION OBJECTIVES

- Describe some human genetic disorders.
- Indicate methods of detecting and treating human genetic disorders.

tri- (three)

mono- (one)

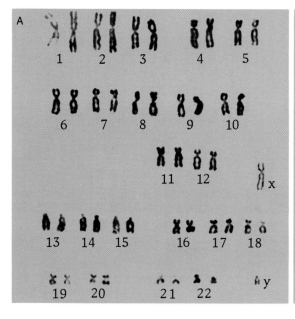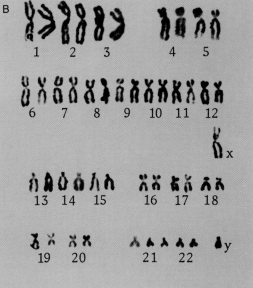

FIGURE 8·5 Karyotypes for a normal male (A) and a male with Down's syndrome (B). Note the extra chromosome 21 in B.

drome occurs in only about 1 birth in 1000 for women in their twenties. For women in their forties the chance of giving birth to a child with Down's syndrome is about 1 in 40. These risks are not affected by the number of times the mother has been pregnant.

Some human genetic disorders are the result of nondisjunction of the sex chromosomes. One of these disorders is Klinefelter's (KLĪN-fehl tuhrz) syndrome. **Klinefelter's syndrome** is a genetic disorder in which the sex chromosome makeup is XXY. A person with this disorder is a male who is sterile. Klinefelter's syndrome occurs once in every 1000 male births. What type of nondisjunction event would lead to an XXY sex chromosome makeup?

In **Turner's syndrome**, another disorder resulting from nondisjunction of the sex chromosomes, the sex chromosome makeup is XO. The O indicates the absence of a sex chromosome. The affected persons are females who are short and have characteristic neck folds. Turner's syndrome occurs once in every 10 000 female births. Why does Turner's syndrome occur less often than Klinefelter's syndrome? Turner's syndrome occurs because of the absence of a chromosome. Sometimes this lack of a chromosome interferes with embryonic development and development is never completed.

In many cases chromosomal alterations are harmful to the health of the organism. The presence or absence of a chromosome may alter the normal functioning of the organism. If nondisjunction occurs in any of the larger chromosomes, embryonic development is seldom completed. Such types of nondisjunction are said to be lethal.

FIGURE 8·6 Persons with Down's syndrome often participate in organized athletic events.

8·5 DISORDERS RESULTING FROM GENE MUTATIONS

Human gene mutations are chance events. The effects of these mutations are sometimes visible, as in the case of albinism. *Albinism* is due to a gene mutation that results in lack of a pigment called *melanin*. An *albino* lacks pigment in the hair and skin and has either pink or pale blue eyes. Most often, gene mutations produce no visible effects but still remain a part of the genotype of the organism. Gene mutations present in gametes are passed on from generation to generation.

Some gene mutations are dominant; only one of these mutant genes is needed for the mutant trait to be expressed. Most gene mutations are recessive and are expressed only when two mutant genes are present in the genotype of the organism.

Sickle cell anemia, a genetic disorder that affects the ability of red blood cells to carry oxygen, is most common among persons of West African ancestry. The disorder affects about 1 in 350 black Americans. Affected persons have two mutant genes for the production of hemoglobin. Hemoglobin is the oxygen-carrying chemical in red blood cells. The abnormal hemoglobin, called hemoglobin *S*, differs from normal hemoglobin, called hemoglobin *A*, by one amino acid. The mutant gene differs from the normal gene in one base.

The red blood cells of a person with sickle cell anemia have a sickle shape, as shown in Figure 8·7. Compare these cells with normal red blood cells. The sickle-shaped red blood cells are poor oxygen carriers and are very fragile. The sickle-shaped cells tend to block small blood vessels.

Persons that are heterozygous for sickle cell anemia have one gene for normal hemoglobin and one gene for hemoglobin *S*. These persons are called *carriers*. A carrier has only mild symptoms of the disorder and some sickling of red blood cells. However, carriers are more resistant to malaria than are noncarriers.

Figure 8·7 shows the possible offspring of two individuals who carry the gene for sickle cell anemia. What proportion of their offspring would be expected to be homozygous for sickle cell anemia? What proportion would be expected to be carriers of the disorder? What proportion of offspring would be expected to show the symptoms of sickle cell anemia?

Research to find a cure for sickle cell anemia is going on. It is hoped that a technique will be developed to replace the mutant gene with a normal gene. This would wipe out the disease.

Phenylketonuria (fehn uhl kee tuh NYUR ee uh), or PKU, involves the inability of a mutant gene to code for an enzyme needed for changing the amino acid phenylalanine to tyrosine. The abnormal products of phenylalanine metabolism build up in the bloodstream. These products are harmful to brain cells and eventually cause mental retardation. PKU may be found in infants by a simple blood test. Mental retardation can be prevented if the affected baby is placed on a special diet that limits phenylalanine intake. PKU is a homozygous recessive condition.

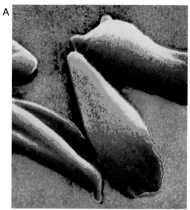

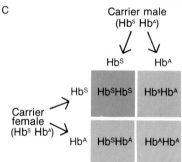

Hb^A = dominant gene for normal

Hb^S = recessive gene for sickle cell hemoglobin

▪ Disease
▪ Carrier
▪ Normal

FIGURE 8·7 The red blood cells of a person with sickle cell anemia (A) are crescent shaped, while normal red blood cells (B) are round. One-fourth of the children of two carriers will probably have sickle cell anemia. (C)

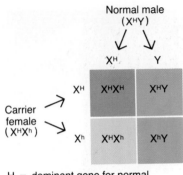

FIGURE 8·8 Hemophilia is a sex-linked disorder. It is possible that one-half of the sons of a normal father and a carrier mother will have hemophilia.

Tay-Sachs (tay saks) **disease** is a genetic disorder that involves the inability to synthesize an enzyme that prevents lipid buildup in brain cells. As the lipids collect, the nervous system breaks down. The disease is caused by a recessive gene. A child has the condition only when he or she receives the recessive gene from both parents. Children with Tay-Sachs disease die at an early age. Carriers of Tay-Sachs disease show no symptoms of the disorder. Blood tests can now determine if a person is a carrier of Tay-Sachs disease.

Sometimes mutant genes are located on the sex chromosomes. Disorders caused by such genes are known as sex-linked disorders. **Hemophilia** (hee muh FIHL ee uh) is a sex-linked disorder caused by a recessive gene. In hemophilia, one of the proteins needed for the blood to clot is missing or is present only in very small amounts. Thus, the blood tends to clot slowly or not at all. A person with hemophilia may bleed severely from small cuts and bruises. Hemophilia was once common in the royal families of Europe. Queen Victoria of England had a son, three grandsons, and six great-grandsons who were hemophiliacs.

The mutant gene for hemophilia is carried only on the X chromosome. It does not appear to have a corresponding allele on the Y chromosome. Hemophilia occurs more often in males than in females. Can you explain why? Carriers, who are always female, have one normal gene and one gene for hemophilia. Figure 8·8 shows the possible offspring of a normal male and a carrier female. What proportion of sons would be expected to be hemophiliacs? What proportion of daughters would be expected to be carriers? Hemophilia is treated by giving the victim a chemical needed for the blood to clot properly.

8·6 FINDING AND TREATING GENETIC DISORDERS

Some genetic disorders can be detected by a study of a person's karyotype. A person who is a carrier of sickle cell disease or Tay-Sachs disease can be found through a blood test. A genetic counselor is a person who can let couples know their probability of having children with these disorders.

Some genetic disorders can be found before birth or just after birth. A fetus can be studied for chromosomal and gene disorders by means of a procedure called **amniocentesis** (am nee oh sehn TEE-sihs). Amniocentesis involves the insertion of a needle through the abdominal wall and uterus of the mother and into the amniotic fluid that surrounds the fetus. Some amniotic fluid, which contains cells of the fetus, is removed. The cells are studied for chromosomal alterations and for chemicals that indicate genetic disorders.

ultra (beyond)
sonus (sound)

Additional techniques that allow physicians to examine fetuses have been developed. One such technique is called ultrasonography (uhl truh sahn AHG ruh fee). **Ultrasonography** involves the use of high-frequency sound waves to produce an image of a fetus on a monitor. This procedure allows a doctor to determine the position of the fetus. Ultrasonography is also used to detect fetal abnormalities.

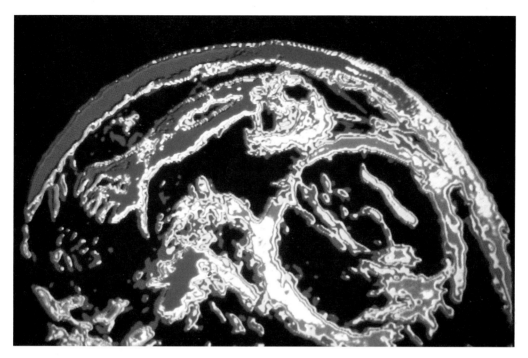

FIGURE 8·9 Using sound waves, doctors can produce an image of a fetus within its mother's body. You can even see the tiny fingers on the fetus's hand in this photograph.

Fetoscopy (fee TAHS kuh pee) is another technique that lets the doctor view the fetus through an instrument called an endoscope while the fetus is still inside the uterus. It is used to detect certain fetal problems. Through the scope doctors can drain excess fluids from the brain, perform surgery, and even do blood transfusions while the fetus remains in the mother's uterus.

Researchers continue to develop new techniques for use in the early detection and treatment of genetic disorders. Although at present there are no cures for genetic disorders, recent advances in genetic engineering suggest that there may be possible breakthroughs in the future.

feto- (fetus)
-scopy (observation)

REVIEW QUESTIONS

6. Name three human chromosomal disorders and give their causes.
7. Name one human disorder caused by a gene mutation and describe its symptoms.
8. What treatment is available for phenylketonuria?
9. What is amniocentesis?
10. A couple in their thirties do not yet have any children. They would like to start a family. What kinds of information do you think a genetic counselor would discuss with them?

Investigation

What Is the Effect of Albinism on Corn Plants?

Goals
After completing this activity, you will be able to
- Describe the effect of albinism on corn plants.
- State the Mendelian ratio of normal plants to albino plants in the offspring.

Materials (for groups of 2)
30 F_2 corn seeds with genes for albinism
potting medium (vermiculite + soil)
planting flat   marking pencil
twist-tie   plastic bag

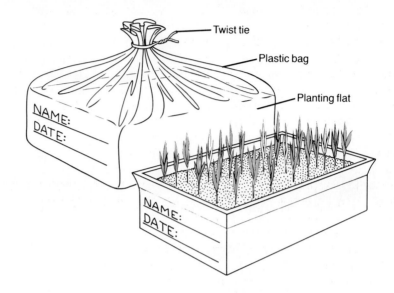

Procedure
A. Partly fill a planting flat with potting medium.

B. Scatter the 30 F_2 corn seeds on the surface of the potting medium. Cover the seeds with a layer of potting medium. Water the seeds thoroughly. Label the flat with your name and the date.

C. Place the flat into a plastic bag. Close the plastic bag with a twist-tie. Place the flat in a warm place.

D. Check the flat after a week for any evidence of sprouting. When you do see some seeds sprouting, open the plastic bag and set the flat in a sunny place. Add water when needed to prevent the soil from drying out.

E. Make a sketch of the position of the seeds in the flat. Note the color of the seedlings. Record the number of seedlings of each color. Find out from other groups in your class the number of seedlings of each color that are sprouting in their flats. Record this information.

F. Observe the growth of the seedlings for several weeks. Note which seedlings thrive and which ones die. Record this information from each group in your class.

G. You may wish to transplant healthy seedlings to a larger area outdoors so that the corn plants can grow and reach maturity.

H. Follow your teacher's instructions for the proper disposal of all materials.

Questions and Conclusions
1. What was the total number of seeds that germinated in your planting flat?

2. Of the seeds that germinated, were there differences in the color of the leaves? Describe what you observed.

3. In a previous generation, some of the corn plants had undergone a *germ mutation*. Seeds from these plants carried the gene for albinism, or absence of chlorophyll. What was the ratio of normal seedlings (with green leaves) to mutant seedlings (with white leaves)?

4. You planted seeds from the F_2 generation. Assume these seedlings were heterozygous and carried the recessive trait for albinism. Devise a genotype for the seeds, and then set up a Punnett square showing the predicted offspring from a monohybrid cross.

5. What is the predicted ratio of corn plants that are phenotypically green-leaved to those that are phenotypically white-leaved? How does this ratio compare to the ratio you observed in this activity?

6. Explain why the gene for albinism is considered a *lethal* gene. Why did the white seedlings grow for a certain period of time and then die?

156   APPLIED GENETICS

Applications of Genetics

8·7 SELECTIVE BREEDING

For centuries farmers collected and planted seeds from the healthiest and highest yielding varieties of plants to improve plant varieties. Early farmers also recognized certain single animals having desirable characteristics. Selecting such plants and animals and crossing them with others having similar desirable characteristics was the beginning of selective breeding. **Selective breeding** is the crossing of animals or plants that have desirable characteristics to produce offspring that have a combination of the parents' desirable characteristics. For example, breeders used wild cabbage to develop green and red head cabbage, broccoli, and cauliflower.

Through breeding, pure strains of various kinds of plants and animals were developed. Such pure strains are still being maintained today by inbreeding. **Inbreeding** is a type of breeding that involves crossing closely related individuals to maintain desirable characteristics. Such breeding keeps homozygous conditions in the genetic make-up of the strain. The various breeds of dogs, such as the boxer and collie, are pure strains maintained by inbreeding. Many strains of vegetables have been carried on by inbreeding.

Another approach to selective breeding is to develop new varieties by crossing individuals of distantly related strains. In this type of breeding, strains having different but desirable characteristics are crossed to produce a variety that has all the desirable characteristics. The crossing of distantly related strains is called **outbreeding** and results in *hybrid* offspring. Hybrids may show hybrid vigor. **Hybrid vigor** is the superiority reached in offspring having the desirable characteristics of both parents. Such offspring also benefit from the absence of undesirable characteristics found in the parent stock.

An example of a hybrid produced by outbreeding involved two breeds of dogs, the bulldog and the mastiff. European landowners needed a dog to patrol and protect their land and game animals. The bulldog had the needed courage and aggressiveness but was too small to guard a large area and stop trespassers. On the other hand, the large mastiff was fast but gentle. A cross between these two breeds, however, produced a new breed, the bull-mastiff. The bull-mastiff, shown in Figure 8·10, joined the desirable characteristics of a bulldog and a mastiff and proved to be a good guard dog.

Outbreeding was used in the development of the Liberty apple. The Liberty apple resulted from a cross between a Macoun and a crab apple known as PRI 54-12. The Macoun is known for its taste, crispness, and long storage life. PRI 54-12 had been developed because it resisted disease. The Liberty apple has desirable characteristics of both parents. It resists four major fungus diseases and has a desirable taste, crispness, and long storage life.

SECTION OBJECTIVES

- Define selective breeding and discuss its usefulness.
- Describe the methods used by genetic engineers to alter genes.

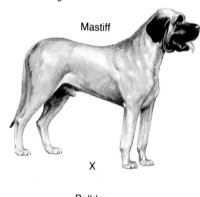

FIGURE 8·10 The bull-mastiff breed was produced from two other breeds—the bulldog and the mastiff.

8·8 GENETIC ENGINEERING

A greater understanding of the gene and its action has led scientists into the development of techniques that allow them to alter the structure of DNA. Altering the structure of a DNA molecule by substituting genes from other DNA molecules is called **genetic engineering.** The DNA molecule that forms from the combining of two different DNA molecules is referred to as **recombinant** (rih KAHM buh nuhnt) **DNA.**

Scientists use the bacterium *Escherichia coli* (*E. coli*) (ehsh uh-RIHK ee uh   KOH lī) to study recombinant DNA. *E. coli* contains one large chromosome, which is found in the cytoplasm. This bacterium may also contain **plasmids,** which are small ring-shaped pieces of DNA. The chromosome and plasmids replicate each time the cell divides.

Genetic engineers can introduce genes into a plasmid so that the chemical reactions inside the cell are changed. As shown in Figure 8·12, the plasmid is removed from the bacterium. Then the DNA of this plasmid is broken at a certain point by a special type of enzyme. At the same time the plasmid is broken, a section of another DNA molecule is broken in the same way. This section of DNA is then introduced into the plasmid. The broken plasmid is joined with the new DNA segment and returns to its original ringed structure. This technique is referred to as **DNA splicing.** The new plasmid is a combination of the original DNA molecule and a newly introduced piece of DNA. This plasmid is inserted back into the bacterium. When the bacterium replicates, the new DNA also replicates.

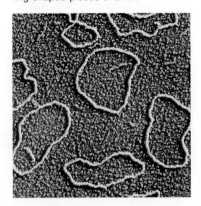

FIGURE 8·11 This transmission electronmicrograph has been specially colored to make the plasmids within the cell easier to see. Plasmids are small ring-shaped pieces of DNA.

FIGURE 8·12 DNA splicing produces recombinant DNA. New DNA is spliced into a plasmid. This plasmid is then inserted into a bacterial cell. When the bacterial cell divides, the plasmid and the new DNA are replicated.

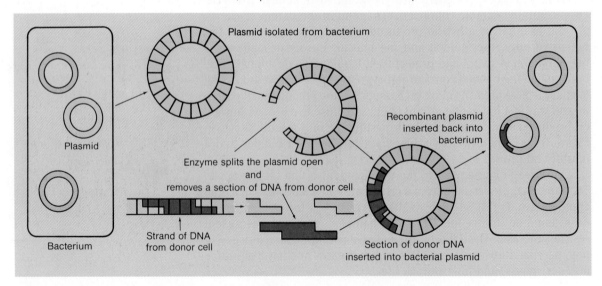

Computers in Biology

Computer Use in Genetics

A T A C C G A flashes across the computer screen, and the "word" continues to grow. What language is this? It is part of the biochemical language that spells out the genetic blueprint of living things. *A, C, G*, and *T* stand for the bases adenine, cytosine, guanine, and thymine, which are found in DNA. By learning the order in which the bases occur in genes—a process geneticists call *sequencing*—a clearer picture of the molecular basis of living things can be obtained.

Today, computers are being used to manage the vast amount of information resulting from the sequencing of DNA in many organisms. While researchers are still far from knowing the 3.5 billion-base sequence that makes up one set of human chromosomes, they are using computers for help in identifying and cataloging gene sequences in simpler organisms, such as bacteria and viruses. Computers also help scientists examine the "misspellings" in gene codes, which may be responsible for certain genetic diseases.

Large bodies of information that can be reached by computer are called *databases*. Databases such as GenBank, located in Massachusetts, and the Nucleic Acid Sequence Database, operated by the National Biomedical Research Foundation in Washington, D.C., have been created to store genetic information. This information is made available to scientists across the country. Newly identified DNA and RNA sequences that appear in scientific journals are typed into a computer at data headquarters. Scientists across the country, hooked into the database by phone, can use their own computers to gain access to this genetic library. Long searches through scientific papers are unnecessary when the computer is used to identify codes or compare codes with a database.

Additional computer programs can be used with sequence databases. For example, one program searches gene sequences for similarities. Some programs "know" details of molecular biology. When the programs are used with the database, sequencing experiments can be conducted. Also, follow-up studies can be suggested.

When genetic databases were first begun in the late 1970s, they contained less than 100 gene sequences, and the sequences were only a few thousand bases in length. Now the database in Washington, D.C., has thousands of sequences, some of which are more than a million bases in length. With resources like this available in their own labs, scientists are identifying sequences at record speeds. Some are studying makeup of viruses, bacteria, and cell parts such as mitochondria. Others are using the information to analyze differences between normal and diseased cells. Such databases are not only of value in genetics; they are also a vital resource in fields of biochemistry, pharmacology, agriculture, and genetic engineering.

Computers are being used in genetics in other ways. These uses include measuring chromosome lengths directly from photographs, doing genetic crosses without living materials, and studying long-term genetic effects.

160 APPLIED GENETICS

FIGURE 8·13 An African violet leaf (top) is able to produce new plants when it is rooted in water. The new plants are identical to the parent plant. Using special cloning techniques (bottom) many more identical plants can be produced.

DNA splicing has led to a new method for making the hormone *insulin.* Insulin regulates the body's use of sugar. Some diabetics lack or do not make enough insulin. Most diabetics take insulin taken from cows and sheep. Some diabetics have allergic reactions to animal insulin. Recently, the gene that produces human insulin was spliced into *E. coli.* A group of bacteria that can make human insulin was obtained. This insulin does not seem to cause the bad effects some diabetics have to animal insulin. *Interferon,* an antivirus drug, has been made in the same way. *Human growth hormone* has also been made by using bacteria into which the proper gene has been added.

Genetic engineers are hopeful that they will be able to correct inherited defective genes by introducing genes to replace the defective ones. Someday defective genes may be corrected in the sex cells, thus preventing their transfer from generation to generation.

Genetic engineering may also improve agriculture. Plants need nitrogen compounds to make proteins. Most plants are unable to use free nitrogen found in the atmosphere. Thus, most nitrogen for plants must come from the soil. Farmers add nitrogen compounds to the soil to obtain the greatest yield from their crops. Commercial nitrogen-compound fertilizers are costly, however. Some bacteria, known as the nitrogen-fixing bacteria, can make enzymes that change free atmospheric nitrogen to usable nitrogen compounds. Genetic engineers hope to improve crop plants, such as corn, by the method of DNA splicing. The crop plant would then have the nitrogen-fixing gene. This plant would be able to make its own nitrogen compounds from free atmospheric nitrogen. At present, genetic engineers have been able to splice the nitrogen-fixing gene from the bacterium *Klebsiella* into a plasmid of *E. coli.* The *E. coli* produced the enzyme necessary to fix nitrogen. In the future, genetic engineers hope to be able to splice the nitrogen-fixing gene directly into the chromosomes of commercial crops, such as corn and wheat.

Scientists must be sure that no genes are added to bacteria that might give them undesirable traits. Such traits might cause new diseases that could spread without control in the environment.

Some methods of genetic engineering are based on natural processes that occur during the reproduction of organisms. For example, you have probably seen someone remove a cutting from a plant and place it in water to root. Cuttings are a form of asexual reproduction called vegetative propagation. Plants formed by vegetative propagation are identical to the parent. This method of reproduction keeps the desirable characteristics of the parent in the offspring. Some varieties of food crops are produced through vegetative propagation and other methods of asexual reproduction.

Cloning is the process of forming a group of genetically identical offspring from a single cell. The cloning of plants, shown in Figure 8·13, is a modern method of vegetative propagation. Cloning has potential economic importance in producing large enough supplies of important crops.

APPLIED GENETICS 161

Cloning has been less successful in animals than in plants. Cloning in animals involves removing the nucleus of an unfertilized egg and replacing it with a diploid nucleus from a body cell. The egg then has a complete set of chromosomes. It is then stimulated to divide, after which it develops into a complete organism. Cloning could be used to produce genetically identical animals, such as frogs, mice, or rats, that are needed in research.

REVIEW QUESTIONS

11. How is selective breeding of commercial value to humans?
12. What is hybrid vigor?
13. Describe the steps in the formation of recombinant DNA.
14. What is cloning? How is cloning of value to someone working in a greenhouse?
15. How might gene splicing be applied to treating some of the disorders caused by gene mutations that are discussed in Section 7·5?

Gene Expression

8·9 GENES AND THE ENVIRONMENT

Environmental factors often influence the action, or expression, of genes. For example, the Himalayan rabbit is homozygous for black fur. But the gene for black fur is only expressed at body temperatures below 33° C. The feet, tail, ears, and nose are usually cooler than 33° C in the rabbit's natural habitat. The rabbit expresses the homozygous gene for black fur. Only the feet, tail, ears, and nose are black. An experiment involving the genes in the hair-producing cells of these rabbits is described in Figure 8·15.

SECTION OBJECTIVES

- Compare gene control systems in prokaryotic eukaryotic cells.
- Discuss how gene expression is influenced by the environment.

FIGURE 8·15 An experiment involving Himalayan rabbits which caused genes to direct the production of black fur.

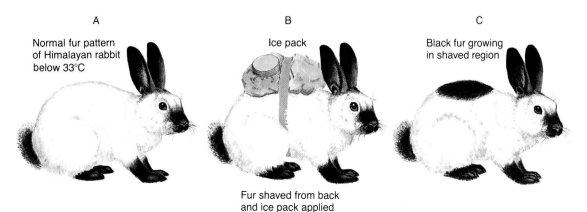

A Normal fur pattern of Himalayan rabbit below 33°C

B Ice pack — Fur shaved from back and ice pack applied

C Black fur growing in shaved region

Reared at 25 °C

Reared at 16 °C

FIGURE 8·16 Temperature influences wing development in *Drosophila*. At cool temperatures, the gene for curly wings is not expressed.

Environmental factors also influence gene expression in the fruit fly. Look at the fruit flies in Figure 8·16. Both flies are homozygous for curly wings. The fly with curly wings was reared at 25° C. The fly with straight wings was reared at 16° C. As you can see, temperature can affect gene expression.

Humans also show the effect of environment on gene expression. Identical twins are genetically identical individuals. Studies have been carried out on twins who, from birth, were raised in different environments. The studies have helped to prove that people are a product of both their genetic makeup and their environment. Although identical twins are very similar, it is clear that they are not identical people.

The environment may limit gene expression in humans. For example, many people have genes for tallness that are never expressed. Malnutrition is one condition that may limit the expression of the genes for tallness.

The internal environment may also affect the expression of genes. In humans the genes for beard growth are found on autosomes and are present in both males and females. Males express the genes, but females do not. The expression of the genes is influenced by an individual's internal environment, namely the concentration of the male hormone *testosterone* (tehs TAHS tuh rohn). The interactions of genes and hormones also govern the expression of other secondary sex characteristics. These types of gene expressions are said to be *sex-influenced characteristics.*

8·10 CONTROL SYSTEMS IN PROKARYOTIC CELLS

Scientists have found that enzymes are not continuously being made in a bacterial cell. They have also found that some cells show differences in the amounts of enzymes present. Such observations lead to the question of what turns genes "on" and "off" in cells. Let us look at one mechanism for genetic control systems in prokaryotic cells.

In 1965, a Nobel Prize was awarded to three biochemists. François Jacob, Jacques Monod, and André Lwoff received the award for their work in developing a model to explain how genes control enzyme production in prokaryotic cells. According to their model, a substrate must be present before the enzyme is made. The substrate involved is known as an *inducer,* and the enzyme formed is called an *inducible enzyme.*

The section of the DNA molecule involved in enzyme production has three regions. Figure 8·17A shows these regions of the DNA molecule. The structural genes and the operator make up the *operon* (AHP uhr ahn). A *structural gene* is that part of the DNA nucleotide sequence that codes for the synthesis of a particular protein. Next to the structural gene is the *operator,* the part of the DNA that controls expression of the structural genes. The third region is the regulator gene. A *regulator gene* is a gene that codes for a protein that controls the expression of another gene.

How is the production of an enzyme induced? As an example, consider the digestion of the sugar lactose. Lactose is broken down to form glucose and galactose. Enzymes that digest lactose are made only when lactose is present. Lactose acts as the inducer. Thus lactose triggers the processes that make the needed enzymes. Jacob, Monod, and Lwoff proposed that three structural genes code for the enzymes needed for lactose digestion. What turns the structural genes "on" so that the enzymes can be made?

The operator is the "on-off" switch of an operon. When lactose is present, the switch is on, and the enzymes are made. How does the presence of lactose control the operator gene?

Part of the answer lies in the region of the chromosome where the regulator gene is found. Part B shows the regulator gene coding for the synthesis of a protein that functions as a *repressor*. When this repressor binds to the operator, the operator is turned off, and no more enzymes for lactose digestion are made.

Part C shows how the operator is turned back on. When lactose is again present, lactose binds with the repressor and blocks the repressor from binding with the operator. When the operator is free, the system is on. Thus the structural genes will be transcribed as mRNA to form enzymes to digest lactose.

FIGURE 8·17 The Jacob-Monod-Lwoff operon model explains how genes control enzyme production in prokaryotic cells.

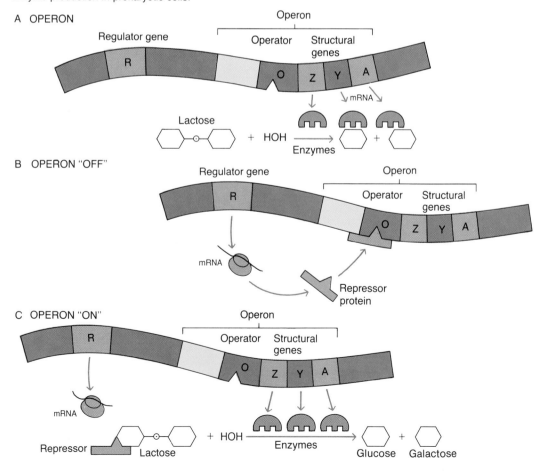

8·11 CONTROL SYSTEMS IN EUKARYOTIC CELLS

Control systems in eukaryotes and prokaryotes are not alike. In eukaryotic cells the expression of the gene seems to be subject to many controls. The control can be in the activation of the gene, or in the processes of transcription and translation, or in the action of the enzyme in chemical reactions.

Eukaryotic chromosomes are different from the chromosomes found in prokaryotic cells. Prokaryotic chromosomes have very little protein material. Eukaryotic chromosomes have large amounts of protein associated with their DNA.

The *histones* comprise a major type of protein that is similar in plant and animal cells. The histones are thought to be important in the structure of the chromosomes. A second type of protein is *nonhistone protein*. Nonhistone proteins vary from species to species and from one type of tissue to another. These proteins even vary within the same tissue at different times. These nonhistone proteins may have a regulatory function. This regulatory function may be affected by the presence or absence of such substances as chemicals or hormones that in some way help turn genes "on" or "off."

FIGURE 8·18 This electronmicrograph shows a strand of DNA combined with a strand of mRNA. The "loops" in the photograph are DNA introns. When the strand of mRNA (dotted line) leaves the nucleus, it will leave the introns behind. The mRNA will consist of a strand of extrons.

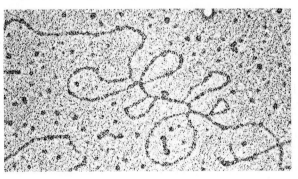

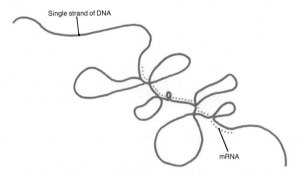

Eukaryotic cells have many split genes, whereas split genes are rare in prokaryotes. *Split genes* are genes in which the region of DNA used to code for a single protein is not continuous. Instead, split genes, as shown in Figure 8·19, have two parts. *Exons* are the sections of DNA that code for a protein. *Introns* are DNA sequences that do not code for protein and whose function is not fully understood.

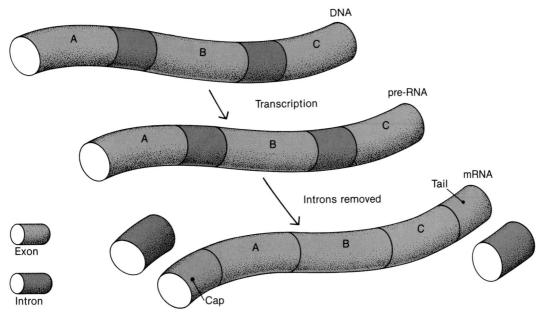

FIGURE 8·19 Before the mRNA from a split gene can leave the nucleus, the introns must be removed and a cap and tail added. Only the mRNAs for needed proteins are processed. This controls which proteins are made in the cell.

The long strand of mRNA transcribed from a split gene is called precursor RNA, or preRNA. The introns must be removed, as shown in Figure 8·19, before the mRNA can leave the nucleus. The exons are joined after the introns are removed. Adenine nucleotides are added as a "tail" to one end of the mRNA. Another base, methylguanosine, is added to the other end as a "cap." The cap is probably involved with initiating translation on ribosomes. Now the mRNA can move through the nuclear membrane. Remember, mRNA must leave the nucleus and move to a ribosome before proteins can be made.

Other controls can be exerted on the expression of a gene after transcription takes place. The mRNA can be inactivated after it leaves the nucleus. What causes this inactivation is still unclear. When mRNA is translated on the ribosome, a protein is made that is also subject to controls. As you can see, the expression of a gene in eukaryotic cells is subject to many controls.

REVIEW QUESTIONS

16. Describe how temperature influences gene expression in *Drosophila*.
17. Describe two factors that may influence gene expression in humans.
18. Describe how the regulator gene controls the operator in prokaryotic cells.
19. What effect would the breakdown of all available lactose have on synthesis of the enzyme lactase?
20. How does an inducible enzyme affect the repressor protein?

Chapter Review

SUMMARY

8·1 Mutations are changes in the hereditary material. The two types of mutations are chromosomal alterations and gene mutations.

8·2 Chromosomal alterations may involve changes in the number of chromosomes or changes in the structure of chromosomes.

8·3 Gene mutations are changes in the sequence of the nitrogenous bases of DNA.

8·4 Human genetic disorders resulting from nondisjunction include Down's syndrome, Klinefelter's syndrome, and Turner's syndrome.

8·5 Human genetic disorders resulting from gene mutations include sickle cell anemia, Tay-Sachs disease, phenylketonuria, and hemophilia.

8·6 Certain genetic disorders can be detected before birth by procedures such as karyotyping, amniocentesis, ultrasonography, and fetoscopy.

8·7 Through selective breeding, pure strains of plants and animals with desirable characteristics can be maintained, and new species can be developed.

8·8 Genetic engineers have developed DNA-splicing techniques, in which segments of DNA are joined to produce recombinant DNA.

8·9 The phenotype of an organism is the product of the interactions of the organism's heredity and the environment.

8·10 A mechanism that may regulate the activity of genes in prokaryotic cells is the inducible-enzyme mechanism.

8·11 The expression of genes in eukaryotic cells involves many internal controls.

BIOLOGICAL TERMS

amniocentesis
chromosomal alteration
cloning
deletion
DNA splicing
Down's syndrome
duplication
fetoscopy
gene mutation
genetic engineering
germ mutations
hemophilia
hybrid vigor
inbreeding
inversion
karyotype
Klinefelter's syndrome
monosomy
mutagen
mutations
nondisjunction
outbreeding
phenylketonuria
plasmids
polyploidy
recombinant DNA
selective breeding
sickle cell anemia
somatic mutations
Tay-Sachs disease
translocation
trisomy
Turner's syndrome
ultrasonography

USING BIOLOGICAL TERMS

1. Define genetic engineering.
2. Distinguish between a mutation and a mutagen.
3. What is a karyotype?
4. Distinguish between the terms *inbreeding* and *outbreeding*.
5. What is selective breeding?
6. What is a chromosomal alteration?
7. Distinguish between a deletion and a duplication.
8. Define the term *cloning*.

Choose the term that does not belong and explain why it does not belong. In your explanation, define the terms.

9. ultrasonography, fetoscopy, monosomy, amniocentesis
10. Down's syndrome, ultrasonography, hemophilia, Klinefelter's syndrome
11. deletion, inversion, translocation, inbreeding
12. Klinefelter's syndrome, sickle cell anemia, Tay-Sachs disease, phenylketonuria

UNDERSTANDING CONCEPTS

1. Distinguish between somatic mutations and germ mutations and indicate which type of mutation may be passed on to offspring. (8·1)

2. Explain the differences between translocation, duplication, deletion, and inversion. How do all these processes differ from nondisjunction? (8·2)

3. Describe three changes that can occur in genes to cause gene mutations. (8·3)

4. Compare Klinefelter's syndrome and Turner's syndrome. (8·4)

5. Using a Punnett square, show the possible offspring of a man and woman who are carriers of the sickle cell trait. Indicate what proportion of their offspring would probably be carriers and what proportion would probably have the sickle cell disease. (8·5)

6. Explain how amniocentesis is used in the detection of genetic disorders in humans. (8·6)

7. How has selective breeding been used to improve plants and animals of value to people? (8·7)

8. What are inbreeding and outbreeding? Give an example of each. (8·7)

9. Explain how DNA splicing is used to produce new plasmids. (8·8)

10. Describe how temperature influences gene expression in the Himalayan rabbit. (8·9)

11. Describe the mechanism by which structural genes are turned "on" in prokaryotic cells. (8·10)

12. List the various influences that can modify gene expression in eukaryotic cells. (8·11)

APPLYING CONCEPTS

1. Albinism is a homozygous recessive trait. A woman with normal pigmentation is the daughter of a woman who is an albino. The normally pigmented woman marries a normally pigmented man whose father is an albino. What are the possible genotypes of the couple's parents? The couple have four children. What percentage would you expect to be albinos?

2. Your future spouse has a sister with phenylketonuria (PKU). You have no record of PKU in your family, but you have been told that your grandfather's brother was retarded. Describe the kind of discussion you and your future spouse might have with a genetic counselor regarding your concerns about having healthy children.

3. You are a genetic counselor. A young couple comes in to ask for information on genetic disorders. Determine the kinds of information you would give the couple. List the questions you would need to ask them.

4. A normal male is married to a female who is a carrier of hemophilia. If the couple has four children, what percentage is likely to have no genes for hemophilia?

5. In what ways could a plant breeder use genetic engineering to improve crops?

EXTENDING CONCEPTS

1. Construct a sequence of models that could be used to demonstrate genetic control systems in prokaryotic cells.

2. Many human disorders are inherited. In addition to those described in this chapter, examples include neurofibromatosis (the condition that afflicted the main character in the film *Elephant Man*), Huntington's chorea, and cystic fibrosis. Select one of these inherited conditions and report on the following: (a) the type of trait (dominant or recessive) that produces the condition, (b) the symptoms, (c) the possibility of detection before birth, and (d) treatment.

3. There has been some controversy over guidelines from the National Institute of Health that govern experiments using recombinant DNA. Set up a committee to do research on the nature of the controversy. Have a debate on the pros and cons of government intervention in genetic research.

4. Abraham Lincoln is believed to have suffered from the genetic disorder Marfan's syndrome. Do research on the condition. Find out the symptoms and treatment. Prepare a chart showing Lincoln's family tree and indicate how he might have inherited the disorder.

READINGS

Chambon, P. "Split Genes." *Scientific American,* May 1981, p. 60.

Langone, J. "Treating the Littlest Patient." *Discover,* January 1981.

Lawn, R.M. and G.A. Vehar, "The Molecular Genetics of Hemophilia," *Scientific American,* March 1986.

9. Evolution

This iguana spends much time in the sea. Other kinds of iguanas live exclusively on land. Many people believe that the similarities between the two kinds of iguanas indicate that they are related. What do you think? In this chapter, you will study evidence indicating that many modern plants and animals are similar to, yet different from, plants and animals of the past.

CHAPTER OBJECTIVES

After completing this chapter, you will be able to
- **Restate** the evidence that supports the theory of evolution.
- **Compare** the theories of evolution as presented by Lamarck and Darwin.
- **Compare** gradualism and punctuated equilibrium.
- **Describe** the modern theory of evolution.
- **Explain** how speciation is thought to occur.

Evidence for Evolution

9•1 WHAT IS EVOLUTION?

Do you think that horses and zebras are related in some way? Is a snowy owl related to a barn owl? The animals in each pair are distinct yet similar. The animals in each pair belong to different species. A *species* is made up of organisms of a certain kind that naturally interbreed and produce fertile offspring.

You may make observations about likenesses among species many times. Scientists also make these observations. Certain species of animals and plants seem to be related to other species. Organisms alive today seem to be related to similar but different species of living things of the past. Why do these apparent relationships occur?

There is a scientific explanation for the apparent relationships among living things. The explanation states that organisms alive today are related to extinct organisms of the past. The explanation for both the great diversity and yet apparent relatedness of organisms is called the *theory of evolution*.

The main idea of the theory of evolution is that populations of organisms may undergo biological change over generations. A *population* consists of those members of a species living in a certain region. According to the theory of evolution, horses of today are related to and have descended from horselike animals of the past. However, horses of today are not of the same species as horselike animals of, say, 10 million years ago. Further, horses and zebras both have descended from some species of horselike animal of the past. That is why we believe that horses and zebras are related.

Evidence shows that life has been on the earth for about 3.5 billion years. The history of life is a history of change. **Evolution** is the process of change by which new species come about from preexisting species. Evolution explains how different life forms have developed and why living things show both similarities and differences.

SECTION OBJECTIVES

- Explain how fossil evidence supports the theory of evolution.
- Describe how evidence from living organisms supports the theory of evolution.

FIGURE 9·1 Fossils of a woolly mammoth (top left), an insect in amber (bottom left), and a fern (top right).

9·2 FOSSILS

The history of life is recorded in the rock layers of the earth. In these layers of rock, fossils preserve a history of change. **Fossils** are the remains and traces of organisms that once lived.

Fossils are formed in many ways. The nearly complete remains of living things have sometimes been preserved. For example, woolly mammoths that have been preserved in ice for 40 000 years have been found. The remains of insects and other small organisms have sometimes been preserved in amber, as shown in Figure 9·1. The remains of actual animal skeletons, such as those of the saber-toothed tiger in the La Brea tar pits in Los Angeles, have been found.

Some fossils are the *imprints* made by parts of animals and plants in mud. As the organism breaks down, its impression is left in the mud, which slowly hardens into rock. Other fossils are formed when spaces in the shells and skeletons of animals and in the hard parts of plants are filled by mineral matter.

sedere (to settle)

Most fossils are found in sedimentary (sehd uh MEHN tuhr ee) rock. **Sedimentary rock** is rock that forms from buildup of sediment, tiny grains of eroded rock and other materials that settle to the bottom of a body of water. The sediment builds up, layer upon layer upon layer, as it settles from the water. As sediment compacts and hardens, layers of sedimentary rock are formed. These layers of sedimentary rock are called **strata** (STRAY tuh). In undisturbed strata (sing., *stratum*) the oldest layers are the deepest and have the oldest fossils. Look at Figure 9·2, which shows a cross section of undisturbed sedimentary rock. Which stratum would have the oldest fossils?

The relative age of a fossil can be found by comparing the stratum of sedimentary rock in which it is located with other strata. The actual age of a fossil can be estimated by using radioactive dating. Some of the elements that make up fossils and sedimentary rock are radioactive. Radioactive elements break down into nonradioactive elements at constant rates. The time in which half the radioactive atoms of a certain kind in rocks or fossils will break down into atoms of another element is called the **half-life.**

Carbon-14 is a naturally occurring radioactive isotope of carbon. Both carbon-12, which is nonradioactive, and carbon-14 are found in the atmosphere in a constant ratio. Living organisms are always taking in both forms of carbon, thus keeping this constant ratio within their tissues. When an organism dies, it stops absorbing carbon. As radioactive carbon-14 decays, the ratio changes. Carbon-14 has a half-life of 5710 years. By studying the proportion of carbon-14 to carbon-12, scientists can estimate the age of fossils up to about 75 000 years old.

To date older fossils, other radioactive isotopes, such as potassium-40 and uranium-238, can be used. Potassium-40 (half-life, 1.3 billion years) is used to date fossils more than 100 000 years old. Uranium-238 (half-life, 4.5 billion years) is used to date rocks in which fossils are found.

The presence of certain fossils in a particular layer of rock can be used to tell the age of the rock layer compared with other layers. The fossils give the rock layer a relative place in geologic time. Because of such fossil evidence, including radioactive dating, scientists have been able to organize fossils by their ages.

The oldest strata contain fossils of simple organisms. Newer strata have fossils of more complex organisms. Fossils of organisms found in the newest layers of rock show more similarities to organisms alive today than do fossils of organisms of more distant times. This pattern would be expected if evolution has taken place.

The fossil record also suggests certain similarities among groups of organisms. These similarities have led scientists who study ancient life to infer that certain similar groups may have evolved from a common ancestor. Look at Figure 9·3. The fossil record suggests that an animal called *Hyracotherium* (hī ra koh THEER ee uhm) is the common ancestor of the horse. Note the changes that have taken place in the toes and leg bones of the many forms apparently descended from *Hyracotherium*. An organism intermediate in characteristics between two other kinds of organisms and showing a link between these organisms is called a **transitional form**.

FIGURE 9·2 The river at the bottom of this canyon has cut through the sedimentary rock layers over thousands of years, revealing the strata. Where would you expect to find the youngest fossils?

trans- (across)

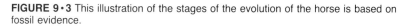

FIGURE 9·3 This illustration of the stages of the evolution of the horse is based on fossil evidence.

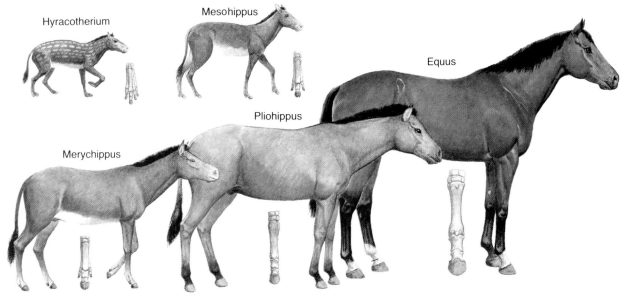

A well-known example of a transitional form is *Archaeopteryx* (ahr-kee AHP tuhr ihks). Several fossilized skeletons of *Archaeopteryx* have been found. Based on the skeleton alone, *Archaeopteryx* would be classified as a small reptile with two legs. In fact it looks much like certain small two-legged dinosaurs. The skeleton includes a long, bony tail, a mouth full of teeth, and clawed fingers. However, the imprints of feathers are also part of the remains of *Archaeopteryx*. Feathers are characteristic of birds. Thus *Archaeopteryx* is usually classified as a primitive bird. But it has teeth and clawed fingers on its wings! *Archaeopteryx* is a transitional form between reptiles and birds. The presence of transitional forms is evidence that evolution has taken place.

9·3 COMPARATIVE ANATOMY

Evidence of evolution is not limited to the remains of organisms that lived in the past. By studying the similarities and differences in structure among living things, scientists have found additional evidence of evolution.

Figure 9·5 shows the structure of a dog's leg, a whale's flipper, a bird's wing, and a lizard's leg. At first glance, these forelimbs seem to have little in common. However, closer examination reveals that their internal structure is surprisingly similar.

Most scientists believe that certain structural similarities indicate that these organisms are descended from a common ancestor. In this case, the ancestor had a particular arrangement of bones in its forelimb. As groups of organisms evolved from the ancestral form, the bones changed. However, the basic bone pattern remained the same. The forelimbs shown in Figure 9·5 are an example of homologous structures. **Homologous structures** are parts of different organisms that are similar in structure but serve different functions. Homologous structures suggest evolutionary descent from a common ancestor.

FIGURE 9·4 A fossil of *Archaeopteryx*. Note the feather impressions.

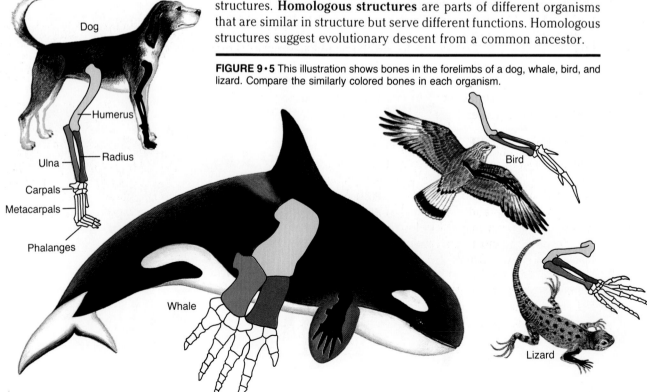

FIGURE 9·5 This illustration shows bones in the forelimbs of a dog, whale, bird, and lizard. Compare the similarly colored bones in each organism.

The presence of vestigial organs in many animals is additional evidence for evolution. **Vestigial organs** serve little or no function. They are thought to be the remnants of organs that were once functional in an ancestral form. The tiny hindlimb bones in whales indicate that their ancestors once walked on four legs. Vestigial organs are often homologous to functional organs in other species. This suggests that species evolved from a common ancestor.

Alternatively, similarities in organ structure and function may indicate that organisms have evolved along different lines. Butterflies and birds have wings. Although these organs share the same function, their internal structures are very different. The wings of a butterfly and a bird are analogous structures. **Analogous structures** are similar in function but differ in internal structure. Analogous structures do not suggest evolution from a common ancestor.

vestigium (footprint)

9·4 COMPARATIVE EMBRYOLOGY

Certain organisms show strong similarities in the development of their embryos (EHM bree ohz). An *embryo* is an organism in the earliest stages of development. Figure 9·6 shows the embryos of various kinds of vertebrates, or animals with backbones. It is somewhat difficult to tell these vertebrate embryos apart.

All vertebrates have a notochord at some time. The *notochord* is a rod of tissue along the back. In most adult vertebrates the notochord is replaced by the backbone. However, in some kinds of fish the notochord remains in the adult. The existence of this structure in all vertebrates indicates a relationship between different vertebrates.

All vertebrate embryos form *gill pouches* at some time. In fish and amphibians, working gills develop from these pouches. In reptiles, birds, and mammals, the gill pouches disappear during development. However, the presence of gill pouches in all vertebrates shows a relationship between those with gills and those that lack them.

Many scientists think that similarities in patterns of development show that similar genes are controlling these early stages of development. Such similarities in embryological development suggest that organisms have descended from a common ancestral form.

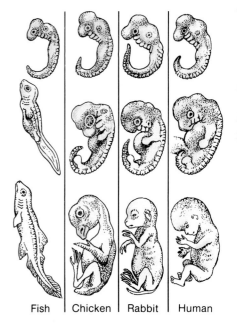

FIGURE 9·6 Vertebrate embryos are very similar in early stages of development. In time the embryos become more and more dissimilar.

Fish | Chicken | Rabbit | Human

9·5 COMPARATIVE BIOCHEMISTRY

You have already learned about some very basic similarities in the chemical processes that take place in cells. Energy is stored in cells in ATP. Cellular respiration is similar in most organisms. Protein synthesis, hydrolysis, and dehydration synthesis are other similar processes. Similar enzymes control these processes.

Perhaps the most remarkable similarity is the fact that organisms have DNA. Recall from Chapter 8 that bacteria are able to make human insulin and human growth hormone. It is possible for a bacterial cell to translate the DNA sequence of a human gene because the DNA code is the same in both organisms. The universal nature of the genetic code suggests evolutionary relationships among organisms.

Scientists can analyze proteins in many different organisms. They can compare the order of amino acids of these proteins. Also, scientists can now analyze genes and learn the order of bases in DNA. The close evolutionary relationships thought to exist between organisms can be confirmed in this way. Similarities in the sequences of bases in DNA result in similarities in the proteins made. The greater the biochemical similarities among different groups of organisms, the closer the evolutionary relationships are thought to be. Biochemical evidence often is used to show that species thought to be closely related are, in fact, closely related.

REVIEW QUESTIONS

1. What is evolution?
2. How does the arrangement of fossils in strata provide evidence to support the idea that evolution has occurred?
3. What are homologous structures? How do they provide evidence that evolution has occurred?
4. Give an example of close biochemical similarities among organisms that suggest evolutionary relationships.
5. Scientists often find fossils of the same kind of organism on different continents. Explain how such widely separated finds could have occurred.

Theories of Evolution

SECTION OBJECTIVES

- Contrast Lamarck's and Darwin's theories of evolution.
- Describe how organisms adapt to their environment.
- Distinguish between gradualism and punctuated equilibrium.

9·6 LAMARCK'S IDEAS

There is much evidence that evolution does take place. But *how* do organisms evolve? There have been several hypotheses to explain the mechanism of evolution. Early in the nineteenth century Jean Baptiste Lamarck, a French scientist, proposed a hypothesis to explain the mechanism of evolution.

Lamarck suggested that events in an organism's life could produce changes in that organism. Organs that were used would develop more than those that were not used. This is called the *hypothesis of use and disuse*. Lamarck explained the long neck of the giraffe in this way: Giraffes evolved from animals with short necks. As these ancestral short-necked animals stretched their necks to reach leaves on the upper branches of trees, their necks grew long. Lamarck believed these changes could be passed on to the organisms' offspring. Lamarck called this the *inheritance of acquired characteristics*. As generations passed, the giraffes acquired longer and longer necks. Lamarck's hypothesis is shown in Figure 9·7.

If acquired characteristics can be inherited, then changes in body cells would have to be accompanied by changes in the genetic infor-

FIGURE 9·7 According to Lamarck, short-necked forms of giraffes (left) stretched their necks to reach leaves, causing their necks to grow long.

mation in the sex cells. Some mechanism for changing the genetic information in the sex cells would have to exist. There is no evidence for such a mechanism.

A great many scientists have done experiments to test Lamarck's hypothesis. However, there is no evidence to support the idea of the inheritance of acquired characteristics. Thus Lamarck's hypothesis of evolution has been rejected by scientists.

9·7 DARWIN'S IDEAS

In 1859, Charles Darwin, an Englishman, published *On the Origin of Species by Means of Natural Selection*. This book presented a theory of evolution that totally changed biology. Many years earlier, in the 1830s, Darwin had spent 5 years on a trip around the world aboard the *H.M.S. Beagle*. The *Beagle* was a British naval ship on a scientific journey to map the coast of South America and gather specimens. At each stop on this voyage, Darwin, who was the ship's naturalist, took notes as he studied the local animals and plants. He noticed that each species seemed well adapted to its environment. In addition to observing organisms, Darwin found rich fossil beds. Studying the fossils, Darwin became aware of forms of life that no longer existed.

One region that Darwin studied was an isolated group of volcanic islands called the Galápagos. This Pacific island group, located some 950 km off the coast of Ecuador, had plants and animals that were clearly related to, but different from, those on the mainland of South America. Darwin also noticed that slightly different environments often were home to similar, but different, species.

Darwin studied 14 species of finches on the Galápagos Islands. Finches are birds that are found all over the world. However, the finches Darwin saw on the Galápagos did not live anywhere else. Most

FIGURE 9·8 Four of the 14 species of finches studied by Darwin on the Galápagos Islands and the type of food they ate.

finches eat seeds. Some of the finch species on the Galápagos eat seeds. However, there are also fruit-eating and insect-eating finches. Yet all the species of finches on the Galápagos are similar in many ways. And they are similar to the seed-eating finches found in South America.

It seemed to Darwin that the Galápagos Islands had been colonized in the past by finches from South America. Over many generations, different populations of finches developed. Eventually, 14 related species existed, each with a different lifestyle. However, the question of how such evolutionary changes happened had not been answered.

After returning to England, Darwin spent the next 20 years organizing the data he had gathered on his voyage. Darwin also read a great deal. One of the books that he read was an essay on population growth. It had been written in 1798 by the Reverend Thomas Malthus. Malthus predicted that the human population was growing so quickly that it would someday outgrow the available food supply.

Malthus warned that the population growth would in time be stopped by starvation, diseases, and war. Darwin found that Malthus's prediction of population growth and decline was likely to apply to all groups of organisms. All organisms produce more offspring than can survive. Environmental factors, such as limited food supply and available breeding sites, keep the quick growth of populations in check. Darwin realized, as Malthus had before him, that among organisms there is a competition to survive.

What determines which organisms live? Darwin lived in the country in England. He had watched farmers selectively breeding plants and animals. Selective breeding involves choosing those living things with desirable traits and breeding them. By selecting which organisms would be bred, farmers produced offspring with the traits they thought desirable. Over many generations, selective breeding produced many varieties of livestock and crops.

Darwin suggested selection takes place in nature as well. During his travels Darwin saw that there was much variation in traits among the members of a species. For example, some were stronger or faster than others. Darwin believed that these differences happen by chance. He thought that some variations make an organism better adapted to live in a certain environment. Those organisms that lived could pass their traits on to their offspring. Darwin called the process resulting in survival of those organisms best adapted to the environment **natural selection.**

The evolution of the giraffe's long neck can be explained by natural selection. Some of the giraffes in a group have necks that are longer than others. Those giraffes with the longer necks can feed on leaves that cannot be reached by the shorter-necked giraffes. The long-necked giraffes are more likely to live and reproduce than are shorter-necked giraffes, thereby passing the trait for long necks on to their offspring.

Darwin's **theory of evolution by natural selection** can be summarized as follows.

- *Overproduction* Favorable conditions allow a population to increase in size. Over time environmental pressures limit the number that can survive.
- *Variation* No two organisms in a population are exactly alike. They differ in size, behavior, and other features.
- *Competition* Due to the environmental pressures, the organisms within a population must compete with each other to survive.
- *Survival of the Fittest* The individuals who are best adapted to that environment are the ones most likely to survive. They possess variations that give them a selective advantage.
- *Reproduction* Individuals that survive and reproduce can pass their traits on to their offspring.
- *Speciation* Over time, the population changes because some traits are passed on and others are not. When a population differs enough from the original population, it is a new species.

Darwin could not explain how variations were produced or how variations were passed to offspring. However, the science of genetics can explain how these occur. Recall that genes determine the traits of an organism. The variations in a population result from the different combinations of genes in the organisms in the population. In each generation new combinations of genes are produced. Thus, more variations may appear. Random changes in genes, called mutations, can occur. Mutations can result in changed traits. A change in the order of genes on a chromosome also will introduce variations. As you will see, the modern theory of evolution includes principles of genetics.

FIGURE 9·9 This population of elephant seals is basking on a beach in California. Favorable conditions allow a population to increase in size. Competition between individuals will eventually limit the population's size.

9·8 NATURAL SELECTION OBSERVED

Darwin never observed evolution occurring. While no one has seen the evolution of a new species, scientists have seen the effects of natural selection on populations. One of the most famous of such observations involves the peppered moth. In England in the mid-1800s there were two kinds of peppered moths. The light variety was the most common; the dark-colored variety was very rare. These night-flying moths rested during the day on light-colored tree trunks. The moths were preyed upon by birds while on the trees. Look at Figure 9·10A. It shows a dark-colored peppered moth on a light-colored tree trunk. Why is this moth likely to be preyed upon?

As a result of the industrial revolution during the nineteenth century, soot darkened the trunks. Look at Figure 9·10B. Notice the light-colored moth on the dark-colored tree trunk. What do you predict would happen to the light-colored moths in an industrial area? By the end of the nineteenth century, the population of peppered moths in England was made up mostly of dark individuals.

In 1950 an experiment was performed to test the hypothesis that natural selection favored the survival of the dark-colored moths in industrial areas and the light-colored moths in rural areas. Equal numbers of the two varieties of moths were placed on tree trunks in an industrial area and in a rural area. A hidden camera recorded the birds eating the released moths. Figure 9·11 summarizes the data of this

FIGURE 9·10 Black moths are conspicuous against light-colored trees, while white moths are easier to see on sooty trees. Birds eat the moths that are easier to find.

experiment. Which kind of moth was better adapted to survive in the industrial area? What can you predict about the color of future generations of peppered moths in industrial areas that enforce strict air-pollution guidelines?

This experiment showed that organisms whose color provides better camouflage are more likely to survive and reproduce. It showed a population changing as a result of natural selection. Thus there is evidence to support Charles Darwin's theory of evolution by natural selection.

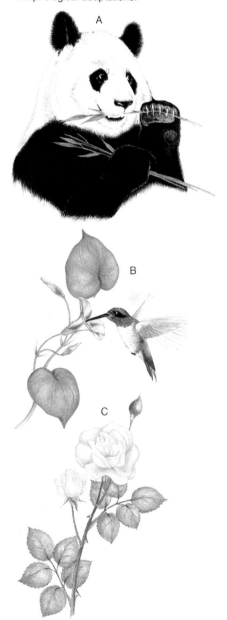

FIGURE 9·12 The panda's thumb is actually a modified wrist bone which is used to gather food. The hummingbird's long bill allows the bird to reach the nectar at the bottom of a flower. The rose's thorns are modified leaves that help protect the rose against plant-eating animals. These are all examples of morphological adaptations.

FIGURE 9·11 The peppered moth population changed over time as a result of natural selection.

VARIETY OF PEPPERED MOTH	NUMBER OF MOTHS EATEN	
	RURAL AREA	INDUSTRIAL AREA
Light moth	26	43
Dark moth	164	15

9·9 TYPES OF ADAPTATIONS

Natural selection can result in a species that is better adapted to a particular environment. An **adaptation** is an inherited trait that makes an organism better able to survive in a certain environment. A species that is better adapted than another to an environment is said to have an adaptive advantage. There are three types of adaptations—morphological (mor fuh LAHJ uh kuhl), physiological (fihz ee uh LAHJ uh kuhl), and behavioral (bih HAYV yuhr uhl).

Morphological adaptations are structural traits that make an organism better able to live in a given environment. Some morphological adaptations are related to food getting. Examples are the stinging cells of jellyfish, the sticky tongues of anteaters, and the opposable "thumbs" of pandas. How does a panda's "thumb" provide an adaptive advantage? The panda feeds on bamboo shoots. It strips the bamboo shoots by using the "thumb" and five fingers on each forepaw. The "thumb" is an elongated wrist bone that opposes the five fingers and allows the panda to strip the bamboo. How do the other adaptations listed make the organisms better able to survive?

Other morphological adaptations include the hollow bones of birds (an adaptation for flight) and the specialized flower parts of certain plants (adaptations that are specific for pollination by insects or wind). Think of other morphological adaptations.

Physiological adaptations are those adaptations that involve the metabolic processes of an organism. The production of certain enzymes for digestion and of blood chemicals such as hemoglobin for oxygen transport are physiological adaptations. The release by egg cells of chemicals that attract only the sperm cells of possible mates is a physiological adaptation for reproduction.

FIGURE 9·13 The viceroy butterfly (bottom) resembles the bitter-tasting monarch butterfly (top).

Behaviorial adaptations are inherited behaviors that help an organism to survive in a given environment. The woodpecker finch sometimes uses a cactus spine to remove insects from a tree. Many birds, fish, reptiles, and insects show courtship displays that allow organisms within the group to recognize and attract members of the opposite sex. Male combat in certain groups of animals determines which males are the strongest and will mate with the females. This behavior helps ensure that the most fit organisms mate. What other behavioral adaptations can you think of?

It is often difficult to describe an adaptation as morphological, physiological, or behaviorial because many adaptations include all of these. Monarch butterflies, shown in Figure 9·13, feed on milkweed plants that contain bitter-tasting chemicals. Monarchs take these chemicals into their bodies and become bitter-tasting themselves. A bird that eats a monarch will become sick and will then avoid monarchs. Although viceroy butterflies are not bitter-tasting, they also are avoided by birds. Viceroys have structural and behavioral adaptations that aid in their survival. They look like the monarchs, and they copy the monarchs' behavior. How do these adaptations aid the survival of the viceroys?

9·10 THE RATE OF EVOLUTION

Nearly all biologists accept the theory of evolution by natural selection. However, they still disagree over some of its details. Perhaps the most important dispute concerns the rate of evolution. Darwin believed that the rate is very slow. He knew that no major changes in species had been observed or recorded throughout human history. He believed that any such change must take place slowly.

If evolution is a gradual process, there should be many intermediate stages in the history of every species. However, scientists have found very few complete fossil records showing the gradual change of one species to another. Instead, there are large gaps in the fossil record. Some scientists have explained the lack of such intermediate stages by pointing out that fossils are formed by chance. Soft-bodied organisms decay too quickly to produce fossils, and the remains of land organisms are not likely to become encased in sediment. Many fossils have been destroyed by erosion. Since the remains of very few organisms become fossilized, the gaps are not surprising.

In 1972, Niles Eldredge and Stephen Jay Gould proposed a theory of evolution that explains the gaps in the fossil record. They suggested that these gaps are real and that there are very few intermediate forms as species evolve. In their view, a species may show little or no change for millions of years. This long period of time favors the chance that a fossil record of this species would form. At some time, however, a very quick change may take place, with a new species arising in only a few thousand years. In terms of geologic time, such a change would be very quick. There would be little chance that the intermediate stages would leave a fossil record.

Eldredge and Gould called this pattern of evolution *punctuated equilibrium*. According to the **theory of punctuated equilibrium,** species change very little or not at all for long periods of time. This equilibrium sometimes is punctuated, or interrupted, by sudden change. The result is a new species. This new species, in turn, may remain unchanged for millions of years.

The theory of punctuated equilibrium does not contradict Darwin's view of natural selection as the process directing evolution. It simply alters Darwin's idea of the way in which evolution occurs. Instead of long, slow change, Eldredge and Gould see quick jumps separated by long periods in which there is little or no change at all.

Biologists are divided in their reactions to this theory. Some prefer Darwin's view, which is called gradualism (GRAJ u uh lihz uhm); others have accepted the idea of punctuated equilibrium. Figure 9·14 compares gradualism with punctuated equilibrium. In this chart, the red species evolves into the blue species in a time period of 1 million years. According to **gradualism,** the process of evolutionary change is slow and steady. Over the period of 1 million years, intermediate forms evolve. According to punctuated equilibrium, the red species remains relatively unchanged for most of the 1 million years. In about 100 000 years, rapid evolution occurs. During this period of change, few fossils of intermediate species form. In both theories, species evolve by the process of natural selection.

FIGURE 9·14 You can see that punctuated equilibrium produces a new species in less time than gradualism.

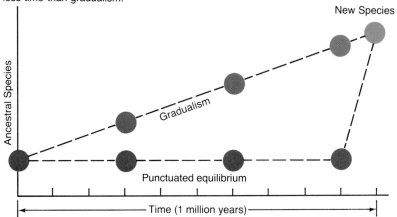

REVIEW QUESTIONS

6. Describe Lamarck's idea of how evolution occurs.
7. Describe Darwin's theory of evolution by natural selection.
8. Give examples of the three types of adaptations.
9. Explain how gradualism and punctuated equilibrium differ.
10. Propose a hypothesis to explain why the rate of evolution is sometimes rapid, as suggested in the theory of punctuated equilibrium.

Investigation

How Can Natural Selection Be Simulated?

Goals

After completing this activity, you will be able to
- Demonstrate how coloration can be a factor in natural selection.
- Predict which individuals will survive in a population of animals.
- Predict changes in a population over time.

Materials (for groups of 4)

two sheets of newspaper
sheet of black construction paper
sheet of white construction paper
scissors
jar or other large container

Procedure

A. Cut a section of the newspaper so that it is the same size and shape as the construction paper. Fold the sheet of newspaper into 64 rectangles. Cut the rectangles apart. (Each rectangle will measure about 2.7 cm by 3.3 cm.) In like manner, fold and then cut the sheet of black construction paper into 64 rectangles. Do the same with the white paper.

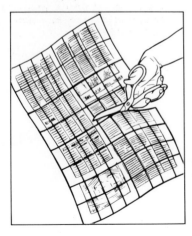

B. The black, white, and newspaper rectangles represent animals in an environment. The uncut sheet of newspaper represents the environment.

C. Place all the rectangles in the jar and shake the jar thoroughly. Dump all the rectangles on the open newspaper so that they are scattered.

D. You and your lab partners are going to be "predators." You will "prey" on the animals in the newspaper environment. Each person in your group must "capture" five paper animals. Prey on the animals in this way: Quickly choose an "animal." Look away for a moment. Turn back and choose another animal *as quickly as possible.* Continue until five animals have been captured.

E. On a sheet of paper copy the chart shown below. Record the number *of each type* of paper animal captured.

F. Compute the number of each type of paper animal surviving. (Remember that you began with 64 of each color.)

G. Compare the data you and your partners have gathered with that of others in the class. Compute the total number of survivors of each type of animal.

Questions and Conclusions

1. From the data gathered from the entire class, which of the paper animals survived in greatest numbers? Why do you think these animals survived?
2. Predict the kinds of changes you might be likely to find in this population of paper animals over a period of time.
3. Suppose that the environment were to change. For example, suppose that the environment gradually darkened. How might this change affect the numbers and kinds of animals in the population?
4. Describe a situation in the real world that illustrates what was just simulated in your classroom.

	WHITE PAPER ANIMALS	BLACK PAPER ANIMALS	NEWSPAPER ANIMALS
STUDENT 1			
STUDENT 2			
STUDENT 3			
STUDENT 4			

Mechanisms of Evolution

9·11 SPECIATION

Recall that a species is made up of organisms that naturally breed among themselves. Through natural selection, a species may become better adapted to a certain environment. It might seem that the species in time becomes so well adapted that evolution stops. However, environments change. The changes may be small or they may be great. A species may change, or adapt, as the environment changes. **Speciation** (spee shee AY shuhn) is the process by which new species evolve from ancestral species.

The members of a species are alike in many ways. Yet if evolution is to occur there must be inheritable variations among the members. These variations, which may be expressed in the organisms' phenotypes, come from different alleles in the species. However, although there are differences within a species, the members still can mate and produce fertile offspring. How does a new species that can no longer mate with the original group evolve? Scientists think that for a new species to evolve from an ancestral species, some of the members of the group must become isolated, or separated, from the rest of the group.

Geographic isolation happens when a physical barrier, such as a mountain range, stops members of two populations of a species from coming in contact with each other. Thus, interbreeding between the two populations is not possible. Different environmental pressures may then favor different adaptations. In time the populations become

SECTION OBJECTIVES

- Describe how isolation leads to speciation.
- Identify the conditions that result in genetic equilibrium.

FIGURE 9·15 At one time the population of animals on both sides of the mountain range could interbreed. A landslide closed off the pass. Now the animals cannot interbreed because they are isolated from each other by a geographic barrier.

183

FIGURE 9·16 The dodo bird, isolated on the island of Mauritius, survived because it had no natural enemies. When people brought domestic animals to the island, the dodo bird was no longer able to survive. Without its protective isolation, this species became extinct.

different enough that they can no longer interbreed, even if the geographic barrier disappears. They now may have slightly different breeding seasons, different courtship rituals, or different chromosomes. If they cannot interbreed and produce fertile offspring, **reproductive isolation** has taken place.

A probable example of speciation brought about by geographic isolation involves the dodo bird. A fierce storm may have blown several pigeons from Asia to Mauritius, a remote island in the Indian Ocean. On this island, where there was little competition, a small population of the birds probably thrived. Mutations in this isolated population were different from those on the mainland. Also, the environmental conditions were different. Natural selection favored traits different from those on the mainland. In time a new species, the dodo, evolved. Figure 9·16 shows the now-extinct dodo. This bird evolved in an area where there were few predators. It was thus unable to live when humans came into its land. What was the physical barrier that led to geographic isolation in this case?

On land, populations may become separated by rivers, or even by the building of a highway. A newly-formed meadow, for example, may be a barrier to two populations of forest-living animals.

Natural selection may affect organisms in different parts of a range in different ways. The population in the north may differ in some ways from the population in the south. For example, the white sheep and the stone sheep are members of the same species. But natural selection has favored the survival of sheep with white fur in the north and those with black fur (stone sheep) in the south. However, these two populations still interbreed. If a barrier isolates these two populations, two distinct species may evolve.

Recall that Darwin saw 14 species of finches on the Galápagos Islands. These 15 small islands are isolated from the coast of South America, the nearest landmass. Darwin observed that the Galápagos finches seemed to be related to each other. They were also similar to finches found in South America. Form a hypothesis to explain how speciation resulted in the existence of the Galápagos finches.

9·12 POPULATION CHANGES

In 1937, Theodosius Dobzhansky wrote the book *Genetics and the Origin of Species*. This book modified the theory of evolution to include genetic principles. The **modern theory of evolution** is a joining of Darwin's idea of natural selection with the concept of population genetics. **Population genetics** is the application of genetic principles to groups of organisms rather than to single organisms. In the modern theory of evolution, evolution is defined as a change in the genetic makeup of a population.

Discovery

Mass Extinction

Throughout the history of the earth, some species of plants and animals have flourished and others have died out. Usually the process of extinction involves only a few species over a very long time. However, the fossil record shows several instances of mass extinction, in which many species—sometimes more than 1000—became extinct in a relatively short period.

Scientists have proposed various hypotheses to explain these events, but no single explanation is conclusive. There is strong disagreement about the causes of the most recent mass extinction, at the end of the last Ice Age. Fifty-seven species of large land mammals disappeared from the American continents about 11 000 years ago. Some scientists believe this extinction was caused by environmental changes; others insist that human hunters were responsible for killing off so many animals.

According to the environmental hypothesis, the melting of glacial ice at the end of the Pleistocene Epoch changed the climate and other environmental conditions. Because of these changes, many plant species died, and the animals that were dependent on them had to search for new food supplies. As a result, some animal species that had previously coexisted were forced to compete with each other for food. The less successful animals died out. With the loss of the herbivores, many of the large predators and scavengers also probably died out.

The hunter hypothesis suggests that the mass extinction was caused by excessive hunting by early human societies. To support this idea, scientists have compared the approximate time of the extinction with the time that hunting people arrived in the Americas. The extinct large mammals were last in existence around 11 000 years ago, and the big-game hunters arrived at least 12 000 years ago. The rapid extinction over roughly 1000 years would therefore support the hunting hypothesis.

The disagreement among scientists over the cause of the last mass extinction continues. New information may provide the answer, or perhaps a new hypothesis will be proposed. It is unlikely, though, that the question will be settled soon.

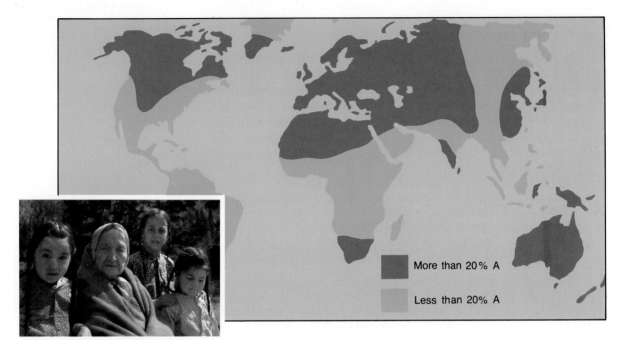

FIGURE 9·17 A map showing the frequency of blood group A. The frequency of this blood group is high among the Blackfoot Indians of North America.

The whole genetic makeup of a population is the population's gene pool. The **gene pool** is made up of all the genes of all the members in the population. Within a gene pool may be one or several alleles for a particular gene. In the human gene pool there are three alleles for the major blood types. Figure 9·17 shows the frequency of blood group A. The percent of each allele in the gene pool is the **gene frequency**. If the gene frequencies remain the same from one generation to the next, the population is in **genetic equilibrium**. If the gene frequencies change, evolution is taking place. Changes in gene frequencies do not take place automatically. They occur when something disrupts the genetic equilibrium. Thus, the factors that cause changes in gene frequencies are the factors that promote evolution.

In 1908, G. H. Hardy and G. Weinberg showed mathematically that under certain ideal conditions genetic equilibrium is maintained. They stated their idea in the Hardy-Weinberg Law. According to the **Hardy-Weinberg Law**, the gene pool in a sexually reproducing population tends to remain stable when all the following conditions are met.

- The population must be large enough so that chance alone will not change gene frequencies.
- There can be no movement into or out of the population.
- There can be no mutations.
- Reproduction must be totally random, meaning natural selection must not favor any one allele over any other.

The population stability suggested by the Hardy-Weinberg Law rarely takes place, since some or all of the conditions for stability usually are not met.

In a small population, chance alone may change the frequency of certain alleles. In such an instance, the gene pool of the population would be changed. For example, in a small population of frogs in a pond, the chance death of a few frogs with identical alleles will change the gene frequency of the population. The change that takes place in a gene pool, caused by chance is called **genetic drift**. Genetic drift may work for or against natural selection. For example, it may decrease the frequency of a helpful gene that is present in a low frequency.

Gene flow is the movement of alleles into or out of a population. Gene flow results from immigration and emigration. As individuals that can reproduce enter or leave a population, changes in gene frequency can occur. These changes can disrupt genetic equilibrium and produce changes in the gene pool. Few populations are totally isolated. Therefore, immigration and emigration usually occur.

Genetic equilibrium requires that there be no mutations. But mutations are always taking place. They are chance events, and they cannot be predicted. Mutations produce changes in the gene pool. These changes are hard to detect, since mutant genes are often recessive and not expressed in the phenotype. Given enough time, a homozygous recessive genotype with the mutant alleles may appear. Only when the mutation is expressed can it be acted upon by natural selection. Since mutations are random changes in the genes, they do not control the direction of evolution but rather provide the raw material upon which natural selection acts.

One requirement for a stable gene pool is that reproduction be totally random. In most populations, however, reproduction is not random. Natural selection plays a large part in the reproductive success of organisms. Certain alleles make an organism better able to survive, attract a mate, and produce many fertile offspring. The genes that increase reproductive success will probably be present in greater frequency in the gene pool of the next generation.

The Hardy-Weinberg Law describes the conditions needed to keep genetic equilibrium. Since all of these conditions are seldom met, equilibrium is often disrupted. Evolution occurs when one or more of the conditions described by the Hardy-Weinberg Law are not met.

REVIEW QUESTIONS

11. Describe how a geographic barrier may promote speciation.
12. What is the connection between reproductive isolation and speciation?
13. What two ideas are integrated in the modern theory of evolution?
14. According to the Hardy-Weinberg Law, what conditions are necessary for a gene pool to remain stable?
15. A and A' are alleles for a trait. In a certain population the gene frequency of A is 60% and of A' is 40%. In what way will the frequencies change in a large, isolated population in which natural selection favors allele A' over allele A?

Chapter Review

SUMMARY

9·1 Evolution is a natural process of change by which new species develop from species that already exist.

9·2 The fossil record provides evidence that evolution has taken place.

9·3 Homologous structures are similar structures that differ in function and indicate descent from a common ancestral form.

9·4 Similarities in the embryos of different organisms suggest an evolutionary relationship between the organisms.

9·5 The more biochemical similarities there are among different groups of organisms, the more closely related the groups of organisms are thought to be.

9·6 Lamarck proposed a theory of evolution based on the idea of the use and disuse of body parts and on the inheritance of acquired characteristics.

9·7 Darwin proposed a theory of evolution based on the action of natural selection on variations in traits among individuals within a species.

9·8 Natural selection has been observed in populations of peppered moths.

9·9 Species adaptations to the environment can be morphological, physiological, and behavioral.

9·10 Gradualism is the theory that the rate of evolution is slow and steady. Punctuated equilibrium is the theory that a species undergoes little change for long periods of time interrupted by periods of rapid change.

9·11 Speciation may often involve geographic isolation of populations followed by reproductive isolation.

9·12 The modern theory of evolution is a combination of the concept of natural selection and the principles of genetics. Evolution is a change in the gene frequencies of a population's gene pool.

BIOLOGICAL TERMS

adaptation
analogous structures
behavioral adaptations
evolution
fossils
gene flow
gene frequency
gene pool
genetic drift
genetic equilibrium
geographic isolation
gradualism
half-life
Hardy-Weinberg Law
homologous structures
modern theory of evolution
morphological adaptations
natural selection
physiological adaptations
population genetics
reproductive isolation
sedimentary rock
speciation
strata
theory of evolution by natural selection
theory of punctuated equilibrium
transitional form
vestigial organs

USING BIOLOGICAL TERMS

1. What is evolution?
2. Distinguish between analagous structures and homologous structures.
3. Distinguish between geographic isolation and reproductive isolation.
4. The Hardy-Weinberg Law describes the factors under which a particular condition is maintained. Which term names this condition?
5. Which term identifies Charles Darwin's main contribution to our concept of evolution?
6. Distinguish between morphological adaptations, behavioral adaptations, and physiological adaptations.
7. Which term refers to all the genes in a population?
8. Define the term *speciation*.
9. Which term is a measurement of the decay time of radioactive isotopes?
10. What is a transitional form?

UNDERSTANDING CONCEPTS

1. What is the central idea of the theory of evolution? (9·1)
2. What are fossils? Give examples of different kinds of fossils. (9·2)
3. How are fossils evidence that evolution has occurred? (9·2)
4. What are homologous structures? Using examples, explain how homologous structures provide evidence that evolution has occurred. (9·3)
5. What are vestigial organs? Using examples, explain how vestigial organs provide evidence that evolution has occurred. (9·3)
6. What structural similarities in embryos indicate relationships between all groups of vertebrates? (9·4)
7. List some of the biochemical evidence that indicates that evolution has occurred. (9·5)
8. What were the two main ideas in Lamarck's hypothesis of evolution? (9·6)
9. Why has Lamarck's hypothesis of how evolution occurs been rejected by scientists? (9·6)
10. Summarize the main points of Darwin's theory of evolution by natural selection. (9·7).
11. How do Larmarck's ideas on evolution differ from Darwin's ideas? (9·6, 9·7)
12. How did observations of populations of peppered moths provide direct evidence of evolution? (9·8)
13. What is an adaptation? Distinguish between the three types of adaptations. (9·9)
14. Compare the ideas of gradualism and punctuated equilibrium. (9·10)
15. Describe the conditions under which scientists think speciation occurs. (9·11)
16. What is the Hardy-Weinberg Law? (9·12)
17. Explain how population genetics has led to the modern theory of evolution. (9·12)

APPLYING CONCEPTS

1. Use your knowledge of genetics to explain why acquired characteristics are not passed on to offspring.
2. Use of insecticides often leads to the development of insecticide-resistant populations. Explain what is occurring in terms of natural selection and evolution.
3. What type of fossil evidence would support the hypothesis of punctuated equilibrium?
4. Do you think that evolution is occurring today? Give reasons to support your answer.
5. Most major industrial countries are trying to reduce environmental pollution. Describe what might happen to the peppered moth population if this effort is successful in England.

EXTENDING CONCEPTS

1. The fossil record indicates that organisms called therapsids are an evolutionary link between reptiles and mammals. Prepare a report on therapsids. Present your report to the class.
2. The various breeds of dogs all belong to the same species. Prepare a "species tree" to show how the various breeds are thought to be related.
3. Prepare an account of Darwin's voyage on the *H.M.S. Beagle*. Include a map of the journey.
4. Choose one species of animal or plant and observe the organism and read about it. List as many of its adaptations as you can, identifying each as a morphological, physiological, or behavioral adaptation. Use your knowledge of natural selection to explain how each adaptation might have evolved.
5. Ask an animal breeder or a plant breeder to describe how a desirable mutation can be preserved through artificial selection.

READINGS

Ayala, F.J. *Origin of Species*. Carolina Biology Reader. Carolina Biological Supply Company, Burlington, North Carolina, 1983.

Dixon, Dougal. *After Man: A Zoology of the Future*. St. Martin's Press, 1981.

Gould, Stephen Jay. *The Panda's Thumb*. W.W. Norton, 1980.

Herbert, Sandra. "Darwin as a Geologist." *Scientific American,* May 1986, p. 116.

Miller, Jonathan, and Borin Van Loon. *Darwin for Beginners.* Pantheon, 1982.

Moore, Ruth. *Evolution*. Time-Life Books, 1980.

10. History of Life

Imagine what an exciting discovery it was for the scientist who found this fossil of a dinosaur hatching from its egg. Fossils like this one reveal a great deal about life in the earth's past. In this chapter you will explore some theories that explain how life began on earth and developed over time.

CHAPTER OBJECTIVES

After completing this chapter, you will be able to

- **Describe** how conditions on the earth contributed to the development of life forms.
- **Discuss** the origin of eukaryotes.
- **Explain** how life on earth has changed over time.
- **Identify** major patterns of evolution.
- **Discuss** human evolution.

Origin of Life

10·1 THE EARLY EARTH

Astronomers believe that about 5 billion years ago the solar system began to form from a huge, very hot cloud of gas and dust. It is thought that the largest part of this cloud slowly condensed to form the sun. The nine planets, including the earth, formed from smaller portions of gas and dust revolving around the newly formed sun. The earth took form as its matter slowly cooled over many millions of years. Heavier materials collected to form a dense core. Lighter materials toward the outside formed the crust and a primitive atmosphere. Volcanic eruptions poured out lava from the hot inner region, adding to the material of the crust. Steam that was given out from these volcanoes condensed and fell back to the surface as rain. As the earth continued to cool, more water condensed to form oceans.

The earth's early atmosphere was not made up of the same mixture of gases as is the air we breathe. Today's atmosphere is mostly nitrogen (N_2) and oxygen (O_2). Scientists agree that the early atmosphere was a different mixture, but they disagree about what that mixture was. Figure 10·1 shows two possible mixtures for the early atmosphere. Most scientists think that the early atmosphere consisted mainly of ammonia (NH_3) and methane (CH_4). However, some think that it consisted mainly of nitrogen and carbon dioxide. There was probably also some hydrogen (H_2) and water vapor (H_2O).

The gases of the early atmosphere probably contained the elements that are found in living things: carbon, hydrogen, oxygen, and nitrogen. Large amounts of the gases of the atmosphere were dissolved in the oceans, just as they are today. Ocean water also contained dissolved minerals, such as sodium (Na), calcium (Ca), phosphorus (P), potassium (K), and magnesium (Mg), as it does today. Organisms contain all these elements too. Could these inorganic substances give rise to living things, and if so, how?

SECTION OBJECTIVES

- Explain how life on earth may have begun.
- Describe a hypothesis explaining how eukaryotic cells may have evolved.

FIGURE 10·1 As you can see in this table, the atmosphere of the early earth was very different from the atmosphere of the earth today.

TODAY'S ATMOSPHERE	POSSIBLE EARLY ATMOSPHERES	
N_2 (78%)	NH_3	N_2
O_2 (21%)	CH_4	CO_2
CO_2 (0.03%)	H_2	H_2
H_2O (0 to 4%)	H_2O	H_2O

10·2 CHEMICAL EVOLUTION

Darwin described biological evolution—the way species change as the result of biological processes. Hypotheses about the origin of life draw upon Darwin's ideas to describe chemical evolution. **Chemical evolution** is evolution based on nonbiological chemical processes that involve changing inorganic and simple organic compounds into more complex organic compounds.

Earlier in this century, A. I. Oparin, a Russian scientist, wrote about how chemical processes in the ancient seas might have given rise to life. The results of such processes, Oparin realized, would have had to bring about two important traits of living things. First, living things are made up largely of complex organic molecules. Chemical evolution would have had to produce such molecules from smaller building blocks. Second, systems of organic molecules in living things are *encapsulated* (ehn KAP suh layt ihd), or enclosed, as separate units. Such units are the cells that make up all living things. This encapsulation of materials also would have had to occur for life to arise.

According to Oparin's hypothesis many chemical reactions took place among the ingredients of the ancient seas. Over millions of years the molecules produced in these reactions probably included amino acids and the nucleotides that make up DNA or RNA. Because of the presence of these molecules, scientists sometimes call the solution that made up the early oceans the **primordial** (prī MOR dee uhl) **soup**.

primordius (original)

Suppose, however, that once in trillions of reactions, several amino acids joined to form a protein that could function as an enzyme. In turn this enzyme might have linked several other amino acids. In this case there would have been several amino acid chains and perhaps even complete proteins. Such rare accidents might also have produced short pieces of nucleic acids with the ability to replicate themselves. In each of these cases, the appearance of one organic molecule could have led to the production of many more.

The formation of complex organic molecules from smaller building blocks would have needed energy. Oparin suggested that there were several possible sources of energy. These sources were electrical energy from lightning, radiant energy from the sun, heat energy from volcanoes, and energy from the decay of radioactive substances.

In 1954 the American scientists Harold Urey and Stanley Miller tested the hypothesis that random processes could produce complex organic molecules. Their apparatus is shown in Figure 10·2. A mixture of ammonia, methane, hydrogen, and water vapor was circulated through the apparatus. These four gases were believed to make up the ancient atmosphere. An electric discharge, much like lightning, was passed through the gases from time to time. When the results were reported, biologists were amazed. After running the apparatus for a week, Urey and Miller found that several amino acids had been produced. If scientists were right about the makeup of the early atmosphere, the organic compounds found in living things could have appeared very soon after the earth was formed.

FIGURE 10·2 The atmosphere of the early earth experienced many severe electrical storms. The Urey and Miller apparatus simulated the conditions on the early earth.

Later experiments have given similar results. Some experiments used mixes of gases similar to Urey and Miller's mixture and others used solutions representing the primordial soup. Some of these studies have formed many different complex organic compounds, including ATP. American scientist Cyril Ponnamperuma and his co-workers found adenine, one of the DNA bases, among the products of their experiments. Other experimenters have formed compounds such as hydrogen cyanide (HCN) and formaldehyde (H_2CO). These two compounds can lead to the formation of other compounds found in living things. When hydrogen cyanide is dissolved in water and exposed to various forms of energy, it forms proteins. Reactions involving formaldehyde (for MAL duh hīd) can form sugars similar to those found in DNA and RNA. When phosphorus compounds are added to a mixture of bases and sugars, short pieces of nucleic acids are formed.

The experiments by these and other scientists do not prove that the events which happened in the laboratory were the same as those that led to the development of life. However, they do give evidence of some possible ways in which the complex molecules found in organisms could have formed from the materials thought to have existed on the primitive earth.

HISTORY OF LIFE   193

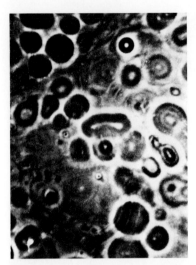

FIGURE 10·3 Liposomes, like coacervates, can form membranelike structures that may be similar to primitive cellular membranes.

micro- (small)

Oparin described how the complex compounds may have formed. He also described how the original compounds of life could have been encapsulated. He pointed out that mixtures of organic compounds can form clumps, which he called coacervates (koh AS uhr-vayts). A **coacervate** is a microscopic group of droplets formed by attractions between molecules. A coacervate can form from a mixture of protein and sugar in water. The droplets on the inside are protein molecules. The boundary around these droplets is formed of water molecules. This boundary layer acts somewhat like a cell membrane. Coacervates can exchange materials with their environment through this boundary in a manner similar to the way a cell does. To Oparin, these droplets suggested the shapes of cells. Like a single cell, each droplet is separate and distinct.

Experiments done since Oparin's time have shown that many different mixtures can form coacervates and similar kinds of droplets called *microspheres*. Microspheres called liposomes are shown in Figure 10·3. These studies also have shown that these droplets grow by absorbing more of the surrounding material. The droplets may even form buds that grow and break off and become separate droplets.

The studies of Oparin's hypothesis have shown that the kinds of molecules found in living things could have formed in the earth's early history. They have also shown that groups of these molecules could have become encapsulated. Such encapsulated groups of molecules—containing water, proteins, sugars, and nucleic acids—could have grown by taking in materials from their surroundings. The molecules may have been able to duplicate themselves by using the materials that were taken in. Finally, droplets that broke off could have produced exact copies of entire encapsulated groups of molecules.

Could such units have evolved and become the first cells? Many biologists think so. However, they point out that this hypothesis describes only what *could* have happened.

10·3 THE FIRST ORGANISMS

The rocklike formation shown in Figure 10·4 is a stromatolite (stroh-MAT uh līt). A **stromatolite** is a mound of limestone formed by the activities of one-celled organisms. Scientists have found a few modern stromatolites. However, some stromatolites are more than 800 million years old. Fossils of small organisms, such as those found in stromatolites, are called **microfossils.**

Microfossils have been found in rock formations in different parts of the world. One such formation, in South Africa, is called the Fig Tree series. Many of the organisms in the Fig Tree rocks were in the process of dividing when they died and were preserved.

Modern organisms that are similar in appearance to these ancient organisms are the prokaryotes, such as bacteria. As you know, prokaryotic cells are the simplest types of cells. The earliest cells on earth must have been simple prokaryotic cells. If a prokaryote were described simply as a bag of water and chemicals, would that description seem to apply to a coacervate? Some biologists think that coacervatelike structures evolved, by a process involving many steps, into the first prokaryotic cells.

The fossil evidence shows that prokaryotes had appeared by about 3.5 billion years ago. It is hard to pinpoint their first appearance, or to find out the nature of the first kind of organisms. However, some prokaryotes seem more likely than others to have appeared first. Most prokaryotes and eukaryotes are aerobic. An aerobic organism is an organism that needs oxygen. Clearly, if the early atmosphere did not include oxygen, it is unlikely that the earliest organisms were aerobic. However, a few prokaryotes are anaerobic—that is, they do not need oxygen. Many scientists think the first organisms to appear were anaerobic.

FIGURE 10·4 These stromatolites are still being built in Shark Bay near Australia. They are similar to fossil stromatolites built more than 800 million years ago.

Examples of present-day anaerobic prokaryotes are the *methanogens* (muh THAN uh juhnz). A methanogen is an anaerobic prokaryote that changes hydrogen, carbon dioxide, or certain organic compounds into methane. Methanogens live in the mud beneath bodies of water, where there is little oxygen. Because methanogens are anaerobic, they could have evolved in an atmosphere without oxygen. They could be similar to the first kinds of living things to appear.

Prokaryotes that could carry on photosynthesis probably evolved next. As you have learned, the process of photosynthesis uses carbon dioxide. If the early atmosphere included carbon dioxide, photosynthetic organisms could have thrived. Biologists think that many early microfossils were made by photosynthetic organisms.

The evolution of photosynthetic organisms was probably a required step before a wider variety of organisms could appear. As you have learned, oxygen is a by-product of photosynthesis. If the early atmosphere lacked oxygen, photosynthetic organisms may have added oxygen to the atmosphere.

196 HISTORY OF LIFE

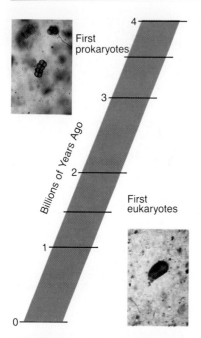

FIGURE 10·5 A time line comparing the appearance of the first prokaryotes with the appearance of the first eukaryotes. A cyanobacteria fossil (top) and an early protist fossil (bottom).

10·4 ORIGIN OF EUKARYOTES

Figure 10·5 shows a time line comparing when the first prokaryotic and eukaryotic organisms appeared. There is evidence that the first prokaryotes appeared about 3.5 billion years ago. You can see that eukaryotes, or cells with organelles, developed much later. Due to the lack of fossil evidence, it is hard to find out exactly when the first eukaryotic cells appeared. However, many biologists believe that eukaryotes evolved from prokaryotes between 1 and 2 billion years ago. Thus the time line shows that for the first 2 billion years after the appearance of life, only prokaryotes were on the earth.

How did eukaryotes evolve? Most biologists think eukaryotes evolved from prokaryotes. Several hypotheses have been proposed to explain how this happened. One explanation is called the symbiotic (sihm bī AHT ihk) hypothesis. The **symbiotic hypothesis** states that eukaryotic cells evolved from prokaryotic cells when certain prokaryotes began to live inside other cells. *Symbiosis* is an association among organisms of different kinds, often with mutual benefit.

Many biologists believe eukaryotes arose from a symbiosis in which some cells simply absorbed others. Or perhaps smaller cells moved into larger ones. If both kinds of cells were helped, they could have continued to live in that way. The smaller cells would have continued to grow and divide within their larger hosts. The larger cells also would have continued to grow. When the larger cells divided, each of the daughter cells would have received some "guest" cells. This symbiotic hypothesis of evolution is shown in Figure 10·6.

In support of the symbiotic hypothesis, biologists have pointed out similarities between some organelles in eukaryotic cells and some prokaryotic cells. For example, Figure 10·7 shows that a mitochondrion is similar in some ways to certain bacterial cells. Chloroplasts are similar to blue-green bacteria. Prokaryotic organisms are about the same size and have about the same shape and internal structure as the mitochondria and chloroplasts found in eukaryotic cells. It has also been shown that mitochondria and chloroplasts have their own DNA

FIGURE 10·6 Organelles like mitochondria and chloroplasts may have been free-living organisms that were taken in by ancestral eukaryotic cells.

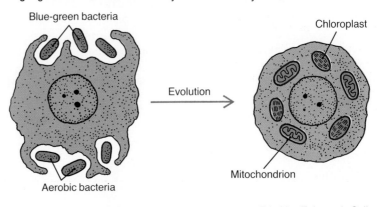

and ribosomes, similar to the DNA and ribosomes in bacteria, and can carry on reproduction and protein synthesis.

What advantages would symbiosis have for prokaryotic cells? As you know, mitochondria provide a cell's energy. Perhaps they have descended from bacteria that were able to produce so much energy that they had an extra amount of energy. A host cell could have used this energy. In turn, perhaps bacteria were protected in their new environment from extreme heat or cold, from water loss, or from being eaten by other cells. The ancestors of chloroplasts might have gained similar advantages while providing food for their host through the process of photosynthesis.

Supporters of the symbiotic hypothesis point to examples of symbiosis among simple organisms in the modern world as further support of the theory. For example, amoebas (one-celled eukaryotes) infected with bacteria have been studied. Some of these amoebas survive with the bacteria continuing to live inside them. In some cases a bacterium and an amoeba come to depend on each other. Neither can live alone, just as eukaryotes and their organelles cannot live without each other.

Not all biologists accept the symbiotic theory. A major objection to the theory is the fact that the DNA in the chloroplasts and mitochondria cannot make all the proteins needed by these structures. DNA from the nucleus is needed to make some of their proteins. If bacteria evolved and formed mitochondria and chloroplasts, some process would have had to happen to transfer hereditary traits to what was to become a nucleus. Some biologists have offered another explanation for the development of eukaryotic cells. They believe that prokaryotes developed directly into eukaryotes. Organelles in the first eukaryotes were formed by the infolding and pinching off of regions of the cell membrane.

Evidence can be found to support both the symbiotic hypothesis and the hypothesis that organelles of eukaryotes are the result of infolding of the cell membrane. Perhaps both processes played a role in the evolution of eukaryotic cells. Some organelles may have come about by one process, and others may have developed by another process.

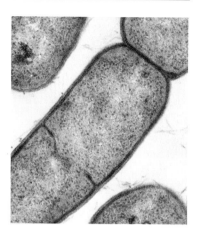
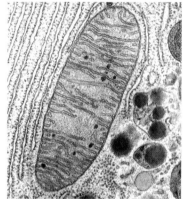

FIGURE 10·7 Mitochondria (bottom) are similar in size, shape, and structure to bacterial cells (top).

REVIEW QUESTIONS

1. Briefly describe Urey and Miller's experiment that simulated conditions on the early earth.
2. What is Oparin's hypothesis of the encapsulation of the first compounds of life?
3. Why is it likely that the first prokaryotes were anaerobic?
4. According to the symbiotic hypothesis, how did eukaryotic cells originate?
5. Using Figure 10·1, identify atmospheric sources of the elements in carbohydrates, lipids, and proteins.

Patterns of Evolution

SECTION OBJECTIVES

- Give examples of periods of mass extinction and rapid evolution in the history of living organisms.
- Compare divergent and convergent evolution.

10·5 DIFFERING RATES OF EVOLUTION

Scientists estimate that from the time life first appeared on earth, there have been from about 1.5 billion species to more than 15 billion. Most of these species are extinct. In fact, it has been estimated that 99 percent of the species that ever lived no longer exist today.

Evidence suggests that species do not appear or become extinct at a steady rate. Instead, there have been times in the earth's history when many new kinds of organisms have appeared suddenly. There have been other times when many species or whole groups of related species have disappeared quickly.

A geologic time chart is shown on page 199. A **geologic time chart** is a chart that summarizes the history of life. It is based on the fossil record in the earth's crust. Notice that the chart shows the oldest divisions of time at the bottom and the most recent divisions at the top. Since new rocks usually are deposited over existing ones, the lowest rocks and fossils in a deposit are the oldest. The divisions of time within the geologic time chart are based on fossils found in rocks of different ages. Notice that the largest intervals of time are called **eras,** and these are subdivided into **periods** and **epochs** (EHP uhks). As you study the chart, you can also see some of the kinds of organisms that flourished at different times in the earth's history.

The fossil record shows evidence of many variations in the rate of evolutionary change. In the geologic time chart you can see that two periods of mass extinction have taken place in the history of the earth. **Mass extinction** is the sudden disappearance of many species. Mass extinctions are sometimes followed by periods of adaptive radiation. **Adaptive radiation** is the rapid evolution of many new species, usually within a few million years. This term reflects the idea that the new species radiate, or branch outward, from a small number of species. Note the periods of adaptive radiation in the time chart.

10·6 ADAPTIVE RADIATION

Fossil evidence shows that a great period of adaptive radiation began at the start of the Cambrian (KAM bree uhn) Period, about 600 million years ago. There is fossil evidence that some kinds of multicellular animals had evolved before the Cambrian Period. These animals were early kinds of worms, jellyfish, and other soft-bodied invertebrates (animals without backbones). They left very few fossils.

Cambrian rocks contain many marine fossils. They record the existence of a great variety of animal life. The first invertebrates with shells appeared during the Cambrian Period. Trilobites (TRI luh bīts), distant relatives of insects, also appeared. A trilobite fossil is shown in Figure 10·8. In fact, representatives of all the major kinds of invertebrates lived during the Cambrian Period. There is fossil evidence that

FIGURE 10·8 Trilobites were among the first invertebrates with shells. They are distant relatives of insects.

ERA	PERIOD	EPOCH	BEGINNING (millions of years ago)	IMPORTANT EVENTS
Cenozoic	Quaternary	Recent	Less than .01	Modern humans appear
		Pleistocene	2.5	Ice Age
	Tertiary	Pliocene	6	Mammals dominant
		Miocene	25	
		Oligocene	38	
		Eocene	55	
		Paleocene	65	**Adaptive Radiation** (mammals)
Mesozoic	Cretaceous		140	**Mass Extinction** (reptiles and others); **Adaptive Radiation** (flowering plants)
	Jurassic		195	First birds appear; dinosaurs dominant
	Triassic		230	First dinosaurs and first mammals appear; cone-bearing plants abundant
Paleozoic	Permian		290	**Mass Extinction** (marine invertebrates); Mammallike reptiles dominant
	Carboniferous		345	First insects and first reptiles appear; club mosses and horsetails abundant
	Devonian		400	Fish abundant; first amphibians appear
	Silurian		440	Land plants appear and first jawed fish appear
	Ordovician		500	Shelled organisms abundant; jawless fish appear
	Cambrian		600	**Adaptive Radiation** (marine invertebrates)
Pre-Cambrian			1.5 billion	First eukaryotes
			3.5 billion	First prokaryotes

HISTORY OF LIFE

primitive fish, the first vertebrates (animals with backbones), appeared near the end of the Cambrian Period. Some biologists call the great increase in the variety of life forms during this period the Cambrian Explosion.

What caused such quick evolution? Some biologists believe that there may have been a change in the kinds or amounts of chemicals, such as calcium, in the ocean. Calcium is a major ingredient of shells and bones. However, no one has a definite answer to this question.

Another period of adaptive radiation involved plants. It took place about 120 million years ago, at the beginning of the Cretaceous (krih-TAY shuhs) Period. At that time, ancestors of all the modern kinds of flowering plants evolved. Since their development in the Cretaceous, flowering plants have made up most of the species of land plants.

This great increase in species took place in a period of perhaps 10 million years. How did it happen in so short a time? In many flowering plants, reproduction depends on birds and insects to carry pollen (male gametes) from one plant to another. Fossil evidence shows that the evolution of birds and insects took place before the adaptive radiation of flowering plants. Perhaps they had increased in number to a point that helped the success of the flowering plants.

A third period of adaptive radiation took place about 65 million years ago, at the beginning of the Cenozoic (see nuh ZOH ihk) Era. During a period of about 10 million years, mammals spread and became the dominant animals on earth. The first mammals had appeared at least 100 million years earlier. These early mammals were small and were probably insect eaters. The early mammals lived at the time of the dinosaurs. The dinosaurs died out suddenly about 65 million years ago. It was then that the number of species of mammals began to grow quickly.

Some scientists think that the adaptive radiation of the mammals is related to the extinction of the dinosaurs. During the time that dinosaurs lived, mammals were small creatures. Most of the large land animals were dinosaurs. It seems that the mammals were unable to compete successfully with the dinosaurs and did not evolve into large forms. With the extinction of the dinosaurs, mammals evolved and filled the roles formerly held by the dinosaurs. Thus the death of one group of animals cleared the way for the rise of another.

10·7 MASS EXTINCTIONS

Evidence shows that many species of organisms have evolved, thrived, and died out during the earth's history. Few species of organisms die out during any period of time. However, during mass extinctions thousands of species have disappeared within a short time. A few species that died out during mass extinctions are shown in Figures 10·9 and 10·10.

Mass extinctions have long been mysteries to biologists. What could cause death on such a large scale? Many hypotheses have been suggested to explain these events. Perhaps oceans retreated, becom-

FIGURE 10·9 The organisms that became extinct at the end of the Permian Period included eurypterids and trilobites.

FIGURE 10·10 The organisms that became extinct at the end of the Cretaceous Period included dinosaurs, pterosaurs, and mosasaurs.

ing shallower and reducing the space and food available for marine species. Perhaps oceans advanced, reducing the land area. Perhaps mountains rose, changing land environments. There may not have been enough organisms with an adaptive advantage to survive under these new conditions. Perhaps climates changed. Perhaps new predators appeared that were such successful hunters that they wiped out their own food supply. Any of these hypotheses may be true. Each mass extinction may have had a different cause or several causes operating at once.

In 1979, two American scientists made a discovery that may help to explain the dinosaurs' disappearance. Luís and Walter Alvarez studied sedimentary rock that had formed at the end of the Cretaceous Period. They found a layer containing large amounts of the element iridium (ih RIHD ee uhm). Iridium (Ir) is very rare on the earth, but it is often found in meteorites. Moreover, the Alvarezes found the same evidence in rocks of the same age in several places around the world. They concluded that a gigantic meteorite, perhaps 10 km in diameter, may have hit the earth. This meteorite may have produced the large amounts of iridium found in these sediments.

HISTORY OF LIFE   201

FIGURE 10·11 According to the meteorite impact hypothesis, a meteorite caused the mass extinction in the late Cretaceous Period.

The Alvarezes believe the meteorite would have hit the earth late in the Cretaceous Period. This event would have caused the mass extinction that killed the dinosaurs and other organisms. They hypothesized that upon impact this meteorite would have exploded, throwing huge amounts of dust into the air. It would have taken months or years for the dust to settle. In the meantime, the wind would have carried it around the globe. The dust would have blocked out the sunlight, leaving the earth in continuous twilight or darkness.

One result would have been the death of much of the earth's plant life, which depends on sunlight for photosynthesis. Another result would have been the death of many plant-eating animals when their food supply was destroyed. A further result, the death of the meat-eating animals would have followed the death of their food supply, the plant-eating animals. The only animals to live would have been the small species, such as mammals. Mammals could have fed on food items such as seeds, which can survive for an extended time. The hypothesis that the mass extinction in the late Cretaceous Period was caused by a meteorite is called the **meteorite impact hypothesis.**

Not all scientists accept the meteorite impact hypothesis as the cause for the extinction of dinosaurs. Some scientists say that fossil evidence in the western United States suggests that dinosaurs were extinct before the supposed meteorite hit the earth. Those scientists who disagree with the hypothesis of meteorite impact often believe that changes in climate brought about the mass extinction of the late Cretaceous Period.

10·8 DIVERGENCE AND CONVERGENCE

The evolution of species has followed some general patterns. The most common pattern is called divergence (duh VER juhns), or divergent evolution. **Divergence** is an increase in the differences among descendants of a single ancestral species as time passes. (As you may have guessed, adaptive radiation is simply divergence on a large scale.)

Remember that horses of today are thought to have evolved from an ancestral species named *Hyracotherium*. All members of the horse family probably evolved from this species. Horses and their living relatives are shown in Figure 10·12. They are thought to be the result of divergence from a common ancestor within the past 5 million years. About 10 million years ago, *Pliohippus* (plī uh HIHP uhs), an ancestor of the horse, donkey, and zebra, roamed the plains of North America. As populations of these horselike animals spread out, they became adapted in different ways to new environments. As populations became separated from each other, they evolved into separate species. While they are alike in many ways, you can see in Figure 10·12 that there are definite differences in their appearance. Donkeys are smaller than horses. Donkeys also have larger ears. Zebras differ from these other two members of the horse family because of their distinctive black and white stripes.

HISTORY OF LIFE **203**

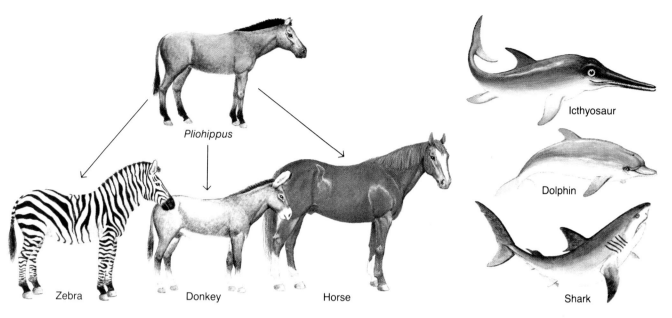

FIGURE 10·12 Horses, donkeys, and zebras are all thought to be the result of divergence from a common ancestor, *Pliohippus*. Convergence as illustrated by a shark, a dolphin, and an icthyosaur.

A second pattern of evolution is called convergence (kuhn VER-juhns), or convergent evolution. **Convergence** is an increase in similarities among species derived from different ancestors as a result of similar adaptations to similar environments. For example, Figure 10·12 shows the result of convergence in an ichthyosaur (IHK thee uh sor) (an extinct sea-dwelling reptile), a shark (a fish), and a dolphin (a mammal). What similarities do you see in these three animals? Notice that all three are adapted for swimming in the ocean. Each has a long, streamlined body, which allows easy movement through the water. Note, too, that their paddle-shaped limbs are adapted for an existence in water.

REVIEW QUESTIONS

6. Explain the difference between mass extinction and adaptive radiation.
7. Describe the adaptive radiation of the Cambrian Period.
8. List some of the possible causes for mass extinction at various times in the earth's history.
9. Distinguish between divergence and convergence.
10. After the extinction of the dinosaurs, birds underwent a period of adaptive radiation that included the evolution of large, flightless, carnivorous birds. However, such large flightless forms were largely unsuccessful. Propose a hypothesis to explain why large flightless birds did not successfully fill the "roles" left by the dinosaurs.

Investigation

What Are the Major Characteristics of Some Fossilized Animals?

Goals
After completing this activity, you will be able to
- Describe the major features of a fossil gastropod.
- Identify fossil specimens of gastropods by using a field guide to fossils

Materials (for groups of 4)
assorted specimens of fossil gastropods
hand lens   metric ruler
field guide to fossils (for class)
field guide to shells (for class)

Procedure
A. Study the drawing below, which shows the major features of the shell of a fossil gastropod. This drawing represents a typical gastropod. Fossil shells show much variation, but they do have certain features in common. Note the two main parts of the shell: the body whorl **(1)** and the spire **(2)**. The spire may have several separate whorls, or complete coils. Look for the suture **(3)**, the line along which the whorls meet. The place where the shell turns inward toward the suture is called the shoulder **(4)**.

B. Locate the aperture **(5)**, which is the opening of the shell. The flaring portion of the aperture is known as the outer lip **(6)**. Often the other side of the opening curls back to form an inner lip **(7)**. The columella **(8)** is the central column of the shell. At the uppermost point of the shell is the apex **(9)**. At the lower end of the shell is a channel called the siphonal canal **(10)**.

C. Shells have various decorative features that aid in identification. The shell may have spiral lines **(11)** and vertical ridges known as varices (sing., *varix*) **(12)**. Note the decorative features in the drawing of the gastropod.

D. Compare each of your fossil specimens with the drawing of the typical fossil gastropod. Locate each of the features described in steps A–C.

E. With your metric ruler, measure the length of the shell from the apex to the siphonal canal. Count the number of whorls in the spire. Note any decorative features, such as spines, bumps, varices, and spiral lines. On a separate sheet of paper, prepare a chart like the one shown below. Record your findings.

F. Use the pictures and descriptions in a field guide to fossils to identify your specimens. Determine the following information: (1) scientific name of organism, (2) geologic period in which it lived, (3) habitat, (4) where it was most abundant.

G. Look through a field guide to shells. Find out whether any of your fossil specimens look like the shells of modern gastropods. Make drawings of the present-day gastropods that are most like your specimens.

H. Follow your teacher's instructions for the proper cleanup of all materials.

Questions and Conclusions
1. Do any of your fossil specimens resemble the drawing on this page? If they differ, tell specifically how they are different.
2. Were any of your fossil specimens incomplete? What part of the shell could you see?
3. List the names of the fossils you could identify through the field guide.
4. Why is a fossil-hunter more likely to find a fossilized snail than a fossilized worm?
5. Why are water-dwelling animals more likely to be fossilized than land-dwelling animals?

FEATURES OF FOSSIL GASTROPODS		
	Specimen #1	Specimen #2
Length		
Number of whorls		
Decorative features		
Sketch		

204   HISTORY OF LIFE

Human Ancestry

10·9 PALEOANTHROPOLOGY

When did humans first appear on the earth? Fossil evidence shows that **Homo sapiens** (HOH moh SAY pee uhnz), the species to which we belong, existed about 50 000 years ago in the same form as it does today. It also shows that a few other hominid (HAHM uh nihd) species lived in earlier times. **Hominids** are humanlike or human species. These hominid species were similar to modern humans in some ways. However, the fossil record does not show exactly when and how *Homo sapiens* evolved.

The study of human and prehuman fossils is called **paleoanthropology** (pay lee oh an thruh PAHL uh jee). This word is made up of *paleontology* (pay lee ahn TAHL uh jee), the study of past life, and *anthropology*, the study of humans and human societies. In studying the fossil record of our origins, paleoanthropologists often focus on three important human traits.

First, humans are *bipedal* (BĪ puh duhl) organisms. That is, humans walk upright on two legs. Few other mammals are bipedal. Therefore, the question of when this trait evolved is important.

Second, humans are intelligent. Intelligence is hard to measure. Only indirect evidence can give clues to the intelligence of past species. One such kind of indirect evidence is the use of tools. A second clue to intelligence is brain size, which can be measured when fossil skulls are found. However, care must be exercised in using brain size as a guide to intelligence.

Third, the teeth of humans differ from those of apes, monkeys, and other species thought to be related to humans. The differences include smaller front teeth in humans and curving, rather than straight, rows of teeth at the back of the jaws, as shown in Figure 10·13. Therefore, the shapes, sizes, and arrangements of fossil teeth all are clues to the kind of organism to which they belonged.

10·10 HOMINIDS

It is hard for paleoanthropologists to piece together the early history of humans. Most fossils of hominids consist of a few teeth and fragments of bones. Complete fossil skulls or bones of ancestors of humans are rare. Still, newly found fossil remains of early humans are giving a clearer picture of the story of human evolution.

The first hominids probably appeared in what is now eastern Africa. The oldest fossils that biologists generally accept as true hominids have been found in Tanzania and Ethiopia. These fossils record the existence of several species of the genus *Australopithecus* (aws truhloh PIHTH uh kuhs). A *genus* (pl., *genera*) is a group of related species. Organisms in the genus *Australopithecus* are called **australopithecines** (aws truh loh PIHTH uh seenz).

SECTION OBJECTIVE
- Describe some early hominids.

paleo- (old)

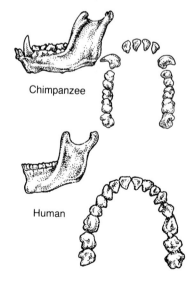

FIGURE 10·13 The structure of a human jaw differs from the jaw of a chimpanzee. The chimpanzee's canine teeth are much larger than a human's.

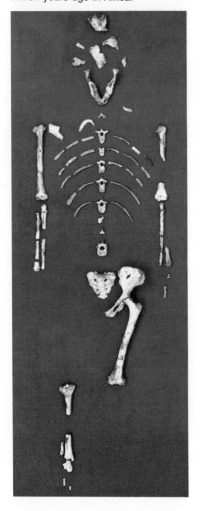

FIGURE 10·14 Lucy is the most complete skeleton of *Australopithecus afarensis* ever found. She lived about four million years ago in Africa.

The oldest of these species, **Australopithecus afarensis** (af ahr-EHN sihs), or *A. afarensis,* lived nearly 4 million years ago. The most complete *A. afarensis* skeleton is nicknamed Lucy. Lucy was a female who died at about age 20. Her teeth and jaws are more apelike than human, and her brain was small. Yet one important characteristic separates her from earlier primates: She was bipedal.

You may wonder how a few bones can provide this kind of information. First, both the pelvis, or hip bones, of a human and Lucy's pelvis are bowl-shaped. Both are adapted to support the internal organs. A four-legged primate, such as a baboon, has a differently shaped pelvis. Second, in all vertebrates there is a hole, called the *foramen magnum* (fuh RAY muhn MAG nuhm), at the base of the skull. The spinal cord, which is connected to the brain, passes through this hole. A bipedal organism carries its skull on top of the spine. Therefore, its foramen magnum must be located on the bottom of the skull, as it is in both modern humans and in Lucy. A four-legged walker carries its skull in front of its body and therefore has its foramen magnum at the back of the skull.

Scientists have pieced together fossil remains of three other australopithecines. **Australopithecus africanus** (af rih KAN uhs), about the same size as Lucy, probably lived from more than 3 million years ago until about 1.5 million years ago. **Australopithecus robustus** (roh BUHS tuhs), a larger species than *A. africanus,* lived from about 2.5 million years ago until a little over 1 million years ago. **Australopithecus boisei** (BOI zee ī), known from just a few incomplete fossils, lived about 2 million years ago.

Some of the australopithecines lived at the same time. For this reason, not all scientists agree that they were four separate species. For example, some think that *A. afarensis* and *A. africanus* were the same species. Others place *A. robustus* in a separate genus. Whether or not *A. robustus* was an australopithecine, it was probably a side branch of hominid evolution. It is possible that either or both *A. afarensis* and *A. africanus* are ancestors of humans.

The next hominid genus was *Homo,* the genus to which modern humans belong. Early members of this genus had larger brains than those of australopithecines, and their teeth are more like those of modern humans. Figure 10·15 shows a fossil skull of the species **Homo habilis** (HAB uh luhs). There is evidence that *H. habilis* used tools. This hominid lived nearly 2 million years ago. Therefore, it appeared while *A. africanus* still lived. Perhaps members of this genus descended from the australopithecines, or perhaps they evolved separately from some other ancestor.

Scientists have given the name **Homo erectus** (ih REHK tuhs) to hominid fossils ranging in age from 1.5 million to 0.5 million years. This species is the earliest hominid found outside Africa. Fossils have been found in Java and China. Its brain size varied from a volume of 800 cm^3 up to 1300 cm^3. The high end of this range is nearly as large as the average brain size of modern humans—about 1400 cm^3.

HISTORY OF LIFE **207**

Along with fossil remains of *H. erectus* are the remains of campfires. The controlled use of fire is an indication of intelligence. In addition, *H. erectus* made stone axes and other tools.

There is no evidence to show what finally happened to *H. erectus*, whose last known fossils are about 500 000 years old. At some time between then and 100 000 years ago, a new species appeared. This species, **Homo sapiens neanderthalensis** (nee an der thawl EHN-sihs), or the Neanderthals, spread throughout what is now the Middle East, North Africa, Central Asia, and Europe.

Neanderthals are classified as a variety of *Homo sapiens*, and they had many features in common with modern human beings. They stood erect, and they used simple stone tools. In the climates in which they lived, they almost certainly wore clothing, probably made of animal hides. Neanderthals were also different in several respects from present humans. They had slightly larger brains than ours, and they had thick skulls with low foreheads, as seen in Figure 10·15. These early humans gathered plant foods and hunted. They lived in family groups in caves. There is evidence that they buried their dead.

The Neanderthals became extinct about 35 000 years ago, shortly after the first appearance of modern human beings. The two species may have existed side by side for several thousand years. However, for some reason the Neanderthals died out.

As the Neanderthals disappeared, they were replaced by a group known as the **Cro-Magnons** (kroh MAG nahnz). This group of *Homo sapiens* cannot be physically distinguished from modern humans. The Cro-Magnons were hunter-gatherers and they made a variety of stone tools. In the caves where they lived, they produced artwork showing the animals they hunted.

Collected fossil evidence suggests that humans developed from primitive primate ancestors. However, the exact path from these ancestors to modern humans is not known. Several models have been presented, but scientists do not agree on their accuracy. Two different current schemes are shown in Figure 10·16.

FIGURE 10·15 Hominid skulls: *Homo habilis* (A), Neanderthal (B), and Cro-Magnon (C).

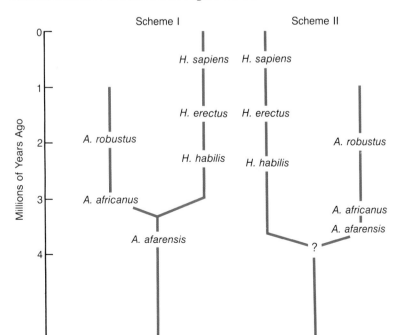

FIGURE 10·16 Pathways outlining the development of humans as proposed by paleoanthropologists Donald Johanson (Scheme I) and Richard Leakey (Scheme II).

Insights

The Case of the Phony Fossil

From the mid-1800s to the early 1900s, Europeans were actively searching for early ancestors. The Germans dug up the fossilized remnants of Neanderthal man and Heidelberg man. The French discovered not only ancient bones but also cave paintings done by early humans.

England, Charles Darwin's home, had no evidence of ancient ancestors. English scientists—both professionals and amateurs—began searching for fossils. Scarcely a cave was left unexplored, scarcely a stone was left unturned. Many scientists asked workers in gravel pits to watch for fossils.

In 1912, Charles Dawson, a part-time collector of fossils for the British Museum, wrote to Dr. Arthur Smith Woodward, keeper of the Natural History Department at the British Museum. Dawson claimed that a human skull he had found in a gravel pit in Piltdown Common, Sussex, "would rival Heidelberg man." Soon Woodward was digging in the gravel pit with Dawson and other eager volunteers. They found a separate jaw that, though apelike, included a canine tooth and two molars, worn down as if by human-type chewing. Flints and nonhuman fossils found at the same dig indicated that the finds were very old.

Despite arguments by some scientists that the jaw came from a chimpanzee or an orangutan, the discoverers reconstructed the skull and connected the jaw to it. They named the fossil *Eoanthropus dawsoni,* Dawson's "Dawn man," and said that it was much older than Neanderthal man. The find came to be known as Piltdown man.

The finds were X-rayed. One dental authority was suspicious of the canine; he said it was too young a tooth to show such wear. However, such contrary evidence was ignored in the general surge of enthusiasm. So the Piltdown man was acclaimed as an important find, a human in which the brain had evolved more quickly than the jaw.

Beginning in the 1940s, the bones were subjected to modern tests. It is now believed that the skull was from a modern human and the jaw was from a modern ape, probably an orangutan. The animal fossils and flints were found to be very old, but not the types that have been found in England. Apparently they had been placed in the gravel pit to make the finds more convincing. Why were scientists and others fooled so easily? Perhaps the desire to find an "ancestor" may have interfered with careful scientific observation.

10•11 EVIDENCE OF PRIMATE RELATIONSHIPS

How many of these earlier hominids were directly related to *Homo sapiens?* No one knows for sure, because the fossil record is too scarce. However, geneticists have recently developed a method that may show how closely various existing primates are related.

This method depends on the ability to find out the exact order of the amino acids that make up a protein. Because cells build proteins under the control of DNA, the amino acid sequence is a clue to the order of the bases in the DNA.

Some biologists have suggested that as different species evolve from a common ancestor, their DNA becomes different at a fairly steady rate. That is, the differences increase as time passes. Thus changes in DNA serve as a genetic clock. If this suggestion is true, then measuring the difference between similar proteins in two related species should provide a clue to the length of time since they diverged from a common ancestor. For example, horses, donkeys, and zebras probably evolved from an ancestral species only a few million years ago. As a result, there should be far less difference between the proteins of these animals than there is between the proteins of any one of them and some other mammal, such as a camel.

For all living things whose proteins have been studied in detail and whose evolutionary history is known as well, this genetic clock has proved to be fairly correct. Therefore, biologists applied this same method to some of the primates.

Biologists discovered that human proteins are more similar to those of gorillas and chimpanzees than to those of any other existing primates. Therefore, humans shared a common ancestor with gorillas and chimpanzees more recently than they shared a common ancestor with, for example, orangutans. Biologists can calculate how long ago the primate groups diverged from one another by using the protein differences of other organisms as a measuring standard. If the rate of the genetic clock has been correctly determined hominids and apes diverged from a common ancestor about 5 million years ago. However, some scientists think that the divergence of homonids and apes may have taken place further back in time.

FIGURE 10·17 Protein studies have shown that gorillas (center) and chimpanzees (right) are closely related to humans. The same studies have shown that the orangutan (left) is a more distant relative.

REVIEW QUESTIONS

11. List the characteristics considered in studying the fossil record of human evolution.
12. In what way was Lucy different from earlier primates?
13. Compare the schemes of hominid evolution shown in Figure 10·16.
14. How is a genetic clock used to provide clues as to when two species may have diverged from a common ancestor?
15. Fossil finds in the Middle East indicate that populations that had characteristics of both Neanderthals and modern humans lived there about 40 000 years ago. What hypothesis can you suggest for these populations having traits of both Neanderthals and modern humans?

Chapter Review

SUMMARY

10·1 The early earth's atmosphere and oceans probably contained the elements that are found in living things.

10·2 Random chemical reactions among materials found on the primitive earth may have produced the molecules of which organisms are composed.

10·3 The first organisms were probably prokaryotes.

10·4 Eukaryotes appear to have evolved from prokaryotes between 1 and 2 billion years ago. They may have evolved as the result of symbiosis among prokaryotes or as a result of direct evolution of prokaryotes.

10·5 The fossil record shows evidence of many variations in the rate of evolutionary change.

10·6 Many new species have evolved rapidly during several periods of adaptive radiation in the earth's history.

10·7 At various times in the earth's history, mass extinctions resulted in the disappearance of many species.

10·8 Evolution has followed different patterns. Divergence is an increase in the differences among species descended from a single ancestral species. Convergence is an increase in similarities among species derived from different ancestors.

10·9 Paleoanthropology is a field of science that studies human and prehuman fossils.

10·10 Fossil evidence indicates that hominids evolved from primitive primate ancestors.

10·11 Protein similarities and differences are being used to determine relationships among similar groups of organisms, including primates.

BIOLOGICAL TERMS

adaptive radiation
australopithecines
Australopithecus afarensis
Australopithecus africanus
Australopithecus boisei
Australopithecus robustus
chemical evolution
coacervate
convergence
Cro-Magnons
divergence
epochs
eras
geologic time chart
hominids
Homo erectus
Homo habilis
Homo sapiens
Homo sapiens neanderthalensis
mass extinction
meteorite impact hypothesis
microfossils
Neanderthals
paleoanthropology
periods
primordial soup
stromatolite
symbiotic hypothesis

USING BIOLOGICAL TERMS

1. What are australopithecines?
2. Distinguish between epochs, eras, and periods.
3. How are the meteorite impact hypothesis and mass extinction related?
4. How are chemical evolution and primordial soup related?
5. What term refers to the rapid development of many species?
6. Define the term *coacervate*.
7. To what genus and species do modern humans belong?
8. Distinguish between Cro-Magnons and Neanderthals.
9. Define the term *stromatolite*.
10. What is paleoanthropology?

UNDERSTANDING CONCEPTS

1. Describe the probable conditions on the primitive earth. (10·1)
2. How might the accidental formation of enzymes have led to the formation of many more organic compounds? (10·2)

3. How might organic materials in a primordial soup become encapsulated? (10·2)

4. Describe a present-day prokaryote that may be similar to the earliest organisms on the earth. (10·3)

5. Why do most biologists think that organisms capable of carrying on photosynthesis must have existed before most other kinds of organisms could have evolved? (10·3)

6. Compare the symbiotic hypothesis of the origin of eukaryotes with the hypothesis of membrane infolding. (10·4)

7. What kind of information is contained in a geologic time chart? (10·5)

8. Give two examples of adaptive radiation, including the times when they occurred. (10·6)

9. How might the disappearance of the dinosaurs have been related to the adaptive radiation of the mammals? (10·6)

10. How could a collision of a large meteorite with Earth cause a mass extinction? (10·7)

11. Give evidence for and evidence against the meteorite impact hypothesis. (10·7)

12. Compare convergence and divergence. (10·8)

13. Briefly explain the nature of the work done by paleoanthropologists. (10·9)

14. Briefly, name and describe hominid fossils that may be ancestral to modern humans. Include a description of the possible sequence of these hominids in time. (10·10)

15. What does the term *genetic clock* mean? How can the genetic clock provide clues to relationships among organisms? (10·11)

APPLYING CONCEPTS

1. Some scientists feel strongly that life may have evolved in other parts of the universe as well as on the earth. Why do they think so?

2. Which of the evolutionary events described in this chapter might support the theory of punctuated equilibrium?

3. You have learned about two features of hominid skeletons that accompany bipedal walking. Describe one other feature of the skeleton that might accompany bipedalism.

4. Many primates use tools. For example, chimpanzees often use sticks to probe ant-hills when searching for food. They also use leaves as sponges to collect drinking water. How is the use of tools by humans different from that of apes?

5. Scientists have discovered a few cases in which photosynthetic prokaryotes live within eukaryotic cells. How do such findings affect the symbiotic hypothesis?

EXTENDING CONCEPTS

1. A few serious scientists have suggested that living organisms might have been "planted" here, arriving from space either accidentally or purposefully. Write a report about one of these hypotheses and discuss it with your class.

2. Do some reading about Peking man, an earlier name for members of *Homo erectus*. Write a report about the unsolved mystery connected with the Peking man fossils.

3. Do a report about the lives and work of Louis and Mary Leakey, paleoanthropologists who have discovered and named several of the fossils discussed in this chapter.

4. Linus Pauling received a Nobel Prize for his work on enzymes. Prepare a report on how his research on enzymes supports the idea of a genetic clock.

READINGS

Colbert, Edwin H. *Dinosaurs, An Illustrated History.* Hammond, Inc., 1983.

Johanson, Donald C., and Maitland A. Edey. *Lucy: The Beginnings of Human Kind.* Warner Books, 1982.

Leakey, Richard E., *The Making of Mankind.* E.P. Dutton, 1981.

McMenamin, Mark A.S. "The Emergence of Animals." *Scientific American,* April 1987, p. 94.

Pilbeam, D. "The Descent of Hominoids and Hominids." *Scientific American,* March 1984, p. 84.

Rensberger, Boyce. "Bones of Our Ancestors." *Science 84,* April 1984, p. 29.

Sattler, Helen R. *Dinosaurs of North America.* Lothrop, Lee & Shepard Books, 1981.

11. Classification

The jewel sea anemone is an organism that remains permanently attached to one spot on the ocean floor. Is a sea anemone a plant or an animal? Years ago, people believed that sea anemones were plants because they do not move around the way most animals do. In fact, the sea anemone is named after a garden flower. However, modern scientists classify sea anemones as animals. In this chapter, you will find out why it is necessary to classify living things. You will also learn about the system that is used in classification.

CHAPTER OBJECTIVES

After completing this chapter, you will be able to

- **Explain** how scientists classify and name organisms.
- **Give examples** of species names using binomial nomenclature.
- **Identify** the classification groups.
- **Apply** a modern classification system to organisms.
- **Describe** the type of evidence used to determine relationships between organisms.
- **Classify** organisms using a taxonomic key.

History of Classification

11·1 WHY A CLASSIFICATION SYSTEM IS NEEDED

Imagine that you are entering a library. You are looking for a particular book. However, there are problems. All the books on the shelves have blank covers, and there is no organization to the placement of the books. You look around for a card catalog, but there is none. As a last resort you approach the reference desk for help, only to find that the librarian speaks a language that you do not understand. This strange situation is similar in some ways to the problem that early biologists faced when they tried to study the living world.

Many forms of life inhabit the earth. Over a million species of animals and 325 000 species of plants have already been discovered. The list of organisms that have been discovered grows longer each year. Biologists suggest that there may be several million different species of organisms sharing this planet.

One task of a scientist is to find order where disorder appears to exist. To bring order to this vast array of life forms, biologists have developed systems for grouping, or classifying, the organisms. **Taxonomy** (tak SAHN uh mee) is the science of classification. It is a science that involves more than identifying and naming organisms. The science of taxonomy is concerned with finding order in diversity. A **taxonomist** (tak SAHN uh mihst), or scientist who specializes in taxonomy, tries to understand relationships among organisms as well as to identify and name organisms. A good classification system allows biologists to know a great deal about an organism simply by knowing the characteristics of the group to which it belongs.

The need to organize is not unique to the scientist. How many things in your daily life do you group into specific categories? Think about the things in your home or in your school. You group your clothes when you decide to hang some in your closet and to place others in your dresser drawers. You further classify them by deciding

SECTION OBJECTIVES

- Describe the function of a classification system.
- Identify the seven major categories in the modern biological classification system.
- Define binomial nomenclature and give some examples of species names.

taxis (arrangement)

213

which items will go into certain drawers. Think about the grouping system you use. You may separate your clothes according to color, or you may separate them according to type of garment—shirts in one drawer, sweaters in a second, and so on. Whatever system you use, grouping saves time and effort when you have to find an item. For the biologist a classification system provides a convenient way of keeping track of all the known life forms.

Biologists classify organisms to provide a basis for accurately naming organisms. By having an accurate naming system, scientists are better able to communicate with one another. Common names for organisms can be misleading. Did you know a seahorse is a fish, a sea cucumber is an animal, and ringworm is a fungus? Think about this list of organisms: jellyfish, cuttlefish, and silverfish. From these names it would appear that the organisms have much in common. You might think that they all live in the water and have fins. Now look at the pictures of these organisms in Figure 11·1. You can see that not one of them is actually a fish. Scientists avoid the use of common names. A good classification system provides a basis for a universally accepted naming system that can be used by scientists of all nations. A standard naming system can help eliminate the confusion that can occur when common names are used.

FIGURE 11·1 A cuttlefish, jellyfish, and silverfish. None of these organisms is actually a fish.

11·2 EARLY CLASSIFICATION SYSTEMS

Since primitive times, people have attempted to classify living things. People developed very simple methods of naming organisms when they began to recognize similarities and differences among organisms in the natural world. The first real effort to develop a classification system began with the ancient Greeks. About 350 B.C. the Greek philosopher Aristotle divided all organisms into two groups, which he called the plant kingdom and the animal kingdom. He then introduced the term *species* to mean "similar life forms." Today the term **species** means "a group of closely related organisms of a particular kind that can naturally interbreed and produce fertile offspring."

Aristotle devised a system in which he divided all animals into three groups based on their natural habitats. The three groups were those animals that lived on land, those that lived in the sea, and those that lived in the air. The chief problem with this system was that it grouped many animals that were structurally very different and that it separated many that were similar.

Theophrastus, a Greek botanist, was one of Aristotle's students. He developed a system for classifying plants that was based on their growth habits. Plants were divided into three groups: herbs (no woody stems), shrubs (many woody stems), and trees (one main woody stem). Theophrastus's system introduced the idea of classification based on structural similarities.

The classification systems put forth by Aristotle and Theophrastus remained in use for almost 2000 years. In the sixteenth and seventeenth centuries, scientists again turned their attention to classification. What caused this new interest in classification? European explorers brought many new, unidentified plant and animal samples back to Europe from other lands. Biologists of the time realized that a better system would help to keep track of all these finds. Sixteenth-century biologists made lists of these plants and animals and organized them according to their structural characteristics and their medicinal value. Some plants having medicinal value are shown in Figure 11·2.

In the seventeenth century the English botanist John Ray developed an improved classification system. He devised a method for classifying seed plants according to the structure of the seed. This method is still used today. John Ray lived 200 years before Darwin and Mendel, yet he was the first scientist to realize that a species is a group of organisms that are capable of interbreeding. He also recognized that the variations in a species are a natural result of this interbreeding.

John Ray was one of the first scientists to understand the need for giving scientific names to organisms. Ray devised a naming system by which each organism was given a Latin name. This Latin name consisted of a long scientific description of that organism. According to Ray's system, the scientific name for the carnation was *dianthus floribus solitariis, squamis calycinis subovatis brevissimis, corollis crenatis.* What was the disadvantage of this naming system?

11·3 LINNAEAN SYSTEM

The system of classification used today had its beginnings in the eighteenth century with the work of Carolus Linnaeus (lih NEE uhs). Linnaeus assigned each organism to a large category—either the plant kingdom or the animal kingdom. Then he successively subdivided each category into progressively smaller categories. The smaller the grouping, the more similar were the members of that group. In the time of Linnaeus, three categories were recognized: species, genus, and kingdom. According to the Linnaean system, a genus was a group of similar species.

In 1753, Linnaeus published a classification system for plants; in 1758, he published a classification system for animals. The species was (and is) the basic unit of the Linnaean classification system. Members of a species are more closely related to each other than they are to members of other groups. In addition, members of a certain species will breed only with other members of the same species. The Linnaean system was based on similarities of body structure. A modified form of this system is used today. For his work in classification, Linnaeus has been called a founder of modern taxonomy.

11·4 BINOMIAL NOMENCLATURE

The development of a naming system for all living things was another contribution Linnaeus made to the science of taxonomy. Linnaeus knew the problems caused by the use of common names. To solve this problem he developed a naming system that is still used by

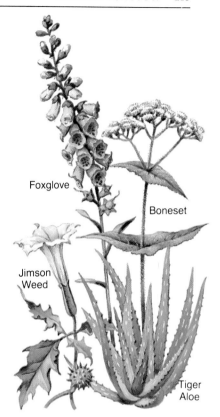

FIGURE 11·2 Biologists searched the New World for plants that had medicinal value. Some plants brought back to Europe are shown here.

216 CLASSIFICATION

bi- (two)
nomen (name)

scientists. His naming system, **binomial nomenclature** (bī NOH mee-uhl NOH muhn klay chuhr), is a system by which each species is given a two-word Latin name. Some examples of names of organisms based on binomial nomenclature are *Rana pipiens* (leopard frog), *Ursus americanus* (black bear), and *Quercus alba* (white oak). The rules for binomial nomenclature are given below.

- The first word of the name tells the genus to which the organism belongs. The first letter of the genus name is always capitalized.
- The second word of the name is a specific, and often descriptive word that indicates the particular species.
- Latin is the language used.
- When the name is handwritten or typed, it is underlined. When the name is printed, it is italicized.
- A species name can be abbreviated by use of the first letter of the genus name as well as the species name, such as *R. pipiens* and *U. americanus*.
- If a subspecies or a variety of a species is identified, a third word is added to the name.

Compare the Latin names of the animals in Figure 11·3. Two animals belong to the genus name *Felis* (FEE lihs) and two belong to the genus name *Canis* (KAY nihs). The similarity in the names is based on structural similarities between the animals. From the scientific names, which animal is most closely related to the house cat—the dog or the mountain lion?

The following are some of the advantages to the use of the Linnaean system of binomial nomenclature.

- Latin is accepted as the language of classification by scientists all over the world.

FIGURE 11·3 Both dogs and wolves belong to the genus *Canis*. House cats and cougars belong to the genus *Felis*.

Canis lupus

Canis familiaris

Felis concolor

Felis domesticus

- Since Latin is not a commonly used language, the language is stable and not subject to change.
- The system shows relationships of species within a particular genus. For example, from their scientific names you can tell that the dog *(Canis familiaris)* is more closely related to the wolf *(Canis lupus)* than to the house cat *(Felis domesticus)*.
- The second word of the Latin name is an adjective. Often this term helps to describe the species. For example, *Acer rubrum* is the name for the red maple. *Rubrum* means "red." *Acer saccharium* is the sugar maple. *Saccharium* refers to sugar. Not all adjectives in the species name are descriptions of the organisms. Some adjectives recognize the scientist who first discovered or developed the organism. For example, the Latin name for the glory lily is *Gloriosa rothschildiana*, named after the developer of the lily, Rothschild.

REVIEW QUESTIONS

1. What is the importance of classification to scientists?
2. Describe the contributions of Aristotle, Theophrastus, Ray, and Linnaeus to the science of taxonomy.
3. What are the strengths of the Linnaean system of binomial nomenclature?
4. What are the two parts of a scientific name?
5. Consider the following list of items found in a desk drawer: pencils, paper clips, rubber bands, thumbtacks, pens, paper, and cellophane tape. Describe a method for classifying these objects into groups.

Classifying Organisms

11•5 BASIS OF MODERN CLASSIFICATION

Linnaeus based his system of classification on similarities in body structure. For example, Linnaeus considered a bat to be a mammal because it had many structural similarities to other mammals. Just because it had wings did not qualify it as a bird. Other evidence shows that even though bats have wings, they are mammals, not birds.

Linnaeus's work on taxonomy was done more than a century before the work of Darwin or Mendel. After Darwin presented his theory of evolution, the similarities and differences in organisms were viewed by scientists as products of evolution. Today, the major emphasis in taxonomy is the study of evolutionary relationships.

The classification of a species is based on the evolutionary history of the species. For example, at one time rabbits and squirrels were both classified as rodents. Studies of the earliest fossils of rabbits show

SECTION OBJECTIVES

- Relate the theory of evolution to modern classification systems.
- List the classification groups in order.
- Describe the main characteristics of organisms in the five kingdoms.

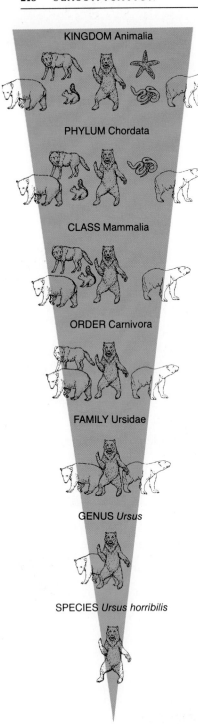

FIGURE 11·4 As you move down the classification groups for the grizzly bear, each category becomes smaller until you focus on an individual species.

that they evolved from an ancestor separate from that of rodents. For this reason rabbits are now classified in a group separate from rodents.

Many of the ideas that support the theory of evolution give a useful basis for classifying an organism as a certain species. Today, taxonomists who classify organisms study them in several ways.

1. The overall structure of the organism is studied to try to find *homologous structures.* Remember that homologous structures are structures that have a basic similarity and have developed in organisms that may have a common ancestor. For example, a bird's wing and a turtle's leg are homologous structures.
2. The life cycle of the species is studied to find *embryological likenesses* to other groups of organisms.
3. The *fossil record,* if available, is studied to show relationships among organisms over time.
4. The degree of *biochemical likeness* between species is determined. The amino acid order of similar proteins differs from one species to another. Similarities in the order of amino acids of certain proteins can help determine how to classify organisms.
5. *Genetic likenesses* between the chromosomes of different species are studied. Information on such likenesses can give very important clues for classification. Relationships between different species can be found by studying and comparing segments of DNA. The more alike the order of nucleotide bases, the more closely related are the species.

11·6 CLASSIFICATION GROUPS

The classification system devised by Linnaeus and still used today is a *hierarchical* (hī uh RAHR kuh kuhl) *system.* It consists of a series of smaller groups that are organized into larger groups. You are probably familiar with other hierarchical systems. For example, if you wanted to organize all the people in the United States according to where they live, you could use a hierarchical system. You could begin by dividing the people into the states in which they live. You could then divide each state into counties, each county into cities, each city into streets, and each street into specific buildings. If a certain building were an apartment building, you could further subdivide it into single apartments. Using this system, you could place every person into a hierarchy of categories as follows: state/county/city/street/building/apartment.

The hierarchical system for the classification of organisms now includes seven major categories and several subcategories. The seven major groups are shown in Figure 11·4. Each category in this hierarchy contains one or more groups from the next lower level. In other words, a **kingdom** is a group of closely related phyla (sing., *phylum*); a **phylum** (sometimes called a *division* in the naming of plants) is a group of closely related classes; a **class** is a group of closely related orders; an **order** is a group of closely related families; a **family** is a group of closely related genera (sing., *genus*); and a **genus** is a group of closely related species. A **species** is a group of organisms of a particular kind that can naturally interbreed and produce fertile young. As you study the hierarchy and look at the organisms within each group, the similarities at each lower level become clearer.

Figure 11·5 traces the classification of three of the organisms shown in *A*. Two kingdoms—**Plantae** (PLAN tee) and **Animalia** (an uh-MAYL yuh)—are represented in *B* and *C*. All the organisms in *C* belong to kingdom Animalia and are more similar to each other than they are to the organism in *B*.

Now look at *D* and *E*. Why has the housefly been separated from the other animals? All the animals in *E* are more closely related to each other than they are to the animal in *D*. All the animals in *E* belong to the same phylum. What phylum is that?

The animals in *E* can now be grouped as shown in *F* and *G*. Why was the robin separated? What similarities are shared by the three animals in *G*? They all happen to be fur-covered animals that feed their young milk. Thus they all belong to the same class—class Mammalia. The dog, wolf, and cat share another characteristic. They are all meat-eaters. Thus they all also belong to the same order—order Carnivora. The animals in *H* can be grouped as shown in *I* and *J*—into families. The dog and wolf are in the same family. By following the chart through to the end, you will find that the dog, wolf, and cat each belong to separate species.

There is one difficulty with this classification system. There are no definite rules for determining whether two species are related enough to be included in the same genus. The classification of the different species in the cat family illustrates this problem.

FIGURE 11·5 Several organisms are classified in this illustration. You can see the relationships between some of the organisms.

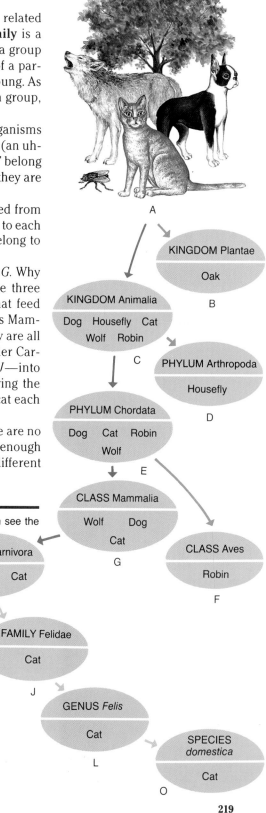

219

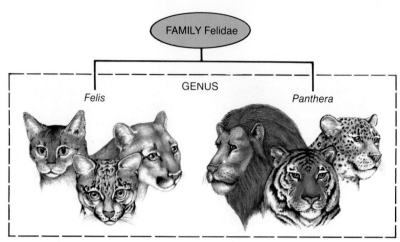

FIGURE 11·6 The classification of cats into two genera. The small cats are placed in the genus *Felis,* while the larger cats are in the genus *Panthera.*

Some biologists lump most cats into one genus—*Felis.* Other biologists split them into separate genera. For example, some scientists classify the small cats—including the house cat, the cougar, and the ocelot—as *Felis.* They classify the larger cats—including the lion, tiger, and leopard—as *Panthera.* Thus, you can see that a classification system, while very useful, is not perfect.

11·7 MODERN CLASSIFICATION SYSTEM

In early classification systems all organisms were divided into two major groups. Organisms that were green and lacked the ability to move were classified as plants; those that were capable of locomotion and that fed on other living things were considered to be animals. At the time of Linnaeus, each of the known organisms could be placed in one of these two kingdoms.

With the development of the microscope, microorganisms could be observed. When biologists tried to classify these organisms, they found that many did not fit neatly into either kingdom. The microorganism *Euglena* best illustrates this problem. Look at Figure 11·7. *Euglena* is a unicellular organism. It contains chloroplasts and carries on photosynthesis. Both of these features are characteristics of plants. However, *Euglena* has animal characteristics also. It moves, lacks a cell wall, and can change shape. In laboratory experiments *Euglena* can lose its chloroplasts under certain conditions. Then it can absorb organic food sources if they are supplied. Some biologists wanted to classify *Euglena* as a plant, and others insisted it was an animal.

FIGURE 11·7 *Euglena* is grouped in the Kingdom Protista. Like an animal, *Euglena* moves and lacks a cell wall. However, like a plant, it carries on photosynthesis within its chloroplasts.

Eventually, an expanded classification scheme was suggested. A new category—Protista (pruh TIHS tuh)—was formed. Kingdom **Protista** would contain all those organisms that did not fit into the plant and animal kingdoms. However, the addition of kingdom Protista also presented a major problem. Classification is based on groupings of similar organisms. The similarity that all protists (members of Protista) share is that they are neither plants nor animals. In time this kingdom became the catchall for organisms that taxonomists could not place in either kingdom Plantae or kingdom Animalia.

As microscopes improved and cell study continued, it became clear that there are two very different kinds of cells. As you know, prokaryotic cells are those that lack a nucleus. Eukaryotic cells are those that have a nucleus. Most organisms have cells that are eukaryotic. The bacteria and the blue-green bacteria are the only living things that are prokaryotes. Most biologists believe that this difference is very important. To provide for the difference in prokaryotes and eukaryotes, taxonomists made another kingdom, called Monera (muh NIHR uh). Kingdom **Monera** contains only those organisms that have prokaryotic cells.

Although many taxonomists accepted this classification system, others thought there was still a problem. Molds and mushrooms are fungi (FUHN jī). Fungi are organisms that depend on other organisms for their food. They reproduce by spores. The cells of fungi are eukaryotic, but the organisms are neither plants nor animals. Most fungi are multicellular organisms and are very different from protists. Many biologists think that fungi should be placed in a fifth category, called kingdom **Fungi.**

The major characteristics of the five kingdoms are shown in Figure 11·8. In order to place an organism into one of the five kingdoms, a biologist asks the following questions.

1. Is a cell of the organism prokaryotic or eukaryotic?
2. Is the organism multicellular or unicellular?
3. Is the organism an autotroph, which makes its own food, or is it a heterotroph, which feeds on other organisms?
4. If the organism is a heterotroph, is it an ingestive heterotroph or an absorptive heterotroph? An **ingestive heterotroph** is a type of heterotroph that takes in food and then digests it. Animals are examples of ingestive heterotrophs. An **absorptive heterotroph** is a heterotroph that digests food before it absorbs it into its cells from the environment. Fungi are examples of organisms that are absorptive heterotrophs.

hetero- (other)
trophus (feeder)

FIGURE 11·8 The major characteristics of the five kingdoms are shown here. You should be able to place an organism in the correct kingdom by comparing it to this table.

	MONERA	PROTISTA	FUNGI	PLANTAE	ANIMALIA
Cell Type	Prokaryotic	Eukaryotic	Eukaryotic	Eukaryotic	Eukaryotic
Multicellular or Unicellular Body Plan	Unicellular	Unicellular	Multicellular	Multicellular	Multicellular
Mode of Nutrition	Some autotrophic; some heterotrophic	Some autotrophic; some heterotrophic	Absorptive heterotrophic	Autotrophic	Ingestive heterotrophic
Examples	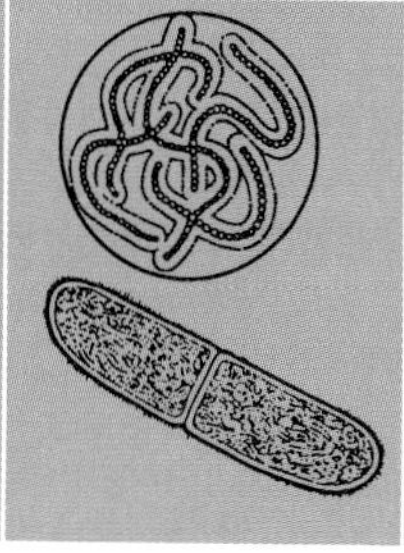	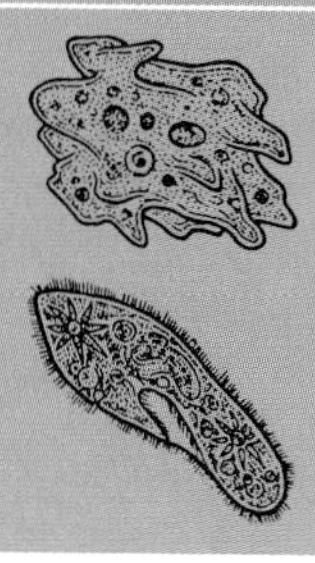	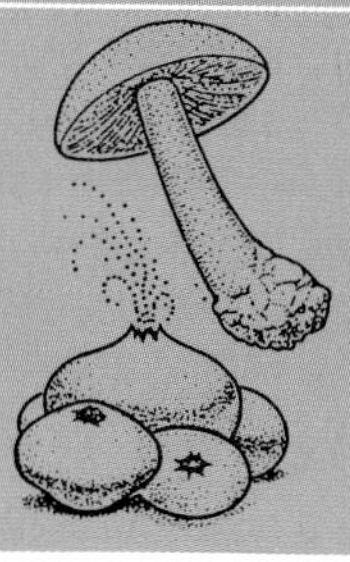		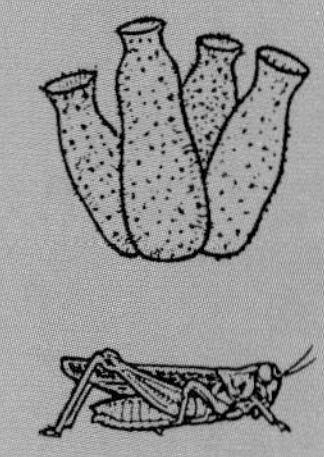

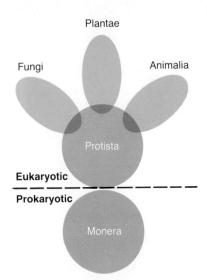

FIGURE 11·9 This diagram shows how the characteristics of organisms in the five kingdoms overlap. Some scientists feel that the overlap represents evolutionary relationships of the five kingdoms.

11·8 MODERN CLASSIFICATION SYSTEM PROBLEMS

In the chapters to come, you will learn about the organisms that are classified within these five kingdoms: Monera, Protista, Fungi, Plantae, and Animalia. You have seen that there has been debate over the number of kingdoms in the classification system. Although many biologists now accept the five-kingdom system, there is still much controversy about what should be included in each kingdom.

In the five-kingdom system of classification, kingdom Monera includes only prokaryotic organisms. Kingdoms Protista, Fungi, Plantae, and Animalia include only eukaryotic organisms. This system places unicellular eukaryotes in kingdom Protista and multicellular eukaryotes in kingdoms Fungi, Plantae, and Animalia. However, the system is not perfect. There are some unicellular species that are closely related to multicellular groups. The green algae, classified as plants, are an example of organisms that do not fit neatly into any one kingdom. Some green algae are multicellular and some are unicellular. Still, the green algae have been placed in kingdom Plantae. There are similar problems with multicellular organisms in kingdoms Animalia and Fungi that are closely related to unicellular organisms. Figure 11·9 represents the views of some scientists on the evolutionary relationships of the five kingdoms.

Scientists studying monerans have recently suggested that the five-kingdom classification system may need to be replaced. Scientists have found a few genera that differ a great deal in their cell chemistry from the other monerans. The **methanogens** are very primitive organisms that are now placed in kingdom Monera. They seem to have a biochemical makeup that is very different from the other members of the kingdom. It has been proposed that these organisms be placed in a new kingdom called Archaebacteria, meaning "primitive bacteria."

All classification systems are made by humans. No human invention is completely perfect. The value of a given system is judged by the help it gives people in learning about the life forms on the earth. Distinctions between the classification categories are not clear-cut. As more genetic, evolutionary, and biochemical data are gathered and analyzed, taxonomists may change the positions of some organisms within the classification system.

REVIEW QUESTIONS

6. List the types of evidence used by taxonomists to determine the relationships between organisms.
7. List the groups in the hierarchical system of classification from largest to smallest.
8. List the five kingdoms in the modern classification system. Give two characteristics of each kingdom.
9. Briefly describe one problem with the five-kingdom system.
10. Birds and bats belong to the same phylum but different classes. What do they have in common that places them in the same phylum? What differences might cause them to be placed in separate classes?

Identifying Living Things

11·9 AIDS FOR IDENTIFYING ORGANISMS

The main value of a classification system is that it provides an organization for the study of living things. Over a million organisms have been classified and named. This vast catalog of scientific names may seem very confusing to the person who observes plants and animals in the field and then wants to identify them by name. The task of identifying plants and animals need not be overwhelming. There are many aids available, such as field guides and taxonomic keys.

A field guide is a book that contains information about a group of things found in nature. There are specific field guides for the identification of organisms such as birds, insects, mammals, reptiles, trees, ferns, and wild flowers. (There are also field guides for identifying nonliving things, such as rocks and minerals.)

How is a field guide used? Suppose you want to identify a certain red bird. You have a field guide that contains many pictures and silhouettes of birds, both perched and in flight.

Figure 11·10 shows four red birds as they might be pictured in a field guide. The drawings are labeled with arrows that indicate identifying features. Since you have a pair of field glasses, you can check on such features as color of wings, shape of head, and type of bill. You can also check a map that tells you where the bird would be likely to live at different times of the year.

The bird you are observing does not have a pointed crest on its head. Therefore you rule out the cardinal. You also note that the wings are not as black as those of the scarlet tanager. Thus you also rule out the scarlet tanager. You think that the bird is either a red crossbill or a purple finch. You check the map in your field guide to determine the bird's range. You are currently visiting friends in Louisiana. According to the map, the crossbill is unlikely to be seen in that area. Thus you identify the bird as a purple finch.

SECTION OBJECTIVES
- Differentiate between a field guide and a taxonomic key.
- Explain how to use field guides and taxonomic keys.

FIGURE 11·10 A northern cardinal, red crossbill, scarlet tanager, and purple finch as each might be seen in a field guide.

Many animals make sounds that you can recognize when you are trying to identify them. Birds, whales, dolphins, toads, frogs, and insects are some of the animals that produce distinctive sounds. Recordings of animal sounds along with descriptions of the animals are available. Some of these recordings are keyed to field guides of the animals. Learning to recognize animals by the sounds they make is helpful for identification.

Another identification tool is the taxonomic key. A **taxonomic key** is a guide to the identification of plants and animals based on certain distinguishing characteristics. These characteristics are arranged in a systematic way so that organisms can be identified.

11·10 USING A TAXONOMIC KEY

Most taxonomic keys are based on pairs of opposing statements. Only one statement in each pair of statements can be true for a given organism. A choice must be made as to which one describes the organism. The statement you choose directs you to another pair of statements in which another choice must be made. Each succeeding pair of statements is more specific in describing the organism. You continue making choices until the organism is identified.

Look at the drawings of the leaves of pine trees in Figure 11·11. You may not know the names of all the different types of pines shown. A taxonomic key in branching form is also shown. Suppose you want to identify the type of pine tree that has leaves like those in drawing C. After examining the leaves, decide between the first two statements.

1. Needles in fives. White Pine.
1. Needles in twos or threes. Go to 2.

You decide that the pine needles are in clusters of twos, so you follow the path to choices number *2*.

2. Needles in twos. Go to 3.
2. Needles in threes. Go to 5.

Since you know that the needles are in clusters of twos, you follow the path to number *3*.

3. Needles 13 cm long. Red Pine.
3. Needles less than 8 cm long. Go to 4.

Since you see that the needles are less than 8 cm long, you follow the path to number *4*.

4. Needles sharp-pointed; mostly blue-green. Scotch Pine.
4. Needles sharp or dull; yellow green. Table Mountain Pine.

At this point you can identify the tree as a Scotch pine.

Generally, taxonomic keys call your attention to small details of the organism you want to identify. The key may refer to the internal structures of the organism or to its seasonal color changes. It may refer to parts of the organism that may not be present when the organism is studied. For these reasons it is helpful if you are familiar with the organisms you are trying to identify.

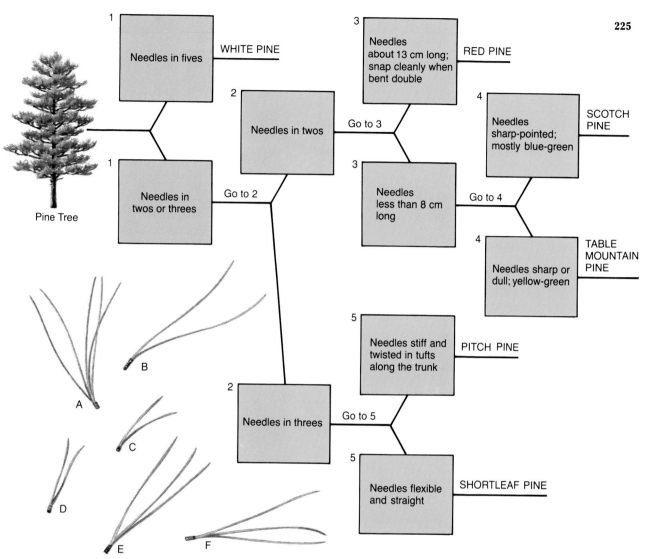

FIGURE 11·11 This illustration represents a taxonomic key for pine trees. You can identify an unknown specimen by comparing it to descriptions in the key.

REVIEW QUESTIONS

11. Under what conditions would you choose to use a field guide instead of a taxonomic key?
12. Under what conditions would you choose to use a taxonomic key instead of a field guide?
13. When would recordings be helpful for identifying organisms? Give some examples.
14. Give a brief outline of the steps to be followed in using a taxonomic key.
15. Field guides generally contain pictures of organisms along with descriptions of their shapes, sizes, and colors. What other types of information do you think would be useful for identifying organisms in the field?

Investigation

How Is a Taxonomic Key Used?

Goals
After completing this activity, you will be able to
- Identify the names of several insects by using a taxonomic key.
- Design a simple taxonomic key of your own.

Materials (For 1)
taxonomic key on this page

Procedure

A. Study the drawing below of the structural features of insects in the order Lepidoptera. Be sure you can identify the antenna, forewing, hindwing, eyespot, and wing margin.

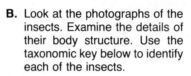

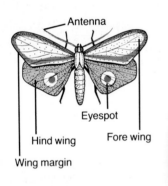

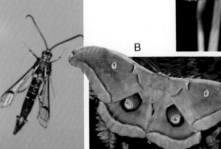

B. Look at the photographs of the insects. Examine the details of their body structure. Use the taxonomic key below to identify each of the insects.

1. **a.** Antennae long and slender, always knobbed at tips (butterflies) . . . . . . Go to statement 2
   b. Antennae featherlike or threadlike, not knobbed at tips (moths) . . . Go to statement 3

2. **a.** Fore wing smooth-edged; hind wing scalloped, without taillike extension . . . . Go to statement 4
   b. Fore wing smooth-edges; hind wing scalloped with taillike extensions . . . . Go to statement 5

3. **a.** Antennae featherlike Go to statement 6
   b. Antennae threadlike Go to statement 7

4. **a.** Wings orange and black . . . . Monarch butterfly
   b. Wings brown; eyespots in fore wing and hind wing . . . . . Wood nymph butterfly

5. **a.** Fore wing black with green margins; hind wing green with white spots . . . . Pipevine swallowtail butterfly
   b. Fore wing and hind wing yellow and black . . . . Palamedes swallowtail butterfly

6. **a.** Wings broad, light brown, with eyespots . . Polyphemus moth
   b. Wings broad, without eyespots Notch-winged geometer

7. **a.** Body heavy, wide-winged . . . . Go to statement 8
   b. Body heavy, narrow-winged . . . . . Peach tree borer moth

8. **a.** Fore wing and hind wing of similar colors . . . . . Gypsy moth
   b. Fore wing and hind wing of different colors . . . . Ilia underwing moth

Questions and Conclusions

1. What are the names of insects A, B, C, and D?
2. Describe how a taxonomic key is used.
3. What structural features of the insects did you use to identify them?
4. Choose three or four closely related living things. Design a taxonomic key to allow another person to identify these things.

Careers in Biology

Genetic Counselor

Genetic counselors offer guidance to people who have a history of genetic disorder in their families. The genetic counselor analyzes family history and determines the risk that a harmful trait will be inherited. With this information, genetic counselors help people decide whether to have children.

To become a genetic counselor, a person must have a bachelor's degree in biology and a master's degree in genetic counseling. In addition, all genetic counselors must be certified by the American Board of Medical Genetics. For more information contact the American Society of Human Genetics, Medical College of Virginia, Box 33, Richmond, VA 23298.

Plant Breeder

Plant breeders conduct experiments to develop new varieties of plants. They use their understanding of genetics, selective breeding, and plant growth to try to improve such features of crops as yield per acre and resistance to disease. Many plant breeders conduct research and teach in the agricultural science division of a university. Others work for the federal government.

To work as a plant breeder, at least a bachelor's degree in biology or horticulture is required. Many plant breeders also have a master's degree in plant breeding or plant pathology. For more information write to the American Society for Horticultural Science, 701 North Saint Asaph Street, Alexandria, VA 22314.

Museum Technician

Museum technicians assist museum professionals in preparing, maintaining, and shipping exhibits. They may work closely with the curator and other experts to preserve specimens, interpret historical research, and help construct exhibits.

Most museums require technicians to have a bachelor's degree in a specialized area. For example, a technician who works with the exhibit designer usually has a degree in design.

Museum technicians work for museums, historical societies, and universities and colleges. For more information write to the American Association of Museums, 1055 Thomas Jefferson Street NW, Washington, DC 20007.

How to Use the Want Ads

Many people find jobs by answering help-wanted ads. Such ads are a good source of information about the kinds of jobs that are available, what salaries are being offered, and what qualifications are required.

Want ads are found in a variety of publications. Newspapers usually have a special classified section that lists job openings. In the Sunday edition of most newspapers, larger ads for better-paying technical and professional jobs are also found in other sections.

When you look through the want ads, be sure to check all the possible job categories that apply to you. Try to answer ads that give specific information about the job along with the name of the company or a person to contact. Also make sure that you follow the instructions in the ad.

Chapter Review

SUMMARY

11·1 Taxonomy is the science of classification. Classification provides an organization that aids people in understanding the relationships among organisms.

11·2 Since early times, people have attempted to classify living things. Their systems used such criteria as habitats, methods of growth, and structural characteristics.

11·3 Carolus Linnaeus developed a classification system based on structural similarities. He is best known for the development of a universally accepted classification system based on binomial nomenclature.

11·4 According to the system of binomial nomenclature, every species is given a name consisting of two Latin words. The first word is the name of the genus; the second word is a descriptive word that names the species.

11·5 Modern classification is based on evolutionary relationships.

11·6 The main groupings in the modern system of classification are kingdom, phylum, class, order, family, genus, and species.

11·7 The modern classification system groups all organisms into five kingdoms: Monera, Protista, Fungi, Plantae, and Animalia.

11·8 Classification systems are inventions of humans and thus are not perfect. Scientists do not always agree on the classification of organisms.

11·9 Field guides and taxonomic keys are aids in the identification of organisms. Field guides help in the quick identification of organisms in their natural environments.

11·10 Taxonomic keys, which are used in the identification of organisms, are based on the distinguishing characteristics of organisms.

BIOLOGICAL TERMS

absorptive heterotroph
Animalia
binomial nomenclature
class
family
field guide
Fungi
genus
ingestive heterotroph
kingdom
methanogens
Monera
order
phylum
Plantae
Protista
species
taxonomic key
taxonomist
taxonomy

USING BIOLOGICAL TERMS

1. Which term means "a group of organisms of a particular kind that interbreed and produce fertile young"?
2. Which term means "a person who specializes in naming organisms"?
3. What is a taxonomic key?
4. Distinguish between Animalia and Plantae.
5. Use a dictionary to find out the meanings of the word parts in the term binomial nomenclature. Explain why this name is appropriate.
6. Distinguish between kingdom Protista and kingdom Monera.
7. What are methanogens?
8. Distinguish between an absorptive heterotroph and an ingestive heterotroph.
9. List these classification catagories in order from broadest to narrowest: phylum, family, kingdom, species.
10. List these classification catagories in order from broadest to narrowest: genus, class, order, phylum.

UNDERSTANDING CONCEPTS

1. State the functions of a classification system. (11·1)

2. What was the basis of Aristotle's classification system? (11·2)

3. What problems did Aristotle's classification system present? (11·2)

4. Briefly describe the Linnaean classification system. What is the basic unit of this system? (11·3)

5. Compare the naming system of John Ray with that of Carolus Linnaeus. (11·2, 11·4)

6. Explain why Latin is a good language for a scientific naming system. (11·4)

7. What is meant by *binomial nomenclature?* Give two examples of organisms named according to this system. (11·4)

8. Describe the kinds of information taxonomists use in classifying organisms. (11·5)

9. Indicate the categories in the classification hierarchy that fit the following descriptions: (a) contains the most diversity of organisms and (b) contains only organisms capable of interbreeding. (11·6)

10. How did the development of the microscope cause problems for taxonomists? (11·7)

11. Summarize the kinds of questions a biologist must ask in order to place an organism into one of the five kingdoms. (11·7)

12. Give reasons why no classification system is perfect. (11·8)

13. Explain what a field guide is. Describe when and how you might use a field guide. (11·9)

14. Explain the differences between a field guide and a taxonomic key. (11·9, 11·10)

APPLYING CONCEPTS

1. You have been given pine leaves to identify. There are three needles to a bundle and the leaves are flexible and straight. Based on the information in Figure 11·11, what type of pine tree are the leaves from?

2. Devise a taxonomic key to distinguish among the following: soccer ball, tennis ball, golf ball, table-tennis ball, basketball, football, baseball, tetherball, and billiard ball.

3. Explain what is meant by the statement "Classification systems are the inventions of humans; diversity is the product of evolution."

EXTENDING CONCEPTS

1. Write a short play in which you explore the life and times of Carolus Linnaeus. Include an interview in which Linnaeus discusses the merits of his system of classification.

2. Find out the basis of the Library of Congress classification system. Compare this system with the Dewey decimal system.

3. Collect a sample of pond water. Using a taxonomic key for the protists, try to classify some of the organisms in your sample.

4. Go to a library or bookstore and find out how many field guides are available. Check out or buy the field guide of greatest interest to you and use it to identify living things in your area.

5. The following are the Latin names of several common organisms: *Musca domestica, Periplaneta americana, Haliaetus leucocephalus, Pinus ponderosa,* and *Rhus radicans.* Use an unabridged dictionary to find out the common names of these organisms. Which names do you think are most appropriate?

6. In your classroom or at home, identify all available organisms, such as plants and pets, with their scientific names. If possible, attach the names to the organisms.

7. The modern classification system represents the evolutionary relationship among groups of organisms. Use this book and reference books to identify the kingdoms from which it is thought the following kingdoms evolved: (a) Animalia, (b) Fungi, (c) Plantae, and (d) Protista.

READINGS

Margulis, L., and Karlene Schwartz. *Five Kingdoms: An Illustrated Guide to the Phyla of Life on Earth.* W. H. Freeman and Company, 1982.

"New Phylum Discovered, Named." *Science,* October 14, 1983, p. 149.

Peterson, R.T. *A Field Guide to the Birds.* Houghton Mifflin Company, 1980.

Vidal, Gonzalo. "The Oldest Eukaryotic Cells." *Scientific American,* February 1984, p. 48.

Woese, Carl R. "Archaebacteria." *Scientific American,* June 1981, p. 98.

Unit II Skills

READING SCIENTIFIC WRITING FOR MEANING

The nature of science is discovery and the formation of new ideas. In order for an idea to gain recognition and acceptance, it must be communicated. Scientists publish papers to present their hypotheses and evidence supporting those hypotheses. Below are excerpts from the writings of two of the notable scientists you have read about in Unit II. Although some concepts and terms may be unfamiliar to you, careful reading and context clues will help you understand the excerpts.

Read each excerpt. On a separate sheet of paper, answer the questions; do not write in this book.

> **A.** Following is an excerpt from Gregor Mendel's paper "Experiments on Plant Hybrids," originally published in 1866.
>
> "The experiments demonstrated throughout that in such a case the hybrids always resemble more closely that one of the two parental plants which possesses the greater number of dominating traits. If, for instance, the seed plant has a short stem, terminal white flowers, and smoothly arched pods, and the pollen plant has a long stem, violet-red lateral flowers, and constricted pods, then the hybrid reminds one of the seed plant only in pod shape, and the remaining traits resemble those of the pollen plant. Should one of the two parental types possess only dominating traits, then the hybrid is hardly or not at all distinguishable from it."

1. What two parental plants did Mendel use in his experiment?
2. What three kinds of plant characteristics are described?
3. What do you think the terms *terminal* and *lateral* mean?
4. Which trait of the parental seed plant appears in the hybrid?
5. Which traits of the parental pollen plant appear in the hybrid?
6. If one of the two parental plants possesses only dominating traits, how will the hybrid plant look?
7. What is the author's main idea?

> **B.** The following is an excerpt from *On the Origin of Species,* by Charles Darwin, which first appeared in 1859.
>
> "The illustration of the swimbladder in fishes is a good one, because it shows us clearly the highly important fact that an organ originally constructed for one purpose, namely, flotation, may be converted into one for a widely different purpose, namely, respiration. The swimbladder has, also, been worked in as an accessory to the auditory [hearing] organs of certain fishes. All physiologists admit that the swimbladder is homologous, or 'ideally similar' in position and structure with the lungs of the higher vertebrate animals: hence there is no reason to doubt that the swimbladder has actually been converted into lungs, or an organ used exclusively for respiration."

8. Which phrase most accurately describes the content of this paragraph?
   a. the functions of the swimbladder in fish
   b. respiration in fish
   c. homologous structures in fish
   d. the conversion of the swimbladder into the lungs
9. What were the two original functions of a fish's swimbladder?
10. *Homologous* is used to describe the similarity in structure and position between a fish's swimbladder and our lungs. Which of our appendages is homologous to the forelegs of a horse?
11. According to the reading, what type of scientist studies the structure and function of the organs of animals?
12. "Something of secondary importance" could be used to describe which term from this reading?
    a. accessory   c. ideally similar
    b. homologous   d. conversion
13. Darwin described the change of the swimbladder into lungs. What concept was he trying to illustrate?

UNIT III

Microbiology

Bacteria reproduce at a rapid rate and can live in a variety of different environments. They are a very abundant form of life on earth. But it was not until the microscope was invented that bacteria and other microscopic forms of life were studied. In this unit you will learn more about bacteria, viruses, protists, and fungi. You will explore the world of tiny viruses and organisms that affect our lives.

Chapters

12 *Viruses and Monerans*

13 *The Protists*

14 *The Fungi*

15 *Microbial Diseases*

12. Viruses and Monerans

Blue-green bacteria are among the most primitive of all life forms. Blue-green bacteria were once called blue-green algae and classified with plants. However they are now grouped with bacteria because of their cellular structure. Blue-green bacteria are grouped in the Kingdom Monera. In this chapter, you will explore organisms in the Kingdom Monera. You will also learn about viruses.

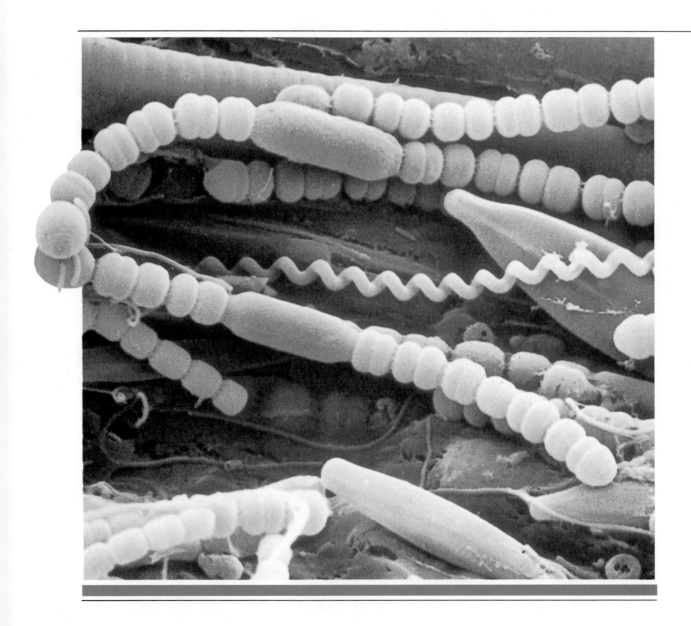

CHAPTER OBJECTIVES

After completing this chapter, you will be able to
- **Recognize** the characteristics of viruses.
- **Describe** some traits used to classify viruses.
- **List** the steps in the life cycle of a virus.
- **Explain** why prokaryotes are the simplest living cells.
- **Identify** the characteristics of monerans.
- **Describe** factors that influence the growth of bacteria.
- **Discuss** ways bacteria are beneficial.
- **Compare** the ways viruses and bacteria reproduce.

Characteristics of Viruses

12•1 WHAT ARE VIRUSES?

Are viruses living organisms? This question has long intrigued scientists. Viruses have some of the properties of living things but lack others. A **virus** is a very small particle usually made of nucleic acid and protein. Viral structure is not cellular. However, viruses are made of nucleic acids and proteins which are substances usually associated with life. A virus can reproduce but only when it is inside a living cell. A virus must get its chemical building blocks and energy from a living cell. When it is not inside a living cell, a virus cannot carry on metabolic processes associated with living things.

Viruses were first identified as the cause of certain diseases. Not until the invention of the electron microscope, however, were scientists able to *see* viruses. Viruses are too small to be seen with a light microscope. Poliovirus, one of the smallest known viruses, is so small that it takes 2500 viruses to span the tip of a pin! Viruses are so small that they will pass through filters that can trap bacteria.

Although viruses are very different in shape and size, they share many traits. The core of a virus is a strand of nucleic acid. The nucleic acid is enclosed within a protein shell, or jacket, known as the **capsid.** The nucleic acid and the capsid together in a virus are called the **nucleocapsid.**

Some viruses that infect animal cells have a nucleocapsid within an envelope. The membranelike envelope is made of protein and lipid. It is thought to come from the membrane of the host cell. A **host cell** is the cell in which a virus or an organism lives. Viral envelopes often have spikes coming out from their surfaces. These spikes are not found on the membrane of the host cell. Scientists believe that the virus changes the membrane of the host cell, forming the spikes. Figure 12•1 shows the structure of the herpesvirus. This type of virus causes blisters and fever sores.

SECTION OBJECTIVES
- List the characteristics of viruses.
- Describe the method by which viruses reproduce.

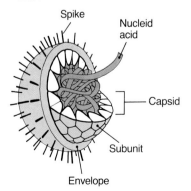

FIGURE 12•1 This illustration shows the shape and some parts of a typical virus.

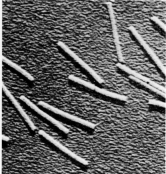

FIGURE 12·2 Tobacco mosaic virus on tobacco leaf (left) and an EM of tobacco mosaic virus magnified about 46 400X (right).

FIGURE 12·3 Viruses vary in shape and size. This variation became obvious after the invention of the electron microscope, a tool that permitted viruses to be observed for the first time.

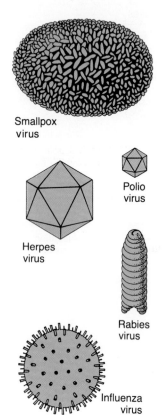

Each kind of virus has its own distinct features. In each kind of virus, the proteins of the capsid have a different makeup. The proteins are arranged in building blocks called capsomeres. A **capsomere** (KAP suh mihr) is a single protein subunit of a capsid. The capsomeres are stacked to form the capsid, much as stones are stacked to form a wall. The arrangement of capsomeres differs in different viruses. In a tobacco mosaic virus (TMV) the stacking of capsomeres produces a rodlike shape. The **tobacco mosaic virus** is a kind of virus that infects plant leaves, causing patches of green and yellow.

12·2 CLASSIFICATION OF VIRUSES

You have learned that a virus is made mostly of nucleic acid and protein. The nucleic acid inside a virus is either RNA or DNA but not both. Thus, a virus differs from a living cell in that it has only one type of nucleic acid. The TMV, for example, contains only RNA. Some viruses that infect bacteria, on the other hand, contain only DNA. Most living cells, however, have DNA and RNA. As with living cells, the RNA or DNA of a virus determines the structure of the virus's protein. In this way the protein shell of each virus is specific to its nucleic acid. The kind of nucleic acid in a virus is one of several features used to classify viruses.

The shape of the capsid is another feature used to classify viruses. In many viruses the capsid has the form of a many-sided figure. The herpesvirus and poliovirus, shown in Figure 12·3, have this many-sided shape. Other viruses, such as TMV and the rabies virus, may be rod-shaped. A virus that infects bacteria (shown in Figure 12·4) has a many-sided capsid, a rod-shaped tail, and several strawlike objects that stick out from the base of the tail.

Size is another factor used to classify viruses. Viruses range in size from just a few nanometers (nm) to over 300 nm. (A nanometer is one billionth of a meter.) Viruses of the pox group, such as the cowpox virus or the smallpox virus, are 250 nm to 275 nm across. Polioviruses are about 27 nm across.

Viruses may be classified by whether they infect plant, animal, or bacterial cells. Viruses are specific in the hosts they may infect. For example, TMV will not infect corn or wheat. Some viruses are even more specific in that they will infect only certain tissues in certain organisms. The rabies viruses, for example, will infect only the nervous tissues of mammals.

Some other features also are used to classify viruses. Animal viruses may or may not have an envelope. The chicken pox virus has an envelope; the poliovirus does not. The number of capsomeres in viruses is another feature used in classification. Figure 12·4 lists some of the features used to classify viruses and gives some examples.

FIGURE 12·4 Scientists use physical characteristics to classify viruses. As you can see, there are two main groups of viruses; one group contains RNA, the other group contains DNA.

NUCLEIC ACID	CAPSID SHAPE	ENVELOPE	HOST TYPE	EXAMPLES
RNA	Many-sided	Absent	Animal	Poliovirus
		Present	Animal	AIDS virus
	Rod	Absent	Plant	Tobacco mosaic virus
		Present	Animal	Flu virus
DNA	Many-sided	Absent	Animal	Cold viruses
		Present	Animal	Herpesviruses
	Rod	Absent	Bacteria	Some bacterial viruses
	Complex	Absent	Bacteria	Some bacterial viruses
		Present	Animal	Smallpox virus, chicken pox virus

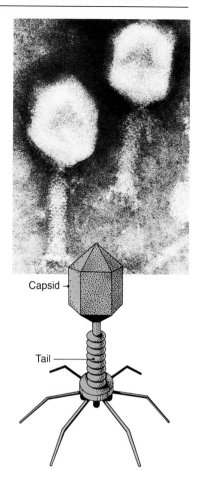

FIGURE 12·5 An EM of T4 bacteriophage magnified 250 800X (top) and a diagram of T4 bacteriophage (bottom).

12·3 BACTERIOPHAGES

Viruses can infect most living things. **Bacteriophages** (bak TIHR-ee uh fayj uhz), also called **phages,** are viruses that infect bacteria. Figure 12·5 shows bacteriophage T4. Notice the many-sided head and the tail built around a rod. This virus has a complex shape. The T4 phage belongs to a family of phages called *T-even phages*. All the viruses in this family have a complex shape and double-stranded DNA. Other bacteriophages carry single-stranded DNA.

Phages have been important in the study of viruses because both the phages and the bacteria are easy to work with in the laboratory. Experiments with phages can be completed in days, because phages reproduce quickly. Experiments with animal and plant viruses, which reproduce more slowly, may take months to complete. Much of what is known about viruses was found first in experiments with phages.

bacterio- (bacteria)
phage (to eat)

236 VIRUSES AND MONERANS

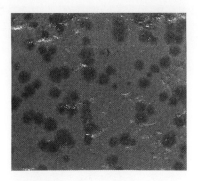

FIGURE 12·6 The clear areas on this Petri dish are called viral plaques. These are areas where bacteria have been killed by bacteriophages.

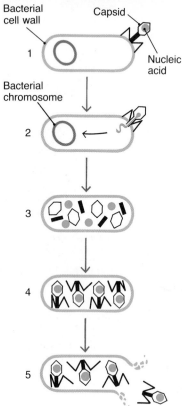

FIGURE 12·7 Viruses can only reproduce when they are inside a cell. This illustration shows viral replication.

As with all viruses, phages are seemingly lifeless particles when they are outside a living cell. Remember, viruses have no metabolic processes. Viruses multiply only when they are inside a living cell. The reproduction of viruses often is called **viral replication** to distinguish it from the reproduction of cells.

A virus can be compared to a safe with a set of master plans inside. The capsid is the safe; the nucleic acid is the set of plans. The plans contain all the instructions for producing more safes containing more sets of master plans. Master plans are not enough, however. Raw materials, tools, and energy are needed to make more safes and plans. Where can all these things be found? The materials can be found in workshops, and workers provide the energy. Living cells are the workshops in which viruses are copied. Amino acids and nucleotides are among the raw materials. Ribosomes and tRNA are among the tools. ATP provides the energy.

When phages enter bacterial cells, they replicate by using the cells' materials and energy. What happens to the bacteria? In most cases, bacteria die as a result of phage infection. Figure 12·6 shows a film of bacterial growth on the surface of a dish. There are also several clear spots called *plaques* (plaks) in the film of bacterial growth. Each of these plaques represents an area in which viruses have replicated, killing the bacteria.

12·4 VIRAL REPLICATION

Most viruses replicate in a similar way. The replication of a bacteriophage is presented here in detail. The process in the phage is similar to replication in most viruses. Refer to Figure 12·7 as you read the numbered steps.

1. When a phage comes in contact with its host bacterium, it sticks to the surface of the bacterial cell by a process called **adsorption.** Adsorption takes place because molecules on the viral tail form a close fit with molecules on the host bacterium. This matching fit is one of the reasons why a virus usually infects a specific host.

2. After adsorption, a reaction or series of reactions take place at the point of attachment. This creates a hole in the cell wall of the bacterium. The tail of the virus shrinks, injecting the viral nucleic acid into the bacterium. The empty capsid stays outside.

3. The viral nucleic acid takes control when it enters the bacterium. The DNA of the host cell is either destroyed or made useless. The normal activities and metabolism of the cell stop. Using the resources of the host, the viral nucleic acid makes copies of itself and of its proteins.

4. New viral particles are assembled.

5. When enough new viruses have collected, the cell bursts, freeing complete viruses. The bursting of the bacterium is called **lysis** (LĪ-sihs). After lysis, the newly formed viruses are free to infect other host cells. They move through their environment until they meet other host bacteria. Then the process begins again. This entire process, from adsorption to lysis, is called a **lytic** (LIHT ihk) **cycle**.

A phage may complete a lytic cycle in 25 to 45 minutes. Up to several hundred new phages may be given off in each lytic cycle. Phages that cause lysis of their hosts are called **virulent** (VIHR yuh-luhnt) **phages**. (*Virulent* means "very poisonous.")

Not all phages are virulent. Phages that stay inside their host cells for a long time without causing lysis are called **temperate phages**. A temperate phage begins its activities in the same way as a virulent phage. It sticks to the surface of its host. Then it injects its nucleic acid into the cell. There the similarity ends. The DNA of the temperate phage then becomes part of the DNA of the host chromosome. The viral DNA acts as an extra piece of information on the bacterial chromosome. Viral DNA attached to a bacterial chromosome is called a **prophage**. When the bacterium copies its DNA before division, it copies the prophage too. The prophage may be carried through many generations of division without apparent harm to the bacteria. Each new bacterial cell that is formed carries a copy of the prophage on its chromosome. Bacteria that carry prophages are called **lysogenic** (lī-suh JEHN ihk) **bacteria**. The process by which viral DNA is attached to and carried by a bacterial chromosome is called **lysogeny** (lī SAHJ uh-nee). Lysogeny is shown in Figure 12·8A.

Even with the prophage present, normal metabolism goes on in the bacterial cell. The prophage may later separate from the host chromosome. If this takes place, the phage will become virulent. It will replicate and cause lysis of the host, as shown in Figure 12·8B.

FIGURE 12·8 This illustration shows stages of lysogeny. After the virus has replicated and is released from the cell it can infect other cells.

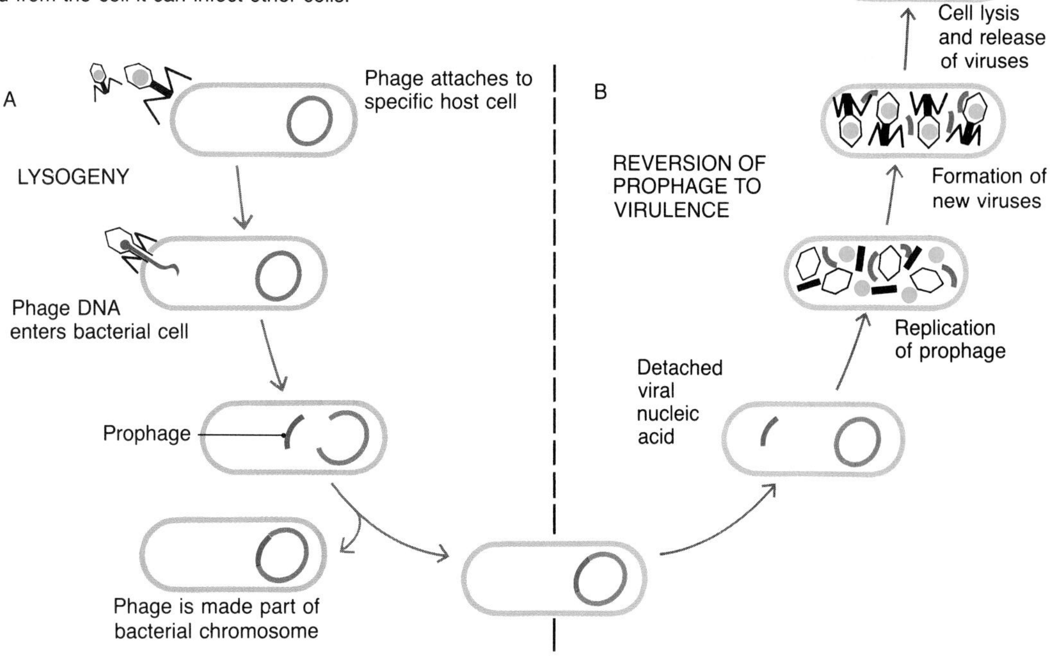

A LYSOGENY
- Phage attaches to specific host cell
- Phage DNA enters bacterial cell
- Prophage
- Phage is made part of bacterial chromosome

B REVERSION OF PROPHAGE TO VIRULENCE
- Detached viral nucleic acid
- Replication of prophage
- Formation of new viruses
- Cell lysis and release of viruses

Animal and plant viruses carry on activities like those of virulent bacteriophages. During replication, animal and plant viruses use the materials and energy of their hosts. As with virulent phages, animal and plant viruses often cause lysis of host cells. Lysis, and the resulting loss of materials and energy, leads to disease.

Animal and plant viruses also can behave as temperate phages do. Viral nucleic acid may be present in the host cell without causing lysis. The presence of this extra nucleic acid can lead to other chemical changes in the host cell, again resulting in disease.

Animal viruses and phages are somewhat different in their methods of infection and replication. A phage injects only its nucleic acid into a bacterium; the capsid remains outside. After an animal virus sticks to the surface of a host cell the whole virus is taken in. Once inside, enzymes of the animal cell dissolve the viral capsid. The viral nucleic acid may then replicate or attach to a host chromosome. An animal virus replicates more slowly than does a phage. It takes 6 or more hours to complete replication. As a newly formed virus leaves an animal cell, it may take part of the cell membrane of the host. This membrane material becomes the envelope of the virus.

12·5 VIRAL TRANSDUCTION

Because they can attach to host chromosomes, temperate phages sometimes carry segments of DNA to other bacteria. Viral DNA may remain attached to its host chromosome for a long time, through many divisions. It is not known why a prophage may later separate from the chromosome and begin a lytic cycle. When a prophage separates from a host chromosome, it sometimes carries a piece of the original host's chromosome with it to a new host. The transfer of host DNA to another organism by a virus is called **viral transduction**. Figure 12·9 shows the events of this process.

1. The prophage separates from a host chromosome and carries a piece of the chromosome with it.
2. The viral DNA, with its piece of host DNA, replicates. New viruses are assembled, each carrying a piece of the bacterial DNA.
3. Lysis of the host bacterium releases the virulent viruses to start a new cycle of infection.
4. As these modified viruses invade other bacteria, they inject the extra bacterial DNA as well. Thus bacterial DNA has been carried from one bacterium to another.

You can see in Figure 12·9 that the new host bacterium gains an extra section of DNA. DNA carries genetic information. Viral transduction is one of many ways that bacteria gain new genetic information. It is one reason for the many inherited variations seen among some living things. It is believed that resistance to antibiotics can be transferred from one bacterium to another by viral transduction.

Viral transduction of DNA has been seen again and again among bacteria. Scientists are not certain that viral transduction can take

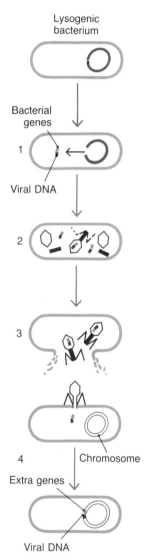

FIGURE 12·9 Viruses can sometimes be used to change the genetic material of a cell. Thus viruses can be used to transfer DNA from one bacterium to another.

place among animals or plants. It does, however, seem likely that the process does take place. A number of animal and plant viruses are known to behave as temperate phages do. They attach their DNA to host chromosomes and may later carry off sections of host DNA to other cells.

12·6 VIROIDS

Viruses were once thought to be the smallest possible agents of disease. This idea has changed. Several plant diseases have been found to be caused by something smaller and simpler than viruses. These diseases are caused by short naked strands of RNA. Because they behave similarly to viruses, the infective strands of naked RNA are called **viroids** (VĪ roidz). Viroids are only a fraction of the size of the smallest viruses. Figure 12·10 shows the sizes of a cell nucleus, a bacterium, a virus, and a viroid. Viroids lack the protective protein capsids of viruses. As with viruses, viroids can only replicate within a living cell.

-oid (like)

Biologists consider viroids to be surprising for two reasons: (1) Viroids can remain powerful agents of disease even though they are single naked strands of RNA in the environment. (2) Viroids are extremely short; the information content of a viroid is less than that needed to control the formation of a single enzyme! With so little genetic information, how can a viroid take control of its host cell? How can the viroid control its own replication? To answer these questions, scientists must do further research.

FIGURE 12·10 This figure shows the relative sizes of a smallpox virus, a bacterial cell, a cell nucleus, and a viroid.

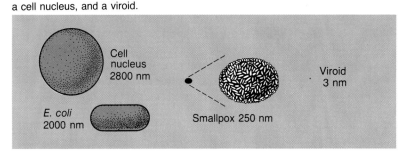

REVIEW QUESTIONS

1. What characteristics are used to classify viruses?
2. Describe the steps in the replication of a bacteriophage.
3. What structure is found only in some animal viruses?
4. Compare virulent phages and temperate phages.
5. A species of bacterium that does not normally cause disease is suddenly found to be able to cause disease. Scientists suspect that a virus may be involved in the change. Explain how a virus could cause such a change.

Bacteria and Other Monerans

SECTION OBJECTIVES

- Identify the characteristics of prokaryotic cells.
- Recognize the varieties of bacteria, their activities, and their methods of reproduction.
- Identify the processes by which genetic material is exchanged between bacteria.

12·7 CHARACTERISTICS OF PROKARYOTES

Prokaryotes are the simplest living cells. They lack many of the organelles found in the more complex cells. For example, the prokaryote does not have a nucleus with its nuclear membrane. Also lacking are mitochondria, chloroplasts, endoplasmic reticulum, and Golgi complex. Can you name a common factor among these structures? They all have membranes. Most prokaryotes contain no membrane other than the cell membrane. Prokaryotes may have some organelles. The enzymes needed for metabolism are in the cytoplasm or on the cell membrane rather than in complex organelles.

Prokaryotes comprise the kingdom *Monera*. Members of kingdom Monera are called **monerans.** Monerans are divided into two main groups: blue-green bacteria (at one time called blue-green algae) and other bacteria.

Monerans exist either as single cells or as colonies. A **colony** is a group of similar cells that are attached to each other. A colony of moneran cells is different from a multicellular organism. Each cell in a colony can live on its own. The cells of a multicellular organism cannot carry out all the functions of the whole organism alone because they are dependent on each other.

12·8 BLUE-GREEN BACTERIA

Blue-green bacteria are autotrophs. For many years the blue-green bacteria were called blue-green algae and were classified as plants. Like plants, blue-green bacteria are photosynthetic. However, their photosynthetic pigments are not found within chloroplasts. Their genetic material is not enclosed in a nuclear membrane. Mitochondria, large vacuoles, and other membrane-bound organelles are not present in blue-green bacteria cells. Unlike algae, blue-green bacteria are prokaryotic. For these reasons blue-green bacteria are classified as monerans.

Within kingdom Monera, blue-green bacteria make up phylum Cyanophyta (sī uh NAHF uht uh). *Cyanophyta* means "blue plants" and refers to the blue pigments found in these bacteria. All blue-green bacteria have at least two pigments. These pigments are *phycocyanin* (blue) and *chlorophyll* (green).

Blue-green bacteria also may have other pigments, especially red and orange ones. The color of any given species depends on the amount of each pigment present. It is thought that the Red Sea is so called because of the presence of red pigments in blue-green bacteria. At certain times of the year, these bacteria increase greatly in number, giving the sea a red color.

Blue-green bacteria can grow almost any place where moisture is present. Though often found in ponds, they also live in salt water, on

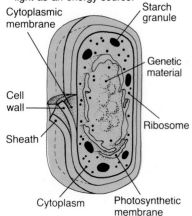

FIGURE 12·11 A blue-green bacterium is able to make its own food using light as an energy source.

FIGURE 12·12 Hot springs in Yellowstone National Park are colored because of the presence of blue-green bacteria.

damp rocks, and in moist soils. Some blue-green bacteria live on snow. Others live in springs that are as hot as 80°C. Colorful forms of blue-green bacteria live in the hot springs of Yellowstone National Park, as shown in Figure 12·12.

Blue-green bacteria have few internal structures. They have photosynthetic membranes that are inward folds of the cell membrane. Chlorophyll and the enyzmes of photosynthesis are found on these membranes. The photosynthetic membranes are not separate organelles, as are the more complex chloroplasts of eukaryotic cells.

The DNA of blue-green bacteria is found in the cytoplasm. Even though this DNA often is called "nuclear material," it is not contained in a nucleus. Blue-green bacteria also have starch granules, which store food for the organisms, and ribosomes in the cytoplasm.

A blue-green bacterial cell has a cell wall outside the cell membrane. As in a plant cell, the cell wall gives support to the bacterial cell. But the cell wall of a blue-green bacterium is different from that of a plant cell. It is made up of a mixture of amino acids and sugars and is chemically similar to the cell walls of other prokaryotes. This similarity is another reason for these organisms being classified as monerans rather than as plants. Outside the cell wall is a jellylike sheath that holds the cell to others in a colony. A few blue-green bacteria exist as single cells, but most form colonies. Colonies may take the form of spheres, sheets, or threads.

Sexual reproduction has not been observed in blue-green bacteria. Reproduction is by the simple asexual process of splitting, called *binary fission.* The bacterial cell makes a copy of its DNA, and the cell pinches into two new cells. When division takes place, the newly formed cells stay together, held by their sheaths.

FIGURE 12·13 Some blue-green bacteria, like *Anabaena* shown here, form threadlike colonies.

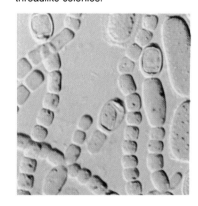

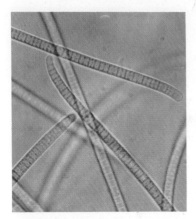

FIGURE 12·14 A photomicrograph of *Oscillatoria*, a blue-green bacterium, magnified 530X.

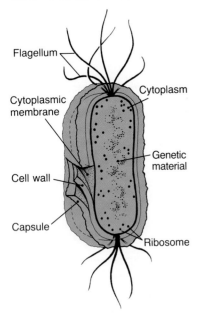

FIGURE 12·15 This illustration shows a bacteria cell. As you can observe, bacteria cells lack a nucleus.

Besides serving as food for water animals, many blue-green bacteria also have a special function. They take nitrogen gas from the air and change it to ammonia. Other kinds of bacteria change the ammonia to nitrates that can dissolve in water. Plants, which use nitrates to make protein, cannot make nitrates. The supply of nitrates from monerans limits the natural growth of many plants in both water and soil. Blue-green bacteria also produce oxygen as a by-product of their photosynthesis. This oxygen is used by many water animals during respiration.

The organic matter and detergents in sewage promote the growth of bacteria. Lakes polluted with sewage often show an increased number of blue-green bacteria. Blue-green bacteria grow quickly under such polluted conditions. Biologists sometimes measure the amount of pollution in a lake by counting the number of blue-green bacteria. Members of the genus *Oscillatoria*, shown in Figure 12·14, are counted for this reason.

12·9 BACTERIA—STRUCTURE AND FUNCTION

Bacteria include many different monerans, most of which are heterotrophs. Bacteria are the most numerous monerans and are found in almost every environment. Bacteria live in soil, water, and air. They may be on things you touch or eat and may be on and in the bodies of most living things. Bacteria are classified in phylum *Schizomycetes* (skihz oh mī SEET eez). *Schizo-* means "split" and refers to the simple division process by which bacteria multiply. *Mycete* means "fungus" and is a holdover from the time when bacteria and fungi were grouped together.

People often think of bacteria as only "germs" of disease, but this belief is not completely correct. Of the more than 1500 species of bacteria, only about 250 are known to cause disease. The activities of most bacteria are useful and needed. People have used many species of bacteria in the making of foods and medicine.

The cells of bacteria are like those of other monerans. They are prokaryotic and lack nuclei and other membrane-bound organelles. Figure 12·15 shows the structure of a representative bacterial cell. How is this bacterial cell different from the blue-green bacterial cell shown in Figure 12·11? Notice that this bacterial cell lacks the photosynthetic membranes found in the blue-green bacterial cell. Observe also that this bacterial cell has flagella, which are not present in the blue-green bacteria shown. Many bacteria move by means of flagella. Some bacteria have one flagellum, others have many, and some have none at all.

Outside the cell membrane of a bacterial cell is a cell wall that is a mixture of amino acids and sugars linked in flat, stiff sheets. The cell wall gives the bacterial cell its shape and holds it together. Some bacteria also have an outer layer called a **capsule.** Even though the makeup of the capsule varies, it is usually made of a sticky material that allows bacteria to stick to surfaces. Removing the capsules from some

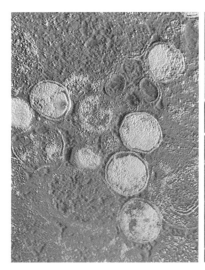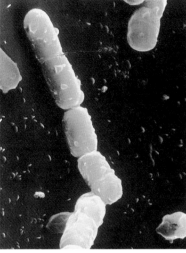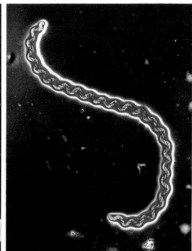

FIGURE 12·16 Photomicrographs of cocci, *Chlamydia* (left) magnified 17 600X; bacilli, *Pseudomonas* (center) magnified 5 880X; and spiral shaped bacteria, *Leptospira* (right) magnified 15 000X.

disease-causing bacteria causes them to be harmless. It is believed that the capsule may "disguise" bacteria and thus they escape the defenses of the host.

Some bacteria grow as single cells. Others remain attached to each other after dividing. These types form colonies shaped like chains or clusters. Figure 12·16 shows some forms of bacteria. Based on their shapes, most bacteria are divided into three groups: cocci (KAHK sī), bacilli (buh SIHL ī), and spirilla (spī RIHL uh).

Cocci (sing., *coccus*) are bacteria that are shaped like spheres. Cocci are generally 0.5 to 1.0 μm in diameter. Some are found in pairs. Paired cocci are called *diplococci*. An example is *Diplococcus pneumoniae*, the cause of bacterial pneumonia. *Streptococci* are cocci that form chains. *Staphylococci* form grapelike clusters. *Staphylococcus aureus* is a coccus that lives on human skin. Occasionally, it gets *in* the skin and causes boils.

Bacilli (sing., *bacillus*) are bacteria shaped like rods. Bacilli range in size from 0.5 to 20 μm in length. Bacilli that grow in chains are called *streptobacilli*. They look like strings of sausage. *Escherichia coli* (*E. coli*) is a bacillus that lives in the intestines of humans. It is one of the most studied organisms. *E. coli* has been used in many thousands of experiments in genetics and biochemistry. Species of *Lactobacillus* are used in the production of yogurt.

Spirilla (sing., *spirillum*) are bacteria shaped like spirals. They vary in length from just a few micrometers to about 500 μm. Spirilla grow only as single cells, not as chains or clusters. Members of the genus *Spirillum* are found in soil and water. Their activities help the soil by forming chemicals that are used by plants in their growth. Another group of spiral-shaped bacteria, called *spirochetes* (SPĪ ruh-keets), includes the organism that causes syphilis.

Discovery

Archaebacteria: a Second Major Group of Bacteria

Prokaryotes have many structural features that distinguish them from eukaryotes. In the past 15 years, many bacteriologists have studied unusual bacteria that are found in very harsh environments, such as salt marshes and hot springs. Detailed research on these bacteria indicates that they belong to a distinct group of prokaryotes. That group is now called *archaebacteria* (archaeic, or ancient, bacteria) to distinguish them from all other bacteria, which are called eubacteria (true bacteria).

Studies of the sequence and structure of one kind of ribosomal RNA in many types of bacteria have provided strong support for the separate grouping of archaebacteria. Such data show clear differences between organisms in the two groups. Also, the cell membranes of archaebacteria have unusual lipids not found in eubacteria or in eukaryotes.

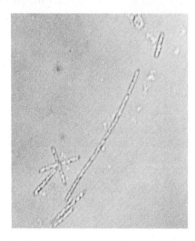

Archaebacteria may be more like the ancestors of eukaryotes than are eubacteria. Studies show that another kind of ribosomal RNA in archaebacteria is like that of eukaryotes. Also, some features of protein synthesis in archaebacteria resemble those of eukaryotes and not eubacteria. However, recent studies have shown clear similarities between the genes of archaebacteria and eubacteria.

These data do not prove that archaebacteria are more ancient than eubacteria or that they are ancestors of eukaryotes. However, it is clear that archaebacteria are a special group of prokaryotes.

Most bacteria are in the form of cocci, bacilli, and spirilla. Two small groups of bacterialike organisms are set apart because of special features. **Mycoplasmas** are tiny monerans that are smaller than most bacteria and lack cell walls. Mycoplasmas are about 0.1 to 0.25 μm in size. Because they lack cell walls, they vary in shape. Mycoplasmas are probably the smallest known living cells. They cause lung diseases in humans and other animals.

Rickettsias (rih KEHT see uhz) are bacterialike organisms that cannot live outside of living tissue. They are slightly larger than mycoplasmas, averaging about 0.45 μm. They are the cause of several serious illnesses in humans. Rickettsial diseases are generally spread by ticks and lice.

12·10 BACTERIAL GROWTH CONDITIONS

Bacteria grow and reproduce actively when conditions are favorable. If conditions become unfavorable, bacteria stop their activities and enter an inactive stage. Bacteria can switch back and forth from the active to the inactive stage as conditions change. What conditions are needed for bacterial growth?

Bacteria need moisture. Like all living cells, bacteria contain water. In dry surroundings, bacteria lose water and become inactive. Bacteria get moisture by living in water, in damp places, or in the moist bodies of animals or plants. Many bacteria live in the thin film of water found on particles of soil.

Each species of bacteria lives best within a certain temperature range. A few bacteria can grow at temperatures near 0°C. Bacteria have even been found growing in Antarctica at −7°C. Most bacteria grow at temperatures between 20°C and 40°C. A few species grow at temperatures as high as 75°C. Bacteria that cause disease in humans grow best at body temperature (37°C). Figure 12·7 shows bacteria about to be grown in an incubator, a device that controls the temperature. Most bacteria grow best in darkness. The blue, violet, and ultraviolet light in direct sunlight kill most bacteria within a few hours or cause bacterial mutation.

The oxygen needs of bacteria vary. Some bacteria need oxygen to live. Other bacteria die in the presence of oxygen. Still other bacteria can live with or without oxygen. More will be said of this later (Section 12·11).

Bacteria need a source of energy. Bacteria have many different ways of obtaining food. Some bacteria obtain nourishment as other absorptive heterotrophs such as fungi do—they feed on the tissues of other organisms. These bacteria secrete through their cell walls digestive enzymes that break down food materials into smaller molecules outside the bacteria. Then the digested food can be absorbed by the bacteria.

Bacteria that harm the living organisms upon which they feed are parasitic bacteria. A **parasite** is an organism that lives in or on a host organism and does harm to this host. When parasitic bacteria feed on organic substances inside living things, this feeding often causes disease in the living host.

Some bacteria feed on dead organic matter. These bacteria are saprobes. **Saprobes** are organisms that use dead organic matter as food. Saprobic bacteria break down dead plants and animals or waste, returning minerals and other nutrients to the soil. Plants need these nutrients for growth. Some saprobes, however, may cause disease if they enter a wound. *Pseudomonas aeruginosa* is a saprobe found almost everywhere in nature. This bacterium can cause disease in both plants and animals. Usually, pseudomonads are not a serious threat to a healthy person. Infants, hospital patients, and the aged are most likely to be infected by these organisms.

Certain bacteria make their own food by photosynthesis. Bacterial photosynthesis is different from photosynthesis in plants in several ways. Some photosynthetic bacteria do not use chlorophyll to capture the energy of sunlight. They may use other pigments instead. Another difference from plant photosynthesis is that photosynthesis in some bacteria (the purple and green bacteria) does not produce oxygen as a by-product. The overall reaction for one such kind of bacterial photosynthesis follows.

$$CO_2 + H_2S + 2H_2O \xrightarrow{\text{light energy}} (CH_2O)_n + H_2SO_4$$
$$\text{(carbon dioxide)} \quad \text{(hydrogen sulfide)} \quad \text{(water)} \quad \text{(carbohydrate)} \quad \text{(sulfuric acid)}$$

sapros (rotten)

FIGURE 12·17 Certain bacteria, especially those that cause disease in humans, need a warm temperature to grow well. A bacteriologist is placing the bacteria in an incubator to encourage their growth.

Compare this equation with the following equation for the process of plant photosynthesis.

$$6CO_2 + 6H_2O \xrightarrow{\text{light energy}} C_6H_{12}O_6 + 6O_2$$
$$\text{(carbon dioxide)} \quad \text{(water)} \quad \text{(glucose)} \quad \text{(oxygen)}$$

Some bacteria obtain nourishment by chemosynthesis (kee moh-SIHN thuh sihs). **Chemosynthesis** is a kind of autotrophic nutrition in which chemical energy is used to make carbohydrates. Bacteria of the genus *Thiobacillus* are chemosynthetic organisms found in soil and water. Energy is taken from chemical reactions with inorganic materials from the soil or water. The energy is then used to build sugars and other food molecules. The following is an example of a reaction carried out by chemosynthetic bacteria.

$$2S + 3O_2 + 2H_2O \longrightarrow 2H_2SO_4 + \text{energy}$$
$$\text{(sulfur)} \quad \text{(oxygen)} \quad \text{(water)} \quad \text{(sulfuric acid)}$$

The energy is used to produce sugar from carbon dioxide and hydrogen. Chemosynthetic bacteria have no chlorophyll; they can obtain energy and grow without sunlight. The ability to obtain energy by breaking down inorganic compounds is unique to a few types of prokaryotes.

12•11 ENERGY PRODUCTION IN BACTERIA

Bacteria may obtain energy either by respiration or by fermentation. Remember that there are two kinds of respiration: aerobic and anaerobic. Respiration in which the final electron acceptor is oxygen is called *aerobic respiration*. Respiration in which the electron acceptor is an inorganic substance other than oxygen is called *anaerobic respiration*. Remember that fermentation is also an anaerobic method of making usable energy. *Fermentation* is the breakdown of glucose and the release of energy in which organic substances are the electron acceptors.

Bacteria that require oxygen for respiration are called **obligate aerobes.** The bacteria that cause tuberculosis and diphtheria are obligate aerobes. Bacteria that grow only in the absence of oxygen are called **obligate anaerobes.** These organisms do not need oxygen to produce ATP. Bacteria of the genus *Clostridium* are obligate anaerobes. Several species of *Clostridium* are found in soil. These bacteria can cause serious diseases in humans if they get into human tissues from improperly sterilized canned foods. Figure 12•18 shows cells and spores of *Clostridium*. Tetanus, gangrene, and botulism (a deadly type of food poisoning) are a few diseases caused by bacilli of the genus *Clostridium*. **Facultative anaerobes** are bacteria that can produce ATP with or without oxygen. *E. coli* as well as many bacteria of the soil and water are bacteria of this kind.

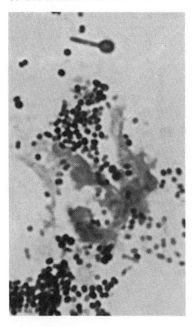

FIGURE 12•18 This photograph shows the cells and spores of a *Clostridium* species magnified 1925X. The spores of this bacteria are resistant to drying out. Several *Clostridium* species cause serious disease in humans.

Investigation

How is the Gram's Stain Used to Identify Bacteria?

Goals
After completing this activity, you will be able to
- Identify a few specific groups of bacteria.
- State what the terms *Gram-positive* and *Gram-negative* mean.

Materials (for groups of 2)
4 or 5 slides of Gram-stained bacteria with examples of Gram-positive and Gram-negative organisms
microscope (with oil-immersion lens, if possible)
oil for oil-immersion lens
pink pencil
purple pencil

Procedure

A. On a separate sheet of paper, make a chart similar to the one below. Leave enough room to fill in data on all the bacteria you observe.

GRAM REACTION	SHAPE	
	Bacilli (rods)	Cocci (spheres)
Positive		
Negative		

Some genera of bacteria—the cocci and the bacilli—can be identified by their shape and by the way they pick up certain stains. A staining technique known as the Gram's stain is used to make bacteria visible under the light microscope. This technique uses two stains—a dark purple stain and a light pink stain. Bacteria that are Gram-positive retain the dark purple stain. Bacteria that lose the dark purple stain and show only the pink stain are called Gram-negative. The difference in staining characteristics is due to differences in the bacterial cell wall. Many bacteria are therefore classified as either Gram-positive or Gram-negative.

Diplococci (pairs of cocci)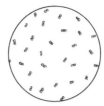

Streptococci (chains of cocci)

Staphylococci (grapelike clusters of cocci)

Bacilli (rod-shaped bacteria)

Bacilli (rod-shaped bacteria)

B. Examine a stained slide of bacteria under the microscope. Look at it first under low power, then under high. If your microscope has an oil-immersion lens, look at the bacteria under this lens. Ask your teacher to demonstrate the correct use of the oil-immersion lens.

C. Write the genus name of the organisms on your chart. (The name will probably be on the slide.) Place the name of the organism in the proper box according to its Gram reaction and its shape. In parentheses after each name note any additional characteristic you might find useful in identifying the bacteria (for example, "in clusters").

D. On a separate sheet of paper, draw a few representative bacteria from each slide. Label the drawings. Color them with either the pink or purple pencil to indicate their Gram reaction.

E. Repeat for each slide.

F. Follow your teacher's instructions for the proper disposal of materials.

Questions and Conclusions
1. What are two reasons for staining bacteria?
2. Explain the meaning of the terms *Gram-positive* and *Gram-negative.*
3. Is an organism's Gram reaction directly related to its shape? Explain.
4. Strep throat is caused by streptococci. Suppose a culture of bacteria from a person with strep throat were given a Gram's stain. What shape and color would you expect the organism to be?

VIRUSES AND MONERANS 247

12·12 BACTERIAL REPRODUCTION

Binary fission is the process by which most bacteria divide. *Binary fission* means "a splitting into two." It is a process that is simpler than mitosis. Several of the features of mitosis are not seen in binary fission. Figure 12·19 compares these two forms of division. How does bacterial binary fission differ from animal cell mitosis? There are no centrioles or spindle fibers in binary fission. During binary fission the bacterial nucleic acid duplicates. Then a membrane begins to grow between the duplicated nucleic acid, and the cell wall pinches in to form two separate bacterial cells. Bacteria can complete binary fission in as little as 20 minutes.

FIGURE 12·19 Binary fission in bacteria differs from mitosis in animal cells in several important ways.

PROCESS	BINARY FISSION	ANIMAL CELL MITOSIS
DNA replicates before division	Yes	Yes
Chromosomes become visible during division	No	Yes
Centrioles and spindle fibers	Absent	Present
Number of daughter cells formed	2	2
Daughter cells receive identical copies of DNA	Yes	Yes

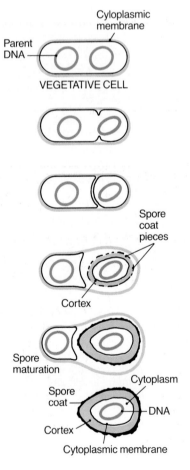

FIGURE 12·20 Endospore formation insures that bacteria will be able to survive conditions that are not favorable for growth. When conditions improve, the spore coat breaks open and a bacteria cell emerges.

12·13 ENDOSPORE FORMATION

Bacteria become inactive when conditions are not favorable. Many bacteria of the bacillus type form a tough structure called an endospore during poor conditions. An **endospore** is a spore that forms inside a cell. See Figure 12·20 for the stages of endospore formation. Endospore formation begins in the way binary fission begins. The nuclear material replicates and then splits into two parts. But a dense spore coat forms around one of the parts. The spore coat is chemically different from a bacterial cell wall. When the coat is fully formed, the bacillus breaks down, freeing the endospore. The free endospore remains inactive as long as the conditions are not favorable. It is very resistant to being killed by heating, freezing, drying, radiation, and some toxic chemicals.

When favorable growing conditions return, the endospore germinates. The spore coat breaks open and a single bacillus comes out. Endospore formation is not a form of bacterial reproduction. The bacillus that is breaking apart forms only one endospore. The endospore forms only one bacillus.

Only a few species of bacteria produce spores, but their spores are nearly everywhere. Methods of killing spores are needed before and during surgery, in food canning, and in laboratory procedures. Moist heat under pressure is a widely used method of sterilization. Pressurized steam at 121°C to 126°C will kill all bacteria and most spores within 15 to 45 minutes. Bacteria are generally killed in an autoclave. This device places the organisms under moist heat and pressure for the required period of time.

12·14 CONJUGATION AND TRANSFORMATION

Earlier in the chapter you learned that transduction is a process by which viruses transfer genes from one bacterium to another. There are other ways in which genes or genetic material can be transferred between bacteria. **Conjugation** is the transferring of genetic material from one bacterium to another by direct cell-to-cell contact. Bacterial conjugation is shown in Figure 12·21. The steps in bacterial conjugation are shown as follows.

1. The chromosome in the donor bacillus replicates.
2. Inside the donor cell the newly formed chromosome opens up.
3. A tube forms between the two organisms. One of the two chromosomes in the donor bacillus travels through the tube into the second bacillus, the recipient.
4. The recipient now has two different chromosomes—its own and the one passed through the tube from the donor.
5. These two chromosomes exchange pieces.
6. When the recipient divides, each new cell will get one of the reorganized chromosomes. Each of the new cells has genes from both of the original bacilli.

Conjugation is similar to sexual reproduction in higher organisms. The offspring get genetic information from both parents.

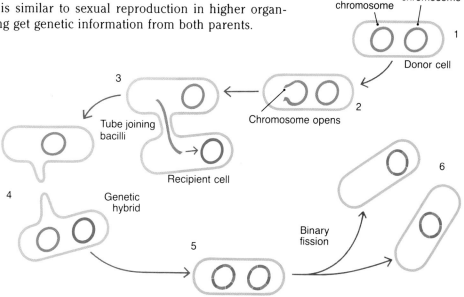

FIGURE 12·21 Bacterial cells exchange genetic material in direct cell-to-cell contact. This exchange contributes to genetic diversity and may lead to the development of a new strain of bacteria.

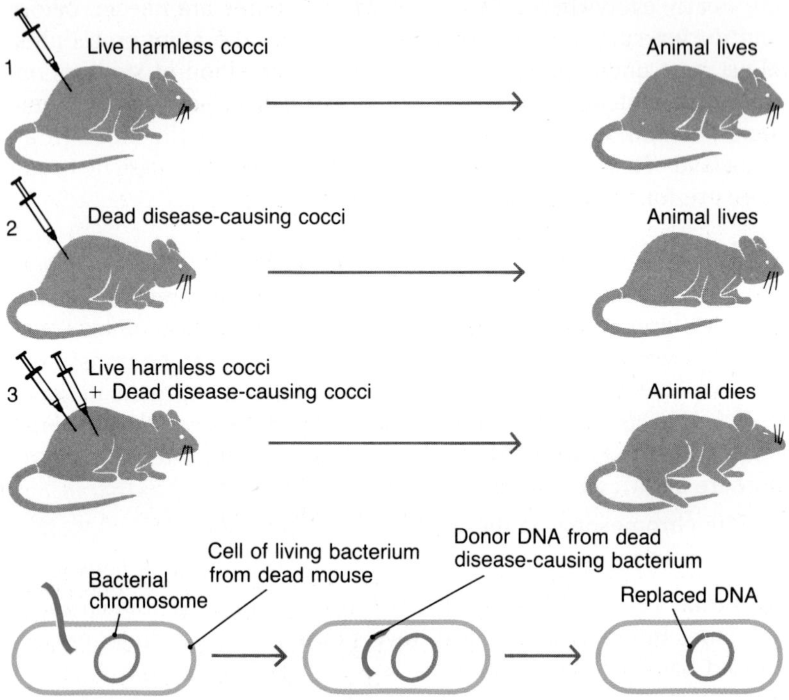

FIGURE 12·22 Bacteria can acquire genetic material from the environment. This genetic material can become a part of the genetic material of the cell.

Transformation is another way in which genes are moved between bacteria. **Transformation** is the process by which genetic material absorbed from the environment replaces part of a bacterium's DNA. A bacterium can cause disease because, by transformation, it has inherited the genes to cause disease.

Figure 12·22 shows an experiment that demonstrates transformation. Study steps 1 and 2 of Figure 12·22. What results do you see? Notice that neither the live harmless bacteria alone nor the dead disease-causing bacteria alone harm the mouse. However, in step 3 a combination of live harmless bacteria and dead disease-causing bacteria kills the mouse. Why did this combination of bacteria kill the mouse? The live harmless bacteria were transformed. They acquired disease-causing properties.

If you could look at the genetic material of a living bacterium from the dead mouse, how might the material be different from its usual form? The diagram at the bottom of Figure 12·22 shows what you might find in the cell of a living bacterium removed from the dead mouse. The donor DNA has been taken into the bacterial chromosome. The transformed cell can cause disease. In fact the transformed cocci killed the mouse.

Some bacteria give off DNA into the environment in which they are growing. They secrete it in a mixture of substances called slime. A few bacteria, called transformable bacteria, can absorb the DNA. Exactly how bacteria absorb DNA remains unknown.

12·15 BENEFICIAL BACTERIA

Bacteria function in many ways that help other living things. Without the bacteria that break down materials, some of the chemicals that plants and animals need to live would not be recycled and available for reuse. Carbon, sulfur, nitrogen, and phosphorus are elements that bacteria continually recycle.

Earlier in this chapter you learned that many blue-green bacteria change atmospheric nitrogen into ammonia. Some nonphotosynthetic bacteria also change atmospheric nitrogen to ammonia. The changing of atmospheric nitrogen (N_2) to ammonia is called **nitrogen fixation.** Only certain monerans can perform nitrogen fixation. Remember that certain kinds of bacteria change ammonia to nitrates, which are needed by plants.

Some bacteria live in and help other organisms. *E. coli,* bacteria that live in the intestines of humans, make vitamins needed by their human hosts. Cows benefit from the bacteria in their stomachs. These bacteria digest the cellulose of grasses. Cellulose is a complex carbohydrate of the cell walls of plants. Animals cannot digest cellulose, while some bacteria can.

Bacteria are helpful in other ways. A number of drug companies use bacteria to make antibiotics and other medicines. Genetic engineering with bacteria has led to the production of human insulin and other substances (Section 8·8). Some toxic waste areas may be cleaned up by bacteria. The protein of certain bacilli acts as an insecticide against certain caterpillars harmful to crops. Bacteria can break down the cellulose of flax and hemp to produce the fiber used to make linen and rope. Many foods—including yogurt, cheese, buttermilk, sauerkraut, and vinegar—are products of bacterial action.

FIGURE 12·23 Relatively few species of bacteria are harmful to people. In fact, bacteria contribute to the manufacture of many common foods.

REVIEW QUESTIONS

6. Bacteria and blue-green bacteria are both monerans. What are some differences between them? What are some similarities?
7. What are several ways in which bacteria are beneficial?
8. How does binary fission differ from mitosis?
9. How do some bacilli survive during unfavorable conditions?
10. Suppose some organisms in a lake were carrying out photosynthesis. How could you determine whether they were plants, blue-green bacteria, or bacteria?

Chapter Review

SUMMARY

12·1 A virus is a particle with a core of nucleic acid surrounded by a coat, or capsid, of protein.

12·2 Viruses are classified by type of host, type of nucleic acid, shape of capsid, size, and other structural features.

12·3 Bacteriophages, which are viruses that infect bacteria, have a variety of capsid shapes and vary in the DNA they contain.

12·4 Viruses can replicate only inside the cells of their hosts.

12·5 Viral transduction, which is the process by which viruses transfer DNA from one bacterium to another, is one way in which bacteria gain new genetic information.

12·6 Infective strands of naked RNA, called viroids, are smaller than the smallest viruses.

12·7 Prokaryotes make up the kingdom Monera. Prokaryotes consist of simple cells that lack nuclei and membrane-bound organelles.

12·8 Blue-green bacteria produce oxygen as a by-product of photosynthesis, and some convert free nitrogen to water-soluble nitrates.

12·9 Bacteria are the most numerous monerans and are found in every environment. They occur as cocci, bacilli, and spirilla.

12·10 While many bacteria are saprobes or parasites, some bacteria carry out photosynthesis or chemosynthesis.

12·11 Bacteria can obtain energy by respiration and fermentation.

12·12 Bacteria reproduce by binary fission.

12·13 Some bacilli form endospores to survive unfavorable conditions.

12·14 Bacteria may exchange genes through conjugation or through transformation.

12·15 Bacteria are beneficial to other living organisms in many important ways.

BIOLOGICAL TERMS

adsorption
bacilli
bacteria
bacteriophage
binary fission
blue-green bacteria
capsid
capsomere
capsule
chemosynthesis
cocci
colony
conjugation
endospore
facultative anaerobes
host cell
lysis
lysogenic bacteria
lysogeny
lytic cycle
monerans
mycoplasmas
nitrogen fixation
nucleocapsid
obligate aerobes
obligate anaerobes
parasite
phage
prophage
rickettsias
saprobes
spirilla
temperate phages
tobacco mosaic virus
transformation
viral replication
viral transduction
viroids
virulent phages
virus

USING BIOLOGICAL TERMS

1. What is a phage?
2. Which term includes the following events: adsorption, penetration, replication, assembly, and lysis?
3. Distinguish between the terms *bacilli, cocci,* and *spirilla.*
4. Distinguish between the terms *binary fission* and *conjugation.*
5. Distinguish between viral transduction and viral replication.
6. When referring to viruses, what is a capsomere and a nucleocapsid?
7. Define the term *endospore.*
8. Distinguish between bacteria and blue-green bacteria.
9. Distinguish between a viroid and a virus.
10. What is an obligate anaerobe?

UNDERSTANDING CONCEPTS

1. Describe the structure of a typical virus. (12·1)
2. Explain why viruses cannot multiply on their own. (12·1)
3. List four characteristics used in the classification of viruses. (12·2)
4. Identify two reasons bacteriophages are used in the study of viruses. (12·3)
5. Describe the stages in the lytic cycle. (12·4)
6. Describe the process of viral transduction. (12·5)
7. What is a viroid? (12·6)
8. Describe the characteristics of organisms found in kingdom Monera. (12·7)
9. Name the green and blue pigments found in all blue-green bacteria. (12·8)
10. Describe the major characteristics of a bacterial cell. (12·9)
11. List the conditions most bacteria require for growth. (12·10)
12. Identify two main ways bacteria obtain energy. (12·11)
13. How do bacterial binary fission and animal cell mitosis differ? (12·12)
14. Describe spore formation in bacteria and explain its significance. (12·13)
15. Explain why conjugation in bacteria is similar to sexual reproduction in higher organisms. (12·14)
16. Give examples of several food products that require the use of bacteria in their manufacture. (12·15)

APPLYING CONCEPTS

1. A good deal of what is known about biochemistry has been learned from experiments with *E. coli*. Information about such topics as metabolism, DNA, RNA, and protein synthesis has come from work with this bacillus. Give some reasons why this bacterium has been studied so much.
2. Viruses can be taken apart and put back together in the laboratory. Their nucleic acids can be modified. Explain how this information might be used in the genetic engineering of plants or animals.
3. Bacteria were classified as plants through much of the history of biology. Why, do you think, were they classified in this way? Give several reasons why bacteria should not be classified as plants.
4. Viruses must adsorb to the surface of a host cell before penetrating. Molecules on the outer surface of a virus have a shape that matches, or "fits" onto, some molecule on the host cell. This matching of shapes seems to be the reason that each type of virus infects a certain type of host cell. What model of chemical activity is this matching of shapes similar to?

EXTENDING CONCEPTS

1. Consult your family physician or your school nurse to find out what diseases students are expected to be vaccinated against by the time they enter school. Which of these diseases are caused by bacteria? Which are caused by viruses?
2. Only a few of the beneficial activities of bacteria have been mentioned in this chapter. Go to the library and investigate this topic. Make as long a list as you can of the beneficial activities of bacteria.
3. The specific identification of bacteria is an important task in a medical laboratory. Two factors that help to identify bacteria are their shape and whether or not they need oxygen. Many other factors must be considered to identify exactly what species a bacterium belongs to. Get information on this from the library. Write a report on how to identify bacteria.
4. Arrange with your teacher and your classmates to organize a debate on this topic: Are viruses living organisms? There are more arguments for and against this question than have been given in this chapter. Consult references for more information.

READINGS

Epps, G. "Viroids Among Us." *Science 81,* September 1981, p. 70.

Gallo, Robert C. "The AIDS Virus." *Scientific American,* January 1987, p. 46.

Novick, R. "Plasmids." *Scientific American,* December 1980, p. 102.

13. The Protists

This is the skeleton of a radiolarian, a single-celled organism. This delicate, elaborate structure is formed of a glasslike substance. The fossilized skeletons of radiolaria are studied to determine the age of the earth and to detect changes in the earth's climate in the past. Radiolarians are not classified as plants or animals. They are classified as protists. What are some characteristics of protists? In what environments are protists found? These and other questions you might have will be answered as you read this chapter.

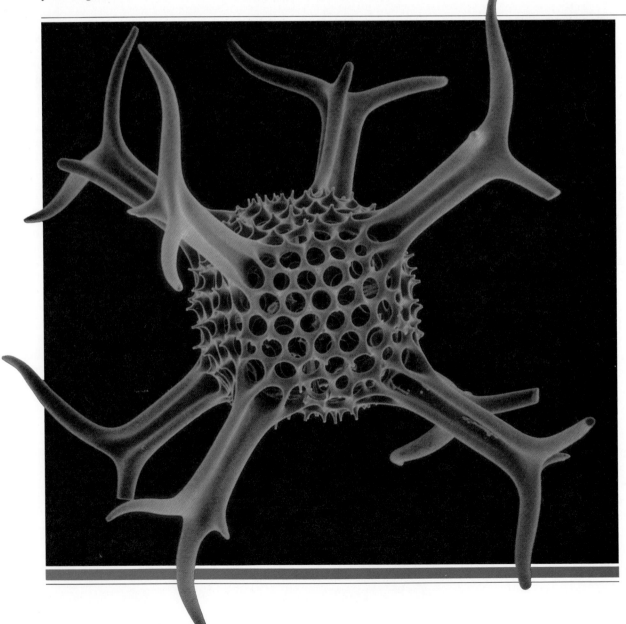

CHAPTER OBJECTIVES

After completing this chapter, you will be able to

- **List** important traits of some organisms in the kingdom Protista.
- **Discuss** the classification system used for protists.
- **Give examples** of some organisms classed in kingdom Protista.
- **Describe** protist reproduction.
- **Explain** how protists affect other organisms.

Introduction to Protists

13·1 THE NATURE OF PROTISTS

Kingdom Protista is a kingdom made up of simple eukaryotic organisms. Organisms grouped under kingdom Protista are called **protists.** The study of protists started in 1675. It was in that year that the Dutch merchant Anton van Leeuwenhoek used one of his simple microscopes to view thousands of microorganisms swimming in a drop of water. Leeuwenhoek called these newly discovered organisms "animalcules," or "little animals." In describing his discovery, Leeuwenhoek wrote, "This was for me, among all the marvels that I have discovered in nature, the most marvelous of all." Leeuwenhoek kept careful records over many years on the tiny organisms he saw. Since Leeuwenhoek's time, hundreds of biologists have contributed to the study of protists. Their studies have shown that there is a great diversity of life forms within kingdom Protista. However, protists are alike in many ways.

Most protists are unicellular. Those that are multicellular have a very simple structure with little specialization of cells. Protists are different from monerans. Protists are eukaryotes, whereas monerans are prokaryotes. The cells of protists contain membrane-bound organelles. Many biologists think that complex eukaryotic organisms, such as plants and animals, came from protists.

Most protists live in the oceans or in fresh water. Many of these protists are autotrophs and serve as a major food source for other organisms. These autotrophic organisms also produce much of the earth's oxygen. Kingdom Protista also includes some heterotrophs, organisms that eat other organisms or dead organic material. Among the heterotrophic protists are some parasites that cause diseases in animals. Other heterotrophs live on decaying matter. There are also a few protists that are autotrophic at some times and heterotrophic at other times.

SECTION OBJECTIVES

- Describe some traits of protists.
- Recognize the diversity of organisms in the kingdom Protista.

13·2 TRAITS AND CLASSIFICATION OF PROTISTS

Figure 13·1 shows a system for classifying organisms of the kingdom Protista. The organisms grouped in subkingdom *Protophyta* (proh TAHF uh tuh) are called **algal protists.** A subkingdom is a division of a kingdom. Many of these organisms are called algae. Algal protists include many different unicellular, plantlike organisms, most of which are autotrophs. The algal protists are found near the surface of the oceans as well as in bodies of fresh water. Since most algal protists have chlorophyll, they can make their own food.

Subkingdom *Gymnomycota* is made up of a group of organisms called **fungal protists.** Fungal protists, also called slime molds, are heterotrophic. Slime molds can be found growing on decaying leaves or logs in cool, shady places. Slime molds show traits of both fungi and animallike protists. They may be multicellular in some stages of their development. Slime molds are classified as protists mainly because in the reproductive stage of their development they look like certain animallike protists.

FIGURE 13·1 This table lists some of the characteristics of organisms in the Kingdom Protista. What is a characteristic of slime molds that is not shared by other protists?

SUBKINGDOM	PHYLUM	EXAMPLES	CHARACTERISTICS
Protophyta —the algal protists	Euglenophyta —euglenoids	Euglenas	Found mostly in fresh water, move with flagella, no cell wall
	Chrysophyta —golden algae	Diatoms	Mostly marine, have yellow-brown pigment, many have silica in cell wall
	Pyrrophyta —fire algae	Dinoflagellates	Mostly marine, usually have two flagella, cellulose in cell wall
Gymnomycota —the fungal protists	Myxomycota —true slime molds	Slime molds	Multicellular in some stages
Protozoa —the animallike protists	Sarcodina —sarcodines	Amoebas and their relatives	Move with pseudopodia (cell extensions)
	Ciliata —ciliates	Paramecia	Move with cilia
	Sporozoa —sporozoans	Plasmodia	No locomotion, all parasites
	Mastigophora —flagellates	Trypanosomes	Move with one or more flagella

Animallike protists are classified in subkingdom *Protozoa*. The animallike protists are called **protozoans**. Protozoans are unicellular, heterotrophic organisms. Like animals, protozoans get nourishment by taking in food. Many protozoans are highly specialized. Within their single cells may be found special structures for food getting, digestion, and reproduction. Figure 13·1 shows protozoans are subdivided into four phyla, mainly on the basis of their means of locomotion.

proto- (first)
-zoa (animals)

REVIEW QUESTIONS

1. How do protists differ from monerans?
2. Name two ways protists may be of significance to other organisms.
3. Distinguish among algal protists, fungal protists, and protozoans.
4. In Figure 13·1 you can see that three phyla of kingdom Protista include organisms that have flagella. To which of these phyla would a flagellated parasite living in a human's bloodstream likely belong? Why?
5. Suppose you are looking through a microscope at various protists in a drop of pond water. The largest protists that you view appear to be using cilia to move about. In what protist phylum would you classify them?

Algal and Fungal Protists

13·3 EUGLENOIDS

The three phyla of algal protists are made up of organisms that have chloroplasts. Phylum *Euglenophyta* (yoo glee NAHF uht uh) is a group of about 800 species of algal protists having both plant and animal traits. Organisms in phylum Euglenophyta are **euglenoids** (yoo GLEE noidz). Euglenoids generally are found in fresh water.

These organisms make their own food, like plants. However, unlike a plant cell, a euglenoid has no cell wall. Like most animals, euglenoids can move about. For locomotion, euglenoids have one or two flagella. *Flagella* are taillike structures used in movement.

SECTION OBJECTIVES

- List the traits and give examples of algal and fungal protists.
- Describe the life cycle of a slime mold.

FIGURE 13·2 As you can see, these three euglenoids vary in size. However, they all possess chloroplasts.

 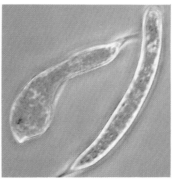 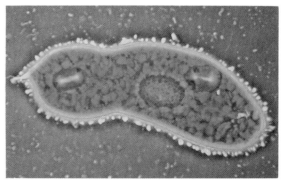

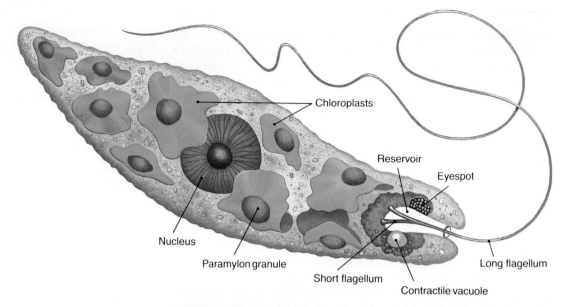

FIGURE 13·3 This illustration shows some of the structures in a euglena. The eyespot is sensitive to light and helps keep the euglena in a lighted area where it can make food though the process of photosynthesis.

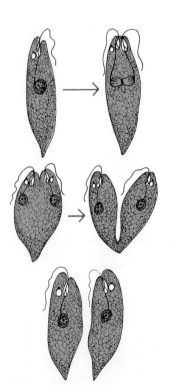

FIGURE 13·4 Euglenas divide longitudinally during cell division when they undergo asexual reproduction.

Euglena is a common genus in the phylum Euglenophyta. The structure of a euglena is shown in Figure 13·3. Locate the structures of a euglena in the figure as you read about this euglenoid. The flagella are attached at the base of a groove called the reservoir. The **reservoir** is a flask-shaped groove at the front of the cell. Notice that there is one long flagellum that sticks out from the reservoir. If you look carefully, you will see a shorter flagellum that does not stick out from the reservoir. A euglena can swim by moving its long flagellum.

Locate the contractile vacuole in the figure. The **contractile vacuole** is a structure that collects extra water and discharges it from the cell. This extra water is sent out through the reservoir.

The **eyespot,** which is near the reservoir, is a red, light-sensitive structure. The eyespot allows a euglena to detect light. A euglena responds to light by moving toward it. Since light is needed for food-making, the eyespot is an adaptation that helps the organism make food.

A euglena has no cell wall. However, inside the cell membrane are strips of protein called **contractile fibers.** These contractile fibers can stretch and contract, causing the shape of the cell to change. In addition to moving by using its flagellum, a euglena also can move in a squirming motion by changing its shape.

Euglenoids reproduce asexually by cell division, as shown in Figure 13·4. The nuclear membrane stays intact during mitosis, and the cell divides lengthwise. Sexual reproduction does not occur among the euglenoids.

Because of its many chloroplasts, a euglena looks bright green. The granules associated with the chloroplasts are believed to be food-storage bodies. Other food storage granules are also scattered through the cytoplasm. These granules store food as *paramylum*, a starchlike carbohydrate.

As long as euglenas are exposed to light, they carry on photosynthesis. However, laboratory experiments have shown that when they are placed in darkness in a nutrient-rich solution, their chloroplasts break apart. The euglenas then become heterotrophic, absorbing nutrients through the cell membrane. When they are returned to the light, they still live as heterotrophs. Euglenas lacking chloroplasts always remain heterotrophs. A euglena without chlorophyll is much like the flagellated animallike protist called *Astasia* (uh STAY zhuh). *Astasia* does not have chlorophyll. How might *Astasia* have come about naturally?

13·4 GOLDEN ALGAE

Phylum *Chrysophyta* (kruh SAHF uht uh) is made up of algal protists that include the diatoms, the golden-brown algae, and the yellow-green algae. All members of this group, sometimes called golden algae, are single-celled photosynthetic organisms. Golden algae are found in both marine and freshwater environments. Their cells contain chlorophyll, but the presence of other pigments causes their colors to range from brown to yellow-green.

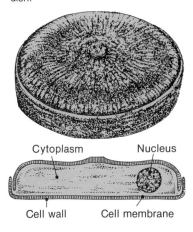

FIGURE 13·5 As you can see, diatoms are made of two parts that fit together much like the two plates in a petri dish.

Biologists believe that there are 6000 to 10 000 species of golden algae, most of which are diatoms. **Diatoms** are algal protists, each living within an overlapping two-part shell. This shell is the cell wall. Notice in Figure 13·5 that the two parts of the shell fit snugly together. Observe the radiating lines in the shell. These lines are grooves with pores that connect the cell on the inside to the outside environment.

The cell walls of diatoms have no cellulose. Instead, they are made of silica. *Silica* is a compound used in making glass. Silica gives the cell walls of diatoms their stiffness and glassy appearance.

The shells of several species of diatoms are shown in Figure 13·6. You can see that diatoms come in a wide variety of shapes and patterns. But in every case, one part of the shell overlaps the other.

Diatoms generally reproduce by mitosis. The two parts of the cell wall separate during mitosis. Each daughter cell then forms a new cell wall that fits inside the one that came from the parent cell.

FIGURE 13·6 These photographs show some of the variation present in diatoms. These organisms vary in size as well as shape.

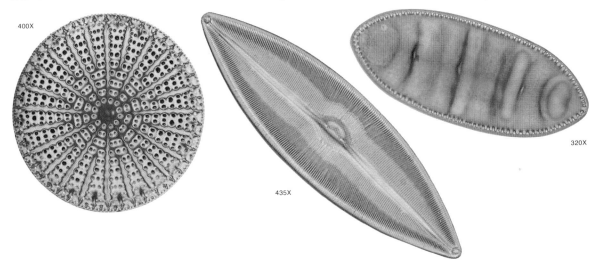

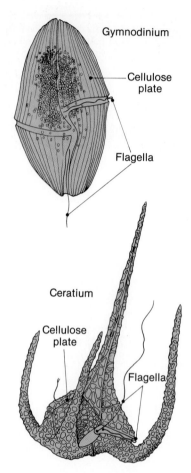

FIGURE 13·7 Dinoflagellates are found in the Phylum Pyrrophyta. Most of the dinoflagellates have two flagella. The flagella help move the organism through the water.

Diatoms lack flagella, but many species are able to move. This movement is produced by the streaming of cytoplasm extended through the grooves on the surface of the cell wall. Other members of phylum Chrysophyta have one or two flagella, which are used to carry on locomotion.

13·5 DINOFLAGELLATES

The organisms shown in Figure 13·7 belong to phylum *Pyrrophyta* (puh RAHF uht uh). Phylum Pyrrophyta is made up of about 1000 species of one-celled algal protists. The cell walls of these protists are made of cellulose plates. Each cell usually has two flagella. Members of phylum Pyrrophyta are called **dinoflagellates** (dī nuh FLAJ uh-layts). Most species are marine, but a few are found in fresh water.

In many dinoflagellates, such as those shown, the cellulose plates are thick structures that cover much of the cell. The flagella are at right angles to each other. The beating of the flagella causes the organism to move with a spinning motion.

Dinoflagellates have chlorophyll, but the chlorophyll is usually hidden by other pigments. Some species have pigments that give the cells a reddish color. A few species lack chlorophyll, and they get their food by taking in other small cells. However, they are still placed in phylum Pyrrophyta because they have so many other features of autotrophic dinoflagellates.

Most dinoflagellates reproduce asexually by cell division, but sexual reproduction takes place in a few species. Sometimes red dinoflagellates reproduce in such great numbers that the seawater is turned a bright color, usually red. This condition is called a *red tide*.

Mitosis in dinoflagellates occurs differently from the way it occurs in other eukaryotes. Remember that in eukaryotes the DNA exists in chromosomes enclosed within the nucleus. During mitosis, the nuclear membrane breaks apart. The replicated chromosomes become attached to the spindle, which separates them into two cells. In dinoflagellates the chromosomes are attached to the nuclear membrane. During mitosis, the membrane remains whole. The chromosomes are separated into two cells as the nuclear membrane and cell membrane pinch into two cells.

13·6 SLIME MOLDS

The organism shown in Figure 13·8 is a slime mold. **Slime molds** are fungal protists, having stages in their life cycle that are like those of both fungi and protozoans. Slime molds are found in cool, shaded areas, growing on damp organic matter such as rotting logs or decaying leaves.

The slime molds are classified in phylum *Myxomycota* (mihks uh-mī KOH tuh). Slime molds also are known as plasmodial slime molds because their structure during the growing stage is known as a plasmodium (plaz MOH dee uhm). A **plasmodium** is a mass of cytoplasm that contains many nuclei. The nuclei of the plasmodium undergo

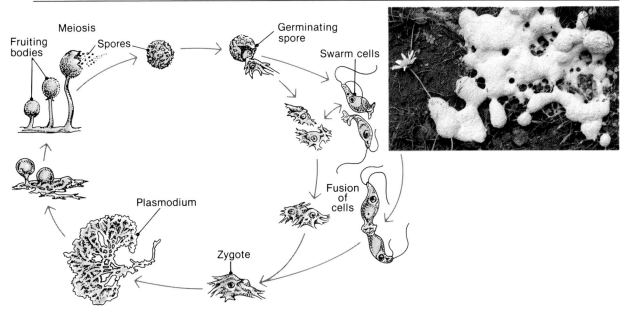

FIGURE 13·8 Slime molds are likely to be found in cool, damp woods. They obtain food from decaying plant material. The life cycle of a typical slime mold and a yellow slime mold (inset) are shown in this illustration.

mitosis, but no division of the cytoplasm occurs. As a result, a large, fan-shaped plasmodium containing thousands of nuclei develops. As the plasmodium gets bigger, it spreads slowly over the organic matter on which it is growing. Plasmodia up to 1 m in diameter have been found. As a plasmodium grows, it surrounds and digests small cells and particles of decayed plant and animal material.

Figure 13·8 shows a slime mold's fruiting bodies. These structures develop from the plasmodium when moisture or food becomes scarce. **Fruiting bodies** are reproductive structures that produce many spores. Spores are reproductive cells that are able to survive unfavorable environmental conditions. Figure 13·8 summarizes the life cycle of a slime mold. Notice the many spores being given off from a fruiting body. If spores land where conditions are favorable, they will germinate. Notice that the germinating spores may develop either into flagellated cells, called *swarm cells,* or into cells that look like small plasmodia. Both of these stages in the life cycle closely resemble protozoans. You can see that these protozoanlike cells serve as gametes, for they join to form a zygote. The zygote then develops into a new plasmodium, thus completing the life cycle.

13·7 IMPORTANCE OF ALGAL AND FUNGAL PROTISTS

Algal and fungal protists often are not noticed by people. Yet most of them are important organisms in the environments in which they are found. Diatoms and dinoflagellates are present in large numbers, and they are a main source of food for microscopic water animals. These microscopic animals then are food for clams, shrimp, and other

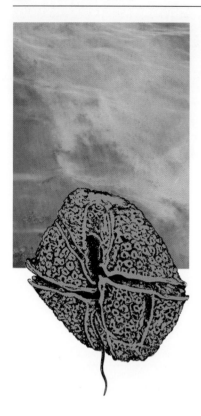

FIGURE 13·9 The green pigment chlorophyll is present in dinoflagellates, but it is usually masked by a red pigment. It is the red pigment that causes the water to appear red when dinoflagellates are present in large numbers (top).

small water animals. In turn, larger organisms, including humans, use these water animals for food. A food chain is thereby established. A *food chain* is a grouping of the living things in a community according to which organism is used by each as food. Organisms in the food chain depend, directly or indirectly, on the algal protists for food. Also, because of their great numbers in the oceans of the earth, the food-making algal protists supply large amounts of oxygen to the earth's oceans and atmosphere.

Diatoms are the source of an important commercial product. After a diatom dies, its glassy cell wall settles to the bottom of the lake or ocean in which the organism lived. Over a long period of time, billions of these diatom shells build up. In some places, where ancient seabeds are now above sea level, thick deposits of diatom shells can be found. The piled-up shells of diatoms form a fine, powdery substance known as **diatomaceous** (dī uh tuh MAY shuhs) **earth.** Diatomaceous earth is used as a gritty substance in toothpaste and scouring powders. Diatomaceous earth is added to the paints used to mark lines on highways. Its glasslike properties cause the paint to reflect light.

Not all activities of algal protists are helpful. Outbreaks of red tide caused by certain dinoflagellates have been a problem for the seafood and tourist industries. Figure 13·9 shows a red tide and one of the dinoflagellates that can cause this condition in coastal waters. The poison produced by the dinoflagellates kills thousands of fish. Many of the dead fish are washed ashore, where they litter beaches in vacation areas. Shellfish, such as clams and oysters, are not killed by the poison. However, the poison builds up in their tissues. Such shellfish are then dangerous for humans to eat. In 1980, clams, mussels, and oysters off the coast of Maine could not be gathered due to red tide. The fishing industry lost about $7 million as a result. As yet, biologists are not certain of the factors that cause outbreaks of red tide.

Slime molds are of special interest to biologists because of the different forms that appear in their life cycle. The plasmodium looks like a fungus, and it produces spores in much the same way fungi do. However, the flagellated cells that fuse to form a zygote are similar to some animallike protists. Research biologists are studying these organisms to gain a better understanding of the processes by which certain cells develop specialized functions.

REVIEW QUESTIONS

6. Describe the structure of a euglena.
7. Distinguish between a diatom and a dinoflagellate.
8. How are slime molds like both fungi and protozoans?
9. Identify some ways in which algal and fungal protists are important.
10. While viewing a sample of seawater under a microscope, a marine biologist observes some unicellular algal protists. Their cells are red and covered with thick cellulose plates. They spin around like tops in the water. To what phylum do these protists likely belong?

Protozoans: Animallike Protists

13·8 SARCODINES

The animallike organisms of kingdom Protista are called protozoans. Like animals, all protozoans are heterotrophic. Even though protozoans are single-celled, they carry out all the processes needed for life.

Most protozoans can move about, and their classification is based generally on their means of locomotion. Phylum *Sarcodina* (sahr kuh-DĪN uh) consists of protozoans that move by means of extensions of the cell called pseudopodia (soo duh POH dee uh), which means "false feet." **Pseudopodia** (sing., *pseudopodium*) are fingerlike projections of a cell that function in locomotion and feeding. The pseudopodium is formed by a flowing action of the cytoplasm against the cell membrane.

Members of phylum Sarcodina are called **sarcodines** (SAHR kuh-dīnz). Sarcodines are found in freshwater puddles, ponds, and lakes. Some species are found in the oceans. Many biologists think that sarcodines may have developed from flagellated forms of protists.

Both asexual and sexual reproduction take place among sarcodines. Asexual reproduction takes place by cell division. In sexual reproduction, cells undergo meiosis to form gametes. These gametes then join to form a zygote, which develops into a new sarcodine.

Sarcodines exist in many different forms. Some of these forms are shown in Figure 13·10. Amoebas move about with broad, flat pseudopodia. A **radiolarian** (ray dee oh LAIR ee uhn) is a sarcodine that has a lacy, glassy shell composed of silica. Radiolarians have sticky, needlelike pseudopodia that trap microscopic organisms for food. Radiolarians are found only in warm oceans. A **foraminiferan** (fuh-ram uh NIHF er ehn) is a sarcodine that has a shell of calcium carbonate. Foraminiferans have a network of branched pseudopodia that come out from holes in their shells. These sarcodines are found mostly in seawater.

SECTION OBJECTIVES

- Name three phyla of Protozoa and give examples of organisms in each phyla.
- Compare the methods of locomotion in protozoa.
- Describe protozoa reproduction.

FIGURE 13·10 Notice the variations in these three sarcodines. A radiolarian is shown on the left, an amoeba in shown on the right, and a foraminiferan is shown below.

13·9 AMOEBA—A SIMPLE PROTOZOAN

Perhaps the most familiar sarcodine is the amoeba. An **amoeba** is a freshwater sarcodine. Amoebas are most often bottom dwellers in freshwater ponds. They can often be gathered from the surface of submerged vegetation or from decayed organic matter from the bottom of a pond. When you look for an amoeba under a microscope, you may have trouble finding it at first. The organism is a dull gray-brown and looks much like the organic matter on which it is found.

Amoeba proteus, a common amoeba, is shown in Figure 13·11. *A. proteus* is a large protozoan, ranging in size from 250 μm to 500 μm. Look at the drawing to identify the structures in *A. proteus* as you read about this familiar protozoan.

The cytoplasm of *A. proteus* contains two regions. The clear outer layer of the amoeba's cytoplasm, just inside the cell membrane, is the **ectoplasm.** The granular form of cytoplasm in the interior, containing the cell organelles, is the **endoplasm.** The endoplasm is a watery fluid that also has fat droplets, minerals, and food granules.

An amoeba moves by means of pseudopodia. An amoeba has no head end or tail end. It can send out a pseudopodium at any point along its cell membrane. As more cytoplasm flows into a pseudopodium, the organism slowly flows along in the direction in which the pseudopodium is extended. The kind of movement shown by an amoeba, in which it moves by changing the shape of its cell, is called **amoeboid movement.** As new pseudopodia form, old ones flow back into the main part of the cell.

ecto- (on the outside)
endo- (inner)

FIGURE 13·11 A live amoeba is shown in the photograph (left). Some of the structures found in an amoeba are shown in the drawing (right). What is the function of the contractile vacuole?

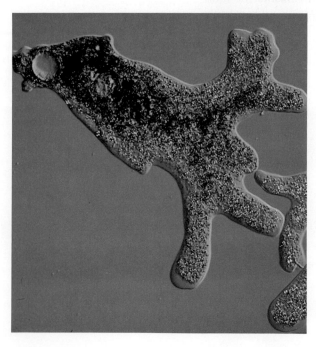
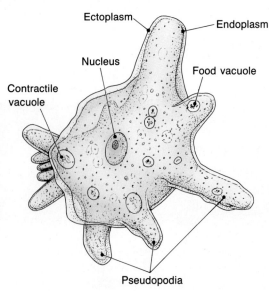

An amoeba uses its pseudopodia not only for locomotion but also for engulfing food. An amoeba takes large, solid materials into its body by a process called *phagocytosis*. The photograph shows an amoeba extending its pseudopodia around a smaller protozoan. The pseudopodia extend around the sides and over the top of the prey. The captured protist is thus surrounded by the amoeba's cell membrane, and a food vacuole forms. A **food vacuole** is an organelle in which food is digested. Enzymes from a lysosome are secreted into the food vacuole. The digested food then is absorbed into the cytoplasm, and the food vacuole disappears. Undigested food particles are moved to the cell membrane, where they are freed from the cell as the organism flows away.

Notice in Figure 13·11 that an amoeba has a large contractile vacuole in its cell. Since the amoeba lives in fresh water, water will usually be entering the cell by osmosis. As in a euglena, the contractile vacuole maintains water balance by gathering extra water and pumping it from the cell.

An amoeba has no special organelles for gas exchange. There is simply a free exchange of oxygen and carbon dioxide across the cell membrane. Oxygen needed for cellular respiration diffuses in, and carbon dioxide diffuses out.

Though an amoeba has no nervous system, it does react to certain stimuli in the environment. For example, an amoeba reacts negatively to light. If light is shone on one side of an amoeba, the organism will move away from the source of light. However, it responds positively to sources of food in its environment.

An amoeba reproduces by cell division. As you can see in Figure 13·12, the amoeba simply divides into two amoebas. An amoeba may live through unfavorable conditions by forming a cyst. A **cyst** is a protective shell secreted around a protozoan cell that allows the cell to survive unfavorable conditions. An amoeba may form a cyst when food or moisture becomes scarce or during periods of extreme heat or extreme cold. Within the cyst, the amoeba's metabolism falls to a level just high enough for survival. When conditions become more favorable, the organism comes out of the cyst and again becomes active. Perhaps you have wondered how protozoan life can appear in a newly dug pond. Cysts of protozoans can be carried from pond to pond by the wind.

13·10 CILIATES

Phylum *Ciliata* is made up of one-celled organisms that have many cilia. *Cilia* are short hairlike extensions from a cell. Members of phylum Ciliata are called **ciliates.** Ciliates are found in both fresh water and seawater. Their food consists mostly of bacteria.

When a drop of pond water is viewed under a microscope, ciliates are generally the most noticeable protozoans. They move rapidly about the field of view, and they often outnumber all the other protozoans in the water drop.

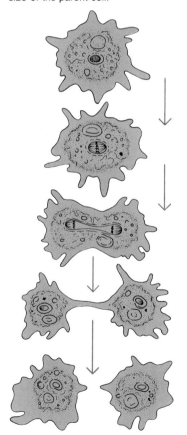

FIGURE 13·12 Amoebas reproduce by fission, or dividing in half. Each of the cells that result from fission is half the size of the parent cell.

macro- (large)

micro- (small)

Ciliates have two kinds of nuclei. Each ciliate has one or more large nuclei, called macronuclei, and one or more small nuclei, called micronuclei. The **macronucleus** is the active control center of the cell. It is made by the micronucleus, and it may contain 50 to 100 copies of the DNA found within the micronucleus. The **micronucleus** is the nucleus that undergoes meiosis before the form of sexual reproduction that takes place in ciliates. A ciliate cell cannot live without a macronucleus, even if the cell still has a micronucleus. On the other hand, a cell can live without a micronucleus.

There are many different kinds of ciliates, with a variety of feeding structures and modifications for locomotion. A few of these varieties can be seen in Figure 13·13. The best-known genus of the phylum is *Paramecium,* which will be discussed in detail in the next section. The whole cell of a paramecium is covered with short cilia. However, *Vorticella* (vor tuh SEHL uh) has cilia only in the end of the cup-shaped feeding structure. These cilia sweep food particles into the cell. Notice that the feeding structure is at the top of a coiled stalk. When something disturbs *Vorticella,* the stalk contracts, lowering the organism away from danger. Now look at *Stentor. Stentor* is a huge trumpet-shaped ciliate that sticks to objects by the narrow end of the trumpet when it is feeding. *Stentor* sucks in its food by setting up water currents with the large fused cilia around its mouth. When *Stentor* is not feeding, it becomes detached and swims about freely. In some ciliates, the cilia are modified for different functions. In *Euplotes* (yoo PLOHT eez), the cilia have been modified to aid movement. In other species of ciliates, the cilia may be joined to form brushlike structures that sweep food into the mouth of the cell.

FIGURE 13·13 Each of the protists in this figure are ciliates. The cilia are used for movement or to create water currents that bring food to the ciliate.

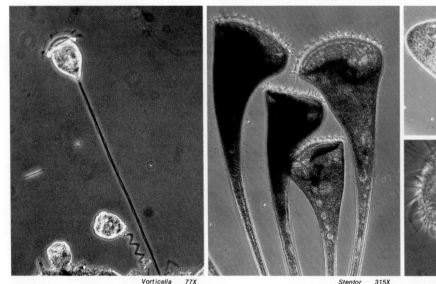
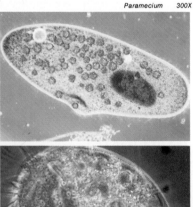

13·11 PARAMECIUM—A COMPLEX PROTOZOAN

The ciliate shown in Figure 13·14 is *Paramecium caudatum*. A **paramecium** is a ciliate often found in pond water. Unlike an amoeba, a paramecium (pl., *paramecia*) has a definite shape, resembling a slipper. A paramecium's shape is kept by the firm but flexible outer covering called the **pellicle.** A paramecium has a definite front end and a definite tail end.

A paramecium is covered with about 25 000 cilia arranged in rows. These cilia do not all beat at once. The beating is a wave motion along the body of the organism. The way in which the cilia beat causes a paramecium to rotate lengthwise as it swims. Beneath the pellicle, each cilium is joined to a network of connecting fibers. These fibers function in regulating the beating of the cilia. A paramecium may move as fast as 2–3 mm/s. This is about equal to a car traveling 160 km/h. Paramecia can also reverse the direction of movement of their cilia and move backward.

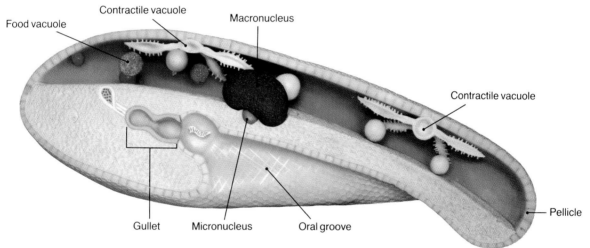

FIGURE 13·14 Some parts of a paramecium's cell are shown in this illustration. Just beneath the pellicle, tiny structures that may contain poison, help protect this protist.

A paramecium is far more specialized than an amoeba for catching and digesting its food. Its food is usually bacteria. The **oral groove,** shown in Figure 13·14, is a tube through which beating cilia drive food toward the mouth opening. A funnellike tube called the **gullet** leads from the mouth opening into the cell. Food vacuoles form at the lower end of the gullet. As the food is being digested, the vacuoles move through the cell, and nutrients are absorbed into the cytoplasm. Any undigested food particles are let out of the cell through a fixed structure called the **anal pore.**

Most species of paramecia have two contractile vacuoles, at opposite ends of the cell. Each of these contractile vacuoles is surrounded by a set of canals that lead into the vacuole. As in amoebas, the contractile vacuoles control the amount of water in the cell. Also, there are no special structures for gas exchange. Oxygen and carbon dioxide are exchanged across the cell membrane by diffusion.

268 THE PROTISTS

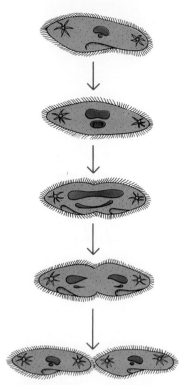

FIGURE 13·15 Unlike the amoeba, a paramecium divides across the middle of its cell.

FIGURE 13·16 During conjugation, two paramecia exchange genetic material. This exchange leads to the development of new strains of paramecia.

Two methods of reproduction have been observed in paramecia. In asexual reproduction the cell divides in two. This is pictured in Figure 13·15. You can see that the cell divides across the middle. During the dividing process, the micronucleus divides by mitosis. However, the macronucleus does not undergo mitotic division. Instead, it lengthens and breaks into two equal parts.

A kind of sexual reproduction known as conjugation takes place in paramecia. **Conjugation** is a process in which two one-celled organisms join and exchange genetic material. A paramecium cannot be distinguished as being either male or female. However, there are a number of different mating strains. Two individuals of the proper mating strains can unite at their oral grooves and transfer genetic material. The process of conjugation in paramecia is summarized in Figure 13·16. Follow the steps in Figure 13·16 as the process is discussed.

1. Two paramecia join at their oral grooves.
2. The macronuclei break apart.
3. The micronuclei divide by meiosis to form four micronuclei in each paramecium.
4. Three of the four micronuclei in each cell break apart.
5. The remaining micronuclei now divide by mitosis. The nuclei formed are of unequal size.
6. The smaller micronuclei are exchanged between cells, and they fuse with the larger micronuclei. The fused nucleus in each cell now has genes from two sources.
7. The two paramecia separate, and several nuclear divisions result in the formation of new macronuclei and micronuclei.
8. Each paramecium divides twice to produce eight new paramecia (four from each original parent).

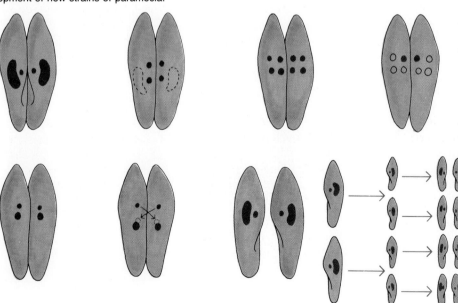

A paramecium responds to a number of changes in its environment. For example, it may respond to certain stimuli by sending out structures called trichocysts (TRIHK uh sihsts). **Trichocysts** are sharp-pointed structures beneath the pellicle that are sent out for defense or for catching prey. The trichocysts may be poisonous.

tricho- (hair)
cystis (pouch)

A paramecium also will show a negative response to light. It will try to move away from a source of bright light. A paramecium will also respond to objects in its path in a trial-and-error way. If a paramecium meets an object, it backs up and swings to one side. It then moves forward again. It continues this behavior until it finds a clear path.

13·12 SPOROZOANS

Phylum *Sporozoa* is made up of parasitic protozoans that form many small cells called spores during some part of their life cycle. Members of phylum Sporozoa are often called **sporozoans.** All sporozoans lack cilia and flagella as adults. Thus they have no external structures for locomotion. Many sporozoans go through complex life cycles that involve moving from one host to another. The spore is the stage in which the sporozoan is transferred to a new host. During sexual reproduction, sex cells join to form a zygote. The nucleus of the zygote then divides many times until a number of nuclei have been formed. Each nucleus becomes enclosed in a small amount of cytoplasm to become a spore. In many sporozoans, the sexual, or spore-producing, stage of reproduction alternates with an asexual stage. The best known sporozoans are members of the genus *Plasmodium,* which includes the organism that causes malaria (Section 15·9).

Malaria — *Discovery*

Malaria is a disease caused by a protist in phylum Sporozoa. The sporozoans that cause malaria are spread from one human to another by mosquitoes. For many years, people have been battling malaria by killing the mosquitoes that spread the disease and by treating infected humans with quinine drugs to fight the disease.

However, the mosquitoes have developed resistance to the chemical sprays, and the sporozoans that cause malaria have developed resistance to the quinine drugs. Malaria, once almost stamped out, is again a major killer and serious disease.

Scientists are now working to develop a vaccine that can stimulate the human body to produce chemicals to protect itself against the malaria-caus-

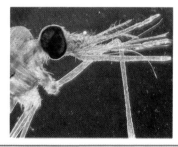

ing parasites. The job is complex because the sporozoan quickly changes into a different form when it enters the human body. Soon it changes again. Thus the body needs to produce three types of chemicals instead of one.

Research teams recently have discovered how to make a specific protein associated with the sporozoan. The protein causes the body to produce a chemical against the sporozoan. Such a protein could be used to make a vaccine that may immunize humans against malaria.

Investigation

Are the Same Kinds of Protozoans Present in Different Water Supplies?

Goals
After completing this activity, you will be able to
- Identify protozoans by type from observation of living forms.
- Describe some characteristics and habits unique to each type of protozoan.

Materials (for groups of 2)
microscope   3 slides
3 coverslips   3 droppers
glass-marking pencil
water samples from 3 different sources in labeled beakers

Procedure

A. With a glass-marking pencil, label each slide with a letter to identify the source of each sample.

B. Prepare a wet mount from each water sample provided. See the figure below for the method for preparing the slide. If debris is present in the water, include some in the drop of water on your slide. Use a different dropper for each source.

C. On a separate sheet of paper, prepare a data chart like the one shown below. Leave enough room in the chart for drawings. In the proper columns draw protozoans observed in each water source. Under each drawing identify the type of protozoan. Use S for sarcodine, C for ciliate, and F for flagellate.

PROTOZOANS OBSERVED		
SOURCE A	SOURCE B	SOURCE C

D. Scan each slide under low power to find protozoans. Watch for movement and search the pieces of debris. Reduce the light slightly to help in seeing clear forms, especially the slow-moving sarcodines.

E. When you locate protozoans that are not moving across the field of view, switch to high power. In the proper column of the data chart sketch the forms observed under high power.

F. Identify the type of protozoan observed by looking for cilia, flagella, or pseudopodia. Refer to the text for descriptions and pictures to aid in identification.

G. Choose one protozoan that is moving. On your data chart describe its method of locomotion.

H. Draw up a class list of all the types of protozoans observed in the three water sources. From this list determine which types occur most often and which occur least often.

I. Follow your teacher's instructions for the proper disposal of all materials.

Questions and Conclusions

1. Which type of protozoan is present in largest numbers in each source?
2. Which type of protozoan occurs in all three sources? Which type occurs least often?
3. Which water source contains the greatest variety of protozoans?
4. Describe the environmental conditions for each water source. (Consider light, temperature, location, currents, and any special factors.)
5. What are some factors that may account for variations in the number and kinds of protozoans in the three sources?
6. Why are sarcodines more difficult to detect than other protozoan types?
7. Is a flagellate pushed or pulled by the flagellum?

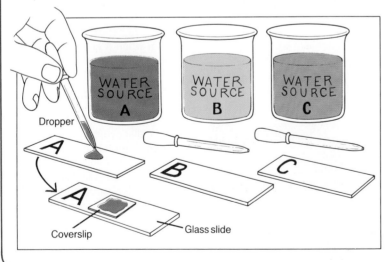

13·13 FLAGELLATES

There are several groups of algal protists whose cells bear one or more flagella. There is also a group of animallike protists whose cells may have flagella. These flagellated animallike protists are classified in phylum *Mastigophora* (mas tuh GAHF uhr uh). Members of phylum Mastigophora are commonly called **flagellates.** The flagellates look much like certain flagellated algal protists, such as *Euglena*. Because of this resemblance, many biologists believe these flagellates developed from algal protists.

Some flagellates are free-living, and some of them are parasites. Members of the genus *Trypanosoma* (trihp uh nuh SOH muh) cause African sleeping sickness in humans. Another flagellate, of the genus *Giardia* (jee AHR dee uh), causes certain intestinal disorders in humans.

A flagellate of special interest to biologists is the *choanoflagellate* (koh uh nuh FLAJ uh layt), shown in Figure 13·17. This protozoan consists of a cell having a high transparent collar that has a single flagellum. Water currents produced by the flagellum bring food particles into the collar, where they are taken in by the cell. These choanoflagellate cells are like the feeding cells that line the inside of a sponge. Some biologists think that ancient protists similar to choanoflagellates may be the ancestors from which simple animals, like sponges, arose.

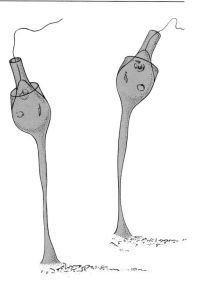

FIGURE 13·17 This flagellate, a choanoflagellate, resembles cells that line the inside of a sponge.

13·14 THE IMPORTANCE OF PROTOZOANS

The many species making up subkingdom Protozoa vary greatly in structure and complexity. Billions of these unicellular organisms can be found throughout the world. Protists live in almost every kind of environment.

Some of the protozoans cause disease. Diseases such as malaria and dysentery cause much human suffering. However, most protozoans are not disease-causing parasites. Some protozoans live in the bodies of humans and other animals without causing disease.

Like the algal protists the protozoans are important sources of food for animals living in the earth's waters. Some kinds of protozoans feed on bacteria. Such protozoans may be important in controlling bacterial populations.

REVIEW QUESTIONS

11. Compare methods of locomotion in these protists: sarcodines, ciliates, and flagellates.
12. Describe the process of conjugation in paramecia.
13. How are radiolarians and foraminiferans alike? How are these organisms different?
14. How are protozoans important to other organisms?
15. Suggest some of the reasons why biologists consider the ciliates to be the most highly specialized and complicated protozoans.

Chapter Review

SUMMARY

13·1 Protists are simple eukaryotic organisms.

13·2 Because of the diversity among members of kingdom Protista, the organisms are subdivided into three subkingdoms.

13·3 Euglenoids are flagellated algal protists that lack cell walls. *Euglena* is a well-known euglenoid.

13·4 The golden algae are unicellular algal protists with pigments that range from brown to yellow-green. Diatoms are the most numerous golden algae.

13·5 Dinoflagellates are flagellated algal protists with cell walls made up of cellulose plates.

13·6 Slime molds are composed of a funguslike plasmodium during part of their life cycle. During the other part of their life cycle, they produce cells resembling protozoans.

13·7 Algal protists are a major source of food for many water organisms. A few, such as the diatoms, are a source of commercial products. Dinoflagellates may produce red tides.

13·8 Sarcodines are protozoans that have pseudopodia for locomotion and capturing food.

13·9 An amoeba is a simple sarcodine commonly found in freshwater ponds.

13·10 Ciliates have cilia for locomotion.

13·11 A paramecium is a complex ciliate with many specialized cell structures.

13·12 Sporozoans are parasitic protozoans that have no external means of locomotion.

13·13 Flagellates are protozoans that use one or more flagella for locomotion.

13·14 Protozoans are single-celled protists that live in nearly every kind of environment. Some cause disease, but others are an important source of food for aquatic animals.

BIOLOGICAL TERMS

algal protists
amoeba
amoeboid movement
anal pore
ciliates
conjugation
contractile fibers
contractile vacuole
cyst
diatomaceous earth
diatoms
dinoflagellates
ectoplasm
endoplasm
euglenoids
eyespot
flagellates
food vacuole
foraminiferan
fruiting bodies
fungal protists
gullet
macronucleus
micronucleus
oral groove
paramecium
pellicle
plasmodium
protists
protozoans
pseudopodia
radiolarian
reservoir
sarcodines
slime molds
sporozoans
trichocysts

USING BIOLOGICAL TERMS

Identify the term that does not belong and explain why it does not belong. Include a definition of each term.

1. amoeba, foraminiferan, radiolarian, paramecium
2. flagellates, sarcodines, sporozoans, euglenoids
3. euglenoids, diatoms, slime molds, dinoflagellates
4. reservoir, cilia, oral groove, pellicle
5. macronucleus, pseudopodia, micronucleus, anal pore
6. eyespot, flagellum, contractile fibers, trichocysts

Study each of the terms below. Based on the origins, explain why each term is appropriate.

Biological Terms	Origin
7. macronucleus	(macro-, large)
8. pseudopodia	(pseudo-, false; -pod, foot)
9. ectoplasm	(ecto-, outer; -plasm, form)
10. protozoan	(proto-, first; -zoa, animals)

11. What are trichocysts?
12. What is amoeboid movement?

13. Distinguish between a micronucleus and a macronucleus.
14. What are fruiting bodies?
15. What is a pellicle?

UNDERSTANDING CONCEPTS

1. How do protists differ from monerans? (13·1)
2. How are protists that live in the oceans important to other organisms? (13·1)
3. Name the three subkingdoms of protists and give the characteristics of each. (13·2)
4. In what ways are euglenoids like both plants and animals? (13·3)
5. What is the function of the contractile vacuole in a euglena? (13·3)
6. Describe the structure of a diatom. (13·4)
7. In what ways do members of phylum Chrysophyta move? (13·4)
8. Describe the structure of a typical dinoflagellate. (13·5)
9. Explain the life cycle of a slime mold. (13·6)
10. Name two ways in which algal protists are important to humans. (13·7)
11. What is a red tide? (13·7)
12. What characteristic is common to organisms in phylum Sarcodina? (13·8)
13. Compare radiolarians and foraminiferans. (13·8)
14. Sketch a typical amoeba and label its main parts. (13·9)
15. How does a cyst help an amoeba survive unfavorable conditions? (13·9)
16. What characteristic is common to organisms in phylum Ciliata? (13·10)
17. Explain how a paramecium takes in and digests its food. (13·11)
18. Describe the two ways in which paramecia reproduce. (13·11)
19. Name two characteristics of protozoans in phylum Sporozoa. (13·12)
20. What characteristic is common to organisms in phylum Mastigophora? (13·13)
21. List several ways in which protozoans may be important to humans. (13·14)

APPLYING CONCEPTS

1. Explain why some algal protists are not colored green even though all of them have chlorophyll.
2. Of what value might flagella be to euglenoids and dinoflagellates?
3. A paramecium is generally considered to be a more advanced protozoan than an amoeba. Give some reasons why this idea is so.
4. Suppose all diatoms suddenly disappeared from the oceans. What effect, if any, would their disappearance have on other kinds of organisms?
5. A biologist proposes placing the algal protists and the protozoans in separate kingdoms. Give reasons why you would support or oppose this idea. If you support, how would you classify the slime molds?

EXTENDING CONCEPTS

1. Using at least two references, prepare a written report on the contributions of one of the following biologists to our knowledge of protists.
   a. Anton van Leeuwenhoek (1632–1723)
   b. Félix Dujardin (1801–1860)
   c. Karl T.E. von Siebold (1804–1885)
2. Make a collection of algal protists. Diatoms may be found on the surface of submerged rocks and stems in ponds. Euglenas may be found in the scum at the edges of shallow ponds or in puddles in a pasture. Bring to class any specimens you find, and examine them with a microscope.
3. Obtain a culture of paramecia from your teacher. Determine the responses of paramecia to a variety of chemical substances. Dip a small piece of thread in the substance to be tested. Place the thread in a drop of the paramecium culture on a microscope slide and observe the behavior of the organisms.

READINGS

Bonner, John Tyler. "Chemical Signals of Social Amoebae." *Scientific American,* April 1983, p. 114.

Douglas, Lee. "Slime Mold—The Fungus That Walks." *National Geographic,* July 1981, p. 130.

Teasdale, Jim. *Microbes.* Silver Burdett, 1985.

14. The Fungi

Mushrooms play a unique role in the ecology of the damp forest or the grassy field in which they grow. Most kinds of mushrooms feed on the remains of dead plant material. In doing this, mushrooms obtain nourishment and also recycle certain materials that green plants can use again. In what way are mushrooms like plants? How are mushrooms different from plants? How do mushrooms reproduce?

CHAPTER OBJECTIVES

After completing this chapter, you will be able to
- **Identify** the reasons fungi are classified in a separate kingdom.
- **List** the characteristics of the major phyla of fungi.
- **Describe** ways the major phyla of fungi reproduce.
- **Discuss** ways fungi are important to plants and animals.
- **Explain** the importance of fungi to humans.

Characteristics of Fungi

14•1 WHY A SEPARATE KINGDOM?

Yeasts, molds, mildews, rusts, mushrooms, and toadstools are among the many different organisms in kingdom *Fungi*. A **fungus** is an organism made of eukaryotic cells with cell walls and that gets its nourishment by absorbing organic substances. Fungi range in size from microscopic, one-celled yeasts to large mushrooms that may be 25 cm or more in diameter.

Have you ever seen a mushroom growing in a shady area after a period of warm, damp weather? If so, you will have noticed that mushrooms are like plants in some ways. Mushrooms, like plants, grow upward from the soil. The cells of a mushroom have cell walls, as do the cells of plants. For these and other reasons, biologists once classified mushrooms and other fungi as plants.

Many fungi grow as microscopic tubes, or filaments. These filaments are like those of some kinds of algae and are surrounded by cell walls. However, the cell walls of fungi are different from those of algae and plants. The cell walls of most fungi contain *chitin* (KĪ tuhn), a substance found in the shells of crabs and the outer coverings of insects.

Unlike plants, fungi have no chlorophyll or chloroplasts. Fungi cannot make their own food. Like animals, fungi use food substances made by other organisms. Fungi also must digest their food. However, fungi do not digest food within their bodies as animals do. Fungi digest food outside their bodies and then absorb the products of digestion. Fungi secrete digestive enzymes into their surroundings. The enzymes break down the organic matter near the fungus into small molecules, which are then absorbed.

You can see that the fungi are organisms with a special mixture of traits. Most biologists now think of fungi as making up a separate kingdom.

SECTION OBJECTIVES

- Discuss the reasons fungi are grouped in a separate kingdom.
- Describe the basic structure of a fungus.

14•2 THE NATURE OF FUNGI

Fungi use organic matter as a source of nourishment. For this reason, fungi generally live close to other life forms or near dead organic material. Fungi need moisture, and many cannot live in strong sunlight. The damp, shady floor of a forest with its covering of dead leaves gives a suitable habitat for fungi. Fungi are often found in forests, but they may be found in many other places as well. Fungi may live inside other organisms. The organisms provide sources of organic matter and the dark, moist conditions in which fungi thrive.

Most fungi are *saprobes*. Remember from Chapter 12 that saprobes consume the dead remains and waste products of other organisms. Saprobic fungi, such as those shown in Figure 14•1, break down dead organic matter. They help to put back into the soil minerals needed for the growth of plants.

Some fungi are *parasites*. Remember that parasites are organisms that get organic nutrients from a living host. A parasite usually harms or injures its host. Fungi may be parasitic on plants, animals, or even other fungi. For example, fungi of the genus *Trichophyton* (trih kuh FĪ tuhn) are parasitic on humans and cause athlete's foot. Wheat rust and corn smut are fungi that are plant parasites that harm important food crops.

FIGURE 14•1 Five examples of saprobic fungi: bracket fungi (A), morels (B), mushrooms (C and D), and cup fungi (E). All of these fungi consume the remains and waste products of other organisms.

A third group of fungi are neither saprobes nor parasites. These are fungi that get organic matter from living tissues but that do no harm. Such fungi live closely with other life forms. Several examples of these associations are presented in Section 14·10.

14·3 GENERAL STRUCTURE OF A FUNGUS

Most fungi are made up of branching filaments, or tubes, called **hyphae** (HĪ fee). As you can see in Figure 14·2, the hyphae (sing., *hypha*) contain cytoplasm. Nuclei are found throughout the cytoplasm. In some fungi the hyphae are divided into cells by cross walls called **septa.** In other fungi, however, no septa appear. Even in fungi with septa, the cytoplasm is continuous. As you can see, each septum has a pore in it.

The hyphae of a fungus grow and branch into a complex, tangled network. Such a network of fungal hyphae is called a **mycelium** (mī SEE lee uhm). The mycelium spreads over the surface or into the food source on which the fungus is growing. Often, the mycelium is hidden from view. It may be within the soil or beneath the bark of a decaying tree.

A fungal mycelium spreads in all directions through its food source. The mycelium forms the body of the fungus. When a fungus is mature, part of the mycelium may develop into a structure specialized for reproduction. Look at the bracket fungus growing on the tree trunk in Figure 14·1. The mycelium of this fungus has spread throughout the bark of the tree. The part of the fungus that you can see in the photograph is the reproductive structure, or *fruiting body*. The parts of a fungus other than the fruiting body often are called *vegetative hyphae* or *vegetative mycelium.*

Most fungi can reproduce both sexually and asexually. Asexual reproduction is often by *fragmentation.* Fragments of broken hyphae may be carried to new places by wind or water. If conditions are suitable, these fragments will grow into new fungi.

Fungi most often reproduce by producing *spores.* The air may be so full of fungal spores in late summer that the spores may cause problems for people with allergies. Spores are formed in the fruiting bodies of fungi by either a sexual or an asexual process. In some fungi the fruiting bodies are microscopic. In others, such as mushrooms and bracket fungi, the fruiting bodies can easily be seen with the unaided eye. The method of producing spores and the kind of fruiting body formed are features used to classify fungi.

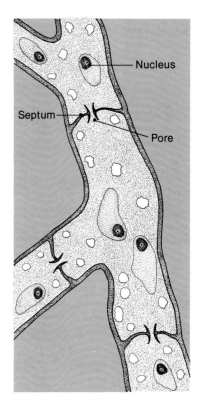

FIGURE 14·2 Most fungi are made up of branching hyphae. As you can observe, the cytoplasm in fungi is continuous. Each septa has a pore in it.

REVIEW QUESTIONS

1. Why are fungi placed in a separate kingdom?
2. How do fungi obtain nourishment?
3. What is a hypha?
4. What features are used in classifying fungi?
5. What similarities are there between fungi and fungal protists?

Classification and Reproductive Patterns

SECTION OBJECTIVES

- Identify characteristics of four major phyla of fungi.
- Discuss ways the four major phyla of fungi reproduce.

14·4 CLASSIFICATION OF FUNGI

Kingdom Fungi includes organisms that vary greatly in form and method of reproduction. Many ways of classifying fungi have been suggested. The classification of fungi in this chapter is based, in part, on how the organisms reproduce. This classification is shown in Figure 14·3. At present you do not know the names of the spores listed in the table. You will be reading about them in the following sections. Refer back to the table as you read about the characteristics and life cycles of the different kinds of fungi.

FIGURE 14·3 This table shows characteristics of four major phyla of fungi. Scientists use the characteristics to classify fungi.

PHYLUM	CROSS WALLS OF HYPHAE	CELL WALL	SEXUAL SPORE	ASEXUAL SPORE
Oomycota	Absent	Cellulose	Diploid oospore	Diploid zoospore
Zygomycota	Absent	Chitin	Diploid zygospore	Monoploid spore
Ascomycota	Present	Chitin	Monoploid ascospore	Monoploid conidia
Basidiomycota	Present	Chitin	Monoploid basidiospore	Varies

14·5 PHYLUM OOMYCOTA

Phylum *Oomycota* (oh eh mī KOHT uh) is a large and varied group that includes the water molds, white rusts, and downy mildews. Organisms in phylum Oomycota are called **oomycotes** (oh eh mī-KOHTS). The water molds are water saprobes. The white rusts and downy mildews are parasitic on plants. Other organisms in this phylum live as parasites in, or on, algal protists, plants, small water organisms, and other fungi.

Most fungi have cell walls made of chitin. Most oomycotes have cell walls containing cellulose, a substance that is found in the cell walls of plants.

Few traits are shared by all oomycotes. One shared trait, however, is the formation of asexual spores called **zoospores** (ZOH uh sporz), each of which has two flagella. The flagella allow the zoospores to swim. Water is needed for asexual reproduction in most oomycotes. Rainwater, the water in soil, or even dew is enough water for the zoospores to swim in.

Sexual reproduction also takes place among the oomycotes. The fertilized egg of an oomycote develops into an **oospore** (OH uh spor). Figure 14·4 shows the life cycle of *Saprolegnia* (sap ruh LEHG nee uh), an oomycote that reproduces both asexually and sexually. *Saprolegnia*, a common freshwater mold, can often be seen as a fuzzy, whitish growth on organisms that have died. This water mold usually lives as a saprobe, but it may be a parasite on injured fish and on fish eggs.

Look at the life cycle of *Saprolegnia* as shown in Figure 14·4. Notice that asexual reproduction is by means of diploid zoospores that have flagella. These zoospores are produced by mitosis within a spore case called a **sporangium** (spuh RAN jee uhm). When the sporangium is ripe, it opens and lets out the zoospores into the water. After a few hours, the zoospores become enclosed within walls to become resting spores, or *cysts*. The cysts later change into zoospores again.

After a few more hours of swimming, these zoospores form cysts again. These cysts change into vegetative hyphae. After a period of growth, the vegetative mycelium may again form sporangia to reproduce asexually, or it may enter into a cycle of sexual reproduction.

The sexual reproduction of *Saprolegnia* is also shown in Figure 14·4. This cycle begins with the development of reproductive structures. The male reproductive structure is called the **antheridium** (an-thuh RIHD ee uhm). The female reproductive structure is called the **oogonium** (oh uh GOH nee uhm). Meiosis takes place inside the antheridium and results in monoploid sperm nuclei. Monoploid eggs are formed within the oogonium by meiosis.

The antheridium develops fertilization tubes that go into the oogonium. Sperm nuclei pass through these tubes to reach the eggs. The sperm nuclei join with the egg nuclei to form diploid *zygotes*. Each zygote forms a thick wall and is released from the oogonium. In this form it is called an oospore. When conditions are favorable, the oospore will form a new vegetative mycelium.

spora (seed)
angeion (vessel)

oo- (egg)
gonos (producing)

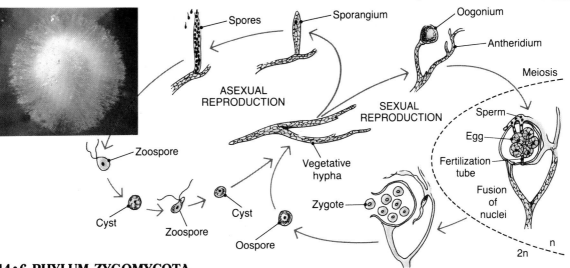

FIGURE 14·4 *Saprolegnia* is a common fungi that lives on dead organisms in fresh water. The life cycle of this organism is shown in this illustration.

14·6 PHYLUM ZYGOMYCOTA

Phylum *Zygomycota* (zī guh mī KOHT uh) is made up of land-dwelling fungi often called molds and blights. Members of phylum Zygomycota are called **zygomycotes** (zī guh mī KOHTS). Most zygomycotes are saprobes. However, some are parasites on plants, animals, or other fungi. All zygomycotes reproduce sexually by means of thick-walled diploid spores called **zygospores** (zī guh sporz).

280 THE FUNGI

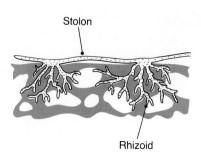

FIGURE 14·5 *Rhizopus* is a common mold that grows on the surface of breads and cakes. You will certainly find *Rhizopus* growing on old bread. Special chemicals are often added to commercial breads to prevent the growth of molds.

FIGURE 14·6 The life cycle of *Rhizopus* and an example of this mold growing on bread (inset).

Rhizopus stolonifer (RĪ zuh puhs stoh LAH nih fer) is a zygomycote mold that often grows on bread and starchy fruits. It is whitish when young; it is black when it reaches maturity and forms spores. Look at the structure of the hyphae of *Rhizopus* shown in Figure 14·5. The hyphae that are parallel to the fungus' growth medium are called **stolons** (STOH lahnz). The many smaller, branching hyphae growing downward from the stolons are called **rhizoids** (RĪ zoidz). The rhizoids go into the food source the fungus is growing on and give out enzymes that digest the food.

Rhizopus reproduces asexually by forming stalklike hyphae called **sporangiophores** (spuh RAN jee uh forz). These structures stick up into the air from the stolons, as shown in Figure 14·6. At the top of each sporangiophore is a sporangium that has a very large number of asexual, monoploid spores. Note that *-phore* means "carrier" and that the sporangiophore carries the spore case, or sporangium. The spores of zygomycotes, unlike those of oomycotes, have no flagella. When the spores are ripe, the sporangia burst and release the spores. These spores can then germinate into new vegetative hyphae.

Rhizopus also can reproduce sexually through the mating of hyphae of genetically different strains. Figure 14·6 shows how mating takes place. Stolons from two strains of *Rhizopus* grow next to each other. A swelling from each stolon grows toward the other. The two swellings meet and nuclei from each side come together and join so that a diploid (2n) cell is formed. This diploid cell forms a thick wall and becomes a *zygospore*.

Meiosis takes place during the germination of a *Rhizopus* zygospore. The germinating zygospore forms a single sporangium having monoploid (n) spores. These spores appear to be the same as those formed in asexual reproduction. When these spores are released, they grow into vegetative hyphae. They may later reproduce either sexually or asexually.

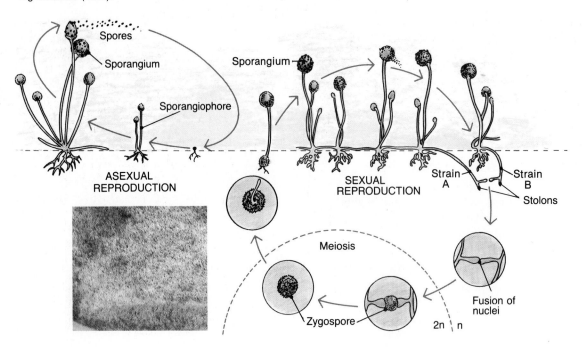

14·7 PHYLUM ASCOMYCOTA

Phylum *Ascomycota* (as kuh mī KOHT uh) is the largest group of fungi, with about 2000 genera. Members of phylum Ascomycota are called **ascomycotes** (as kuh mī KOHTS). The yeasts, the powdery mildews, the cup fungi, truffles, and morels are part of this phylum. Also included are a number of molds that cause food to spoil. Most ascomycotes are saprobes, living on dead plant material. However, other fungi are very destructive plant parasites. Some cause diseases such as Dutch elm disease, chestnut blight, and peach leaf curl. Most ascomycotes are adapted to land habitats.

Ascomycotes are sometimes called sac fungi because spores develop in a sac called an **ascus** (AS kuhs). These monoploid, nonmoving spores are called **ascospores** (AS kuh sporz). The ascospores are formed by sexual reproduction followed by meiosis.

The pink bread mold *Neurospora crassa* (nuhr oh SPOR uh KRA-sah) is an ascomycote that has been used in experiments in genetics and biochemistry. Its life cycle is shown in Figure 14·7.

Neurospora and many other ascomycotes reproduce asexually by forming spores called **conidia** (kuh NIHD ee uh). The conidia are formed on specialized hyphae called **conidiophores** (kuh NIHD ee uh-forz). After the spores fall from the conidiophores, they germinate to form new vegetative hyphae.

Sexual reproduction in *Neurospora* begins with the mating of hyphae of different types. The **ascogonia** (as kuh GOH nee uh) are female reproductive structures. They form tubes that fuse with antheridia on the male hyphae. Male nuclei from the antheridia cross through the tubes into the ascogonia. One male nucleus and one female nucleus become isolated in an extension of each ascogonium. Look at the hook-shaped extension in the diagram. The male and female nuclei join to form a zygote. The zygote goes first through meiosis and then mitosis to produce eight monoploid (n) cells inside an

FIGURE 14·7 *Neurospora* is a pink mold that grows on breads. Geneticists use this mold because it is easy to study the genotype frequencies by crossing one strain of *Neurospora* with another.

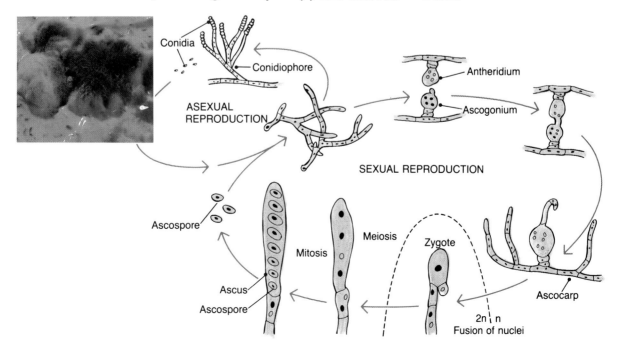

282 THE FUNGI

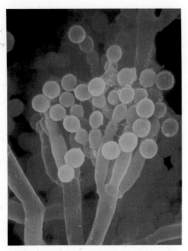

FIGURE 14·8 Conidia are spores that form new hyphae when they begin to grow. These conidia are produced by a *Penicillium* mold.

ascus. The eight cells form thick walls as they ripen into ascospores. Hyphae grow around the developing asci to form a fruiting body called an **ascocarp** (AS kuh kahrp). Each ascospore is able to develop into fungal hyphae.

Yeasts, one-celled ascomycotes of the genus *Saccharomyces,* are used in baking and brewing. *Saccharomyces* means "sugar-fungus." These fungi live on sugary food sources, such as grape juice.

Saccharomyces yeasts reproduce asexually by budding. Budding is a method of reproduction that involves an outgrowth, called a bud, from the cell. The nucleus finally divides and one of the new nuclei moves into the bud. The bud later breaks off, becoming a separate yeast cell.

Saccharomyces yeasts can also reproduce sexually by forming ascospores. In such cases the yeast cell becomes an ascus. Such an ascus may contain four or eight ascospores.

A number of very common molds of the genera *Penicillium* and *Aspergillus* have no known sexual reproduction. These molds have been classified in phylum Ascomycota because they form conidiophores and reproduce asexually by means of conidia. *Penicillium* molds often are seen as bluish-green growths on oranges and other citrus fruits and sometimes on bread.

Insights

Penicillin

The story of how the antibiotic penicillin was discovered is often used as an example of scientific advancement that resulted from good luck. Dr. Alexander Fleming, a British scientist, found a mold growing on a culture of bacteria. The mold, *Penicillium notatum,* seemed to be killing the bacteria. A chemical extract from the mold proved to be the substance, and it was called penicillin.

While serving in the Royal Army Medical Corps during World War I, Dr. Fleming saw many infected wounds. He also observed that the methods used to treat these infections were not very effective.

After the war, Dr. Fleming returned to research, hoping to find a way to kill bacteria without harming the cells of the host. He conducted many carefully planned and controlled experiments on bacterial cultures. At the same time, he kept watch for any unusual occurrences that could provide clues in his search for a way to kill bacteria.

Dr. Fleming discovered penicillin in 1928 and published his findings the following year. At this stage, penicillin was not usable because it had not been purified

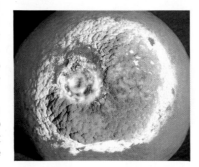

and concentrated. These processes were completed by two other British scientists, Dr. Ernst Chain and Dr. Howard Florey. In 1945 these three scientists shared the Nobel Prize in Physiology and Medicine.

Luck was not the only factor that led to this great discovery. The mold that contaminated the bacterial culture had developed there by chance. Some scientists would have discarded the contaminated culture immediately. Because Dr. Fleming was willing to follow any lead, he kept the culture. This open-mindedness, along with good scientific observation, allowed Dr. Fleming to notice that the mold killed bacteria. Dr. Fleming's training and interest enabled him to recognize the value of this observation.

14·8 PHYLUM BASIDIOMYCOTA

Phylum *Basidiomycota* (buh sihd ee uh mī KOHT uh) includes some of the largest fungi, such as mushrooms, puffballs, and bracket fungi. A puffball is shown in Figure 14·9. Puffballs are similar to mushrooms, but the cap is roundish. Bracket fungi look like shelves growing on trees. Members of phylum Basidiomycota are called **basidiomycotes** (buh sihd ee uh mī KOHTS).

Basidiomycotes produce spores in club-shaped structures called **basidia** (buh SIHD ee uh). The monoploid spores, called **basidiospores** (buh SIHD ee uh sporz), are formed by meiosis of a zygote formed by sexual reproduction. In most basidiomycotes the structure that bears the basidia is called a **basidiocarp** (buh SIHD ee uh kahrp). Because of the shape of the basidia, members of this phylum are called club fungi.

The object often called a mushroom is only a small part of the complete fungus. Most of a mushroom fungus is made of an underground mycelium. The mycelium absorbs nutrients from organic matter in the soil. The part of the mushroom seen above the ground is the basidiocarp, or fruiting body.

Mushrooms do not form asexual spores. The basidiospores, or sexual spores, of mushrooms change into hyphae that develop into underground mycelia. Figure 14·10 shows how hyphae from two different mycelia meet and join to form a hypha with two sets of nuclei. It is these hyphae that grow into the basidiocarp of the mushroom.

The young basidiocarps of some mushrooms look like little rounded balls. A sheetlike *veil* holds the developing mushroom together. Inside the veil may be another sheetlike tissue, the **annulus**, which draws the edge of the mushroom cap toward the stem, or **stipe**. When the basidiocarp grows a bit larger, the veil and annulus tear. This tearing allows the mushroom to open like an umbrella. The cap of the mature mushroom has gills on its undersurface. The **gills** of a mushroom are where the basidiospores are formed.

FIGURE 14·9 These "gem-studded" puffballs are growing on the trunk of a tree. You should never taste or eat mushrooms from the wild. Many are extremely dangerous.

anulus (ring)

FIGURE 14·10 Mushrooms are perhaps the most well-known fungi. This illustration shows the life cycle of a mushroom.

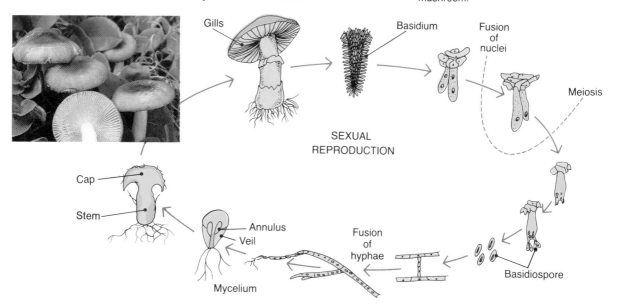

Club-shaped basidia develop from the surface of each gill of the mushroom cap. Within each basidium, two nuclei join to form a zygote nucleus. The zygote nucleus then undergoes meiosis, forming four monoploid nuclei within the basidium. These four nuclei then travel into projections at the outer part of the basidium. The nuclei develop into four monoploid basidiospores. The basidiospores are released from the basidiocarp when they are mature.

Rusts and smuts are basidiomycotes that do not form basidiocarps. They are small and nearly always parasitic on plants. Many rusts have complex life cycles, infecting different hosts at different stages of the life cycle. Unlike mushrooms, many of these basidiomycotes produce asexual spores as well as basidiospores.

Wheat rust, *Puccinia graminis,* infects wheat and barberry plants. Basidiospores develop on the infected wheat plant and are carried by wind to barberry plants. Asexual spores develop from hyphae on barberry plants. These spores are carried by wind to wheat plants.

FIGURE 14·11 Many important crop plants are damaged by fungi each year. Farmers often spray their crops with special chemicals called fungicides. This photograph shows wheat rust, caused by a basidiomycete, on the stems of wheat plants.

REVIEW QUESTIONS

6. Describe the life cycle of a common water mold.
7. Describe the structure of bread mold and explain the life cycle of this organism.
8. What similarities are there in the sexual reproduction of oomycotes and ascomycotes?
9. Compare and contrast sexual spore formation in ascomycotes and basidiomycotes.
10. Suppose you are investigating an unknown fungus. The asexual spores are produced in chambers on stalklike hyphae. The sexual spores are diploid and lack flagella. Classify the fungus.

Investigation

What Are the Characteristics of Bread Mold?

Goals

After completing this activity, you will be able to
- Set up conditions suitable for the growth of bread mold.
- Describe the appearance of bread mold.
- Draw bread mold and include the details of the reproductive structures.

Materials (for groups of 2)

slice of bread   water
wetting agent   2 slides
forceps   dropper
hand lens or stereomicroscope
petri dish
glass-marking pencil

Procedure

A. Place a piece of wet bread in the petri dish. Expose the bread to air for several hours. Wet the bread again, but do not make it soggy. Cover the dish.

B. With the glass-marking pencil, write your name and date on the cover of the dish. Place the dish in a warm place to incubate.

C. Observe the bread daily for 5 to 7 days. Look for cottony spots of mold growth.

D. On the seventh day, observe the individual mold colonies. First use a hand lens or stereomicroscope to look at the overall structure of the mold.

E. With your forceps remove a small quantity of material from the mold. Place the bit of mold in a drop of wetting agent on a slide.

F. Examine the mold first under low power. Then switch to high power.

G. Identify the following structures: sporangiophores, spores, sporangia, stolons, and rhizoids.

H. On a separate sheet of paper, draw the structure of the bread mold and label the structures listed in step G.

I. Compare the density of mold growth on your bread with that of the mold grown by other students in the class.

J. Follow your teacher's instructions for the proper disposal of all materials.

Questions and Conclusions

1. Describe the appearance of the mycelium of the mold.
2. Are sporangia present on the mold? If so, what colors do you observe?
3. Where do you think the mold spores that started the mold growth came from?
4. What would have happened to the mold growth if the bread had been kept in the refrigerator rather than in a warm place?
5. The procedure that you followed encouraged mold growth. How could you discourage mold growth?

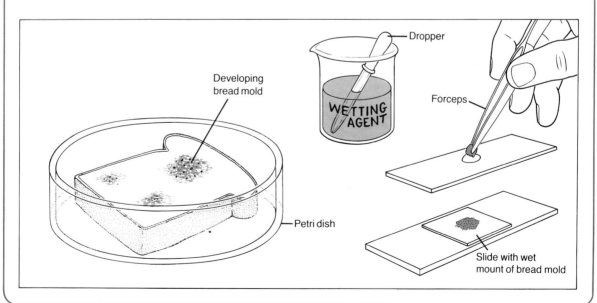

THE FUNGI   285

Activities of Fungi

SECTION OBJECTIVES

- Describe the processes by which fungi get nourishment.
- Identify several types of relationships between fungi and other organisms.
- Give examples of the ecological and economic importance of fungi.

14·9 NUTRITION

A fungus is in direct contact with its food. The fungus secretes enzymes into the food. These enzymes break large food molecules into molecules that are small enough to be absorbed by the hyphae. Different enzymes act on different kinds of nutrients.

After food has been absorbed, it travels to all parts of the fungus. Think about a mushroom, a saprobic fungus. The underground mycelium of the mushroom is in contact with organic matter in the soil. Above the ground the fruiting body, or basidiocarp, has no direct source of food. Food travels from the underground mycelium to the mushroom through the hyphae. Remember that some fungi lack septa in their hyphae. Thus cytoplasm is free to move back and forth through the hyphae, carrying food with it. Even in those fungi with hyphae that have septa, pores in the septa allow the cytoplasm to flow.

Although most saprobic fungi live in the soil, some live in other environments. Bracket fungi are found on both living and dead trees. Those bracket fungi that grow on living trees are believed to be saprobes rather than parasites. They usually consume only dead cells in the bark or in the dead heartwood of the tree.

Parasitic fungi obtain nutrients directly from the living cells and tissues of their hosts. Parasitic fungi often form special hyphae called **haustoria** (haw STOR ee uh). These hyphae are highly branched and specialized for absorbing nutrients. Haustoria pass into the tissues and often even enter the cells of the host.

A few filamentous fungi are predators on such small organisms as protozoans, rotifers, and roundworms. These fungi have special sticky pads or loops that trap the prey. Some types of soil-dwelling fungi form rings along the hyphae. Figure 14·12 shows how these rings are used to catch small roundworms. After trapping a worm, the fungus forms haustoria, which pass into the prey. Digestion of the host's tissues takes place, followed by absorption of the nutrients.

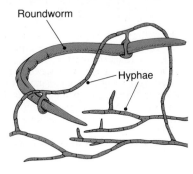

FIGURE 14·12 Some fungi get nutrients from living organisms. This soil fungus has trapped a small roundworm in a special sticky loop on its hyphae.

14·10 FUNGI AS SYMBIONTS

A close relationship between two species in which at least one species benefits is called **symbiosis** (sihm bī OH sihs). A symbiotic relationship in which one organism lives on or in another and harms it is called **parasitism.** If both species are helped by symbiosis, the relationship is called **mutualism.** Some fungi are mutualistic.

A **lichen** (LĪ kuhn) is a symbiotic association between a fungus and a blue-green bacterium, an algal protist, or a green alga (a simple plant). A lichen is shown in Figure 14·13. The fungus may give structural support and supply certain nutrients for the partner. The bacterium, protist, or plant—unlike the fungus—is able to make its own food. It provides food for the fungus. Lichens live in many habitats, from tropical heat to arctic cold, from deserts to rain forests.

The photosynthetic species found in lichens include about 8 species of blue-green bacteria, 20 species of green algae, and 1 algal protist. The blue-green bacterium *Nostoc* and the green alga *Trebouxia* are the most common. All of the photosynthetic species can probably live without the fungi. But the fungi usually need a partner for survival. Lichens that contain *Nostoc* can use nitrogen from the air as a nutrient.

A **mycorrhiza** (mī kuh RĪ zuh) is a structure formed by the mutualistic association of a fungus and the roots of a plant. The fungus grows on the roots, with some hyphae reaching into the tissues of the host plant. Some plants, including elm, maple, and willow, can grow well with or without fungi. Other plants, such as some species of pine, spruce, and oak, need a growth of fungi on their roots. The fungus receives sugars from the roots of the host plant and gives growth-stimulating chemicals to the plant. The fungus also aids the plant in taking in water and minerals.

Orchids have a very close relationship with their mycorrhizal fungi. The fungus invades the cells of the orchid roots and provides organic nutrients to the orchid. Some orchids are so dependent on the fungus that they can live without chlorophyll. Coralroot, shown in Figure 14·13, is an orchid without chlorophyll. Most of the organic matter flows from fungus to orchid. But the fungus may be helped by receiving some amino acids and vitamins from the orchid.

Symbiotic relationships between insects and fungi are fairly common. Many insects have yeasts, along with bacteria, in their digestive tracts. The fungi aid the insect in digesting its food and are helped by receiving a share of that food.

Some kinds of burrowing beetles, termites, and ants keep fungal gardens. The insects bring food to the growing fungi. The fungi serve as food for the insects. Several species of ants of the southern and southwestern United States show such relationships with fungi.

Fungi are common parasites of insects. Sometimes the fungus helps to keep the insect population under control. The citrus mealybug in Florida is controlled by the fungus *Entomophthora fumosa,* which grows well in the warm, rainy season. Yet the citrus mealybug grows well in California, which has a cool, rainy season.

FIGURE 14·13 Some kinds of fungi live in a relationship with other organisms. The British soldiers lichen (top) is an example of a symbiotic association between a fungi and an algae. The Western coralroot (bottom) cannot make its own food. It depends upon the fungi that grow on its roots to supply it with nutrients.

14•11 IMPORTANCE OF FUNGI

A great amount of dead organic material is deposited on the earth's surface each year. If it were not for saprobes, including many fungi, this waste would build up to massive proportions. The buildup of such wastes would exclude most plants and animals from their natural habitats. Saprobes remove this material by breaking it down. They also return inorganic nutrients and minerals to the soil, where these materials can be used for more plant growth.

Saprobic fungi are able to break down organic material. In some cases this ability causes problems. Mildews and molds form quickly in moist environments. Clothing and other cloth products, such as tents, are often attacked by molds and mildew. Cotton is very often attacked because it is made of plant cellulose. Many molds can break down cellulose. These fungi often cause musty odors, stains, and weakened fabric. Wool, which is a protein fiber, is less likely to be attacked by fungi. There are fewer fungi that can digest protein.

Many foods are attacked by molds. You have probably seen moldy fruits, vegetables, and bread. The downy or cottony molds *Penicillium* and *Rhizopus* grow quickly on foods that are in even moderately humid environments. Many of these fungi affect the smell, taste, and appearance of the products they grow on.

Some fungi that attack foods produce poisons that may cause illness in people and livestock. Corn that becomes infected with certain species of *Penicillium* and *Aspergillus* may cause poisoning of pigs and cattle. Horses and other livestock may be poisoned by moldy hay. Death may take place because of the effects of the poison on the nervous and circulatory systems.

FIGURE 14•14 The fungus that causes ergotism grows on rye plants, a grain often used to make bread. The fungus causes severe mental disturbances when it is eaten.

Ergotism is a disease caused by an ascomycote that is a parasite on rye. The developing ascospores invade the seeds and replace the seed material with hardened, purple-black masses of fungal hyphae. The fungus forms several poisons. People who eat the contaminated grain may have convulsions and other mental disorders.

Although some fungi destroy useful products, many valuable products are made as a result of fungal activity. These include bread, alcohol, cheese, drugs, and enzymes. Fermentation of sugar by yeast (usually *Saccharomyces cerevisiae*) forms carbon dioxide and ethyl alcohol. The carbon dioxide causes bread dough to rise. Brewers and winemakers use the ethyl alcohol to make beer and wine.

Cheese is formed from milk that has been coagulated to form semisolid curds and a watery liquid called whey. The curds are the newly-formed cheese, sometimes sold as cottage cheese. The curds are ripened, either by bacteria or fungi, to make other kinds of cheese. Most soft cheeses are ripened by fungi. Camembert cheese is soft but has a stiff rind containing the white mold *Penicillium camemberti.* This fungus gives off an enzyme that makes the cheese soft and buttery and mild in flavor. Roquefort cheese is made from sheep's milk. The curds are broken up, drained until fairly dry, and inoculated with *Penicillium roqueforti,* which grows over the curds.

FIGURE 14·15 This person looks more like a miner than a mushroom grower. Since most mushrooms prefer cool, moist growing conditions, mushrooms are often grown commercially in caves.

Other important substances are produced by fungi. Antibiotics, including penicillin, are formed by *Penicillium*. The drug *ephedrine*, used in treating hay fever and asthma, and the B-vitamins are commercially made by yeasts. Some enzymes and organic acids, including citric acid, are also made using fungi.

Some fungi, such as mushrooms, have been thought of as delicacies since the time of the ancient Greeks and Romans. In North America, mushrooms are grown commercially in large mushroom farms. The commercial mushroom is an excellent source of vitamins and the minerals phosphorus and iron. These mushrooms also contain some proteins. Mushrooms are grown in trays of a mixture of straw and fertilizer. The trays are kept in a moist, dark environment which prevents the growth of soil algae and weeds. The fungal mycelium grows throughout the mixture, and the edible basidiocarps rise up above the surface.

There are many species of wild mushrooms that are edible, but some are deadly poisonous. It is difficult to tell nonpoisonous and poisonous varieties apart. About 50 people a year die from mushroom poisoning in the United States. Usually the killer mushroom is a species of *Amanita*, the "death angel." Even one cubic centimeter of this mushroom can cause death. Since only an expert can identify wild mushrooms, you should never collect and eat wild mushrooms.

REVIEW QUESTIONS

11. What hyphal adaptation for entering host tissue is found in many parasitic fungi?
12. What is a lichen?
13. Describe the structure of a mycorrhiza.
14. What useful products are produced by fungi?
15. Imagine a forest from which all of the fungi had somehow been removed. What would be the short-term effects of the removal of fungi? What would be the long-term effects?

Chapter Review

SUMMARY

14·1 Fungi are organisms with eukaryotic cells that get nourishment by absorbing organic substances. Fungi have a mixture of plantlike and animallike characteristics and are now classified as a separate kingdom.

14·2 Most fungi are saprobes. Some fungi exist in symbiotic associations with other organisms.

14·3 The fungal body is composed of filamentous hyphae, usually divided into a reproductive part, the fruiting body, and a nutrient-gathering part, the mycelium.

14·4 Fungi are classified on the basis of their patterns of reproduction.

14·5 The oomycotes, which include the water molds, have flagellated asexual spores. They form antheridia and oogonia as sexual reproductive structures.

14·6 The zygomycotes, which include the common bread mold, have zygospores that form by fusion of cells from genetically different hyphae.

14·7 The ascomycotes include yeasts, cup fungi, truffles, morels, and molds. Sexual reproduction in ascomycotes produces ascospores, while asexual reproduction produces conidiospores.

14·8 The basidiomycotes produce sexual spores called basidiospores on fruiting bodies called basidiocarps.

14·9 Fungi secrete enzymes that digest external organic matter, which is then absorbed. Nutrients are circulated through the cytoplasm of the hyphae.

14·10 Fungi participate in important symbiotic associations. Lichens are symbiotic associations of fungi and algae; mycorrhizae are mutualistic associations between fungi and plant roots.

14·11 Fungi play an important role as decomposers, removing dead organic matter and making minerals available for use by plants. Fungi cause damage when they decompose food, clothing, and products valued by people.

BIOLOGICAL TERMS

annulus
antheridium
ascocarp
ascogonia
ascomycotes
ascospores
ascus
basidia
basidiocarp
basidiomycotes
basidiospores
conidia
conidiophores
fungus
gills
haustoria
hyphae
lichen
mutualism
mycelium
mycorrhiza
oogonium
oomycotes
oospore
parasitism
rhizoids
septa
sporangiophores
sporangium
stipe
stolons
symbiosis
zoospores
zygomycotes
zygospores

USING BIOLOGICAL TERMS

1. Distinguish between oospores, zoospores, zygospores, ascospores, and basidiospores.
2. What is a mycorrhiza?
3. Which terms apply to sac fungi? What does each of these terms refer to?
4. Which terms name parts of a mushroom? Describe each term.
5. What is a mycelium?
6. Distinguish between an antheridium and an oogonium.
7. What is a lichen?
8. Distinguish between mutualism, parasitism, and symbiosis.

9. Which terms refer to specialized hyphae formed by the zygomycote *Rhizopus?* Describe each term.
10. Distinguish between a basidium, a basidiocarp, and a basidiospore.
11. Distinguish between hyphae, mycelium, and haustoria.
12. Define the term *sporangium.*

UNDERSTANDING CONCEPTS

1. What are some resemblances between fungi and plants? Why are fungi placed in a separate kingdom? (14·1)
2. How do fungi obtain nourishment? (14·2)
3. The body of a fungus generally has two types of parts. What are they? (14·3)
4. How are fungi classified? (14·4)
5. What kinds of organisms are classed as oomycotes? (14·5)
6. Describe the structure and life cycle of the common bread mold. (14·6)
7. How are ascospores formed? (14·7)
8. Describe asexual reproductive methods among ascomycotes. (14·7)
9. Describe the life cycle of a typical basidiomycote. (14·8)
10. How are water molds different from the other phyla of fungi? (14·2–8)
11. Describe a type of predatory behavior found in certain fungi. (14·9)
12. Why is a lichen an example of a mutualistic relationship? (14·10)
13. How does a mycorrhiza benefit a fungus and a tree? (14·10)
14. How do some orchids survive without any chlorophyll? (14·10)
15. Give an example of parasitism in fungi. (14·10)
16. How are fungi important in recycling materials in the environment? (14·11)
17. Why are cotton fabrics more likely than wool to be attacked by mold and mildew? (14·11)
18. Describe how fungi are involved in the making of cheese. (14·11)

APPLYING CONCEPTS

1. Give reasons for and against the proposition that fungi should be classed in kingdom Protista rather than as a separate kingdom.
2. Yeasts do not form hyphae. Why are they classified as fungi?
3. In what way are the roles of many bacteria and fungi similar in the environment?
4. How can you prevent the growth of molds in the home?
5. How do rusts and smuts differ from other basidiomycotes, such as mushrooms?

EXTENDING CONCEPTS

1. Find a mushroom or other large fungus in your yard or in the woods. Mark its location and observe it once a day for 5 to 7 days. Record any changes in its size and appearance.
2. How many kinds of fungi can you find growing in and around your home? Make a list and briefly describe each fungus. It may be helpful to check out a library book on identifying mushrooms and other fungi.
3. Make a list of all the food items in your home that were made partly or wholly with fungi.
4. Look at a piece of moldy bread through a magnifying lens or low-powered microscope. Do you see hyphae? Sporangia? Describe them.
5. If you live near a pond, look for water molds on dead leaves, fish, or other organisms. Describe the mold, its food, and its environment.

READINGS

Aharanowitz, Y., and G. Cohen. "The Microbial Production of Pharmaceuticals." *Scientific American,* March 1981, p. 140.

Arora, David. *Mushrooms Demystified.* Ten Speed Press, 1986.

Miller, Orson, and Hope Miller. *Mushrooms in Color.* N.Y., Elsevier-Dutton, 1980.

Moore-Landecker, Elizabeth. *Fundamentals of the Fungi,* 2nd ed. Prentice-Hall, 1982.

Robb, E.J., and G.L. Barron. "Nature's Ballistic Missile." *Science,* December 17, 1982, p. 1221.

15. Microbial Disease

Since most diseases are caused by microorganisms, it is essential for scientists to understand how these organisms grow, reproduce, and are transmitted. To do this, scientists often inject eggs with microbes and then observe how the developing tissues inside the egg are affected. Which diseases are caused by microorganisms? How do these diseases affect the body? These questions will be answered in this chapter.

CHAPTER OBJECTIVES

After completing this chapter, you will be able to
- **Explain** the nature of disease.
- **Describe** how microorganisms cause disease.
- **List** ways diseases are spread.
- **Recognize** the different agents of disease.
- **Explain** how the human body fights disease.

The Nature of Disease

15•1 WHAT IS DISEASE?

Diseases have probably been on the earth as long as life itself. Fossils of dinosaur bones show evidence of disease. Ancient human fossils show evidence of tuberculosis (too ber kyuh LOH sihs). In many cases, scientists have become interested in an organism because it was found to cause disease.

What is a disease? A **disease** is a condition that interferes with the normal functioning of a living thing but that is not the result of an injury. Generally a disease has certain symptoms. The disease can affect the entire body or only parts of the body. People can be born with certain diseases, such as genetic disorders. Other diseases can develop during a person's lifetime. An **infectious** (ihn FEHK shuhs) **disease** is one that is caused by organisms or viruses that enter the body. Such a disease can be transferred from one person to another by transfer of the organisms. The diseases discussed in this chapter are infectious diseases.

Organisms that cause disease are called **pathogens** (PATH uh-jehnz). Pathogens can be bacteria, fungi, protists, or other organisms. Scientists may debate whether viruses are living or not. But they generally agree that viruses are pathogens, since they are known to cause many types of disease.

Diseases that are transmitted or transferred from one organism to another are called **contagious** (kuhn TAY juhs) **diseases,** or **communicable diseases.** Sometimes distinctions are made between infections and infestations. *Infections* are diseases caused by microbes. *Infestations* are diseases caused by larger organisms, such as tapeworms. Most of the diseases discussed in this chapter are those caused by pathogenic microorganisms, or microbes. Infectious diseases that are caused by pathogenic microscopic organisms are called **microbial diseases.**

SECTION OBJECTIVES
- Define the term disease.
- Explain how diseases are spread.
- Describe the contributions of Koch to the development of the germ theory of disease.

pathos (suffering)
-gen (producing)

A pathogen has a parasitic relationship with the organism in which it is living. The pathogen is a *parasite,* an organism that lives in or on another living thing and obtains its nutrients from that organism while harming it. The living thing on which the parasite lives is called the *host.* In the disease tuberculosis the parasite is a bacterium called *Mycobacterium tuberculosis;* the host is often a human. The bacteria live in and get their nourishment from the human.

15·2 PATHOGENS AND HOSTS

Organisms capable of causing disease are around us all the time. They are present in the air we breathe, in the soil, in the food and water we take in, and on objects that we come into contact with. Some microbes that normally live within us in great numbers are potential pathogens. Others enter the body from time to time through openings such as the nose and ears or through cuts in the skin. Often these organisms have no effect on us, but sometimes they cause disease. What determines if an organism becomes a pathogen and if a parasitic relationship develops?

Certain factors determine if an organism will become pathogenic once inside a potential host. One of these factors is host resistance. *Host resistance* is the degree to which the host is able to fight infection by the disease organism. Different hosts can be resistant to different pathogens. For example, humans are generally resistant to cat distemper. This is a viral disease that can be fatal in cats. The health of the host can affect its resistance to disease organisms. If the health is poor, due to such factors as poor diet, lack of sleep, or stress, the body can be less resistant than when in good health. An organism that has poor resistance to a disease can get that disease.

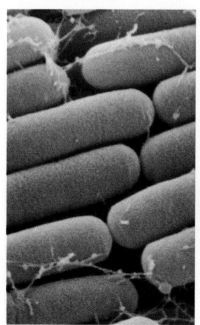 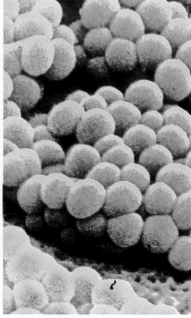

FIGURE 15·1 Scanning electron micrographs of two types of bacteria found in the human body are shown here. The rod-shaped *Escherichia coli* (left) is found in the intestine. The spherical *Staphylococcus aureus* (right) is commonly present on the skin. It can cause boils on the skin and internal infections.

Virulence, the ability of a pathogen to cause disease in a certain host, also is an important factor. Different strains, or subspecies, of an organism may differ in virulence. For example, some strains of *Corynebacterium diphtheriae,* which causes diphtheria in humans, are more virulent than others.

A third factor that affects a pathogenic relationship is the location of the pathogen within the host. Many microbes live within our bodies and either cause no harm to us or are beneficial. Those microbes that are normally found in or on an organism without causing harm make up what is called the normal flora. For example, the bacterium *Escherichia coli* is part of the normal flora of the human intestine. However, if *E. coli* spreads beyond the intestine, such as into the body cavity or into the urinary tract, it can cause infection.

Infection can cause direct damage to the cells of the host by the pathogen itself. Viruses, for example, enter the cells of the host, multiply, and finally cause the cell to undergo lysis. *Plasmodium,* the protozoan that causes malaria, multiplies within red blood cells, finally causing them to break apart (Section 15•9).

Many of the effects of infection are not due to direct damage by the pathogen. Instead, the pathogen may give off poisons that damage the cells or that interfere with the normal functioning of the organism. A poison released by a pathogen is called a **toxin** (TAHK suhn).

Different toxins have different effects on the host. For example, the toxin made by a cholera bacterium interferes with the control of ion balance in the cells lining the intestine. An infection that is limited to a certain area is said to be a localized infection. The toxin made by the diphtheria bacterium hinders protein synthesis. The toxin can affect cells throughout the body. An infection that affects many tissues is said to be a systemic infection.

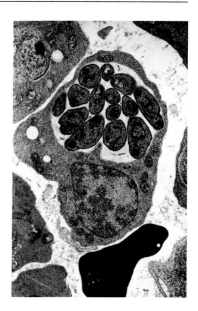

FIGURE 15•2 *Plasmodia* are protozoans that cause malaria. This photograph shows plasmodia inside a human red blood cell. Later, these plasmodia will break out of the cell and infect new blood cells.

15•3 HOW DISEASE IS SPREAD

Diseases spread when pathogens move from one host to another. Diseases can spread from one organism to another either directly or indirectly. Direct spread of a disease refers to the direct passing of organisms from one living thing to another. Indirect spread of a disease refers to the passing of a disease by an intermediate agent.

Direct spread of a disease can occur through sneezing, coughing, physical contact, or blood transfusion. With sneezing and coughing, the microbes are in the droplets of moisture that are forced from the nose and mouth. Cold viruses often are spread in this way.

Physical contact can spread certain diseases. Chicken pox, for example, can spread through skin contact. Among the most serious of diseases are those that are spread by sexual contact. Diseases spread by sexual contact are called **sexually transmitted diseases (STD).** Genital herpes, syphilis, and gonorrhea are STDs.

Some diseases are spread by animal bites. Dogs and bats, for example, can pass on rabies to other mammals, including humans, in saliva that enters the bite wound.

If disease organisms are in the bloodstream of a person who donates blood, those organisms can cause disease in the recipient. Before a person can be accepted as a blood donor, he or she is asked several health questions. Screening donors in this way reduces the number of diseases transmitted through blood transfusions.

Indirect spread of disease takes place when pathogens are passed from person to person by some intermediate agent. Disease organisms present on nonliving objects can be passed on to living things by contact with those objects. The spores of *Clostridium tetani,* which causes tetanus (lockjaw), are found in the soil and on many objects, such as nails. If a person steps on a nail and spores enter the body, he or she may get the disease. Hepatitis is a liver disease that can be spread by contact with used hypodermic needles.

Indirect spread of disease can take place when contaminated food or water is consumed. Typhoid fever, which is caused by bacteria, is spread by drinking water that contains the pathogens. Generally the water has been contaminated with human feces containing the bacteria. Thus typhoid fever is spread under conditions of improper sanitation. Cholera and infectious hepatitis also are spread by consuming food or water that is contaminated with certain pathogens.

Animals such as insects are another means of indirect spread of disease. Animals that spread disease but are not affected by the disease are called **vectors** (VEHK tuhrz). A vector can mechanically transfer pathogens, such as when a fly carries microbes to a human. In many cases, however, the pathogens multiply within the vector and are passed to a host through the bite of the vector.

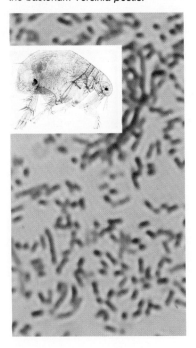

FIGURE 15·3 The rat flea is the vector for bubonic plague, which is caused by the bacterium *Yersinia pestis.*

Fleas, ticks, flies, and mosquitoes have been found to be vectors of certain diseases. Bubonic plague, a bacterial disease caused by *Yersinia pestis,* is spread from rats and other rodents, the usual hosts, to humans by fleas. Yellow fever, a disease caused by a virus, is passed along through the bite of an infected mosquito. Viral encephalitis, a disease of the nervous system, can be spread by several different insect vectors.

When pathogens spread quickly from host to host, a disease travels quickly through a population. A condition in which a disease is widespread in a population is an **epidemic.** A bubonic plague epidemic killed 25 to 40 million people in Europe in the fourteenth century.

In some cases a certain disease is present in a population at all times, but the number of cases is low. A disease known to be present in a population at all times is said to be **endemic.** Measles and flu are examples of endemic diseases. Under certain conditions, endemic diseases can become epidemic.

15·4 THE GERM THEORY OF DISEASE

The theory that microorganisms cause disease was not widely accepted until the 1870s. Remember that Redi disproved the idea that flies arose from decaying meat. In spite of Redi's experiments, many scientists still believed that microbes arose spontaneously. Not until

1864 did Pasteur finally prove that microbes do *not* arise by spontaneous generation. Pasteur's experiments and research also gave support to the **germ theory of disease,** the theory that living microorganisms are the cause of many diseases.

Robert Koch, a German physician, showed that anthrax, a disease of livestock, is caused by a bacillus (a rod-shaped bacterium). In proving that anthrax is caused by a bacillus, Koch applied certain principles for identifying a particular microbe as the cause of a disease. A slightly modified version of the procedure he used is still used today. The way to identify an organism as a disease agent was described in statements called **Koch's postulates** (PAHS chuh lihts). These postulates are as follows.

1. The specific microorganism in question is always present in the body fluids or tissues of the diseased hosts.
2. The microorganism can be isolated from the diseased host and grown in pure culture.
3. The microorganisms obtained from the pure culture injected into a healthy susceptible host will produce the disease in that host.
4. The microorganisms from the experimentally infected host can be recovered in pure culture.

Koch used the procedure to identify the bacteria that cause tuberculosis and cholera. The procedure is still used today to identify not only bacteria but also viruses, viroids, rickettsiae, and other organisms as the agents of certain diseases.

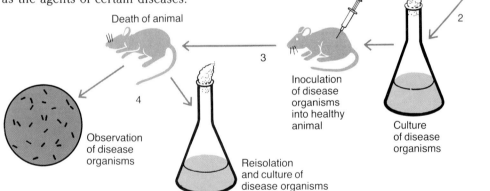

FIGURE 15·4 Scientists can identify agents of disease by applying Koch's postulates. The work of Pasteur and Koch provided the foundation for our knowledge about the causes of human disease.

REVIEW QUESTIONS

1. Describe the relationship that occurs between a pathogen and the organism it affects.
2. What determines whether an organism will be affected by a pathogen?
3. How do disease organisms harm the host?
4. List five ways diseases are directly and indirectly spread.
5. Suppose a certain disease in humans is found to be epidemic during summer but uncommon during the rest of the year. Suggest how a vector may be involved in producing this pattern.

Agents of Disease

SECTION OBJECTIVES

- Compare viral, bacterial, rickettsial, fungal, and protozoan diseases.
- Describe various ways diseases are spread.

15·5 VIRAL DISEASE

Recall that viruses reproduce within living cells (Section 12·4). The viral nucleic acid directs the production of new virus particles at the expense of the cell's normal activities. Viruses are pathogenic because they infect cells and disrupt their normal activities.

Viruses affect cells in many ways. They can change both the structure and function of the cell. For example, viruses can cause infected cells to clump together. They interfere with the production of cellular enzymes and other chemicals made by the cell. Viruses can affect the permeability of the cell membrane, changing the normal movement of materials into and out of the cell. Finally, viruses can cause the cell to lyse, or break open. The freed viruses can then spread to other cells in the host.

Viral diseases often are hard to control or cure. Antibiotics (Section 15·13) are drugs that act against many bacterial diseases but usually do not act against viral diseases. Vaccines (Section 15·14) are available against some viral diseases.

Viruses cause many common diseases in humans, including the childhood diseases of measles, mumps, and chicken pox. The symptoms of both measles and chicken pox are fever and skin rash. Chicken pox is usually mild in children but can be a serious and painful disease in adults. Measles begins with coughing, sneezing, and fever. A red rash later develops. In most cases the disease is not serious, but complications and death sometimes occur.

Another common viral disease is rubella, also called German measles. The disease is generally mild in children. But it can cause a serious problem if a woman has the disease during the fourth through twelfth week of pregnancy. The developing embryo can suffer damage to the heart, the lens of the eye, the inner ear, and the brain. The type of damage depends on the stage of development of the fetus at the time the woman has the disease. There is a vaccine against rubella. Why is it important that all children are vaccinated against German measles?

Genital herpes is a sexually transmitted disease that is caused by a virus. The number of cases of genital herpes is rising. The disease is painful, and the skin sores can return again and again. Infants born to women with the disease can contract the disease. Although research is being done to find a cure, as yet none is available.

Viruses may attack organisms other than humans. The tobacco mosaic virus attacks and damages the leaves of the tobacco plant (Section 12·2). Encephalitis, a viral disease affecting the brain, occurs in several forms and may affect most mammals. Other viral diseases have been found to occur in horses, cows, pigs, and many kinds of insects.

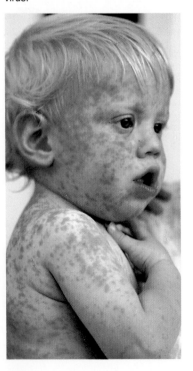

FIGURE 15·5 Measles is a common childhood disease. It is caused by a virus.

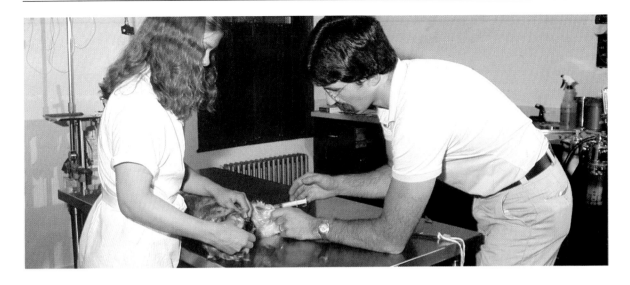

FIGURE 15·6 Dogs may be vaccinated against the rabies virus. This vaccination also protects humans, since a vaccinated dog or cat will not get rabies and transmit the disease to humans.

Besides the infectious diseases discussed, there is evidence linking viruses to certain kinds of cancer. The evidence is strong that some cancers in rodents and other animals are caused by viruses. Viruses are thought to play a part in the development of Hodgkin's disease and Burkitt's lymphoma. Both of these diseases are cancers of the lymph system in humans.

15·6 BACTERIAL DISEASE

Bacteria are another cause of many of the infectious diseases that occur in humans. Tuberculosis, plague, cholera, gonorrhea, and bacterial pneumonia are just some of the serious diseases caused by bacteria.

Recall that bacteria may affect the host by entering and damaging cells or by producing toxins. There are two types of bacterial toxins, exotoxins and endotoxins. An **exotoxin** is a toxic substance made of a protein and given off by a living bacterium. An **endotoxin** is a toxic substance made up of lipid, saccharide, and protein. It is part of the cell wall and usually is given off when the bacterium dies. Many bacteria release more than one toxin.

exo- (outside)
endo- (inside)
toxicum (poison)

Different exotoxins may affect different tissues and functions in the host. *Clostridium tetani,* the bacterium causing tetanus, produces an exotoxin that affects nerve cells. This exotoxin causes the nerve cells to send continuous messages to the muscles to contract. Death of the host results from paralysis of the muscles involved in breathing.

Cholera and bacterial dysentery are both caused by organisms that make exotoxins affecting the cells lining the small intestine. The toxins interfere with the balance of ions and water in the cells. Diarrhea results. The severe loss of ions and water from the diarrhea can result in death.

MICROBIAL DISEASE

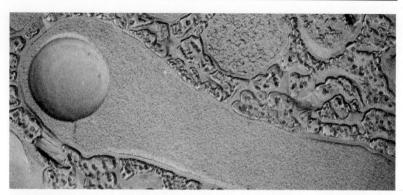

FIGURE 15·7 Under unfavorable conditions endospores, such as those shown here, are formed by some bacteria. Endospores are resistant to heat, drying, and some chemicals. When conditions become favorable, the endospores begin to grow.

Plague affects the blood, lymph glands, and sometimes the lungs. The death rate is high. *Yersinia pestis* has been found to be the cause of plague. However, the exact way in which the exotoxin damages the host's tissues is not yet known.

The effects of endotoxins usually are not as specific as those of exotoxins. Most endotoxins, for example, cause fever in humans. Some endotoxins increase the permeability of cell membranes, causing tissues to lose fluids and sometimes causing bleeding.

Many infectious diseases result not from active bacteria but from their endospores. An endospore is a protective structure formed by certain kinds of bacteria when conditions are unfavorable. Endospores are more resistant to heat and dry conditions than are active bacteria. Endospores change back to active bacteria when conditions are favorable. Tetanus is an example of a disease that can be contracted when endospores penetrate body tissues.

Food poisoning is a condition caused by eating food that is contaminated by bacteria or their toxins. There are two kinds of food poisoning. In the first kind, bacteria within food are eaten. Once inside the small intestine these bacteria multiply and give off toxins that damage the host. Members of the genus *Salmonella* cause most cases of this type of food poisoning. Symptoms include abdominal pain and diarrhea. Most victims recover in a few days. In foods prepared under high enough temperatures, *Salmonella* organisms are killed. However, in foods cooked at lower temperatures, the organisms live and cause illness. Figure 15·8 shows a species of *Salmonella* bacteria.

In the second kind of food poisoning, bacterial toxins within the food are eaten. Any bacteria also eaten are killed in the digestive tract. However, the toxin remains active. Food poisoning caused by organisms of the genus *Staphylococcus* is of this second kind. Symptoms are like those seen in food poisoning due to *Salmonella*.

Botulism is food poisoning caused by the exotoxin of the bacterium *Clostridium botulinum* (Section 12·11). The organism is anaerobic and can grow in improperly canned food. If ingested, the organism

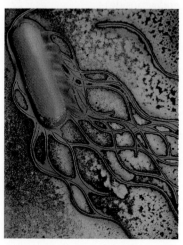

FIGURE 15·8 *Salmonella* bacteria, shown here, are a cause of certain types of food poisoning. Symptoms of this disease may occur several days after food contaminated with *Salmonella* is eaten.

cannot survive in the human digestive tract. Yet the toxin produced by the organism remains. The toxin prevents the passing of signals from nerve cells to muscles. Death results from paralysis of muscles involved in breathing.

Bacterial diseases affect plants and animals as well as humans. Crop losses judged to be in the hundreds of millions of dollars take place yearly due to bacterial infections of plants. One of the first bacterial diseases of plants to be found was fire blight on apples and pears. The bacteria can be spread by rain. Whole orchards can be infected in a single rainy night. The bacteria enter the tree through cracks in the bark or through openings in the leaves. They live between the cells and secrete enzymes that destroy the plant's tissues.

Another kind of plant disease results when bacteria cause increased cell growth, forming what is called crown gall. Crown gall has been studied a great deal because of its similarity to cancer. The cause of the increased growth rate that forms the galls is not known.

Animals serve as hosts for many pathogens. Some bacterial diseases that occur in animals can be passed on to humans. Anthrax, a disease that occurs in cows and sheep, can also attack humans. It is usually fatal. Tularemia, a disease of wild rodents and rabbits, often is contracted by people who have open cuts and who handle wild game. Tularemia also can be passed on by ticks and deer flies.

15·7 VIROID AND RICKETTSIAL DISEASES

Viroids are particles that are the smallest known agents of disease. Recall that a viroid is made of a small strand of RNA with no protein covering (Section 12·6). Like viruses, viroids can replicate only within living cells.

Viroids are known to cause certain diseases of higher plants. Such diseases may involve spotting of the leaves and stunted growth. Viroid diseases have caused much damage to potatoes on farms in the United States and to coconuts on plantations in the Philippines. Viroids also cause stunted growth in chrysanthemums and damage the fruit of cucumber plants. It is thought that viroids are spread from plant to plant by tractors, pruning shears, and other tools. The spread of viroids by such tools allows these organisms to infect whole farms quickly. Viroids have not been identified as agents of disease in animals.

Rickettsiae are bacterialike organisms known to cause a number of diseases in humans and other animals (Section 12·9). Almost all rickettsial diseases are spread by vectors—mainly mites, ticks, and lice. One example of a rickettsial disease is Rocky Mountain spotted fever, which is caused by *Rickettsia rickettsii*. The rickettsia is carried by ticks to dogs, humans, and other animals. The transmission of Rocky Mountain spotted fever is shown in Figure 15·9. Epidemic typhus is caused by another rickettsia that is spread among humans by human body lice or head lice. Getting rid of the vector (the louse) by practicing proper body hygiene and washing clothing and bedding helps to prevent typhus.

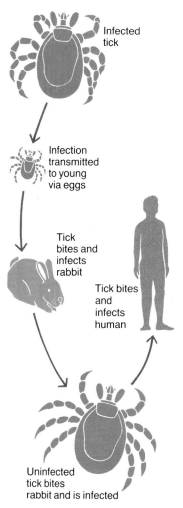

FIGURE 15·9 Rocky Mountain spotted fever, a rickettsial disease, is spread by ticks. Ths disease, in spite of its name, is not limited to the Rocky Mountain area.

15·8 FUNGAL DISEASE

Fungi are among the organisms that can be agents of disease. Fungi do not make their own food. They have parasitic and saprobic relationships with other living or once-living organisms. As parasites, many cause disease symptoms in the hosts.

Human fungal diseases are of three kinds: infections of the skin, infections deep beneath the skin, and infections of the internal organs and systems. An example of a fungal skin infection is athlete's foot. Spores of the athlete's foot fungus enter a person's foot through breaks in the skin. As the fungus grows, the skin becomes scaly and itchy. Scalp ringworm is a skin infection of the scalp caused by another fungus. Both athlete's foot and scalp ringworm are spread by direct contact or by contact with skin scales or hair clippings containing spores. Both are treated with antifungal chemicals.

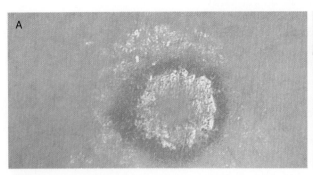 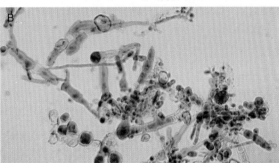

FIGURE 15·10 Ringworm is not caused by a worm, but by a fungus. You can see the characteristic shape of this skin infection (left) and the fungus that causes this disease (right).

Histoplasmosis (hihs tuh plaz MOH sihs) is an example of a fungal disease of the internal organs. The causative agent lives in the soil but can become a parasite in the lungs if spores or hyphae are inhaled. Often the spores can be found in birds' nests. The symptoms of the disease are those of a respiratory infection. The disease is rarely fatal.

Fungi also cause diseases in other animals and plants. For example, rusts and smuts do extensive damage to cereal crops by weakening the plants. Fungal diseases of plants are discussed in greater detail in Chapter 14 (Sections 14·7 and 14·11).

15·9 PROTOZOAN DISEASE

Of the over 28 000 species of protozoans, only about 35 are known to cause diseases in humans and other animals. However, the effect of this small number of pathogens on world health is great. Many protozoan diseases are easily spread and cause many deaths.

One protozoan disease that is a serious world health problem is malaria. Malaria is a disease caused by several species of the genus *Plasmodium*. The protozoans are carried by female *Anopheles* (uh-NAHF uh leez) mosquitoes. Thus the disease is more common in tropical areas where *Anopheles* mosquitoes thrive. The protozoans are passed to humans when humans are bitten by infected mosquitoes.

Figure 15·11 shows the life cycle of *Plasmodium.* Follow the steps in the life cycle as you read about the disease.

1. A female *Anopheles* mosquito bites a person who has malaria. The mosquito ingests not only the blood but also gametocytes, which are the cells that can become gametes.
2. In the digestive system of the mosquito, the gametocytes develop into gametes. The male and female gametes join, forming a zygote.
3. This zygote forms a cyst in the wall of the digestive tract.
4. Within the cysts many spore cells are produced. Then the cysts break open, freeing the spore cells, which move to the salivary glands of the mosquito.
5. The mosquito carrying the spore cells bites an uninfected human, and the spore cells enter the person's bloodstream.
6. In the bloodstream of the new human host, the spore cells travel to the liver.
7. The spore cells divide asexually into many cells.
8. These cells enter red blood cells and keep dividing.
9. Sometimes these cells will enter one red blood cell and reproduce. The daughter cells then enter another red blood cell and reproduce again. Each time the red blood cells break open and release more of these cells, the human host has chills followed by high fever.
10. Finally some of these cells will form gametocytes in the blood. If this human host is then bitten by an *Anopheles* mosquito, the cycle continues.

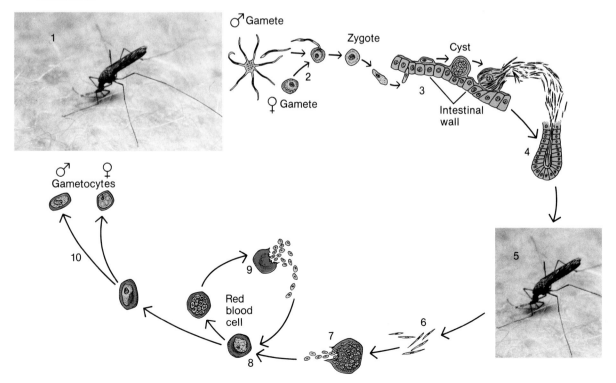

FIGURE 15·11 This illustration shows the life cycle of *Plasmodium,* the protozoan that causes malaria. Although it is relatively rare in the United States, malaria is one of the most common infectious diseases in the world today.

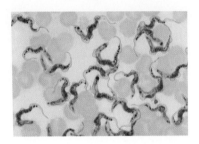

FIGURE 15·12 African sleeping sickness is caused by the protozoan *Trypanosoma*. In this photograph you can see the Trypanosomes (dark purple) among red blood cells (light purple). Trypanosomes are spread by the bite of the tsetse fly.

Another serious disease, African sleeping sickness, is caused by a protozoan called *Trypanosoma*. The organism infects herd animals. It is transmitted from herd animals to humans by the tsetse fly. Control of this disease is difficult because of the huge numbers of animals that can carry the parasite. The parasites live in their host's bloodstream but finally attack the nervous sytem, usually causing death.

The most common intestinal protozoan parasite is an amoeba. Amoebic dysentery is caused by this parasite. The pathogen can be passed on in water and in uncooked vegetables. It can also be carried in materials contaminated by feces from infected hosts. The organism can infect the host's liver, spleen, and brain. Drugs are sometimes helpful in treating the disease, but the disease is sometimes fatal. Proper sanitation prevents the spread of amoebic dysentery.

REVIEW QUESTIONS

6. Why are viruses considered to be parasites?
7. Explain why bacteria that form spores can be more dangerous than those that do not form spores.
8. How is botulism different from a viral disease such as measles?
9. Describe the life cycle of the protozoan that causes malaria.
10. The exotoxin that is produced by the cholera bacterium results in loss of water and ions from the lining of the intestine. What cell structure and what cell function are probably being affected by the exotoxin?

Defenses Against Disease

SECTION OBJECTIVES

- Describe the defenses of the body against disease.
- Explain how the human immune system functions.
- Describe some of the medical treatments used to control and prevent disease.

15·10 BARRIERS TO PATHOGENS

Organisms are not helpless against invading pathogens. The human body, for example, responds to the presence of pathogens and often destroys them. In fact, most pathogens never get inside the body. The skin is the first line of defense against invading organisms. Although the skin is only 0.5 cm thick, microbes cannot easily penetrate it. Microbes can enter the skin only through a cut or wound. Even though clotting of the blood usually seals those openings quickly, infections often occur with such injuries.

Most pathogens enter the body through natural openings, such as the nose and mouth. Mucous membranes line the nose, mouth, and eyes, as well as the digestive, respiratory, excretory, and reproductive systems. These membranes secrete mucus. Microbes that reach these membranes usually are trapped in the mucus and discharged from the body. Sometimes microbial diseases begin as infections of the mucous membranes.

There are other barriers to invading microbes. Tears in the eyes help wash away pathogens. Ciliated cells that line the respiratory system sweep mucus, with its trapped microbes, up toward the mouth. Coughing also moves the mucus up so that it can be gotten rid of. Hairs in the nose can trap dust and microbes, which are then expelled by sneezing. Hydrochloric acid produced by the stomach kills most microbes that are swallowed. Enzymes in digestive juices also can destroy microbes.

Pathogens that enter the skin or mucous membranes can multiply in different body tissues. However, the body has defenses against microbes. **Leukocytes** (LOO kuh sīts) are white blood cells. Many kinds of leukocytes destroy microbes by engulfing them by phagocytosis. Leukocytes are found in the blood, but they can leave the blood vessels and gather at the site of an infection. For example, leukocytes can enter the *intercellular fluid* (fluid that bathes the cells) and destroy pathogens.

Intercellular fluid can enter vessels called lymph vessels. While in the lymph vessels this fluid is called **lymph** (lihmf). As it moves through the lymph vessels, lymph passes through enlarged regions called lymph nodes. Phagocytic white blood cells called **macrophages** (MAK ruh fayj uhz), are found in the lymph nodes. Pathogens can be destroyed in the lymph nodes by these cells. Lymph finally enters the bloodstream (Section 31·10).

Sometimes pathogens get past the body's barriers and are not destroyed by white blood cells. If the pathogens multiply, the infection can become more widespread. The body has yet another infection-fighting response.

leuco- (white)
-cyte (cell)

FIGURE 15·13 The body has many defenses against disease. The ways the body fights disease are shown here.

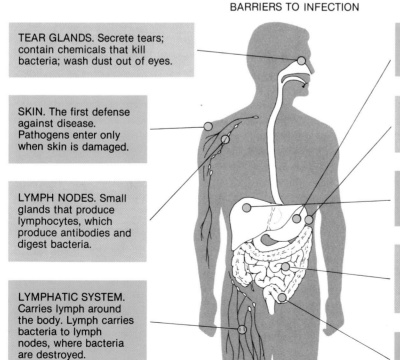

15•11 THE IMMUNE RESPONSE

A person rarely has a disease such as measles more than once. What protects the body against later infections? When substances enter the body, the body forms chemicals. These chemicals, which protect the body by helping to destroy pathogens and neutralizing their toxins, are called **antibodies** (AN tee bahd eez). Any substance that causes the body to make antibodies is known as an **antigen** (AN-tuh juhn). Examples of antigens include proteins on the surfaces of viruses and bacteria, as well as the toxins produced by bacteria.

When a person has an infectious disease, the body makes antibodies against the pathogens. When the disease is over, some antibodies against those pathogens stay in the blood. The cells that make that particular antibody are also found in great numbers. If the pathogen that first caused the disease enters the body again, it is usually quickly destroyed by the antibodies. The body's ability to prevent an infection from developing by having antibodies against the pathogen is called **immunity** (ih MYOO nuh tee).

Immunity that develops as a result of exposure to pathogens or their toxins is called active immunity. In **active immunity** the body makes its own antibodies as a result of exposure to the antigens. The body is still able to quickly produce these antibodies again. **Passive immunity** is a form of immunity caused by receiving antibodies from the blood of humans or other animals that have been exposed to the antigen. The antibody-containing substance is called a **serum** (SIHR uhm). This kind of immunity lasts no more than a few weeks. However, its effect is almost immediate.

The cells in the body that respond to the presence of antigens are white blood cells called **lymphocytes** (LIHM fuh sīts). There are two kinds of lymphocytes: B-cells and T-cells. **B-cells** are lymphocytes that make antibodies and give off the antibodies into the bloodstream. **T-cells** are lymphocytes that join with foreign antigens, such as those on a bacterial cell, and then destroy that cell.

Antibody synthesis by a B-cell starts after the surface of the B-cell comes in contact with an antigen. The B-cell then grows and reproduces, forming a population of cells called plasma cells. A **plasma cell** is an enlarged cell that can make large amounts of an antibody. All of the plasma cells that come from one B-cell make the same kind of antibody. After an infection is over, special plasma cells called **memory cells** stay in the bloodstream. Memory cells can produce this particular antibody. If the same antigen enters the body again, the memory cells will instantly make antibodies to destroy it. Humans and other mammals can produce thousands of kinds of antibodies. Each kind of antibody acts against one type, or a few types, of antigen.

Antibodies are proteins that are shaped like the letter Y. Two of the ends of the Y-shaped molecule have pockets. These pockets, or combining sites, are different on each kind of antibody. It is believed that antibodies attach to antigens in a manner similar to that in which an enzyme attaches to its substrate.

Antibodies can attach to more than one antigen. Antigens that are held together in this way are readily ingested by phagocytes. Antibodies that act against antigens on the surface of a pathogen can cause enzymatic reactions that puncture the cell membrane of the pathogen and kill it.

A T-cell produces only a few antibodies. These antibodies are carried on the surface of the T-cell. When a surface antibody attaches to an antigen on a pathogen, the T-cell kills the pathogen. T-cells can kill body cells that have been infected with viruses because the body cells have foreign antigens on their surfaces.

The cells and tissues involved in the body's response to infection make up the **immune system.** The immune system consists of lymphocytes and other leukocytes, lymph nodes, and those tissues that make white cells, including the bone marrow and the spleen.

An **inflammation** (ihn fluh MAY shuhn) is a tissue response to the presence of microbes, other foreign antigens, or an injury. The symptoms of inflammation are redness, swelling, heat, and pain. During inflammation the injured cells give off chemicals that increase the flow of blood to the area and cause blood vessels to become leaky. White blood cells and antibodies enter the affected area at an increased rate. They destroy microbes and other foreign matter. The lymphocytes make more antibodies, while the macrophages take in dead white cells and other debris.

Increased blood flow to an infected area causes it to become warm and red. The leakage of fluid from blood vessels causes swelling. These are signs, however, of a normal defensive reaction.

Acquired Immune Deficiency Syndrome (AIDS) is a disease that interferes with the normal functioning of the immune system. The cause of AIDS is a virus, the Human Immunodeficiency Virus or HIV.

FIGURE 15·14 This illustration shows the immune response in humans. Two kinds of lymphocytes, B-cells and T-cells, are needed for the immune response to successfully fight disease.

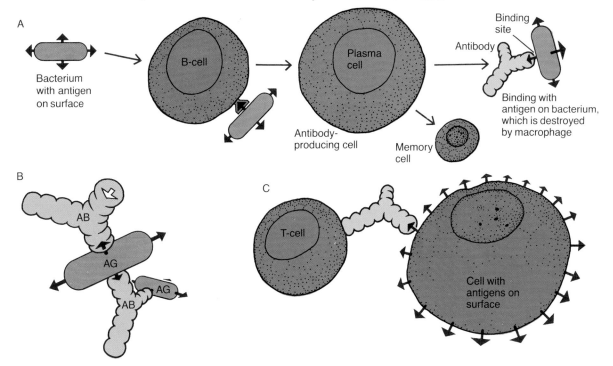

This virus destroys lymphocytes. Thus, a person with AIDS has a much lowered resistance to infection. A person with this disease can die as a result of infections from which a healthy person would recover.

An **allergy** (AL uhr jee) is an extreme reaction by the immune system to the presence of antigens. Many substances, including pollen and drugs, can produce allergic reactions. The allergic reaction involves the release by the immune system of larger than normal amounts of chemicals called *histamines*. Symptoms of an allergic reaction can include watery eyes, sneezing, and intestinal problems.

15·12 INTERFERON

Interferon is a cell-produced protein that helps protect the body against viral infection. Interferon was so named because it "interferes" with the reproduction of viruses. Interferon is of interest because of its protection against viral infections and because it may aid in fighting some kinds of cancer.

Interferon is produced by cells that are infected by viruses. The interferon is given off and causes other cells to make a protein that inhibits viral reproduction. Human interferon is now being produced by genetic engineering (Section 8·8). Interferon is effective in small amounts and its commercial production may lead to improved treatment of viral infections and cancer.

15·13 CHEMOTHERAPY

The body's immune system sometimes cannot control an infection. Medical treatment in the form of drugs is useful against certain kinds of infections. The use of drugs to treat disease is called **chemotherapy** (kee moh THER uh pee).

The use of drugs to fight infectious disease was limited before the discovery of sulfa drugs in the 1930s. The sulfa drugs are effective against certain kinds of bacterial infection. These drugs interfere with the metabolic activities of the bacteria, thus inhibiting their growth. However, sulfa drugs are toxic and must be used with care.

Today antibiotics are more widely used than sulfa drugs. **Antibiotics** (an tee bī AHT ihks) are substances that are formed by microbes or fungi and that inhibit the growth of bacteria. The best-known antibiotic, penicillin, is produced by the fungus *Penicillium notatum*. Penicillin was discovered in 1929 by Dr. Alexander Fleming. Penicillin interferes with the development of the cell wall in bacteria, thus inhibiting their reproduction. Another antibiotic, streptomycin, which is produced by a soil microbe, *Streptomyces griseus*, inhibits protein synthesis. Figure 15·15 shows some antibiotic-saturated disks in a dish in which bacteria are growing. Why are some of the disks surrounded by clear zones?

Some strains of bacteria are now resistant to certain antibiotics. These bacteria have developed as a result of mutations. Researchers are always working to find new antibiotics that are useful against such mutant populations.

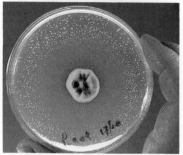

FIGURE 15·15 Antibiotic-saturated disks on a culture of *E. coli* bacteria (top) Notice the clear areas where bacteria have been killed. The mold in the bottom photograph has produced an antibiotic, killing nearby bacteria.

Investigation

What is the Effect of Alcohol on the Growth of Organisms?

Goals
After completing this activity, you will be able to
- Describe the kinds of colonies present after objects have been placed on a growth medium and incubated.
- Compare the number and kinds of colonies present after the objects have been treated with alcohol.
- Explain why alcohol is used as an antiseptic.

Materials (for groups of 2)
2 petri dishes with sterile nutrient agar
forceps soaking in alcohol
2 paper clips   2 thumbtacks
2 small pencil erasers
glass-marking pencil
tape   graduate
70% alcohol solution
100-mL beaker

Procedure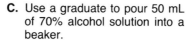

A. Obtain two petri dishes containing sterile nutrient agar.

B. Use a glass-marking pencil to label the lids of the dishes. Label one dish *Control* and the other dish *Soaked in alcohol*. Also write your name and the date on the cover of each dish. *Be sure to keep the dishes closed while labeling them.*

C. Use a graduate to pour 50 mL of 70% alcohol solution into a beaker.

D. Place a paper clip, a thumbtack, and an eraser into the alcohol solution. Allow these objects to remain in the solution 10 minutes.

E. Using a clean pair of forceps, place a paper clip, a thumbtack, and an eraser into the dish marked *Control*. Remember, these objects must *not* be soaked in alcohol. Immediately cover the dish and tape it closed.

F. Using a clean pair of forceps, remove the paper clip, thumbtack, and eraser from the beaker with alcohol. Place these objects into the dish marked *Soaked in alcohol*. Immediately cover the dish and tape it closed.

G. Incubate the petri dishes for a week in your desk drawer or in a place indicated by your teacher.

H. After 1 week examine the two petri dishes. On a separate sheet of paper, draw what you observe in each petri dish. Label one drawing *Control* and the other drawing *Soaked in alcohol*.

I. Follow your teacher's instructions for the proper disposal of all materials.

Questions and Conclusions

1. After 1 week of incubation, what did you observe in the dish marked *Control*? What did you observe in the dish marked *Soaked in alcohol*?
2. How did the growth in these two dishes compare?
3. What effect did the alcohol have on the microorganisms? Give a reason for your answer.
4. Why, do you think, did you use forceps, rather than your fingers, to place objects in the petri dishes?
5. Why was it necessary to *immediately* close the petri dishes after you added the objects?
6. Why do dentists often soak their instruments in alcohol?

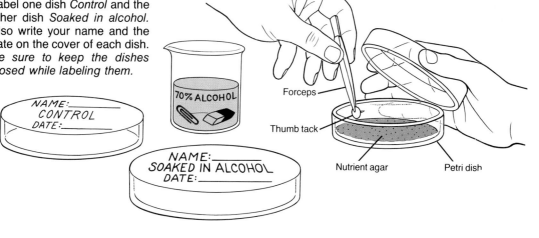

15·14 PREVENTING THE SPREAD OF DISEASE

Stopping the transmission of pathogens prevents a disease from spreading among people. Diseases such as malaria and yellow fever, which are transmitted by insect vectors, can be prevented from spreading by getting rid of the insects.

Quarantine is the isolation of diseased persons. It is one of the oldest methods of preventing diseases from spreading. Crews on ships were often kept on board until health officers confirmed the absence of disease. People can also be quarantined at home. This method is not very effective because it is difficult to enforce.

Before doctors understood the connection between microbes and disease, they often transmitted microbes from infected people to healthy people. **Sterilization** is the process of killing microbes by heat and pressure. This process ensures that infectious microbes carried on instruments and equipment are not spread to patients. **Disinfectants** are chemicals that kill microbes on contact. They are also effective in preventing the spread of disease.

People can acquire active immunity, which prevents their getting a disease, by being exposed to a disease or by being vaccinated. **Vaccination** is the injection of weakened or dead pathogens or their toxins into a healthy person. The person's body reacts by making antibodies, which gives the person immunity to the disease. The substance used in the vaccination is called a *vaccine*. Figure 15·16 shows the major steps in the development and use of a vaccine. The captions under each picture explain the steps in the process.

Infants and children are vaccinated to prevent their getting infectious diseases and to prevent epidemics. The vaccinations are begun in infancy and continue until about age 16. In many states the DPT (diphtheria, pertussis, tetanus) immunization is required by law before children can enter grade school. Vaccines are also available against measles, mumps, and rubella and are usually given to children. There are also certain flu vaccines. These vaccines are often suggested for elderly people, who are at high risk for getting the flu.

Passive immunity can be acquired by the injection of a serum. Recall that such a serum contains antibodies from the blood of humans or other animals. The antibodies will kill the disease-causing microbe.

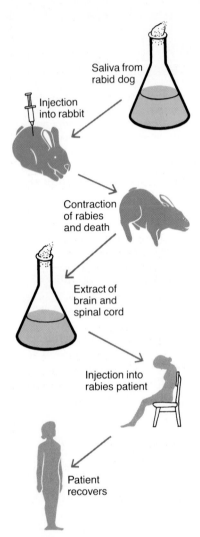

FIGURE 15·16 Rabies is a disease that is almost always fatal. You can see why the development of a vaccine to fight this disease was so important. This illustration shows the steps in the production of this vaccine.

REVIEW QUESTIONS

11. How do the skin and mucous membranes protect against infection?
12. Describe the body's responses to the presence of an antigen.
13. What is the difference between active and passive immunity?
14. What is interferon? Why does it offer hope in the treatment of viral diseases?
15. An individual receives a vaccination. Within 24 hours the person shows mild symptoms of the disease. What was probably the nature of the vaccine that was used?

Careers in Biology

Microbiologist

Microbiologists study the growth, structure, and functions of bacteria, viruses, and other microorganisms. They isolate and culture different species of these microorganisms. Microbiologists work with computers, electron microscopes, and other sophisticated laboratory tools.

Because there are so many kinds of microorganisms, microbiologists usually specialize in one particular area. Medical microbiologists study disease-causing microbes. Food microbiologists develop ways to prevent the growth of microbes that spoil food.

Microbiologists must have at least a bachelor's degree in biology, microbiology, or a closely related field. Many microbiologists have advanced degrees in highly specialized areas of study. For additional information write to the American Society of Microbiologists, 1913 I Street NW, Washington, DC 20006.

Medical Technologist

The job of a medical technologist involves many different skills. Medical technologists conduct specialized laboratory tests that help physicians to diagnose and treat disease. They analyze specimens of urine, blood, and tissues from patients in order to identify disease-causing microorganisms. They prepare tissue for microscopic examination and perform blood tests for transfusions.

To become a medical technologist, a person must have a degree from either a two-year or four-year college. A degree from a two-year college, however, does not allow the technologist to do as many tasks as does a degree from a four-year college. Most states require that medical technologists be licensed or certified. For additional information write to the American Medical Technologists, 710 Higgins Road, Park Ridge, IL 60068.

Mushroom Grower

Mushroom growers raise mushrooms to be sold to stores and restaurants. Many mushroom growers have small farms where they grow only small amounts of mushrooms for sale. Other growers have large farms and sell mushrooms to national distributors.

Mushroom farming requires no formal education, but an understanding of the growth and reproduction of mushrooms is necessary. A solid background in high school biology, chemistry, and some business courses also are helpful. Most libraries have books describing how to start small mushroom farms. Some large commercial farms provide on-the-job training. For additional information write to the Phillips Mushroom Farm, P.O. Box 190, Kennett Square, PA 19348.

How to Find a Summer Job

A wide variety of summer jobs are available for high school students. One way to find out about summer jobs is to read the want ads in your local newspaper. Part-time and full-time summer jobs are usually advertised in separate sections of the want ads.

Many local and state governments provide summer work programs. They offer jobs ranging from office and restaurant work to outdoor work in parks. Most of these programs are advertised in local newspapers, and on local television and radio news or community service programs. Usually there are minimum requirements that potential employees must meet. Check with your local government to find out if such programs are available in your area.

Information about summer jobs may also be available in your school or local library. Many libraries maintain files of local businesses that hire students to work part-time or full-time during the summer.

Chapter Review

SUMMARY

15·1 Parasitic organisms may enter a body, causing disruption of the normal functioning of the body. The resulting condition is called a disease.

15·2 Whether an organism will cause disease in a particular host is related to host resistance as well as to the virulence and location of the entering pathogens.

15·3 Diseases can be spread by various means: air, water, physical contact with an infected host or an intermediate object, and vectors.

15·4 By applying postulates first used by Robert Koch, scientists are able to isolate and identify disease-causing agents.

15·5 Viruses cause diseases in humans and other organisms by altering the structure and functioning of cells.

15·6 Bacteria cause diseases in humans and other organisms by invading host cells or by releasing exotoxins and endotoxins, which disrupt cell functioning.

15·7 Viroids and rickettsiae cause disease in a number of plants and animals.

15·8 Human fungal disease includes diseases of the skin and of the internal organs and systems.

15·9 Protozoans cause some human diseases, such as malaria, African sleeping sickness, and amoebic dysentery.

15·10 The body's barrier to pathogens include the skin, tears, mucus of the mucous membranes, and enzymes of the digestive juices. Once inside the body, pathogens can be destroyed by leukocytes.

15·11 The production of antibodies by white blood cells produces active immunity, a long-lasting chemical defense against the invasion of the body by pathogens.

15·12 Interferon, a cell-produced protein, prevents viral infection by inhibiting the reproduction of viruses.

15·13 Antibiotics and drugs such as sulfa may be used to treat microbial disease.

15·14 Diseases can be prevented from spreading by interfering with the transmission of the microbes or killing them.

BIOLOGICAL TERMS

active immunity
allergy
antibiotics
antibodies
antigen
B-cells
chemotherapy
communicable diseases
contagious diseases
disease
disinfectants
endemic
endotoxin
epidemic
exotoxin
germ theory of disease
immune system
immunity
infectious disease
inflammation
interferon
Koch's postulates
leukocytes
lymph
lymphocytes
macrophages
memory cells
microbial diseases
passive immunity
pathogens
plasma cell
serum
sexually transmitted diseases (STD)
sterilization
T-cells
toxin
vaccination
vectors

USING BIOLOGICAL TERMS

1. Use a dictionary to find the meaning of the word parts *endo-* and *exo-* in the terms *endotoxin* and *exotoxin*. Explain why these terms are appropriately named.
2. List all terms that refer to methods that help prevent microbial diseases. Define the terms.
3. List all terms that refer to any methods by which microbial diseases can be treated. Define the terms.
4. Identify all terms that refer to parts of the body's immune system. Define the terms.
5. Distinguish between the terms *epidemic* and *endemic*.
6. Distinguish between the terms *antigen* and *antibody*.
7. Distinguish between B-cells and T-cells.

8. Explain the major differences between *active immunity* and *passive immunity*.
9. Using a dictionary, find the meaning of the word parts that comprise the term *antibiotics*. Explain why the term is appropriately named.

UNDERSTANDING CONCEPTS

1. Why is an infectious disease an example of a parasitic relationship? (15·1)
2. Describe the major conditions affecting whether or not an organism will become pathogenic. (15·2)
3. Explain what is meant by "direct spread of a disease." Give several examples of the ways in which a disease is spread directly. (15·3)
4. Describe the role of a vector in spreading disease. Give an example of a disease involving an insect vector. (15·3)
5. List the steps a microbiologist would follow in isolating and identifying a pathogen. (15·4)
6. Explain why German measles, a disease that is usually mild, is dangerous when contracted by a pregnant woman. (15·5)
7. How do bacterial endotoxins and exotoxins affect host cells? (15·6)
8. What is a viroid? How are viroids thought to be spread from plant to plant? (15·7)
9. Summarize the stages in the transmission of Rocky Mountain spotted fever. Name the microbe that causes this disease. (15·7)
10. Give an example of a fungal disease in humans. (15·8)
11. Describe two methods for the control of malaria. (15·9)
12. Describe four of the body's natural defenses against invasion by pathogens. (15·10)
13. List the various body responses that accompany an inflammation. How are these responses beneficial to the body? (15·11)
14. What is interferon? (15·12)
15. Briefly describe the ways in which antibiotics control bacterial disease. (15·13)
16. Name some diseases against which children must be immunized before they enter school. Why is immunization required by law? (15·14)

APPLYING CONCEPTS

1. Many microorganisms produce antibiotics in their natural environment. What do you think the function of the antibiotics might be in nature?
2. Parasites that do not kill their host are considered by biologists to be more "successful" than those parasites that do kill their host. Explain why.
3. Athlete's foot is a condition caused by a fungus that lives on dead organic matter. Explain how a fungus can live on a human foot.
4. Suppose you step on a piece of sharp, stiff wire. The wire punctures your foot. What disease should you be most concerned about? What aspect of your medical history would be useful in deciding whether you should seek immediate medical attention?

EXTENDING CONCEPTS

1. Write a research paper on how interferon works and why it offers hope for a treatment against viruses.
2. Write a report on the construction of the Panama Canal and how the construction relates to what you have studied in this chapter.
3. Your vaccination with tetanus vaccine helps only you. Your vaccination with polio vaccine or diphtheria vaccine helps your neighbors as well. Write a report in which you explain the above statement.
4. Research the development of the polio vaccine. Find out how mass vaccinations were conducted during the 1960s. What is an iron lung? Find out its role in the treatment of polio before vaccinations became available.

READINGS

Goodfield, J. "Vaccine on Trial." *Science 84,* March 1984, p. 78.

Habicht, G.S., Gregory Beck, and J.L. Benach. "Lyme Disease." *Scientific American,* July 1987, p. 78.

Hall, S.S. "The Flu." *Science 83,* November 1983, p. 56.

Hirsch, Martin S. and Joan C. Kaplan. "Antiviral Therapy." *Scientific American,* April 1987, p. 76.

"Interferon by Bacteria, On Trial." *Science News,* January 24, 1983, p. 54.

Unit III Skills

USING SCIENTIFIC ABBREVIATIONS

You already have seen many shortcuts that biologists take in writing chemical and biological information. You have seen abbreviations used for measurements and symbols used for chemical elements. You also have seen letters used to represent genes and chromosomes. In the following exercises, you will review the meanings of some of these biological abbreviations. You will examine the ways in which abbreviations and symbols are created. Finally, you will predict the meaning of some abbreviations with which you may not be familiar.

As you work through the exercises, use the page references given if you need help from the text. Sometimes you may have to refer to pages you have already studied. Sometimes you may need to look at pages that you have not yet read in the book. Work on a separate sheet of paper; do not write in this book.

1. Some of the abbreviations or symbols from the first three units are listed below. Write the word or words for which each abbreviation or symbol stands.
   a. Cl
   b. ER
   c. kg
   d. ATP
   e. O_2
   f. $2n$
   g. tRNA
   h. coA
   i. E_1
   j. SI

As you might have noticed, abbreviations represent many different kinds of words. Abbreviations also take many different forms. For example, in some abbreviations, one or more letters represent the words. Other abbreviations are shortened forms of the words. Some shortened forms eventually become part of everyday English usage.

2. Write the word for which each of these shortened word forms stands.
   a. flu
   b. lab
   c. biped
   d. dicot
   e. forams
   f. phage

Sometimes letters that abbreviate words spell "words" themselves. Abbreviations that are used as other words are called *acronyms* (AK roh nihmz).

3. Using this book and other references, determine the meaning of each of the acronyms below.
   a. scuba
   b. AIDS

Not only are there different kinds of abbreviations, there are specific rules about writing certain abbreviations. For example, the genus given in a scientific name often is abbreviated. It is correctly abbreviated by its first letter followed by a period. (Remember that scientific names are expressed in italics or are underlined.) For example, *Canis familiaris* is abbreviated correctly as *C. familiaris*.

4. How would you correctly abbreviate the following names?
   a. *Neurospora crassa*
   b. *Pseudomonas aeruginosa*

5. What do the following abbreviations represent?
   a. *E. coli*
   b. *H. sapiens*
   c. *A. proteus*
   d. *D. melanogaster*

6. Find two other scientific names in this book. For each, write the full scientific name, the correct abbreviation, and the page on which you found the name.

Some science abbreviations represent things other than science words. For example, the names of certain governmental agencies are represented by initials. EPA stands for the Environmental Protection Agency.

7. Find out what governmental agencies are represented by the following abbreviations.
   a. NIH
   b. NSF

Names of scientific journals also may be represented by abbreviations. Many library reference books use abbreviated journal names.

8. Find out what the following journal abbreviations represent.
   a. *Nat Hist*
   b. *Sci Am*
   c. *Pop Sci*
   d. *Natl Wildl*

9. Find out what the abbreviations for the following journals are.
   a. *National Geographic*
   b. *Science Digest*

UNIT IV

The Plants

Plants seem silent and passive, incapable of the complex behaviors seen in animals. But, in fact, without the activities of green plants most other kinds of living things could not survive. The activities of plants and the many kinds of behaviors of which plants are capable will be discussed in this unit.

Chapters

16 *Algae to Ferns*
17 *The Seed Plants*
18 *Transport in Plants*
19 *Plant Reproduction*
20 *Growth and Behavior*

16. Algae to Ferns

Have you ever noticed that there are many plants growing in ponds, lakes, streams, and oceans? You might have seen a layer of green scum on the surface of a pond or some seaweed that had washed up on a beach. How are these plants like land plants? How are they different from land plants? What are the economic and ecological values of these plants?

CHAPTER OBJECTIVES

After completing this chapter, you will be able to
- **Describe** the structures and life cycles of algae.
- **Identify** primitive land plants.
- **Describe** the life cycles of ferns.
- **Recognize** ways vascular plants are adapted to a life on land.

The Simplest Plants

16•1 VASCULAR PLANTS AND NONVASCULAR PLANTS

The plant kingdom contains about 350 000 known species. Members of the plant kingdom range from small single-celled organisms to complex trees. Although they differ in many ways, all plants have some things in common. Plants are grouped into five phyla. The classification of plants illustrates the evolutionary relationships between groups of plants. Classification also allows us to study plants in an organized manner.

The most complex members of the plant kingdom make up phylum *Tracheophyta* (tray kee AHF uht uh). This phylum includes all the vascular plants. **Vascular plants** are plants that have specialized tissues for carrying water and food throughout the plant body. Water and minerals move from the soil up to all parts of a vascular plant in specialized conducting tissue called *xylem* (ZĪ lehm). Food and other substances are carried throughout the plant in tissue called *phloem* (FLOH ehm). Both the xylem and the phloem are made of rigid cells. As a result, they serve as structural support for the plant. Vascular plants, such as ferns, conifers, and flowering plants, are adapted to life on land. Most of the plants with which you are familiar are vascular plants.

Plants without vascular systems are simpler and much older in origin than vascular plants. Plants without vascular tissue are called **nonvascular plants.** Nonvascular plants are classified into several phyla. These plants lack true roots, stems, and leaves. They originally developed in the sea. Most nonvascular plants are aquatic. Water and the materials dissolved in it are readily available to all parts of the plants. Water also helps to support the plants. Thus both the transport of materials to cells and the structural support of these plants are accomplished without vascular tissue. You have probably seen nonvascular plants floating on ponds.

SECTION OBJECTIVES
- Compare vascular and nonvascular plants.
- Describe different types of aquatic nonvascular plants.
- Compare the life cycles of green, brown, and red algae.

16·2 GREEN ALGAE

The simplest types of nonvascular plants are the **algae** (AL jee). Algae are mostly aquatic plants, occurring in both salt water and fresh water. Some also grow in moist soils, on tree bark, and on wood.

In water, floating algae often are found with other simple organisms. The algae, bacteria, fungi, and small animals together are called *plankton*. Algae in the plankton are called **phytoplankton** (fī toh-PLANGK tuhn). Plankton forms the basis of food chains in the sea.

phyto- (plant)
planktos (drifting)

Much phytoplankton is unicellular algae—simple plants made of one cell. Other slightly more complex algae in phytoplankton are colonial algae. Algal colonies are groups of cells. Each cell is similar in structure and function to a single-celled alga.

In more complex algae, cells stay together and form specialized structures. Algae with many cells, some of which are specialized for certain functions, are called multicellular algae. Long stalks, branches, and leaflike structures can form in these algae. Many of these plants grow attached to objects such as stones, shells, wood, and other plants in the water. Figure 16·1 shows some different kinds of algae.

Algae are classified according to the pigments they contain, the manner in which they store food, the chemicals in their cell walls, and the presence or absence of specialized structures for movement. The major groups of algae described in this chapter are phylum *Chlorophyta* (kloh RAH fuht uh), phylum *Phaeophyta* (fee AHF uht uh), and phylum *Rhodophyta* (roh DAHF uht uh).

Members of phylum Chlorophyta are called **green algae.** While most green algae live in fresh water, others live in salt water, in snow, in the soil, and on trees. The pigments and the chemicals in the cell walls of green algae are very similar to those in the vascular plants. Most green algae store food as starch.

FIGURE 16·1 Green algae belong to the family Chlorophyta. Examples of green algae include (left to right) *Colenkinia*, *Closterium*, and *Cladophora*.

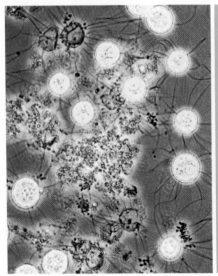
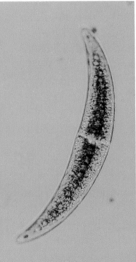
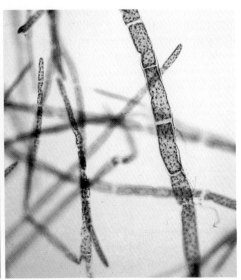

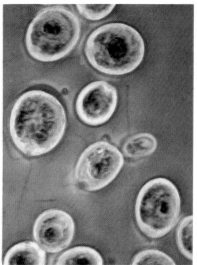 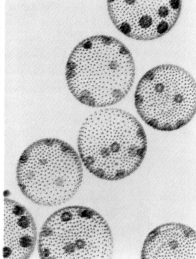 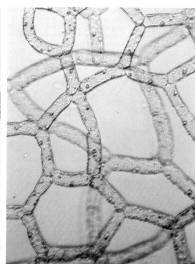

FIGURE 16·2 The unicellular green alga *Chlamydomonas,* the motile colonial green alga *Volvox,* and the nonmotile colonial green alga *Hydrodictyon* (left to right).

Classification of green algae is based on whether the plants are unicellular, colonial, or multicellular, and whether or not the algae move. Motile algae move by means of *flagella*.

A typical motile unicellular alga is the common freshwater alga *Chlamydomonas* (klam uh DAH muh nuhs). As you can see in Figure 16·2, each *Chlamydomonas* is a pear-shaped cell with a large chloroplast. In some species the chloroplast has a red "eyespot" that can detect light. Surrounding the chloroplast are starch deposits, the cell's food supply. Extending from the smaller end of the cell are two flagella. *Chlamydomonas* moves rapidly by beating these two flagella. Recent studies have shown that some *Chlamydomonas* species move in response to a magnetic field. Others move in response to light.

Volvox is a motile colonial green alga. A *Volvox* colony is shown in Figure 16·2. It is a hollow ball made of a single layer of between 500 and 600 cells. The cells are held together by cytoplasmic strands. Each cell has two flagella that beat rapidly. The *Volvox* ball spins on its axis as it moves through the water. How are *Volvox* and *Chlamydomonas* alike? How do they differ?

An example of a nonmotile unicellular green alga is *Chlorella*, a common saltwater and freshwater plant. *Chlorella* is round and has a cup-shaped chloroplast. Nonmotile green algae also form colonies. *Hydrodictyon* is a nonmotile colonial alga common to ponds, lakes, and quiet streams. Each colony is made of cylindrical cells that form a larger hollow cylinder. Patterns formed by *Hydrodictyon* cells give this alga the common name "water net."

The multicellular algae are a more complex group of plants. Such plants often have cells that form specialized structures. In the green algae, three types of multicellular plants exist: simple unbranched filaments; branched or membranous plants; and tubular plants.

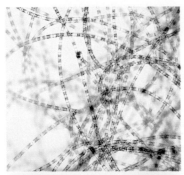

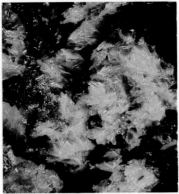

FIGURE 16·3 Multicellular algae may have cells arranged as filaments like *Ulothrix* (top). The cells of *Ulva* (bottom) are bladelike.

A common unbranched filamentous, or threadlike, green alga is *Ulothrix* (YOO luh thrihks). This alga can be found in both fresh water and salt water. The cells of each filament are alike except for the cell at the base. This cell is modified to hold the plant to stones and other objects in streams and lakes, and is called the **holdfast.** Holdfasts are found in many multicellular algae.

Another common unbranched threadlike green alga is *Spirogyra* (spī ruh JĪ ruh). Whereas *Ulothrix* grows attached to objects, *Spirogyra* usually floats. It forms bright-green slimy masses in small bodies of fresh water. The green ribbonlike structure is the chloroplast.

Ulva, or sea lettuce, is a common membranous green alga found in salt water. Its flat bladelike body is adapted to float on the water's surface. How might this adaptation increase the level of photosynthesis in *Ulva*? In some species of *Ulva,* the blade can be as long as 1 m. Specialized cells are adapted to hold the plant to rocks, wood, and larger algae. Thick cell walls allow *Ulva* to survive drying during low tide.

16·3 LIFE CYCLES OF GREEN ALGAE

Reproduction in the green algae can occur in several ways. Some green algae, like *Chlorella,* reproduce only asexually. Many others reproduce both asexually and sexually.

Spirogyra reproduces by a process called conjugation. **Conjugation** is a form of sexual reproduction in which the genetic material of one cell combines with the genetic material of another cell. As shown in Figure 16·4, different filaments of *Spirogyra* become connected by structures called *conjugation tubes.* The cell contents from each cell in one filament, referred to as the "−" filament, flow through the conjugation tube and join with the cell contents in the "+" filament. The nuclei fuse. The diploid cells that result from conjugation are called **zygospores.** Meiosis occurs in the diploid zygospore, which then germinates to produce a monoploid filament of *Spirogyra.*

FIGURE 16·4 *Spirogyra* is a filamentous green algae. *Spirogyra* reproduce by conjugation. During conjugation, the cell contents of one filament are passed to cells in another filament.

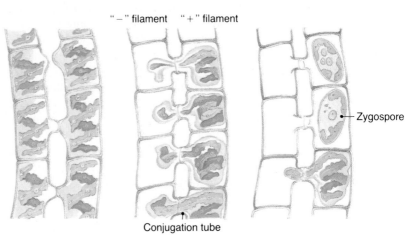

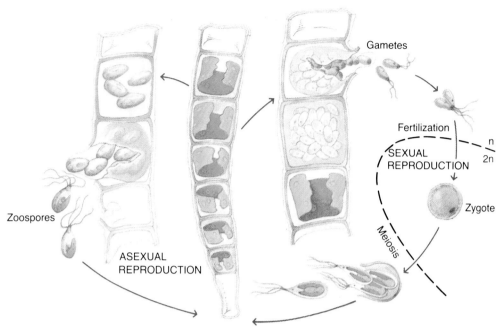

FIGURE 16·5 *Ulothrix* reproduces asexually as well as sexually. The life cycle of *Ulothrix* is shown here.

Asexual reproduction in *Ulothrix* is shown in Figure 16·5. Any of the cells other than the basal cell can undergo asexual reproduction. The cell divides several times and forms four or eight cells. Each of these cells has four flagella. These asexual reproductive cells are called **zoospores.** When the wall of the original plant cell breaks, the zoospores are released. The zoospores swim for a period of time. Then each one settles and attaches to an object. The zoospore loses its flagella, divides, and develops into a new *Ulothrix* filament.

Ulothrix is also able to reproduce sexually as shown in Figure 16·5. Under certain conditions, cells in the filament undergo repeated cell divisions to form flagellated cells smaller in size and more numerous than zoospores. These cells, with two flagella each, are the gametes of *Ulothrix*. All the gametes produced by *Ulothrix* are identical. A single gamete cannot develop into a new *Ulothrix*. It must first unite with another gamete, usually one from a different filament. This union results in the formation of a zygote. The zygote does not have flagella. It develops a hard thick wall that protects the zygote during a dormant period of several months. Once the dormant period ends, the diploid zygote undergoes a meiotic division that results in the formation of four monoploid flagellated cells called **meiospores.** Each meiospore can then develop into a new *Ulothrix* filament.

Compare the formation of zoospores and meiospores in Figure 16·5. How are these reproductive spores different? Most of the *Ulothrix* life cycle is monoploid. The filament, zoospores, gametes, and meiospores are all monoploid. The only diploid stage is the zygote.

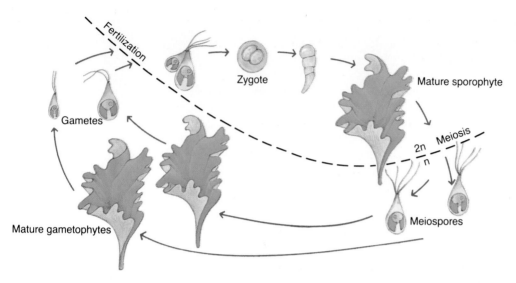

FIGURE 16·6 *Ulva* alternates between a diploid stage and a monoploid stage. This alternation between stages during one complete life cycle is called alternation of generations.

In *Ulva* the plant is monoploid for part of the life cycle and diploid for part. For sexual reproduction to occur, certain cells on edges of the diploid plant undergo meiosis. Each cell produces four monoploid meiospores. These flagellated meiospores develop into new plants which look like the diploid plants, although they are monoploid.

The monoploid *Ulva* plants produce gametes. In some species of *Ulva* the gametes are of two sizes. Some of the monoploid *Ulva* plants produce small gametes called male gametes. Other plants produce larger gametes called female gametes. When a female and a male gamete unite, a zygote is formed. The zygote then develops into a diploid *Ulva*. The cycle is then complete.

As you can see, half the *Ulva* life cycle is spent in the diploid state. The other half is spent in the monoploid state. Compare this with the sexual cycle of *Ulothrix* in Figure 16·5. The diploid plant that produces meiospores is called the **sporophyte** (SPOR uh fīt). The monoploid plant that produces gametes is called the **gametophyte** (guh MEE tuh fīt). This alternation between diploid and monoploid stages during one complete life cycle is called **alternation of generations**. This alternation of generations is typical of the life cycles of many algae, as well as the mosses and vascular plants.

16·4 BROWN ALGAE

Members of phylum Phaeophyta are called **brown algae.** Whereas most of the green algae are freshwater plants, most of the brown algae are marine. The brown algae are found most abundantly in cold water along rocky coastlines, such as the Pacific coast of North America. Some grow in shallow waters and are exposed to air during low tide. Others grow deep enough to be constantly submerged. Brown algae grow attached to rocks, shells, or to other marine algae.

The color of the brown algae comes from the pigment *fucoxanthin* (fyoo koh ZAN theen). The chlorophyll in brown algae is masked by the brown of fucoxanthin. In the cell wall there is a gummy substance called *algin*. This substance helps the plants to withstand the action of waves. Algin is used in making ice cream, medicine, and varnish.

Brown algae are multicellular. Some are simple branching filaments. But the most common brown algae are the large complex *kelps*. These are the plants most often called seaweed.

Laminaria (lam uh NAIR ee uh), shown in Figure 16·7, is a typical kelp. At its base a branched holdfast secures the plant to rocks, shells, or other objects. From the holdfast extends a *stipe*, a stemlike part. Above the stipe is the leaflike *blade*. Cells on the surface of the blade and stipe make food.

Many of the cells in some kelps are too far below the surface of the water to receive enough light to make food. Such kelp plants have specialized cells to carry food throughout the plants. The cells are similar in form and function to phloem tissue.

The growth of the kelps is very rapid and great. *Macrocystis* (mak-ruh SIHS tihs), sometimes called the "sequoia of the sea," can grow 30 cm to 65 cm a day. It reaches its mature length of 50 to 100 m in about 6 months. Air pockets in the blades of *Macrocystis* and other large kelps keep the blades afloat. What effect would the air pockets have on the kelps' ability to perform photosynthesis?

Many of the kelps have life cycles that are unlike those of the green algae. For example, notice in Figure 16·7 that the kelp *Laminaria* has both diploid and monoploid plants. However, the two types of plants look very different. The diploid plant, or sporophyte, is the large complex long-lived plant. Meiosis occurs inside special pockets in the diploid plant. Monoploid spores are produced. The spores develop into male and female gametophytes. What differences can you see between the male and female gametophytes of *Laminaria?*

FIGURE 16·7 This illustration shows the life cycle of *Laminaria*, a typical kelp. The holdfast anchors this plant to rocks, shells or other objects.

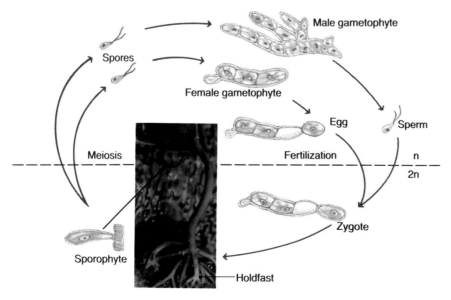

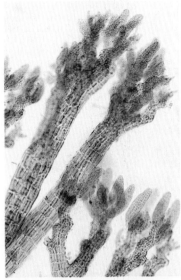

FIGURE 16·8 Photosynthesis in red algae can occur in deep water. Pigments in the plants can absorb energy from those light waves that penetrate deep water. Examples of red algae are (top to bottom) *Polysiphonia, Porphyra,* and *Chondrus.*

The female gametophyte produces eggs. The male gametophyte produces motile sperm that unite with eggs to form diploid zygotes. Each zygote can develop into a new diploid plant. How does the life cycle of *Laminaria* differ from that of *Ulva?*

16·5 RED ALGAE

Plants in phylum Rhodophyta are called **red algae.** Like brown algae, red algae are mostly marine plants. Unlike brown algae, red algae grow well in tropical waters, often at great depths.

Photosynthesis in the red algae can occur at great depths because of certain pigments in the plants. *Phycoerythrin* (fī koh uh RIHTH rihn) is a red pigment commonly found in the red algae. Other pigments in red algae are blue. All these pigments can absorb green, violet, and blue light waves. These light waves penetrate to great depths in the sea. The pigments transfer the absorbed energy to chlorophyll. As a result, red algae can grow at much greater depths than can the green or brown algae. Some red algae have been found at depths of 175 m.

The cell walls of red algae are often surrounded by a layer of slime. Some red algae are covered by additional gelatinlike materials called *agars, carrageenans,* and *gelans.* These materials often are harvested for commercial use. Agar is used as a medium in laboratories for growing microbes. Carrageenan is used as a thickener in ice cream, yogurt, and other foods.

A few unicellular and colonial red algae exist, but the majority of red algae are multicellular. Most of the multicellular red algae are filamentous or membranous.

Several of the red algae produce calcium carbonate on the surfaces of their cells. As a result, the plants become hard and crusty. These algae can form major parts of warm water coral reefs. In some cases, algae form three times as much of the reef as do coral animals. Thus the rate of reef growth is controlled mostly by the algae.

The life cycles of most red algae are complex. Alternation of generations exists in most of the plants. In some of the red algae, diploid and monoploid plants are identical in appearance. In other red algae, such as *Porphyra,* the two types of plants are not alike. The diploid stage of *Porphyra* is a simple filamentous plant. The monoploid stage is a large membranous plant.

Most of the red algae produce eggs and sperm in specialized structures on their branches. Neither sperm nor eggs are motile. Sperm are carried to the eggs by water currents.

16·6 SIGNIFICANCE OF ALGAE

Though most of the algae are small, their importance in the world and to the study of biology is considerable. For example, the pigments, the composition of the cell wall, and the compounds that serve as food storage are the same in both green algae and in land plants. This evidence indicates that the ancestors of land plants were similar to, if not the same as, some of the green algae.

FIGURE 16·9 A kelp bed may provide hiding places for many kinds of fish and other animals. Mature kelp plants can be as long as 100 meters, about the length of a football field.

Green algae form part of the basis of the food chain in the sea. They are eaten by small and large organisms. Algae also release oxygen into the water. The oxygen is used by animals. Observe the forest of brown kelps in Figure 16·9. How do animals benefit from the environment of the kelp bed?

Because of their rapid growth, kelps are being studied as a source of methane gas. Experimental farms off the California coast raise large beds of kelps. Once these kelps are harvested and put under anaerobic conditions, they may prove to be a valuable source of methane.

Algae have for centuries been a food source for humans. The Japanese cultivate and harvest the red alga *Porphyra* for food. Recently, green algae have been cultivated experimentally as a possible high-protein food for humans.

The gelatinlike substances in the cell walls of red and brown algae are the major economically important algal products. Agar for bacterial cultures is made from red algae. Algin from brown algae and carrageenan from red algae are used as thickeners in many products.

REVIEW QUESTIONS

1. Why are algae not able to grow on land to any great extent?
2. How does *Ulva* demonstrate alternation of generations?
3. Why can red algae photosynthesize at greater depths than green and brown algae?
4. What are the adaptive advantages of a holdfast to algae that live in coastal waters?
5. Which of the two *Laminaria* generations, monoploid or diploid, is better adapted for photosynthesis? Explain your answer.

Investigation

What Are the Characteristics of Some Green Algae?

Goals
After completing this activity, you will be able to
- Recognize three species of green algae.
- Compare unicellular, colonial, and multicellular green algae.

Materials (for groups of 2)
Chlamydomonas culture
Volvox culture
Spirogyra culture   microscope
3 slides   3 coverslips
3 droppers   iodine solution

Procedure

A. Remove a sample of the *Chlamydomonas* culture with a dropper. Place it on a slide and cover it with a coverslip.

B. Observe the sample under low power to find cells that are motile. When you find a motile cell, switch to high power and observe the cell under high power.

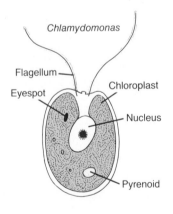

C. On a separate sheet of paper, make a drawing of a single cell. Label the chloroplast, the cell wall, the flagellum, and the pyrenoid. A pyrenoid is a structure on a chloroplast that stores excess food. Label the eye spot if it is visible.

D. Make a wet mount of the *Volvox* culture following the procedure you used in step A.

E. Observe the culture under low power. On a separate sheet of paper, make a drawing of the colony showing several cells in detail. Label the cell wall, the chloroplast, the cytoplasm, and the flagellum of one of the cells.

Volvox

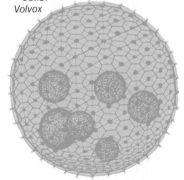

F. Look for differences in the cells of the colony. If you see any, show these differences in your drawing.

G. Make a wet mount of the *Spirogyra* culture.

H. Observe the culture under low power. Switch to high power and make a drawing of two cells showing how they are joined together. Label the cell wall, the cytoplasm, a pyrenoid, and the chloroplast of one of the cells.

I. Remove the coverslip. Place a drop of iodine on the culture. Dry the coverslip and put it back on the *Spirogyra* culture. (Iodine turns blue-black in the presence of starch.) Look for evidence of starch in the cells.

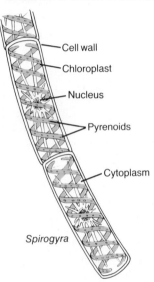

Spirogyra

J. Follow your teacher's instructions for the proper disposal of all materials.

Questions and Conclusions

1. Describe the shape of the chloroplast in *Chlamydomonas* and *Spirogyra*.
2. Compare the structure of *Volvox* and *Spirogyra*. How are they alike? How are they different?
3. In what way are *Chlamydomonas* and *Volvox* cells animal-like? In what ways are they plantlike?
4. Were any of the cells in the *Volvox* colony different from the others? If so, how?
5. Are unicellular organisms necessarily simple organisms? Explain your answer.

Nonvascular Land Plants

16•7 THE BRYOPHYTES

Most algae live in water. Water and the nutrients it carries are readily available to all parts of these plants. Physical support is provided by water. Transport of gametes during sexual reproduction occurs by means of water.

Plants that live on land do not have these advantages. Parts of a land plant are not in contact with water and its nutrients. The plant is surrounded by air, which causes the plant to lose moisture. The air does not provide physical support for a land plant. And water is not as readily available as a medium for sexual reproduction.

Only one phylum of nonvascular plants is adapted for life on land. Members of this group, phylum *Bryophyta* (brī AHF uht uh), are called **bryophytes** (BRĪ uh fītz) and include the liverworts and mosses. All bryophytes are small plants, rarely more than several centimeters long and rising no more than several centimeters high. Some bryophytes are shown in Figure 16•10. Bryophytes are often found in moist environments.

In all bryophytes it is the gametophyte that is the main plant. The gametophyte does not have true roots, stems, or leaves, since it does not have a vascular system. Rootlike structures called **rhizoids** anchor the bryophyte to the ground. The leaflike structures of bryophytes are usually one cell thick. They are adapted for rapid absorption of water. Bryophytes living in somewhat dry environments can absorb enough moisture from sources such as rain and morning dew. Most bryophytes live in clusters. This growth pattern helps retain moisture.

Liverworts are bryophytes that grow flat along a surface, rather than upright. Liverworts grow in extremely moist environments. Some liverworts, such as *Marchantia* (mahr KAIN shih uh), have simple undifferentiated gametophytes that are fleshy. Other liverworts, such as *Porella*, have thin, delicate structures that grow out of the plant's main axis.

Compare the moss in Figure 16•10 to the liverworts. How are they different? **Mosses** are bryophytes whose gametophytes grow erect, up to several centimeters high. Mosses have leaflike parts arranged in a spiral around their stemlike parts. Many mosses have simple food- and water-conducting cells. As a result, many mosses can live in drier places than can liverworts.

The bryophytes are not as important commercially as the algae. *Sphagnum* moss, which is highly absorptive, is added to soil to increase the amount of water the soil can hold. It also is used to form a moist base for flower arrangements. *Sphagnum* forms the underlying layers of bogs. It fills in small bodies of water by forming floating mats. More moss and, eventually, larger plants grow on these mats. Long after a bog has been formed, layers of decomposed and compacted

SECTION OBJECTIVES

- List some characteristics of the bryophytes.
- Describe the life cycle of a typical moss.

FIGURE 16•10 Both mosses (top) and liverworts (bottom) are nonvascular land plants. These plants live in moist areas.

FIGURE 16·11 *Sphagnum* moss (top) is often used by gardeners to improve soil quality. *Sphagnum* can form peat after it dies and decomposes (bottom).

Sphagnum form peat. In some areas, such as Ireland, peat is cut, dried, and burned as fuel. The "peat moss" you may have used in your garden is dried *Sphagnum* moss.

Many mosses, such as *Polytrichum* (puh LIH truh kuhm), can grow in areas where no other vegetation exists. Plants that are the first to grow in an area are called *pioneers*. Mosses growing over newly exposed soil can prevent soil erosion. Mosses growing on bare rock can create a layer of soil in which other plants can then grow. Small spores and seeds that land in this layer can germinate. The pioneer role of mosses is extremely important in areas destroyed by landslides, floods, and fires.

It is likely that the bryophytes, like other land plants, originated from a green algalike ancestor. But the bryophytes probably did not give rise to the vascular plants. Phylum Bryophyta forms only one of the branches that probably developed from algal ancestors.

16·8 LIFE CYCLE OF BRYOPHYTES

In both mosses and liverworts, the sperm are produced in specialized structures called **antheridia** (an thuh RIHD ee uh). Structures called **archegonia** (ahr kuh GOH nee uh) produce the eggs. In some bryophytes one plant can grow both archegonia and antheridia. In other bryophytes there are separate male and female gametophytes.

The transfer of the sperm to the egg is dependent on water. In some cases a film of dew or a single raindrop is enough for this transfer. The short height of bryophytes is related to the need for water for sexual reproduction. Sexual organs must be close to the ground, where water is available. The closely woven clusters of bryophytes allow easy transfer of sperm to egg.

The sporophyte, or diploid phase, results from the union of sperm and egg. The sporophyte is not a separate plant. It is a small, usually unbranched stalk that grows out of the gametophyte. In some bryophytes the sporophyte is green and can make food. In others the sporophyte must depend on the gametophyte for food.

Monoploid spores are produced by meiosis in the sporophyte. These spores are released from the plant and become new monoploid gametophytes. The spores of bryophytes have thick walls. These thick walls enable them to withstand drying.

The life cycle of *Polytrichum,* the common haircap moss, can serve as a model for the reproductive cycle of bryophytes. The life cycle of *Polytrichum* is summarized in Figure 16·12. Follow the events in the life cycle as you read about them.

proto- (first)
nema (thread)

1. The gametophyte of *Polytrichum* begins as a simple branching filament called a **protonema** (proh tuh NEE muh). What kind of nonvascular plant does a protonema resemble? The protonema develops branching rhizoids that penetrate the soil. Though the rest of the protonema is green and photosynthetic, the rhizoids are usually brown and not able to make food.

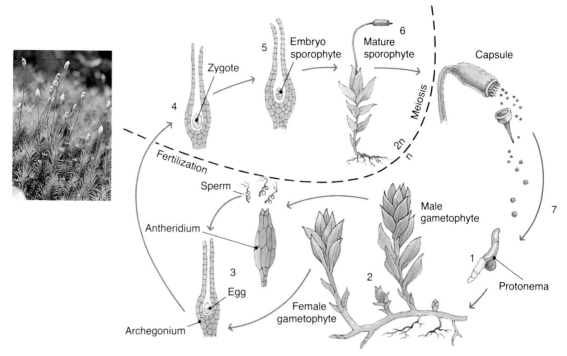

FIGURE 16·12 The life cycle of *Polytrichum*, the haircap moss, is shown here. This moss undergoes alternation of generations when it completes its life cycle.

2. The protonema develops into a mature gametophyte. *Polytrichum* gametophytes are either male or female. Both the antheridia and archegonia of *Polytrichum* are held in clusters at the tips of the branches of the mature gametophytes.

3. Eggs are produced at the bases of the long necks of the archegonia. Sperm are produced in the long vase-shaped antheridia. The antheridium takes in water, opens, and releases its sperm. The sperm move out of the antheridium. They swim through water to the archegonium.

4. Fertilization takes place in the archegonium. The cell that results from the union of sperm and egg is the zygote.

5. The zygote develops into a sporophyte within the archegonium. Though the sporophyte of *Polytrichum* is photosynthetic, it gets much of its food from the gametophyte. Food and water are transferred from the gametophyte to the developing sporophyte.

6. At the top of the sporophyte is a structure called the **capsule,** in which monoploid spores are formed as a result of meiosis. The spores mature in the capsule and eventually are released.

7. Spores landing in favorable places develop into new protonema.

REVIEW QUESTIONS

6. What factors limit the size of liverworts and mosses?
7. How are bryophytes better adapted for life on land than algae?
8. How is water a critical factor in the reproduction of bryophytes?
9. What are the sexual reproductive structures of bryophytes?
10. Mosses are often plant pioneers. In what kinds of environments are mosses unlikely to be able to be pioneers?

Vascular Plants

SECTION OBJECTIVES

- Compare the types of life cycles in simple, vascular plants.
- Discuss the adaptations of land plants that enabled them to survive outside of water.

16·9 CLUB MOSSES AND HORSETAILS

Bryophytes and a few algae are the only land plants that are nonvascular. The majority of land plants are much better adapted to dry conditions. These plants, which are in phylum Tracheophyta, are vascular plants with specialized conducting tissues, xylem and phloem. Vascular plants have tough leaves that are adapted to retain water. Waxy materials, such as **cutin** (KYOO tihn), cover the leaves of vascular plants.

Two groups of primitive vascular plants are the subphylum *Lycopsida* (lī KAHP suh duh) and the subphylum *Sphenopsida* (sfee-NAHP suh duh). Though not numerous today, both groups date back to the Devonian Period. During this period, about 390 million years ago, both groups were among the dominant land plants.

Today, subphylum Lycopsida is made up of small plants called **club mosses.** Though not real mosses, their leaves look like the leaves of bryophytes. *Lycopodium* (lī kuh POH dee uhm) is one of the most common club mosses. Sometimes called ground pine, it usually grows on the floors of temperate zone forests. Erect branches, usually 10 to 20 cm tall, grow from horizontal underground stems. Small tough leaves grow in spirals around the branches.

Alternation of generations in *Lycopodium* is shown in Figure 16·13. The main plant in the life cycle of *Lycopodium* is the sporophyte. Monoploid spores are borne on top of specialized leaves called *sporophylls*. The gametophyte formed from a *Lycopodium* spore is very small and in some species grows underground. The gametophyte develops archegonia and antheridia. Water is necessary for the flagellated sperm to swim to the eggs. The zygote that forms develops into a new *Lycopodium* sporophyte, while the old gametophyte dies.

FIGURE 16·13 *Lycopodium* is sometimes called ground pine, although it is not related to the large pine trees that grow in the forest. The life cycle of *Lycopodium* is show here.

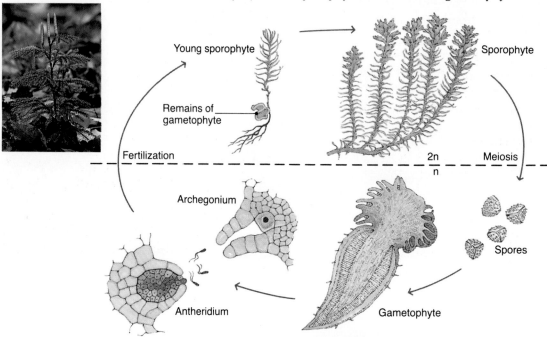

Subphylum Sphenopsida includes only one living genus, *Equisetum* (ehk wuh SEE tuhm), commonly called the **horsetail.** It grows in moist or damp areas, by streams, and along the edge of woods. In temperate climates *Equisetum* is small, usually no taller than 1 m. In the tropics *Equisetum giganteum* may grow up to 5 m tall.

As you can see in Figure 16·14, the major part of the horsetail plant is the stem. The stem is jointed and ribbed. The tough ribbed stems earned *Equisetum* the name "scouring rush" in colonial and pioneer times, when the plants were used to clean pots. Small scaly leaves grow in clusters around the stem at the joints. Judging from the figure, in which structure would you expect most photosynthesis in *Equisetum* to take place?

Spores of *Equisetum* are borne in umbrella-shaped structures at the tips of branches. In some species, only special nonphotosynthetic branches bear spores. The gametophytes that develop when spores land on damp soil are about the size of a pinhead.

FIGURE 16·14 Horsetails grow mostly in moist or damp regions. Their ribbed stems contain the element silicon.

Though most of the modern representatives of subphylum Lycopsida and subphylum Sphenopsida are small nonwoody plants, many of the earlier species were large wood-producing trees. These trees were dominant some 325 million years ago in great swamp forests. The decayed plant material from those forests became the sources of today's coal, natural gas, and petroleum.

16·10 FERNS

More abundant than the club mosses and the horsetails are the **ferns,** plants belonging to subphylum *Pteropsida* (tuhr AHP suh duh). Ferns are easily distinguished from the horsetails and club mosses by their leaves, which contain branching veins. The leaves of club mosses and horsetails are simpler and contain veins that do not branch.

FIGURE 16·15 Before they uncoil, the fronds of ferns resemble the tops of string instruments like the violin. Thus the uncurled fronds are called "fiddleheads."

Ferns grow in a wide variety of environments. They can be found in wetlands, forests, and fields and on trees and cliffs. Most ferns grow in tropical rain forests. There, some species called tree ferns grow as tall as 25 m. In the United States, ferns most often are found in moist woodlands.

The leaves of ferns are familiar to most people. A fern leaf is called a **frond.** Each frond grows out of a horizontal underground stem called a **rhizome** (RĪ zohm). Rhizomes store food and give rise to new fronds. Small roots grow out of the rhizomes. These roots anchor the plant and absorb water and minerals.

Many ferns lose their leaves at the end of the growing season. Their rhizomes and roots remain underground through the winter. They produce new fronds in the spring. Other ferns, like the Christmas fern, keep their fronds through the winter.

In most ferns each frond develops from a **fiddlehead,** a tightly coiled frond, shown in Figure 16·15. As it grows, a fiddlehead uncoils and expands, giving rise to a mature frond. Even the fronds of the tree ferns of tropical rain forests begin as fiddleheads. You may have noticed fiddleheads in the soil during early spring.

16·11 LIFE CYCLE OF FERNS

The life cycle of a typical fern is shown in Figure 16·16. The dominant, conspicuous fern plant is the sporophyte. A fern's spores usually are borne on the undersides of fronds in structures called **sori** (SOR ee). Sori (sing., *sorus*) are masses of spore cases. Each spore case usually holds between 48 and 64 monoploid spores.

When the spores are mature, the outer ring of the spore case dries out. It then breaks apart in such a way as to fling the spores out of the sorus and away from the fern plant.

Wind carries the spores farther away from the plant. Only those spores that land in suitable environments have a chance to germinate. The gametophyte that develops out of a fern spore is called a **prothallus** (proh THAL uhs). It is small and it has no specialized conducting tissue. Rhizoids anchor the heart-shaped prothallus to the ground and absorb water and minerals. Prothalli tend to grow in moist areas, such as wet soil, moss beds, and cracks in rocks.

pro- (before)
thallus (green shoot)

Antheridia develop at the base of the prothallus. Flagellated sperm are produced in the antheridia. Flask-shaped archegonia are usually located close to the upper notch of the prothallus. At the base of each archegonium, an egg is produced. The sperm swim through water from an antheridium to an archegonium. They swim down the archegonium neck, where one unites with the egg. The resulting zygote develops into a young sporophyte fern. The young plant grows out of the archegonium neck, still attached to the gametophyte. After the young fern develops its first root and leaf, the prothallus dies.

Ferns are well adapted to life on land. They can live in a wide range of environments. Some ferns, like the bracken fern, are able to live in extremely harsh areas in which few or no other plants can live. For example, bracken ferns can grow over an area after a forest fire.

FIGURE 16·16 The life cycle of a fern. Notice the sori on the underside of the frond.

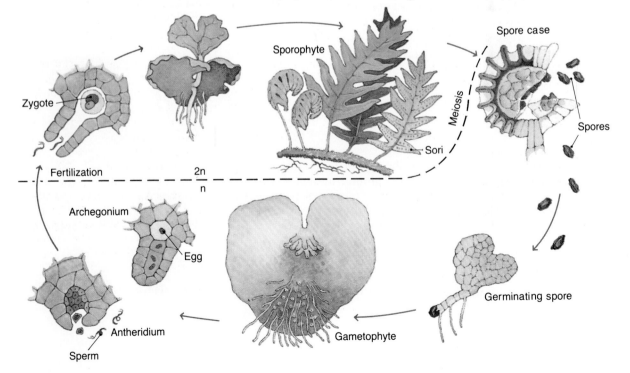

The relationship of ferns to higher plants is not a direct one. The fossil record shows that many of the trees that lived some 345 million years ago had the spores and fronds of ferns, but they also had characteristics of early seed plants. Some biologists consider these plants ferns; others put them in a separate division. Whether these seedlike ferns are related to seed plants is not known.

REVIEW QUESTIONS

11. Describe alternation of generations in *Lycopodium*.
12. How do horsetails and club mosses differ from ferns?
13. Describe alternation of generations in a fern.
14. How do rhizomes help ferns to spread in the environment?
15. How does alternation of generations in bryophytes differ from that in vascular plants studied thus far?

Peter H. Raven, (1936–)

Peter H. Raven began his international travels at an extremely early age. Only a year after his birth in Shanghai, China, in 1936, his family moved to California. It was during his early life in California that Peter Raven developed his interest in plants. He graduated in 1957 with highest honors from the University of California, Berkeley. He earned his doctorate in 1960 from the University of California, Los Angeles. After receiving his doctorate, Peter Raven continued his research and education in California, England, and New Zealand.

Throughout his career in botany, Dr. Raven has received several honorary degrees and awards, including an International Environmental Leadership Award in 1982. He has also been a member of many professional organizations, including the California Botanical Society, the American Society of Plant Taxonomists, and the Botanical Society of America.

Currently, Dr. Raven holds several positions. He is a professor of botany at Washington University in St. Louis, Missouri. He is also an adjunct professor of botany at both St. Louis University and the University of Missouri at St. Louis. In addition to his teaching positions, Dr. Raven is Director of the Missouri Botanical Garden in St. Louis—the botanical garden with the largest membership in the world. As director, Dr. Raven oversees the operation of the entire botanical garden, supervises at least 25 doctoral students in St. Louis and abroad, and directs the garden's extensive research programs. Current research programs under Dr. Raven's direction include the identification and cataloging of tropical plants and the development of new food sources for the growing human population. As Dr. Raven has said, "We're really a botanical research institute that happens to be surrounded by a magnificent garden."

In recent years, Dr. Raven has written and published over 300 journal articles and reports. His nine books include a very successful college botany textbook. Dr. Raven recently traveled to the People's Republic of China, where he continued his study of botany on a worldwide basis.

Chapter Review

SUMMARY

16·1 The vascular plants, phylum Tracheophyta, are better adapted to life on land than are the nonvascular plants.

16·2 Algae are simple, mostly aquatic nonvascular plants. Phylum Chlorophyta is the group of algae most similar to higher plants.

16·3 Life cycles of green algae differ from species to species and involve both sexual and asexual means of reproduction.

16·4 Brown algae are mostly large, complex, cold-water marine plants that are commonly called kelps.

16·5 Red algae are typically multicellular warm-water marine plants with pigments that enable photosynthesis to occur at great depths.

16·6 Algae are important biologically in their role as food sources for aquatic organisms and important economically as a source of chemicals.

16·7 Bryophytes, including liverworts and mosses, are nonvascular plants with simple adaptations for life on land.

16·8 The life cycle of mosses includes a dominant gametophyte upon which the sporophyte is dependent.

16·9 Horsetails and club mosses are primitive vascular land plants whose ancestors were dominant in early forests.

16·10 Ferns are more advanced land plants with conspicuous fronds and underground rhizomes.

16·11 The life cycle of ferns includes separate diploid and monoploid plants.

BIOLOGICAL TERMS

algae
alternation of generations
antheridia
archegonia
brown algae
bryophytes
capsule
club mosses
conjugation
cutin
ferns
fiddlehead
frond
gametophyte
green algae
holdfast
horsetail
liverworts
meiospores
mosses
nonvascular plants
phytoplankton
prothallus
protonema
red algae
rhizoids
rhizome
sori
sporophyte
vascular plants
zoospores
zygospores

USING BIOLOGICAL TERMS

1. Identify the type of algae in each of these three phyla: Chlorophyta, Phaeophyta, Rhodophyta.
2. Protonema is to a moss as _____ is to a fern.
3. Which plants belong in the division Bryophyta?
4. Define the term *vascular plants?*
5. Which terms name structures that are part of the mature sporophyte of a fern? Define each term.
6. Which terms name structures that are part of the gametophyte of a fern? Define each term.
7. Which terms refer to specialized structures for producing sex cells? Define each term.
8. What are phytoplankton?
9. Write a definition of alternation of generations using the words sporophyte and gametophyte.
10. Bryophyta is to mosses and liverworts as Tracheophyta is to _____.

UNDERSTANDING CONCEPTS

1. What are the purposes of a plant classification system? (16·1)
2. How is phytoplankton important to life on the earth? (16·2)
3. What is a colonial alga? Give an example of one. (16·2)

4. What is the function of a holdfast? (16·2)
5. Both zoospores and meiospores are monoploid flagellated cells that can develop into new plants. What is the difference between them? (16·3)
6. In what ways are giant kelps similar to vascular plants? (16·4)
7. In what type of environment are red algae found? (16·5)
8. What characteristics of green algae make them likely ancestors to the higher land plants? (16·6)
9. What are some of the reasons red and brown algae are harvested? (16·6)
10. How do mosses differ from liverworts? (16·7)
11. How can mosses pave the way for other plants to grow on rock? (16·7)
12. How is water important to sexual reproduction in mosses? (16·8)
13. Why are club mosses not really mosses? (16·9)
14. How are ferns better adapted for life on land than mosses? (16·9, 16·10)
15. What is a fiddlehead? (16·10)
16. Describe the life cycle of a fern, beginning with the fern sporophyte. (16·11)
17. In what type of environment would you expect fern prothallia to grow? (16·11)
18. How do rhizomes enable some ferns to survive the effects of fire? (16·11)

APPLYING CONCEPTS

1. Giant kelps are sometimes called the "sequoias of the sea." Communities of kelps are often referred to as "kelp forests." In what ways are such names accurate? In what ways are they inaccurate?
2. Many flowering plants have leaves that are very similar in appearance to those of ferns. Some of these plants are even called ferns, such as the asparagus fern, though they are not true ferns. How would you determine whether or not such a plant was a fern?
3. Why is dependence on water for sexual reproduction a disadvantage for land plants?
4. *Ulothrix* is a nonmotile filamentous green alga whose asexual and sexual spores are flagellated motile cells. What might this indicate about *Ulothrix*'s ancestry?
5. What characteristics found in fossils from the swamp forests would indicate that trees from that time were members of the subphylum Sphenopsida or Lycopsida?

EXTENDING CONCEPTS

1. Prepare an illustrated report on bogs. How and where do they form? What role does moss play in bog development? What characteristics of mosses allow them to grow in bogs? What other types of plants grow in bogs? Find out if there are any bogs nearby that you can visit.
2. Collect several types of fern fronds. Set up a demonstration showing how ferns are identified. How are fronds classified? What are the names of the leaflets of fronds? Demonstrate the different locations of sori. If possible, show the locations of fertile and nonfertile leaflets.
3. Do a study of the kelp forests that grow off the California coast. What types of kelps grow there? How large are they? What other organisms live within the kelp community? What advantages do the kelps provide to other organisms? What effects would harvesting the kelps have on this environment? Illustrate your report with photographs or your own drawings.
4. Do a survey of the food in your local supermarket. What foods contain products from algae? What are the products? What types of algae are they from? What role do the algal products play? Write up your findings.

READINGS

Earle, Sylvia A. "Undersea World of a Kelp Forest." *National Geographic,* September 1980, p. 410.

Frankel, Edward. *Ferns: A Natural History.* The Stephen Green Press, 1981.

Hodge, Walter H. "Where a Heavy Body Is Likely to Sink." *Audubon Magazine,* September 1981, p. 98.

Thomas, Barry. *Evolution of Plants and Flowers.* St. Martin's Press, 1981.

Wexler, Jerome. *From Spore to Spore: Ferns and How They Grow.* Dodd, Mead, 1985.

17. The Seed Plants

These beautiful flowers are lady-slipper orchids. They grow wild in parts of the United States. They grow in shady woods where enough dappled sunlight can break through the leafy tree cover to provide the orchids with the light they need to make food. The lady-slipper orchid is only one kind of plant that makes seeds. What are some characteristics of seed plants? How do they survive in many different environments on earth?

CHAPTER OBJECTIVES

After completing this chapter, you will be able to
- **Distinguish** between seed plants and all other types of plants.
- **Distinguish** between the two groups of seed plants.
- **Describe** the structures and functions of stems, roots, and leaves.
- **Give examples** of the ways in which seed plants are adapted to life in a variety of environments.
- **Infer** a plant's native environment, based on the plant's adaptations.
- **List** reasons for the success of the seed plants.

Classifying Seed Plants

17·1 TRAITS OF SEED PLANTS

The **seed plants** are vascular plants that reproduce by forming seeds. These plants, which make up subphylum *Spermopsida* (sper-MAHP suhd uh), are the most abundant plants. There are over 200 000 species of seed plants. They are found in most types of land environments and in many water environments.

The seed plants are divided into two groups, based on where the seeds develop. **Angiosperms** (AN jee uh spermz) are the plants whose seeds develop within structures called ovaries. The angiosperms also are called flowering plants. **Gymnosperms** (JIHM nuh spermz) are the plants whose seeds do not develop within ovaries. The most familiar gymnosperms are cone-bearing trees, such as pines and spruces.

Recall that seed plants reproduce by forming seeds. A *seed* is a structure made up of a plant embryo, food for the embryo, and an outer covering. Reproduction in seed plants is discussed in detail in Chapter 19.

Most seed plants are adapted to life on land. Many of their adaptations to land involve reproduction. For example, water is not needed for fertilization to take place in seed plants. The seed itself is adapted to the dry conditions of land. The covering prevents the young plant inside from drying out. The food stored in the seed supplies energy for the early stages of growth, after which photosynthesis begins. Seeds may also be adapted for being spread throughout the land by wind, animals, or other means.

Some of the adaptations of seed plants to living on land are also found in nonseed plants. As in other vascular plants, food moves through the plant in phloem tissue. Water is carried in xylem tissue. The vascular tissues also give structural support in seed plants as well as in ferns, club mosses, and horsetails. Thus vascular tissue is an adaptation to living on land that occurs in several plant groups.

SECTION OBJECTIVES

- Describe the characteristics of seed plants.
- Compare gymnosperms and angiosperms.
- Distinguish between monocots and dicots by their characteristics.

angeion (vessel)
sperma (seed)
gymnos (naked)

17·2 GYMNOSPERMS

Although there are many kinds of gymnosperms, they share one characteristic. These plants all have naked seeds. Naked seeds are seeds that are not enclosed in an ovary. The gymnosperms include cycads (SĪ kadz), ginkgoes (GING kohz), gnetales (nuh TAY leez), and conifers.

The **cycads** are a small group of tropical plants that look like large ferns or palm trees. Cycads were much more common on the earth in the Mesozoic Era, about 200 million years ago. The approximately 100 species that exist today occur in Australia, Africa, the West Indies, Mexico, and the southern United States. Cycads often are grown in tropical gardens and greenhouses.

The only cycad native to the United States is *Zamia* (ZAY mee uh), shown in Figure 17·1. The trunk of *Zamia* is covered with the bases of leaves that have been shed. The cluster of leaves at the top of the plant gives *Zamia* its palmlike appearance. Reproductive organs, on separate male and female plants, form in conelike structures at the top of the plant.

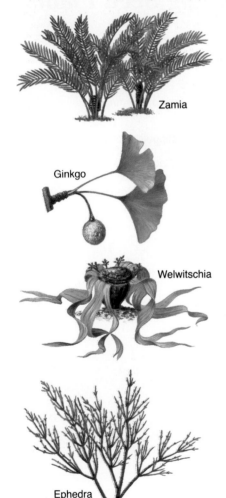

FIGURE 17·1 Gymnosperms vary greatly in appearance. Four examples of gymnosperms are: *Zamia* (A), ginkgo (B), *Welwitschia* (C), and *Ephedra* (D).

Another type of gymnosperm that was more common in the Mesozoic Era is the ginkgo, also shown in Figure 17·1. **Ginkgoes** are trees with small fan-shaped leaves. Unlike most gymnosperms, ginkgoes lose their leaves each winter. Ginkgo seeds develop on short stalks.

Gnetales are a group of small, cone-bearing plants containing three genera. The three genera are dissimilar, and all the plants are relatively uncommon. The rarest gnetale is *Welwitschia* (wehl WIHCH ee uh). It is found in the dry regions of southwestern Africa. Most of a *Welwitschia* plant is under the ground, in the form of a long root. Its long leaves, visible in Figure 17·1, remain on the plant throughout the plant's life. The largest genus is *Ephedra* (ih FEE druh). It is made up of shrubby plants with small scalelike leaves that are found in arid areas. Notice in Figure 17·1 how much *Ephedra* resembles a horsetail. The genus *Gnetum* (NEET uhm) is made up of tropical trees and vines with broad, tough leaves.

The most familiar and important of the gymnosperms are the conifers. **Conifers** are the group of plants that usually produce cones, the structures that contain the seeds of the plants. The sizes and shapes of cones vary according to species. A few conifers, such as the yew and the juniper, do not have cones but have individual seeds that develop on their branches.

Conifers grow as trees or shrubs. The tallest of these vascular plants, the redwoods of the California and Oregon coasts, are conifers that can attain a height of about 120 m. Other conifers include pines, firs, spruces, and hemlocks. These trees grow in large, dense groups, forming vast forests in parts of North America and parts of northern Europe and Asia. Because of the value of conifers for lumber and paper production, many of these forests have been severely reduced in size by logging.

FIGURE 17·2 Three conifers: yew showing seed cups (A), blue spruce and cones (B), and bald cypress, a non-evergreen gymnosperm (C).

Leaves of most conifers are small scalelike needles. A few Southern Hemisphere conifers, however, have broad leaves rather than needles. The tough water-resistant covering of conifer needles is one of many adaptations of conifers to dry conditions. Because conifers have so many adaptations to dry conditions, it is thought that they developed during an arid period. This was probably the Permian Period, about 280 million years ago (Section 10·5).

Most conifers are evergreen. They do not lose all their leaves each year. They shed some needles each year but hold most of them. A pine tree may retain some of its needles for 2 to 14 years. Two nonevergreen gymnosperms are the larch and the bald cypress (Figure 17·2), which grows in the swamps of the southern United States.

17·3 ANGIOSPERMS

More diverse than the gymnosperms are the angiosperms, or flowering plants. There are approximately 235 000 species of these plants living today. Angiosperms make up the largest plant group. They are found in more environments than any other plants.

The flower is a trait of the angiosperm. It greatly increases a plant's chances for successful reproduction. One part of the flower is the ovary. This structure encloses the seed. The role of the flower in reproduction is discussed in detail in Chapter 19.

Whereas the gymnosperms are mostly trees and shrubs, the flowering plants also include vines, herbs, floating plants, and what are typically called flowers. The vegetative, or nonreproductive, organs of angiosperms vary in form. The stem of a cactus, for example, is quite different from the stem of an oak tree. The leaves of a palm tree differ greatly from the leaves on a sprig of parsley. Angiosperms vary in the habitats to which they are adapted. They are found in deserts, tropical rain forests, freshwater ponds, and even in salt water. Many of the plants with which you are familiar are angiosperms.

340 THE SEED PLANTS

The angiosperms are the most recent of plant groups to develop. It is estimated that angiosperms first developed in the Cretaceous Period, about 123 million years ago.

17·4 MONOCOTS AND DICOTS

Angiosperms are divided into two groups: dicots and monocots. **Dicots** (dicotyledons) are angiosperms in which each seed contains two primary leaves. Primary leaves are called cotyledons. There are about 170 000 species of dicots, among which are most familiar trees and shrubs (other than conifers) and many of the small nonwoody plants. Some familiar dicots are maple trees, roses and cactus. **Monocots** (monocotyledons) are angiosperms in which each seed contains one primary leaf. There are approximately 65 000 species of monocots. These species include grasses and orchids.

In addition to seed type, there are other differences between dicots and monocots. Many of these differences can be seen with the unaided eye. One difference is in the patterns made by veins in the plants' leaves, as shown in Figure 17·3. A pattern made by veins is called **venation.** In general, dicots have *net venation.* Monocots, for the most part, have leaves with *parallel venation.* This means there are several main veins running the length of each leaf.

Another difference between monocots and dicots is in the number of flower parts, such as petals, found in the plants. Generally, dicot flowers have parts in multiples of four or five. Most monocots have flower parts in multiples of three. Based on number of flower parts, classify each plant in Figure 17·4 as a monocot or a dicot.

Other differences between dicots and monocots require careful

FIGURE 17·3 The leaves of monocots and dicots have different patterns of veins. The birch (left), a dicot, has net venation. The daffodil (right), a monocot, has parallel venation.

FIGURE 17·4 The number of flower parts differs in monocots and dicots. The three flowers below, a day lily (A), a red trillium (B), and a hibiscus (C) have different numbers of flower parts.

examination of plant parts. For example, a cross section of a stem under a microscope shows an arrangement of the vascular tissues that differs in monocots and dicots (Section 17•9). A stem cross section also shows whether or not rings of wood are present. In dicots, stem thickening often involves the production of woody tissue. In monocots, wood production is rare. Microscopic views of pollen grains show differences between the two groups as well.

Exceptions to these differences exist. For example, the leaves of the jack-in-the-pulpit, a monocot, have net venation. Palms, which are monocots, have woody stems. Several characteristics, rather than only one, should be used to identify a plant as a monocot or a dicot.

REVIEW QUESTIONS

1. In what ways are seed plants adapted for life on land?
2. What are conifers?
3. List some of the diverse growth forms found among angiosperms.
4. How do dicots differ from monocots?
5. A fossil plant is found to have eight petals. Predict the venation and the seed type.

Roots and Stems

17•5 TISSUES OF SEED PLANTS

The seed plants and other vascular plants are made of groups of cells of different types. A group of structurally similar cells performing a particular function is called a *tissue*. The body of a seed plant is made of many kinds of tissues.

The body of a seed plant is highly specialized. The life-supporting activities are carried out by the plant's organs—its roots, stems, and leaves. Each organ is specialized to carry out certain functions and is made of two or more tissues.

Epidermal tissue is a layer of cells that covers the entire plant, much as your skin covers your entire body. Epidermal cells are specialized to protect the plant against physical injury and to control water loss. Epidermal cells contain a layer of waxy material called cutin in their cell walls. The outer layers of those stems and roots that produce wood are made up of a tissue called *periderm*. For example, the bark of a tree trunk is part of its periderm.

Parenchyma (puh REHNG kuh muh) **tissue** consists of thin-walled unspecialized cells found in roots, stems, and leaves. Parenchyma cells can function in food making, food storage, and water storage. **Sclerenchyma** (sklih REHNG kuh muh) **tissue** consists of thick-walled cells specialized to strengthen parts throughout the plant body.

SECTION OBJECTIVES

- List the tissues found in seed plants.
- Discuss the functions of roots.
- Describe the function and structure of stems.
- List adaptations of roots and stems.

epi- (on)
derma (skin)

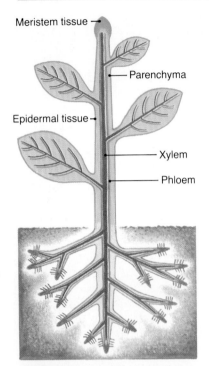

FIGURE 17·5 The tissues that make up a seed plant are shown here. What is the function of meristem tissue?

Vascular tissue consists of cells that conduct water and food through the plant. Because this tissue runs through the entire plant, it is often called the vascular system. The vascular system is made of two types of tissue: xylem and phloem. Xylem is a tissue that conducts water. It also provides support for the plant. In **woody plants,** xylem cells develop hard, thickened cell walls that give strength and support to the plant. In **herbaceous** (her BAY shuhs) **plants,** the cell walls of the xylem are not thickened, and the plant is not as stiff as a woody plant. The xylem forms a continuous path through the plant, from the roots through the stem to the leaves. Phloem is a tissue that conducts food throughout the plant. The structures of xylem and phloem are discussed in Chapter 18, along with details of plant transport.

Some plant tissues are made of highly specialized cells that cannot reproduce. Less specialized cells that are capable of frequent cell division make up **meristem** (MEHR uh stehm) **tissue.** Meristem tissue provides for growth of specialized plant tissues. It is found in all areas of the plant that grow either in length or in thickness.

Though it is convenient to study the plant in terms of distinct organs—roots, stems, and leaves—it is important to remember that the plant functions as a whole. The same basic types of tissues may be found throughout the plant body. However, tissues are specialized to carry out specific functions in specific organs. For example, the epidermal tissue in roots is specialized for the intake of water. In leaves, this tissue is specialized to control the outflow of water.

17·6 ROOTS

The underground organ of most seed plants is the root. The **root** is the organ that anchors a plant in the ground or other substrate. It also absorbs water and minerals from the soil, and conducts water and minerals to the aboveground parts of the plant. Food that is made in the leaves is sometimes stored in the root.

The first root to develop in a seedling is called the *primary root.* In gymnosperms and dicots, the primary root develops into the largest root of the plant. This large main root is called the *taproot.* The taproot grows far into the ground and develops branches called *lateral roots.* Lateral roots are also called secondary roots because they develop from the primary root.

In monocots, the primary root lives only a short time. Other roots, called adventitious (ad vehn TIHSH uhs) roots, arise from the monocot stem. *Adventitious roots* are roots that develop from structures other than roots themselves. The adventitious roots of monocots are all about the same size, and they do not grow as far into the ground as taproots. The shallow, highly branched roots of monocots are called *fibrous roots.* Compare the fibrous roots and the taproot in Figure 17·6. Though not as long as the taproots of dicots, the many branches of fibrous roots anchor a plant firmly to the soil. A plant's fibrous roots also have a great amount of surface area, which increases the amount of water absorption possible.

Root growth occurs in a region near the tip of the root called the root-tip meristematic region. Growth that occurs in such regions is called *primary growth* and results in the elongation of the plant part. Tissues produced in these regions are called primary tissues.

The structure of a typical root is shown in Figure 17·7. The root cap is located beneath the meristematic region. The **root cap** covers the root tip, giving protection and aiding in water and mineral absorption. The root cap is made of parenchyma cells. During root growth, some of these cells are cast off and release a slimy substance. This substance is thought to protect the root from harmful materials in the soil. It may also keep the root tip from drying out, making it easier for materials to be absorbed by the root. The root cap also is thought to play a role in the growth response of the root to gravity.

The epidermis of the root is specialized to absorb water and minerals from the soil. As shown in Figure 17·7, many epidermal cells have tubular extensions called root hairs. **Root hairs** are epidermal structures that increase the absorbing surface of the root.

The vascular system forms a cylinder in the center of the root. This vascular cylinder is made of primary xylem, primary phloem, and parenchyma. The vascular cylinder is called the **stele**. Notice that the xylem makes up the core of the stele. The phloem occurs in bundles around the core of xylem. The layer of parenchyma cells at the edge of the stele is called the **pericycle**. The pericycle cells undergo cell division to form lateral roots.

The **cortex** of a root is a region of parenchyma tissue outside the stele. The cortex is encircled by the epidermis. The inside layer of the cortex forms a ring around the stele and is called the **endodermis**. All materials moving from the cortex to the vascular system pass through the cytoplasm of the endodermal cells. These cells regulate the movement of substances into the vascular system.

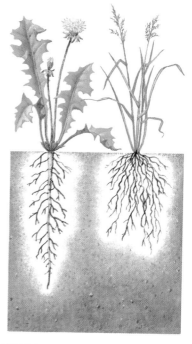

FIGURE 17·6 Compare the taproot of a dicot (left) and the fibrous root of a monocot (right). If even a small piece of the taproot remains in the ground when the top of the plant is removed, the plant can regrow.

peri- (around)
kyklos (circle)

endo- (inner)

FIGURE 17·7 Examine the longitudinal section (left) and the cross section (right) of a typical root. Notice that the xylem and phloem are found in the center of the root.

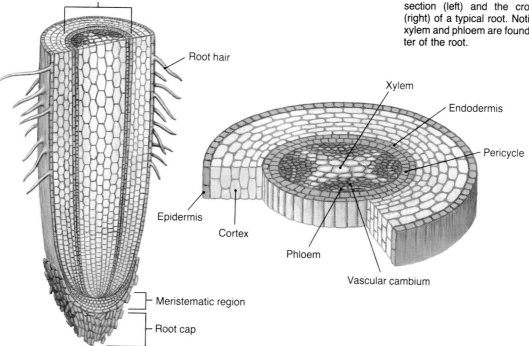

Discovery

Plant Carcinogens

In recent years much attention has been paid to high levels of *carcinogens* in the environment. Carcinogens are chemicals that can cause cancer. Some are natural parts of the environment. In fact, many plants contain chemicals that can cause mutations or cancer in animals. In some cases these same chemicals are useful to the plant because they protect it from infection by various organisms. If plants could be engineered to produce increased amounts of substances toxic to disease organisms, then those plants might be grown more economically.

However, there is a potential problem. The benefits of decreased food costs would have to be balanced against the dangers of increased exposure to carcinogens in food crops. There is a great need for more research on this topic so that we can learn how to minimize the risks.

In primary growth, an increase in cells makes the root longer. *Secondary growth* causes plant parts, such as the stems and roots of gymnosperms and dicots, to become thicker. Most of the tissue made in secondary growth is vascular tissue. The tissue that produces secondary growth is called the vascular cambium. In the root, the vascular cambium forms a ring between the phloem and the xylem of the stele. Thus the **vascular cambium** is a band of meristematic cells that has phloem cells on its outer side and xylem cells on its inner side.

Secondary growth results from the production of new phloem and xylem cells by the vascular cambium. The new xylem cells, called secondary xylem, are formed from the inner side of the vascular cambium. Rings of secondary xylem accumulate, forming wood. Each ring forms farther from the primary xylem in the center of the root. In most woody plants the older xylem tissue does not transport water. As a result, water transport is carried out by the most recently formed cells. These cells are located in the outermost rings of the xylem.

New phloem cells are formed from the outer side of the vascular cambium. The new phloem cells push the older phloem cells toward the outside of the root. The outward growth of newly formed xylem and phloem eventually causes the epidermis, cortex, and older phloem to split and fall off the root. Thus the root has only a thin layer of functioning phloem arranged around the many layers of xylem.

What happens in the root to replace the epidermal cells that fall off the root? Near the surface of the vascular cylinder, a meristematic layer called the **cork cambium** gives rise to a protective layer called cork. **Cork** is a tissue made up of thick cells, and it forms the outer protective tissue of woody plants.

Since cork is a tough, water-resistant layer, the older, woody root parts thickened by secondary growth are not suited for water absorption. Most water absorption occurs in the tips of roots and lateral roots. In these areas the thin epidermis, with its root hairs, is better adapted for water absorption than is cork tissue.

17·7 ADAPTATIONS OF ROOTS

The structure and development of a typical seed plant root have been described. However, different plants vary in root structure, depending on the environments to which the plants are adapted. One adaptation of roots that has been mentioned is the adventitious root. Adventitious roots develop from plant parts other than roots themselves. In many trees, an adventitious root forms if a root has been injured. After an adventitious root arises, the tree functions normally despite the injury.

Adventitious roots that form above the ground are called aerial roots. Many tropical trees, such as palms and mangroves, have aerial roots called prop roots. A mangrove's prop roots, shown in Figure 17·8, extend out from the base of the stem, giving added support to the plant. Prop roots are an adaptation for growth in environments, such as swamps, where ordinary roots would not give enough support to the

THE SEED PLANTS

plant. The mangrove's prop roots also absorb oxygen. The mangrove's primary roots are below water and have no access to air.

Some vines have aerial roots called climbing roots, as shown in Figure 17·8. Climbing roots grow out of a vine's stem and cling to the surface of objects. These roots provide support for the vine.

Many roots are specialized for food storage. The roots of the carrot and beet plants are examples of storage roots. Such roots have a great many storage parenchyma cells among the secondary xylem and phloem. Many storage roots, such as those of carrot plants, store food made in the first year of a plant's life. In the second year, the food provides energy and building materials to produce flowers.

The roots of most trees and shrubs have an adaptation that involves fungi. A certain species of fungus lives in or on the roots of a tree. The fungus takes food from the roots, but it aids the tree as well. It does this by increasing the capacity of the tree's roots to absorb water and minerals. The structure formed by a fungus in a relationship with a plant root is called a mycorrhiza (Section 14·10).

The roots of certain plants are adapted so that bacteria grow within them. Members of the legume (LEHG yoom) family—peas, beans, clover, and alfalfa—have growths called nodules on their roots in which bacteria live. The bacteria are capable of nitrogen fixation. In this process nitrogen is taken from the atmosphere and put into a chemical form that can be used by the plant. Since nitrogen is an important component of proteins, the plant benefits from this adaptation.

FIGURE 17·8 The prop roots of mangroves anchor this plant in wet ground (left). These roots also absorb oxygen from the air. The climbing roots of English ivy hold the plant to the surface it climbs on (right).

17·8 EXTERNAL STRUCTURE OF STEMS

In most seed plants the stem is the organ that continues aboveground from the root. The **stem** is the structure that conducts water and minerals, absorbed by the roots, upward into the leaves. It also conducts food, produced in the leaves, downward, to the roots. The stem gives the plant structural support and produces leaves and flowers.

Figure 17·9 shows a twig of a woody dicot. This twig shows many of the external structures common to seed-plant stems. As the external features of the stem are discussed, find them in Figure 17·9.

Each place where one or more leaves are attached to the stem is called a *node*. The region of a stem between two nodes is called an *internode*. In the winter, *leaf scars* show where leaves were attached at the nodes. Notice that bundle scars can be seen in each leaf scar. *Bundle scars* are scars that show where vascular tissue from the stem was connected to vascular tissue in the leaf.

Meristematic regions of the stem are located in structures called buds. A *bud* is a meristematic region that is enclosed by newly formed unopened leaves. Stem meristematic regions are found in the *terminal bud*, located at the tip of the stem, and in *lateral buds*, located along the sides of the stem. Notice that lateral buds develop above the nodes. Meristematic growth in the buds produces new leaves and new sections of stem. Some buds also produce flowers. *Terminal bud scars* show where a terminal bud had been before a new stem section grew from the bud.

The **bark** on the twig is the outermost layer of protective tissue produced by the stem. *Lenticels* (LEHN tuh sehlz) in the twig's bark are openings through which air can enter the stem.

17·9 INTERNAL STRUCTURE OF STEMS

Recall that primary growth causes plant structures to grow longer. Primary growth of the stem occurs, as it does in roots, in meristematic regions. However, the meristematic regions of stems are located in buds and give rise to both stem and leaf tissues.

Figure 17·10 shows cross sections of the stems of a monocot and a nonwoody dicot. Since the stem, in most cases, is above the ground, its epidermal tissue is adapted to control the loss of water to the atmosphere. The single layer of epidermal cells is covered with cutin, a fatty, water-resistant substance.

The vascular tissues of a stem occur in bundles, with the phloem on the epidermal side. Such bundles of xylem and phloem are called **vascular bundles.** In dicots, the vascular bundles form a ring within the stem. A layer of vascular cambium may occur between the xylem and phloem. In some monocots, the vascular bundles are arranged in two rings within the stem. In other monocots, the vascular bundles are scattered throughout the stem. In both monocots and dicots, at each node in the stem, one or more vascular bundles in the stem connect with the vascular system of the leaf.

In a stem in which the vascular bundles form a ring, parenchyma cells may be found within as well as outside the ring of vascular tissues. As in the root, the layer of tissues surrounding the vascular tissue is called the cortex. The cortex is made mostly of parenchyma cells. The parenchyma cells in the cortex of many plants are photosynthetic. Often the cortex contains sclerenchyma cells, which provide support for the stem.

FIGURE 17·9 Scientists can often identify plants from a twig like this. The pattern of structures on woody twigs is different in different species.

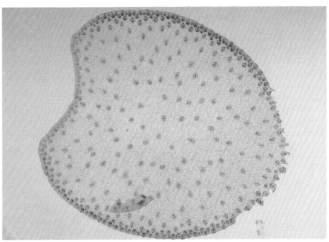
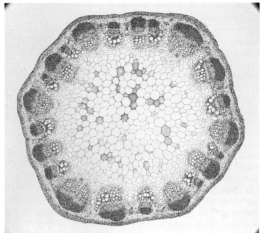

FIGURE 17·10 Vascular tissues are scattered throughout a cross section of a monocot stem (left). These tissues are in a ring in a nonwoody dicot stem (right).

Parenchyma tissue found within the vascular ring make up the **pith.** Parenchyma cells of the pith may be photosynthetic. *Intercellular spaces,* or air spaces are common in the pith. In some stems, the pith breaks down during growth. This results in hollow stems.

Secondary growth can result in plants of large size. In stems, as in roots, secondary growth comes from the growth of new vascular tissues by a layer of vascular cambium. Most monocots and some dicots do not have vascular cambium in their stems. Thus they are incapable of secondary growth. Many gymnosperms and dicots do have vascular cambium in their stems. Of these plants, those that live long and continue secondary growth become woody. The large woody stems of most common trees result from secondary growth.

As in a root, new xylem cells form on the inner side of the vascular cambium, and new phloem forms on the outer side. As the stem increases in diameter, layers of epidermis, cortex, and old phloem eventually split and fall off the plant. At each split in the stem, a layer of cork cambium develops. The cork cambium produces cork cells that protect the stem. The cork usually forms wherever a split occurs, rather than as a continuous layer around the stem. This results in the ridged appearance of most bark on mature tree stems. The term *bark* refers to the two outer layers of the woody stem. The outer bark consists of cork and the older, dead phloem cells just beneath the cork. The inner bark is the newer, live phloem.

The wood of trees is xylem tissue. The differences in cell size in the wood produced at different times of the year form noticeable rings, called *annual rings,* in the stem. One ring is produced each year.

The outermost, or youngest, xylem cells are the ones that conduct water. These functioning secondary xylem cells make up the *sapwood* of a tree. The older, nonconducting xylem cells in the center of the tree form the *heartwood.* The heartwood supports the tree.

FIGURE 17·11 You can see the annual rings that are formed during each growing season in this cross section of a woody dicot stem.

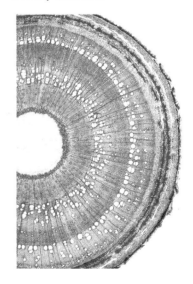

Investigation

What Are the Structural Features of a Dicot Stem?

Goals
After completing this activity, you will be able to
- Identify the developing leaves of a stem tip.
- Locate the meristem tissue of a stem tip.
- Describe the arrangement of vascular bundles in a dicot stem.
- Draw and label the epidermis, cortex, vascular bundles, and pith in dicot stems.

Materials (for groups of two)
prepared slide of *Coleus* stem tip
prepared slide of sunflower (*Helianthus annuus*) stem cross section
microscope   colored pencils

Procedure

A. Using the low power of your microscope, look at the slide of *Coleus* stem tip. Focus on the tip of the stem.

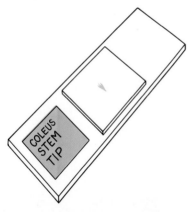

B. Switch to high power. Look for signs of cell divisions in the tip. Recall that the area of frequent cell division is called meristem tissue.

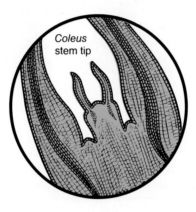

Coleus stem tip

C. Switch back to low power. Look for the developing leaves that extend out from the sides of the stem tip. The youngest leaves are closest to the stem tip.

D. Look for columns of cells in the center part of the developing leaves. These columns are developing vascular tissue.

E. Focus on the angle between the larger developing leaves and the stem. Look for small mounds of tissue that will develop into new branches as the plant grows.

F. On a separate sheet of paper, make a drawing of the *Coleus* stem tip viewed under low power. Label the stem tip and developing leaves. If vascular tissue is present, label it as well.

G. Under low power look at the slide of the cross section of the sunflower stem. On a separate sheet of paper, draw a cross section of the stem.

H. Locate the outermost layer of cells, the epidermis. Epidermal cells often become thickened and secrete a waxy substance called cutin on their outer surface. Locate the cutin, if it is present.

I. Between the epidermis and the vascular bundles is the cortex. Draw and label several cells of the cortex.

J. Inside the cortex is a ring of wedge-shaped vascular bundles. In your drawing show the arrangement of these vascular bundles.

K. Each vascular bundle consists of three main tissues. Phloem is in the outer part of the bundle, toward the epidermis. Xylem is toward the center. Between the xylem and the phloem is the cambium. Locate these tissues in one of the vascular bundles. Label the phloem, xylem, and cambium in your drawing.

L. The center part of the stem consists of thin-walled cells that make up the pith. Draw and label several cells of the pith.

M. Follow your teacher's instructions for the proper disposal of all materials.

Questions and Conclusions

1. Where is the meristem tissue in a stem tip? Describe the appearance of this tissue.
2. Suggest a function for cutin.
3. List at least two characteristics that distinguish epidermal cells from pith cells.
4. Which of the three tissues located in a vascular bundle forms new tissue that increases stem diameter?
5. Suggest how the arrangement and location of vascular bundles may be related to stem function.

17·10 ADAPTATIONS OF STEMS

The woody stem of a tree is typical of the kind of stem found in many seed plants. But the stems of some plants differ in appearance and structure from the stem of a tree. Many of the differences result from adaptations for functions other than support and the transport of water and food. Many differences are due to adaptations of the stems of the different plants to their different environments.

Notice that the stem of the cactus shown in Figure 17·12 is a thick green structure that, in places, resembles leaves. The spines of the cactus are leaves that are adapted as protective organs. The spines are not photosynthetic. For what function might the stem of this cactus be adapted?

The stem of the asparagus plant also is adapted as a photosynthetic organ. The leaves of the asparagus are the small brown scales that grow out of the edible stem. The remaining aboveground portion of the plant is photosynthetic stem.

Stems may also be adapted for water storage. Nonphotosynthetic parenchyma cells in the stem of the cactus store water. This adaptation to the dry desert environment accounts for the thick juicy flesh of the cactus stem.

Other stems are adapted for food storage. The familiar white potato is a modified stem called a *tuber*. A tuber is an underground stem that is adapted for food storage. The food is stored in the form of starch. The eyes of the white potato are buds, which can give rise to new leafing branches.

In some plants, stems are adapted to grow horizontally. Horizontal stems are called *stolons*. In most cases, stolons give rise to new plants. Thus stolons are adaptations that allow for reproduction of a plant over a wide area. The strawberry plant reproduces by means of aboveground stolons. An iris reproduces by means of thick, food-storing, underground horizontal stems called *rhizomes*.

The stems of many vines are adapted for climbing. The stem develops modified branches called tendrils that coil around structures as the plant climbs. The tendrils of Virginia creeper develop cups at their tips that hold onto the structure that the plant climbs. What types of environments can seed plants with tendrils inhabit?

FIGURE 17·12 The thick green structures of this prickly pear cactus are stems. The leaves of this plant are its spines.

REVIEW QUESTIONS

6. Name four types of tissue found in the stele of a root.
7. What are two main functions of roots?
8. Describe a root adaptation to an aquatic environment, such as a swamp.
9. Describe secondary growth in woody stems.
10. Sweet potatoes are underground storage organs. Unlike white potatoes they do not have eyes. What type of plant organ do you think the sweet potato is?

Leaves: Structure and Function

SECTION OBJECTIVES

- Describe the external and internal structure of leaves.
- Discuss the function of stomata.
- List the adaptations of leaves.

17·11 EXTERNAL STRUCTURE OF LEAVES

The **leaf** is the plant organ that absorbs sunlight and carries out photosynthesis. Most leaves are broad and flat, exposing a large surface area to the sun. The more sunlight that strikes the plant, the more energy the plant can absorb for photosynthesis.

Find each of the following structures in Figure 17·13. The broad flat part of the leaf is called the **blade.** The blade is where most photosynthesis takes place. Connecting the blade to the stem is a stalk called the **petiole** (PEHT ee ohl). Inside the petiole are the vascular bundles that connect the blade's vascular system to the stem's system.

The veins of the leaf usually can be seen with the unaided eye. Vascular bundles in the leaf are called **veins.** The veins form a pattern in the blade (Section 17·4). In dicot leaves, one large vein usually runs the length of the blade, forming a prominent ridge. This large central vein of a dicot leaf is called the midrib. From the midrib, a series of smaller and smaller veins branch out into all parts of the blade.

In many dicots, the leaf blade is not one single structure; it is made of several distinct parts. Leaves with one whole blade, such as the oak leaf shown in Figure 17·13, are called simple leaves. Leaves with blades made of several divisions are called compound leaves. The parts of a compound leaf blade are called *leaflets.* Compound leaves are classified according to how their leaflets are arranged along the leaf's petiole. *Pinnately compound* leaves have leaflets that run along opposite sides of an extension of the petiole. *Palmately compound* leaves have leaflets that arise from the tip of the petiole.

17·12 INTERNAL STRUCTURE OF LEAVES

The leaf is the organ in which the greatest amount of photosynthesis occurs in most plants. The leaf is also the organ from which most of a plant's water is lost to the atmosphere. The structure of the leaf is adapted for these two functions: food production and control of water loss.

The internal parts of a leaf are shown in Figure 17·14. The top layer of a leaf is called the upper epidermis. The bottom layer is called the lower epidermis. The walls of the epidermal cells contain the water-resistant substance cutin. Thus the epidermal layers help to control water loss. Also in the epidermis are structures called stomata (STOH muh tuh). **Stomata** (sing., *stoma*) are openings in the leaf epidermis through which water vapor and oxygen pass out of the leaf and carbon dioxide passes into it.

Between the epidermal layers is the *mesophyll* (MEHS uh fihl), a tissue made of parenchyma cells that contain chloroplasts. In the mesophyll, carbon dioxide is used for food-making. In between the

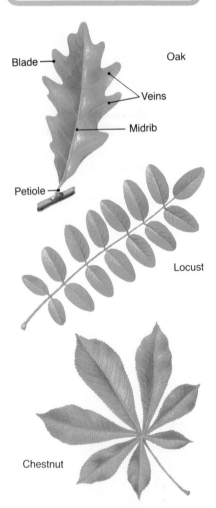

FIGURE 17·13 Dicot leaves may be simple, like an oak leaf, or compound. A Locust leaf is pinnately compound. A chestnut leaf is palmately compound.

350 THE SEED PLANTS

parenchyma cells is a network of air spaces that connect with the stomata. Thus carbon dioxide that passes into the stomata enters the network of intercellular spaces between the parenchyma cells of the mesophyll. This arrangement allows for rapid movement of carbon dioxide from the atmosphere to the mesophyll cells.

In most plants that live in temperate climates, the mesophyll is made of two distinct types of parenchyma cells. The **palisade mesophyll** is a layer of rectangular parenchyma cells, elongated at a right angle to the surface of the leaf. The palisade mesophyll is near the upper side of the leaf in Figure 17·14. In some leaves such a layer is located near both the upper side and the lower side. Palisade mesophyll cells contain many chloroplasts. It is in the palisade mesophyll that most photosynthesis occurs.

The other type of parenchyma cell in the mesophyll is in the spongy mesophyll. **Spongy mesophyll** is a layer of irregularly shaped parenchyma cells around the intercellular spaces in the mesophyll. Notice in Figure 17·14 that spongy mesophyll contains fewer chloroplasts than the palisade layer.

Food that is made in the mesophyll is moved into the phloem of the leaf. The food then is moved through the phloem of the blade and the petiole and into the phloem of the stem. From here the food is carried to all parts of the plant. It is used for growth, development, or is stored.

mesos (middle)
phyllon (leaf)

FIGURE 17·14 Study this model of the internal structure of a leaf. Most of the food-making in a plant takes place in the palisade mesophyll.

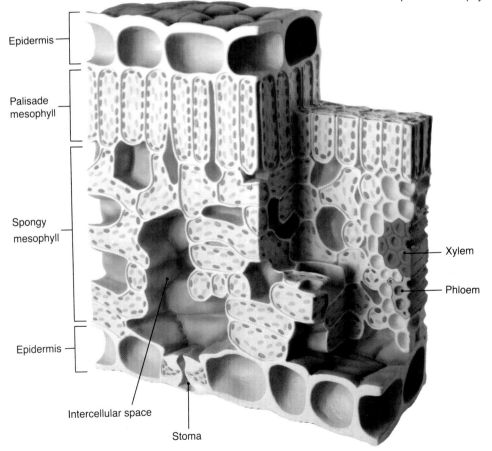

The xylem of the leaf carries water that was absorbed by the roots and transported through the stem into the leaf. Some of this water is used in photosynthesis. Much of it is lost through the stomata.

Recall that the phloem and xylem of the leaf form vascular bundles called veins. Surrounding the veins are continuous sheaths of parenchyma and/or sclerenchyma cells. These *bundle sheaths* keep the vascular bundles from having direct contact with the intercellular spaces of the mesophyll. All materials traveling into the vascular system from the mesophyll must pass through the living cytoplasm of the bundle sheaths. The cytoplasm of the bundle sheath cells may regulate what materials can go into the phloem.

17·13 FUNCTION OF STOMATA

More than 90 percent of the water taken in by a plant is lost through the leaves. Water evaporates from the cell walls that border the many intercellular spaces in the mesophyll. The water vapor moves by diffusion through the spaces of the mesophyll to the stomata. Water then diffuses through the stomata directly into the atmosphere.

While water vapor passes out of the stomata, carbon dioxide from the atmosphere enters the leaf through the stomata. If the stomata were always open, the plant would have a constant source of carbon dioxide, but it might become dehydrated. If the stomata remained closed, water loss would be at a minimum, but photosynthesis would stop for lack of carbon dioxide.

Each stoma is bordered by two specialized epidermal cells called **guard cells.** Guard cells control the opening and closing of a stoma by changing shape. The shape of a guard cell depends on how much water is inside the cell. Recall that a large amount of water in a cell increases the water pressure, or turgor pressure, in the cell. When the two guard cells are full of water, turgor pressure causes them to become stiff. The thick cell walls of the guard cells become arched, and the stoma opens. When the two guard cells lose turgor, they become limp and the arch collapses, closing the stoma. Which pair of guard cells in Figure 17·15 has the greater turgor pressure?

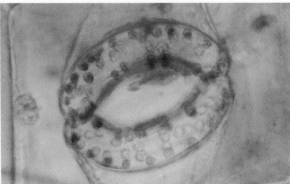

FIGURE 17·15 Stoma open and close to regulate the amount of gases that pass into and out of a plant. Notice the closed stoma (left) and the open stoma (right).

One of the most important environmental factors affecting the opening and closing of stomata is water loss. If the amount of water in a leaf goes below a certain point, the guard cells lose turgor and the stomata close. As a plant wilts because of water loss, the closing of the stomata decreases further water loss.

Under normal conditions, the stomata of most plants are open during the day and closed at night. The graph in Figure 17·16 compares time of day with stomatal opening and water loss in a plant. At what time of day is water loss from the plant the greatest? How does the closing of the stomata in the plant affect water loss?

17·14 ADAPTATIONS OF LEAVES

As you have seen, the leaves of a plant play a vital role in light absorption, food-making, and water loss. But availability of light and water differ greatly among environments. How can plants carry out photosynthesis in places that are shady? How can they retain moisture in dry places? Adaptations of the leaves to these different environmental factors are needed for a plant to stay alive.

In some leaves that are adapted to dry climates, the stomata lie below the surface of the epidermis (Section 18·10). Plants in dry environments also tend to have fewer stomata than do plants in moist environments. Why would this adaptation help reduce water loss?

Another adaptation to dry environments is storage of water in the leaf's mesophyll or epidermis. Plants with water-storing mesophyll or epidermis can be recognized by the juiciness of their leaves. One example is the century plant.

One type of nonphotosynthetic leaf is that adapted for underground food storage. Such leaves form the main part of a bulb, a small underground stem surrounded by modified storage leaves. A familiar example of a bulb is the onion. The parts of an onion that we eat are thick food-storing leaves. The food stored in the bulb is used to produce a new plant in the spring.

Another familiar type of leaf adaptation is that found in the garden pea plant. The pea plant has a pinnately compound leaf whose end leaflet is adapted for physical support of the plant. This end leaflet is a tendril, a modified plant organ that coils around structures as the plant climbs a support. What other organ in some seed plants can be modified as a tendril?

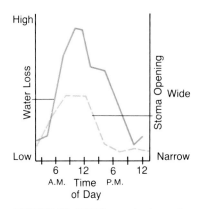

FIGURE 17·16 This graph shows the amount of water loss and stomatal opening versus the time of day. Water loss is greatest during the late morning hours.

REVIEW QUESTIONS

11. Most leaves are flat and broad. How is this structure adaptive?
12. What gases travel into and out of a leaf via the intercellular spaces in the mesophyll?
13. How is the leaf mesophyll adapted for photosynthesis?
14. Describe what happens to a pair of guard cells that causes a stoma to open.
15. Devise an experiment to show that plants with fleshy, juicy leaves are better adapted to arid environments than are other plants.

Chapter Review

SUMMARY

17·1 Both the vegetative and reproductive organs of seed plants are adapted to life on land.

17·2 The older of the two seed plant groups, the gymnosperms, includes cycads, ginkgoes, the gnetales, and conifers.

17·3 The angiosperms, or flowering plants, are more diverse and successful than the gymnosperms.

17·4 Angiosperms are classified as monocots and dicots, according to differences in plant and seed structures and growth patterns.

17·5 The organs of the plant body are made of various types of tissues.

17·6 Roots are adapted to absorb water and minerals, anchor a plant, and transport food and water.

17·7 Adaptations of roots to different environments enable plants to grow when regular root growth would not be sufficient.

17·8 The stem produces buds, from which leaves and flowers develop.

17·9 The stem supports a plant and transports water and food. The secondary thickening of stems with wood occurs in many dicots.

17·10 Adaptations of stems for water and food storage, climbing, and photosynthesis allow plant growth in a variety of environments.

17·11 The leaf's main function as a photosynthetic organ is evident in its external appearance. Leaf characteristics are important clues to a plant's identity.

17·12 In addition to light absorption and photosynthesis, control of water loss is an important function of the leaf.

17·13 The balancing of water loss and carbon dioxide intake is controlled by the opening and closing of stomata in the leaf epidermis.

17·14 Leaves show more adaptations to environmental factors than do stems and roots. Leaves may be specially adapted to dry environments. Some leaves are adapted for food storage.

BIOLOGICAL TERMS

angiosperms
bark
blade
conifers
cork
cork cambium
cortex
cycads
dicots
endodermis
epidermal
  tissue
ginkgoes
gnetales
guard cells
gymnosperms
herbaceous
  plants
leaf
meristem tissue
monocots
palisade
  mesophyll
parenchyma
  tissue
pericycle
petiole
pith
root cap
root hairs
roots
sclerenchyma
  tissue
seed plants
spongy
  mesophyll
stele
stem
stomata
vascular
  bundles
vascular
  cambium
vascular tissue
veins
venation
woody plants

USING BIOLOGICAL TERMS

1. Which of the terms refer to structures or layers of a stem? Define each term.
2. Distinguish between monocots and dicots.
3. Distinguish between angiosperms and gymnosperms.
4. What is the relationship between guard cells and stomata?
5. How are meristem tissue and vascular cambium related?
6. What is a vascular bundle?
7. Distinguish between palisade mesophyll and spongy mesophyll.
8. Which terms refer to external parts of a leaf? Explain the function of each part.
9. What is venation?
10. What are root hairs?

UNDERSTANDING CONCEPTS

1. How are seed plants similar to other vascular plants? How do they differ from other vascular plants? (17·1)
2. Identify the four main groups of gymnosperms. (17·2)
3. What distinguishes gymnosperms from angiosperms? (17·2)
4. Compare the variety of life forms of gymnosperms with those of angiosperms. What conclusion can you draw about the diversity of adaptations in the two groups? (17·3)
5. The lima bean is the seed of the lima bean plant. Each half of the bean is a primary leaf, or cotyledon. Based on this seed structure, what would you predict the venation of lima bean leaves to be? (17·4)
6. What type of tissue is wood? (17·5)
7. To which function or functions of roots are root hairs adapted? (17·6)
8. Explain how secondary growth occurs in a root. (17·6)
9. Why doesn't secondary phloem accumulate during secondary growth? (17·6)
10. How do adventitious roots increase the ability of a plant to adapt to the environment? (17·7)
11. What are bundle scars? What do they indicate about the vascular systems of leaves and stems? (17·8)
12. How does the internal structure of a monocot stem differ from that of a dicot stem? (17·9)
13. Give an example of a plant in which the stem is the main photosynthetic organ. (17·10)
14. Distinguish between simple and compound leaves. How are compound leaves classified? (17·11)
15. Describe the internal structure of a leaf and discuss the function of each part in relation to photosynthesis. (17·12)
16. Describe what would happen to guard cells and stomata during drought. What effect would this change have on water and carbon dioxide movement into and out of a leaf? (17·13)
17. Identify two adaptations of the leaves of plants that live in dry environments. (17·14)

APPLYING CONCEPTS

1. How does the epidermis of leaves and stems differ in function from that of roots? How is this difference an indication of the location of the three organs in relation to the atmosphere?
2. The production of annual rings during secondary growth is a result of the differences in rates of growth, determined by season. Why are the annual rings in one tree not always the same size from year to year?
3. If a strip of bark is removed in a complete encircling band from the trunk of a tree, the tree may die. Why might loss of bark kill a tree? Under what conditions would it not kill a tree?

EXTENDING CONCEPTS

1. Choose one of the following environments: arctic, swamp, desert, temperate forest. Prepare an illustrated report on the seed plants that live in the environment and the special adaptations they have in order to survive there. Include the following: unusual environmental factors to which plants must adapt; your own predictions of how plants might adapt to these factors; actual ways in which plants are adapted. Focus on adaptations of nonreproductive organs.
2. Devise an experiment to show what effects temperature, time of day, light, and soil moisture have on stomatal opening and closing. What will indicate whether stomata are open or closed?
3. Prepare an illustrated guide to tree leaves in your geographical area. Use both angiosperm and gymnosperm leaves. Be sure to include whether leaves are evergreen or not, simple or compound, as well as venation and arrangement of leaflets.

READINGS

Brown, Lauren. *Weeds in Winter.* Norton, 1986.

Burton, Maurice, and others, editors. *The International Book of the Forest.* Simon and Schuster, 1981.

Gordon, Bonnie Bilyeu. "Cactus Rustling." *Science 80,* June 1980, p. 52.

Peterson, Russell. *The Pine Tree Book.* Brandywine Press, 1980.

18. Transport in Plants

The giant saguaro cactus is one of the most successful desert plants. In spite of the desert's intense heat and the scarcity of water, some of these plants live to be 200 years old. Like other green plants, the cactus collects water through its roots. How do you think the roots of a cactus differ from the roots of other plants? How does water travel from the roots upward to all parts of this plant? Does the same method of transporting water in this cactus also occur in giant redwood trees?

CHAPTER OBJECTIVES

After completing this chapter, you will be able to
- **Describe** how water is used by plants.
- **Describe** how water enters plants.
- **Compare** translocation in simple plants and in vascular plants.
- **Compare** xylem and phloem.
- **State** the transpiration-cohesion theory of water translocation.
- **Describe** the translocation of food in plants.
- **List** several plant adaptations for water retention.

Water Balance

18•1 WATER AND PLANTS

Life on the earth is dependent on water. The simplest plants are single-celled and live in water. In plants, water is the medium for transporting minerals, proteins, sugars, and fats. Water is also one of the basic substances needed to produce those materials.

Plants depend on water in several ways. (1) Water is the medium that carries most of the materials needed for life. Materials dissolved in water are transported throughout the plant. This movement of dissolved materials from one part of the plant to another is called **translocation**. (2) Water plays a very important role in the plant during the process of photosynthesis. In that reaction the hydrogen of the water molecule combines with carbon dioxide to form sugars. (3) Water in the plant cells gives the cells their shape and support.

Plants living on land have adaptations to obtain water and to prevent water loss. Some plants can live in very dry areas, such as deserts. Other plants grow only where their roots are constantly covered by water. All of these plants, however, use water as the medium of transport.

Plants can affect the amount of water in the atmosphere and alter the climate. As water moves into the roots from the soil and is conducted upward to the leaves, water vapor is released to the atmosphere. The process by which water vapor escapes from plant surfaces, generally through the stomata of the leaves, is called **transpiration**. The transpired water enters the atmosphere and is later condensed and returned to the earth in the form of precipitation. In areas of the earth where much vegetation has been removed, the amount of rainfall has decreased. Such a change in the rain cycle in turn alters the kinds of plants that can live in the area. Different kinds of plants support different animal populations, and soon both the living environment of an area and the climate have changed.

SECTION OBJECTIVES
- List several reasons plants need water.
- Describe how water enters the roots of plants.

trans- (across)
locus (place)

spirare (breathe)

Transpiration does more than affect the amount of water in the atmosphere. Evaporation from the leaves of a plant pulls water up from the roots. This process also allows minerals dissolved in the water to be brought into the plant. The plant uses some minerals, such as nitrates, phosphates, and sulfates, to synthesize proteins and nucleic acids. Other minerals, such as potassium, magnesium, and iron, are essential parts of the enzyme systems of plant cells. Also, as a result of evaporation the leaves are cooled. This cooling effect is important because only about 3 percent of the sunlight absorbed by a leaf is used in photosynthesis. The remainder is transformed into heat. If this heat were not removed, it would cause the death of the plant cells.

18·2 HOW WATER ENTERS PLANTS

For the single-celled plant living in water, the need for water is easily met. The difference in concentration of dissolved materials within the cell and within the environment results in water diffusing into the cell. Since the cell's interior is not far from the surface of the cell, no system is needed to distribute the water.

As large plants that lived on land developed, means of distributing water also developed. Plant bodies became modified with different tissues performing different roles. You will recall that mosses and liverworts (bryophytes) are held in place by slender rootlike structures called *rhizoids.* Like roots, rhizoids absorb water and minerals from the soil. However, the rhizoids and other structures of bryophytes lack vascular tissue to distribute water through the plant. The water simply diffuses from cell to cell throughout the short body. Because water can move upward in this manner only short distances, mosses and liverworts are limited to a height of about 15 cm.

FIGURE 18·1 Notice the root hairs on the roots of these seedlings. Water enters a plant through these tiny root hairs.

Water is obtained by most taller plants through the roots. Root hairs, which are extensions of epidermal cells, are found near the tips of roots and extend into the soil. Root hairs often give a young root a fuzzy appearance. Note the large number of root hairs produced by the young roots of the germinating seeds in Figure 18·1. Studies have shown that the numerous root hairs greatly increase the absorbing surface of the root. The length of a root hair may be hundreds of times greater than its diameter, giving it a large surface area and increasing the absorbing surface of the root system. Because root hairs live only a few days, new root hairs are constantly being produced near the growing tip of the elongating root. Most water absorption occurs near the tips of roots.

In general, root systems are more extensive than aboveground parts of plants. For example, tree roots generally spread horizontally below the ground well beyond the tips of the branches above the ground, as shown in Figure 18·2. The extensive root system provides both adequate water absorption and anchorage. In a study of a rye plant (a grass), scientists found that the roots had 50 times more surface area than the leaves. There were over 600 km of roots with a surface area of 230 m^2. The leaf surface area was only 4.7 m^2.

Nearly all ground water exists in the spaces between grains in the soil and in the spaces in rocks. Most of the water and minerals needed for the various activities of the plant are absorbed through the plant's root system. The part of the root through which most absorption takes place is the root hair zone. This absorption of water and minerals by the root hairs involves several processes.

Scientists believe that most water enters the root by the process of osmosis, which you read about in Chapter 4. Recall that in the process of osmosis, water moves across a selectively permeable membrane from a region of higher concentration to a region of lower concentration. At least two factors appear to be involved in maintaining a difference between the concentration, or pressure, of water in the soil and the water pressure inside the root. One reason for reduced water pressure in the root is the loss of water from the plant by transpiration. During the day when the stomata of the leaves are open, water is transpired rapidly. Water moves up from the root to the leaves and is evaporated. This loss of water reduces the water pressure in the root. The pressure of water in the soil remains higher than that in the root. This difference in water pressure between the inside and the outside of the root causes water to move into the root hairs.

FIGURE 18·2 The root systems of plants are very extensive. Often the roots spread over a greater area than the above ground parts of plants.

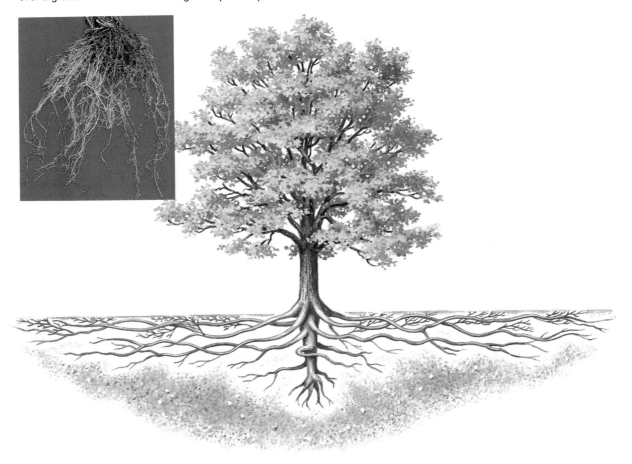

360 TRANSPORT IN PLANTS

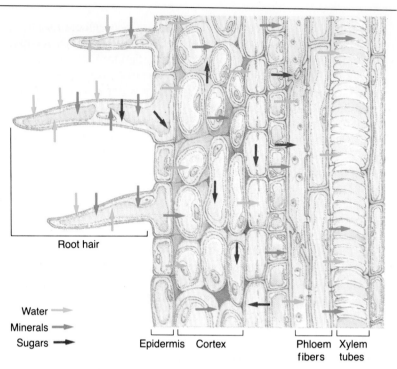

FIGURE 18·3 Water and minerals in the soil enter a plant by passing through the root hairs. The water and minerals then enter the xylem at the center of the root.

A second reason for the difference in water concentration between the root and the soil is the dissolved materials in the water of root cells. A living root contains many dissolved substances, such as sugars and salts. Because of these dissolved substances, there is a greater concentration of water outside the root hair than inside. Thus water molecules move by osmosis from the soil into the root hair. The resulting increase in water concentration in the root hair causes water molecules to move into the next cell and so on into the center of the root. Thus water moves from outside the root hair into the center of the root by osmosis from cell to cell. In Figure 18·3 you can trace the movement of water from the soil into the root.

Roots take in minerals as well as water. The minerals needed by plants for growth are the ions of calcium and potassium as well as nitrates, sulfates, and phosphates. The concentration of these minerals is greater inside the root hair cells than in the surrounding water of the soil. Therefore some energy must be expended to bring these ions into the root hairs. The energy to move the minerals into the root hairs is supplied by ATP.

Minerals are brought into the root by the process of active transport. Recall that *active transport* requires the use of energy by the cell for the movement of materials across the cell membrane. The active transport of minerals increases the movement of water into the root hairs. Increasing the concentration of minerals inside the cell decreases the water concentration. Thus water diffuses into the cell. This active accumulation of materials takes place mainly in rapidly growing and dividing cells.

Water enters the plant through the root hairs. It then passes from cell to cell to the vascular cylinder of the root. This core has two transport tissues, xylem and phloem. Recall that xylem is vascular tissue that carries water and minerals to the leaves. Phloem is vascular tissue that carries materials produced in the leaves to other parts of the plant. A layer of cells surrounds the vascular cylinder and helps to keep the water within the core.

REVIEW QUESTIONS

1. List three reasons plants need water.
2. Where are root hairs found? What is their function?
3. Name two factors that help to maintain a difference in water concentration between soil water and water inside a root.
4. Explain how water moves from the root hairs to the center of the root.
5. Explain why a houseplant grown in a closed terrarium does not need to be watered as often as when it is grown indoors in a clay flowerpot.

Movement of Fluids Within Plants

18·3 DIFFUSION IN SIMPLE PLANTS

In single-celled plants, substances such as water and minerals are distributed by diffusion and by streaming movements of the cytoplasm. Diffusion is a relatively slow process. However, this process can deliver materials throughout a cell. In those multicellular plants in which water moves only by diffusion, the size of the plant is limited by the rate of diffusion.

The multicellular algae lack specialized transporting tissues. Yet some of these plants grow to a very large size. These large algae are aquatic, and so water is always available. The plants are usually flat or filamentous. The alga is bathed in water and is thereby supplied with the minerals and water needed for life. In Figure 18·4 you can see the structure of the brown alga *Fucus*. Notice that this marine alga has a branching body. This body form allows the individual cells to be in direct contact with the environment.

18·4 TRANSPORT IN VASCULAR PLANTS

Vascular plants have specialized tissues that transport materials throughout the plant. However, plants do not have a circulatory system in the way that animals do. The fluid in plants is not pumped throughout the plant as blood is pumped throughout an animal. There is, however, upward and downward movement of materials. The flow of water in plants is similar to the flow of liquid in a pipeline. Water

SECTION OBJECTIVES

- Identify parts of a plant's vascular system.
- Discuss ways water and food are moved throughout a plant.

FIGURE 18·4 The round objects found on the brown algae, *Fucus,* are gas-filled bladders that help this plant float near the water surface.

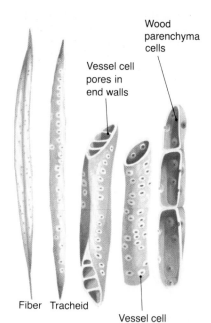

FIGURE 18·5 These different cells make up the xylem tissue in vascular plants. These cells help to move water throughout a plant.

FIGURE 18·6 A photomicrograph of xylem cells in cross section (left). A scanning electron micrograph of xylem cells (right). The dark area is the space within a large xylem vessel.

moves through a system of tubes that connects the roots to the leaves. Recall that the water vessels in the plant are composed of xylem. **Xylem** is a vascular tissue that provides mechanical support of the plant and conducts water and minerals from the roots to the leaves.

Several types of cells make up the xylem, as shown in Figure 18·5. All are fairly long, narrow, tubelike cells. **Tracheids** (TRAY-kee ihdz) are elongated and thickened xylem cells that conduct water. Rather than being open like sections of pipe, tracheids have tapered ends and thin pits in their walls to allow for the passage of water. **Vessel elements,** another type of xylem cell, are open-ended cells that conduct water. Vessel elements are joined end-to-end to form continuous tubes called **vessels.** In angiosperms the xylem consists of both tracheids and vessels. In gymnosperms most of the water-conducting tissue consists of tracheids.

Most xylem cells are elongated and are arranged parallel to the main axis of the stem. Often the cellulose walls of xylem cells become thickened. As a result the xylem cells strengthen the stem as well as conduct materials. As soon as one of the xylem cells becomes fully developed, it dies and its cytoplasm disappears. What remains is a thick, woody cell wall. The result is a series of dead cells, arranged end-to-end, that form a system of tubes throughout the length of the stem. Figure 18·6 (right) shows an electron micrograph of xylem cells arranged as tubes. It also shows large xylem vessels in cross section (left).

Once water has entered the root hairs, it passes into the root xylem in the vascular cylinder. From the root it passes upward into the stem. In Figure 18·7 you can see a cross section of a dicot stem. You will recall that in dicots the vascular bundles are found in a ring around the outside of the stem. Identify the vascular bundles in the figure. In the labeled vascular bundle, locate the xylem tubes. Tracheids and three large vessels make up the xylem tissue of the bundles. Water and minerals pass upward from the roots to the leaves through these conducting tubes. You can see that another vascular tissue, phloem, is also found in the vascular bundles.

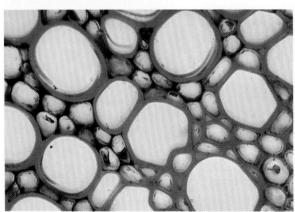
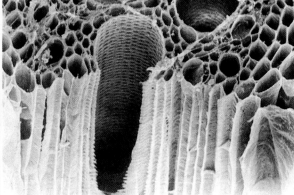

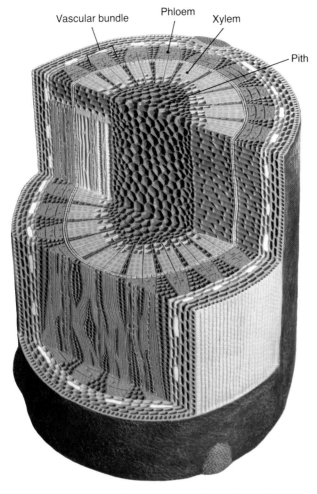

FIGURE 18·7 The structure of a typical dicot stem is shown here. Notice that the vascular bundles are arranged in a ring around the outside of the stem.

Phloem is a vascular tissue that conducts food throughout the plant and also provides storage and support. As you can see from Figure 18·8, phloem is made up of sieve-tube elements that form sieve tubes. **Sieve-tube elements** are cells that are joined end-to-end to form continuous tubes throughout the plant. Sieve-tube elements do not have rigid cell walls; unlike mature xylem cells, these phloem cells contain living protoplasm. The cells of sieve-tube elements join end-to-end to form **sieve tubes.** In angiosperms the sieve-tube elements have sieve plates in the ends of their cells. **Sieve plates** contain holes called *sieve pits,* which provide connections between the sieve-tube elements. Sieve-tube elements are unique in that they have no nuclei. In gymnosperms the sieve-tube elements have no sieve plates.

Next to the sieve-tube elements are parenchyma cells called companion cells. **Companion cells** are specialized cells that are believed to provide some materials and energy used by the sieve-tube elements. A companion cell has a nucleus and abundant cytoplasm. Notice the positions of the sieve-tube element and its companion cell. The smaller companion cell lacks holes.

FIGURE 18·8 Phloem is made up of sieve tube elements and companion cells. Phloem conducts food throughout the plant and also provides storage and support.

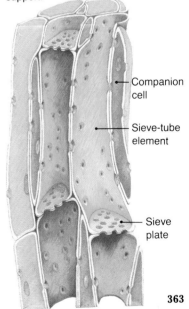

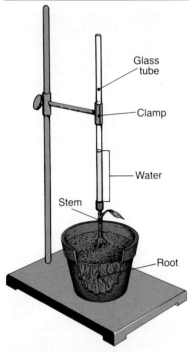

FIGURE 18·9 This apparatus shows that root pressure can push water up in the glass tube. But root pressure alone cannot push water to the top of a very tall plant stem.

FIGURE 18·10 Capillarity will cause water to rise in a thin tube. The narrower the tube, the higher the water will rise.

18·5 CAPILLARITY AND ROOT PRESSURE

At sea level air pressure can support a column of water 10.4 m high. How, then, can water rise to the top of a giant tree that may be 40 m or taller? Is water somehow pumped up to the leaves? As early as 1727, Stephen Hales thought that the roots pumped water upward. In that year he sealed a glass tube into the stump of a living tree, and he found that water rose in the tube. As shown in Figure 18·9, this action can be demonstrated in the laboratory by connecting the cut stem of a rooted and well-watered plant to a section of glass tubing. Water will rise in the tubing.

Root pressure, which is the pressure in the xylem resulting from the inward movement of water, contributes to the movement of water and minerals up the plant. Root pressure can be partially explained by osmosis. Recall that osmosis is the diffusion of water moving across membranes from areas of higher concentration of water molecules to areas of lower concentration of water molecules. As water moves from the soil into the root cells, pressure is created within the root. The pressure is passed on to the xylem tissues. This pressure contributes to the rise of water in the stem.

Since a root that is dead no longer exerts the same force on the water column, metabolic activity of the living root must also contribute to root pressure. However, scientists have found that root pressure alone is not great enough to raise water to the tops of tall trees.

Water can rise in a stem by capillarity (kap uh LAR uh tee). **Capillarity** is the tendency of a liquid to rise inside a tube of very small diameter. It is caused by the combination of two forces: **cohesion,** or the attraction of like molecules for each other, and **adhesion,** the attraction of unlike molecules for each other. The principle of capillarity can be shown by placing several tubes of differing diameters into a dish of water, as seen in Figure 18·10. Because of the force of adhesion, water molecules cling to the sides of the glass tubes and move upward. As the water molecules move upward, the force of cohesion causes water molecules at the top of the water column to pull other water molecules up. The smaller a tube's diameter, the higher the water rises in the tube.

Xylem cells in plants form a system of tubes of very small diameters. However, even in very small tubes, such as xylem, capillarity can raise water only 1 or 2 m above the ground. Water will not rise in xylem tubes by capillarity to the height of the tallest trees.

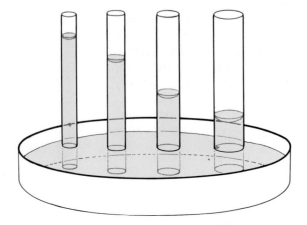

18·6 TRANSPIRATION-COHESION THEORY

The pulling power of evaporation combined with the attractive force of like molecules for each other can be demonstrated by an apparatus such as the one shown in Figure 18·11. In the apparatus a glass tube is filled with water and then set in a beaker of mercury. A stem cutting with attached leaves is connected to the top of the glass tube. The water begins to evaporate from the leaves as a result of transpiration. As it does, the mercury rises in the tube. If the evaporation continues, the water will eventually pull the column of mercury above 76 cm, the point it would reach by the force of atmospheric pressure alone. This is shown in the diagram.

Evidently, in addition to air pressure, some force or forces act on the mercury. Scientists have identified two forces involved in causing the mercury to rise: (1) the evaporation of water from the leaves and (2) the cohesion of water molecules.

The pull of evaporation and the cohesion of water molecules form the basis for the transpiration-cohesion theory of water conduction in plants. The **transpiration-cohesion theory** states that evaporation from leaves (transpiration) *pulls* the water up from the roots.

According to the theory, the column of water is not broken because of the cohesion of the water molecules. As long as the leaves are transpiring, a decreasing amount of water will exist in the leaves. There will be a tendency for water to flow into the leaves to replace the transpired water. Water will flow from the leaf veins into the leaf mesophyll. Water from the stem xylem flows into the veins in the leaves. Finally, water flows up the stem from the roots. A continuous column of water is pulled along by the loss of water evaporating from the leaves. The more quickly water is transpired from the leaves, the more quickly water is pulled up from the roots. The transpiration-cohesion theory has been supported by many experiments, observations of living plants, and mechanical models that demonstrate the principle.

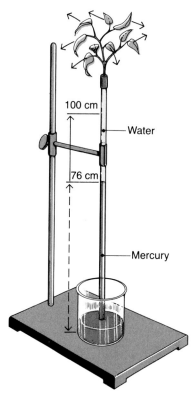

FIGURE 18·11 Transpiration from leaves causes water to rise in the tube. Air pressure alone will cause the mercury to rise to the 76 cm mark. Transpiration causes the mercury to rise even higher.

The transpiration-cohesion theory relies on evaporation from leaf surfaces. How does water move up in a tree that has lost its leaves, such as a deciduous tree in spring before new leaves have appeared? At the time when a tree is leafless, root pressure and capillarity provide a pushing force. Water does rise in leafless trees, so the movement of water in plants involves a combination of root pressure, capillarity, and transpiration-cohesion.

18·7 TRANSLOCATION OF FOOD

Water and the minerals dissolved in it can be shown to flow from the roots to the leaves. What happens to the sugars manufactured in the leaves during photosynthesis? The sugars are needed not only by the living cells of the leaves but also by living cells in the stems and roots. What transport system distributes the food needed by all cells for metabolism?

Translocation of the products of photosynthesis is by phloem. Phloem transports food from one part of the plant to another. Through

FIGURE 18·12 Sap collected by putting a hollow tube into the phloem of certain maple trees is used to make maple syrup.

the use of radioactive elements, phloem has been shown to carry food and water both up and down stems. Unlike the water-carrying vessels of xylem, the conducting tubes of phloem are alive. Recall that phloem is made up of cells called sieve-tube elements, which are cylindrical cells with cellulose walls. Sieve-tube elements joined end-to-end form sieve tubes.

Since the sieve-tube elements are alive and have holes in the ends and sides, it has been suggested that sieve tubes may act as living pumps. It has been shown that food transport is stopped or slowed down when oxygen is eliminated. These results strongly suggest that food transport is a metabolic activity that uses energy (ATP) from respiration.

Phloem also contains groups of very long and thick-walled cells called **phloem fibers.** These cells serve as protection for the thin-walled sieve tubes. Some parenchyma cells may be present, especially in herbaceous plants. These cells appear to serve as temporary storage sites for food.

The fluid inside phloem tissue is **sap,** which is made up mostly of water and dissolved substances, mainly the sugar sucrose. Sap obtained from some types of maple trees is used to make maple syrup. The sap is obtained by putting a hollow tube through the bark of the tree and into the phloem. A bucket is hung from the end of this hollow tube. The sap flows into the bucket. The tapping of maple trees and other evidence indicate that the sap is under pressure. However, the mechanism behind the movement of sap remains unknown.

Evidence that food moves through phloem tissue is provided by trees that have been girdled. **Girdling** occurs when a groove is cut so that it encircles the trunk and cuts through the phloem tubes. Animals such as rabbits and beavers often girdle trees by gnawing through the bark. In most trees the phloem tissue forms a thin inner layer beneath the bark. The tree will continue to live for a time after girdling. Eventually, however, it will die as the roots and tissues below the cut use up their stored food. If a cut is made deep enough during the girdling process, the leaves will wilt and the tree will die rapidly. What tissue has been cut through in that case?

REVIEW QUESTIONS

6. Name the processes that are involved in the transport of water within simple plants.
7. Why are nonvascular land plants small in size compared with vascular plants?
8. What three factors are involved in the transport of water through xylem?
9. What evidence is there that phloem tissue is active in the transport of sap?
10. Suppose a substance that breaks cohesive forces enters ground water. Predict the effect on nearby plant life.

Investigation

What Tissue Transports Water in a Celery Stalk?

Goals
After completing this activity, you will be able to
- Describe the location of water-conducting tissue in a celery stalk.
- Name at least two distinguishing features of the water-conducting tissue.

Materials (for groups of 2)
celery stalk with leaves in beaker of colored water
scalpel slide
coverslip dropper
microscope cutting board

Procedure

A. From your teacher obtain a celery stalk that has been standing overnight in a beaker of water to which food coloring has been added. Look for signs of movement of the colored water into the celery stalks. Observe the leaves carefully to determine if the colored water is present in the leaf veins.

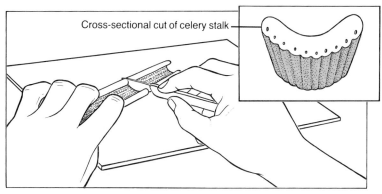

Cross-sectional cut of celery stalk

B. With the scalpel make a cross-sectional cut of the celery stalk about 4 cm from the bottom.

CAUTION: Use great care in handling the scalpel. Direct the edge of the blade *away* from your body. Be sure to work on a firm surface.

C. Examine the surface of the cross-sectional cut. Look for areas containing the colored water. On a separate sheet of paper, make a drawing of the cross section of the stalk. Show the location of the water-conducting tissue.

D. Cut off a 3-cm section of the celery stalk. Cut lengthwise into the section and remove one of the colored tubes from the stalk. With the scalpel make a very thin cross-sectional cut of the tube. Make a wet mount of this piece of tissue. Observe the slide under low power. Try to locate thinner areas that are a few cells to one cell thick. Switch to high power and observe the cells. Draw several cells.

E. Remove another colored tube. Split the tube lengthwise and cut a wedge-shaped section from the tube. Make a wet mount of this piece of tissue. Observe it under low power and high power. Draw some of the cells you see.

F. Follow your teacher's instructions for the proper disposal of all materials.

Questions and Conclusions

1. Where in the stalk is the water-conducting tissue located?
2. From your examination of the water-conducting tissue, describe two structural features of the cells that relate to their function of support.
3. Name at least three environmental factors that may affect the height to which the colored water rises in the celery stalk.
4. Liquids generally move further up the stalk if the bottom of the stalk is cut under water, and the stalk is quickly transferred to the colored solution. Suggest a reason for this response.
5. Suppose one celery stalk was placed in a beaker of colored water in a sunny area. Another similar stalk was placed in a beaker located in a cool, shady area. In which stalk would you expect the colored solution to rise higher in 3 hours? Why?

TRANSPORT IN PLANTS 367

Plant Responses to Water

SECTION OBJECTIVES

- Identify the effects of a lack of water on plant tissues.
- Discuss ways plants are adapted to living under different environmental conditions.
- Describe ways plants get rid of excess water.

18·8 TURGOR PRESSURE AND WILTING

You have probably noticed that a well-watered plant is firm. Plant cells that absorb water expand, and the plant's shape is maintained. The expanded condition caused by the absorption of water is called *turgor* (TER guhr). **Turgor pressure** is the pressure exerted by water and the materials dissolved in it against the cell membranes and walls of a plant. In a well-watered plant the turgor pressure causes the entire plant to become firm, or turgid. You might think the increasing pressure of water inside a plant cell would soon cause the cell membrane to burst. However, the cell wall is strong enough to prevent the bursting from occurring. A turgid plant cell can be compared to a bicycle tire and its inflated inner tube. When filled with air, the inner tube presses tightly against the tire. The inner tube is limited in its expansion by the firm covering of the tire.

In soft plant parts, such as leaf blades and flower parts, turgor pressure maintains the form, firmness, and proper functioning of the plant part. Turgor pressure provides the force needed to push seedlings up through the soil and the force needed to open flower buds.

Picture how a plant looks if it has not been watered in some time. The leaves, stem, and flowers droop. The plant wilts. **Wilting** is the loss of turgor in leaves, stems, and flowers of a plant due to the loss of water from its tissues.

Temporary wilting can occur when the transpiration rate is greater than the rate of water absorption by the roots. Both high temperatures and low humidity will speed the rate of evaporation from the leaves. If these two conditions occur during the day when the stomata are open, the plant may wilt. On a hot, dry day, absorption of water is not rapid enough to match the loss of water by transpiration. At night the stomata close, and the plant can regain normal turgidity in its cells. Temporary wilting does no permanent damage to a plant.

Permanent wilting occurs if sufficient water is not available for an extended time, such as during a drought. It also can occur when the roots or root hairs are destroyed by chemicals or damaged during transplanting. Permanent wilting is followed by death.

18·9 WATER AND PLANT ADAPTATIONS

Many plants have adaptations to obtain water, to prevent water loss, or to store water. Plants inhabit most of the land surface of the earth. Thus they have adapted to conditions ranging from the most arid deserts to saltwater marshes.

Hydrotropism (hī DROHT ruh pihz uhm) is the growth response of plants to moisture. This response usually involves the roots. The root tip is sensitive to variations in the water content of the soil and will turn from a region of low water content to a region of higher water

FIGURE 18·13 The stems of the well-watered plant (top) are held above the pot. The stems of the wilted plant (bottom) hang limply over the pot's sides.

content. Roots will grow a great distance toward water, especially when the surface layers of soil are drying out. The roots of certain trees, such as willows, often grow into drainage pipes.

Some plants, such as those in a tropical rain forest, have few adaptations to prevent water loss. These plants, therefore, grow best only in places where evaporation is low. Evaporation from the leaves is affected by temperature, movement of the air, and light intensity, as well as by relative humidity. Shady, moist habitats where the humidity of the air remains high are needed by many types of plants.

Computers in Biology

The Branching of Plants

The first plants to live on dry land arose 400 million years ago. As they spread, covering the land, groups of plants developed different patterns in the arrangement of their branches. When plant growth becomes dense, it is more difficult for plants to get adequate sunlight for photosynthesis. The branches of trees reach high and wide, exposing the leaves to the sun. But there is a limit to how long the branches can grow and still be able to support their weight.

Growth may place a great deal of stress on trees unless there is *balance* in the branching patterns. A tall tree, therefore, may produce layers of branches, with the longest branches at the bottom and the shortest at the top.

Are the balanced branching patterns the result of chance, or are the patterns the result of natural selection? One approach to answering this question is to map out all the possible branching patterns of a plant. Can you imagine trying to do that by hand? Such a project might take an entire lifetime!

Computers, however, make this huge task manageable. Karl Niklas at Cornell University has made a two-dimensional statistical program for a computer to map the possible branching patterns of plants. The computers also can evaluate which patterns of development are most probable.

Niklas has run hundreds of computer simulations of possible branching patterns. The simulations indicate a steady decrease in variation in the branching patterns of major species over time. It would seem that some patterns are favored by natural selection and others are eliminated.

The two-dimensional model, however, cannot account for a variable such as the rotation angle—that is, how each branch is positioned around the trunk of the tree. The solution to this research question will be approached through the three-dimensional model that Niklas and his colleagues are developing. A three-dimensional simulation could map 10 000 patterns in about 8 days. Imagine having to build these patterns by hand.

FIGURE 18·14 Xerophytes are adapted to living in areas where water is scarce. Two examples of xerophytes are: cactuses (left) and blooming aloes (right).

xero- (dry)
-phyte (plant)

Many plants grow in regions where the air is hot, dry, and constantly moving. These plants have structural modifications to prevent water loss. A **xerophyte** (ZIHR uh fīt) is a plant that grows where water is scarce. Two different types of xerophytes, cactuses and aloes, are shown in Figure 18·14. Many xerophytes have modifications that limit water loss through transpiration. In some plants the stomata of the leaf are reduced in number and may be sunken in pits. These stomata may be surrounded by hairs or protected by special guard cells. In some xerophytes, leaf surface area is greatly reduced. These plants may produce scalelike or needlelike leaves, such as some conifers and cactuses do. Some xerophytes have leaves with a heavy coating of wax to retard water loss. Other plants produce resins and other antidrying agents in the tissues.

In some xerophytes the stems and leaves are fleshy and juicy, or succulent. A **succulent** (SUHK yuh luhnt) is a plant that has large fleshy leaves or stems in which water is stored. Many plants of desert and semiarid regions, such as cacti, stonecrops, yuccas, and spurges, are succulents. The chief characteristic of succulent plants is that the bulk of their bodies is made up of water-storage cells. These cells enable the plants to survive periods of drought. In dry areas, plants can also be adapted to reducing the competition for water. Some plants in dry areas produce chemicals that inhibit the growth of seedlings around them. These plants are thus widely spaced. The wide spacing gives each plant a large area from which to absorb the water that is available. With less competition from other plants, the plants are better adapted for survival in their dry environment.

halo- (salt)

A **halophyte** (HAL uh fīt) is a plant adapted to growing in salty soils. In these environments, there may be large amounts of soil water. However, the larger amount of dissolved salts results in slowed diffusion of water into the roots. Many halophytes have structural modifications, such as fleshy or hairy stems and leaves, similar to those of succulents. The scarcity of water available to halophytes is due to problems of absorption rather than to an actual lack of water. In some of the world's agricultural regions, the only water readily available for irrigation contains slight amounts of salt. Where such water has been used for many years, the buildup of salt in the soil prevents normal crops from growing. Researchers are looking for halophytic strains of crops that might grow in these soils.

In different parts of the world, many plants are deciduous. Recall that a *deciduous* plant is one that drops its leaves during certain seasons of the year. In temperate climates broad-leaved trees, such as the maple, oak, hickory, and beech, shed their leaves and become dormant as winter approaches. In the spring when favorable conditions return, they grow new ones. In parts of the world where there is a rainy season and a dry season, deciduous plants lose their leaves during the dry season. The deciduous condition is an adaptation that prevents excessive water loss by transpiration when water is scarce. In plants without leaves, transpiration is dramatically reduced in dry or cold seasons.

18•10 GUTTATION

Under conditions where transpiration is low but water absorption by the roots is high, excessive pressure can be set up within the plant. This pressure is the result of root pressure, and it can cause certain tissues to expand rapidly and rupture. Tomatoes may burst after a series of rainy days during their growing season. This pressure from excess water is prevented in some plants by guttation. **Guttation** (guh TAY shuhn) is the process by which a plant rids itself of excess water through special pores on the leaf. These water pores, or **hydathodes** (HĪ duh thohdz), are near the ends of certain veins, especially at the tip of the leaf or its marginal teeth. An example of guttation is shown in Figure 18•15. This process of guttation is the source of much of the dew on grass in the early mornings in summer. In the morning a droplet of water may come out from the tip of each leaf. The dew that forms from condensation coats the entire leaf. Besides grasses, some other plants that display guttation are strawberries, tomatoes, and potatoes.

FIGURE 18•15 Excess water may harm plant tissues. Excess water may be forced out of a plant by a process called guttation.

REVIEW QUESTIONS

11. Distinguish between permanent wilting and temporary wilting.
12. Name three plant adaptations that help prevent water loss.
13. Some plants produce chemicals that inhibit the growth of nearby seedlings. How are these chemicals an adaptation to dry conditions?
14. Compare the processes of guttation and temporary wilting.
15. Both conifers and cactuses have needlelike leaves. Would you expect these leaves to have hydathodes? Explain your answer.

Chapter Review

SUMMARY

18·1 Plants need water to produce food, to transport materials, and to provide support for tissues.

18·2 Water enters plants by the process of osmosis. Minerals enter by active transport.

18·3 In single-celled aquatic plants and in simple multicellular aquatic plants, water and dissolved materials are transported by diffusion.

18·4 In vascular plants, water and dissolved minerals are transported from the roots to the stems and leaves through the xylem.

18·5 Water is moved upward to some extent in xylem by the combined processes of root pressure and capillarity.

18·6 The transpiration-cohesion theory accounts for the rise of water in tall plants. This theory suggests that evaporation of water in the leaves combined with the cohesion of water molecules pulls water up from the roots.

18·7 Food is translocated throughout a plant by phloem.

18·8 Well-watered plants remain firm and erect because of turgor pressure, or the pressure exerted by water and dissolved minerals against a plant's cell membranes and walls. High temperatures, low humidity, high transpiration rates, and insufficient uptake of water are factors in wilting.

18·9 Different plants have different adaptations to scarcity of water, including the reduction of leaf surface area and the storage of water in the leaves and stems.

18·10 Guttation is an adaptation by some plants that eliminates excess water.

BIOLOGICAL TERMS

adhesion
capillarity
cohesion
companion cells
girdling
guttation
halophyte
hydathodes
hydrotropism
phloem
phloem fibers
root pressure
sap
sieve plates
sieve-tube elements
sieve tubes
succulent
tracheids
translocation
transpiration
transpiration-cohesion theory
turgor pressure
vessel elements
vessels
wilting
xerophyte
xylem

USING BIOLOGICAL TERMS

1. Using a dictionary, find the meaning of the word part *halo-* in *halo*phyte and *xero-* in *xero*phyte. Explain why these terms are appropriately named.
2. Compare the meanings of the terms *adhesion* and *cohesion*.
3. Distinguish between the terms *translocation* and *transpiration*.
4. List all terms referring to the structures for conducting water upward through the stem. Define each term.
5. List all terms referring to the structures for conducting food throughout the plant. Define each term.

In the statements that follow, some terms are italicized. Write TRUE if the underlined term is used correctly. If it is not used correctly, from the list of terms choose a term that will make the statement correct.

6. The upward movement of water in the vascular system of a plant may be explained by *root pressure, capillarity,* and the *transpiration-cohesion theory.*
7. Within phloem tissue, *companion cells* provide connections between sieve-tube elements.
8. Water rises in tubes of small diameter due to the attractive forces of *adhesion* and *girdling*.

9. Loss of stiffness in plant parts due to a lack of water is called *turgor pressure.*
10. *Xerophytes* and *halophytes* are plants that are adapted to live where water is scarce or difficult for the roots to absorb.

UNDERSTANDING CONCEPTS

1. List several ways in which plants are dependent on water. (18·1)
2. Explain how transpiration can affect climate. (18·1)
3. How do root hairs increase the absorption of water by plants? (18·2)
4. Describe the processes by which water and minerals are absorbed by roots. (18·2)
5. Suggest a reason why mosses and liverworts never grow very large. (18·3)
6. Compare the structure and function of xylem in angiosperms with the structure and function of phloem in angiosperms. (18·4)
7. Explain how each of the following functions in plant transport: (a) root pressure, (b) capillarity, (c) transpiration, and (d) cohesion. (18·5, 18·6)
8. State the transpiration-cohesion theory in your own words. (18·6)
9. Describe the transport system of plants. (18·6, 18·7)
10. How does girdling provide evidence that food is transported in phloem tissue? (18·7)
11. Explain the mechanism by which plant parts become firm after watering. (18·8)
12. Compare temporary wilting and permanent wilting. (18·8)
13. Explain how a deciduous condition can be an adaptation to control water loss. (18·9)
14. Identify some adaptations of xerophytes to dry conditions. (18·9)
15. Explain what is meant by *guttation.* Give examples of plants that exhibit this phenomenon. (18·10)

APPLYING CONCEPTS

1. Growing roots and stems can exert tremendous pressures, often sufficient to split rocks. What causes this power?
2. Root hairs are usually absent in water plants. Explain why this is so.
3. Celery, carrots, sliced cucumbers, and similar vegetables are often placed in water for a time before being served. Suggest some reasons why this is done.
4. Why will some fertilizers, if applied abundantly, often kill plants?
5. Give reasons why some plants can live in salt marshes and on beaches while others cannot.

EXTENDING CONCEPTS

1. You can observe transpiration in plants growing outdoors or indoors. Place thin plastic bags, such as those used for storing food, over leaves of shrubs or trees. (Fill the bags with water first to check for leaks, and then wipe dry.) Secure the bags with rubber bands or twist ties. Be sure to add a control, using a similar bag tied to a part of the stem with no leaves inside. Observe the bags after an hour. Which bag collected moisture? Where did it come from?
2. Perform the activity in Extension 1, but measure the collected water by finding the mass of the bags before and after placing them on the plants. Compare the mass of the water collected on different plants, on different days, in different locations, at different times of day. Suggest reasons for any differences you note.
3. Read about some of the current research on salt-tolerant plants that may be raised for food crops. Report to the class on (a) adaptations of these plants, (b) where they may be grown, and (c) the economic advantage gained by growing these plants.

READINGS

Ayensu, Edward S., ed. *Jungles.* Crown Publishers, 1980.

Boyle, Robert H., and Rose Mary Mecham. "Anatomy of a Man-Made Drought." *Sports Illustrated,* March 15, 1982, p. 46.

Carolina Biological Reader. *Stomata.* 1981.

Galston, Arthur W. *The Life of the Green Plant,* 3rd ed. Prentice-Hall, Inc., 1980.

Raven, Peter, and others. *Biology of Plants,* 4th ed. Worth Publishers, 1986.

19. Plant Reproduction

The coconut is the fruit of the coconut palm tree. Inside each fruit is a large nut which contains the coconut seed. Like other seeds, the coconut seed contains an embryo plant and stored food that will be used by the developing plant after it germinates. How does the formation of a seed occur? How are seeds distributed in the environment? What conditions are needed for the germination of a seed?

CHAPTER OBJECTIVES

After completing this chapter, you will be able to

- **Define** vegetative propagation and **give examples** of the natural and artificial processes.
- **Describe** the stages in the life cycles of a gymnosperm and an angiosperm.
- **Compare** the function of the cone in the gymnosperm with that of the flower in the angiosperm.
- **Describe** the dispersal and germination of seeds in flowering plants.
- **Describe** two hypotheses about the success of the angiosperms.

Asexual Reproduction

19·1 NATURAL VEGETATIVE PROPAGATION

Many seed-bearing plants can reproduce in two ways: sexually and asexually. In sexual reproduction, offspring are produced by the union of male and female sex cells. In asexual reproduction, offspring are produced by a single parent, without the union of male and female sex cells.

When a seed plant reproduces asexually, structures other than seeds are involved in producing the new plants. Such asexual reproduction is called vegetative propagation. **Vegetative propagation** is the production of new plants from a nonreproductive part—such as a root, stem, or leaf—of the parent plant. There are two types of vegetative propagation: natural and artificial. *Natural vegetative propagation* includes all the asexual reproductive processes that naturally occur in plants. *Artificial vegetative propagation* includes all the methods used by humans to bring about asexual reproduction in plants. All of these methods are based on the ability of plants to reproduce vegetatively.

Stems are the plant parts most often involved in natural vegetative propagation. Stems or parts of stems can form new plants by first growing adventitious roots. *Adventitious roots* are roots that grow from structures other than roots. For example, the strawberry plant produces long horizontal stems called **runners,** or stolons, that grow above the ground. Adventitious roots are produced at various points along these runners. Where adventitious roots form, stems and leaves develop. In this way, new plants are produced.

Kentucky bluegrass plants are produced vegetatively by rhizomes (RĪ zohmz). **Rhizomes** are horizontal stems that grow under the ground. Adventitious roots and shoots can develop from rhizomes. The white potato plant has a thick food-storing rhizome called a tuber. Have you ever noticed shoots and roots sprouting from a tuber?

SECTION OBJECTIVES

- Compare natural and artificial vegetative propagation.
- Discuss ways plants can be propagated by artificial means.

376 PLANT REPRODUCTION

Vegetative propagation takes place in gladiolus plants by means of corms. A **corm** is a rounded, thick underground stem that stores food. In the spring a corm develops adventitious roots and sends up a new plant. Figure 19·1 shows a corm. Notice the adventitious roots on the corm. Some of the food produced by the leaves will be used to produce buds on the original corm. These buds will grow into new corms. In many areas corms can survive the winter underground. The following spring these new corms will develop into new plants, completing the cycle.

Tulips, daffodils, hyacinths, and onions are examples of plants that can be propagated by bulbs. A *bulb* is a short underground stem that is surrounded by thick, fleshy food-storing leaves. Compare the bulb and the corm shown in Figure 19·1. Both are underground storage stems. Notice the fleshy leaves of the bulb.

Leaves are sometimes involved in vegetative propagation. The kalanchoe (kal uhn KOH ee), commonly grown as a house plant, develops new plants from plantlets. A *plantlet* is a tiny plant that develops on a larger plant. Plantlets eventually fall to the ground, where they can grow adventitious roots and develop into new plants. Identify the plantlets on the leaf shown in Figure 19·1.

Roots can be the means of vegetative propagation in some plants. A sweet potato is an example of a root that can develop into a new plant. Each plant produces several fleshy roots. Each root can grow new roots and shoots and develop into a new plant.

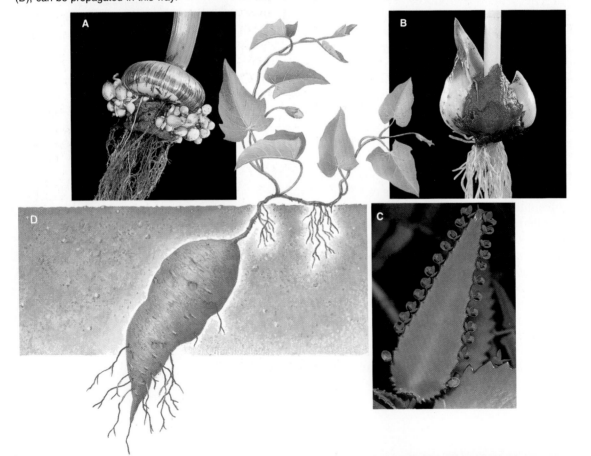

FIGURE 19·1 Various plant parts can be propagated vegetatively. A gladiolus corm (A), a tulip bulb (B), plantlets on a kalanchoe leaf (C), and a sweet potato (D), can be propagated in this way.

19·2 ARTIFICIAL VEGETATIVE PROPAGATION

Gardeners, landscapers, house-plant growers, and greenhouse workers all may grow new plants by asexual methods. They use artificial vegetative propagation.

The use of cuttings is the simplest method of artificial vegetative propagation. A **cutting** is a plant part, most often a leaf and an attached stem or a leaf alone, that is used for growing a new plant. Figure 19·2 shows a cutting taken from a coleus plant. Notice that new roots have grown at the base of the stem.

African violets and begonias can be propagated from leaves. New plants may be developed by placing a begonia leaf on sand. Cuts are made in the veins. Later a plant develops at each cut.

Plants such as black raspberry and forsythia reproduce vegetatively by a process called layering. **Layering** is a method of vegetative propagation in which a branch of the parent plant is bent to the ground and its tip is covered with soil. Adventitious roots develop from the buried tip, vertical shoots grow upward, and a new plant develops.

Some methods of artificial vegetative propagation, such as grafting, can produce offspring with traits of the parent as well as desired traits of another plant. **Grafting** is the joining of a twig of one plant onto the body of a different plant. Grafting requires two parts: the **scion** (SĪ uhn), which is a twig of the plant to be grafted, and the **stock,** which is the rooted stem of the parent plant. Generally, grafting is done with trees.

What is the purpose of grafting? This method can be used to produce a tree that has the most desirable features of two plants. For example, one kind of tree may produce large tasty fruit but be susceptible to disease. Another kind may produce small tasteless fruit but be resistant to disease. If the scion of a tree with tasty fruit is grafted onto the stock of the disease-resistant tree, tasty fruit will be produced by a disease-resistant tree.

Budding is a method of vegetative reproduction that involves grafting buds rather than twigs. A bud is placed inside a slit in the bark of the stock. Several shoots can then develop from each bud.

Artificial vegetative propagation is a quick and efficient way of producing plants with traits identical to parent plants. It is an important method of propagating plants without the use of seeds.

FIGURE 19·2 Cuttings are frequently used to propagate plants. If you look closely you can see the roots at the base of this cutting.

cion (a cutting)
stocc (stump)

REVIEW QUESTIONS

1. What is the major difference between asexual and sexual reproduction in plants?
2. Why is vegetative propagation a form of asexual reproduction?
3. Compare vegetative propagation by runners and by corms.
4. Describe how a new plant can develop from layering.
5. Suppose a citrus farmer wants to propagate a variety of seedless orange. Suggest a method that can be used to reproduce this seedless variety.

Sexual Reproduction

SECTION OBJECTIVES

- Compare methods of reproduction in gymnosperms and angiosperms.
- Discuss the function of various flower parts.

19·3 REPRODUCTION IN SEED PLANTS

Sexual reproduction is the production of young by the joining of male and female sex cells. In the seed-bearing plants, sexual reproduction is the main way of producing young. Sexual reproduction results in the production of young that have a variety of traits. This can help the species survive if environmental conditions change.

Fossil evidence of seed-bearing plants goes back 360 million years. The early seed-bearing plants are called *seed ferns* because of their fernlike leaves. Figure 19·3 shows a fossil of a seed fern. Seed ferns developed at a time when changes in climate caused many places on the earth to get drier. Since the early seed plants had a reproductive cycle that did not rely on water, they did well in the dry environment. Seed ferns no longer exist on the earth. As more fossil evidence is found, the link between seed ferns and modern seed-bearing plants may become clearer.

Recall that there are two modern groups of seed-bearing plants. The more primitive group of plants makes up the class Gymnospermae. Gymnosperms produce seeds that are naked—that is, the seeds are not enclosed in an ovary. Four classes of seed plants make up the gymnosperms: the familiar conifers—such as pines, spruces, and hemlocks—and the less familiar ginkgoes, gnetales, and cycads. Ginkgoes and cycads were common on the earth 150 to 200 million years ago. The conifers also first appeared on the earth at that time.

The more advanced group of seed-bearing plants belongs to the class Angiospermae, the flowering plants. The *angiosperms* produce seeds that are enclosed in special structures called ovaries.

FIGURE 19·3 In this fossil you can see an early seed fern. Seed ferns are thought to be the ancestors of the seed-bearing plants alive today.

Alternation of generations occurs in seed-bearing plants, as it does in the more primitive plants, such as the algae, mosses, and ferns (Section 16·3). Figure 19·4 gives an overview of alternation of generations in seed plants. The stages are described below.

1. The sporophyte generation is the form of the plant that is most familiar. It is the diploid plant with roots, stems, and leaves.
2. The sporophyte plant produces two types of monoploid spores. One type of spore develops into a male gametophyte. The other type of spore develops into a female gametophyte.
3. The male gametophyte produces sperm, and the female gametophyte produces an egg.
4. Following fertilization, a diploid zygote is produced. The zygote is the first cell of the sporophyte plant.
5. A sporophyte embryo forms from the zygote. This embryo, enclosed within a protective structure, is the seed.
6. Under proper conditions, the embryo develops into the adult sporophyte plant.

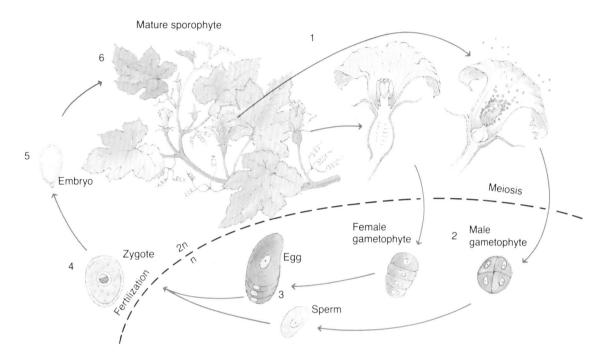

FIGURE 19·4 The sporophyte generation is the most obvious generation in seed plants. This diagram shows the general pattern of alternation of generations in seed plants.

How does alternation of generations in seed-bearing plants differ from this cycle in simpler plants? One major difference is that in seed-bearing plants the gametophyte is very small and is inside the mature sporophyte plant. The gametophyte is not a separate plant, as it usually is in mosses or ferns. Another difference is that in seed-bearing plants the male and female gametophytes develop from different types of spores.

19·4 REPRODUCTION IN CONIFERS

A conifer, or cone-bearing plant, is a shrub or tree whose seeds are carried in cones. Most of the conifers are evergreen—that is, they have leaves throughout the year. Pines, spruces, firs, redwoods, junipers, and hemlocks are evergreen conifers. A few of the conifers are deciduous—that is, they drop their leaves during part of the year. The larch and bald cypress are two examples of deciduous conifers.

The distinguishing feature of the conifers is the cone. The **cones** are the structures that bear the male and female gametophytes. Cones are compact structures made of scales. Figure 19·5 shows a cone from a spruce. When you think of a cone, you probably think of a hard, dry brown structure. Actually, such a cone is a mature female cone. Immature female cones and male cones are smaller.

FIGURE 19·5 This is the cone of a white spruce. Inside of the cone you will find the developing seeds.

An examination of the life cycle of a familiar conifer, the pine, will show the role of the cone in the reproductive cycle of the conifer. Refer to Figure 19·7 to help you follow the cycle of alternation of generations in the pine.

The tree, including the cones, is the sporophyte. In winter the immature male cone is covered with brown bud scales. On each scale, diploid cells called *microsporocytes* divide by meiosis to produce monoploid microspores. Each **microspore** is a cell that produces a four-celled gametophyte. This male gametophyte is the **pollen grain.** In the spring the bud scales fall off and the male cone grows in size, allowing the pollen grains to leave the cone. You can see that the pollen grains have thin wings, which aid in dispersal of the pollen by wind. The dispersal of the pollen grains by wind occurs in most kinds of conifers.

When they first develop in the spring, female cones are soft and green. Inside these cones specialized reproductive structures called **ovules** form. Each ovule contains a diploid *megasporocyte* that divides by meiosis to produce four monoploid nuclei. Three of these nuclei disintegrate and one nucleus becomes the megaspore. The **megaspore** is a cell that produces a multicellular female gametophyte. This female gametophyte remains within the ovule in the cone on the pine tree. When the female gametophyte matures, several egg cells are produced.

FIGURE 19·6 Three conifers and photographs of their cones: sequoia (A), pine (B), and hemlock (C). As you can see, the cones vary greatly in appearance.

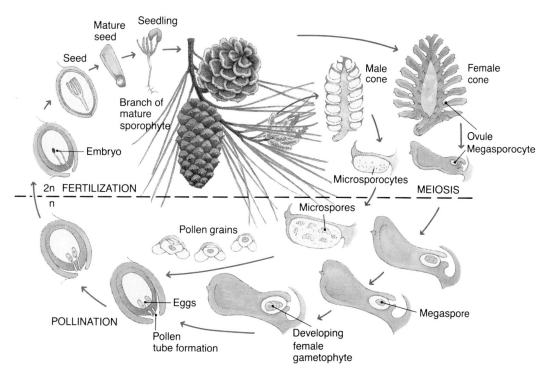

FIGURE 19·7 This diagram shows the life cycle of a pine, a familiar conifer. The pine tree is the sporophyte generation of this plant.

At a certain time of the year, when pollen grains are released from the male cones, the female cone secretes a sticky fluid. The fluid traps the pollen grains, drawing them toward the female gametophyte. Once pollen grains are trapped inside, the female cone enlarges, closing the spaces between its scales.

After the pollen grains are trapped in the female cone, many processes occur that lead to the formation of a seed. A tube called a pollen tube grows from the pollen grain into the ovule. One of the nuclei within the pollen grain divides by mitosis to form two sperm cells. Sperm cells are carried through the pollen tube and reach the egg inside the ovule. Inside the ovule one sperm unites with the egg to form a zygote, the first cell of the sporophyte generation. As the zygote develops into an embryonic plant, the ovule develops into a seed.

A **seed** is a structure containing three main parts: an embryo, which is a developing plant; food for the embryo, which is tissue of the female gametophyte; and a seed coat, which is the wall of the ovule. About four months are required for the seeds to mature and the female cone to open, releasing the seeds. Pine seeds are winged and are carried easily by the wind. The seeds will not grow into seedlings until the following spring. About two years are needed for the mature pine tree to complete its reproductive cycle.

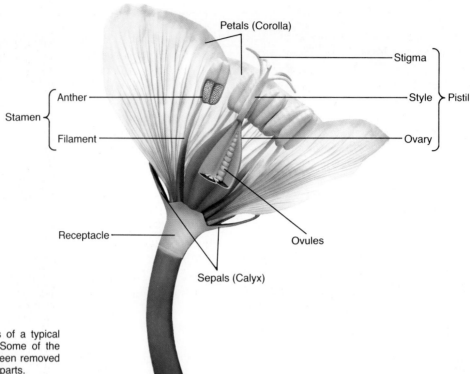

FIGURE 19·8 The parts of a typical flower are shown here. Some of the petals and sepals have been removed to show the reproductive parts.

19·5 FLOWERS

The reproductive structure of an angiosperm is the **flower.** Flowers differ in their shapes, colors, and sizes. They range in form from the simple buttercup to the complex orchid. Although there is great diversity in their form, most flowers have the same basic parts.

Figure 19·8 shows a flower in cross section. The **receptacle** (rih-SEHP tuh kuhl), which is located at the base of the flower, is the structure to which all flower parts are attached. The receptacle also connects the flower to the rest of the plant. The small leaflike structures above the receptacle are the **sepals** (SEE puhlz). All the sepals together form the **calyx** (KAY lihks). Inside the sepals is a ring of brightly colored structures, the **petals** (PEHT uhlz). All the petals together form the **corolla** (kuh RAHL uh). The corolla is usually the most noticeable part of the flower. Inside the corolla are the male reproductive structures, the **stamens** (STAY muhnz). Notice that each stamen contains an anther (AN thuhr) and a filament (FIHL uh muhnt). The **anther** is the structure in which the male gametophytes, or pollen grains, are produced. The **filament** is a long thin stalk that attaches the anther to the receptacle. In the center of the flower is the **pistil** (PIHS tuhl), the female reproductive structure. The pistil contains three major parts: (1) the **ovary** (OH vuhr ee), the structure that contains one or more ovules and developing gametophytes; (2) the **stigma** (STIHG muh), the upper part of the pistil upon which pollen grains land; and (3) the **style** (stīl), the connecting stalk between the stigma and the ovary.

ovum (egg)

PLANT REPRODUCTION 383

The ovary is the distinguishing feature of the angiosperm. It is the part that encloses first the developing gametophyte and later the developing seed. Recall that the gymnosperms have ovules, which later develop into seeds. Only in the angiosperms are the ovules enclosed inside the ovary.

The appearance of the basic flower parts can vary so much from one plant to another that the parts are sometimes difficult to identify. Look at Figure 19·9. Notice that in the tulip the sepals are brightly colored, rather than green. In the flower called butter-and-eggs, the sepals join with the petals, forming a structure called a spur.

In some species of plants there are male and female flowers. Such flowers are called *imperfect flowers.* Male flowers are called *staminate* (STAM uh niht) *flowers,* because they contain only stamens. Female flowers, called *pistillate* (PIHS tuh liht) *flowers,* contain only pistils. The flowers of a corn plant are imperfect flowers. The tassels on top of the plant are the staminate flowers, and the young ears of corn are the pistillate flowers. Other plants, such as the tulip, contain both stamens and pistils in one flower and are called *perfect flowers.*

19·6 POLLEN GRAINS AND OVULES

In flowering plants, male and female sex cells develop within the flower. Figure 19·10 shows the development of pollen grains in the anther. Within the anther are millions of diploid microsporocytes. Each microsporocyte divides by meiosis and produces four monoploid microspores. The nucleus of each microspore divides by mitosis to form two nuclei. One of these nuclei becomes the generative nucleus and the other becomes the tube nucleus. The structure that contains the two nuclei is the pollen grain, or male gametophyte. In Figure 19·10, you can see the mature anther releasing pollen grains. The pattern of spines and ridges on pollen grains is unique to each species.

FIGURE 19·9 In a tulip (top), the sepals are brightly colored. In a butter-and-eggs (bottom) the sepals and petals are joined to form a spur.

FIGURE 19·10 The anther and the filament make up the stamen. The development of pollen grains, the male gametophytes, occurs in the anthers.

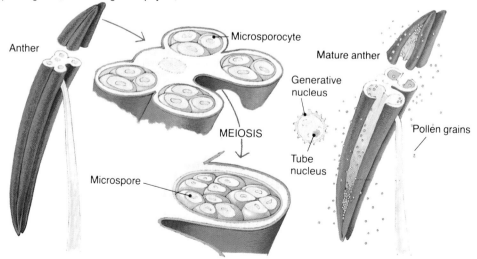

The female gametophyte of a flowering plant forms inside the ovule in the ovary. Refer to Figure 19·11 as you read through the steps that explain how the female gametophyte forms.

Find the ovule in the drawing. Inside an ovule is a cell called a megasporocyte, which is $2n$. The nucleus of this cell divides by meiosis, and four megaspores are formed. Three of these die. The last one divides by mitosis to form two nuclei. Mitosis continues until a female gametophyte with eight nuclei forms. The female gametophyte within the ovule is called the **embryo sac.** Three of the nuclei move to one end of the embryo sac and three move to the other end. The two remaining nuclei move to the center and are called the *polar nuclei.* One of the three nuclei at the lower end of the embryo sac becomes the egg cell. Look closely at Figure 19·11. Notice that there is an opening at the end of the ovule in which the egg is located. This opening in the ovule is called the **micropyle** (MĪ kruh pīl). It is through the micropyle that sperm later enter the embryo sac.

FIGURE 19·11 The ovary is located at the base of the pistil. The development of the female gametophyte occurs in an ovule in the ovary.

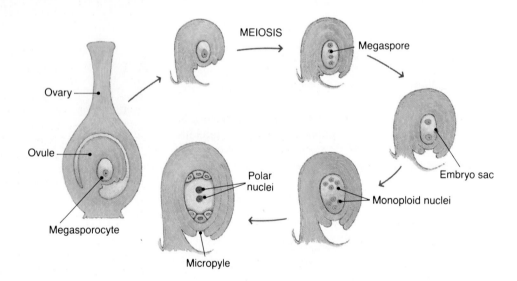

19·7 POLLINATION AND FERTILIZATION

Pollination (pahl uh NAY shuhn) is the transfer of pollen from the anther of a stamen to the stigma of a pistil. This transfer can be done by wind, insects, birds, bats, or by other living things. Most flowers are specifically adapted to particular methods of pollination. These adaptations for pollination account for the great variety in flowers.

Plants such as grass are adapted for pollination by wind. Grass flowers have no sepals or petals covering the anthers and stigmas. As a result, the anthers and stigmas are exposed to the wind. This adaptation increases the chance of pollination. Plants that are pollinated by insects attract the insects by means of showy petals and/or sepals, sweet-smelling nectar, or the pollen itself, which is a rich food for

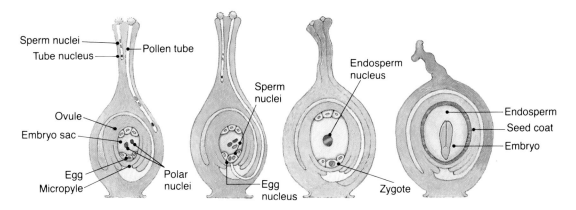

FIGURE 19·12 In flowering plants, pollination is followed by double fertilization. One sperm nucleus joins with an egg to form a zygote. The other sperm nucleus joins with two polar nuclei to form the endosperm nucleus.

bees. The banana plant is adapted to pollination by bats. The banana flowers are open at night, the time when bats are active. Bat-pollinated flowers tend to hang down from the rest of the plant. This adaptation makes contact with the bats easy. Generally, bat-pollinated flowers have sour, musty odors that are unattractive to people but extremely attractive to bats.

If a pollen grain lands on the stigma of a flower of the same species, a certain chain of events occurs. Special proteins located on the wall of the pollen grain become activated by the chemicals on the stigma. Figure 19·12 shows the next few events that occur.

1. The tube nucleus becomes activated, and the pollen tube begins to grow down the style. As the pollen tube grows through the style, the generative nucleus divides to form two sperm nuclei.
2. The pollen tube enters the ovule through the micropyle.
3. Inside the ovule fertilization occurs. During fertilization one sperm nucleus joins with the egg to form a zygote. The other sperm nucleus unites with the two polar nuclei, forming the endosperm nucleus. Because both sperm nuclei unite with nuclei inside the ovule, the flowering plant has a *double fertilization.*

19·8 SEED AND FRUIT FORMATION

After fertilization occurs, the ovule develops into a seed. In Figure 19·12 compare the ovule in which fertilization has occurred with the seed that is developing from it. You should see that certain structures in the ovule become structures in the seed. The endosperm nucleus divides by mitosis and develops into the endosperm (EHN doh sperm). The **endosperm** is rich food-storage tissue, filled with starch grains, fats, and proteins. The endosperm nourishes the developing embryo.

Insights

Hay Fever

There are millions of hay fever sufferers who probably wish that pollen did not exist. *Hay fever* is a misnomer for *seasonal pollen allergy.* This ailment is an allergy, not a fever, and it is rarely caused by hay. The real culprits are plants that produce wind-carried pollen.

When pollen is inhaled by an allergy-prone person, the body produces certain antibodies. The combination of the pollen and these antibodies causes substances called *histamines* to be released. Histamines increase the activity of mucous glands and cause fluids to leak from small blood vessels. The effects of histamine production are the itchy and runny nose, teary eyes, and coughs associated with hay fever.

Most hay fever sufferers are allergic to a specific kind of pollen. Their hay fever "season" depends on the time of year in which the particular plant produces pollen.

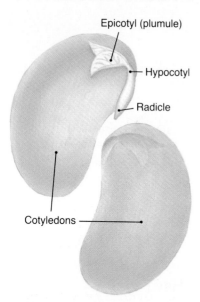

FIGURE 19·13 The parts of a bean seed, are shown here. The largest part of the seed, the cotyledons, contains food for the germinating plant.

Soon after the endosperm nucleus begins dividing, the zygote undergoes cell division and develops into the plant embryo. Around the developing endosperm and embryo, the outer layer of the ovule thickens into a protective, usually hard, **seed coat.**

The embryo within the seed has four main parts. The *radicle* (RAD-uh kuhl) is the beginning root. Above the radicle is a structure called the *hypocotyl* (hī puh KAHT uhl), which connects the tissue between the radicle and cotyledon(s). The hypocotyl later develops into the stem. An embryo has one or two cotyledons, also called seed leaves, which contain food stored for the developing plant. Above the cotyledons is the *epicotyl* (ehp uh KAHT uhl), also called *plumule,* or terminal bud. The epicotyl consists of a stem tip and several tiny leaves.

A mature seed may or may not have an endosperm. If the endosperm is absent, it may already have been used by the developing embryo in the formation of large, thick, fleshy cotyledons. These cotyledons, rather than the endosperm, then act as the food source for the germinating seedling. A corn kernel is a seed made up mainly of endosperm. A pea seed, on the other hand, has no endosperm at all. It consists mainly of fleshy cotyledons. Thus a mature seed consists of an embryo, food for the embryo, and a seed coat.

While the fertilized ovule develops into a seed, the surrounding ovary matures into a fruit. A **fruit** is an enlarged, ripened ovary that contains one or more seeds. Fruits are the seed-enclosing structures that distinguish flowering plants from gymnosperms.

Many fruits, such as the grape and the orange, are formed just from the enlarged ovaries. A fruit such as the apple is formed from the ovary along with other parts of the flower, which fuse together. The apple's ovary wall is the hard, thin membrane around the seeds called the core. The rest of the apple—the fleshy edible part—is formed from the fused bases of sepals, petals, and stamens.

When you hear the word *fruit,* you probably think of a sweet, juicy structure such as a peach or a plum. In common usage, a peach or plum is a fruit. However, the scientific meaning of *fruit* is much broader. It includes anything that is a ripened ovary with its enclosed seeds. Cucumbers, green beans, tomatoes and walnuts all are fruits. All are ripened ovaries that enclose seeds.

REVIEW QUESTIONS

6. List the four main parts of the plant embryo and explain what each embryonic part develops into.
7. What are two major differences between alternation of generations in seed-bearing plants and in primitive plants?
8. Give the function of each of the three parts of a mature seed.
9. Describe the sequence of events occurring after a pollen grain falls onto the stigma of a flower of the same species.
10. Flower structure varies from species to species. Structurally, what would be the simplest type of flower?

Investigation

What Are the Parts of a Seed?

Goals

After completing this activity, you will be able to

- Identify the structures in a monocot and a dicot seed.
- Describe the parts of the embryo from which the parts of a mature plant develop.
- Locate the stored starch in a monocot and a dicot seed.

Materials (for groups of 2)

water-soaked bean seeds and corn kernels
7-day-old bean and corn seedlings
forceps   iodine solution
hand lens   scalpel
dropper   metric ruler
cutting board

Procedure

A. With the forceps carefully remove the outer seed coat from the seed. Look for the cotyledons, two thick fleshy structures that fill most of the space inside the seed coat.

B. Remove one of the cotyledons so that you can see the embryo. The embryo consists of an epicotyl and a hypocotyl. The epicotyl is a pair of small folded leaves enclosing a shoot tip. The hypocotyl is a small shaft of tissue below the epicotyl. The lower tip of the hypocotyl is the radicle, or young root. Suggest what each of these structures will be in an older plant. On a separate sheet of paper, make a drawing of the bean seed showing the embryo attached to one of the cotyledons. Label the following: seed coat, cotyledon, epicotyl, hypocotyl, and radicle.

C. Place a drop of iodine solution on one of the cotyledons. Note any color change. A blue-black color indicates a positive test for starch.

D. Carefully remove a 7-day-old bean seedling from the soil and shake the soil from the roots. Measure the length of the stem and the root. Record your measurements on a separate sheet of paper. Identify the part of the bean seed from which the stem and the root grew. On a separate sheet of paper, draw the bean seedling and label the root, hypocotyl, young leaves, cotyledon, and seed coat.

E. Examine the soaked corn kernel. Identify the silk scar, a pointed structure at the top of the kernel. Using your scalpel, cut the kernel in half from front to back. Cut through the center of the light-colored area.

CAUTION: Be very careful in handling the scalpel. Direct the edge of the blade *away* from your body. Work on a firm surface.

F. Refer to the figure on this page. Locate the following structures:
(1) the embryo in the center of the light-colored area near the front of the kernel; **(2)** the cotyledon, which is the whitish area in which the embryo is embedded; **(3)** the endosperm, which is the remainder of the kernel outside the cotyledon; **(4)** the epicotyl (or plumule), which is the part of the embryo formed of two leaves rolled into a spear; and **(5)** the radicle, or young root.

G. Make a drawing of the corn kernel. Label the embryo, cotyledon, endosperm, epicotyl, and radicle.

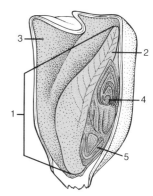

H. Place a drop of iodine solution on the endosperm area. Note any color changes.

I. Remove a 7-day-old corn seedling from the soil. Measure and record the length of the leaves and the root.

Questions and Conclusions

1. How are the dicot seed and the monocot seed alike? In what ways are they different?
2. Based on the results of the iodine tests, explain the function of the cotyledon in the bean seed and the function of the endosperm in the corn kernel.
3. Describe what each of these structures becomes in the bean seedling: epicotyl, hypocotyl, and radicle.
4. You observed the 7-day-old bean seedling. Based on your measurements, which part of the plant embryo grew the most during the first 7 days? Why might one part grow more quickly than others? Answer the same questions for the corn seedling.
5. Describe the changes in the cotyledons in the bean seedling. Describe the changes in the endosperm of the corn.

From Seed to Plant

SECTION OBJECTIVES

- List ways seeds are dispersed.
- Discuss the conditions needed for seeds to germinate.
- Give reasons for the success of flowering plants.

19·9 SEED DISPERSAL

Fruits protect the enclosed seeds and also aid in seed dispersal. **Seed dispersal** is the transport of seeds away from the parent plant. When seeds are dispersed to places far from the parent plant, the developing seedlings face less competition from the parent and other offspring for light, water, and minerals. Good seed dispersal increases the *range* of a species, or the area in which a species lives.

One method of seed dispersal is by animals. Fleshy fruits—such as apples, cherries, and oranges—ripen when their seeds are ready to be dispersed. Attracted by these fruits, birds and mammals eat the fleshy parts and release the indigestible seeds.

As you can see from Figure 19·14, some dry fruits have sharp, curved hooks that tend to cling to clothing and animal fur. Cocklebur, burdock, and thistle are examples of plants that have fruits with such hooks. Animals disperse these fruits as they move from place to place. Eventually, the fruits fall apart, and the seeds fall to the ground.

Wind and water also aid seed dispersal. Maples, ashes, sycamores, and elms have light, winged fruits that are adapted for wind dispersal. The fruit of the coconut tree can float in the ocean for up to two years. Thus, a coconut seed may germinate on an island far from the parent tree. Some fruits are adapted for explosive dispersal. For example, the fruit of the jewelweed, or touch-me-not, springs open when ripe, throwing its seeds far from the parent plant.

In many plants the seed itself is adapted for dispersal. Orchid seeds are so tiny that they float in the air. This adaptation allows the seeds to be dispersed by wind. The milkweed seed, with its silky white plume, is also adapted for wind dispersal. The plume is a modification of the seed coat.

FIGURE 19·14 The seeds of burdock (top) are dispersed by animals. The seeds of maple (bottom) are dispersed by wind.

19·10 SEED GERMINATION

When conditions are suitable, a seed undergoes germination (jer-muh NAY shuhn). **Germination** is the development of an embryo into a seedling. For germination to occur, each of the following must be available in the proper amounts: water, heat, and oxygen. The proper amount of each of these varies from species to species.

Germination can occur only in a seed that is viable. A *viable seed* is one in which the embryo is alive. The length of time a seed can remain viable varies from species to species. The seeds of many plants can remain viable for 10 to 50 years. Lotus seeds have been found to be viable after 1000 years.

Not all viable seeds will germinate even when given proper amounts of water, heat, and oxygen. Many seeds must go through a period of *dormancy,* which is a period of reduced activity during which growth does not occur. Dormancy is an adaptation that prevents germination of a seed until certain conditions occur. For example, apple seeds, which form in late summer or in autumn, must be exposed to several weeks of cold before they germinate. The apple's requirement for a prolonged period of cold to end dormancy ensures that the seeds will germinate after winter. If the seeds were to germinate as soon as they formed in the summer or fall, the seedlings would probably die at the first frost. Since germination is delayed until spring, the seeds and seedlings have a greater chance for survival.

In some species of plants, seeds may remain dormant for several years. Dormancy ensures that not all seeds germinate at the same time. Suppose the conditions in the environment during a given year are not good. If all the seeds were to germinate in that one year, all the new seedlings would be killed. With germination spread over several years, however, there is a greater chance that some of the seeds will germinate when conditions are good.

Germination begins once dormancy is broken and all the needed environmental conditions are met. Figure 19·15 shows the events in the germination of a bean seed. Observe these diagrams as you read about germination.

1. The dry seed absorbs water and swells. The radicle grows, and the seed coat breaks. The embryo is digesting food from the cotyledons, the food source. The cotyledons are the embryo's only source of nutrition until the seedling's leaves begin to photosynthesize.

2. The radicle continues to grow, and primary and secondary roots develop. These roots anchor the plant and absorb water. The hypocotyl and the cotyledons push through the surface of the soil. Both the cotyledons and the young leaves begin to make sugar.

3. The stem lengthens and the young leaves enlarge. As food is used up, the cotyledons begin to shrivel. The seed coat drops off.

4. Germination is completed with the development of the seedling. The seedling has leaves that are actively making sugar. The seedling is self-sufficient and is capable of developing into a mature plant.

FIGURE 19·15 The germinating bean plant uses food stored in the cotyledons until its true leaves develop. Once its leaves develop, it can make food on its own. The cotyledons shrivel and fall off.

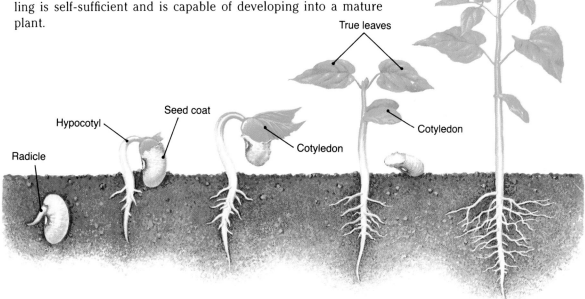

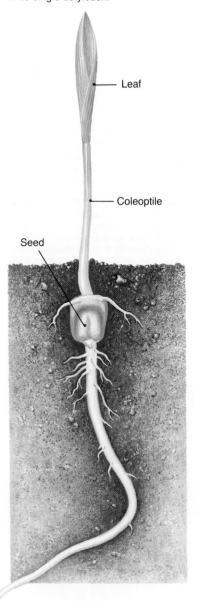

FIGURE 19·16 Corn is a monocot. The food for the germinating seed is stored in its single cotyledon.

The germination of the bean seed is fairly typical for many dicot seeds. Differences occur in the position of the cotyledons in different species. In beans the cotyledons break through the surface of the soil and begin to make sugar. In a plant such as the pea, the cotyledons remain below ground, and only the true leaves emerge above ground and carry on photosynthesis.

Germination in monocots is similar to that in dicots. Figure 19·16 shows a germinating corn seedling, a monocot. The stored food in a corn seed is in the single cotyledon and in the large endosperm. The seed remains below ground as the shoot and the root develop. The first part of the seed to emerge from the soil is the coleoptile (koh lee AHP-tuhl). The *coleoptile* is the sheath that covers the newly developing shoot. Above ground, the shoot becomes photosynthetic.

19·11 SUCCESS OF THE FLOWERING PLANTS

Angiosperms evolved later than all the other major plant groups. However, the angiosperm group contains more species than all the other plant groups combined. What accounts for this dramatic success of the angiosperms? The possible reasons for their success are hypothetical. The hypotheses presented here are those that are generally accepted by botanists who have studied the relative success of angiosperms.

Notice in Figure 19·17 that some of the adaptations of angiosperms are shared with the gymnosperms. For example, the transfer of male gametes to the female reproductive organs does not depend on the availability of water. How does this adaptation contribute to the success of angiosperms and gymnosperms?

Male and female gametophytes are very small in angiosperms and gymnosperms. Apart from their movement during pollination, male gametophytes and gametes do not live outside the sporophyte plant. Female gametophytes and gametes never do. The gametophyte generation thus is protected and is much less vulnerable to unfavorable conditions than it is in the life cycles of lower plants. The embryos of the angiosperms and gymnosperms are contained in seeds. The seed protects the embryo and also contains food.

Many flowering plants have much shorter and faster reproductive cycles than do other plants. For example, seeds of desert plants germinate only during rainy seasons. These plants complete their life cycles within a few weeks. This speed is made possible by a number of factors: (1) very small, quickly formed gametophytes; (2) fast-growing pollen tubes that result in a short time between pollination and fertilization; and (3) quickly formed endosperm, leading to fast growth of the young plant. What is the advantage of a rapid life cycle? The sooner a plant can produce seeds, the lesser the chances that the plant will be destroyed before it can reproduce. The faster and more often a plant species produces seeds, the greater the chances that the species will survive. In addition, there is more chance for the species to spread into suitable environments.

PLANT GROUP	GAMETOPHYTE GENERATION	SPOROPHYTE GENERATION	POLLINATION	FERTILIZATION	SEEDS	SEED DISPERSAL
Mosses	Main plant	Small, growing out of gametophyte	—	Water needed	—	—
Ferns	Small independent plant	Main plant	—	Water needed	—	—
Gymnosperms	Greatly reduced, inside and dependent upon sporophyte	Main plant	By wind	No water needed	Naked	Winged seeds for wind dispersal
Angiosperms	Greatly reduced, inside and dependent upon sporophyte	Main plant	By wind, insects, birds, bats, other living things	No water needed	Enclosed in fruit	By wind, animals, water

FIGURE 19·17 This table compares the growth and development of major plant groups. The seed-bearing plants are the dominant form of plants on earth today.

Pollen transfer by animals is unique to the angiosperms. This method is more efficient than the wind pollination of gymnosperms. Adaptations for specific kinds of pollinators have resulted in a great number of different kinds of flowers.

Angiosperms are the only plants to bear their seeds in fruits. The protective and dispersing functions of fruits give angiosperms a great advantage over all other plants. The adaptations of fruits for dispersal are greatly responsible for the many kinds of environments in which angiosperms are found.

The success of the angiosperms can be seen almost everywhere. Flowering plants dominate most landscapes, including those of city, farm, mountains, seashore, desert, and tropical rainforest. Within each of those environments, the variety of flowering plants is great. As sources of food, lumber, shade, oxygen, chemicals, and beauty, the angiosperms greatly enrich the earth.

REVIEW QUESTIONS

11. List three agents which disperse the seeds of angiosperms and give an example of a seed dispersed by each agent.
12. Describe the sequence of events occurring in germination.
13. Describe two adaptations for survival found in seeds.
14. Give examples of three environmental factors that affect germination.
15. Several seeds are placed in a moist, airtight container and are placed in a dark closet. The seeds begin to germinate and then die. Offer an explanation for the death of the seeds.

Chapter Review

SUMMARY

19·1 Natural vegetative propagation, the naturally occurring form of asexual reproduction in seed-bearing plants, is the production of new plants from a nonreproductive plant part.

19·2 Artificial vegetative propagation includes all the methods used by plant growers for the production of new plants without growing them from seeds.

19·3 Sexual reproduction, the main means of reproduction in seed-bearing plants, includes a sporophyte and a gametophyte generation.

19·4 In conifers, the most successful of the gymnosperms, seed development occurs inside the female cone.

19·5 In angiosperms, the flowering plants, the flower is the reproductive structure.

19·6 In flowering plants the male gametophyte is the pollen grain, which is produced in the anther. The female gametophyte is the developing ovule, which is produced in the ovary.

19·7 Pollination, which is the transfer of pollen from the anther to the stigma, is accomplished by such agents as animals, wind, and water. Fertilization occurs in the ovule, when the sperm nucleus joins with the egg to form a zygote.

19·8 After fertilization the ovule develops into the seed, which contains the embryo, the food source, and the seed coat. The ovary develops into the fruit.

19·9 Fruits as well as seeds are adapted for various methods of seed dispersal.

19·10 Germination, which is the development of a plant embryo into a seedling, occurs only when certain environmental conditions are met.

19·11 The adaptive advantages unique to angiosperms account for their success in the plant world.

BIOLOGICAL TERMS

anther	grafting	runners
budding	layering	scion
calyx	megaspore	seed
cones	micropyle	seed coat
corm	microspore	seed dispersal
corolla	ovary	sepals
cutting	ovules	stamens
embryo sac	petals	stigma
endosperm	pistil	stock
filament	pollen grain	style
flower	pollination	vegetative
fruit	receptacle	propagation
germination	rhizomes	

USING BIOLOGICAL TERMS

1. List all the terms that apply to methods of vegetative propagation. Distinguish between the methods.
2. Which two terms refer to reproductive structures that contain either all male parts or all female parts? Distinguish between the two terms.
3. List all the terms that refer to the female organ of the flower. Define each term.
4. What is endosperm?
5. List all the terms that refer to a flower and its parts. Define each term.
6. What is vegetative propagation?
7. Distinguish between rhizomes and runners.
8. Define the term *micropyle*.
9. Microspore is to a male gametophyte as _____ is to a female gametophyte
10. What are cones?
11. Distinguish between a scion and a stock.
12. Define the term *corm*.
13. Distinguish between calyx and corolla.

UNDERSTANDING CONCEPTS

1. Explain why the plants produced by vegetative propagation are genetically identical to the parent plant. (19·1)
2. Explain how it is possible to have a peach tree with several different varieties of peaches on different branches of the tree. (19·2)
3. In your own words, explain what is meant by *alternation of generations* in seed plants. (19·3)
4. Describe the major events occurring in the life cycle of a pine. (19·4)
5. Compare the female reproductive structure of a cone-bearing gymnosperm with the female reproductive structure of an angiosperm. (19·5)
6. Compare the development of the pollen grain with the development of the ovule. (19·6)
7. Describe two methods of pollination. Give examples of plants that are adapted to these two methods. (19·7)
8. Describe the structures formed as a result of double fertilization in angiosperms. (19·7)
9. List the sequence of events occurring between the fertilization of the egg and the development of a mature seed. (19·8)
10. Give examples of the ways that fruits aid in seed dispersal. (19·9)
11. Describe how animals are involved in the dispersal of seeds. (19·9)
12. How is dormancy an adaptation for survival? (19·10)
13. For each characteristic listed, indicate whether it is found in primitive plants or in more advanced plants: (a) seed, (b) fruit, (c) main plant is an independent gametophyte, (d) small, enclosed male or female gametophyte, (e) embryo with a supply of food, (f) sperm requiring water to swim to the egg. (19·11)

APPLYING CONCEPTS

1. Mature female cones of the jack pine are encased in a resin that melts only in the intense heat of a fire. What advantage might there be in such an adaptation?
2. It has been jokingly said that "if dandelions had opposable thumbs, they would take over the world." What is meant by this statement?
3. If certain seeds require six weeks of prolonged cold before dormancy ends, what prevents them from germinating in the middle of January after a particularly cold November and December?
4. Devise a mental model of a type of seed dispersal. Design a fruit or seed that is specifically adapted to that method of dispersal. Draw a picture of your fruit or seed or construct a model out of materials of your own choosing.

EXTENDING CONCEPTS

1. Visit a local nursery, greenhouse, or plant research laboratory. Write and illustrate a short paper describing the types of vegetative propagation used at the site visited. Identify the specific plants used for each method. Try to include at least two methods not described in this chapter. If chemicals are used, explain how their use is applied to the method of propagation.
2. Using house plants, set up a series of experimental cuttings with at least five different plants. From each type of plant, take two 7.5 to 15 cm-long cuttings from the ends of stems. Cut a stem just below a leaf joint. Put the ends of each cutting in slightly moist soil or sand. Place the pots with the cuttings on a window sill. Every two days gently pull each cutting out of the soil and measure root growth. Make a chart comparing root development among the plants during a three-week period. Compare your results with those of other students.
3. Make a collection of plant seeds from the outdoors, from seed packets, and from foods. For each seed, describe the method of dispersal, indicating whether the seed or its fruit is adapted for dispersal. Be sure your collection is of seeds, not fruits.

READINGS

Gibbons, Bob. *How Flowers Work: A Guide to Plant Biology.* Blandford Press, 1984.

Hartman, Hudson T., and Dale E. Kester. *Plant Propagation.* Prentice-Hall, Inc., 1983.

Hohn, Reinhardt. *Curiosities of the Plant World.* Universe Books, 1980.

Mulcahy, David L. "Rise of the Angiosperms." *Natural History.* September, 1981.

Niklas, Karl J. "Aerodynamics of Wind Pollination." *Scientific American.* July 1987, p. 90.

20. Growth and Behavior

Environmental conditions are important factors in the growth and development of plants. In addition, plants exhibit behavioral responses to certain changes in the environment. For example, the blossoms of a day lily live only from sunrise to sunset. What causes this kind of behavior? In this chapter you will begin to identify the cause of different kinds of plant behavior.

CHAPTER OBJECTIVES

After completing this chapter, you will be able to
- **Identify** the environmental factors that affect plant growth.
- **Describe** the mechanisms that allow plants to respond to the environment.
- **Explain** how a plant develops from an embryo into a mature plant.
- **Identify** the internal and external factors that cause flowering in a plant.
- **Define** and **give examples** of types of plant movement that occur in response to environmental factors.

Factors Affecting Plant Growth

20•1 ENVIRONMENTAL INFLUENCES

Multicellular plants, like multicellular animals, grow and develop. As a seed germinates, new cells are produced. The cells grow larger and then differentiate. Plant organs, such as roots, stems, leaves, and flowers, are produced, and a mature plant results.

Just as a seed needs a proper environment to germinate, a plant needs a proper environment to grow and develop during its life. Water, temperature, air movement, chemicals, light, and the presence of other plants are all factors that affect a plant's growth. These conditions may vary from time to time and place to place. In some environments certain factors may be scarce. In other environments such factors may be abundant. This scarcity or abundance has strong effects on the growth of plants. These factors can affect whether or not a certain plant will grow in a certain environment. These factors also affect the rate at which a plant will grow and how much it will grow. For some plants, the scarcity or abundance of certain factors makes a certain environment unsuitable.

Water is one factor that is vital to plant growth. Photosynthesis cannot occur without water. Water accounts for much of the volume of plant cells. It gives plant cells rigidity. You know that a plant without enough water in its cells will wilt.

Temperature is another factor that affects plant growth. The rate of a plant's metabolic processes usually increases, to a point, as the temperature increases. Thus the growth rate will increase as the temperature rises. As the temperature continues to go up, however, growth slows and finally stops. If the temperature gets too high, vital proteins in the plant are destroyed. A temperature that is too low also can damage plants. Plants can have frost injury if ice crystals form inside their cells. A great amount of water can be lost from a plant if it freezes and then thaws.

SECTION OBJECTIVES
- Discuss ways environmental factors affect plant growth.
- Describe how light affects plant growth and development.
- List the effects of plant hormones on plant growth.

The critical temperature at which metabolic rates decrease and at which damage occurs differs from plant to plant. Most plants grow best within certain temperature ranges. The optimal range varies from species to species. Optimal temperature ranges also vary with time of day for each plant. For example, a tomato plant grows best at a temperature near 25°C during the day and from 17°C to 20°C at night. Thus, tomatoes grow best in temperate to warm climates. They do not grow well in regions where temperatures stay either above 25°C or below 17°C.

In addition to temperature, air movement can affect plant growth. Trees growing in high elevations and exposed to high winds often have leaves and branches only on the sheltered side of the plant. Such trees, like the one shown in Figure 20·1, are said to be flagged. Notice how the growth pattern has been affected by the wind.

Certain chemical factors also affect plant growth. Some chemical factors are needed for growth. For example, carbon dioxide is needed for photosynthesis. Oxygen is needed for respiration. Nitrogen is needed to make proteins. Other chemical factors can be harmful to plant growth. Salt in large amounts, either in the soil or the atmosphere, can prevent growth in plants that are not adapted to environments with high levels of salt. If you live where winters are icy, you may have noticed plants that were damaged by salt put down to help melt ice on roads. Air pollutants from certain industries also can be harmful to plants. Areas around some industrial sites have little or no plant growth.

Some chemicals that prevent plant growth are produced by plants themselves. Bare regions around certain shrubs are caused by chemicals produced by the shrubs. What adaptive advantage could the shrubs have by producing these chemicals? Figure 20·2 shows various plants growing in environments containing different chemicals.

In addition, neighboring plants can affect a plant's growth in a number of other ways. Factors needed for growth, such as space, water, and nutrients, may be used by surrounding plants and thus be unavailable to a plant. The light a plant receives may be limited by surrounding plants. Trees in a forest affect the amount of light available to plants growing beneath them. Plants that need much light cannot grow in the shade of a forest. Some plants, such as early spring violets, are adapted for quick growth, before the leaves of trees overhead come out. By the time the leaves of the trees open and block the light, the early spring wildflowers have already grown and completed their life cycles.

Plants growing in environments with limited amounts of light are adapted for getting more light. For example, many plants that grow in shaded areas have leaves that are arranged so that they do not overlap. Light also affects the development of flowers and other structures in plants. The length of daylight is one of the major factors that determines when a plant will flower.

FIGURE 20·1 The wind can distort the normal growth pattern of plants. This tree now points in the same direction as the prevailing winds on this mountain top.

FIGURE 20·2 The growth of the plants in the picture at left has been stimulated by the use of chemicals. Chemicals released by the grass in the picture at right inhibit the growth of other plants.

20·2 LIGHT AND PHOTOPERIODISM

Plants carry out certain processes at specific times each year. The timing of these processes is crucial. For example, in regions with a short growing season, flowers develop quickly. Thus the plant can complete a reproductive cycle before the cold weather arrives. The timing of many processes, such as flower development, is tied to photoperiod (foh toh PIHR ee uhd). **Photoperiod** is the ratio of light to dark in a 24-hour time span. Certain photoperiods may trigger the development of flowers, the growth of bulbs and tubers, dormancy in trees, and the falling of leaves. The response of an organism to the length of light and darkness is called **photoperiodism.**

Photoperiod changes as the seasons change. Changes in daylength occur at the same time each year. However, other factors, such as temperature and humidity, can vary greatly if measured at the same time year after year. There is an adaptive advantage to photoperiodism. Responses, such as flowering in a particular species, occur roughly at the same time each year. The timing coincides with a favorable period for growth and reproduction.

The specific ratio of daylight to dark that is needed for flowering varies from plant to plant. With regard to photoperiodism, there are three types of plants. **Short-day plants** are those plants that develop flowers after being exposed to a light period that is shorter than a critical number of hours. For example, poinsettia, strawberry, cocklebur, goldenrod, and ragweed are short-day plants. In general, short-day plants bloom in the spring or fall. **Long-day plants** are those plants that flower only if daylength is longer than a critical number of hours. Dill, clover, spinach, beet, lettuce, and petunia are long-day plants. Long-day plants usually bloom in the summer. **Day-neutral**

photo- (light)
periodus (cycle)

FIGURE 20·3 Wild strawberry (left) and goldenrod (middle) are short-day plants. Petunias (right) are long-day plants.

plants are those plants in which flowering does not depend on photoperiod. Cucumber, tomato, string bean, corn, and dandelion are examples of day-neutral plants. Day-neutral plants have a long flowering season.

The names *short-day plant* and *long-day plant* are, in a way, misleading. Flowering depends more on the length and continuity of the dark period than on the light period. For example, if the dark period of a short-day plant is interrupted by light, the plant will not flower. The long dark period appears to be the factor that determines flowering in a short-day plant. A long-day plant will flower even if exposed to a long night, if that night is broken by periods of light. This interruption creates two short dark periods. The short dark period appears to be the factor that determines flowering in a long-day plant.

The number of times a plant needs to be exposed to certain lengths of light and dark in order to flower varies from species to species. Cocklebur, for example, needs only 1 day of its critical photoperiod to flower. But soybean requires more than 25 days of its needed photoperiod before it will flower.

Photoperiod requirements of a plant can limit where that plant will live. Most short-day plants can live only where daylength is less than about 12 hours during the growing season. As a result, most short-day plants cannot grow in far northern or far southern latitudes. In these two regions daylength exceeds 12 hours during the growing season. Long-day plants, which usually need more than 12 hours of daylight to flower, cannot grow near the equator, where daylength never exceeds 12 hours.

Commercial plant growers can make plants bloom at any time of the year by controlling the photoperiod. Nurseries often control the lengths of the light and dark periods to which plants are exposed. In this way they can produce colorful poinsettias and mums at various times of the year.

20·3 HORMONES AND PLANT GROWTH

Plant structures that produce stems, leaves, and flowers are called **buds.** Buds that produce stems and leaves are called vegetative buds; buds that produce flowers are called floral buds. Floral buds develop from vegetative buds. The development of floral buds depends greatly on the relative lengths of daylight and darkness.

How photoperiod controls the development of floral buds is not fully understood. A chemical called phytochrome is found in the leaves of plants. This chemical undergoes changes when exposed to certain amounts of light and dark. These changes in phytochrome cause a flower-promoting substance to be made. This flower-promoting substance then moves to plant buds. There it causes vegetative growth to stop and floral growth to start. There are many questions still to be answered about the flowering response to photoperiod. Two of them are what the chemical nature of the flower-promoting substance is and how it works.

The flower-promoting substance is one of many plant-produced chemicals that affect plant growth. The chemicals that affect plant growth are *hormones* (HAWR mohnz). A hormone is a chemical that is produced by organisms in tiny amounts and that regulates certain functions. Hormones are produced by both animals and plants. Plant hormones are called **phytohormones** (fī toh HAWR mohnz).

Phytohormones usually are produced by a plant in response to something in the environment. For example, photoperiod is the factor that triggers the production of a hormone that stimulates flowering. Phytohormones usually are made in one part of a plant and move to another, where they affect growth. Notice in Figure 20·4 that the flowering hormone is made in the leaves. The hormone moves to the buds, where it starts floral development.

All development in plants is influenced by one or more plant hormones. Some hormones promote growth; others inhibit growth. Five major groups of phytohormones are known. Auxins, gibberellins, and cytokinins are mainly growth promoters. Ethylene and abscisic acid are mainly growth inhibitors. Plant growth and development usually result from the interaction of two or more of these hormones.

FIGURE 20·4 The length of daylight and darkness affects the flowering of many plants. A particular photoperiod results in the production of a phytohormone in the leaves. The hormone moves to the bud where it induces flowering.

FIGURE 20·5 This is the leaf of a streptocarpus plant. This plant can reproduce vegetatively from a cut leaf. The leaf at the top received no auxin. The leaf at the bottom was treated with auxin. Notice the increase in the growth of roots in the leaf that was treated with auxins.

20·4 AUXINS

Auxins were the first phytohormones to be discovered. A Dutch biologist, Fritz Went, discovered the presence of a growth-promoting substance in the protective covering, or coleoptile, of grass seedlings. The substance caused the tips of grass seedlings to bend toward the light. He called the substance auxin. **Auxin** (AWK sihn) is a phytohormone that regulates growth.

The bending of a plant toward light is only one of many effects that auxin has on plant growth. Several other hormones, similar in chemical nature to auxin, also have been found in plants. These hormones, taken as a group, are also called auxins. All auxins have effects on plant growth. Auxins are found in all vascular plants. They are produced in buds, in young tissues, in developing embryos, and in pollen.

Usually auxins cause elongation of plant cells. This lengthening of plant cells results in longer stems, roots, leaves, and flower parts of plants. The amount of auxin needed for cell elongation is very small. Only a few parts per million are needed in stems and a few parts per billion in roots. If auxin is present in much larger amounts, growth may in fact be inhibited.

Auxins also stimulate cell division in some plant tissues. For instance, together with other hormones, auxin triggers cell division in the wood-producing parts of trees each spring. Auxin is produced in the buds of the tree. When the buds enlarge in the spring, auxin moves down the trunk, where it promotes cell division in the vascular cambium.

In many plants, auxins aid the development of adventitious roots. Synthetically-produced auxins are used by commercial plant growers to aid in vegetative propagation. Look at the leaf shown in Figure 20·5. Note how roots have developed at the base of cuttings exposed to synthetic auxin.

Auxins also stimulate the development of fruit. After fertilization occurs, auxins from pollen and from the developing embryo cause the ovary to mature into a fruit. In some plants, auxins from other sources in the plant can trigger fruit development, even if fertilization has not occurred. Applications of auxin are used in the commercial production of seedless fruits, such as seedless grapes, figs, and cucumbers.

Auxins can also act as growth inhibitors. In most plants the bud on the shoot tip stops lateral buds from becoming branches. The inhibiting effect of the terminal bud on the lateral buds is called **apical dominance** (AP uh kuhl DAHM uh nuhns). Apical dominance is a result of the auxin made by the terminal bud. If the terminal bud, and thus its auxin, are removed from a plant, the lateral buds will become branches. Bushy plants can be grown by pinching off the terminal buds.

Auxin applied in large amounts can be deadly to some plants. The herbicides 2,4-D and 2,4,5-T are synthetic auxinlike substances that kill some plants by causing uncontrolled growth. When applied to a broadleaf dicot plant 2,4-D causes abnormal growth and death of the plant. The same auxinlike compound has little effect on narrow-leaf monocot plants, such as grasses. Therefore, 2,4-D can be used by landscapers to produce dandelion-free lawns and by farmers to kill broadleaf weeds in corn and wheat fields.

20·5 EFFECTS OF OTHER PHYTOHORMONES

Although auxins usually are growth promoters, they sometimes act as growth inhibitors. In addition, large amounts of auxin in a plant can cause the production of **ethylene,** a phytohormone that can act as a growth inhibitor. It is now believed that in apical dominance, the lateral buds are inhibited by an increased production of ethylene. This ethylene production is triggered by auxin from the terminal bud.

Ethylene also is involved in the falling of leaves from plants. The falling of a plant part is called *abscission* (ab SIHZH uhn). Leaf abscission is triggered when auxin is no longer made by a leaf. In this case, the *decrease* in auxin stimulates the production of ethylene. The ethylene then promotes the formation of an abscission layer. An **abscission layer** is a specialized layer of cells that forms where the petiole is attached to the stem. The cells in this layer become weaker and, in time, the leaf abscises, or falls off.

abscissio (to tear off)

The stimulation of the ripening of fleshy fruits is another important effect of ethylene. Ethylene is usually produced in a plant just before the fruits ripen. Fruit growers use ethylene to control the ripening of fruits. If unripened fruits are kept in well-ventilated places, where ethylene cannot build up, the fruits can be stored longer.

Another group of growth-promoting plant hormones are the **gibberellins** (jihb uh REHL ihnz). The most obvious effect of gibberellins is that they cause quick elongation of stems. Just before some plants flower, production of gibberellins increases, and the stems quickly grow longer. Look at the plants in Figure 20·6. Gibberellin has been applied to one of them. As you can see, the effects of gibberellin are very dramatic.

FIGURE 20·6 Gibberellins make plants grow taller. Compare the control plant (left), with the plant treated with gibberellins (right).

FIGURE 20·7 Gardeners often remove the terminal buds of a plant to make the plants bushier. Removal of the terminal bud results in the development of lateral buds.

In some plants and seeds, gibberellin production increases in the spring in response to increased length of daylight. It is believed that increased amounts of gibberellin in the spring stimulates dormant winter buds. Applying gibberellin to dormant buds can cause sprouting. In monocot seeds, gibberellins are important in starting the process of seed germination. When gibberellin concentration increases, certain enzymes are produced, and digestion of stored carbohydrates and proteins begins. Energy and building materials are available for growth. As a result, the seed germinates. In the brewing industry, gibberellin is used to speed the germination of barley seeds used for malt.

The **cytokinins** (sī toh KĪ nuhnz) make up another group of plant hormones that aid growth. Auxins and gibberellins usually promote growth through cell enlargement. Cytokinins, however, promote growth through cell division. They cause cells to divide by mitosis. Cytokinins cause buds to begin to develop. It is thought that the balance between auxins, ethylene, and cytokinins in a plant's terminal and side buds accounts for the growth of terminal, not lateral, buds. The auxins in terminal buds prevent the cytokinins in the side buds from aiding growth. When the terminal bud is removed, the cytokinins can then cause cell division in the side buds.

Cytokinins may also play a role in keeping leaves alive and functioning. The death of a leaf may result when a plant stops sending cytokinins to the leaf.

Abscisic acid is a plant hormone that acts mainly to inhibit growth. Abscisic acid causes active buds to become dormant winter buds covered by thick scales. When present in buds and seeds, abscisic acid maintains dormancy by keeping buds from sprouting. It also keeps seeds from germinating. Environmental factors such as long exposure to cold cause a decrease in the concentration of abscisic acid. Such factors, as well as an increased production of gibberellins in the spring, break the dormant period. Then a new cycle of growth begins.

Each of the natural roles of phytohormones described is a response to one or more environmental factors. These factors cause plants to produce various kinds of phytohormones. The phytohormones then trigger different types of growth responses. Thus, responses induced by phytohormones are indirect responses to the environment.

REVIEW QUESTIONS

1. Why would limiting a plant's water supply limit the plant's growth?
2. What type of photoperiod is needed to make a short-day plant flower?
3. What are two effects of auxin on plant growth?
4. Why is abscisic acid considered a growth inhibitor?
5. Explain why fruits kept in a sealed container ripen more quickly than fruits kept in an open bowl.

Growth and Development

20·6 DEVELOPMENT OF VEGETATIVE STRUCTURES

Growth and development of plants are the results of three kinds of cell activity: cell division, cell enlargement, and cell differentiation. Cell division involves mitosis and leads to the formation of new cells. Cell enlargement is the permanent increase in the size of a cell. Cell differentiation is the development of generalized cells into cells that are specialized for a particular plant function, such as xylem cells.

The three kinds of cell activities involved in plant growth and development occur mainly at the tips of the shoot and the root. At the tip of each is the meristematic region, which is composed of meristematic cells. The small, dome-shaped area at the tip of the meristematic region is the **apical meristem** (AP uh kuhl MEHR uh stehm).

Figure 20·8 shows a seed and the mature plant that develops from it. In the seed are two zones—the apical meristems—that give rise to the stem, leaves, flowers, and roots. These two zones are the source of all the cells in the mature plant. The apical meristem that gives rise to the stem, leaves, and flowers is called the **shoot apex.** It is located at the top of the seedling's hypocotyl or in its epicotyl. The **root apex** is the apical meristem that gives rise to the developing root. It is found at the bottom of the hypocotyl or in the root. Cells formed in the apical meristems continue to be active throughout a plant's life.

SECTION OBJECTIVES
- Discuss ways apical meristems contribute to plant growth.
- Describe plant cycles of growth and dormancy.

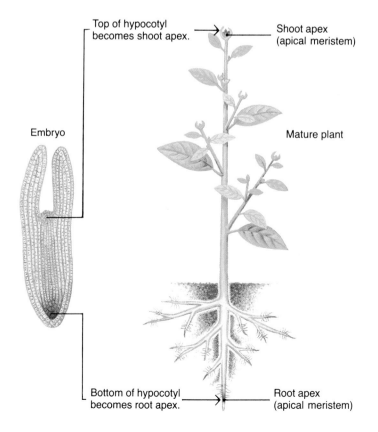

FIGURE 20·8 An embryo (left) and the type of plant that develops from it. In the seed, locate the meristems that give rise to the root and shoot.

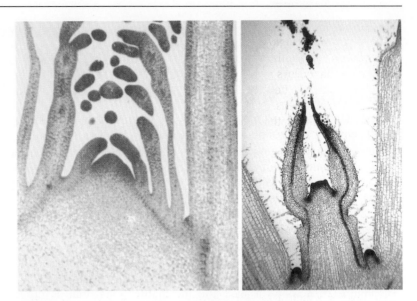

FIGURE 20·9 These are photomicrographs of the apical meristem of two plants. Notice the tiny devloping leaves in these vegetative buds.

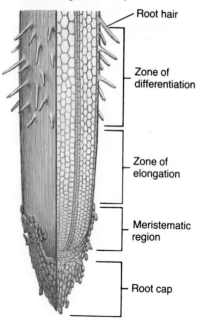

FIGURE 20·10 The apical meristem in this root is located just behind the root cap. The root cap protects the delicate meristematic cells from being damaged as the root grows through the soil.

The process by which meristematic regions produce enlarged, specialized cells is the same in both the shoot tip and the root tip. First the meristematic cells divide. One of the two cells formed during the division remains a meristematic cell and continues to divide. The other cell formed during the division enlarges. It differentiates into a particular type of plant cell, such as a vascular cell. The type of cell it becomes depends mainly on its location in the meristematic region.

Figure 20·9 shows leaves developing from apical meristems. An undeveloped leaf is called a **leaf primordium.** A bud is an apical meristem surrounded by and including a leaf primordium. A plant's highest apical meristem and its leaf primordia together form the terminal bud of the plant. Side buds, formed below the terminal bud, are called lateral buds. Most buds are covered with protective bud scales.

While leaf primordia form around the shoot apex, the cells beneath the apical meristem elongate and then differentiate into specialized cells. Some cells become the xylem and phloem of the vascular system. Others become epidermal cells or structural cells. This differentiation occurs, along with further cell enlargement, a distance below where cell division occurs.

Roots produce no leaves or lateral buds. Their apical meristems are much simpler than those of shoots. Notice in Figure 20·10 that the root apex is found above the root cap. The root cap protects the apical meristem and helps the root push through the soil.

Some cells produced in the meristematic region of a root enlarge. The area where cell enlargement occurs is the **zone of elongation.** Cells elongate because they absorb water, which fills the cells' vacuoles and causes the cells to stretch.

Above the zone of elongation is the **zone of differentiation,** where the newly lengthened cells differentiate. This process results in cells that are specialized in structure and function. The type of specialization that occurs depends on the location of the cell. The cells on the outer surface become epidermal cells. Those in the interior become vascular cells. Those between them form the cells of the cortex. As a result of the activity in these areas, the root grows in length.

In many plants, especially trees, the shoots and roots thicken as a result of secondary growth. Recall that secondary growth results from meristematic tissue called vascular cambium (Sections 17·6 and 17·9).

20·7 DEVELOPMENT OF REPRODUCTIVE STRUCTURES

Before a plant can develop flowers, certain conditions in the environment and in the plant itself must be met. Most plants will flower only when they have reached a certain age or stage of development. The stage at which a plant is ready to flower is called **ripeness to flower.** The conditions needed for ripeness to flower vary from species to species. For example, a rye plant needs to have at least seven leaves before it will flower. Most trees will not flower until they are at least 5 years old.

Factors that affect a plant's ripeness to flower include light and temperature. Some plants must be exposed to several weeks of cold before they can flower. The exposure of plants to a period of cold to induce flowering is called **vernalization** (ver nuh luh ZAY shuhn). The flowering of plants can also be affected by the alternation of day and night temperatures. As you know, many plants need a specific photoperiod in order to flower.

Once a plant is ready to flower, one, some, or all of its shoot apical meristems can become floral buds. The flowering hormone causes an apical meristem to stop producing young leaves and to begin producing immature floral parts. Usually the area of the apical meristem decreases as the parts of the flower emerge. Look at Figure 20·11. It shows the shape of an apical meristem in a floral bud. Notice that the shape of an apical meristem in a floral bud is different from the shape of one in a vegetative bud.

Little is known of how the different floral parts grow from the apical meristem. It is thought that plant hormones have a controlling role. For example, it is known that phytohormones can affect the development of male or female flowers. When auxin or ethylene is applied to a developing cucumber flower bud, the flowers that emerge are all pistillate (have only female reproductive organs). Auxin also is needed for the development of the female gametophyte in orchids.

After a flower bud opens and pollination and fertilization occur, most of the petals and sepals die and fall off. In many plants, nutrients from the dying floral parts move to the ovary and are involved in fruit and seed development.

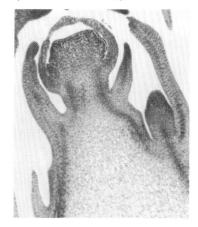

FIGURE 20·11 This is a photomicrograph of a flower bud. Notice that the shape of the apical meristem of this flower bud differs from the shape of the apical meristem of a vegetative bud.

20·12 A developing plant ovary can grow to remarkable size. This gigantic pumpkin couldn't even be lifted by an adult.

Most of what is known about fruit development is known about fleshy fruits. The development of an ovary into a fleshy fruit is probably triggered by auxins in pollen and in the growing embryo. The growth of an ovary into a fleshy fruit is mostly a matter of cell division and enlargement. The shape of the ovary is generally the same as the shape of the mature fruit. Enlargement of ovaries is usually quite rapid, especially in the melons and squashes. Many of these fruits can grow very large in a matter of weeks.

Ripening of fleshy fruits is triggered by the hormone ethylene. Starches in the fruits turn to sugars, flesh softens, and pleasant odors are produced. Ripening makes fruits attractive for seed-dispersal agents, such as animals, which eat the tasty fruits and later excrete the seeds.

20·8 DORMANCY

annus (year)

per- (through)

The development of reproductive structures is usually the last activity of a certain apical meristem. In some plants, flowering and fruit production mark the end of the entire life cycle. Plants that live and reproduce for only one season are called **annuals** (AN yu uhlz). Other plants live longer than one season. These plants go through many reproductive cycles in their lifetime. Plants that live and reproduce for an indefinite number of years are called **perennials** (puh-REHN ee uhlz).

In most regions, there is at least one season each year that is not suitable for plant growth. In temperate climates, winter is too cold for most plants to grow. In some deserts, much of the year is too dry for plants. Annuals, which die before these seasons begin, produce seeds that survive the unfavorable conditions. How do perennials, which live for many years, live through periods of cold or drought?

Most perennials are adapted to greatly reduce their growth before the cold or dry season begins. They remain in a state of reduced growth throughout the unfavorable season. Such plants are said to be in a state of dormancy. **Dormancy** is a condition of decreased growth occurring when environmental conditions are unfavorable. By remaining dormant through an unfavorable season, damage to such delicate tissues as flowers and leaves does not occur. Dormancy ensures that most of a plant's energy and nutrients are used only at times of the year when growth and reproduction will be successful.

Most perennials need specific temperatures or photoperiods to become dormant. These factors stimulate the production of phytohormones that cause buds to greatly reduce growth and, in some cases, cause leaves to die. For example, the shortening photoperiods of autumn cause the increased production of abscisic acid in birch and sycamore trees. The abscisic acid causes the tree buds to become dormant.

In trees, bud dormancy often occurs at the same time as leaf drop. The processes that cause dormancy, however, are often independent of those that cause leaf abscission. In many trees, buds are already dormant in the middle of summer, while leaves are still green.

Leaf death and leaf drop usually occur once temperatures begin to drop. Chlorophyll production stops in the leaf. Other pigments present in the leaf create the bright colors of autumn leaves. The ethylene content of a leaf increases. The ethylene causes enzymes to form at the base of the leaf stalk. Recall that an abscission layer forms at the junction of the leaf and the stem. The enzymes destroy the walls of cells in the abscission layer, and the leaf falls.

Slow growth can still occur even within a dormant plant. If soil temperature and moisture content are high enough, some root growth occurs. At temperatures above 5°C, some photosynthesis can take place, especially in conifers that retain needles. Buds can also increase in size. But growth rates are much slower than normal, and major developmental processes, such as reproduction, do not occur.

Dormancy is broken once proper environmental conditions exist. The change in season is detected by a plant, either through photoperiod, temperature, or a combination of both. Some plants, like the lilac in Figure 20·13, require exposure to a period of cold for dormancy to end. Many seeds and bulbs end dormancy in response to an increase in soil temperature.

Once dormancy ends, growth resumes at a normal rate. Cells in the leaf primordia and flower primordia within the buds enlarge, and the leaves and flowers open.

FIGURE 20·13 The lilac requires a period of exposure to cold temperatures to break dormancy. The part of the plant that remained within the warm room has not begun to grow leaves.

REVIEW QUESTIONS

6. What three activities occur in a meristematic region?
7. Where do meristematic regions form in a mature plant?
8. What determines whether a bud will be vegetative or reproductive?
9. What is ripeness to flower?
10. Imagine that you carve your initials into a tree at a height of 1 m above the ground. After 10 years the tree has grown 1.5 m. How high off the ground will your initials be?

Plant Behavior

SECTION OBJECTIVES

- Describe how tropisms are a plant's reaction to a stimulus.
- Discuss ways turgor and nastic movements affect plants.

20·9 TROPISMS

Most, if not all, of the plant growth processes described thus far occur in response to environmental factors. A factor that causes a response in organisms is called a stimulus. Behavior is the response of an organism to some stimulus or set of stimuli. Though behavior is usually thought of as something unique to animals, plants also show certain kinds of behavior.

Some of the responses induced by stimuli involve plant movement. These responses can appear to be very much like animal behavior. One type of behavioral response in plants is the movement of a plant part toward or away from a stimulus. A movement of a plant part toward or away from a stimulus is called a **tropism** (TROH pihz uhm). The stimulus causes a higher concentration of one or more phytohormones on one side of the plant part. Growth is thus uneven. The more rapid growth on one side results in a curving toward the side of the plant with less rapid growth. Figure 20·14 shows this uneven growth in a stem. Growth toward a stimulus is called a positive tropism. Growth away from a stimulus is called a negative tropism.

Scientists' observation of the tropic response of grass seedlings to light led to the discovery of auxin. The growth response of a plant to light is called **phototropism** (foh toh TROH pihz uhm). Auxin tends to move away from the side of the stem that is closest to the light source. The hormone is thus concentrated on the side of the stem away from the light. This higher auxin level causes the cells to lengthen. The unequal growth on the two sides of the stem causes the stem to grow toward the light. Look at the plant in Figure 20·14. This African violet plant was placed in a sunny window. After a period of time, the stem grew toward the sun. What is the adaptive advantage of phototropism for the plant?

Another tropic response of plants is to gravity. The growth of a plant part toward or away from gravity is called **geotropism** (jee AHT-ruh pihz uhm). Because stems grow away from gravity, they are considered to be negatively geotropic. Roots, on the other hand, grow toward gravity and thus are positively geotropic.

The growth of the pollen tube from a flower's stigma to its ovules is caused by another kind of tropism. The pollen tube grows in response to chemicals made by the pistil. This growth toward chemicals is called **chemotropism** (keh MAHT ruh pihz uhm). Chemotropism directs the growth of the pollen tube straight to the micropyle and thus speeds fertilization.

The growth of vines around objects is the result of a tropic movement. The stems and tendrils of vines curve in response to contact with objects. Tropic response to touch is called **thigmotropism** (thihg MAHT-ruh pihz uhm).

FIGURE 20·14 This African violet is growing on a sunny window sill. As you can see the plant has now started to bend toward the light. This is an example of phototropism.

Investigation

How Do Germinating Seeds Respond to Gravity?

Goals

After completing this activity, you will be able to

- Describe the direction in which the stems and roots of developing seedlings grow.
- Predict the effect of changing the position of the stem and roots on the direction of growth.
- Predict the effect of planting a seed upside down.

Materials (for groups of 2)

1 petri dish
4 water-soaked corn kernels
white filter paper   water
absorbent cotton
glass-marking pencil
masking tape   clay

Procedure

A. Place 4 water-soaked corn kernels on the bottom of a petri dish. Place each kernel with its pointed end toward the center of the dish (see Figure A).

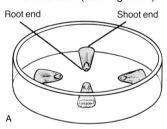

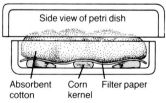

B. Place a piece of filter paper the size of the dish over the seeds. Be careful not to move the seeds.

C. Place cotton on top of the filter paper. Pack the dish so that the seeds will be held firmly in place when the lid is on the dish.

D. Dampen the cotton with tap water. Put the lid on the dish and seal the lid tightly to the bottom with tape.

E. With the glass-marking pencil, draw an arrow on the lid pointing toward the top of the dish (see Figure B). Mark this arrow TOP. Label the lid with your name and the date.

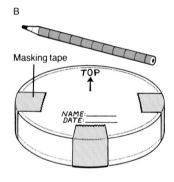

F. Prop the petri dish up (as shown in Figure C) with lumps of clay. Place the dish in a warm place, such as near a sunny window.

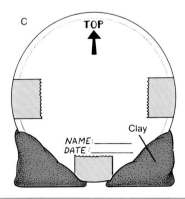

G. Observe the developing roots and shoots each day, if possible. After a week draw the position of the shoots and roots of *each* corn kernel. In your drawing, label the top and bottom of the dish.

H. Turn the dish so that the arrow which was at the top of the dish is now pointing toward the bottom. Observe the dish for several more days. Note any changes in the direction of growth of each root or shoot.

I. Make a second drawing showing the location of the roots and shoots after several days.

J. Follow your teacher's instructions for the proper disposal of all materials.

Questions and Conclusions

1. Describe what part of the seedlings grew up and what part of the seedlings grew down.

2. Find out the meaning of the terms *positive geotropism* and *negative geotropism*. What part of the seedlings showed positive geotropism? What part showed negative geotropism?

3. Describe the changes you observed in the growth of the seedlings after the dish was turned.

4. Suppose a gardener planted all seeds upside down. In what direction would the shoots grow? The roots?

5. What environmental factors other than gravity could a seedling be responding to?

6. Describe an experiment that would allow you to show that the results you observed were not due to a response to light.

GROWTH AND BEHAVIOR

FIGURE 20·15 The snapping shut of a Venus's flytrap is a nastic movement. The trap closes when a fly steps on trigger hairs present on the inside of the trap.

20·10 NASTIC MOVEMENTS

In tropisms, the response of a plant depends on the direction or location of a stimulus. **Nastic movements** are plant responses that do not depend on the direction of the stimulus. Most nastic responses involve different rates of growth on different surfaces of a plant part. When buds open, nastic movements account for the outward growth of the leaves. This outward growth is caused by greater growth on the inside surface of the primordia than on the outside surface.

Nastic movements account for the opening of primrose flowers at dusk and the drooping of touch-me-not leaves at night. Such nastic movements are induced by changes in light. Tulip and crocus flowers open in response to rising temperatures and close in response to falling temperatures.

20·11 TURGOR MOVEMENTS

Many plants have leaves that droop and close at night and reopen in the daylight. In some plants these movements are the results of changes in the amount of water in certain cells.

When cells are full of water, the water pushes against the outer cell wall and causes the cell to be rigid. The pressure of the water against the cell wall is called turgor pressure (Section 18·8). Movements that result from the loss and gain of turgor pressure are called **turgor movements.** These changes in turgor pressure last only a short time and do not cause permanent increases in plant size.

Movements resulting from changes in turgor pressure can be seen in the Venus's-flytrap. The Venus's-flytrap, shown in Figure 20·15, traps insects through nastic movements of its leaves. This plant obtains nitrogen by digesting insects.

Each leaf has two lobes that are hinged in the middle and that can close like a trap. The edges of the lobes are toothed, and the inside surface of each lobe has three hairs. Sweet-smelling liquid produced by the lobes attracts insects to the trap. When an insect touches two of the hairs or touches one of the hairs twice in less than 20 seconds, both lobes of the leaf close. The closing of the leaves is due to changes in turgor pressure triggered by the touch of the insect.

The chemical stimulus of the insect body then triggers further closing of the lobes. Digestion of the insect takes place between the tightly closed lobes. Enlargement of the cells on the inside of the lobes occurs after digestion is completed, and the lobes reopen.

REVIEW QUESTIONS

11. How does auxin cause a plant to grow toward light?
12. What is the difference between a tropism and a nastic movement?
13. What is the adaptive advantage of positive geotropism in roots?
14. Explain how turgor movements are produced.
15. Predict what would happen to seeds that were planted upside down.

Careers in Biology

Plant Physiologist

People who study the way plants function are called plant physiologists. Some plant physiologists specialize in the study of one type of plant such as trees. Others study a particular function.

The work of a plant physiologist may involve extensive field observation and detailed laboratory experiments. For instance, a plant physiologist might identify the factors that cause a plant to produce fruit or flowers. Using this information, a plant physiologist may be able to increase the yield of a plant.

A person must have a bachelor's degree in biology, botany, or physiology to become a plant physiologist. Often a plant physiologist has an advanced degree in an area of botany. For additional information write to the American Society of Plant Physiologists, P.O. Box 1688, Rockville, MD 20850.

Landscape Architect

A landscape architect designs land areas, such as the grounds around schools and offices. Recreational areas, such as parks and the land along highways, are also designed by landscape architects.

Landscape architects analyze information about the physical features of the land. They then prepare plans that show how the land should be developed. The plans include the location of roads, sidewalks, and other structures. The kinds of plants to be used and where they should be planted are also determined by landscape architects.

A person must have a bachelor's degree in landscape architecture and a state license to become a landscape architect. A person with the degree can receive a license by passing the state exam and by working under the direction of a licensed landscape architect. For more information write to the American Society of Landscape Architects, 4401 Connecticut Ave., NW, Washington, DC 20008.

Arborist

An arborist is a person who is responsible for the care and preservation of trees and shrubs. This work includes controlling diseases and insect pests; and selecting, planting, and fertilizing trees and shrubs.

Arborists work anywhere there are trees—in large cities, small communities, rural areas, and forests. They work for government agencies, landscaping firms, and large corporations.

No formal education is necessary, but a high school student interested in becoming an arborist should study biology. Most arborists gain knowledge from on-the-job experience. For additional information write to the National Arborist Association, 1400 Wantagh Avenue, Suite 207, Wantagh, NY 11793.

How to Write a Cover Letter

Often, your first contact with a potential employer is through the mail, by means of a cover letter. In a cover letter you introduce yourself to an employer. Your cover letter should get the employer's attention and convince him or her to seriously consider you.

There are two kinds of cover letters. One kind is used when you are following up on a specific job lead, for example, an ad in a newspaper. This letter should state which position you are applying for and where you learned about the opening. The other kind of cover letter is used when you do not know of a specific job opening.

Your cover letter should be original and interesting, and it should be neatly typed, without errors. You must use correct grammar and spelling. Close the letter with a request for an interview or by mentioning a time when you will call for an appointment.

Chapter Review

SUMMARY

20·1 Environmental factors, such as water, temperature, light, and chemicals, have important effects on plant growth.

20·2 Photoperiodism determines the time of flowering in many plants.

20·3 Photoperiodism and all other growth responses to the environment are controlled by phytohormones.

20·4 Auxin is a phytohormone that usually promotes plant growth through cell enlargement and division.

20·5 Interactions between auxin and other phytohormones, such as gibberellins, ethylene, cytokinin, and abscisic acid, result in both promotion and inhibition of plant growth.

20·6 Cell division, enlargement, and differentiation occur mainly in regions at the tips of shoots and roots.

20·7 Buds can develop into new shoots or flowers, depending on environmental conditions and ripeness to flower of a plant.

20·8 Plants that live and reproduce from year to year undergo a period of dormancy, which is an adaptation for survival during unfavorable seasons or conditions.

20·9 Plant responses to environmental stimuli such as light, gravity, and touch can take the form of tropisms.

20·10 Plant movements that are not dependent on the direction of a stimulus are called nastic movements.

20·11 Turgor movements result from the loss and gain of water from certain plant cells.

BIOLOGICAL TERMS

abscisic acid
abscission layer
annuals
apical dominance
apical meristem
auxin
buds
chemotropism
cytokinins
day-neutral plants
dormancy
ethylene
geotropism
gibberellins
leaf primordium
long-day plants
nastic movements
perennials
photoperiod
photoperiodism
phototropism
phytohormones
ripeness to flower
root apex
shoot apex
short-day plants
thigmotropism
tropism
turgor movements
vernalization
zone of differentiation
zone of elongation

USING BIOLOGICAL TERMS

1. Which of the terms refer to structures in a bud? Define the terms.
2. Which phytohormones promote plant growth?
3. What is the difference between an annual and a perennial?
4. What is the difference between a short-day plant and a long-day plant?
5. What is apical dominance?
6. What are day-neutral plants?
7. What term describes any movement of a plant part toward or away from a stimulus?
8. What is photoperiodism?
9. What term describes a plant response that is not dependent on the direction of the stimulus?
10. What term describes movements in a plant that are due to changes in the water content of the cells?

UNDERSTANDING CONCEPTS

1. What is one way a plant might be adapted for growth on a forest floor, where light is limited? (20·1)
2. What are two environmental sources of growth-inhibiting chemicals? (20·1)

3. Why is too high a temperature harmful to plant growth? (20·1)
4. What is the adaptive advantage of photoperiodism? (20·2)
5. Why would an interruption in the length of day not affect flowering in a long-day plant? (20·2)
6. How are phytohormones involved in flowering? (20·3)
7. How do auxins stimulate plant growth? (20·4)
8. Explain why the lateral buds on a plant usually fail to develop. (20·4)
9. How do large amounts of auxin inhibit growth? (20·4)
10. Why will fruits stored in a well-ventilated place keep for a long time? (20·5)
11. What effect does abscisic acid have on plants? (20·5)
12. What parts of the plant embryo develop into apical meristems? (20·6)
13. Describe the process of growth in a root. (20·6)
14. What environmental factors can determine whether or not a plant flowers? (20·7)
15. Why don't annual plants go through a period of dormancy? (20·8)
16. What is the adaptive advantage of dormancy in perennials? (20·8)
17. Describe the growth changes that cause a positive phototropic response in stems. (20·9)
18. How does chemotropism speed fertilization in flowering plants? (20·10)
19. How are nastic movements different from tropisms? (20·10)
20. Give an example of a turgor movement in a plant. (20·11)

APPLYING CONCEPTS

1. How would commercial agriculture be different if seeds were not geotropic?
2. How could a plant be kept from reproducing if production of its flowering hormone were triggered by temperature only?
3. If mature annuals do not go dormant in winter, how do annuals reappear in the spring and summer? In what form can an annual survive the winter?
4. Under what conditions might it be true that "one rotten apple spoils the bunch"?
5. Invent and name a tropic response that would be advantageous to plant growth.
6. Some ways in which neighboring plants affect a plant's growth have been described. How might neighboring animals affect a plant's growth?

EXTENDING CONCEPTS

1. Make an illustrated report on the alpine environment. What growth-limiting conditions exist there? In what ways are alpine plants adapted for growth under these conditions?
2. Design an experiment that tests the hypothesis that ethylene stimulates fruits to ripen.
3. Do a study of several kinds of insectivorous plants. Where do the plants grow? Why do they need insects? How are they adapted for the trapping and digesting of prey? If movements are involved, are they tropic, nastic, or turgor movements? Illustrate your report.
4. Place several house plants in different locations in one room. Keep daily track of their phototropic responses over a period of two or three weeks. After this period, alter the direction of the light source. How quickly do the plants respond to the change? If one of the plants is turned 90° each day, what effect does that have on its phototropic movement?
5. Using a combination of text and illustrations, describe the entire life history of a tree. Be sure to include the following phases: seed; germination; development of shoot and root apexes; growth of stem, branches, leaves, and roots; production of wood; development of flowers; reproduction; and dormancy. Your description can be of a specific tree or of several trees. Illustrations can include sketches and photographs.

READINGS

Evans, Michael, Randy Moore, and Karl-Heinz Hasenstein. "How Roots Respond to Gravity." *Scientific American,* December 1986, p. 112.

Hendricks, Sterling B. *Phytochrome and Plant Growth,* Carolina Biology Reader, 1980.

Slack, Adrian. *Carnivorous Plants.* The MIT Press, 1980.

Unit IV Skills

PREPARING BIOLOGICAL ABSTRACT

The amount of information written and published about biology in any given year is tremendous. Suppose you wanted to review all the current research on measles outbreaks in the United States. Where and how would you even begin?

One way to start is to review and read biological abstracts. An abstract is a short, self-contained summary of a research paper or a review article. A research paper describes original research and its results. A review article summarizes research and information already published. The abstract highlights only the essential information in an article. A 100-word abstract, for example, can summarize a 50- to 60-page technical research paper. You can clearly see how reviewing an abstract for the information you need can save time.

Abstracts are useful to both authors and researchers. An author who wants to have an article published writes an abstract to summarize and present the purpose and content of his or her article. A researcher uses abstracts to survey quickly the contents of available articles or research papers.

Many people find abstracts difficult to write. Since an abstract must contain only essential information with a minimum number of words, each word must be carefully selected. Many authors write their abstracts only after their papers or articles have been written.

Biological Abstracts is the publication of the biology abstracting service. Abstracts are arranged by topic. Topics are listed alphabetically according to a number of key words in the index. *Biological Abstracts* is updated on a regular basis, usually every two weeks. This publication can be found in some large public libraries and in many college and university libraries.

Below is a summary of guidelines for writing an abstract. Following these guidelines is a sample abstract of Chapter 15, "Microbial Disease," from this text.

- The abstract should be completely self-explanatory and self-contained. The reader of the abstract should not need to refer to the original article.
- Assume that the reader has some previous general knowledge of the topic covered in the original article. Do not include definitions of scientific terms.
- Use grammatically complete and correct sentences.
- Do not use any abbreviations in the title.
- Use abbreviations only for weights and measurements.
- Do not draw any conclusions that are not specifically mentioned in the original article.
- Keep the abstract short—75 to 150 words.
- In a research abstract, include the problem, method, results, and conclusions. Include the design of any new equipment required for any experiments described in the original article.
- In a review abstract, include the topics covered, the central focus of the article, observations, and conclusions.
- At the end of the abstract, select up to 10 key terms that can be used to index the abstract.

DISEASE
MICROBIAL

AUTHORS' NAMES, (Authors' addresses—university/school affiliations), *Biology: The Living World,* Chapter 15: 292–313. 1989. **The germ theory of disease.**—The germ theory of disease states that pathogens such as bacteria, fungi, protists, and viruses cause a variety of microbial diseases. Pathogens act like parasites in host organisms and cause disease by directly damaging host cells or by producing toxins. Diseases are spread when pathogens move from one host to another. Organisms have physical barriers that prevent pathogens from entering their bodies and causing disease. The immune response results in active or passive immunity. Disease is also controlled with the use of antibiotics and other drugs. Diseases can be prevented from spreading through the use of quarantine and vaccination. Key terms: germ theory of disease, microbial disease, pathogens, immune response, active immunity, passive immunity, antibiotics, vaccination

---

Try to write an abstract. Select one of the articles listed in the Readings section at the end of this or another chapter. Write an abstract for the article on a separate sheet of paper. Remember to keep the abstract short and self-explanatory. You also may want to visit a nearby college or university library to examine *Biological Abstracts.*

UNIT V

Invertebrates

A slug is an animal that is classified as an invertebrate because it has no backbone. But this may be the only feature that invertebrates have in common. Groups of invertebrate animals differ greatly from each other in size, form, and way of life. In this unit you will study the major phyla of invertebrates, going from the simplest to the most advanced.

Chapters

21 *Sponges to Worms*
22 *Mollusks and Echinoderms*
23 *The Arthropods*
24 *Comparing Invertebrates*

21. Sponges to Worms

The feather-duster worm and soft coral may look like flowers, but they are actually hungry marine animals. The feather-duster worm filters food particles from the water with its "feathers." The coral captures and kills small animals with special harpoonlike cells. In this chapter, you will learn more about worms, corals, and other simple invertebrates.

CHAPTER OBJECTIVES

After completing this chapter, you will be able to

- **Describe** the different types of symmetry in animals.
- **List** the characteristics of sponges, cnidarians, and worms.
- **Give examples** of sponges, cnidarians, and worms.
- **Compare** the life cycles of sponges, cnidarians and worms.

Simple Invertebrates

21•1 INVERTEBRATE ORGANIZATION

Multicellular animals are called metazoans. Metazoans are adapted to live in many kinds of places and show a variety of forms. Most metazoans have different kinds of tissues in their bodies. The more complex metazoans have distinct organs and organ systems.

Metazoans can be divided into two groups: vertebrates and invertebrates. **Vertebrates** are animals that have a supporting structure called a backbone. Fish, turtles, birds, and humans are all vertebrates. **Invertebrates** are animals that do not have a backbone. Worms, insects, and starfish are invertebrates.

The bodies of most multicellular animals show symmetry. Symmetry is the similar arrangement of parts on opposite sides of a plane or around a central axis. If you divide a symmetrical body in half, you obtain two halves that are mirror-images of one another. Animals with **radial symmetry,** such as the starfish in Figure 21•1, can be cut into mirror-image halves along many lengthwise planes. The body parts of animals with this type of symmetry are arranged in a circle around a central axis. Other animals, such as the insect and flatworm in Figure 21•1, have **bilateral symmetry.** The bodies of these animals can be divided into mirror-image halves along only one lengthwise plane. What type of symmetry do you have?

When studying animals, it is often necessary to refer to different areas of the body. Dorsal refers to the back or upper surface of an animal. Ventral refers to the belly or lower surface. Anterior refers to the head or front end, while posterior refers to the tail or hind end. Locate these different areas on the insect in Figure 21•1.

One way to classify animals is by comparing the number of cell layers that form in an embryo. Animals usually develop from fertilized eggs. In metazoans other than sponges, a fertilized egg forms two or three primary cell layers in the embryo. The **ectoderm** is the outer cell

SECTION OBJECTIVES

- Distinguish between radial and bilateral symmetry and classify animals according to their symmetry.
- Describe the anatomy and physiology of sponges and cnidarians.

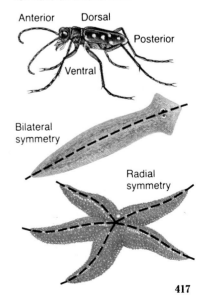

FIGURE 21•1 Compare bilateral symmetry in the flatworm with radial symmetry in the starfish. Where is the anterior end of the flatworm? Does the starfish have an anterior end?

418 SPONGES TO WORMS

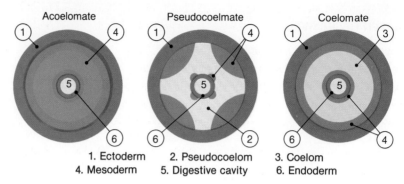

FIGURE 21·2 The body plans of an acoelomate, a pseudocoelomate, and a coelomate.

endo- (inner)
mesos (middle)
derma (skin)

layer, which becomes the epidermis. The **endoderm** is the inner cell layer, which lines the digestive cavity. The **mesoderm** is the middle cell layer, from which the skeletal, muscle, circulatory, and reproductive systems develop. Simple metazoans usually have only two layers. These layers are the ectoderm and the endoderm. Most complex organisms have all three layers.

Animals are also classified as to whether or not a coelom (SEE-luhm) is present. A **coelom** is a body cavity that is surrounded by mesoderm. An animal that has a coelom is called a *coelomate* (SEE-luh mayt). A **pseudocoelom** (soo doh SEE luhm), or false coelom, is a body cavity that is not surrounded entirely by mesoderm. Some organisms have no coelom. Such an organism is called an *acoelomate* (ay-SEE luh mayt). The simplest invertebrates are acoelomates. The more complex invertebrates are coelomates.

21·2 SPONGES

Sponges are simple aquatic animals. They lack many of the features that usually are found in more complex animals. Sponges have neither tissues nor organs. They have no digestive tract and no nervous system. Sponges are able to regenerate. **Regeneration** is the process of regrowing missing or damaged parts.

Sponges belong to phylum *Porifera* (poh RIHF er uh), meaning "pore-bearing." The body of a sponge has a system of canals and pores that water flows through. Unlike most animals, sponges do not move. They are usually found attached to objects under the water. An organism that does not move is said to be *sessile* (SEHS uhl).

A typical sponge has a large opening at the top called an **osculum** (AHS kyuh luhm). Sponges also have pores on the sides. An interior chamber called the *spongocoel* (SPAHN goh seel) is also present.

Sponges feed on microscopic plants and animals called plankton. Water containing plankton flows into the sponge through the pores. It flows out of the sponge through the osculum. Sponges are *filter feeders,* or organisms that get their food by filtering it out of the water around them. This means of food getting allows the sponge to grow and sur-

FIGURE 21·3 Complex sponges, such as this one, have branching chambers.

vive even though it is a sessile organism. Sponges are composed of a variety of cells, as shown in Figure 21·5. Locate each cell type as it is described.

Specialized *pore cells*, which have openings through their middle, are found in some sponges. Water enters the body of a sponge through these cells. Water circulation is maintained by the beating of the flagella on cells called choanocytes (KOH uh nuh sīts). A **choanocyte** is a specialized cell that has a single flagellum surrounded by a collar. The collar serves as a net that captures microscopic food particles which slide down to the base, where the collar joins the cell. The cell then ingests the food particles. Digestion occurs within the food vacuoles of these cells and other cells. Since choanocytes keep water moving through the sponge, undigested foods are swiftly carried away. Each cell obtains oxygen from the water by diffusion. Carbon dioxide and nitrogenous wastes, usually in the form of ammonia, also diffuse from the individual cells into the water. The water containing these wastes exits through the osculum. Stagnant waters may kill sponges. Can you explain why?

A sponge contains many cells that resemble the protozoans called amoebas. These amoebalike cells of a sponge are called **amoebocytes** (uh MEE buh sīts). Amoebocytes are located in the *mesenchyme* (MEHS-ehng kihm), or middle layer, of the sponge. Some amoebocytes digest food. Others are involved in carrying food to other cells.

Other specialized cells in a sponge form reproductive cells—eggs and sperm. Certain cells are flat and serve as the outer covering, or *epithelium* (ehp eh THEE lee uhm), of the sponge.

Support for sponges may be provided by hard structures called **spicules** (SPIHK yoolz) or by a flexible material called **spongin**. Spicules and spongin are secreted by specialized amoebocytes in the mes-

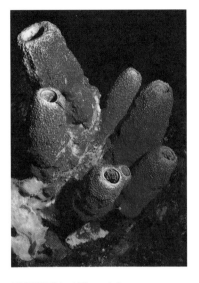

FIGURE 21·4 These tube sponges are growing on a reef. These simple sponges have a vaselike shape and a single chamber. The large hole at the top of each sponge is the osculum.

FIGURE 21·5 The structure of a tube sponge is shown in this figure. Note the different kinds of cells in the sponge's body. Spicules in the mesenchyme support the sponge.

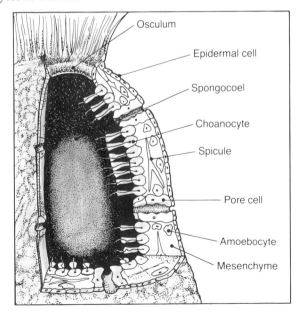

enchyme. Spicules are made up of either calcium carbonate or silicon dioxide. Spongin is made up of branched protein fibers. Dense networks of these fibers provide support in some of the larger sponges. Some sponges contain both spongin and spicules. Some, such as those used as bath sponges, contain only spongin.

Simple sponges have a vaselike shape and a single chamber. In the more complex sponges, many chambers and a branching system of canals are present. Water circulates through the canals and exits through the osculum.

The once large sponge-gathering industry has greatly declined. Less expensive synthethic sponges have been developed. Natural sponges that are sometimes used for bathing are actually the remains of once living sponges.

21·3 REPRODUCTION OF SPONGES

Sponges can reproduce both asexually and sexually. Many sponges reproduce asexually by forming external buds. These buds develop out of the body wall. At maturity the buds may break off and drift to new locations and attach to surfaces. Or the buds may remain attached to the parent sponge and form colonies. Some sponges form internal buds, or *gemmules,* which are clusters of specialized amoebocytes often protected by spicules. When the parent sponge dies, the gemmules that are released drift away. Each gemmule grows into a new sponge.

Some kinds of sponges are hermaphrodites (her MAF ruh dīts). A **hermaphrodite** is an organism that produces both male and female sex cells. Hermaphroditic sponges produce both eggs and sperm. Other sponges produce either eggs or sperm. During sexual reproduction, eggs and sperm form from simple, undifferentiated amoebocytes or from choanocytes. Many sperm are carried out of the sponge with the excurrent water. Some may then enter another sponge with the incurrent water, as shown in Figure 21·6. The sperm move into the mesenchyme, where they fuse with mature eggs. In many sponges the resulting zygote develops into a hollow larva. This larva has flagellated cells at one end and nonflagellated cells at the other end. The larva breaks out of the sponge and swims for a time. Eventually, the flagellated cells push into the hollow center of the larva, forming a spongocoel. The larva then attaches to some object and grows into an adult.

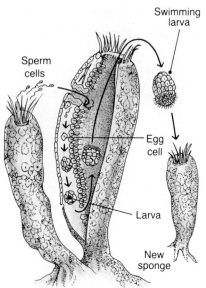

FIGURE 21·6 Sperm cells form one sponge swim in through the pores of another sponge and fertilize its eggs. The fertilized egg develops into a larva which swims out of the parent sponge.

gastro- (stomach)

21·4 CNIDARIANS

Phylum *Cnidaria* (nī DARE ee ah) contains organisms such as hydras, jellyfish, sea anemones (uh NEHM uh nees), sea fans, and sea corals. Members of the phylum Cnidaria are called **cnidarians** (nī-DARE ee ahns). Inside the body of each of these animals there is a cavity called a **gastrovascular cavity.** This cavity has only one opening, the mouth. Food enters through the mouth and is digested in the gastrovascular cavity.

Most cnidarians live in salt water. A few, such as hydras, live in fresh water. Most are sessile. All show radial symmetry—that is, they have body parts arranged around a central axis. Based on differences in body forms and life cycles, members of phylum Cnidaria may be grouped in three classes: class *Hydrozoa,* class *Scyphozoa* (sī fuh-ZOH uh), and class *Anthozoa.*

Some cnidarians have two different body forms during their life cycle. One body form is the polyp (PAHL ihp), shown in Figure 21·7. A **polyp** is a tubelike form with the mouth on the upper surface. The other body form is the **medusa** (pl., *medusae*), a bell-shaped form with the mouth on the lower surface. Both forms have *tentacles,* structures used in food getting, around the mouth. Compare the polyp and medusa in Figure 21·7.

The polyp and the medusa are similar in structure but different in several ways. The polyp is usually attached by a stalk, either branched or unbranched, to a surface such as a rock. By contrast, the medusa is adapted for swimming. In the polyp form, the mouth and tentacles point upward. In the medusa form, the mouth and tentacles point downward.

Unlike sponges, cnidarians have true tissues. The polyps and the medusae have two body layers—an outer **epidermis** and an inner **gastrodermis.** Between these two layers is a gelatinous material called *mesoglea* (mehs uh GLEE uh), in which cells are embedded. Figure 21·8 shows the cells that make up a hydra. Note the two body layers with the mesoglea between.

A cnidarian's tentacles contain special cells called **cnidoblasts** (NĪ duh blasts), each of which has a threadlike stinging structure called a **nematocyst.** The presence of nematocysts is a trait that is unique to cnidarians. When food comes near the tentacles, the nematocysts are discharged. They pierce the prey and paralyze it. The tentacles then pass the food to the mouth.

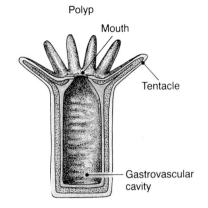
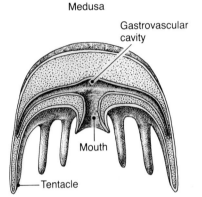

FIGURE 21·7 Compare the medusa form with the polyp form in this figure.

FIGURE 21·8 This is a model of the body wall of the hydra, a cnidarian. Note the two cell layers separated by the mesoglea.

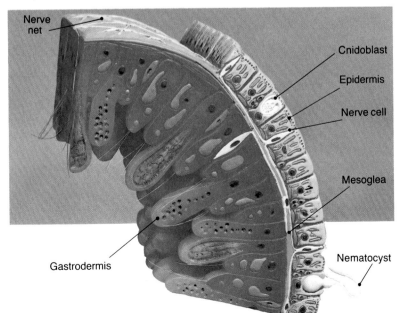

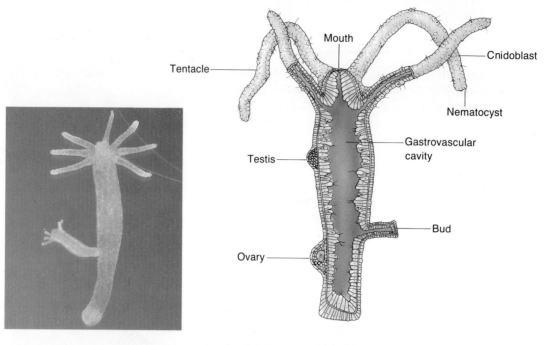

FIGURE 21·9 A hydra is a freshwater cnidarian that lives only as a polyp (left). Its body is composed of two cell layers. The bumps on its tentacles are cnidoblasts, which contain stinging nematocysts. Hydras can reproduce both sexually and asexually. Locate the testis, ovary, and asexual bud in this figure (right).

21·5 HYDRA

A hydra is a cnidarian in the class *Hydrozoa*. Hydras are composed of two layers and have only the polyp form. They are commonly found in ponds. Hydras are only a few millimeters long when contracted but may stretch to a few centimeters when in their food-catching position. They can be seen with the unaided eye and may be found hanging from the undersides of submerged objects, aquatic leaves, and lily pads that float on the surface.

When prey is sensed, a hydra releases its nematocysts. The prey is usually pierced by many nematocysts and is paralyzed by a poison from the nematocysts. The wounds of the prey leak fluids into the surrounding water. These fluids include a protein that is common in animal tissues. This protein stimulates the feeding response of the hydra and causes the mouth to open. The prey is moved toward the mouth by the tentacles. The food is taken in through the mouth into the grastrovascular cavity, where digestion takes place.

Once the food is inside the grastrovascular cavity, cells in the gastrodermis secrete enzymes. These enzymes aid in breaking down the food. Partly digested food may then be engulfed by cells of the gastrodermis. Digestion is then continued in the food vacuoles of these cells. Digested food reaches the body cells by diffusion. Food that cannot be digested leaves the body through the mouth.

Oxygen and carbon dioxide exchange occur in each cell by the process of diffusion. Ammonia, a nitrogenous waste, is also removed by diffusion.

The nervous system of the hydra is made up of a branching *nerve net* of interconnecting nerve cells. Impulses may pass in either direction along any given nerve cell. There is no central brain to coordinate responses. A typical response involves contraction of the hydra's entire body.

The hydra can reproduce asexually or sexually. Asexually, it can form new individuals as *buds* on the surface of the parent. A bud first occurs as a pouch or outgrowth of the parent's body wall. In Figure 21·9 you can see that the gastrovascular cavity of the parent is continuous with that of the bud. In time, tentacles and a mouth form and the new hydra drops off the parent and begins life on its own.

In sexual reproduction, the hydra develops testes and ovaries. A testis is the male reproductive organ; an ovary is the female reproductive organ. In a few species of hydras, both testes and ovaries form on the body of one animal. These species are hermaphroditic. In most other species, testes and ovaries form on separate animals.

Typically, many monoploid sperm mature and are released into the water. Some will reach the ovary of a hydra that is close by. Within the ovary, a single monoploid egg matures. The egg then moves to the outside of the ovary. One sperm joins with the exposed egg. While the fertilized egg is still on the ovary, it undergoes several cell divisions by which it forms the epidermis, the gastrodermis, and the mesoglea. A cyst develops around the embryo. The cyst then breaks away from the parent. While the embryo is inside the cyst, it is protected from unfavorable environmental conditions. When conditions are suitable, a young hydra comes out of the cyst. It is complete with tentacles and looks like a tiny adult hydra.

21·6 OTHER CNIDARIANS

Hydra is an example of a cnidarian having only the polyp stage. Most hydrozoans have a life cycle that includes an attached asexual polyp stage and the free-swimming sexual medusa stage. *Obelia* (oh-BEEL yuh) is a hydrozoan that shows this alternation of body forms. In Figure 21·10, identify the polyp and the medusa forms. In contrast to *Hydra*, which exists as single polyps, *Obelia* is a colonial form. Many polyps are attached by branched stalks. Some of the polyps function only in feeding. Others are specialized for asexual reproduction. A branch that is specialized for reproduction is called a *gonangium* (goh NAN jee uhm). The gonangia produce medusae by budding.

FIGURE 21·10 The life cycle of *Obelia* shows an alternation of body forms. In part of its life cycle, *Obelia* exists as colonial polyps. These polyps are connected to one another by stalks. Some of the polyps are specialized for feeding. Others produce medusae through asexual reproduction.

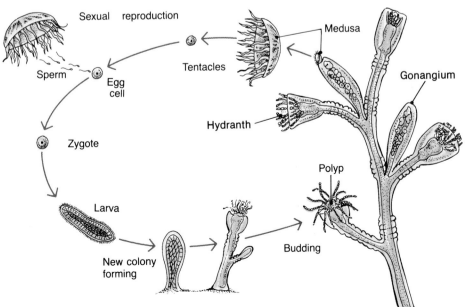

424 SPONGES TO WORMS

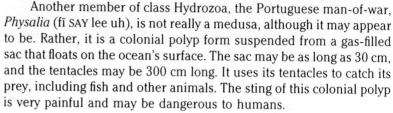

FIGURE 21·11 Like *Obelia*, *Aurelia* shows an alternation of forms in its life cycle. Medusae produce larvae through sexual reproduction. The larvae develop into polyps.

Another member of class Hydrozoa, the Portuguese man-of-war, *Physalia* (fī SAY lee uh), is not really a medusa, although it may appear to be. Rather, it is a colonial polyp form suspended from a gas-filled sac that floats on the ocean's surface. The sac may be as long as 30 cm, and the tentacles may be 300 cm long. It uses its tentacles to catch its prey, including fish and other animals. The sting of this colonial polyp is very painful and may be dangerous to humans.

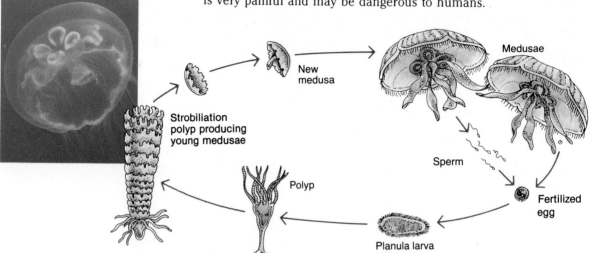

FIGURE 21·12 The sea anemone is a large, complex polyp. It has no medusa stage.

Jellyfish often are seen washed up on beaches. Many are large and have long tentacles. These large jellyfish, called *Aurelia* (aw REEL-yuh), belong to class Scyphozoa. The medusa stage in *Aurelia* is dominant. The polyp stage is very small. The medusae are male or female. Each medusa produces eggs or sperm. The sperm are released into the water and enter a female medusa's gastrovascular cavity. There, fertilization takes place, and the fertilized egg develops into a ciliated larva called a *planula*. The planula swims out of the parent's mouth, drifts about, and eventually becomes attached to some object. It develops into a hydralike organism that undergoes a special type of asexual reproduction, shown in Figure 21·11. Each saucerlike structure eventually separates, inverts, and becomes an adult medusa. Thus the cycle continues.

Sea anemones and corals belong to class Anthozoa. Members of this class have a more complex body structure than members of class Hydrozoa or Scyphozoa. These sea animals have a large polyp stage and no medusa stage. Sea anemones vary in size from 1 centimeter to several centimeters. They eat many kinds of animals, including crabs, clams, and small fish.

Corals are colonial relatives of the sea anemones. They are colonies of polyps. They have a nonliving calcium carbonate exoskeleton secreted by the epidermis of the coral. When the coral dies, the exoskeleton is left behind. Over a long period of time, a great many exoskeletons form a coral reef. You can see a coral reef in Figure 21·13. The reef serves as a habitat for many kinds of living things. The reef-

FIGURE 21·13 Different kinds of coral can be seen in this photograph.

building corals have a symbiotic relationship with single-celled algae. The coral do not eat the algae but depend on the algae to remove carbon dioxide and nitrogen from the water. The algae also play a role in building the calcium carbonate reef.

REVIEW QUESTIONS

1. Describe the body form of a simple sponge.
2. Compare the structure of a polyp with that of a medusa.
3. What is a nematocyst? Explain how it is used to capture prey.
4. Identify the main stages in the life cycle of *Aurelia*.
5. While skin diving, you observe a large purple flowerlike organism eating a small fish. Would you classify this organism as a plant, a sponge, or a cnidarian? Explain your choice.

Flatworms and Roundworms

21·7 FLATWORMS

Phylum *Platyhelminthes* (plat ee hehl MIHN theez) includes the animals commonly called **flatworms**. All flatworms develop from three tissue layers: ectoderm, endoderm, and mesoderm. Another trait of all flatworms is that they are acoelomates.

Flatworms are the simplest animals that have bilateral symmetry. Flatworms have a distinct head end. They show *cephalization* (sehf-uh luh ZAY shuhn); a clustering of nerve cells form a brain.

There are both parasitic and free-living flatworms. All forms have nervous, excretory, and reproductive systems. Some parasitic forms, however, have no distinct digestive system. Phylum Platyhelminthes is divided into three classes. Planarians are in class *Turbellaria*; flukes are in class *Trematoda*; tapeworms are in class *Cestoda*.

SECTION OBJECTIVES

- Discuss the characteristics of the three flatworm classes.
- Identify the major characteristics of roundworms and rotifers.
- Explain why flatworms and roundworms are important to humans.

21·8 PLANARIANS

A planarian is a free-living flatworm. Planarians usually live in water. They are found in ponds, lakes, and the oceans. Some live on land and are found in damp soil.

In Figure 21·14A you can see the gastrovascular cavity of the planarian. Locate the pharynx (FAR ihngks), which is enclosed in a sheath. The **pharynx** is a tubelike muscular structure that joins the mouth with the rest of the digestive tract. When food is taken into the body, the pharynx of the planarian can be extended through the mouth on the ventral surface of the worm. Most planarians are carnivorous; they eat small crustaceans, worms, and insects.

Food is broken down in the gastrovascular cavity with the aid of enzymes secreted by the gland cells that line the cavity. The partially digested food particles are then engulfed by cells of the gastrodermis. Digestion is completed inside these cells. Undigested particles pass out of the body through the pharynx.

The planarian has no special respiratory organs. Exchange of oxygen and carbon dioxide takes place by diffusion through the flattened surface of the body. These flatworms have neither a heart nor blood vessels.

Most of the excretory wastes diffuse from the cells of the planarian into the environment. The excretory system of the planarian, shown in Figure 21·14B, mainly controls the amount of water in the animal. It consists of a series of branching canals called excretory tubules, which connect to flame cells. A *flame cell* removes excess water and possibly other wastes from the body of the planarian. Water is driven into the excretory tubules by cilia located in the flame cells. The tubules open to the outside at pores located along the sides.

The nervous system of the planarian is shown in Figure 21·14A. Note the two main nerve cords that run the length of the body. These nerve cords have shorter nerves connecting them. Near the anterior end of the planarian are two masses of nerve cells, called **ganglia,** that serve as a simple brain. A variety of sensory cells are connected to the nervous system. These cells are sensitive to touch, chemicals, or light and are most abundant toward the anterior end. The light-sensitive cells are organized into a pair of eyespots.

Planarians can reproduce asexually by fission. A worm may split in two behind the pharynx. The anterior and posterior parts regenerate the missing parts. Planarians can also reproduce sexually. They are hermaphrodites, since they contain both ovaries and testes. Even though planarians are hermaphrodites, they cross-fertilize. That is, two planarians mate and exchange sperm. The sperm may be placed in the *oviduct*, the tube that leads to the eggs; or the sperm may be placed into the tissue surrounding the ovary. The sperm move to the eggs, and fertilization takes place. The fertilized eggs pass through the oviduct and leave the worm through the *genital pore*. Find the oviduct and genital pore in Figure 21·14B. In most planarians, the zygote develops directly into a new worm that looks like the adult.

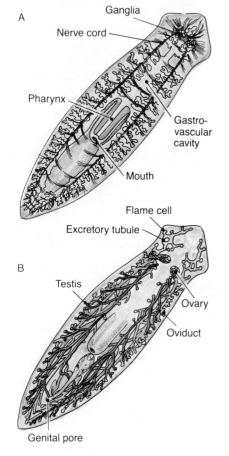

FIGURE 21·14 The digestive system and the nervous system of a planarian (A). The reproductive system and the excretory system of a planarian (B).

Reproduction in the planarian can be stimulated artificially by cutting the worm and allowing the resulting pieces to regenerate. Figure 21·15 shows regeneration in planarians.

21·9 FLUKES

Flukes belong to class Trematoda. Outwardly, flukes resemble the free-living flatworms. However, flukes are parasitic. They cannot live long outside a host's body. The body of the fluke is highly specialized for its parasitic way of life. For example, the adult fluke has structures, called oral and ventral suckers, by which it attaches itself to its host. Its mouth is at the center of the oral sucker. The fluke does not have a highly developed digestive system. It gets its food from its host. Cells and fluids from the host's body are pumped into the intestine by the muscular pharynx found behind the mouth.

Flukes may parasitize many organs of the human body. The life cycle of the liver fluke *Opisthorchis sinensis* (ah puhs THAWR kihs sih-NEHN sihs) is shown in Figure 21·16. As an adult, the liver fluke lives in the liver of a human host. Flukes are hermaphroditic, and sexual reproduction occurs while the fluke is in its human host. The fertilized eggs pass out of the body of the human host in the feces, or solid waste. If the fertilized eggs are released in water, they may be eaten by a secondary host, a snail.

Inside the snail, the fluke develops into a structure called a *sporocyst*. A larval stage develops from the sporocyst. In this stage, the organism has a mouth, intestine, suckers, and a tail. This larva develops into another larval form that looks like a tadpole. This tadpolelike larva leaves the snail and penetrates the skin of another secondary host, a fish. Inside the muscle tissue of the fish the larva develops a cyst, or capsule, around itself. If a human eats the fish without cooking it, the larva will not be killed. The larva leaves its capsule and develops into the adult fluke. The life cycle is then complete. Look at Figure 21·16. Suggest a way to prevent the transfer of the eggs to snails.

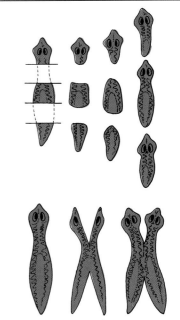

FIGURE 21·15 If a planarian is cut into pieces, the pieces will regenerate their missing parts and become new planarians.

FIGURE 21·16 The life cycle of the liver fluke involves three hosts: snails, fish and humans. If any of these hosts is missing, the fluke cannot complete its life cycle.

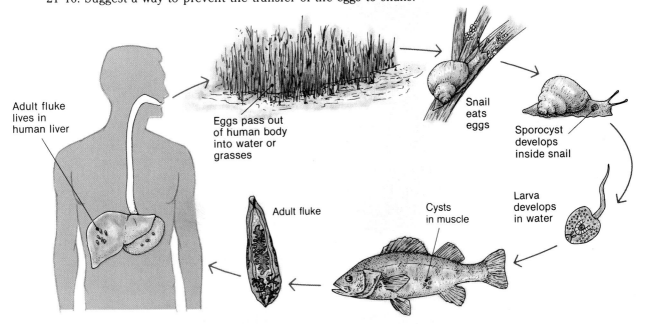

21·10 TAPEWORMS

Tapeworms belong to class Cestoda. Members of this class are specialized for their parasitic way of life. Most adult tapeworms live in the intestine of their vertebrate host. They attach to the intestine by a structure called the scolex. The **scolex** is a knob-shaped head that has hooks and/or suckers. Behind the scolex is a short section called the neck. The rest of the tapeworm's body is made up of body sections called proglottids (proh GLAHT ihds). A **proglottid** is a body section that has male and female reproductive organs. Find the parts of a tapeworm in Figure 21·17.

Tapeworms measuring as much as 18 m have been found. These worms were made up of thousands of proglottids. The tapeworm, lacking a digestive system, obtains its food directly from the intestine of its host. Food is absorbed directly through the body wall of the tapeworm.

Figure 21·17 shows the life cycle of the beef tapeworm. The adult worm lives in the human intestine. A mature proglottid has both male and female reproductive organs. Generally, sperm from one proglottid fertilize the eggs in another proglottid. Sperm can also be transferred from one worm to another. Sperm are stored in a chamber called a *seminal receptacle*. After the eggs are fertilized, they are held in a large saclike structure called a *uterus*. A single mature proglottid may contain as many as 100 000 fertilized eggs. The uterus with its eggs will occupy most of the space in a "ripe" proglottid.

Ripe proglottids break away from the tapeworm and pass out of the host in the feces. Proglottids then rupture, and fertilized eggs may be deposited on the ground and on vegetation. These eggs will develop after they are eaten by a cow.

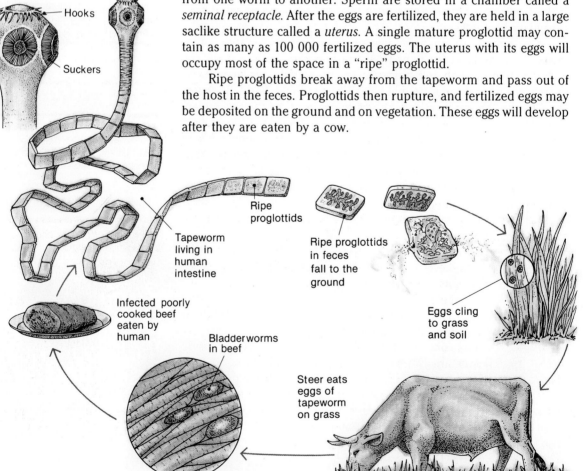

FIGURE 21·17 The life cycle of the beef tapeworm involves two hosts: humans and cows. The closeup of the tapeworm's head shows the hooks and suckers which the tapeworm uses to attach to a human's intestine.

Eggs in the cow's intestine hatch into offspring that bore through the cow's intestinal wall and get into the bloodstream. They then move into muscle tissue, where they form capsules around themselves. At this stage the organism is called a bladder worm. The bladder worm is about 10 mm long. Beef containing bladder worms may be cooked and eaten by humans. If the meat is not fully cooked, bladder worms will not be killed. Once inside a human's intestine, the bladder worm grows into an adult tapeworm. The worm competes with the human for food and may thus cause weight loss and weakness in the human host.

21·11 ROUNDWORMS AND ROTIFERS

Phylum *Aschelminthes* (ask hehl MIHN theez) includes two classes of organisms that appear very different in structure: class *Rotifera* and class *Nematoda*. Class Rotifera includes microscopic aquatic organisms called rotifers (ROHT uh fuhrz). Class Nematoda includes a broad range of worms that as a group are called roundworms. Some roundworms may be free-living, while others are parasitic.

Biologists group roundworms and rotifers together under phylum Aschelminthes because of their body plan. Both rotifers and roundworms derive their tissues from ectoderm, endoderm, and mesoderm. They are the simplest organisms to have a complete digestive system. A complete digestive system has two openings—a mouth at the anterior end and an anus (AY nuhs) at the posterior end. The **anus** is the body opening through which undigested food leaves the body. Both rotifers and roundworms are pseudocoelomates.

Rotifers are multicellular organisms found in freshwater ponds. They are not parasites. They are about the same size as the larger protozoans. Look at Figure 21·18. At the anterior end is a crown of cilia. The rotifer uses the cilia to create water currents that aid in taking in food that is in the water around the rotifer. This crown is often referred to as the wheel organ. Inside the pharynx are jawlike structures that grind up the protozoans and algae which are the rotifer's primary sources of food. On the posterior end is a foot with which a rotifer can cling to a surface.

Roundworms are slender cylindrical worms that are pointed at both ends. Because biologists classify roundworms in the class Nematoda, these organisms are often referred to as *nematodes*. The free-living roundworms are common in soil, mud, and rotting organisms. Roundworms are important in the decomposition of organic matter on land and in mud under lakes and oceans.

Parasitic roundworms include hookworms, pinworms, and filaria worms. Hookworms are common in warm climates in soil that contains feces from infected humans. Hookworms enter the body by burrowing through the soles of the feet. Pinworms live in the intestine. Filaria worms are transmitted by mosquitos. These roundworms live in the lymph system. They often block lymph vessels, causing tremendous swelling and discomfort.

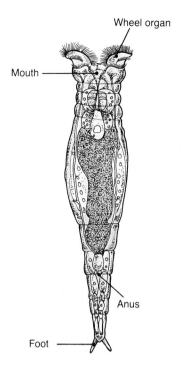

FIGURE 21·18 The multicellular rotifer is approximately the same size as many unicellular protists. The rotifer uses moving cilia to sweep food into its mouth.

FIGURE 21·19 These hookworms are parasitic roundworms.

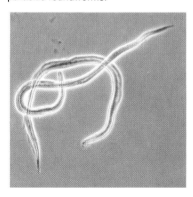

SPONGES TO WORMS 429

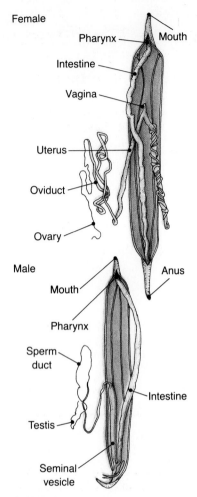

FIGURE 21·20 The structures of a male and a female *Ascaris*.

21·12 ASCARIS

Ascaris lumbricoides (AS kuhr ihs luhm bruh KOID eez) is a roundworm that is a common parasite of humans. A similar species infects pigs and horses. The adult worm lives in the host's intestine. Fertilized eggs pass out of the body in the feces. These eggs are highly resistant to drying and can survive for years in the soil. The eggs may also cling to leaves and stems of plants.

It is fairly easy for the eggs to be transferred to a new host. They may be consumed when unwashed vegetables are eaten or when children who are playing in dirt put their fingers in their mouths. The developing eggs hatch in the intestine, and the worms bore through the intestinal wall into the bloodstream. In the blood they are carried through the heart and into the lungs. There they bore out of the blood vessels and into the air passages. At this stage the worms can cause symptoms similar to those of pneumonia. The worms travel up into the throat and then enter the digestive tract. After reaching the intestine, they mature, completing their life cycle.

As with all nematodes, *Ascaris* is a pseudocoelomate. Look at Figure 21·20. You can see that the digestive system is complete. It is made up of a mouth, pharynx, intestine, and anus. There is no circulatory system. Materials move throughout the animal in the fluid-filled pseudocoelom. There is no respiratory system. The nervous system is made up of dorsal and ventral nerve cords, lateral nerves, and a nerve ring around the pharynx. The muscles in *Ascaris* are longitudinal. They cause the worm to move in a whiplike fashion.

In *Ascaris* the sexes are separate. The female *Ascaris* may be as long as 33 cm and may produce 200 000 eggs per day. The female reproductive system is made up of a pair of ovaries, oviducts, and uteri, and a vagina. The *vagina* is the structure leading from the uteri to the outside of the body. The male *Ascaris* is somewhat smaller than the female. The male has tiny spines at its posterior end that are used to grasp the female during the transfer of sperm. The male reproductive system is made up of a single testis, a sperm duct, and a seminal vesicle. The sperm duct is the tube through which the sperm travel from the testis. Sperm are stored in the seminal vesicle. Fertilization is internal in *Ascaris*.

REVIEW QUESTIONS

6. Describe how a planarian obtains nutrition.
7. Identify the hosts in the life cycle of a fluke.
8. How do tapeworms get nourishment?
9. Explain why rotifers and roundworms are both classified under phylum Aschelminthes.
10. You have just taken a job as a meat inspector in a large meat-packing company. As your first assignment, you must inspect the beef for the presence of tapeworms. What would you look for?

Segmented Worms

21·13 ANNELIDS

Segmented worms, also known as **annelids** (AN uh lihds), are worms in which the body is made up of sections called segments. Some of the largest annelids have hundreds of segments. The youngest segments are at the posterior end. The older segments, containing the brain, hearts, and reproductive organs, are at the anterior end. Annelids can be found in soil and in freshwater streams and ponds. They are also found in the oceans. Segmented worms are classified in phylum *Annelida*. The phylum includes three classes of worms: class *Oligochaeta* (earthworms, freshwater worms); class *Polychaeta* (sandworms, tubeworms); and class *Hirudinea* (leeches).

Annelids have three body layers derived from the ectoderm, endoderm, and mesoderm. Most have a true coelom. The coelom is fluid-filled and is surrounded by a membrane. The coelom aids in support and transport.

The true coelom is one of the most important evolutionary advances in annelids. Another advance is the "tube-within-a-tube" body plan. The outer tube is the muscle and skin of the body wall. The inner tube is the digestive tract.

Organ systems in annelids are well-developed. The digestive tract is complete and thus has two openings—a mouth and an anus. The circulatory system is closed. This means that blood is always contained within vessels. Reproduction in annelids is usually sexual and involves two parents. Although some species of annelids have separate sexes, other species are hermaphroditic.

21·14 THE EARTHWORM

The earthworm, *Lumbricus terrestris* (LUHM bruh kuhs teh REHS-trihs), is a representative annelid. This familiar worm lives in rich, moist soil. The earthworm is found throughout the world. An earthworm's body is divided into numerous segments. Each segment has four pairs of setae (SEE tee). **Setae** are bristles that aid the earthworm in moving.

SECTION OBJECTIVES

- List the three classes of segmented worms and give examples of each class.
- Discuss the physiology and anatomy of the earthworm.

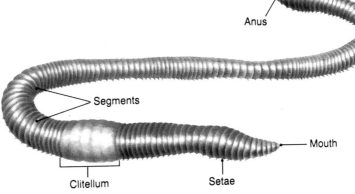

FIGURE 21·21 The major external structures of an earthworm are shown here. You may have seen many earthworms after a heavy rain. The earthworms would drown in their waterfilled burrows if they did not come up to the surface for air.

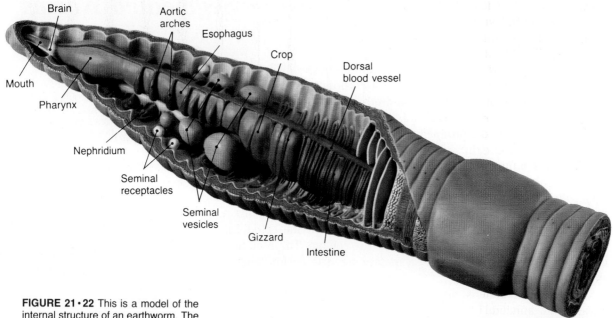

FIGURE 21·22 This is a model of the internal structure of an earthworm. The ovaries cannot be seen in this illustration. They are located in segment 13, just beneath the seminal vesicles.

The anterior, or head, end of the earthworm is closer to the clitellum. The **clitellum** (klih TEHL uhm) is the swollen band around segments 31 through 37 that functions in reproduction. The mouth is found on the ventral side of the first segment. Covering the mouth is a fleshy overhang, the *prostomium*. The prostomium is part of the first segment. The skin of the earthworm secretes a thin protective layer called a cuticle. In addition, mucous glands of the skin secrete a thick coating of mucus, which keeps the surface moist so that gas exchange can occur.

The digestive system of the earthworm has specialized regions. Find each organ in Figure 21·22. The mouth opens into a muscular pharynx that connects the mouth with the rest of the digestive tract. Food is taken in by contraction of the pharynx muscles. Secretions made by glands in the pharynx wet the food. Contractions of the muscles push food into the *esophagus* (ee SAHF uh guhs), a long, mostly straight tube. The contraction and relaxation of the muscles in the walls of the esophagus and other digestive organs is called *peristalsis* (pehr uh STAHL sihs).

In earthworms the posterior end of the esophagus expands into a thin-walled **crop,** which temporarily stores food. A tough, muscular **gizzard,** which is posterior to the crop, is used to grind food. The rest of the digestive tract is the intestine, which secretes enzymes that aid in completing digestion and also absorbs digested food. Undigested food leaves the digestive tract through the anus.

The circulatory system of the earthworm is a closed system. Look at Figure 21·22. The system is made up of a dorsal blood vessel, which pumps blood toward the anterior end, and a ventral blood vessel, which delivers blood toward the posterior end. Branching between the dorsal and ventral vessels are lateral vessels. These vessels carry blood to and away from capillaries in the skin and organs. Capillaries are thin-walled blood vessels. Nutrients and oxygen pass from the blood through the walls of the capillaries to the cells. Wastes and carbon dioxide move in the opposite direction. Five pairs of lateral vessels in the anterior end of the earthworm are expanded into muscular tubes called **aortic arches.** These aortic arches, which help to pump the blood, are often called hearts.

Included in the earthworm's circulatory system is a network of capillaries near the surface of the body. These capillaries allow gas exchange with the environment. There is no respiratory system.

Each segment of the earthworm, except the first three segments and the last segment, contains a pair of nephridia (nih FRIHD ee uh). A **nephridium** is a structure that filters excretory wastes, ammonia, urea, salts, and excess water from the fluid in the coelom. The nephridial tubule is long and coiled. It opens to the exterior through a pore on the ventral surface of the earthworm.

The nervous system of the earthworm has a brain and a main nerve cord with lateral branch nerves. The main nerve cord runs along the ventral surface of the worm. Lateral branches circle the digestive tract. Several pairs of these nerves begin in each segment from bulb-like ganglia in the main nerve cord. The brain consists of two large ganglia that are anterior to the pharynx.

The anterior segments of the worm have many sensory nerves that are connected to the brain. Earthworms show a negative response to light. That is, they move away from bright light. The dorsal and lateral body walls contain light-sensitive nerve endings, or photoreceptors. These photoreceptors detect differences in light intensity.

Earthworms are hermaphroditic. However, they most often reproduce by cross-fertilization. Sperm are produced by paired testes. The sperm mature in large sacs called seminal vesicles. The sperm are later released into the coelom. From there they pass through a funnel into the sperm ducts and leave the worm. Eggs are produced in the paired ovaries but mature in the coelom. Eggs leave the worm through oviducts. The sperm ducts and oviducts open on the ventral surface through separate male and female pores. Mating requires contact.

During mating, the sperm pass from the male pore of each worm into the seminal receptacle of the other worm, where they are stored. A few days later the clitellum secretes a cocoon that contains the protein albumen. The cocoon slips forward toward the anterior end. As it passes the female pore and the seminal receptacle pore, eggs and sperm enter the cocoon. Fertilization takes place within the cocoon.

Annelids are ecologically important. Earthworms consume humus and small organisms in the soil. They also improve drainage and air circulation in the soil by burrowing into, loosening, and mixing the soil. They provide food for a variety of animals.

Discovery

New Phylum Identified

Recently, Dr. Reinhardt Kristensen, of the University of Copenhagen in Denmark, made a rare biological discovery. He found a worm that did not belong to any of the known phyla of animals. A new phylum was created as a result of his discovery.

The name of the new phylum is Loricifera. The newly discovered organism is named *Nanaloricus mysticus*. This worm is one of many small organisms that live in sand below the ocean surface.

Dr. Kristensen first found an organism in phylum Loricifera in 1975. Unfortunately, the single specimen that he found was destroyed while it was being prepared for microscopic study.

Another researcher, Dr. Robert Higgins, had predicted the existence of the new phylum in 1961. Dr. Kristensen and Dr. Higgins began to work together in August 1982. They found several adult forms of the worm off the coast of Florida.

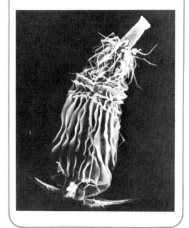

Investigation

How Are Earthworms Adapted for Living in the Soil?

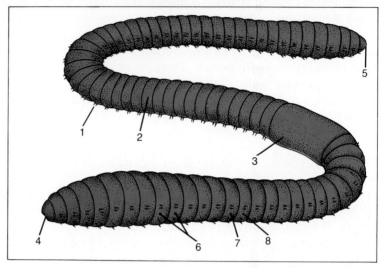

Goals

After completing this activity, you will be able to

- Describe the external features of the earthworm.
- Identify those features that adapt the earthworm to living in soil.
- Describe the responses of earthworms to some external stimuli.

Materials (for groups of 2)

living earthworms in a covered container of soft soil
hand lens
paper towels   vinegar
beaker of water   cotton swab
dropper   flashlight

Procedure

A. Open the container of earthworms suddenly and shine a flashlight on the worms. Note their response. Record your observations on a separate sheet of paper.

B. Remove a worm from the container and place it on a paper towel. Gently touch the earthworm on the anterior, posterior, dorsal, and ventral surfaces of its body. On a separate sheet of paper, record the earthworm's responses.

C. Dip a cotton swab into vinegar. Hold the swab close to the anterior end of the worm. Record the worm's responses. Hold the swab close to the posterior end and record the worm's responses.

D. Hold the earthworm in your hand and rub your finger along its body. Note whether the body is wet or dry. Observe the body shape, especially the ends of the earthworm. On the ventral surface, notice the pairs of bristles, or setae **(1)**.

E. Estimate the total number of segments **(2)** in your worm. Locate the clitellum **(3)**, a swollen band about one third of the way down the body from the anterior end.

NOTE: Do not allow the earthworm to dry out during your observations. If needed, moisten the worm with water. From time to time, return the worm to its container.

F. Use a hand lens to locate the following openings in the earthworm: mouth **(4)**, anus **(5)**, seminal receptacle pores **(6)** on the ninth and tenth segments, oviduct pores **(7)** on the fourteenth segment, and sperm duct pores **(8)** on the fifteenth segment.

G. On a separate sheet of paper, draw the earthworm and label the following parts: mouth, clitellum, anus, oviduct pore, seminal receptacle pore, and sperm duct pore.

H. Follow your teacher's instructions for the proper disposal of all materials.

Questions and Conclusions

1. What was the earthworm's response to bright light? How can this response be an adaptation to life in the soil?
2. Describe the earthworm's reaction to being probed and to coming close to vinegar. How might these responses adapt the worm to life in the soil?
3. How is the anterior end of the earthworm adapted for movement into and through soil?
4. Tell how each of the following features is an adaptation for life in the soil: **(a)** a soft body; **(b)** a long, narrow body.
5. How are the setae arranged on the body? How are setae useful to the earthworm?
6. Why is it unlikely the earthworm would be found in hard soil?
7. What problems would the earthworm have in surviving on the surface of the soil?

21·15 OTHER SEGMENTED WORMS

Most species of annelids belong to class Polychaeta. Polychaetes have a pair of appendages called *parapodia* on each segment. Polychaetes (PAHL ee keets) are abundant in most marine habitats. Many crawl over and hide under rocks or other living things. Some live in burrows lined with mucus, while others are attached to a surface and live as filter feeders. Some polychaetes have a pharynx that can be pushed out through the mouth to catch prey.

Leeches belong to another class of annelid worms, class Hirudinea. They have an anterior and a posterior sucker. Most live in fresh water, but a few live in the sea or on land. Leeches have no heads or parapodia.

Some leeches eat different kinds of small invertebrates. Many species of leeches live on blood. They attach to a vertebrate host and secrete a substance into the wound that prevents the blood from clotting. Blood-sucking leeches do not feed very often. When they do feed, they take in great quantities of blood. Some consume several times their own weight in blood. The water in the blood is quickly extracted and excreted by their nephridia. The blood cells that remain are digested over a long period of time. Some leeches need several months to digest a meal.

FIGURE 21·23 Annelids live in different places and obtain their food in various ways. Polychaetes (left) live in the ocean. Some of them grab their prey with jaws. Others filter their food from the water. Leeches (right) are usually found in fresh water. They are parasites which attach themselves to their host with suckers. They puncture their host's skin and drink the blood that flows from the cut.

REVIEW QUESTIONS

11. Explain why annelids are said to have a "tube-within-a-tube" body plan.
12. State two ways in which segmented worms are more advanced than flatworms or roundworms.
13. Describe the major features of the circulatory system of the earthworm.
14. How do polychaetes differ from oligochaetes?
15. A friend brings you a specimen of an organism and asks you to identify it. You observe that it lacks parapodia and has anterior and posterior suckers. It has no distinct head. How would you classify this organism? Do you think it is a parasite? Why or why not?

Chapter Review

SUMMARY

21·1 Animals vary in their degree of cellular specialization and in body form.

21·2 The sponges show some cellular specialization, such as amoebocytes and choanocytes, but have no organs or organ systems.

21·3 Sponges carry on both asexual and sexual reproduction. Sponges produce buds, or gemmules, in asexual reproduction and gametes in sexual reproduction.

21·4 The cnidarians are metazoans having a body composed of two layers of specialized tissues as well as specialized cells, such as cnidoblasts.

21·5 A hydra is a cnidarian that lives only as a polyp form.

21·6 The hydrozoans—the jellyfish, sea anemones, and corals—carry on life cycles that involve an asexual polyp stage and a sexual medusa stage.

21·7 Platyhelminthes are bilaterally symmetrical flatworms lacking a coelom but having specialized organs.

21·8 Although the planarians are like cnidarians in having a single opening for the digestive sac, they differ in having specialized organs for ingestion, reproduction, and water regulation.

21·9 Parasitic flukes have life cycles involving larval forms and several intermediate hosts, sometimes including humans.

21·10 Parasitic tapeworms are well adapted to live in the intestines of their hosts; their body design consists of a scolex for attachment, a neck, and proglottids that contain reproductive organs.

21·11 The Aschelminthes—rotifers and nematodes—have a tube-within-a-tube body plan and a pseudocoelom, but otherwise they appear quite different from each other.

21·12 Parasitic roundworms have complicated life cycles, including living in a host's circulatory system when immature and then reaching maturity in the host's intestine.

21·13 Segmented worms are bilaterally symmetrical and differ from the roundworms in having a true coelom and a closed circulatory system.

21·14 The earthworm, a representative annelid, has some specialized organs that are repeated in each segment as well as some organ systems that span nearly all the segments.

21·15 The most common annelids, the polychaetes, are abundant in marine habitats. Most leeches are adapted to live on blood.

BIOLOGICAL TERMS

amoebocytes
annelids
anus
aortic arches
bilateral
   symmetry
choanocyte
clitellum
cnidarians
cnidoblast
coelom
crop
ectoderm
endoderm
epidermis
flatworms
ganglia
gastrodermis
gastrovascular
   cavity
gizzard
hermaphrodite
invertebrates
medusa
mesoderm
nematocyst
nephridium
osculum
pharynx
polyp
proglottid
pseudocoelom
radial
   symmetry
regeneration
rotifers
roundworms
scolex
setae
spicules
sponge
spongin
vertebrates

USING BIOLOGICAL TERMS

1. Which terms are used when classifying multicellular organisms by the presence or absence of a backbone?
2. Distinguish between the terms *bilateral symmetry* and *radial symmetry*.
3. What is the difference between spongin and spicules?
4. What is the relationship between cnidoblasts and nematocysts?
5. Which terms refer to structures that are part of the digestive system of the earthworm? Define the terms.
6. Distinguish between the terms *medusa* and *polyp*.

7. What are amoebocytes and choanocytes?
8. Which term refers to any organism that contains both male and female reproductive organs?
9. Name the three primary tissue layers, from the outermost to the innermost layer.
10. What is a nephridium?

UNDERSTANDING CONCEPTS

1. Sketch a flatworm. Label the anterior and posterior ends; label the dorsal and ventral surfaces. (21·1)
2. Give examples of animals having bilateral symmetry and radial symmetry. (21·1)
3. Describe why sponges are known as filter feeders. (21·2)
4. Describe the specialized cells that occur in sponges. (21·2)
5. Outline the stages in the sexual reproduction of the sponge. (21·3)
6. Compare the polyp and medusa with regard to structure and orientation. (21·4)
7. List the characteristics of cnidarians. (21·4)
8. Compare sexual and asexual reproduction in the hydra. (21·5)
9. Explain what is meant by alternation of body forms in *Aurelia*. (21·6)
10. Identify four characteristics common to flatworms. (21·7)
11. Name and describe the main structures in the nervous system of the planarian. (21·8)
12. Outline the stages in the life cycle of the human liver fluke. (21·9)
13. Name and describe the three main parts in the body of the tapeworm. (21·10)
14. Give reasons why Rotifera and Nematoda are different classes under phylum Aschelminthes. (21·11)
15. Explain how *Ascaris lumbricoides* can be transferred from one host to another. (21·12)
16. How do annelids differ from other worms? (21·13)
17. List the features of the earthworm that are typical of segmented worms. (21·14)
18. Give reasons why the circulatory system of the earthworm is called a closed system. (21·14)
19. Distinguish between the polychaete and oligochaete worms. (21·15)

APPLYING CONCEPTS

1. A new species is discovered that is bilaterally symmetrical, shows eyespots, a mouth, and is found to have no coelom. In what phylum would you place this species and why?
2. You receive a live hydra from your teacher to observe and maintain. Several weeks after getting the hydra, you observe smaller hydras living in the same bowl. Since you are sure that no other hydra had been introduced into the bowl, what other explanations might you provide to explain this occurrence?
3. Your friend has just returned from a scuba-diving trip in the ocean and has brought back several large sponges that look and feel just like the expensive kinds sold in stores. However, upon examining them, you both notice that they have a foul odor, unlike the ones bought in stores. What would you tell your friend is the source of the odor? Why is it lacking in the store-bought sponges?

EXTENDING CONCEPTS

1. Animals can change the surface of the earth. Research the formation of the Great Barrier Reef, located along the eastern coast of Australia. Prepare a report for your class, including a map of the area.
2. On poster paper, illustrate and describe the life cycle of the dog heartworm.
3. Prepare a presentation on "The Earthworm as a Food Source for Humans." Research where earthworms are used as food, how they are prepared, and what their economic value is.

READINGS

Brownlee, S. "Jellyfish Aren't Out to Get Us." *Discover,* August 1987, p. 42.

Fichter, G.S. "Animals That Build Islands." *International Wildlife,* September/October 1981, p. 50.

Johnson, S. "Colorful Flatworms." *Sea Frontiers,* January/February 1983, p. 2.

Williams, T. "Leeches in My Breeches." *Audubon,* March 1982, p. 62.

22. Mollusks and Echinoderms

This colorful and unusual mollusk is a nudibranch, or sea slug. Although it lacks the shell that characterizes most other mollusks, it is not defenseless. Each of the finger-like projections on the nudibranch's back is loaded with stinging nematocysts. If a predator bothers this nudibranch, it receives a painful sting. In this chapter, you will discover more about the mollusks. You will also learn about the spiny-skinned animals known as echinoderms.

CHAPTER OBJECTIVES

After completing this chapter, you will be able to
- **List** the characteristics of the three major classes of mollusks.
- **Identify** the distinguishing characteristics of echinoderms.
- **Compare** the anatomy and physiology of mollusks and echinoderms.
- **Describe** the economic importance of mollusks and echinoderms.

The Mollusks

22·1 TRAITS OF MOLLUSKS

Phylum *Mollusca* is a group of soft-bodied invertebrates. They have a muscular foot and a mantle that often secretes a hard shell made of calcium. The animals belonging to this phylum are commonly called **mollusks** (MAHL uhsks). A mollusk's **foot** is a muscular structure adapted for moving from place to place. The thin fleshy tissue surrounding most of the soft body of a mollusk is called the **mantle**. The digestive tract is complete, and there is a structure inside the mouth of many mollusks called a radula (RAJ u luh). The **radula**, which is used to scrape food, is a strip covered with teeth that are like files. Many kinds of mollusks have a hard, external shell. Mollusks are a large group of organisms. They are found in habitats ranging from seas to deserts.

The hard external shells of ancient mollusks have formed excellent fossils. The fossil record indicates that mollusks have been on the earth for at least 600 million years. These ancient mollusks gave rise to the more than 50 000 species of present-day mollusks. In numbers of species, the mollusks make up the second-largest animal phylum. Only phylum Arthropoda is larger.

The soft bodies of most adult mollusks are bilaterally symmetrical, and their bodies are unsegmented. Although the body plan of mollusks is quite different from that of annelids, many biologists believe mollusks and annelids share a common ancestor. One reason for this belief is that some mollusks and annelids develop from a similar kind of larva, seen in Figure 22·1, called a trochophore (TRAHK uh-for). A **trochophore** is a top-shaped, bilaterally symmetrical larva of certain mollusks and marine annelids. You can see that the larva is encircled by a band of short cilia. Notice that there are longer tufts of cilia elsewhere on the body. You can also see that the digestive tract has both a mouth and an anal opening. Since the trochophore larva is

SECTION OBJECTIVES

- Describe the structure and function of the various body parts of mollusks.
- Name some of the ways in which mollusks are important to humans.

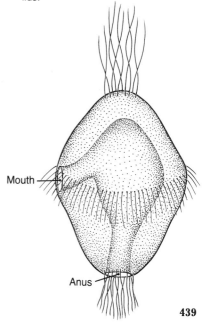

FIGURE 22·1 A trochophore larva, the type found in some mollusks and annelids.

439

440 MOLLUSKS AND ECHINODERMS

FIGURE 22·2 Chitons live in rocky areas at the ocean's edge. Their large muscular foot allows them to cling to rocks without being washed away by waves. If dislodged from a rock, a chiton can roll up into a ball for protection.

common to both annelids and mollusks, these phyla may have arisen from a common ancestor. However, the annelids developed a segmented body plan, while the mollusks developed a nonsegmented body plan.

Phylum Mollusca includes a wide variety of forms. Snails, clams and oysters, squids, and octopuses are all mollusks. These organisms may appear to be too different to belong in the same phylum. However, each form has a mantle and a muscular foot, although the foot is sometimes modified. Four classes of mollusks are described in this chapter.

1. Class *Pelecypoda,* the **pelecypods** (puh LEHS uh pahdz), includes mollusks having a compressed body within a two-part hinged shell.
2. Class *Gastropoda,* the **gastropods** (GAS truh pahdz), consists of mollusks having a well-developed head, a large flat foot, and often a one-part shell.
3. Class *Cephalopoda,* the **cephalopods** (SEHF uh luh pahdz), consists of marine mollusks having a ring of tentacles around the mouth.
4. Class *Amphineura* (am fuh NUR uh) is made up of sluggish, wormlike mollusks with a shell made up of eight plates.

Figure 22·3 presents a summary of the distinguishing characteristics of mollusks. A few representative organisms are listed. To which class do squids and octopuses belong?

The classes Gastropoda, Pelecypoda, and Cephalopoda are the major classes of mollusks. Figure 22·2 shows a chiton (KĪ tuhn), which is a member of the class Amphineura. A chiton is a mollusk that lives

FIGURE 22·3 Major characteristics of four classes of mollusks are summarized here.

CLASS	CHARACTERISTICS	EXAMPLES
Pelecypoda	Compressed body with no head. Wedge-shaped foot for digging. A two-part hinged shell covers the body (bivalve).	Clams, oysters, mussels
Gastropoda	Head with eyes and tentacles. Large flat foot for creeping. Coiled one-part shell is often present (univalve).	Snails, slugs
Cephalopoda	Head with eyes. Foot is divided into arms or tentacles. Shell may be external, internal, or absent.	Squids, octopuses, cuttlefish
Amphineura	Small head without eyes or tentacles. Large flat adhesive foot. Shell is composed of eight dorsal plates.	Chitons

in the sea. It feeds by scraping algae off rocks near the seashore. Chitons cling so tightly to the rocks on which they live that it is almost impossible to pry them loose. Biologists believe that chitons are the mollusks that are most like the wormlike ancestors of the phylum.

Variations in the basic molluscan body plan are shown in Figure 22·4. Refer to this figure as you compare the classes of mollusks. First, compare the different structures of these animals. Recall that a mollusk's foot is a muscular structure adapted for moving from place to place. The snail and chiton crawl on the foot. A clam uses the foot as a wedge for digging in sand or mud. A squid swims about and catches prey with the foot, which is modified into arms.

FIGURE 22·4 A comparison of the basic anatomy of a snail, a clam, a squid, and a chiton.

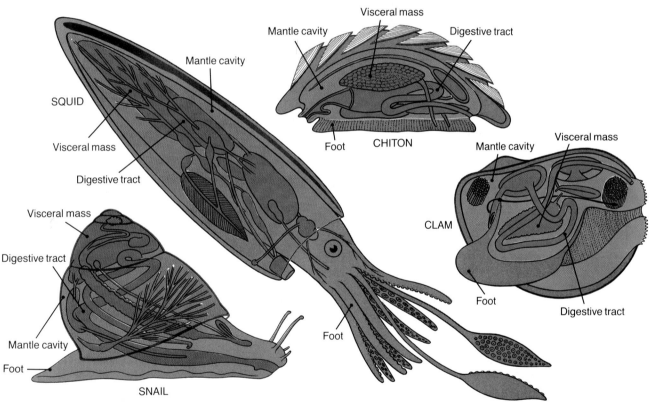

Next, locate the visceral (VIHS uhr uhl) mass in each of the four mollusks. The **visceral mass** is the region of the body that is made of most of the organs of a mollusk. The organs of digestion, excretion, and reproduction are found here. You can see that the digestive tract is arranged differently in each of these mollusks.

Another major characteristic of mollusks is the mantle. The mantle is the tissue that surrounds and covers the visceral mass. A mollusk's shell is secreted by the mantle. In Figure 22·4 notice the differences in the structure and position of the shells. A chiton's shell is made up of eight separate overlapping plates. The clam lives inside a hinged two-part shell. The head and foot of a snail stick out from a one-piece spiral shell. The shell (sometimes called the pen) of a squid is inside the body. This shell does not protect; rather, it supports the body.

The mantle is folded and encloses the mantle cavity. The **mantle cavity** is a space lying between the mantle and the visceral mass. Mollusks that live in water breathe by means of gills found in the mantle cavity. A **gill** is a structure in an aquatic animal through which the respiratory gases are exchanged between the animal's blood and the water in which the animal lives. Gills usually are made up of many filaments having a large surface area and a rich supply of blood. Oxygen diffuses inward, and carbon dioxide diffuses outward through the filaments of the gills. The gills are located within the mantle cavity of the chiton, clam, and squid. In a land snail, the mantle cavity is modified into a lung for air breathing.

22·2 THE CLAM

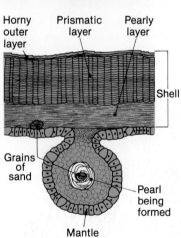

FIGURE 22·5 The top picture is of the outside of a clam's shell. Locate the umbo and growth lines. The diagram shows the formation of a pearl from a grain of sand that has become trapped between the mantle and shell.

Clams and the other members of the class Pelecypoda are sometimes called bivalves. **Bivalves** are mollusks that live within a shell made up of two sections, or valves. A pair of valves is held together by internal muscles and an elastic hinge. A clam's shell is shown in Figure 22·5. In the figure you can see that the shell is oval in shape. Along the upper surface is the structure called the umbo. The **umbo** is the oldest part of a bivalve shell. Surrounding the umbo are *growth lines,* which are added as the clam grows.

Also notice in Figure 22·5 that the shell is made up of three layers. The **horny outer layer** is the layer of a clam shell that is made first. This is a rough layer that protects against the weak acids made by carbon dioxide in water. The thick middle layer of a clam shell is called the **prismatic** (prihz MAT ihk) **layer.** This layer, which is made up of crystals of calcium carbonate, gives strength to the shell. The innermost part of the shell is the **pearly layer.** The pearly layer is made up of thin layers of calcium carbonate lying parallel to the outer layer. The surface of this layer is shiny and smooth. Sometimes foreign matter, such as a grain of sand, gets stuck between the mantle and the pearly layer. The mantle will then deposit more pearly layers around the particle. The buildup of many layers of this protective substance results in the formation of a pearl. You can see a pearl being formed in Figure 22·5. Pearls can be made artificially by putting glass beads between the mantle and shell of clams and oysters. After several years the pearls that form can be removed to be sold as jewelry.

If you removed one of the valves from a clam and dissected the soft body, you could see the structures shown in Figure 22·6. Refer to this drawing as you read. Locate the two powerful muscles that open and close the two-part shell. The muscle near the mouth is the *anterior adductor muscle.* The one at the rear is the *posterior adductor muscle.* These muscles are so powerful it is very difficult to open a clam shell. A clam usually can be opened only by putting a thin-bladed knife between the two parts of the shell and cutting the muscles.

Next, find the incurrent siphon and the excurrent siphon. These tubelike structures are formed from folds of the mantle. The **incurrent siphon** is a structure that regulates the movement of water into the mantle cavity. The **excurrent siphon** is a structure that regulates the movement of water out of the mantle cavity. Within the mantle cavity are four platelike gills. Locate the gills in Figure 22·6. The beating of cilia on the surface of the gills and mantle regulates the circulation of water through the siphon system. As water brought in through the incurrent siphon passes over the gills, oxygen that is dissolved in the water is taken in. Carbon dioxide that is dissolved in the clam's blood diffuses from the blood into the water. The water then passes out through the excurrent siphon.

Water entering through the incurrent siphon also carries small food particles, such as protozoans and algae. Mucus is secreted by the mantle and gills. The mucus traps these small food particles on the surface of the gills. Heavier bits of debris that cannot be used for food fall to the bottom of the mantle cavity. The cilia of the gills and mantle move the mass of mucus and trapped food. The mass is moved forward from the gill surface to the region of the mouth. Particles that cannot be used for food are filtered from the food-mucus mass near the mouth opening. These particles, as well as the heavy debris at the bottom of the mantle cavity, are forced out between the valves by contractions of the adductor muscles.

FIGURE 22·6 In this model of the internal structure of a clam, the mantle, two gills, and part of the foot have been removed so that you can see the internal organs.

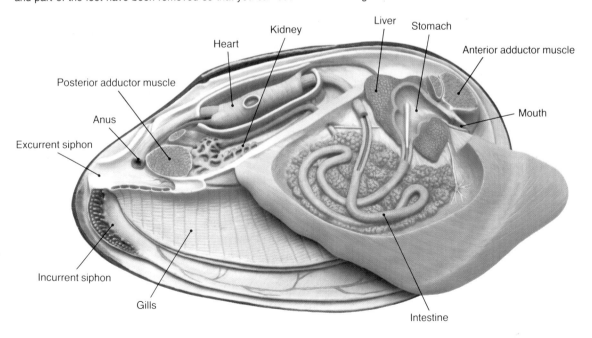

Food swept into the mouth passes through a short esophagus to a saclike stomach. In the figure, notice that the liver, a digestive gland, is joined to the stomach. Digestion occurs in the stomach. The digested food is absorbed in the long intestine. Undigested food passes out the anus and is carried out of the body through the excurrent siphon.

The clam has an open circulatory system. In an open circulatory system, the blood does not always flow within vessels. Recall that the earthworm has a closed circulatory system, in which the blood always flows through vessels. Locate the heart in Figure 22·6. The clam's heart pumps the blood through two arteries. The blood travels toward the posterior and anterior regions of the body at the same time. In some parts of the body, blood is released directly into spaces in the tissues. It is then collected by veins and carried to the kidneys. Wastes are removed from the blood by the two U-shaped kidneys lying below the heart. The kidneys empty the wastes into the excurrent siphon. From the kidneys, blood flows to the gills, where it is oxygenated. Other veins then direct the blood back to the heart.

A clam's simple nervous system is made up of three pairs of ganglia and the attached nerve cords. *Ganglia* are collections of nerve cell bodies. Paired ganglia are found in the mouth region, in the foot, and near the posterior adductor muscle. Each pair of ganglia controls the part of the body in which it is found. The clam has no sense organs. However, along the edge of the mantle it has sensory cells. These cells respond to touch and light. The clam also appears to be sensitive to chemicals.

The sexes are separate in most species of clams. A testis or ovary surrounds the coil of the intestine in the foot. In saltwater clams the eggs and sperm are shed into the water, where fertilization takes place. The zygote then develops into a swimming larva and then into an adult.

22·3 OTHER BIVALVES

Class Pelecypoda is made of mollusks having widely different body forms and habits. Mollusks in this class are found in many different types of aquatic habitats as well. A few of the many types of bivalve mollusks are shown in Figure 22·7.

Though many clams spend their lives burrowing into sand or mud, other bivalves can swim about. The scallop swims by clapping the two valves of its shell together. As the shell opens and closes, water is forced in and out. This propels the scallop through the water. If you look carefully, you can see a row of blue eyes along the edge of the mantle. The scallop uses these eyes to detect moving objects nearby. Mussels do not swim, but they creep by pulling themselves along by means of a threadlike substance they secrete. With these threads they anchor themselves to rocks or to other mussels. Some bivalves, such as oysters, do not move at all as adults. An oyster develops from a ciliated larva that settles to the bottom after about two weeks and remains there.

FIGURE 22·7 The scallop (top) is able to escape from predators. It snaps its valves together and shoots away by jet propulsion. Mussels (center) attach themselves to rocks, piers and other objects by adhesive threads. Young mussels can climb vertical walls using these threads. Oysters (bottom), like scallops and mussels, are valued as food. Some species of oysters produce precious gems—pearls.

22·4 GASTROPODS

Class Gastropoda is made up of the snails and their relatives. These mollusks are commonly known as the univalves. **Univalves** are mollusks that have a one-piece shell. In many univalves the shell is coiled. In others the shell may look like a cone. Some gastropods, such as slugs, have no shell at all. The gastropods are a varied group of mollusks, ranging in size from those that are microscopic to some that may have shells as long as 60 cm. They are found in both fresh water and seawater, and on land.

uni- (one)

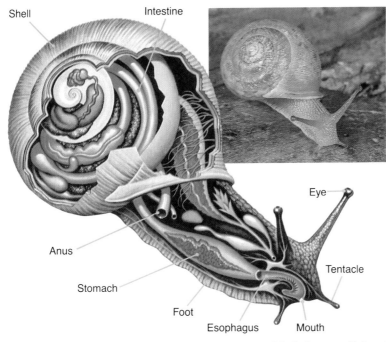

FIGURE 22·8 Notice how the intestine and the rest of the digestive tract is twisted to fit inside the snail's shell. The small drawing is a closeup of the snail's radula, a file-like organ which is used to obtain food.

The body plan of a snail is shown in Figure 22·8. As a snail develops from a larva, the organs in the visceral mass are twisted upward into the shell. As you can see in the figure, the digestive tract is looped so that the anus opens in the front, above the head.

Unlike the bivalves, the snail has a clearly defined head region. Notice the two pairs of tentacles on top of the head. The shorter tentacles, which can be retracted, are thought to be the organs involved in the sense of smell. Light-sensitive eyes are found at the tips of the longer tentacles. The mouth, on the ventral surface of the head, has a structure called a radula. Recall that a radula is a strip covered with filelike teeth found on the tongues of many mollusks. The structure of a snail's radula is shown in Figure 22·8. A snail uses its radula to scrape away the surface of the vegetation on which it feeds.

The snail moves at a very slow pace—only about 5 cm per minute. With its large fleshy foot, a snail can move about in a gliding manner. Mucous glands located at the front of the foot secrete a slimy path just ahead of the snail, which aids in movement.

FIGURE 22·9 Some univalves lack shells. The slug (top) must remain in moist places during the day to avoid drying out. The pteropod (bottom) swims with wing-like extensions of its foot.

Land-dwelling snails breathe by using the mantle cavity as a lung. There are many blood vessels near the surface within the mantle cavity; gas exchange occurs through these vessels at the moist surface of the mantle cavity. Most aquatic snails breathe by means of gills. However, aquatic snails without gills must come to the surface at regular intervals to take bubbles of air into the mantle cavity.

Snails are adapted to live in all parts of the world, from dry deserts to frozen arctic regions. During the winter and during periods of dryness, land snails become dormant. When dormant, they secrete a membrane over the opening of the shell to keep the soft, moist body from drying out. Some water snails also become dormant when the water dries up and become active again when moisture returns.

The slug is a gastropod that lacks a shell. Some slugs feed on plants in gardens and yards, usually at night. By day they are found in moist, sheltered places, such as under rocks and logs. Lacking a shell, they would dry out during the heat of the day. Nudibranchs (NOO duhbrangks) are strikingly beautiful slugs found in shallow coastal waters. The sea hare is a large, fleshy slug found in shallow ocean water. When it is disturbed, it secretes a purplish fluid. This fluid helps to hide the sea hare from predators.

The pteropod (TEHR uh pahd) has its foot modified as a pair of winglike extensions. When it uses the winglike foot, it resembles a butterfly in flight. It is often called the sea butterfly.

22·5 CEPHALOPODS

The most specialized, most active, and largest of the mollusks are those belonging to the class Cephalopoda. Squids, octopuses, and nautiluses are cephalopods. All cephalopods are marine animals, and all of them have a large head region with a well-developed brain. Most of them have two large complex eyes. The foot of a cephalopod is divided into a number of arms or tentacles surrounding the head. The shells of different cephalopods are different in form. The nautilus has a large external coiled shell composed of many chambers. In the squid and cuttlefish, the shells are internal. The octopus has no shell at all.

The squid is a common mollusk found in all oceans. Most squids grow to about 30 cm long, but a few species grow up to 6 m long with tentacles 9 m long. Unlike the univalves and bivalves, squids are active hunters that can move rapidly through the water. In Figure 22·10, notice the large fins at the back of the squid's body. The squid uses these muscular fins for ordinary swimming at slow speeds. To move rapidly it can force water out of the siphon, which is located near the head. The rapid movement occurs by jet propulsion, in much the same way that a balloon moves after it is blown up and released. The squid can turn the siphon in different directions to move either forward or backward. When the siphon is pointed backward, the squid can swim forward rapidly to catch prey. When the siphon is pointed forward, jet propulsion moves the squid backward. A squid will often move backward when fleeing from an enemy.

The squid's mouth is surrounded by ten arms bearing suckers that catch and hold prey, such as fish and other mollusks. Notice that one pair of arms is longer than the others. The longer arms are used in capturing prey, while the shorter arms force it into the mouth. Two heavy jaws crush the prey, which is then shredded into fine pieces by a radula in the mouth. The digestive organs of a squid are similar to those of other mollusks.

The squid has a well-developed circulatory system. Unlike the circulatory system of the clam, it is a closed system. The squid also has two types of hearts. The **arterial heart** pumps blood throughout the body. A gill heart is located at the base of each gill. The **gill hearts** of a squid receive blood from all over the body and pump it through the gills. These separate hearts for pumping blood through the gills allow for a higher rate of gas exchange than is possible in other mollusks.

The nervous system of a squid is highly developed. The squid has a pair of organs located in the head that help it maintain balance. On the head it also has a pair of large eyes. In fact, the structure of a squid's eye is similar to that of a human's eye.

A squid can protect itself from predators in several ways. When bothered by such predators as large fish or whales, the squid squirts an inky fluid from its siphon. This fluid hides the squid and temporarily dulls its pursuer's sense of smell. A squid can also protect itself by camouflage. Numerous pigment cells of blue, purple, yellow, and red are located beneath the skin. These pigment cells lie in elastic sacs surrounded by muscles. As these muscles contract, various amounts of pigment can mix. This changes the squid's color to blend with the surroundings.

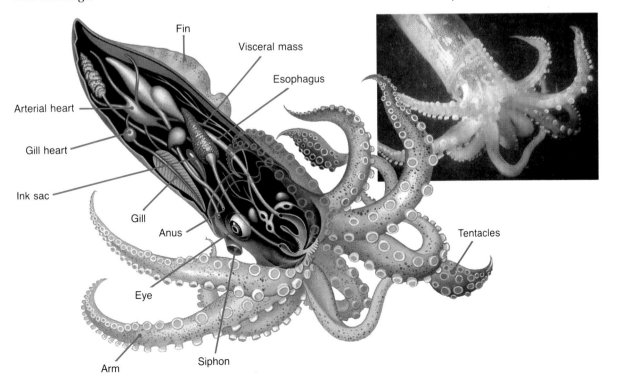

FIGURE 22·10 The major internal structures of the squid are shown here. The squid's ten arms surround its mouth.

FIGURE 22·11 The adult nautilus and young nautilus (left) are cephalopods with shells. Gas-filled chambers in the shell help keep the animal upright. A nautilus has 38 arms arranged in two rings. The octopus (right) does not have a shell. It has eight arms.

Two other cephalopods are shown in Figure 22·11. The octopus has eight long arms that are used for crawling along the ocean floor in search of food. Like the squid, the octopus can propel itself through the water, but usually it crawls slowly over rocks.

The chambered nautilus has a flat coiled shell made up of many chambers, built over time. The animal lives in the largest, outermost chamber in the shell. The nautilus secretes a gas into the other chambers. The gas-filled chambers enable the animal to float.

22·6 ECONOMIC IMPORTANCE OF MOLLUSKS

Many of the aquatic mollusks are used for food by humans all around the world. These mollusks are often called shellfish. The term *shellfish* includes clams, oysters, scallops, and mussels. Oysters are harvested in many places along the Atlantic and Pacific coasts. Abalones (ab uh LOH neez), large marine snails, are also used for food. Some land snails are served in fine restaurants where they are known by their French name—escargots (ehs kahr GOH).

Several commercial products are made from mollusks. Natural pearls formed within clams and oysters have long been collected from the ocean floor by pearl divers. However, many pearls are now artificially cultured by inserting foreign objects between the shell and mantle of an oyster. Bits of clam shells are cut out and polished to make buttons. From a type of squid called the cuttlefish comes sepia, an ink used by artists as a water color.

Unfortunately, many mollusks are harmful. Because of the damage that shipworms do to underwater wooden structures, millions of dollars have been spent to find ways to combat these animals. Also, most snails and slugs are thought of as pests because they cause a great deal of damage to plants. Some snails are dangerous because they serve as intermediate hosts for such parasites as sheep liver flukes and human blood flukes.

Today, mollusks are being used for studies in physiology, learning, and animal behavior. For example, research has shown that octopuses are able to learn. Squids have nerve fibers hundreds of times larger than the nerve fibers of vertebrates. Because of the large size of these nerve fibers, it is possible for researchers to place electrodes into them to determine the nature of the nerve impulses.

REVIEW QUESTIONS

1. Describe the structure and function of the mantle in a mollusk.
2. Describe the way in which a snail uses its foot for locomotion.
3. Describe two ways in which a squid can move.
4. Name three ways in which mollusks are beneficial to humans. Name three ways in which mollusks are harmful.
5. The type of shell is one feature used to classify mollusks, yet some mollusks of classes Gastropoda and Cephalopoda lack shells. Suppose a marine biologist finds an unknown mollusk without a shell. How might the biologist decide whether the mollusk is a gastropod or cephalopod?

Profiles

Libbie Henrietta Hyman (1888–1969)

Libbie Hyman grew up in a poor farming community in Iowa. She often roamed the fields, collecting and identifying plants, flowers, butterflies, and moths.

Libbie Hyman graduated first in her high school class at Fort Dodge, Iowa, in 1905. She continued to take biology and language courses at the high school during the next year while working part-time in a factory. In 1906, Libbie Hyman received a scholarship to attend the University of Chicago. There she began her formal study of invertebrate zoology. She received her bachelor's degree in 1910 and her doctorate in 1915. Dr. Hyman remained at the University of Chicago until 1930. During that time she worked as a laboratory and research assistant and coordinated and taught the undergraduate laboratory courses in comparative invertebrate and vertebrate zoology.

Between 1916 and 1932 Dr. Hyman wrote several articles on her research on invertebrates. In 1919 and 1922 she wrote and published laboratory manuals for invertebrate zoology and comparative anatomy. These laboratory manuals were revised and republished for more than 40 years.

Her growing reputation as an authority on invertebrate anatomy and taxonomy brought increased contact from colleagues around the world. On numerous occasions Dr. Hyman was asked to identify unknown invertebrate organisms. She soon recognized the need for an organized collection of information on the taxonomy and anatomy of invertebrates. In 1931 she began to work on such a collection of information. Between 1940 and 1967, Dr. Hyman published a six-volume survey of the biology of invertebrates entitled *The Invertebrates*.

During the last 10 years of her life, Dr. Hyman was handicapped by Parkinson's disease. As a result she was unable to complete her volumes on the biology of mollusks and arthropods. Although she did not reach her original goal of surveying all invertebrate groups, her work is still one of the most authoritative and accurate surveys of the invertebrates.

Investigation

What Are the Major Structural Features of Mollusks?

Goals
After completing this activity, you will be able to
- Identify the major structural features of the clam, and the squid.
- Compare the major characteristics of the clam and the squid.
- Draw and label the major features of the squid.

Materials (for groups of 3)
preserved clam
preserved squid
dissecting pan
dissecting needle or probe
scalpel
hand lens
stereomicroscope (if available)

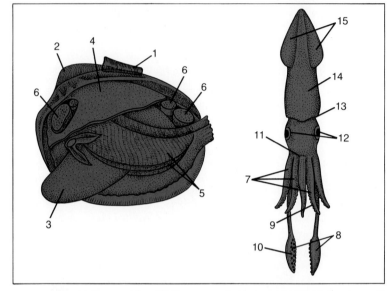

Procedure

A. Place the preserved clam in a dissecting pan. Note the two hard shells joined together by a hinge ligament **(1)** on one side. Identify the swollen hump, or umbo **(2)**, which is the oldest part of the shell. Note the lines of growth around the umbo.

B. Using your scalpel, pry open the two shells. Put an object, such as a pencil, between the shells so that they stay apart. Keeping the blade of the scalpel close to the shell, cut the muscles inside the shell.

CAUTION: Use great care in handling the scalpel. Direct the edge of the blade *away* from your body. Work on a firm surface.

C. Open the clam shell and identify the foot **(3)** and the mantle **(4)**. You will also be able to see the pair of gills **(5)**. Note the remains of the muscles **(6)** that held the shells together. Check to see whether the clam has a head, tentacles, or eyes.

D. Place the squid in the dissecting pan. Use the drawing to identify various structures. Find the 10 arms **(7)**, two of which are longer than the others. **(8)**. Note the adhesive disks **(9)** on the shorter arms. Look for the suckers **(10)** on the longer arms. At the base of the circle of arms is the mouth **(11)**.

E. Below the arms, locate the two large eyes **(12)**. The eyes are located just above the squid's collar **(13)**. Locate the spear-shaped mantle **(14)** with the fins **(15)** on either side. Run your fingers along the length of the mantle. You should be able to feel a long, narrow, hard structure called a pen. It is analogous to the shell in the snail and the clam.

F. On a separate sheet of paper, draw the squid. Label all the features described in steps D and E.

G. On a separate sheet of paper, set up a chart comparing the clam and the squid. List data on the shell, head, foot, mantle, tentacles, and arms.

H. Follow your teacher's instructions for the proper disposal of all materials.

Questions and Conclusions

1. Of the two mollusks studied, which lacks a shell? Does this animal have anything similar to a shell? Explain.
2. Compare the foot in the squid and the clam. What part of the squid is similar to the foot?
3. Which of the mollusks has a head? Describe any sensory organs you saw on the head.
4. Which of the mollusks seemed better adapted to protect itself from its predators? Explain why.

The Echinoderms

22·7 TRAITS OF ECHINODERMS

Unless you live near the ocean or have visited the seashore, you probably have not seen a live starfish. All species of starfish and their relatives are found only in the oceans.

The starfish is a member of phylum *Echinodermata*, the group of spiny-skinned marine invertebrates. The animals classed in this phylum are called **echinoderms** (ih KĪ nuh dermz). Like mollusks, the echinoderms have left fossils that date back to the Paleozoic Era. At present about 6000 species are known. In addition to the starfish, other echinoderms you may know about are brittle stars, sea urchins, sea cucumbers, and sand dollars.

The word *Echinodermata*, meaning "spiny skin," refers to the spines found on most echinoderms. The spines project from bony plates found just below a layer of soft skin. These plates make up an endoskeleton. An **endoskeleton** is a skeleton that lies inside the body of an animal. As adults, echinoderms have radial symmetry. Recall that in animals with *radial symmetry,* the body parts are arranged around one longitudinal axis. What other group of radially symmetrical animals did you study earlier? The radial symmetry of some echinoderms is sometimes described as being *pentaradial,* or arranged in five sections. Echinoderms such as starfish, sea lilies, and brittle stars have five arms.

Though an adult echinoderm has radial symmetry, it develops from a bilaterally symmetrical larva called a **dipleurula** (dī PLUR yuh-luh). This larva is free-living and it swims by means of two bands of cilia. The dipleurula larva of an echinoderm can be seen in Figure 22·12. It is similar to the larval forms of some simple members of the phylum to which vertebrates belong. Because of the similar larvae and the presence of endoskeletons, the echinoderms and vertebrates are believed to have developed from a common ancestor.

A special feature of the echinoderm is the water-vascular system. An echinoderm's **water-vascular system** is a connected system of water-filled canals that serves in locomotion. The internal canals that make up this system are connected to rows of tiny tubelike structures which project from the body. Pressures caused by water entering and leaving these structures enable the echinoderm to pull itself along or attach itself to objects on the ocean floor.

22·8 THE STARFISH

The most familiar echinoderms are the starfish, also known as sea stars. They are found in abundance in coastal waters and along rocky seashores. Starfish vary greatly in size—from about 1 cm to about 1 m in diameter. In their natural environment, starfish are found in a wide variety of colors.

SECTION OBJECTIVES

- Describe the structure and function of the parts of a starfish.
- Explain why echinoderms are important to humans.

endo- (inside)

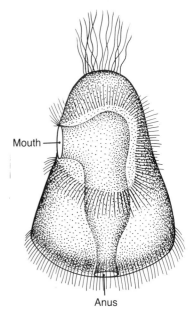

FIGURE 22·12 The dipleurula larva of an echinoderm.

ab- (away from)
oris (mouth)

The typical body plan of a starfish, five arms radiating from a central disk, is shown in Figure 22·13. In the figure, part of the body wall has been removed to show some of the internal organs.

Special terms are used to describe the surfaces of animals with radial symmetry. The **aboral surface** is the surface opposite the mouth. The **oral surface** is the surface on which the mouth is located. The mouth of the starfish cannot be seen in the figure because it is located on the oral, or bottom, surface in the middle of the central disk.

The upper surface of the starfish is covered with spines. The spines project from plates beneath the skin. Because of its spiny covering, a starfish is rarely eaten by other animals. Skin gills and tiny pincers surround the spines. The **skin gills** are hollow tubes that project from the lining of the body cavity. These organs are used for gas exchange. The pincers clean the body surface and protect the skin gills.

The starfish's water-vascular system is made up of internal water-filled canals and tiny tubelike structures that project from the oral surface of the body. The water-vascular system is involved in feeding and locomotion. Located on the aboral surface of the starfish, the **sieve plate** contains tiny pores through which water enters the water-vascular system. From the sieve plate, water passes through a short tube called the **stone canal** to the ring canal. The **ring canal** is a circular tube that surrounds the mouth. Each of the starfish's arms has a **radial canal** to carry water from the ring canal to the tip of the arm. Along each radial canal are many pairs of tube feet. The **tube feet** are

FIGURE 22·13 This aboral view shows some of a starfish's major internal structures. The tube feet and radial canal of the water-vascular system are shown in blue. The digestive system is shown in green, and the reproductive organs are shown in reddish-tan.

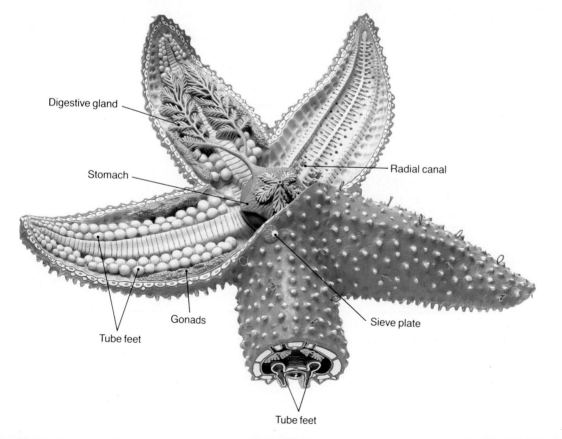

hollow, thin-walled cylinders that are found in grooves on the underside of the arms. The tube feet resemble old-fashioned automobile horns in shape. Each tube foot has a disk-shaped sucker at its tip, and the upper end is attached to a bulblike sac. When these sacs contract, water is forced into the tube feet, causing them to extend. The suckers of the extended tube feet can then attach to a surface. When the muscles in the walls of the tube feet contract, water is forced back into the sac, and the starfish is pulled forward. By attaching and releasing its hundreds of tube feet, the starfish can propel itself along a surface.

A starfish also uses its tube feet when it feeds. Its food consists of oysters, clams, fish, snails, barnacles, and worms. The starfish opens an oyster or clam by spreading its arms over the edge of the shell opposite the hinge. It attaches its tube feet to the two shell halves, and the suction produced by the dozens of tube feet gradually draws the halves apart. At first the mollusk resists. After it becomes exhausted, however, it relaxes and the shell opens. The starfish then turns its stomach inside out and thrusts it through the mouth opening into the space between the halves of the mollusk's shell. A space of about 0.1 mm is all that is needed for the starfish to insert its stomach. Digestive juices from the starfish's stomach begin digestion of the soft parts of the mollusk while it is still inside its own shell. The partially digested food is taken into the stomach, which is then pulled back into the starfish's body. Further digestion occurs within the *digestive gland,* which you can see in Figure 22·13. Digested food passes into the fluid of the body cavity.

A starfish has no circulatory system. Cilia within the body cavity circulate fluid that carries digested food to all body parts. There are no specialized organs for excretion. Amoeboid cells, circulating in the fluid of the body cavity, pick up wastes and then pass through the walls of the skin gills. These amoeboid cells are made by the lining of the body cavity.

A starfish's nervous system is very simple, with no centralization of nerve cells. A nerve ring surrounding the mouth branches off into nerve cords, one extending into each arm. The nerve cords are located below the radial canals. An eyespot sensitive to light and a tentacle sensitive to touch are located at the end of each arm. The tube feet are also sensitive to touch.

The sexes in starfish are separate, and the reproductive system is made of paired gonads (testes or ovaries) in each arm. Very large numbers of eggs and sperm are released into the water, where fertilization takes place. One female starfish may release over 200 million eggs in a single season. The number of sperm produced by the male is far greater than the number of eggs produced by the female. The larva that develops from a zygote is at first bilaterally symmetrical. It later shows the radial form of the adult.

The starfish shows great powers of regeneration. If an arm is broken off, the starfish will regenerate a new arm. A starfish regenerating three arms is shown in Figure 22·14.

FIGURE 22·14 A starfish is able to regenerate missing arms. The two long arms of this starfish are of normal length. The other three are being regenerated. Why can't you kill a starfish by cutting it in half?

454 MOLLUSKS AND ECHINODERMS

22·9 OTHER ECHINODERMS

Various species of echinoderms are shown in Figure 22·15. Some of these are much like the starfish. Others are not.

The most primitive echinoderms are the sea lilies and the feather stars. Both are found in deep water, attached to the ocean floor by stalks. Unlike other echinoderms, their mouths face upward. They use their branched arms to capture food by filtering small plants and animals from the water.

The most active echinoderms are the brittle stars, which have five snakelike arms radiating from a central disc. They can be found all the way from the coastline to at least 6 km below the water's surface. These echinoderms are named brittle stars because parts of their arms break off if they are disturbed. The missing parts are then rapidly regenerated.

A sea urchin is an echinoderm that resembles a pincushion. This creature has an endoskeleton made up of plates that are fused to form a shell called a *test*. The long, sharp spines protrude from pores in the

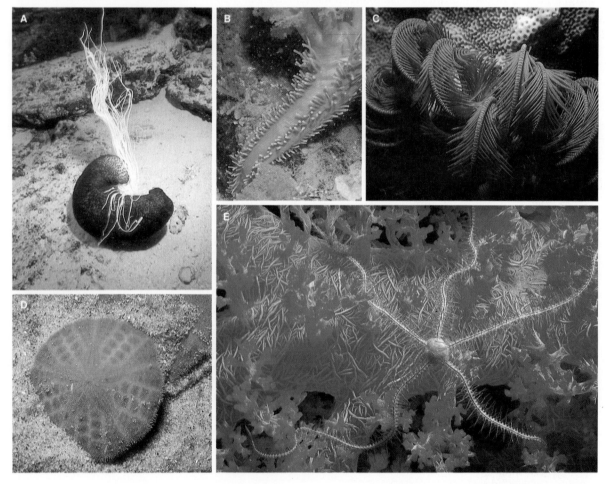

FIGURE 22·15 Echinoderms come in a variety of shapes and colors. (A) A sea cucumber can repel predators by ejecting its sticky intestines through its mouth. Later, the sea cucumber regenerates its internal organs. (B) The tube feet on this orange sea cucumber can be seen clearly. (C) Feather stars live attached to rocks on the ocean bottom. (D) Sand dollars have flattened bodies and short spines, an adaptation to living buried in the sand. (E) Brittle stars are the most active types of echinoderms.

test. In some giant species of sea urchins, the spines may be 30 cm long, and they may be poisonous. Tube feet also extend from between the spines. A sea urchin moves across the sea bottom by moving its spines.

The sand dollar has a flat body with very short spines and tube feet. Notice that there is a starlike pattern on the aboral surface. On the oral surface there are five grooves radiating from a short central region. Sand dollars are found living along seacoasts, where they often bury themselves in sand on the bottom of the sea.

The sea cucumber is an echinoderm that looks like a cucumber with a fringe of tentacles at one end. These tentacles, which surround the mouth, are modified tube feet. The anus is located at the opposite end of the body. A sea cucumber crawls by means of its tube feet and by contractions of muscles in the body wall.

22·10 ECONOMIC IMPORTANCE OF ECHINODERMS

Because of their spiny external covering, echinoderms are not usually a source of food for humans. In the West Indies, however, people seek sea urchin eggs for food during the seasons when the females produce them in great abundance. In parts of Asia, dried sea cucumbers are used to make soup.

An adult starfish can eat nearly a dozen clams or oysters daily. For this reason starfish are considered pests in shellfish beds. At one time oyster harvesters cut starfish into small pieces. They then threw the pieces back into the water. This practice only made the problem worse, because many of the pieces of starfish would regenerate. Now the starfish are caught with nets and destroyed. Sometimes the shellfish beds are treated with the chemical lime. The lime kills the starfish, but it is harmless to the mollusks.

The eggs and sperm of echinoderms, especially those of starfish and sea urchins, are widely used in biological experiments in embryology. *Embryology* is the study of the development of organisms from the time of fertilization to their hatching or birth. Echinoderm eggs and sperm are ideal for studies in embryology because so many of them are produced, and they are easily obtained.

REVIEW QUESTIONS

6. Describe the outer body covering and symmetry of echinoderms.
7. Explain how a starfish uses its water-vascular system to move.
8. List two ways in which sea lilies differ from other echinoderms such as starfish.
9. In what area of scientific research are echinoderms important?
10. Some mollusks, such as octopuses, appear to be more advanced than echinoderms. Yet many biologists believe that vertebrates, which are advanced animals, are more closely related to the echinoderms. Why do biologists believe vertebrates are more closely related to echinoderms than to mollusks?

Chapter Review

SUMMARY

22·1 Mollusks are soft-bodied invertebrates that have a muscular foot and a mantle, which often secretes a shell.

22·2 The clam is a familiar representative of the mollusks. Its soft body is enclosed in a shell made up of two sections.

22·3 Bivalve mollusks have a wide diversity of body forms, and they are found in a wide variety of habitats.

22·4 The gastropods are a group of mollusks that includes the snails, which have a one-piece shell, and slugs, which lack a shell.

22·5 Cephalopods, which are the most active mollusks, have a head region with a brain and eyes and have arms or tentacles surrounding the head.

22·6 Mollusks are a major source of food and of commercial products. Certain mollusks are harmful.

22·7 Echinoderms are spiny-skinned marine invertebrates having radial symmetry as adults.

22·8 The starfish is a well-known representative of the echinoderms. Like other echinoderms, the starfish uses its water-vascular system for feeding and locomotion.

22·9 Other classes of echinoderms are represented by the sea lilies, sea urchins, and sea cucumbers.

22·10 Echinoderms are of limited economic importance; however, their eggs and sperm are widely used in embryological studies.

BIOLOGICAL TERMS

aboral surface
arterial heart
bivalves
cephalopods
dipleurula
echinoderms
endoskeleton
excurrent siphon
foot
gastropods
gill
gill hearts
horny outer layer
incurrent siphon
mantle
mantle cavity
mollusks
oral surface
pearly layer
pelecypods
prismatic layer
radial canals
radula
ring canal
sieve plate
skin gills
stone canal
trochophore
tube feet
umbo
univalves
visceral mass
water-vascular system

USING BIOLOGICAL TERMS

1. Which terms refer to types of mollusks? Distinguish between these terms.
2. Which terms refer to larval forms of mollusks and echinoderms? Distinguish between these terms.
3. Which terms name parts of the water-vascular system? Define the terms.
4. Which terms refer to the sections of a clam's shell? Distinguish between these terms.
5. Distinguish between the mantle cavity and the mantle.
6. What is a radula?
7. What is an umbo?

Explain the difference between the terms in each set.
8. radial canals, ring canal
9. arterial heart, gill hearts
10. incurrent siphon, excurrent siphon
11. bivalve, univalve
12. prismatic layer, pearly layer
13. aboral surface, oral surface

UNDERSTANDING CONCEPTS

1. Explain how the trochophore larva indicates a relationship between annelids and mollusks. (22·1)
2. Name four classes of mollusks and give an example of a member of each class. (22·1)
3. Describe the three layers of a clam's shell. (22·2)
4. Explain the circulation of water through the siphon system of a clam. (22·2)
5. Describe the nervous system of a clam. (22·2)
6. Compare locomotion in adult clams, scallops, and mussels. (22·3)
7. What changes occur in the anatomy of a snail as it develops from a larva? (22·4)
8. What is a radula, and how is it used by a snail? (22·4)
9. How is the circulatory system of a squid different from that of a clam? (22·5)
10. Compare the shells of a squid and a nautilus. (22·5)
11. Why are squids often used to study the nature of the nervous system? (22·6)
12. In what habitat are echinoderms found? (22·7)
13. Compare the symmetry of echinoderms as larvae and as adults. (22·7)
14. How does a starfish use its water-vascular system to feed on clams or oysters? (22·8)
15. Describe the life cycle of a starfish. (22·8)
16. Compare locomotion in starfish, sea urchins, and sea cucumbers. (22·8, 22·9)
17. How do oyster harvesters eliminate starfish from oyster beds? (22·10)

APPLYING CONCEPTS

1. In what ways are the shells of clams and snails both an advantage and disadvantage?
2. Some mollusks seem to be capable of learning; however, echinoderms seem incapable of learning. Why is this so?
3. Why are fossils of mollusks and echinoderms more abundant than those of the worm phyla?
4. Starfish produce large numbers of eggs and sperm. How is this production an adaptive advantage?
5. Squids and octopuses have well-developed eyes, similar in structure to those of vertebrates. Yet mollusks and vertebrates do not appear to be closely related. How might such similar structures have developed in these two unrelated groups of animals?

EXTENDING CONCEPTS

1. Using appropriate references, prepare an oral or written report on one of the following topics.
   a. The Economic Importance of Mollusks
   b. The Giant Squid
   c. Sea Cucumbers
   d. Regeneration in Echinoderms
2. In 1952, a primitive mollusk, *Neopilina,* was discovered in a deep trench of the Pacific Ocean. Using appropriate library resources, prepare a report on the significance of the discovery of *Neopilina.*
3. For the bulletin board, prepare a large drawing showing the internal anatomy of a clam or a starfish.
4. Study the development of snails. Many aquarium snails lay their eggs in masses attached to plants in the aquarium. Remove the developing eggs to aquarium water in a culture dish, and use a stereomicroscope or hand lens to watch the embryos and ciliated larvae develop.
5. Shell collecting is an interesting hobby. Many books on shell collecting and identification are available in book stores and libraries. If you live near the ocean, or if you vacation at the seashore, you may wish to start a shell collection.

READINGS

Fleming, Carroll B. "Snail Pays Dearly for Its Tasty Meat and Beautiful Shell." *Smithsonian,* March 1982, p. 119.

Hedeen, Robert A. *The Oyster: The Life and Lore of the Celebrated Bivalve.* Tidewater, 1986.

Roper, Clyde F.E., and Kenneth J. Boss. "The Giant Squid." *Scientific American,* April 1982, p. 96.

Sumich, James L. *An Introduction to the Biology of Marine Life,* 3rd ed. William C. Brown & Co., 1984.

Ward, Peter D. "Nautilus: Have Shell, Will Float." *Natural History,* October 1982, p. 64.

23. The Arthropods

Last spring, this luna moth was a creeping caterpillar that fed on leaves. Later, while in its cocoon, every molecule in the caterpillar was completely rearranged. The chewing jaws of a caterpillar changed into a strawlike tube through which the moth drinks the sweet nectar of certain flowers. The caterpillar could only crawl along on tiny feet; as a moth it can fly with its large green and purple wings. In this chapter, you will learn more about the strange metamorphosis of moths, as well as about the other kinds of animals that are grouped in the phylum Arthropoda.

CHAPTER OBJECTIVES

After completing this chapter, you will be able to

- **Discuss** the anatomy, physiology, and behavior of arthropods.
- **List** the major subphyla and classes of arthropods.
- **Give examples** of arthropods.
- **Contrast** incomplete and complete metamorphosis.
- **Describe** some ways insects are adapted to their environment.

What Are Arthropods?

23•1 TRAITS OF ARTHROPODS

Phylum *Arthropoda* includes animals such as lobsters, crayfish, ants, bees, spiders, millipedes, and centipedes. Members of phylum Arthropoda are called **arthropods.** Arthropod fossils date back to the Paleozoic Era, over 500 million years ago. The earliest arthropods lived in the sea. The arthropods were the first animals to move successfully from the ocean to dry land.

Arthropods share certain traits that distinguish them from other animal groups. One of the traits is a *segmented body*. In many kinds of arthropods, the body is divided into three distinct parts. The **head** is made up of the most anterior of the segments. The **thorax** contains the middle segments. The **abdomen** is made up of the posterior segments.

Another trait is an exoskeleton. An **exoskeleton** is a skeleton covering the outside of the body. The exoskeleton of arthropods is made of *chitin* (KĪ tihn). Chitin is a fairly hard, waterproof covering made by the cells of the epidermis. The exoskeleton aids the animal in many ways. It protects the organs inside the animal's body. In addition, it prevents the loss of water from the body. This allows arthropods to survive in places where other animals, such as annelids, would dehydrate.

Together, the exoskeleton and muscles produce locomotion. But an exoskeleton that encloses the body limits the growth of the animal. After it is secreted, the exoskeleton hardens, restricting further growth. To grow, the animal must first shed the exoskeleton. The periodic shedding of the exoskeleton is called **molting.** As the old skeleton is shed, a new skeleton is secreted beneath it. At first, the new skeleton is soft and flexible. After shedding the old skeleton, the animal inflates its body, enlarging itself while its new skeleton hardens. The exoskeleton offers some protection from predators. However, while its skeleton is hardening, the arthropod is vulnerable to predators.

SECTION OBJECTIVES

- Describe the characteristics of arthropods.
- Explain the success of the phylum Arthropoda.
- Discuss the importance of the arthropods.

arthron (joint)
podos (foot)
exo- (outside of)

460 THE ARTHROPODS

FIGURE 23·1 Arthropods live in many different environments and show a wide variety of forms. (A) The grasshopper is an insect that eats plants. (B) The swallowtail butterfly is also an insect. It can fly, and feeds on nectar. (C) Spiders are eight-legged arthropods. Many spin webs to trap their prey. (D) This tropical spiny lobster is a predator and scavenger. (E) Ants are social insects.

A third trait of arthropods is *paired jointed appendages*. Generally, these body parts are very specialized. Some appendages, such as legs and wings, allow for ease of movement. They are adapted for walking, hopping, climbing, swimming, and flying. Others, such as claws and mouth parts, are adapted for food getting and feeding. Still others are adapted for gas exchange and reproduction. Some, such as antennae, are highly developed sense organs. Locate some of the distinguishing arthropod characteristics in each of the animals shown in Figure 23·1.

23·2 CLASSIFICATION OF ARTHROPODS

Classification of the arthropods is difficult because of the large number of species. Classification is also difficult because similarities in structure do not always mean that organisms are related. Most scientists divide the phylum Arthropoda into three subphlya: Trilobita (tri luh BĪT uh), Chelicerata (kuh lihs uh RAHT uh), and Mandibulata (man dihb yuh LAHT uh).

All members of the subphylum Trilobita are extinct. These primitive arthropods lived in the oceans about 500 million years ago. Trilobite fossils are abundant in rock formed during the Paleozoic Era. Figure 23·2 shows how a trilobite may have appeared on its ventral surface. What arthropod traits are seen in the trilobite?

The subphylum Chelicerata contains two classes of modern arthropods. The class *Xiphosura* (zihf uh SUR uh) is represented by horseshoe crabs. The class *Arachnida* (uh RAHK nuh duh) is represented by spiders, ticks, mites, and scorpions. Figure 23·3 shows the dorsal and ventral view of the horseshoe crab.

One of the major traits of subphylum Chelicerata is a body made up of two divisions. The anterior division is the **cephalothorax** (sehf-uh loh THOR aks), which is a fused head and thorax. The posterior division is the abdomen.

The second major trait is the specialization of the appendages. Chelicerates have six pairs of appendages, which extend out from the cephalothorax. The first pair of appendages are the **chelicerae** (kuh LIHS uhr ee) (sing., *chelicera*), which are modified as claws or fangs. The second pair are the pedipalps (PEHD ih pahlps). The **pedipalps** function as feeding organs, as walking legs, as sensory organs, or as reproductive organs that transfer sperm from the male to the female. The last four pairs of appendages are walking legs. In the horseshoe crab there are book gills. *Book gills* are aquatic gas-exchange surfaces. In terrestrial arachnids, gas exchange is through internal book lungs.

The subphylum containing the most species and numbers of organisms is Mandibulata. This subphylum is divided into four classes: (1) *Crustacea* (kruhs TAY shuh), represented by shrimp, crayfish, crabs, and lobsters; (2) *Chilopoda* (kī LAH puh duh), represented by centipedes; (3) *Diplopoda* (dih PLAHP uh duh), represented by millipedes; and (4) *Insecta,* represented by grasshoppers, butterflies, bees, ants, and beetles. Members of this subphylum have **antennae,** which are sensory appendages, and **mandibles,** which are mouth parts adapted for chewing and tearing food. More detailed descriptions of these classes will appear in later sections of this chapter.

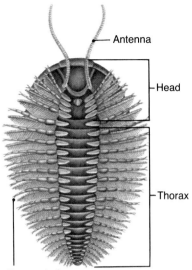

FIGURE 23·2 This drawing shows what the ventral side of a trilobite may have looked like. Note the feathery gills between the legs.

FIGURE 23·3 The dorsal and ventral surfaces of a horseshoe crab are shown here. The long, pointed tail is not a stinger. If a horseshoe crab is turned upside down, it uses its tail to right itself.

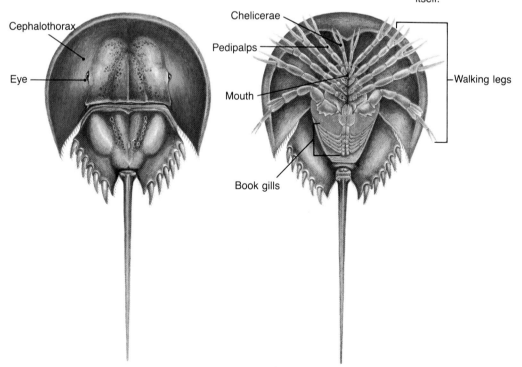

23·3 SUCCESS OF ARTHROPODS

Is the phylum Arthropoda a successful phylum? Success in the biological world can be judged in several ways. A "successful" group or phylum is one that has members living in many different habitats. Arthropods live in marine, freshwater, and land habitats. They have been found in hot springs as well as on the snows of the Arctic.

Another way to judge success is by the number of different species present in a phylum. Figure 23·4 shows a graph representing all the animals on the earth. As you can see, arthropods make up 80 percent of all known animal species.

A successful phylum is also indicated by the numbers of new individuals produced by members of the phylum. Some female arthropods lay a very large number of eggs. These eggs need only a short period of development to reach the adult form. In some insect species, several egg layings can occur in one season. Other species of insects produce large numbers of offspring because there are many more females than males.

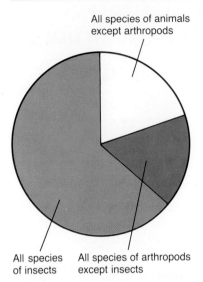

FIGURE 23·4 There about 1,250,000 species of animals on earth. About a million species are arthropods. About three-quarters of the arthropods are insects.

23·4 IMPORTANCE OF ARTHROPODS

Arthropods are important to many other organisms, including humans. Crustaceans are an important part of the diets of many fish, such as herring and mackerel. The blue whale feeds exclusively on floating animal and plant life made up largely of crustaceans called krill. Many other marine organisms depend either directly or indirectly on krill for food. Crustaceans such as lobsters, shrimp, and crabs are considered delicacies by many people.

Insects are an important part of the diets of many animals, especially certain birds and mammals. Insects carry out another important role—pollinating flowers. Fruit growers often place beehives in their orchards. The bees help ensure that a great number of flowers will be pollinated and therefore that more fruit will be produced for sale. Insects are also valuable to humans for certain products that they produce, such as honey, wax, and silk.

Arthropods—insects, in particular—eat more crops than any other group of animals. In fact, insects are our chief competitor for food. For many years scientists have tried to control insect populations by chemical means. Their efforts have resulted in controlling the numbers of insects, not in eliminating any species of insect.

REVIEW QUESTIONS

1. Describe three distinguishing characteristics of arthropods.
2. List three subphyla of arthropods and give an example of an animal in each.
3. Give two indications that the arthropods are a successful phylum.
4. In what ways are arthropods useful to humans?
5. How does the exoskeleton of arthropods aid survival in a terrestrial environment?

Kinds of Arthropods

23·5 ARACHNIDS

Spiders, scorpions, ticks, mites, and daddy longlegs are arthropods in the class Arachnida. A number of arachnids are shown in Figure 23·5. Recall that arachnids are a large class in the subphylum Chelicerata. With about 60 000 species, arachnids are very common animals. Every arachnid has four pairs of walking legs, chelicerae, pedipalps, simple eyes, and no antennae. The simple eyes are called **ocelli** (oh SEHL lī). They detect light but do not form images. The body of an arachnid has two divisions—the cephalothorax and the abdomen.

With over 30 000 species identified, spiders are the most plentiful of the arachnids. The abdomen of the spider is oval and unsegmented. It is joined to the cephalothorax by a narrow waist. The cephalothorax of the spider has four pairs of simple eyes. The chelicerae are modified as fangs that are joined to poison glands. The poison of only a few species of spiders is dangerous to humans. The pedipalps of spiders are short sensory appendages that function mainly in tearing and chewing their prey. In the male spider the pedipalps become specialized for transferring sperm into the genital opening of the female during mating.

Spiders are carnivores and feed mostly on insects. They capture their food by hunting or trapping. Once the prey is caught, the fangs inject both poison and digestive enzymes into the prey. The poison causes paralysis and the enzymes begin digestion. Once the body of the prey has been digested, the spider sucks up the juices.

SECTION OBJECTIVES

- Compare the internal structures of the crayfish and the grasshopper.
- Describe the structures necessary for gas exchange in arthropods.

FIGURE 23·5 (A and B) Ticks are parasites which drink their hosts' blood. (C) A scorpion grabs its prey with its claws and paralyzes or kills it with the poison stinger at the end of its tail. (D) Black widow spiders can be identified by the red hourglass on their ventral side. They have a venomous bite. (E) The daddy-longlegs uses its legs as feelers, as well as for walking.

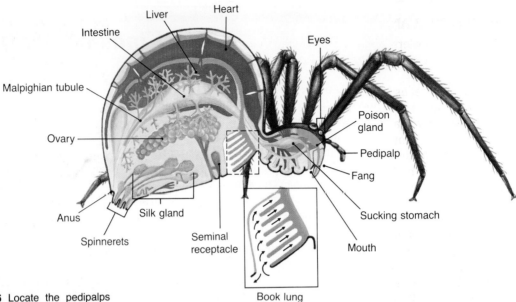

FIGURE 23·6 Locate the pedipalps and poison fangs on this diagram of a female spider. Note the large tubular heart, spinnerets and book lungs. The book lungs in modern-day arachnids evolved from the book gills of aquatic ancestors. Why does the book lung have so many "pages"?

FIGURE 23·7 Find the cephalothorax, waist, and abdomen of this garden spider. How does the spider's silk help it survive?

Spiders have an open circulatory system. The blood is not always contained within vessels and flows freely within the body cavity, or *hemocoele* (HEE muh seel). The blood contains an oxygen-binding pigment, *hemocyanin* (hee muh SĪ uh nihn), which aids the blood in transporting oxygen to the cells of the organism. The blood is pumped to the anterior end of the body by a large dorsal tubular heart. The blood is pumped through several blood vessels and then out into the body cavity. As the blood flows over the body cells, nutrient and gas exchange occurs. Eventually the blood reenters the heart through several openings called **ostia** (sing., *ostium*).

Gas exchange in spiders is carried out by book lungs, by tracheal tubes, or by both. These organs provide an internal gas exchange surface, an adaptation to life on dry land. Air enters the organism through **spiracles** (SPIHR uh kuhlz), which are openings found on the ventral surface of the body. **Book lungs** are a series of air-filled plates with a large surface area. Oxygen and carbon dioxide exchange occurs by diffusion through these air-filled plates. Oxygen, which diffuses into the blood, joins with the hemocyanin. The oxygen is then carried to all cells of the body. Tracheal tubes, which carry oxygen and carbon dioxide, may also be present. **Tracheal tubes** are tubes extending into the body tissue. Oxygen can diffuse directly from the tracheal tubes into the tissue.

At the posterior end of the spider are several specialized appendages called **spinnerets,** from which silk is released. Each spinneret is attached to a silk gland that produces silk. Spiders use silk in several ways. Some spiders spin webs that are used to trap their prey. Spiders also use silk as a line to lower themselves from one place to another. In addition, spiders use silk to encase their eggs in cocoons.

23·6 CRUSTACEANS

Crayfish, lobsters, shrimp, crabs, barnacles, and pill bugs are arthropods in the class Crustacea. All members of this class are called *crustaceans*.

Most crustaceans live in water. Lobsters, shrimp, crabs, and barnacles are animals that live in the oceans. Crayfish and *Daphnia*, a microscopic water flea, live in fresh water. A few crustaceans, such as pill bugs, live in damp places on land. Many different feeding habits are found among crustaceans. Many are free-living predators, scavengers, or filter feeders. A filter feeder is an organism that gets its food by filtering it out of the water. Some crustaceans are parasitic.

There are about 25 000 species of crustaceans. Organisms of this class have a pair of grinding mouth parts called mandibles. They also have two pairs of **maxillae** (mahk SIH lee) (sing., *maxilla*), which are feeding organs, and two pairs of antennae. One pair of antennae, called antennules, is smaller than the other pair. The bodies of most crustaceans are segmented. The segments are made up of two divisions, the cephalothorax and the abdomen. In most crustaceans the exoskeleton is hardened with calcium compounds in addition to chitin. A shield of chitin, the **carapace** (KAR uh pays), covers the dorsal surface of the cephalothorax.

23·7 THE CRAYFISH

The crayfish, *Cambarus,* is a typical crustacean. Various species of crayfish inhabit many freshwater environments. They usually live along the bottom of lakes, ponds, and streams. Crayfish grow to a length of about 15 cm.

The body of the crayfish is divided into two main parts. The first part, the cephalothorax, is the head and thorax fused together. The second part is the abdomen. Both parts are made up of fused segments. The cephalothorax is made up of the five fused segments of the head and the eight fused segments of the thorax. The abdomen is made up of six separate segments.

On the head are compound eyes at the ends of short stalks. The crayfish has a pair of long movable antennae that receive stimuli from the environment. The short antennules also act as sensory organs. They are thought to aid in maintaining the crayfish's balance. The second pair of maxillae have special flaps called *gill bailers.* These flaps draw water into the body and pass it over the gills.

On each of the eight segments of the thorax is a pair of appendages. Each appendage is branched. One branch is partially enclosed in the carapace. This branch is modified as a gill. The other branch is modified to serve various other functions. The first three pairs of thoracic appendages are maxillipeds (mahk SIHL uh pehds). **Maxillipeds** are appendages adapted for feeding. The fourth pair of thoracic appendages, the **chelipeds** (KEE luh pehds), are large pincers adapted for capturing food and for defense. The last four pairs of thoracic appendages are *walking legs,* which function in locomotion.

FIGURE 23·8 Most crustaceans are marine animals that are active predators or scavengers. These two organisms are exceptions to the general rule. The pill bug (top) lives on land. The barnacles (bottom) attach themselves to rocks, piers, and other hard surfaces in the ocean. They filter food from the water with their feathery legs.

maxilla (jaw)
pedis (foot)
chele (claw)

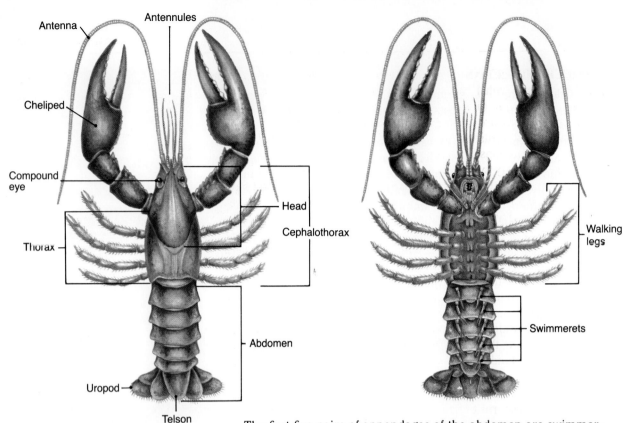

FIGURE 23·9 Dorsal and ventral views of a crayfish are shown here. Locate the major body regions and the appendages. How do the appendages help the crayfish survive?

The first five pairs of appendages of the abdomen are swimmerets. **Swimmerets** are appendages that aid both in gas exchange and in reproduction. Swimmerets help to circulate water over the gills. The first two pairs of swimmerets on males are tubular and transfer sperm to females during mating. The last three pairs of swimmerets in females carry the fertilized eggs and the young. The last pair of abdominal appendages are the uropods. **Uropods** are wide, flat appendages that form part of a posterior paddle enabling the crayfish to swim backward rapidly. The last segment extends to form a flat, taillike structure called the **telson.**

In the crayfish, food that has been ground up by the mandibles is passed into the mouth, and it moves down the esophagus into the stomach. From the stomach, partially digested food passes into the large digestive gland, where digestion is completed and nutrients are absorbed. The undigested part of the food moves into the intestine and out the anus.

The crayfish, like all other arthropods, has an open circulatory system. It has a dorsal tubular heart in a cavity, the pericardial sinus. When the heart pumps, blood is forced through arteries that distribute it throughout the body. The arteries end in body spaces called *sinuses.* The blood fills the sinuses, bathing the cells. Nutrients and oxygen enter the cells. Carbon dioxide and wastes leave the cells and enter the blood. Eventually the blood collects on the ventral side of the thorax, where it bathes the internal surfaces of the gills. From the gills the blood is then directed back to the pericardial sinus and into the heart.

Gas exchange in the crayfish is through the gills. *Gills* are specialized structures with large surface areas that aid in the exchange of oxygen and carbon dioxide. The gills, which are branches of the thoracic appendages, are located in the gill chamber. As water passes over the gills oxygen diffuses from the water into the blood. Carbon dioxide diffuses in the opposite direction.

The excretory system of the crayfish consists of a pair of **green glands** located in the head, near the esophagus. As the blood bathes the green glands, wastes are removed. The wastes are excreted through ducts that open at the bases of the antennae.

The nervous system of the crayfish is highly developed. The crayfish has a large ganglion that functions as a brain. It also has a ventral nerve cord with ganglia in many segments. It has a nervous system with many specialized sense organs that connect to a central nervous system by way of nerves. Much of the body surface is covered by fine hairs that are sensitive to touch.

The antennules and antennae are organs sensitive to taste, touch, and smell. Inside the antennules are **statocysts,** which are small sac-like organs that aid in balance. Each statocyst has many fine sensory hairs that are connected by nerves to the brain.

The sexes in the crayfish are separate. At the time of mating, the male transfers sperm to the seminal receptacle of the female with his first and second pairs of swimmerets. The sperm cells remain in the seminal receptacle until mature eggs pass out of the female's reproductive organ. When the female lays eggs, the eggs are fertilized by these stored sperm. The fertilized eggs stick to the ventral surface of the female's abdomen during development. The young crayfish remain there for several weeks after hatching. Upon hatching, the young feed, develop, and molt while still attached to the adult female. After leaving the female, the young molt several more times during the first year.

FIGURE 23·10 Some of the internal organs of the crayfish are seen in this model. Note how muscular the tail is. Crayfish are very similar to lobsters and shrimp.

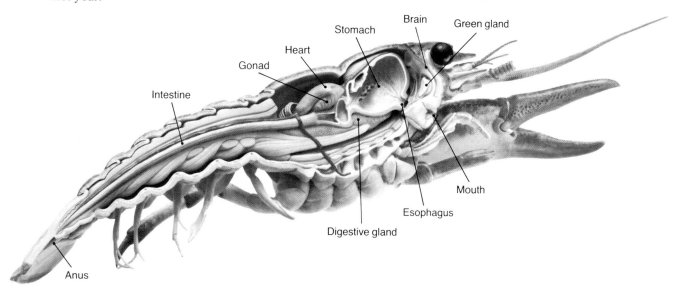

Investigation

What Are the External Features of a Crayfish?

Goals
After completing this activity, you will be able to
- Describe the movements and eating habits of a crayfish.
- Identify the major features of the external anatomy of a crayfish.

Materials (for groups of 2)
live crayfish in partly filled aquarium having a large rock
preserved crayfish
hand lens
probe
small bits of meat
metric ruler

Procedure

Ventral View of Head of Crayfish

Labels: Antennule, Antenna, Mouth, Mandible, First maxilliped, Second maxilliped, Third maxilliped, Cheliped

A. Observe a living crayfish in an aquarium. Describe its method of moving. With a blunt probe gently touch various parts of the animal's body. On a separate sheet of paper, describe its response. Place a few bits of meat on a rock near the crayfish. If it feeds, note how it obtains and eats the food. Record your observations.

B. Examine the preserved crayfish. Identify the cephalothorax, which is made up of the head and thorax. Identify the abdomen. Count the number of segments that make up the abdomen. Look at the last abdominal segment. Describe its shape. This part is the telson. Gently try to bend the two regions of the animal's body. Describe your findings.

C. Turn the crayfish so that you can see the head. Locate the compound eyes, mounted on stalks. Use the drawing below to help in identifying other parts. Look for the long feelers, or antennae. Find the smaller, branched antennules. Locate the large pincers, or chelipeds. Note the part of the body to which they are attached. Open and close the pincers to see how they move. Locate the jaw-like mandibles. Find the first, second, and third pairs of maxillipeds.

D. Turn the crayfish so that its ventral side faces you. Locate the four pairs of walking legs. To which part of the body are they attached? Describe any differences you note in the legs.

E. Examine the ventral side of the first five segments of the abdomen. Describe the appendages on these segments. These appendages are the swimmerets. Look at the appendages on the last segment. These appendages are the uropods.

F. On a separate sheet of paper, make a sketch of the ventral side of the crayfish. Label the following parts: cephalothorax, abdomen, telson, compound eyes, antennae, antennules, chelipeds, mandibles, maxillipeds, walking legs, swimmerets, and uropods.

Questions and Conclusions

1. In obtaining food, which of its appendages does the crayfish use most actively? Describe how the crayfish obtains and eats food.

2. The crayfish is enclosed in an exoskeleton. What are the major features of the exoskeleton? How is it an adaptive advantage to the crayfish?

3. How do you think the parts of the telson are used by the crayfish in locomotion?

4. In what ways does a crayfish use its chelipeds? Can you think of a tool that is similar to the cheliped?

5. Why is it an adaptive advantage to the crayfish to have its eyes on stalks?

6. How many parts does each walking leg have? How do the walking legs differ from one another?

THE ARTHROPODS

23·8 CHILOPODS AND DIPLOPODS

The chilopods and diplopods are arthropods that live on land. They are classified in the subphylum Mandibulata. Figure 23·11 contains photographs of a centipede, class Chilopoda, and a millipede, class Diplopoda.

There are about 3000 known species of centipedes. The body of a centipede is long, made up of a head and many posterior segments. The head appendages are a pair of sensory antennae, a pair of mandibles, and two pairs of maxillae. On the head are two clusters of simple eyes. Each of the segments behind the head has a pair of appendages. The appendages of the first segment are modified as poison claws that are used for capturing prey. Centipedes are predators, feeding mainly on insects. Each of the other segments has one pair of walking legs, which adapts the animal for quick movement. Although the name *centipede* implies that the animal has 100 legs, it may have as few as 30 or as many as 340 legs.

There are about 7500 known species of millipedes. The body of a millipede is long, consisting of a head, a short thorax of four segments, and a long abdomen consisting of many segments. The head appendages are a pair of antennae, a pair of mandibles, and only one pair of maxillae. Two clusters of simple eyes are present. Many of the abdominal segments are fused double segments, each with two pairs of walking legs. The actual number of walking legs varies with the number of segments. Millipedes lack poison claws. They are scavengers, feeding mainly on decaying vegetation.

The internal anatomy of both the chilopods and the diplopods is similar to that of other land arthropods. The circulatory system is an open system with a long tubular heart. They have tracheal tubes for gas exchange and Malpighian (mahl PIHG ee uhn) tubules for excretion. **Malpighian tubules** are excretory structures that remove wastes from the blood. These organs also are present in insects and will be described in greater detail in Section 23·11.

23·9 INSECTS

Insects can be found in most habitable environments on the earth. The only environment in which insects are not present is the ocean, where crustaceans are the dominant arthropods. The class Insecta is the largest group of land-dwelling arthropods. In fact, as you have seen in the graph in Figure 23·4, the insects are the largest class of animals. There are more than 750 000 species of insects. True insects have certain distinguishing characteristics, which are shown in Figure 23·12.

The body of an insect is divided into three main parts—the head, the thorax, and the abdomen. The head has two compound eyes and three simple eyes. The appendages of the head include one pair of antennae and several pairs of mouth parts, including mandibles and maxillae. The thorax is composed of three fused segments. Each segment has a pair of walking legs. In many insects each of the last two

FIGURE 23·11 A millipede (top) defends itself by curling into a ball and secreting unpleasant or poisonous chemicals. The bite of a centipede (bottom) while painful, is not fatal. The bite of some centipedes is like a severe yellow jacket or hornet sting.

FIGURE 23·12 The major body parts of an insect are shown on this ant. Wings are present only on the reproductive females and males before mating. After mating, the males die and the females shed their wings.

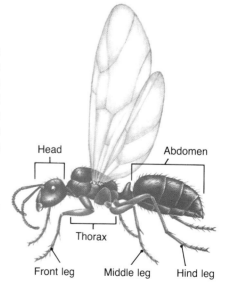

470 THE ARTHROPODS

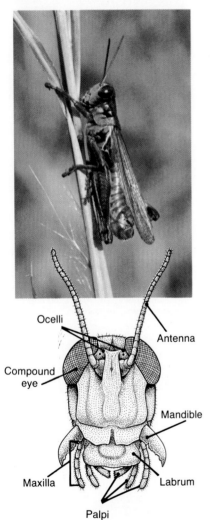

FIGURE 23·13 Locate the grasshopper's well-developed jumping legs in the photograph (top). The drawing (bottom) is a closeup of a grasshopper's head.

segments has a pair of wings. Insects are the only invertebrates that have wings. The abdomen is composed of no more than 11 segments. In adult insects the abdomen lacks walking legs. In many insects the last segments are modified as reproductive structures. As in all arthropods, insects have a chitinous exoskeleton.

23·10 THE GRASSHOPPER: EXTERNAL ANATOMY

The grasshopper is an example of an insect. Grasshoppers are found throughout the world in grasslands, where they feed on leaves. Look carefully at the photograph of a grasshopper in Figure 23·13. Locate the head, thorax, and abdomen. Below the photograph in Figure 23·13 is a drawing of the grasshopper's head. In this drawing you can see the antennae, the mouthparts, the compound eyes, and the ocelli, or simple eyes.

On the head, there are four mouth parts. The **labrum** is the upper lip. The mandibles are the chewing mouth parts and have hardened toothlike surfaces. They move from side to side, cutting and grinding the food. The next two pairs of mouth parts are the maxillae and the labium. The **labium** is the lower lip. Both the maxillae and the labium bear palpi, which are sensory structures that aid in tasting and handling the food.

Look at the thorax, which is composed of three fused segments. Each segment bears a pair of legs. All three pairs of legs are used for walking and climbing. The last pair of legs, which are long and muscular, are adapted for jumping. In addition to legs, the second and third thoracic segments have a pair of wings each. One pair of wings is used for flight; the other hardened pair serve as a protective cover for the flight wings.

The abdomen of the grasshopper contains 11 segments. On the lateral surfaces of each segment is a small opening, the **spiracle,** which leads to the internal gas exchange surface. As in all adult insects, these segments lack appendages. However, the first abdominal segment has a pair of large oval **tympanic membranes,** which are sensory organs for sound. The posterior segments are modified in the male and the female as reproductive structures. In the male the posterior segment forms an organ for internal fertilization. In the female the posterior region is modified as an **ovipositor,** which is an organ used in digging a hole in the ground where eggs can be laid.

23·11 THE GRASSHOPPER: INTERNAL ANATOMY

The internal anatomy of grasshoppers is typical not only of insects but also of other terrestrial members of the subphylum Mandibulata, such as chilopods and diplopods. As you read the descriptions of the various systems, refer to Figure 23·14, which shows the internal anatomy of the grasshopper.

The digestive system of the grasshopper consists of a digestive tract and several accessory glands. Ingestion involves the action of the mandibles, which cut, tear, and crush the food. The maxillae hold and

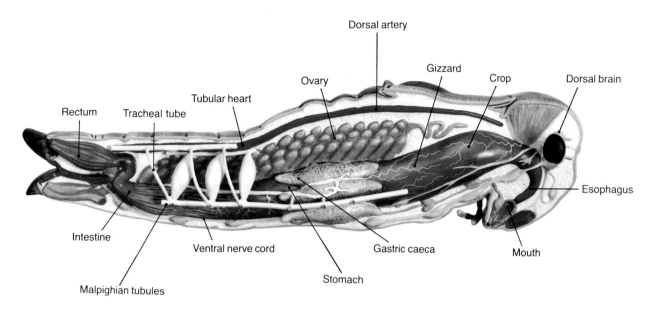

FIGURE 23·14 This model illustrates the internal organs of a grasshopper. The nervous system is shown in yellow, the digestive system in green, circulatory system in red, and respiratory system in white. The Malpighian tubules are shown as blue lines.

cut the food. The labium helps to hold and push the food into the mouth. In the mouth, food mixes with saliva, a digestive fluid secreted by the large salivary glands.

The food is moved through the esophagus into the crop. The *crop* is a thin-walled sac where the food is stored temporarily. While the food is in the crop, some digestion occurs, since the food had been mixed with saliva in the mouth. From the crop the food passes into a muscular organ, the *gizzard*. The interior wall of the gizzard is lined with chitinous plates that mix and grind the food. This food then passes into the stomach. The stomach bears several fingerlike projections, the *gastric ceca* (SEE kuh), which are glands that secrete enzymes into the stomach. Within the stomach, digestion is completed and nutrients are absorbed into the blood. The undigested food moves into the intestine and is forced out through the rectum.

The circulatory system of the grasshopper is an open system typical of all arthropods. It consists of one long vessel located in the dorsal portion of the body. This long vessel, called the heart, is closed at the posterior end. It is extended at the anterior end by a dorsal artery that carries blood into the head. Blood enters the heart through paired ostia, or openings, located in each abdominal segment. The ostia are one-way valves. When the heart contracts, blood is forced frontward into the dorsal artery and then out into the body cavity. As the blood flows over the body cells, nutrients and wastes are exchanged. The insect's blood lacks pigments that bind oxygen. The circulatory system is not involved in the transport of oxygen.

The respiratory system of the grasshopper, which serves for gas exchange and gas transport, is a system of tracheal tubes and air sacs. The tracheal tubes are air-transport tubes that extend to all parts of the body. Air enters through the spiracles, which open and close by means of valves. Air then flows through the branching tracheal tubes. Finally the air passes into the smallest tubes, the *tracheoles* (TRAY kee ohlz). Tracheoles penetrate throughout the body. The tracheoles have moist, permeable membranes. Gas exchange occurs between the tracheoles and the body cells. Oxygen diffuses into the body cells and carbon dioxide diffuses into the tracheoles.

Malpighian tubules are the excretory organs of the grasshopper found in the body cavity at the place where the stomach and the intestine meet. Malpighian tubules are closed tubes that extend into the body cavity, where they are bathed by the blood. Metabolic wastes are absorbed from the blood and emptied into the intestine. These wastes, along with the undigested food wastes, pass out of the body through the anus.

The grasshopper's nervous system, shown in yellow in Figure 23·14, is like that of the crayfish. The nervous system is made up of a dorsal brain, with nerves to the sensory organs of the head. There is also a ventral nerve cord. In most segments the nerve cord contains ganglia with nerves connecting to the body organs. The sense organs of the grasshopper are well developed. They include organs of touch (tactile hairs), organs of smell (antennae), organs of taste (palpi), organs of sight (compound eyes and ocelli), and organs of hearing (tympanic membranes).

Grasshoppers mate in late summer. The posterior end of the male's abdomen is modified as an organ for internal fertilization. During mating, sperm are transferred to the seminal receptacle of the female. In the ovary the female produces up to 200 eggs, each of which is enclosed in a tough shell. With her ovipositor, which is the pointed structure at the end of her abdomen, the female digs a hole in the ground. She then lays the fertilized eggs in this hole. The young hatch in the spring. Development of the young is discussed in the following section.

REVIEW QUESTIONS

6. What is an open circulatory system? What substances does the open circulatory system of the crayfish transport?
7. How do chilopods and diplopods differ?
8. Name three sense organs found in the grasshopper.
9. How do gills, book lungs, and tracheal tubes differ?
10. An unknown arthropod is found. It has the following characteristics: three body divisions, compound eyes, one pair of antennae, and three pairs of walking legs. To what class does this animal belong? Predict what structures it has for gas exchange.

Arthropod Adaptations

23·12 INSECT METAMORPHOSIS

An adaptation among insects is the growth pattern called metamorphosis. **Metamorphosis** is the series of changes through which an organism passes from egg to adult. During metamorphosis, an insect may change in size, body form, feeding habit, and habitat.

In some kinds of insects, metamorphosis is a slow change. The immature animal that hatches from the egg of an insect is called a **nymph.** The grasshopper nymph has no wings. As the nymph grows, it molts a number of times and develops into a sexually mature adult. This process of change from egg to nymph to adult in insects is called **incomplete metamorphosis.** Figure 23·15A shows the incomplete metamorphosis in a grasshopper.

In most insects the life cycle involves a series of distinct changes in form. The larva that hatches from the egg is very different from the adult. An insect **larva** is the wormlike stage that follows the egg stage.

Look at Figure 23·15B, which shows the metamorphosis of the monarch butterfly. The larva of the monarch butterfly feeds on the leaves of the milkweed. The adult butterfly, however, feeds on the nectar in flowers. The larva crawls along the leaves of the plant, but the adult flies from plant to plant. Look carefully at both the young and the full-grown caterpillars. Growth from young to full-grown caterpillar occurs between moltings. The growth periods between molts are called *instars*. During each instar the larva eats a great deal.

SECTION OBJECTIVES

- Discuss the stages of complete and incomplete metamorphosis.
- Describe some behavioral adaptations in social insects.

meta (change)
morphe (form)

FIGURE 23·15 A grasshopper undergoes incomplete metamorphosis (A), while a butterfly undergoes complete metamorphosis (B).

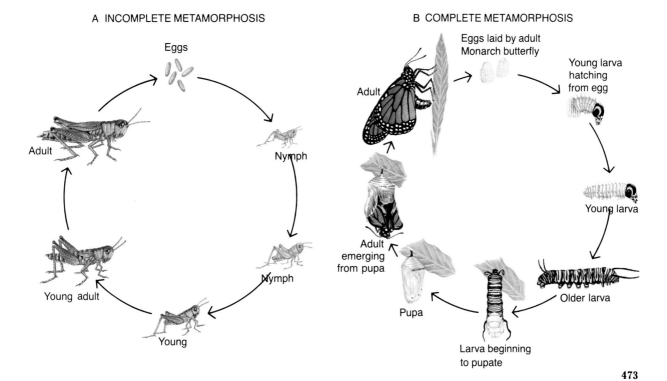

Discovery
Fireflies

How do fireflies make their flashes? In the last segment of the firefly's abdomen, networks of air tubes contain a pigment called luciferin and an enzyme called luciferase. When the firefly lets air into these networks, the luciferase acts as a catalyst in a reaction between the oxygen in the air and the luciferin. The result is energy in the form of light.

Scientists have developed a way to produce light outside the firefly's body by making a mixture of luciferin and luciferase. This mixture glows when exposed to ATP from any living cell. The more ATP present, the brighter the glow. Thus the mixture acts like a light meter, measuring the amount of ATP in a cell.

When the larva completes its growth, it enters a stage of great change called the pupa. A **pupa** is a stage between the larva and the adult. The pupa of a butterfly, shown in Figure 23·15, is called a *chrysalis* (KRIHS uh lihs). The pupa of a moth is encased in a *cocoon*. Although the pupa may appear to be lifeless, within the pupa many changes are taking place. The larval tissues composed of complex molecules are digested. The adult tissues are made from the products obtained. When the changes are complete, the chrysalis breaks open, and the adult emerges. The process of change from egg to larva to pupa to adult is called **complete metamorphosis.**

Insect metamorphosis is controlled by hormones. Three hormones interact to control metamorphosis: a brain hormone, a juvenile hormone, and a molting hormone called *ecdysone* (EHK duh sohn).

Figure 23·16 shows the stages of metamorphosis in the *Cecropia* (sih KROH pee uh) moth. Cells in the brain secrete the brain hormone. This hormone stimulates a specific gland, which then produces ecdysone. Ecdysone is the hormone that must be present if molting is to occur. The juvenile hormone is produced by a part of the brain called the corpus allatum (KOR puhs uh LAY tuhm). If both ecdysone and the juvenile hormone are present in large amounts, the insect will molt from one larval form to another larval form. This type of molt allows for growth.

The amount of juvenile hormone decreases as the insect gets older. When there are only small amounts of juvenile hormone with the ecdysone, the molt will be from larva to pupa stage. After the insect has been in the pupa stage for some time and there is no juvenile hormone present, ecdysone will stimulate the molt from pupa to adult. The length of each stage and the number of larva-to-larva molts varies from one species to another. However, the control mechanism is similar for all insects.

FIGURE 23·16 Hormones control insect metamorphosis, as shown in this illustration of the metamorphosis of the *Cecropia* moth. Scientists are developing ways to control insect populations by using such hormones.

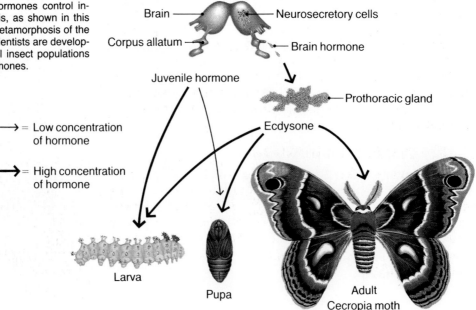

23·13 STRUCTURAL AND PHYSIOLOGICAL ADAPTATIONS

Arthropods have adaptations to a wide variety of habitats. These adaptations give the individual a greater ability to survive and thus increase its chance of producing young. Metamorphosis is a key adaptation that helps account for the success of arthropods. The life cycle of most arthropods is short. For most insects, the sexually mature adult form is reached quickly. Most species are adapted to have large numbers of offspring.

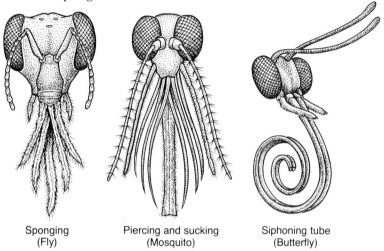

Sponging (Fly)   Piercing and sucking (Mosquito)   Siphoning tube (Butterfly)

FIGURE 23·17 The mouth parts of different insects are adapted to different kinds of feedings. This fly uses its spongy labium to soak up food. The mosquito uses its mouth parts to pierce an animal's skin and suck its blood. The butterfly uses its mouth parts like a straw to siphon nectar from flowers.

The wide variety in the types of mouths has allowed many different insect species to use many different food sources. Look at Figure 23·17. Compare the mouth parts of each insect. The mouth of the housefly is spongy. Compare it with the piercing-sucking mouth of the mosquito, and the siphoning type of mouth of the butterfly.

Some insects show variations in how their bodies function and in the secretions they produce. The salivary glands of grasshoppers produce digestive enzymes. The salivary glands of some caterpillars produce silk for making nests and cocoons. The salivary glands of mosquitoes produce a substance that prevents the clotting of blood in the animal on which they are feeding.

Other adaptations of insects are protective and thus give an insect a greater chance of survival in a certain environment. Can you find the insects in Figure 23·18? The adaptive coloration and form provide very good camouflage.

FIGURE 23·18 The coloration and form of both the treehopper (top) and the dead leaf butterfly (bottom) are adaptations that aid survival.

23·14 BEHAVIORAL ADAPTATIONS

In addition to structural and physiological adaptations, arthropods have behavioral adaptations. They have complex nervous systems with which they detect and respond to changes in the environment. Web building in spiders, for example, is a behavioral adaptation that aids in securing food. Although a spider's sight is poor, it can determine the size of its trapped prey by detecting the intensity of the vibrations of the web.

Other behavioral adaptations involve courtship responses. Courtship behavior is begun by the release of pheromones (FEHR uh-mohnz). **Pheromones** are chemicals released by one animal that cause a response in another animal of that species. Courtship behavior usually involves displays that prepare the animals for mating. In arthropods these displays may serve two functions. They identify members of the same species but of the opposite sex. They also make the female less likely to attack a possible mate. In some spiders, the male presents the female with a small silk-covered package that may contain food. This offering distracts her, identifies him as a proper mate, and allows mating to take place.

23·15 SOCIAL INSECTS

Behavioral adaptations include those found in social insects. Insects that form colonies are called social insects. Honeybees, wasps, ants, and termites are considered social insects. Social organization is needed for survival. It increases defense. It also allows for the division of labor, in that groups of individuals carry out specific functions within the colony. These functions might include defending the colony, getting food, building nests, or rearing young.

The honeybee colony is an example of this social organization. Within the colony are thousands of female worker bees, several drones, and one queen. The drones are the male honeybees that developed from unfertilized eggs. The only role of the drones is to mate with the queen. The drones cannot feed themselves. At the end of the season, they are driven from the hive and left to starve.

FIGURE 23·19 Honeybees are social insects. Their society is highly organized. Some members of the colony serve only in reproduction. Others gather food, take care of the young, or build and repair the hive.

The workers are sterile females, unable to reproduce. They perform all the jobs in the hive except reproduction. The specific job of a worker changes with her age. When a worker emerges from the pupa, she assumes the role of a nurse bee. Nurse bees feed pollen and honey to the larvae, the drones, and the queen. They have special glands on the sides of the head that secrete a nutrient-rich fluid called *royal jelly*. This fluid is fed to the youngest larvae and to the queen. A larva that is being prepared to develop into a queen is fed royal jelly throughout its period of larval development.

After a few days, the workers enter "middle age." The glands that produce royal jelly shrivel up. Wax glands on the abdomen begin to secrete wax. The worker then takes on the new jobs of cleaning, repairing, and building wax cells in the hive. She also guards the entrance to the hive. At about three weeks of age, the worker, now in "old age," becomes a forager. She leaves the hive to gather pollen and nectar. She continues this search for pollen and nectar for a week or two, until she dies.

The queen is the one fertile female in the hive. The queen's only job is to reproduce. Soon after the queen emerges from the pupa, she flies off, followed by the drones. She mates and then returns to the hive. The queen may live for five years. She continuously produces eggs that are fertilized by the sperm she stored during the initial mating.

23·16 INSECT COMMUNICATION

Communication is essential in a social organization as complex as that of a beehive. Bees communicate in several ways. In addition to using pheromones, bees communicate by their movements. When a forager bee finds a source of nectar, she is able to communicate information about her find to the rest of the bees in the hive. She informs the other bees by means of one of two types of dances.

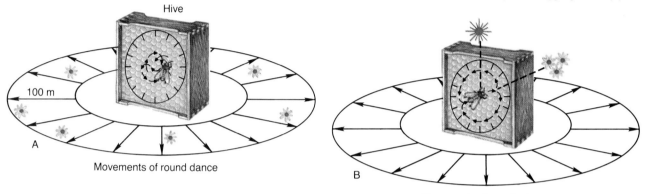

FIGURE 23·20 A worker bee can communicate the location of food to other bees in the hive. The bee dances in circles if the food source is within 100 m of the hive (A). If the food source is beyond a 100 m radius, then the bee will dance in a figure 8, waggling her tail (B).

The simpler dance, the *round dance,* is a dance done by a worker bee communicating that nectar is within 100 m of the hive. Figure 23·20A illustrates the movements of a bee doing the round dance. The bee returns to the hive and releases a drop of nectar. Then she begins to move, circling in one direction and then in the other. While she is moving, the other bees use their antennae to detect the scent from the flower on her body. When they leave the hive, they search for a nearby flower with that scent.

The other type of dance is the waggle dance. The *waggle dance* is a dance done by a worker bee to communicate a source of nectar farther than 100 m from the hive. This dance communicates not only the distance but also the direction of the nectar from the hive with respect to the sun. When the bee returns to the hive, her movements form a figure 8, such as those observed in Figure 23·20B.

REVIEW QUESTIONS

11. What are the stages in incomplete metamorphosis? Give examples of insects that undergo these changes.
12. What are the stages in complete metamorphosis? Give examples of insects that undergo these changes.
13. Identify some structural adaptations of arthropods.
14. Explain the role of each of the following members of a honeybee society: queen, drone, and worker.
15. A student examining a colony of bees found no drones in the hive. What hypothesis might the student make about the queen?

Chapter Review

SUMMARY

23·1 The distinguishing characteristics of an arthropod are a segmented body, a chitinous exoskeleton, and paired jointed appendages.

23·2 Phylum Arthropoda is divided into three subphyla based on the specializations of the body segments and the appendages. The subphyla are Trilobita, Chelicerata, and Mandibulata.

23·3 Arthropods are the most numerous animals in number of different species on the earth, having successfully adapted to most terrestrial and aquatic habitats.

23·4 Arthropods are beneficial to many organisms, including humans, both as sources of food and for the products they produce.

23·5 Arachnids are a class of chelicerates that includes spiders, ticks, and scorpions.

23·6 Crustaceans are aquatic mandibulates. The crustaceans include crayfish, lobsters, shrimp, barnacles, and hundreds of microscopic species.

23·7 The crayfish is a freshwater crustacean that has many typical crustacean features.

23·8 The chilopods and diplopods are terrestrial arthropods. The chilopods are the centipedes, and the diplopods are the millipedes.

23·9 Insects are a class of terrestrial arthropods. The class Insecta represents almost 80 percent of all animal species on the earth.

23·10 The grasshopper is a typical insect. It has three body divisions, three pairs of walking legs, and two pairs of wings on the thorax.

23·11 The grasshopper has a complete digestive system, an open circulatory system, and a complex nervous system. Gas exchange involves a system of tracheal tubes, an adaptation for life on land.

23·12 The life cycle of an insect involves either incomplete or complete metamorphosis. Complete metamorphosis is most common and involves several distinct stages: egg, larva, pupa, and adult.

23·13 The success of the arthropods has been the result of structural and physiological adaptations.

23·14 Behavioral adaptations of the arthropods include food-gathering and courtship behaviors.

23·15 Honeybees, ants, wasps, and termites are social insects that live in complex societies, where division of labor exists.

23·16 For social interactions to occur, communication between individuals is necessary. Communication in arthropods depends on pheromones and movements.

BIOLOGICAL TERMS

abdomen
antennae
arthropods
book lungs
carapace
cephalothorax
chelicerae
chelipeds
complete metamorphosis
exoskeleton
green glands
head
incomplete metamorphosis
labium
labrum
larva
Malpighian tubules
mandibles
maxillae
maxillipeds
metamorphosis
molting
nymph
ocelli
ostia
ovipositor
pedipalps
pheromones
pupa
spinnerets
spiracles
statocysts
swimmerets
telson
thorax
tracheal tubes
tympanic membranes
uropods

USING BIOLOGICAL TERMS

1. Which terms refer to the two stages of complete metamorphosis that follow the egg stage?
2. Mandibles are to insects as _____ are to spiders.
3. Green glands are to Crustacea as _____ are to Insecta.
4. Identify all the structures listed that are involved in gas exchange. Define the terms.
5. Which terms refer to the mouth parts of an insect? Identify their functions.
6. Name the large thoracic appendage used by the crayfish for capturing prey.
7. Distinguish between a nymph and a pupa.

8. What are ocelli?
9. Identify all terms having a word part meaning "foot."
10. Which structure is used by female insects to dig holes and lay eggs?

UNDERSTANDING CONCEPTS

1. Describe the distinguishing characteristics of the phylum Arthropoda. (23·1)
2. Describe two characteristics of subphylum Chelicerata and subphylum Mandibulata. (23·2)
3. List and explain three characteristics of arthropods that account for their success. (23·3)
4. Describe two ways arthropods are beneficial to humans. (23·4)
5. How are spiders adapted to feeding on insects? (23·5)
6. Describe gas exchange in arachnids. (23·5)
7. List three characteristics of the class Crustacea. (23·6)
8. Describe the appendages of a crayfish. (23·7)
9. Describe the circulation of blood in a crayfish. (23·7)
10. Identify two differences between a centipede and a millipede. (23·8)
11. List the body segments of an insect and the appendages of each. (23·9)
12. Identify the thoracic appendages of the grasshopper and tell the function of each. (23·10)
13. Describe the digestive system of a grasshopper. (23·11)
14. Compare gas exchange in the grasshopper with that in a crayfish. (23·11)
15. What is the adaptive advantage of complete metamorphosis to the success of insects? (23·12)
16. How are differences in the structures of mouth parts related to differences in feeding behavior of insects? (23·13)
17. What is the role of pheromones in reproduction among arthropods? (23·14)
18. Describe the division of labor in a beehive. (23·15)
19. What information is communicated by the waggle dance? (23·16)

APPLYING CONCEPTS

1. Discuss the importance of nutrition in the development of the honeybee larvae destined to become workers and the queen.
2. How would a permeable exoskeleton on a terrestrial arthropod affect the type of habitat in which you would find this arthropod?
3. Evaluate the advantages and disadvantages of introducing a species of insect into an area to rid the area of some other type of insect pest.
4. Compare and contrast the social behavior exhibited by insects with the social behavior of humans.
5. Although millions of dollars are spent annually to get rid of insect pests, no pest species has ever been completely destroyed. Offer an explanation to account for this.

EXTENDING CONCEPTS

1. Ants, like honeybees, are social insects. Find out about ant societies, including the roles of members of different social orders.
2. Obtain a taxonomic key to the different orders of insects. Key several insect specimens to the order to which they belong.
3. Devise a method for insect control that would have a minimum number of negative side effects.
4. Design an experiment that would test this hypothesis: Houseflies can detect different colors of light.
5. Prepare a report on the courtship behavior of the black widow spider.

READINGS

Barth, Friedrich G. *Insects and Flowers: The Biology of a Partnership*. Princeton University Press, 1985.

Evans, Howard E. *The Pleasures of Entomology: Portraits of Insects and the People Who Study Them*. Smithsonian Institution Press, 1985.

Pringle, Laurence. *Here Come the Killer Bees*. Morrow, 1986.

Silverstein, Robert M. "Pheromones: Background and Potential for Use in Insect Pest Control." *Science*, September 18, 1981, p. 1326.

24. Comparing Invertebrates

The starfish, coral and murex snail are all marine invertebrates. The starfish has radial symmetry. The murex is protected by a hard, spiral shell. The coral polyps are permanently attached to the other members of their colony. Despite their obvious differences, all these animals share similarities in structure and function. In this chapter, you will compare the various kinds of invertebrates.

CHAPTER OBJECTIVES

After completing this chapter, you will be able to
- **Discuss** how invertebrates perform their life functions.
- **Relate** invertebrate body structure to their functions.
- **Describe** the evolution of invertebrate body systems and body forms.
- **Compare** and **contrast** invertebrate form and function.

Body Plans of Invertebrates

24•1 UNITY IN DIVERSITY

In Chapters 21–23 you studied the traits of animals without backbones. You learned that invertebrates are divided into groups based on their traits. Although the animals in different groups are different, they have common biological needs. This idea can be expressed in the thought "Unity in Diversity." There are common functions, *unity,* in different animals, *diversity.*

The expression "Unity in Diversity" does not just apply to animals. There are certain biological functions common to all living things. All organisms digest food and get it to parts of their bodies. They exchange gases with their surroundings and get rid of wastes. Most organisms also have some form of nervous control.

In this chapter you will examine the structures and functions of invertebrates. In the past chapters you studied examples of invertebrates from different groups one at a time. In this chapter you will study one system, such as the digestive system, at a time. You will see how each system, even though different in different animals, has the same basic function.

24•2 BODY LAYERS AND COELOMS

One of the simplest animal groups is composed of the sponges. Although some specialized cells are present, sponges do not have specialized tissues and organs. There is little coordination of activity among the individual cells. Water moves through channels in the body. Sponges do not appear to be closely related to most other groups of animals.

Cnidarians (nī DAIR ee uhns) are animals whose body plan is made up of two body layers. These layers are the **ectoderm,** or outer body layer, and the **endoderm,** or inner body layer. The endoderm surrounds a *gastrovascular cavity* that has only one opening. Between

SECTION OBJECTIVES
- Compare the body layers of invertebrates.
- Describe the evolution of the coelom.
- Discuss the support system of invertebrates.

ecto- (outer)
derma (skin)
endo- (inner)

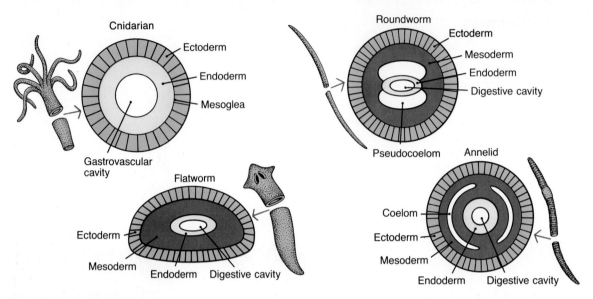

FIGURE 24·1 A cnidarian's body has only two cell layers. Flatworms have three body layers and are also acoelomates. Roundworms are pseudocoelomates. Annelids have a true coelom.

meso- (middle)

koiloma (cavity)

the ectoderm and endoderm is a gelatinous material, the *mesoglea*. These layers are shown in Figure 24·1. The cells of the cnidarians show some degree of specialization. Examples include the endodermal cells, which are specialized for digestion, and nerve cells lying at the bases of ectodermal cells, which coordinate simple body activities such as tentacle movement.

Invertebrates in the remaining major phyla develop a third primary body layer, mesoderm. The **mesoderm** is the middle body layer, which develops from the endoderm at an early stage of embryonic development. The animals with three body layers have specialized tissues and organs. Some of these invertebrates have a **coelom** (SEE-luhm), an internal, usually fluid-containing cavity that is completely surrounded by mesoderm. Sponges, cnidarians, and other groups that have no coelom are called **acoelomates** (ay SEE luh mayts), which means animals without a coelom.

Flatworms differ from sponges and cnidarians in being composed of three body layers, as shown in Figure 24·1. In the flatworms, organ formation involving the mesoderm occurs, but there is no coelom. Like the cnidarians, flatworms are acoelomates.

Roundworms also have three body layers. Roundworms are called **pseudocoelomates** (soo doh SEE luh mayts) because they have a false coelom. As shown in Figure 24·1, the pseudocoelomate has a fluid-filled body cavity between the mesoderm and endoderm. This cavity is not a true coelom because it is not completely surrounded by mesoderm.

Annelids, mollusks, echinoderms, and arthropods have body plans with three primary body layers. In these animals a true coelom forms at some stage. These animals are called **coelomates** (SEE luh-mayts). The true coelom is surrounded by mesoderm, as shown in Figure 24·1. The fluid-filled coelom functions in circulation of substances and is the space in which a circulatory system may develop. Also, specialized reproductive organs form inside the coelom.

24·3 BODY FORMS

Locomotion is the process of moving from one place to another. Although there are exceptions, most invertebrates are able to move around. Organisms that are able to move are said to be motile. Body forms show many adaptations in animals that are able to move. Many motile animals have bilateral symmetry and a body with a definite head region. This streamlining of the body may make moving easier. In addition, the head region has specialized sense organs and a brain. The anterior end of the body is able to detect food and predators.

Many adaptations of animals involve the supporting structures and muscles. The sponges have internal supportive structures in the form of spicules. The spicules may be made of spongin, silica, or calcium carbonate. However, there are no muscles connected with these supporting structures.

Most cnidarians lack any kind of supporting structure. Although the individual polyps of a coral reef live in structures made of limestone, there are no muscles connected with this supporting structure. The limestone serves mostly as protection. In many cnidarians, such as hydras, a small amount of motion is brought about by musclelike fibers located in the ectoderm and endoderm layers.

Though worms can move more than cnidarians can, they lack any special supporting structures. However, they do have muscles that are used in moving. These muscles develop from mesoderm, and they are arranged in layers.

Unlike the worms, some mollusks have muscles that are connected to a supporting structure. Mollusks have a muscular foot adapted for locomotion, and many of them have a calcium carbonate shell. Muscles attach a clam's foot to its shell. By pulling this foot forward and backward, the clam is able to move and to dig into sand or mud. In other mollusks the shell neither supports the body nor works with the muscles in moving. A snail uses its muscles to glide along a surface, but the shell only offers the animal some measure of protection. Arthropods and echinoderms have supportive skeletons that, together with muscles, aid in movement. These skeletons may be made of fibrous proteins or of inorganic compounds of calcium or silicon.

The skeletons of animals are classified as exoskeletons or endoskeletons. An **exoskeleton,** which occurs in arthropods, is an external support system secreted by the epidermal tissue beneath it. The major disadvantage of an exoskeleton is its inability to grow as the body grows. The exoskeleton is periodically molted, or shed. Although another exoskeleton is secreted before the old one is shed, the new covering needs some time to harden after the old one is lost. During this time the animal is vulnerable to attack and injury.

FIGURE 24·2 Muscles pull against the fly's exoskeleton to produce movement. The snail uses its muscular foot to glide along a surface.

exo- (outside)

484 COMPARING INVERTEBRATES

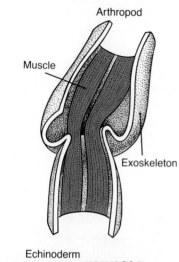

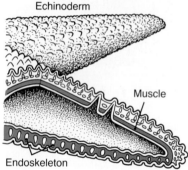

FIGURE 24·3 Arthropods, such as a grasshopper, have exoskeletons. Echinoderms, such as a starfish, have endoskeletons.

The exoskeleton functions in support and locomotion by providing a place of attachment for muscles. The attachment of the muscles of a grasshopper's leg is shown in Figure 24·3. Notice that the muscles are attached to the inner surface of the exoskeleton. In addition, the exoskeleton prevents water loss on land due to evaporation. Since all the organs are contained within the exoskeleton, what other function might the exoskeleton serve?

An **endoskeleton** is an internal support system to which most body muscles are attached. The major functions of an endoskeleton are maintaining the shape of the organism, anchoring the muscles used in locomotion, and protecting the vital internal organs. Unlike an exoskeleton, an endoskeleton usually can grow. Molting does not occur. The echinoderms are the only invertebrates that have an endoskeleton. This skeleton is composed of calcium carbonate plates to which skin and muscles are attached. Figure 24·3 shows an arm of a starfish. Note that the plates that make up the skeleton lie under the skin. Muscles located beneath these plates allow the starfish to bend its body.

REVIEW QUESTIONS

1. Distinguish between the following organisms: acoelomates, pseudocoelomates, and coelomates.
2. Identify some advantages of the presence of a coelom.
3. What adaptations of body form are found in motile organisms that are not found in sessile organisms?
4. How does an exoskeleton differ from an endoskeleton?
5. Suppose a newly discovered organism is found to have three body layers. Scientists think that it is a worm but disagree as to which group it belongs in. How could they assign it to the proper group based on body structure?

Systems for Body Maintenance

SECTION OBJECTIVES

- Describe digestion and excretion in invertebrates.
- Differentiate between open and closed circulatory systems.
- Contrast the gas exchange systems of land and water invertebrates.

24·4 DIGESTION

All animals need the same nutrients. Amino acids, glucose, fatty acids, and glycerol are the building blocks from which animals make larger compounds. However, these simple molecules are not often found in the environment. Instead, animals take in large molecules—proteins, starches, and lipids—and break them down. Animals have many different structures that break down these foods. Although these structures may vary, the enzymes that break down foods are similar in most animals.

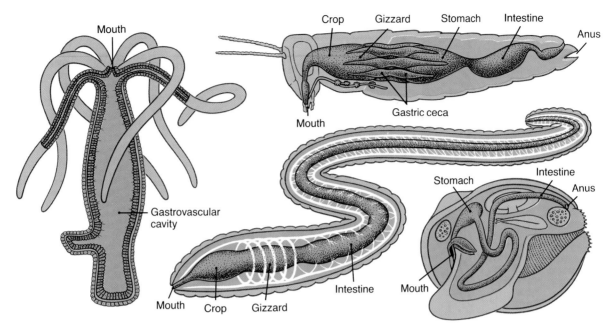

FIGURE 24·4 A hydra has an incomplete digestive system with only one opening. The grasshopper, earthworm and clam all have complete digestive systems.

Like the amoebas and other protists, sponges digest food inside their cells. In **intracellular digestion,** enzymes break down the food inside food vacuoles. As the food vacuoles move in the cytoplasm, the end products of digestion are distributed throughout the cell.

The cnidarians, such as the hydras, possess a saclike digestive space, the gastrovascular cavity. Food is taken in through the mouth and passes into the gastrovascular cavity, where extracellular digestion begins. In **extracellular digestion,** enzymes are secreted into the gastrovascular cavity, where the food is broken down. The partially digested food then is engulfed by specialized cells that line the gastrovascular cavity. Digestion is completed inside the food vacuoles of these cells. Undigested food is released through the mouth. The hydras are said to have a two-way digestive tract because food enters and leaves through a single opening.

The flatworms also have a saclike digestive cavity with only one opening. Both extracellular and intracellular digestion occur in flatworms. The digestive cavity of a flatworm has more branches than that of a cnidarian. What function do these branches serve?

The roundworms, annelids, arthropods, and echinoderms have a "tube-within-a-tube" body plan. The internal tube is the digestive system, and the outer tube is made up of the muscles and skin—and skeleton, if present. All these phyla have a one-way digestive tract with two openings—a mouth and an anus.

The one-way digestive tract includes specialized organs. In the annelids, for example, food is ingested through the mouth and passes through the esophagus to the crop. Food is temporarily stored in the crop. From the crop, food passes to the gizzard, where it is ground. From the gizzard, food passes into the intestine, where the chemical breakdown of food takes place. After this process is completed, the end products are absorbed. Undigested food exits through the anus.

COMPARING INVERTEBRATES 485

The digestive systems of arthropods and some mollusks are similar to the digestive system of the annelids. Observe Figure 24·4. Note the similar organs in each of the organisms. Locate organs in the grasshopper's digestive tract that are not found in the earthworm.

24·5 CIRCULATION

Circulatory systems transport nutrients and, usually, oxygen, to body cells. The simple invertebrates, such as sponges, cnidarians, and flatworms do not have circulatory systems. Each cell of these animals is in close contact with its watery environment. Oxygen and carbon dioxide exchange takes place by diffusion. The many branches of the digestive cavity of flatworms result in the diffusion of nutrients to all cells.

The roundworms have some cells that are not in close contact with the external environment. Yet these worms have no circulatory system. The fluid in the pseudocoelom serves as a liquid that carries nutrients throughout the worm's body.

Echinoderms have no circulatory system. Like the pseudocoelom of roundworms, the fluid-filled coelom of echinoderms carries nutrients and wastes. The other invertebrates have circulatory systems.

There are two types of circulatory systems found in animals. The **closed circulatory system** is that in which the transport medium, blood, is always contained within vessels. The **open circulatory system** is that in which the blood is not always contained within vessels and there is direct contact between the blood and the body cells.

Figure 24·5 shows the closed circulatory system of the earthworm. Note the many small blood vessels, called capillaries, of the intestine. Oxygen diffuses from the environment through the moist skin and the capillaries and into the blood. As the blood moves throughout the body, oxygen diffuses from the blood into the cells.

Arthropods and some mollusks have an open circulatory system. As illustrated in Figure 24·5, the grasshopper's blood collects in the heart and is pumped through the dorsal blood vessel into *sinuses*, or open spaces. As the blood surrounds the body cells, nutrients diffuse into the cells, and wastes diffuse into the blood.

FIGURE 24·5 An earthworm has a closed circulatory system. Materials diffuse through the walls of the blood vessels. A grasshopper has an open circulatory system. Blood bathes the tissues directly.

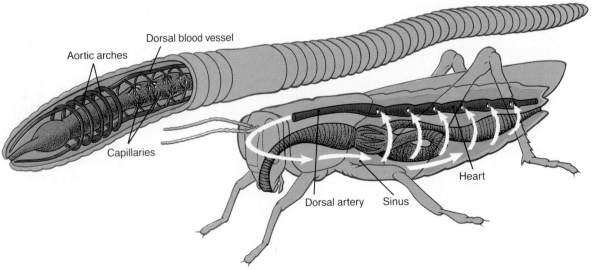

Investigation

What Are the Differences in the Plans and Structures of Two Invertebrates?

Goals

After completing this activity, you will be able to
- Name the body layers in a cnidarian and a segmented worm.
- Identify the differences in the body plans and structures of the two invertebrates studied.

Materials (for groups of 2)

microscope
slide of a cross section of *Hydra*
slide of a cross section of *Lumbricus terrestris*

Procedure

A. You will be looking at slides of the cross sections of two invertebrates. These animals represent two major body plans. As you look at each slide, refer to the drawings on this page. The drawings will help you interpret what you see.

B. Obtain a prepared slide of a cross section of *Hydra*, a cnidarian. Look at the organism under low power. Switch to high power when you wish to see a specific structure in greater detail. Locate the epidermis **(1)**, or outermost layer of cells. Find the gastrodermis **(2)**, or innermost layer of cells. The epidermis develops from the ectoderm. The gastrodermis develops from the endoderm.

C. Locate the mesoglea **(3)**, which is material between the epidermis and the gastrodermis. The epidermis consists of many different types of cells. You may be able to find cells called cnidocytes **(4)**. Each of these special cells contains a stinging structure called a nematocyst

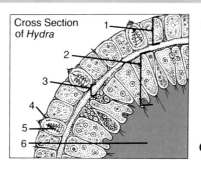

Cross Section of *Hydra*

(5). Locate a cnidoblast with its nematocyst. Look for the gastrovascular cavity **(6)** in the center of the cross section of *Hydra*.

D. On a separate sheet of paper draw what you see and label the epidermis, gastrodermis, mesoglea, and gastrovascular cavity. Label one cnidoblast.

E. Obtain a slide of a cross section of *Lumbricus terrestris,* an earthworm. Find the thin layer called the cuticle **(7)**. Just below the cuticle, look for a layer of cells called the epidermis **(8)**. The epidermis develops from the ectoderm. Below the epidermis are two muscle layers— an outer circular layer **(9)** and an inner longitudinal layer **(10)**. The outer layer runs circularly around the worm. The inner layer runs the length of the worm.

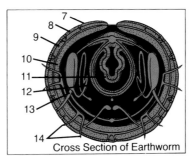

Cross Section of Earthworm

F. Study the center of the slide. You will see an infolded structure, the digestive tract **(11)**. Between the muscle layers and the digestive tract is a body cavity, the coelom **(12)**. The coelom is entirely surrounded by the muscle layers, which develop from the mesoderm.

G. Refer to the drawing of the earthworm. Locate a nephridium **(13)**, which is an excretory structure. Find one nephridium on each side of the worm's body. You should also be able to see the four pairs of setae **(14),** or bristles. Recall that the earthworm uses its setae to cling to the soil.

H. On a separate sheet of paper, draw what you see. Label the cuticle, epidermis, muscle layers, digestive tract, coelom, nephridia and setae.

Questions and Conclusions

1. What are the main body layers of *Hydra?* What is the name of the material between these two layers?

2. Find out the meaning of the term *acoelomate.* Which of the two organisms studied may be called an acoelomate? Give a reason for your answer.

3. Why do you think the cnidarians are sometimes called "animals with stinging cells"? Describe the appearance of a stinging cell.

4. Compare the structure of *Hydra* and the earthworm regarding (a) presence or absence of a coelom and (b) complexity of body layers.

5. Explain the relationship between the digestive tract, the muscle layers, and the coelom in the earthworm.

COMPARING INVERTEBRATES

24·6 GAS EXCHANGE

The process of cellular respiration is similar in all animals. The chemical energy stored in food is changed into a form that the organism can use. This process usually involves oxygen, with carbon dioxide produced as a waste product. Invertebrates have many different structures that are used in the exchange of gases. However, diffusion across a moist, thin membrane always is involved.

In an aquatic environment, membranes for gas exchange stay moist because of the water around them. Land animals have specialized structures that remain moist despite the dry air. The cells of sponges, cnidarians, flatworms, and roundworms are either in contact with the external environment or only a few cells away. How does the exchange of oxygen and carbon dioxide take place?

In annelids such as the earthworm the skin is kept moist by mucous secretions. It is the surface through which gas exchange occurs. The closed circulatory system includes capillaries in the skin. These capillaries, as well as capillaries in other body tissues, are the locations where the exchange of gases between the cells of the body and the blood takes place.

Aquatic mollusks and arthropods, such as the crayfish, have gills. As blood flows through the gills, oxygen from the water enters the blood, and carbon dioxide passes into the water. As the blood flows through the body, oxygen enters the cells, and carbon dioxide is removed.

Like other animals that live on land, the grasshopper does not have gills. While the exoskeleton prevents drying out of the body, it also blocks the exchange of gases through the surface of the body. Instead of gills, the grasshopper has a system of *tracheal tubes,* as shown in Figure 24·6. The many branches of these tubes end in moist membranes, where gas exchange takes place. Although the grasshopper has a circulatory system, the system is not involved in the exchange or transport of gases. Look at Figure 24·6 and compare gas exchange and transport in terrestrial and aquatic arthropods.

FIGURE 24·6 In land arthropods, like this grasshopper, gas exchange occurs in the tracheal tubes. In aquatic arthropods, like this crayfish, gas exchange occurs in the gills.

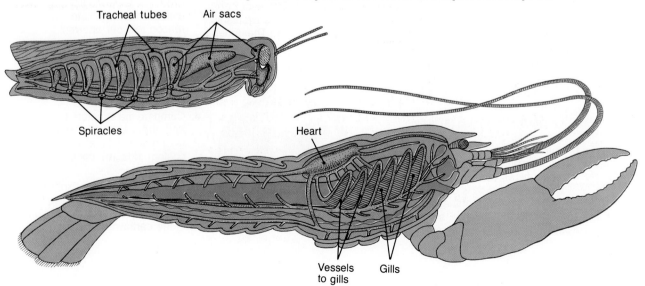

24·7 EXCRETION

The removal of the metabolic waste products of an organism is called excretion. These metabolic wastes include carbon dioxide, water, mineral salts, and some form of nitrogenous waste. Wastes containing nitrogen are produced when amino acids are broken down. These wastes include the highly toxic *ammonia,* the less toxic *urea,* and the insoluble and even less toxic *uric acid.*

Invertebrates that live in water usually excrete nitrogenous waste in the form of ammonia. This waste is toxic, but it is water soluble and diffuses through cell membranes.

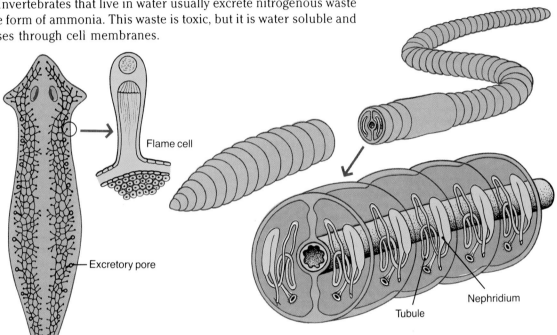

FIGURE 24·7 A planarian's metabolic wastes diffuse through the body surface. Flame cells remove excess water. An earthworm's nephridia filter out wastes and carry the wastes outside the body through pores.

Planarians give off most of their nitrogenous waste as ammonia, which diffuses through the body surface. These worms have a flame cell system as well. This system, as shown in Figure 24·7, is made up of two networks of tubules with many openings called excretory pores. Some branches of these tubules connect with cells that contain clusters of cilia. These cells are called *flame cells.* The cilia move water and possibly some waste materials along the tubules to the outside of the worm's body. Thus flame cells seem to function in excretion as well as water balance.

Land organisms, such as earthworms, excrete urea. The earthworm has a pair of *nephridia* (nih FRIHD ee uh) in most body segments. These organs filter wastes from the coelomic fluid. In Figure 24·7 notice the coiled tubule of the nephridium. Useful salts are reabsorbed from the watery waste in the tubules back into the coelomic fluid. The tubules then carry wastes to the outside of the body. Because urea is a less toxic waste than ammonia, less water is needed to dilute it. Excretion of urea is a major adaptation to life on land, where water conservation is important for survival.

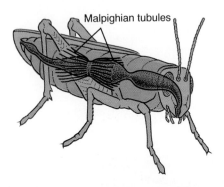

FIGURE 24·8 A grasshopper's Malpighian tubules reabsorb water and useful materials back into the blood, then release wastes into the intestine.

Other invertebrates, such as the terrestrial arthropods, excrete uric acid. In the grasshopper, wastes are removed from the blood by the *Malpighian tubules,* located in the blood sinuses. As the waste-containing solution moves through the Malpighian tubules, some water and useful salts are reabsorbed into the blood. The solution that remains is released from the tubules into the intestine. Still more water is reabsorbed in the rectum. The uric acid crystals that remain then leave the organism in combination with the feces. Uric acid production conserves water, since uric acid is not diluted with water. This adaptation enables insects to live in arid environments.

REVIEW QUESTIONS

6. Distinguish between the two types of circulatory systems in invertebrates.
7. How can the coelenterates and flatworms survive without circulatory systems?
8. Contrast the gas exchange system of a crayfish with that of a grasshopper.
9. What relationship is there between the nitrogenous waste eliminated by an organism and the organism's environment?
10. In what kind of environment would you expect to find an animal with an open circulatory system and tracheal tubes? What is likely to be the form of nitrogenous waste excreted by this animal?

Control Systems

SECTION OBJECTIVES

- Compare nervous systems in invertebrates.
- Explain how hormones regulate body functions in invertebrates.

24·8 NERVOUS SYSTEMS

Most complex animals have a head end with specialized sensory structures, such as eyes. In addition, a "nerve center" localized in the head end interprets stimuli from these sensory structures and quickly coordinates responses. The trend toward having sense organs and brain located in the anterior region is called **cephalization.**

As you might expect, sponges show little coordination among cells. These animals have no sense organs. Some cnidarians are sessile, and some are motile. The mostly sessile hydra has a *nerve net,* but there is no control center. The hydra has cells that respond to certain stimuli, such as touch, by contracting. The hydra also has sensory cells that allow the animal to detect food. The feeding response, begun by the nervous system, involves the release of stinging structures called nematocysts and the contraction of the tentacles. Figure 24·9 shows the nervous system of the hydra. Signals may be carried in either direction along any given nerve cell of the hydra.

The more motile cnidarians, such as the jellyfish, have a somewhat more specialized nerve net than that of the hydra. At the edge of the bell of the jellyfish, nerve cells are organized into nerve rings. These nerve rings coordinate the actions necessary for swimming. These active cnidarians also have more specialized sensory cells than do the sessile forms. Some jellyfish have cells that are specialized to detect light; other cells can detect the pull of gravity.

The flatworms have a simple central nervous system. The anterior region of the organism has *ganglia,* which act together as a control center. Two nerve cords extend along the length of the worm. Various receptors, such as eyespots, detect environmental stimuli. A planarian's nervous system can be seen in Figure 24·9.

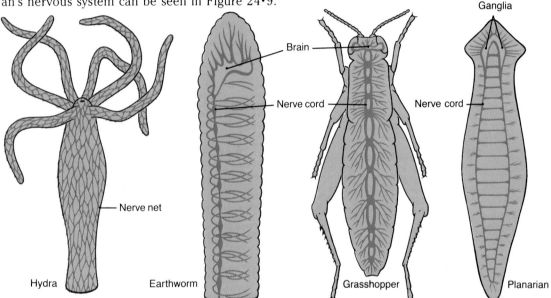

The nervous systems of the annelids and arthropods are similar. In each group, fused ganglia in the anterior end form a simple brain. Each of these animals has a ventral nerve cord and branching nerves from the cord. Each body segment contains small ganglia. Single nerve cells in the nerve cord are specialized to carry nerve signals in one direction, either from the brain to the body or from the body to the brain.

Look at Figure 24·9. Describe the differences in the nervous systems of the hydra, planarian, grasshopper, and earthworm. Note the more numerous and highly developed receptors of the grasshopper. With increased ability to move, animals tend to have more complex nervous systems as well as more numerous and varied sensory structures.

In the mollusks there is great diversity in the development of the nervous system. It is simplest in bivalves, such as clams. These animals lack a brain. Instead, they have three pairs of ganglia, each pair located in the part of the body that it controls. Each pair of ganglia is

FIGURE 24·9 A hydra's nervous system is distributed evenly throughout its body. The nervous systems of an earthworm, a grasshopper, and a planarian show cephalization and the development of nerve cords. The earthworm's nervous system is shown in a side view.

FIGURE 24·10 The eyes of a snail are specialized receptors that are sensitive to changes in light intensity.

FIGURE 24·11 The color of a fiddler crab is determined by pigment-containing cells called chromatophores, which are under the control of hormones.

connected to the other two pairs by two long pairs of nerve cords. A clam's only sense organs consist of sensory cells located near the siphons and on the mantle. These cells are sensitive to touch, chemical changes, and light. The scallop, another bivalve, has a hundred or so light-sensitive eyes located on tentacles at the edge of the mantle.

Snails have a more complex nervous system than do bivalves. Some of them have as many as six pairs of ganglia connected by nerve cords. Also, these animals have eyes that are sensitive to changes in light intensity. The most highly developed molluscan nervous system is found in squids and their relatives. Each of these mollusks has a large well-developed brain. Each has a pair of image-forming eyes on the head, and bundles of giant nerve fibers that control the muscles in its body.

In echinoderms, coordination is centered in the nerve ring located around the mouth. Radial nerves branch off this nerve ring. The radial nerves are connected to a nerve net located on the inside of the oral and aboral surfaces. There is no structure resembling a brain. The only sense organs found in echinoderms are light-sensitive eyespots located at the ends of the arms of starfish.

24·9 HORMONES IN INVERTEBRATES

Hormones are chemicals that are produced in one part of an organism and that affect another part of the organism. Hormones usually move within the circulatory system of the organism. Little is known about the role of some hormones in invertebrates, but in general their function appears to be related to development, reproduction, and regeneration.

Studies have identified some of the hormones that regulate body functions in invertebrates. A hormone from the brain stimulates sexual maturation of the octopus. In some annelids, sexual maturation does not happen until a hormone from the brain stops being produced. Studies of insects have shown that hormones control growth and development. Hormones also are known to influence blood sugar levels, salt and water balance, and protein metabolism in insects.

Most research on hormones in invertebrates has been done on insects and crustaceans. Crustaceans change color and take on the color of their environment. Such changes are under the influence of hormones. In the fiddler crab, shown in Figure 24·11, the exoskeleton has many cells called chromatophores (KROH muh tuh forz). *Chromatophores* are pigment-containing cells that are located on the outer surface of many animals. A crab is pale when the pigment is concentrated in the center of each chromatophore. But the crab becomes darker when the pigment spreads across each cell. A hormone causes the pigment to spread, which can change the color of the crab in a few minutes or even a few seconds. Because of these changes, the crab can protect itself from predators by blending with its surroundings.

In insects, hormones are involved in the control of growth and metamorphosis (meht uh MOR fuh sihs). *Metamorphosis* is the distinct change in body form that an organism undergoes from egg to adult (Section 23•12).

REVIEW QUESTIONS

11. How do the nervous systems of a cnidarian and an annelid differ?
12. Compare the nervous system of an earthworm with that of a grasshopper. What structures are the same in both animals? What structures are found in one animal and not in the other?
13. Name several ways in which a squid, because of its nervous system, is adapted for a more active life than is a clam.
14. Describe the way in which a hormone influences color change in a fiddler crab.
15. Biologists have discovered that hormones stored in the eyestalks of the crayfish prevent the animal from molting. What do you think would happen if a researcher removed the eyestalks of a crayfish?

Reproduction and Development

24•10 REPRODUCTIVE SYSTEMS

There are two types of reproduction: asexual and sexual. Asexual reproduction is any form of reproduction not involving the union of sperm and eggs. The offspring produced by asexual reproduction are genetically identical to the parent.

Three types of asexual reproduction take place in invertebrates: fragmentation, budding, and parthenogenesis (pahr thuh noh JEHN uh-sihs). **Fragmentation** is the separation of an organism into parts, with each part growing into a new individual. Fragmentation happens in some aquatic annelids as well as in sponges, cnidarians, and flatworms. **Budding,** which occurs in hydras, is the formation of an outgrowth on the parent's body. **Parthenogenesis** is the production of monoploid offspring from unfertilized eggs. This process occurs in some insects. Parthenogenesis results in the production of large numbers of offspring in a short period of time.

Sexual reproduction involves the union of two gametes to produce offspring. An offspring receives a new combination of genes and thus is different from its parents.

In some species of cnidarians, flatworms, and annelids, each individual is a hermaphrodite (her MAF ruh dīt). A *hermaphrodite* is an organism that has both male and female reproductive structures. Some hermaphrodites, such as the tapeworm, may self-fertilize. Other hermaphrodites, such as the earthworm, mate and cross-fertilize.

SECTION OBJECTIVES

- Discuss asexual and sexual reproduction in invertebrates.
- Describe the development of invertebrates.

In other invertebrates, such as echinoderms, arthropods, and most mollusks, the sexes are separate. The gonads of a clam surround the intestines. Clams shed their gametes into the mantle cavity. Usually, the gametes then leave the body through the excurrent siphon and are fertilized externally. However, some female bivalves hold their eggs in the space around their gills. Sperm enter the female's mantle cavity through the incurrent siphon and fertilize the eggs. The zygotes develop into free-swimming larvae inside the female, and are later released into the water.

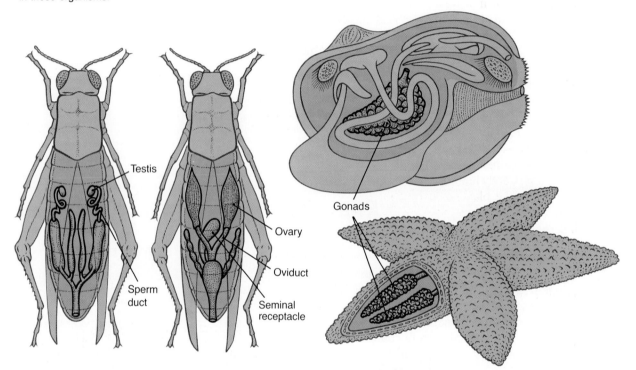

FIGURE 24·12 Compare the reproductive systems of a grasshopper, a clam, and a starfish. The sexes are separate in these organisms.

Unlike the clam, the gonads of a starfish open directly to the outside of its body. Sperm and eggs are released through aboral pores into the sea, where fertilization takes place. The zygotes then develop into swimming larvae.

Look at the views of the reproductive systems of male and female grasshoppers in Figure 24·12. The female produces eggs in two ovaries, which lead to two oviducts. The male produces sperm in two testes, which open into the sperm ducts. Sperm are passed to the female during mating and are stored in the seminal receptacle until eggs pass through the oviducts. Fertilization occurs inside the female's body. The female then uses her pointed ovipositor to deposit the fertilized eggs into the soil, where they will remain through the winter before hatching.

Fertilization is external in most invertebrates that live in aquatic environments. **External fertilization** is the union of sperm and eggs outside the body. Invertebrates that are adapted for external fertilization need only be in the general vicinity of other individuals of the same species for reproduction to be successful. The release of large numbers of sperm and eggs increases the chance that some will unite.

Internal fertilization is the union of sperm and eggs inside a female's body. It occurs in many invertebrates, especially those living on land. Organisms in which internal fertilization occurs usually produce fewer sperm and eggs than do organisms adapted for external fertilization. Though internal fertilization occurs in all land invertebrates, development of the eggs may take place either inside or outside the body of the female.

24•11 EARLY DEVELOPMENT

How does a one-celled zygote become a multicellular animal with different body layers and organs? First, the zygote undergoes a series of rapid cell divisions and becomes an embryo. The series of divisions that follows fertilization is called **cleavage.** Figure 24•13 shows radial cleavage in a starfish. Radial cleavage occurs in echinoderms and vertebrates. Cleavage results in a hollow ball of cells called a **blastula.**

The primary body layers of an animal are formed in the next stage of development. The blastula begins to indent, much like a soft rubber ball when you push your finger into it, forming the **gastrula.** This infolding produces two body layers: the endoderm and the ectoderm.

FIGURE 24•13 A starfish zygote (A) undergoes radial cleavage (B and C). A blastula (D) forms. In the gastrula (E), note the cells moving from the endoderm into the space between the two cell layers. These cells will become the mesoderm. The larva (F) has bilateral symmetry. It will later metamorphize into a young starfish (G) with radial symmetry.

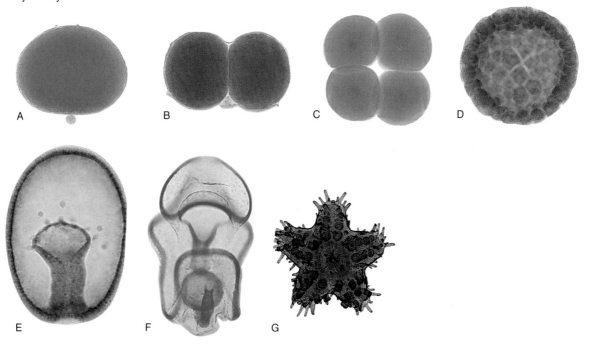

496 COMPARING INVERTEBRATES

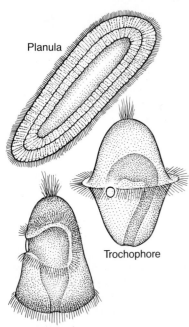

FIGURE 24·14 Many cnidarians have a planula larva (A). Mollusks and annelids have a trochophore larva (B). A dipleurula larva (C) is found in echinoderms.

The third body layer, the mesoderm, forms from cells that move into the space between the endoderm and ectoderm. Figure 24·13E shows the formation of the starfish's mesoderm.

The organs and body systems are formed from folds and pouches of the three body layers. The ectoderm produces the body covering and the nervous system. The endoderm gives rise to the lining of the digestive tract. The mesoderm becomes muscle and also is involved in forming most of the internal organs.

24·12 LARVAL FORMS

In many invertebrates a larval form occurs before the adult stage. *Indirect development* involves the production of an intermediate form, or larva, that does not resemble the adult.

Many adult invertebrates show little or no locomotion. As the more motile larvae of those animals spread out in the environment, they increase the chance of populating a new area. This spreading out also decreases the chance of overcrowding the original area.

Many cnidarians go through a life cycle that has a larval form called a planula. The **planula** type of larva is a ciliated, free-swimming larva that has a body made of two cell layers. Like the adult it becomes, the planula has radial symmetry. Annelids and mollusks have a bilaterally symmetrical, free-swimming larva called a **trochophore** (TRAHK-uh for). Note in Figure 24·14 that the trochophore has a one-way digestive system. Although adult echinoderms show radial symmetry, their larvae show bilateral symmetry. The **dipleurula** (dī PLUR yuh luh) type of larva of echinoderms is a free-swimming larva with a complete digestive system.

Some organisms do not form a larval stage. These organisms undergo *direct development,* in which the immature form is a small copy of the adult. In insects, direct development is called **incomplete metamorphosis,** and the immature form is called a nymph. Other insects show an indirect type of development called **complete metamorphosis,** in which the larva goes through a pupa stage before becoming an adult.

REVIEW QUESTIONS

16. Name three types of asexual reproduction in invertebrates. Why is each type considered to be a form of asexual reproduction?
17. How does external fertilization differ from internal fertilization?
18. What body structures develop from each of the three primary body layers?
19. How is a planula larva different from trochophore and dipleurula larvae?
20. A certain species of insect may lay 200 to 300 eggs within a season of 2 to 3 months. A female starfish may produce over 2.5 million eggs during its mating season. Why does the starfish produce so many more eggs than does the insect?

Careers in Biology

Parasitologist

Parasitologists study the characteristics, habits, and life cycles of parasites such as tapeworms, flukes, and protozoans. Parasitologists use the information they find to develop methods to prevent the spread of parasites or to treat diseases caused by them.

Some parasitologists specialize in only one species or type of parasite. For example, a medical parasitologist studies parasites that attack people. Others are more general in their study. Many parasitologists work with physicians in hospitals or medical research centers. Others work for government health agencies, universities, or private industry.

To become a parasitologist, a person must have an advanced degree. A person with a bachelor's degree in biology may be able to work as a laboratory assistant for a parasitologist. For additional information write to the American Society of Parasitologists, 1041 New Hampshire Street, Lawrence, KS 66044.

Entomologist

Entomologists study insects and their life cycles. They raise insects to use in research and in pest-management programs. Some entomologists study ways to protect food and other products from insect damage during storage.

Entomologists work in many different areas, such as agriculture and human or veterinary medicine. They are employed by federal and state agencies, chemical companies, pest control organizations, and colleges and universities.

A beginning job as an entomologist requires a bachelor's degree with a major in entomology. A strong background in chemistry and mathematics is also important. Because of growing competition for jobs, work experience is also helpful. Many schools offer work-study programs to help students get on-the-job training. For more information write to the Entomological Society of America, 4603 Calvert Road, College Park, MD 20740.

Beekeeper

Beekeepers raise bees to produce honey and pollinate crops. The bees are raised in hives that can be moved from one orchard or field to another. Beekeepers often rent their hives to farmers who need the bees to pollinate crops. Beekeepers remove the honeycombs from the hive in order to collect the honey.

Most beekeepers raise bees as a hobby or as a part-time second job. Opportunities for full-time beekeepers are limited.

No formal education is required for the job of beekeeper. Some knowledge of the habits and behavior of bees, however, is essential. Many high schools and community colleges offer courses on beekeeping. For additional information write to the Bee Industries Association of America, 102 Broadway, Hamilton, IL 62341.

How to Write a Résumé

A résumé is a one-page summary of your education and experience. It tells a potential employer what your qualifications are. Therefore, the résumé should be written in a way that highlights your skills and accomplishments.

The résumé must have your name, address, and phone number. It must also list your work experience, either by date or by job category. For each position you have had, give the employer's name and address and describe your job responsibilities.

Another important section is education, which includes the name of your high school, the courses you took, and any special summer programs related to your career interests. You may also list any honors or awards that you have received. Résumés sometimes include information about hobbies, club memberships, and other activities that may be of interest to a potential employer.

COMPARING INVERTEBRATES

Chapter Review

SUMMARY

24·1 All types of invertebrates have the same basic needs and carry out similar functions. However, a particular system may vary from one invertebrate group to another.

24·2 Except for sponges, invertebrates develop from either two or three primary body layers. While the simplest invertebrates lack a coelom, a coelom exists in annelids, arthropods, mollusks, and echinoderms.

24·3 Although structures involved in body support are found in most invertebrates, only the arthropods and echinoderms have skeletons with muscles attached for locomotion.

24·4 In the simplest invertebrates, the digestive tract has a single opening, and digestion is intracellular. A one-way digestive tract and extracellular digestion are characteristic of complex invertebrates.

24·5 In invertebrates, nutrients and wastes are exchanged between cells and body fluids through diffusion. In annelids, mollusks, and arthropods, specialized circulatory systems distribute nutrients.

24·6 In invertebrates, gas exchange occurs through the moist body surface, through gills, or through tracheal tubes.

24·7 The more complex invertebrates have special excretory organs. The types of nitrogenous wastes that are excreted vary, being adapted to environmental conditions.

24·8 Among the more motile invertebrates there is a general tendency toward greater complexity of nervous systems and to more numerous and varied sensory structures.

24·9 Hormones control many basic changes in the bodies of invertebrates, such as changes associated with metamorphosis.

24·10 Asexual reproduction and sexual reproduction occur among the invertebrates.

24·11 Zygotes become multicellular through the process of cleavage. The three body layers are formed during the gastrula stage.

24·12 Some invertebrate reproductive patterns include a larval stage.

BIOLOGICAL TERMS

acoelomates
blastula
budding
cephalization
cleavage
closed circulatory system
coelom
coelomates
complete metamorphosis
dipleurula
ectoderm
endoderm
endoskeleton
exoskeleton
external fertilization
extracellular digestion
fragmentation
gastrula
incomplete metamorphosis
internal fertilization
intracellular digestion
mesoderm
open circulatory system
parthenogenesis
planula
pseudocoelomates
trochophore

USING BIOLOGICAL TERMS

1. What is parthenogenesis?
2. Distinguish between coelomates and acoelomates.
3. What is a trochophore?
4. Distinguish between an open circulatory system and a closed circulatory system.
5. Which of the three body layers is not found in cnidarians?

Identify the term that does not belong and explain why it does not belong.

6. budding, blastula, gastrula, cleavage
7. ectoderm, endoderm, endoskeleton, mesoderm
8. fragmentation, budding, parthenogenesis, fertilization
9. planula, dipleurula, pseudocoelomate, trochophore
10. complete metamorphosis, incomplete metamorphosis, exoskeleton, intracellular digestion

UNDERSTANDING CONCEPTS

1. Why is it said that invertebrates show unity in diversity? (24·1)
2. Describe the arrangement of body layers in acoelomates, pseudocoelomates, and coelomates. (24·2)
3. What special adaptations are associated with motile organisms? (24·3)
4. What is a major difference between the supporting structures of arthropods and the supporting structures of echinoderms? (24·3)
5. List the differences between the hard outer covering found around many mollusks and the hard covering on insects. (24·3)
6. Identify the type of digestion—extracellular or intracellular—that occurs in sponges; in hydras; in earthworms; and in grasshoppers. How do extracellular digestion and intracellular digestion differ? (24·4)
7. What is a tube-within-a-tube body plan? Give an example of an organism with this body plan. (24·4)
8. How does the closed circulatory system of an earthworm differ from the open circulatory system of an insect? (24·5)
9. Identify adaptations that improve gas exchange between the outside environment and the internal cells of complex invertebrates. (24·6)
10. Explain one reason why the excretion of uric acid by insects represents a valuable adaptation to a terrestrial environment. (24·7)
11. Compare the nervous systems of a cnidarian, a flatworm, and an annelid. (24·8)
12. List three functions of hormones in invertebrates. (24·9)
13. Describe three types of asexual reproduction that occur in invertebrates. (24·10)
14. What is the adaptive advantage of the release of many sperms and eggs by organisms that have external fertilization? (24·10)
15. Describe how a zygote becomes a multicellular embryo. How are the three body layers formed? (24·11)
16. What is a larva? What are the adaptive advantages of producing a larva? (24·12)

APPLYING CONCEPTS

1. Contrast the process of digestion in a cnidarian and an annelid.
2. Design a table that summarizes the control systems of invertebrates.
3. Suppose a new phylum of invertebrates is discovered that has a gastrovascular cavity and a very thin layer of mesodermal cells but shows no organ development. Where would you place this new phylum with respect to the ones studied in this chapter? Explain your placement.
4. Living things often have structures that are designed to increase surface area. Identify some invertebrate structures that are designed for increasing surface area and indicate the function of each.

EXTENDING CONCEPTS

1. Construct or draw a new form of invertebrate having characteristics of one existing phylum or intermediate characteristics of two related phyla. From your completed model or drawing, your class should be able to identify the animal's symmetry, locomotion, nervous system, method of circulation, excretory system, type of digestive system, and based upon its gas exchange structure, whether your new "species" is terrestrial or aquatic.
2. Using colored clay, different colors of yarn, or other items, make three-dimensional models showing the development of the primary body layers and coelom in an organism.
3. Prepare a report on some of the invertebrate phyla not covered in this chapter. Prepare a drawing to be hung in your classroom of a phylogenetic "tree" showing all of these phyla.
4. Using your library, do a report on the *Cecropia* moth experiments that were performed in studies of hormonal control.

READINGS

Blonston, Gary. "To Build a Worm." *Science 84,* March 1984, p. 62.

Epps, Garrett. "The Brain Biologist and the Mud Leech." *Science 82,* January/February 1982, p. 34.

Horsman, Paul V. *Seawatch: The Seafarer's Guide to Marine Life.* Facts on File, 1985.

Unit V Skills

UNDERSTANDING BIOLOGICAL TERMS

Biologists and other scientists use certain terms to describe areas of plants and animals or the position of these areas. Most of these terms are used to describe animals, while a few terms apply to plants.

You have already read about some of these terms in Units IV and V. Some of the terms are directional and come in opposing pairs. For example, *anterior* refers to the forward region of an animal and *posterior* refers to an animal's hind region. The term *dorsal* describes the upper or back area of an animal or one of its parts, while *ventral* refers to the under or front area of an animal or its parts. An additional term, *caudal,* refers to the area near the tail of an animal.

Some terms have specific meanings when used to describe the human body. For example, the terms anterior and posterior describe the front and back of the human body. In humans, the direction toward the head is called *superior* while the direction away from the head is called *inferior.* Sometimes biological terms describe the position of body parts in relation to the whole body. *Medial* refers to the direction toward the midline or center of the body; *lateral* describes the direction away from the midline or center, or to the side of the body. *Proximal* means toward or near the origin of a structure. The direction away from the origin of a structure is termed *distal.* The illustration on this page labels some directional terms on the human body.

You will notice in your biology reading that some terms apply only to plants. For example, *terminal* locates a part at the tip or end of a plant. Some terms apply to both animals and plants. Lateral refers to a plant part that is located at the side or along the side of the plant. The terms superior and inferior can be used to describe the upper and lower parts of plants. For example, a plant's flower is superior to its root.

Directional terms generally refer to the exterior of a plant or animal structure only. Biologists use other special terms to describe the appearance of a structure when it is cut into sections. Two of these terms are longitudinal section and cross section. A *longitudinal section* of a plant or animal is one cut lengthwise, while a *cross section* is made at right angles to the longitudinal axis. You can see both longitudinal and cross sections in Figure 17·7.

To help you become familiar with biological terms and their meanings, answer the following questions on a separate sheet of paper. Sometimes you will need to refer to sections and figures in this book.

1. How do the terms anterior and posterior differ when used to describe human beings and other animals?
2. Look at Section 31·9 in your book. What do the terms superior vena cava and inferior vena cava mean?
3. What is one part of the human body that is most proximal? What part or parts are the most distal?
4. How would you make a lateral movement with your leg?
5. How could you use the terms terminal and lateral to describe the stem of a woody dicot?
6. Look at Figure 19·8 in your book. How would you use the terms superior or inferior to describe the position of the stigma and the receptacle?

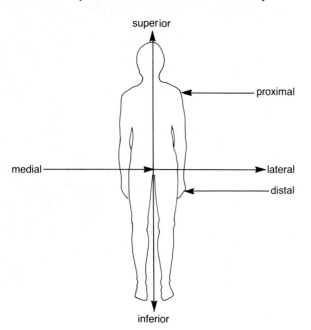

Unit VI

Vertebrates

The African elephant is the largest land animal. An adult male may be 7.5 m long, stand about 3.2 m high, and weigh 6000 kg. In contrast, a dwarf goby fish may be only 8 mm long, shorter than the length of your thumbnail. Despite the enormous differences in size, elephants and gobies are both vertebrates, or animals with backbones. This unit focuses on the specialized systems and behavior of each class of vertebrates.

CHAPTERS

25 *Fish and Amphibians*

26 *Reptiles and Birds*

27 *The Mammals*

28 *Comparing Vertebrates*

25. Fish and Amphibians

This beautiful lionfish uses its outspread pectoral fins to corner smaller fish on which it feeds. Its exotic beauty is prized by aquarium enthusiasts, even though the lionfish is one of the most venomous fish in the world. Its elegant plumelike dorsal fin rays have poison sacs at their base. Although kept as a pet, the lionfish is aggressive and may try to puncture its owner's hand with its poison spines. In this chapter, you will learn more about fish and amphibians.

CHAPTER OBJECTIVES

After completing this chapter, you will be able to
- **Distinguish** between the three subphyla of phylum Chordata.
- **Name** three classes of fish.
- **Describe** the distinguishing characteristics of fish.
- **Identify** the major external and internal organs of a bony fish.
- **Describe** the characteristics of amphibians.
- **Identify** the major external and internal organs of a frog.

The Phylum Chordata

25•1 TRAITS OF CHORDATES

Animals in phylum *Chordata* are called **chordates** (KOR dayts). Chordates have a notochord, a dorsal nerve cord, and gill slits at some time in their life cycle.

A **notochord** is a slim and flexible rod that supports the body. In invertebrate chordates the notochord is the only supporting structure. In vertebrate chordates the notochord most often occurs only in early stages of development. It is then replaced by a supporting column of bone or cartilage.

A **nerve cord** is a tubular bundle of nerves that runs along the dorsal, or upper, surface of the body. In invertebrate chordates the nerve cord lies above the notochord. In vertebrates the nerve cord is enclosed in a column of bone or cartilage.

Paired **gill slits** are structures located behind the mouth in the region of the pharynx, or throat. In the lower chordates the gill slits are involved in food-gathering and, to some extent, gas exchange. In fish the gill slits become the **gills,** which are organs through which oxygen is taken from the water and carbon dioxide is given off. In most larval amphibians external gills develop. In reptiles, birds, and mammals, the gill slits are present only in the early stages of development. These groups develop **lungs,** which are organs in which blood takes in oxygen from the air and gives up carbon dioxide.

All of the lower chordates are marine animals. The lower chordates are divided into two subphyla: *Urochordata* (yoo roh kor DAY-tuh) and *Cephalochordata* (sehf ah loh kor DAY tuh). The higher chordates make up subphylum *Vertebrata* (ver teh BRAY tuh). In this subphylum are the animals with backbones, the **vertebrates**—the fish, amphibians, reptiles, birds, and mammals.

Several traits distinguish the animals in the subphylum Vertebrata. One of these traits is the presence of a *vertebral column,* or back-

> **SECTION OBJECTIVES**
> - Describe the major characteristics of the chordates.
> - Compare and contrast tunicates, lancelets and vertebrates.

noton (back)
chorde (string)

504 FISH AND AMPHIBIANS

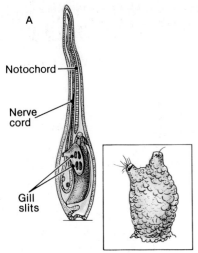
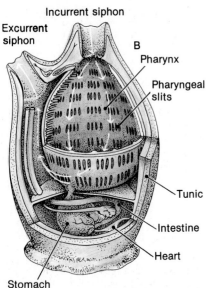

FIGURE 25·1 The tadpolelike larva of a tunicate (A) exists for less than 36 hours. It soon attaches itself to a hard surface and changes into an adult (B). During metamorphosis, the larva loses its notochord, nerve cord, sense organs and tail.

bone. A vertebral column encloses, supports, and protects the spinal cord, or nerve cord. In all vertebrates except the most primitive class of jawless fish the notochord is absent in the adult organism. It is replaced by the backbone in an early embryonic stage. All vertebrates have an *endoskeleton*. This skeleton provides protection for internal organs and is an attachment site for the muscles that aid in locomotion. Unlike an exoskeleton, this internal skeleton contains living cells. The skeleton grows as the organism grows. It does not have to be shed in periodic molts.

25·2 THE LOWER CHORDATES

Organisms in subphylum Urochordata are commonly known as **tunicates** (TOO nuh kihts). Tunicates are invertebrate chordates. Tunicates have a tunic, or tough outer covering, formed of a substance similar to cellulose. They range in size from kinds so small they can be seen only with a hand lens to others as much as 30 cm in diameter. Some tunicates are called sea squirts, because they squirt water from openings in the outer covering when touched. Most of the roughly 2000 species inhabit shallow seas, but some are found as much as 5 km beneath the surface of the sea.

The larval form of a tunicate looks much like a tadpole, as you can see in Figure 25·1. Identify typical chordate characteristics in the drawing. It is only at the larval stage that tunicates show all of their chordate features. Adults lack a notochord. As adults, some continue to be free-swimming, but most become sessile, like sponges. Such tunicates are the only chordates not able to move as adults. Most kinds live as separate individuals, but some live in groups or colonies.

Adult tunicates take in water through the incurrent siphon, or mouth opening. The water passes through gill slits in the pharynx. Oxygen is removed from the water as it passes through the body. The tiny organisms that make up the tunicate's food are strained out of the water in the gill slits, and these enter the digestive tract. Water and wastes are expelled through the excurrent siphon. The body contains a simple circulatory system and nervous system.

Organisms in subphylum Cephalochordata are called **lancelets.** There are about 30 species of these small animals living in warm seas throughout the world. *Amphioxus* (am fee AHK suhs) is the common type of lancelet that is studied in laboratories as a representative of invertebrate chordates. As you can see in Figure 25·2, an *Amphioxus* is slim and fishlike in shape.

Most of the time a lancelet stays in a sand or mud burrow with only its mouth sticking out. Cilia within the lancelet's mouth draw in currents of water. The small organisms making up a lancelet's food are strained from the water as it passes through the gill slits. The digestive tract is a branched tube running straight through the body. The dorsal nerve cord, which extends into the head region, does not enlarge to form a distinctive brain. In a lancelet, the notochord is present in the adult.

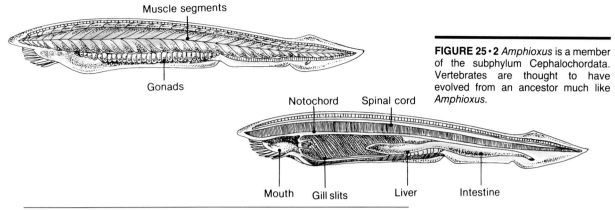

FIGURE 25·2 *Amphioxus* is a member of the subphylum Cephalochordata. Vertebrates are thought to have evolved from an ancestor much like *Amphioxus*.

REVIEW QUESTIONS

1. What are the characteristics of animals in phylum Chordata?
2. How does a urochordate differ from a cephalochordate?
3. How do vertebrates differ from the lower chordates?
4. List some kinds of vertebrates.
5. Tunicates are sometimes called filter feeders. How might this type of food-getting be related to the fact that these two types of organisms have poorly developed systems of locomotion?

The Classes of Fish

25·3 TRAITS OF FISH

Fish are vertebrates that live in water and breathe with gills. Fish are ectotherms, or cold-blooded animals. An **ectotherm** is an animal that obtains most of its body heat from the environment. The body temperature of an ectotherm rises and falls with the environmental temperature. Fish have backbones either of *cartilage,* which is firm but flexible, or of *bone,* which is stronger but less flexible. Most fish are adapted for living in either fresh or salt water. Most fish have **fins,** which are membranes that extend from the body and are supported by rays or spines. Most fish also have **scales,** which cover and protect the body. A fish's body is divided into a distinct head, trunk, and tail. A fish's digestive, circulatory, muscular, and nervous systems are well developed.

There are three classes of fish. The most primitive of the three classes of fish are the jawless fish, *Agnatha* (AG nuh thuh). There are fewer than 50 species in this class. Sharks and rays, of which there are about 600 species, make up class *Chondrichthyes* (kahn DRIHK thee-eez). In members of this class, the skeleton is composed entirely of cartilage. The largest class, with more than 30 000 species, is class *Osteichthyes* (ahs tee IHK thee eez). This class is made up of fish with a skeleton formed largely of bone.

SECTION OBJECTIVES

- List the three classes of fish and give examples of each class.
- Identify the external and internal structures of a bony fish.
- Describe breeding and reproductive behavior in bony fish.

ecto- (outside)
-therm (heat)

FISH AND AMPHIBIANS   505

Profiles

Karel F. Liem (1935–   )

Karel Liem was born in Java, Indonesia. As a high school student, he had a keen interest in biology, especially in plant diseases that affected agricultural crops. His interest in plant diseases lead him to study botany at the University of Indonesia. As an undergraduate, Liem studied under an internationally known Dutch botanist. Liem wanted to continue to study biology but was uncertain of his next step.

While on an early evening field trip with his fellow students, Liem observed an event that led him to become interested in the biology of fish. Liem saw several eel-shaped, air-breathing fish. These "amphibious" fish were traveling across a field, from one body of water to another! Fascinated by the fish, Liem did research on them and wrote his master's thesis on their unique characteristics and behavior. Liem received his master's degree in biology in 1958 from the University of Indonesia.

Liem's research caught the attention of an American biology professor who was visiting at the University of Indonesia. With the help of this professor, Karel Liem was accepted as a graduate student of ichthyology at the University of Illinois. He received his doctoral degree in 1961.

Although he was still very interested in the biology of air-breathing fish, Liem's first position as a professor was in the medical school at the University of Illinois. There he taught human anatomy. While associated with the University of Illinois, Liem was also the head of the division of vertebrate anatomy at the Chicago Natural History Museum.

In 1972, Liem became a professor at Harvard University. Today, Karel Liem teaches introductory biology at Harvard. In addition, he continues his research into the biology of fish. Liem is also involved in research with the New England Aquarium in Boston.

25•4 JAWLESS FISH

cyclo- (circle)
stoma (mouth)

Agnatha, the name of the class to which jawless fish belong, means "without jaws." The living species of jawless fish are called **cyclostomes,** a term that means "circular mouths." There are two types of jawless fish: hagfish and lampreys. Both types have slim round wormlike bodies covered with slime. They lack scales and paired fins. They also lack true teeth, but they do have toothlike structures. There is a single nostril in the middle of the snout. The notochord is present in the adult, but there is also a skeleton of cartilage, including a backbone.

Like all fish, cyclostomes have a two-chambered heart. Blood enters a chamber called the **atrium.** The atrium pumps the blood into a second chamber, called the **ventricle.** The ventricle pumps the blood throughout the body, including the gills. The gills open to the exterior through pores just behind each side of the head. Oxygen and carbon dioxide are exchanged as water passes over the gills.

Reproduction is sexual. Females release their eggs into the water, where they are fertilized by sperm spread over them by the males.

All hagfish live in the sea. They are scavengers. Using the toothlike structures in the circular mouth, a hagfish bores into a dead or dying fish and then feeds on the blood and body fluids.

Lampreys occur in both fresh and salt water. Some are small free-living eellike creatures. The best known lamprey, however, is the parasitic sea lamprey, which may reach a length of 1 m. The sea lamprey has many more toothlike structures in its mouth than does a hagfish. Even its tongue is covered with toothlike structures. The round mouth is a sucker with which the lamprey fastens itself to the outside of its prey. The lamprey then rasps a hole through the body wall with its toothlike structures and feeds on the blood and body fluids until the prey dies.

25·5 CARTILAGINOUS FISH

Sharks and rays make up class Chondrichthyes. Fossil evidence indicates that sharks appeared more than 300 million years ago. Rays did not develop until about 200 million years ago. Both sharks and rays have skeletons of cartilage rather than bone. Paired fins, scales, and jaws distinguish sharks and rays from the more primitive jawless fish. The scales bear toothlike spines, called dermal spines, which are shown in Figure 25·3. Notice that a spine has a coating of enamel, a dentine core, and a pulp cavity. A shark's teeth are similar in structure to the dermal spines. Tooth replacement is continuous in sharks.

Sharks and rays have five to seven gill openings on each side of the body. Sharks take in water through the mouth and expel it through the gill slits. Rays take in water through two openings called *spiracles,* one on each side of the body, just behind the eyes. When the spiracles close, the water is forced out over the gills.

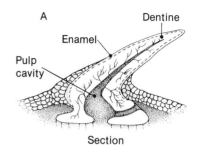

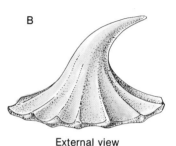

FIGURE 25·3 The spines on a shark's skin are similar to teeth in structure as well as appearance. These dermal spines make the skin so rough that sharkskin is used as sandpaper.

FIGURE 25·4 Sharks and rays belong to the class Chondrichthyes. Sharks are streamlined and fast-moving. Many of them, including the Mako shark, are predators. Rays, such as the manta, have flat bodies. Mantas "fly" through the water by flapping their large pectoral fins. Mantas are harmless to humans.

Most sharks are streamlined and can swim swiftly. Sharks tend to sink if they do not keep swimming, although the large amounts of oil in their livers help keep them afloat. Some sharks eat dead animals. A few kinds feed on plankton, the microscopic and near-microscopic forms of life that are abundant in the sea. One kind that feeds on plankton is the giant whale shark, which may reach a length of 16 m and have a mass of 13 600 kg. It is the largest of all fish. However, most sharks are predators, feeding on fish and other animals in the sea.

A ray is unlike a shark in form. Rays are flat-bodied and have broad winglike pectoral fins. Most rays live on or near the sea floor, feeding on fish and invertebrates. But the largest ray—the manta, or devilfish—inhabits the open sea, where it frequently basks at the surface. A manta may measure over 6 m from wingtip to wingtip and have a mass of 1360 kg. The two "horns" at the sides of its mouth are not weapons but rather are soft flaps of flesh that help channel small fish and other food into the mouth.

In the sharks and rays, the sexes are separate and reproduction involves internal fertilization. Sperm produced in the testes of the male are transferred to the reproductive tract of the female through grooves on special fins called claspers. Some species are oviparous (oh VIHP uhr uhs). The fertilized eggs of **oviparous** organisms develop outside the female's body. The dehydration of these eggs in seawater is prevented by an impermeable coating that surrounds the eggs. Other species are ovoviviparous (oh voh vī VIHP uhr uhs). The eggs and embryos of **ovoviviparous** organisms are kept within the female's body throughout the period of embryonic development.

25•6 BONY FISH

In fish belonging to class Osteichthyes, the skeleton consists mainly of bone. The more than 30 000 species of bony fish—more than all other vertebrates combined—live in fresh and salt waters throughout the world. Like other fish, bony fish breathe with gills. Most bony fish have scales and fins, although the fins are modified in different species.

The swimming motion of a fish is a result of the alternate contractions of the muscle groups on either side of the body. As the muscles on each side of the body contract and relax, they produce wavelike motions that propel the fish forward. The fins also help most fish to steer and maintain balance in moving through the water.

The bony fish have been divided into two major groups: the *ray-finned fish* and the *lobe-finned fish*. Most fish belong to the group called the ray-finned fish. Several traits distinguish this major group of fish. Their fins contain many parallel bony spines and rays but no skeletal muscles. Their nostrils open into a nasal cavity, which is an organ for smell. The nasal cavity does not have an internal opening to the mouth and is not part of the respiratory system. Most of these fish also possess a **swim bladder,** a structure that is filled with gases and helps the fish keep afloat in the water. By controlling the amount of gases in the

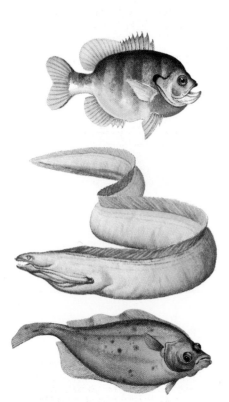

FIGURE 25•5 Bony fish vary greatly in body shape. Compare the sunfish (top), eel (middle), and flounder (bottom). How are their body shapes adapted to the way they live?

swim bladder, ray-finned fish can remain at certain depths in the water without using a lot of energy. The organs for gas exchange in the ray-finned fish are the gills.

The second group of bony fish—the lobe-finned fish—is composed of six living species. One type of lobe-finned fish is the lungfish. A lungfish is shown in Figure 25·6. Lungfish have primitive lungs. Although gills are used for gas exchange, these fish also breathe through their lungs. Unlike the nasal cavity in a ray-finned fish, the nasal cavity in some lungfish opens into the mouth. Therefore, such lungfish can breathe air through their nostrils. Living lungfish are found in rivers and lakes of Africa, Australia, and South America.

Another type of lobe-finned fish, the coelacanth (SEE luh kanth), is of special interest to scientists. Although coelacanths were thought to have been extinct for the last 70 million years, a living specimen, termed *Latimeria* (lat uh MIHR ee uh), was found in the deep waters off the coast of Africa in 1939. Since then, several more living specimens have been found. *Latimeria* is considered to be a "living fossil" because it is the only living species in this kind of lobe-finned fish. Biologists think that amphibians evolved from a now-extinct ancestor of lobe-finned fish.

FIGURE 25·6 A lungfish is a type of lobe-finned fish. Some live in lakes and rivers that dry up during the summer. Lungfish can live up to three years buried in mud. During this time, they live off their own muscle tissue.

25·7 THE PERCH: EXTERNAL ANATOMY

The yellow perch shows the features of a typical ray-finned fish. The body is divided into a head, a trunk, and a tail. There are several fins. On the back are two *dorsal fins,* the anterior with spines and the posterior with soft rays. The *caudal fin,* or tail fin, is broad. There is a single *anal fin* just behind the anus. The *pectoral fins* and the *pelvic fins,* which are along the sides, are paired.

The fins aid in locomotion. The dorsal, caudal, and anal fins help to prevent the fish from rolling; the pectoral, pelvic, and caudal fins are important for steering.

FIGURE 25·7 A yellow perch shows the characteristics of a typical bony fish. Note the different types of fins and the lateral line.

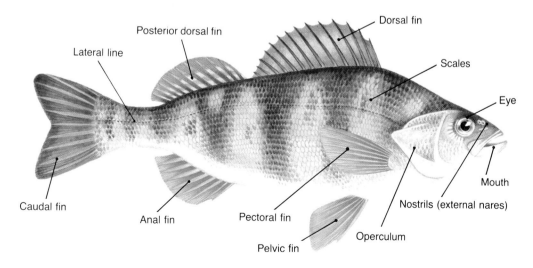

Cycloid scales

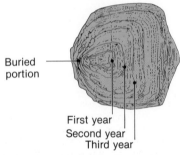

FIGURE 25·8 The flat scales that are found on a perch can be used to identify the age of the fish.

The perch's body is covered with scales that provide a protective covering by overlapping like the shingles on a roof. The scales continue to grow throughout a fish's life. The total number remains the same, but new scales grow in to replace any that are lost. The scales are embedded in the skin. The skin contains special cells, called chromatophores, that give the fish its color. The skin also contains glands that secrete mucus. The mucus lubricates the fish's body, helping the fish to slide through the water smoothly.

The **operculum** is the protective bony covering over the gills. Water taken in through the mouth moves across the gills and out through the opening provided by the operculum. The movement of the operculum and the action of muscles creates a suction that helps draw the water out.

Locate the lateral line on the perch. The **lateral line system** is a series of sensory pits forming a line on each side of the body from the operculum to the tail. This system detects vibrations and changes in water pressure and current direction. The lateral line system allows the perch to find its way through water both in the day and at night. It also helps the fish find food and keep with others, in schools. The lateral line system is found in cartilaginous and bony fish.

25·8 THE PERCH: INTERNAL ANATOMY

In the perch, body scales and the spines and rays in the fins are made of a bony material and comprise the external part of the skeleton. The internal skeleton is a flexible supporting structure for the skeletal muscles and the rays and spines of the fins. It protects the internal organs and also forms the structure of the jaws, the opercula, and the bony arches of the gills.

The skeleton of the perch, shown in Figure 25·9, can be divided into two parts: the axial skeleton and the appendicular skeleton. The **axial skeleton** includes the skull, which houses the brain and the sensory organs of the head, and the backbone, which runs the length of the body. The skull is so firmly attached to the backbone that a perch cannot move its head without moving its body. The axial skeleton also includes many pairs of ribs as well as the spines and rays of the dorsal, caudal, and anal fins. All of these structures are attached to the backbone.

The paired pectoral fins and pelvic fins are not attached to the backbone. They are attached to the bones that form the pectoral girdle and the pelvic girdle. These bones form the **appendicular skeleton.** These girdles provide support for the pectoral and pelvic fins during swimming.

Figure 25·9 illustrates some of the other systems and organs of a perch. Locate the structures of the digestive system as they are discussed. The perch's food—smaller fish and other aquatic organisms—is swallowed whole. The food passes through a short esophagus into a saclike stomach, where some digestion occurs. The liver secretes a fluid that helps to break up fats. The gall bladder stores this secretion

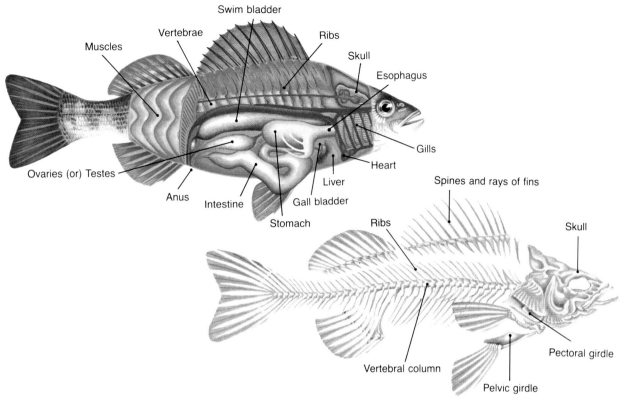

and releases it into the intestine. Food passes from the stomach into the intestine, where digestion is completed and absorption occurs. Undigested material is eliminated through the anus.

Identify the gills, the main respiratory organs, in Figure 25·9. The gills are thin filaments filled with blood vessels. Water taken in through the mouth washes over them. As it does so, carbon dioxide in the blood is exchanged with oxygen in the water.

Figure 25·10 shows the circulatory system of a perch. As in other fish, the heart is composed of two chambers. The arrows in the figure indicate the path of the blood through the perch. Note that the blood is pumped from the ventricle to the gills. From the gills it flows to all parts of the body. The blood returns to the atrium and the cycle is repeated.

FIGURE 25·9 The internal structures of a yellow perch are shown here. The internal organs (top) include the swim bladder, which regulates bouyancy, and a two-chambered heart. The skeletal system (bottom) includes the spines and rays that support the fins.

FIGURE 25·10 This diagram shows the circulatory system of the perch. Blood vessels that carry oxygenated blood are shown in red, and vessels that carry deoxygenated blood are shown in blue.

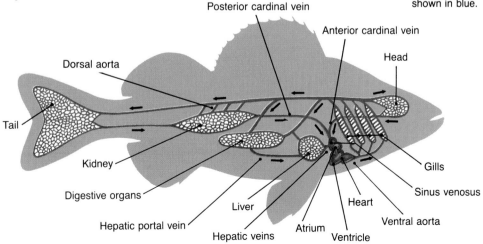

511

Above the stomach, close to the backbone, is a pair of kidneys. Both the kidneys and the gills excrete wastes. Occupying much of the body cavity, especially during the breeding season, are the gonads—ovaries in a female and testes in a male.

Across the top of the body cavity is the gas-filled swim bladder. The swim bladder is an organ of buoyancy, controlling the depth at which the fish swims. As the swim bladder takes in or releases gases to the blood, the fish's level in the water changes. Flatfish do not have swim bladders. Thus they sink to the bottom when not swimming.

The nervous system of the perch includes the brain, spinal cord, and nerves. The brain, located in the head, is protected by the bones of the skull. The spinal cord, which runs along the back, is enclosed in the backbone.

25·9 BREEDING BEHAVIOR AND DEVELOPMENT

While some fish give birth to live young, in most species the young develop externally. Breeding in fish is referred to as spawning. The female lays eggs in the water, and the male deposits sperm on them. Some kinds of fish build nests and lay the eggs in them. In some species the parents guard the eggs until they hatch. In a few cases the parents continue to guard the young fish until they reach a certain size.

Some bony fish, such as guppies, are ovoviviparous, or live-bearers. The female carries the eggs inside her body; then she gives birth to the hatchlings. In most cases, live-bearers produce fewer numbers of young than do fish that lay their eggs in the water.

A newly hatched fish usually carries with it an undigested supply of yolk from the egg. The little fish at first looks much like a tadpole. Its built-in food supply is enough to last for a day or two. Then the little fish must obtain food for itself. How long it takes a little fish to begin looking like an adult varies not only with the species but also with water temperature and the availability of food.

FIGURE 25·11 Some fish, such as these chinook salmon, travel thousands of kilometers to reach their breeding grounds.

25•10 WATER BALANCE IN FISH

In bony fish that live in the sea, the body fluids have only about one third of the concentration of solutes found in seawater. Such fish tend to lose water through osmosis and take in large quantities of salts. Some of the excess salts are removed by the kidneys. The rest are removed by special cells in the gills. These cells take up most of the excess salts and excrete them through the gills.

The body fluids of freshwater fish have a higher concentration of solutes than the water. Thus water tends to enter the body by osmosis. In freshwater fish, the kidneys excrete large amounts of water. The kidneys reabsorb some salts rather than excreting them.

REVIEW QUESTIONS

6. How do members of class Osteichthyes differ from members of class Chondrichthyes and class Agnatha?
7. How do the two groups of bony fish differ?
8. What are the functions of fins in fish?
9. How are the swim bladder and lateral line system adaptations for living in water?
10. Each of the following is an adaptation to the problem of maintaining water balance in fish. Indicate whether the adaptation is characteristic of saltwater fish or freshwater fish.
    a. Drinking large volumes of water to replace water lost
    b. Excreting large volumes of water from the kidneys

The Amphibians

25•11 TRAITS AND CLASSIFICATION

Amphibians are vertebrates that have gills as larvae and, usually, lungs as adults. The amphibians—class *Amphibia*—were the first vertebrates to become adapted to life on land. After hatching from eggs, most young amphibians are aquatic and breathe by means of gills. As adults, most amphibians live on land at least some of the time, and they breathe with lungs.

The most familiar and abundant of the three groups of amphibians are frogs and toads. About 2600 species of frogs and toads make up order *Anura*. Frogs and toads are tailless amphibians. They have two pairs of legs. The hind legs are much larger than the front legs. Most adult frogs spend their lives in and near water, but toads and some other kinds of frogs live mainly on land. With few exceptions, frogs and toads return to bodies of water to reproduce.

SECTION OBJECTIVES

- Identify the external and internal structures of a frog.
- Discuss the life cycle of a typical amphibian.

amphibios (double life)

FIGURE 25·12 The crimson spotted newt (top) is an example of a salamander. Its bright red color warns predators to stay away, since it has poison in its skin. The caecilian (bottom) is an example of a legless amphibian.

FIGURE 25·13 The membranes inside a frog's large mouth are sometimes used for gas exchange. Locate the two kinds of teeth in the frog. Why does a frog have to swallow its prey whole?

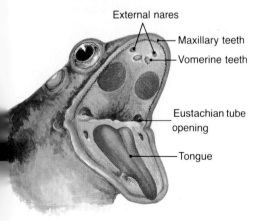

Salamanders are amphibians that have a tail and two pairs of legs approximately the same size. They make up order *Caudata*, which includes about 300 species. In body shape, salamanders resemble lizards, which are reptiles.

Salamanders have thin moist skin, which is characteristic of most amphibians. Much gas exchange occurs through the skin. Salamanders can survive only in moist places. Many live in and on damp debris in the forest. Some kinds live in streams and lakes. Some salamanders are completely aquatic and keep their gills throughout their lives. Internal fertilization occurs in many kinds of salamanders.

The third group of amphibians, caecilians (see SIHL yuhnz), make up order *Apoda*. As you can see in Figure 25·12, caecilians are legless and wormlike. They live in loose soil, where they feed on larvae, worms, and other soil-dwelling creatures. All of the approximately 150 species live in the warm parts of Africa, Asia, and the Americas.

Amphibians are ectotherms. They are active on warm days, but if exposed to very high temperatures, they will die. Too much heat dries out their moist skin, hindering gas exchange. To keep cool and moist, salamanders may bury themselves in loose soil.

Amphibians become dormant during cold weather. Their body temperature drops as the surrounding temperature falls. This period of inactivity during cold weather is called **hibernation.** While hibernating, amphibians need little oxygen, and they eat no food. Their energy requirements are met by the fat stored in their bodies. Many salamanders hibernate in holes in the ground or in loose soil.

25·12 THE FROG: EXTERNAL STRUCTURE

The green frog, *Rana clamitans,* is a common frog. Its skin is moist and smooth. As in salamanders, gas exchange can occur through the skin. Mucous glands in the skin secrete mucus, which helps to keep the skin moist. Chromatophores give color to the frog's skin.

The frog has big eyes on the top of the head. Two openings that lead into the mouth, the **external nares** (NAIR eez), are in front of the eyes. A frog can stay below water with only its eyes and external nares above the surface. The flat circle just behind each eye is a **tympanic membrane,** or eardrum.

The tongue is attached at the front of the lower jaw. When a small organism comes close, the frog flicks out its tongue. With the prey caught on the sticky tip, the tongue is then pulled back into the mouth.

Frogs do not chew their food. Around the inside edge of the frog's upper jaw are tiny **maxillary teeth.** There are two slightly larger teeth, called **vomerine teeth,** on the roof of the mouth. These teeth are used only to hold the prey prior to swallowing it whole.

The **internal nares,** two openings near the vomerine teeth, are connected by channels to the external nares. When the frog breathes, air passes into the mouth through these channels. Then the air passes into the lungs through the glottis, a slit on the floor of the mouth that is the opening to the windpipe.

25·13 THE FROG: INTERNAL STRUCTURE

Figure 25·14 shows the skeleton of a frog. The frog's bony skeleton protects the delicate internal organs, provides places for muscle attachment, and forms the framework for the body shape. The skull, for example, is made up of several bones, some of which protect the brain.

The frog's backbone consists of nine vertebrae and a long slim bone, the urostyle, extending to the tip of the body. The frog does not have ribs. The bones of the pectoral girdle, which form the shoulders, give some protection to the internal organs beneath. The bones of the pectoral girdle also connect to the bones of the front legs. At the rear of the body, the pelvic girdle, or hipbones, connects to the hind legs.

The digestive system of the frog is shown in Figure 25·14. The frog's mouth opens into a short esophagus that connects to a slim whitish stomach. At the far end the stomach narrows and joins the small intestine that in turn connects to the shorter large intestine. The large intestine narrows to a rectum that enters the cloaca (kloh AY-kuh). The **cloaca** is a chamber into which the products of the digestive, excretory, and reproductive systems are emptied.

FIGURE 25·14 Study the ventral views of the skeleton (left) and digestive system (right) of a frog. Note the urostyle and the lack of ribs in the skeleton. One lobe of the frog's liver has been removed to show the gall bladder.

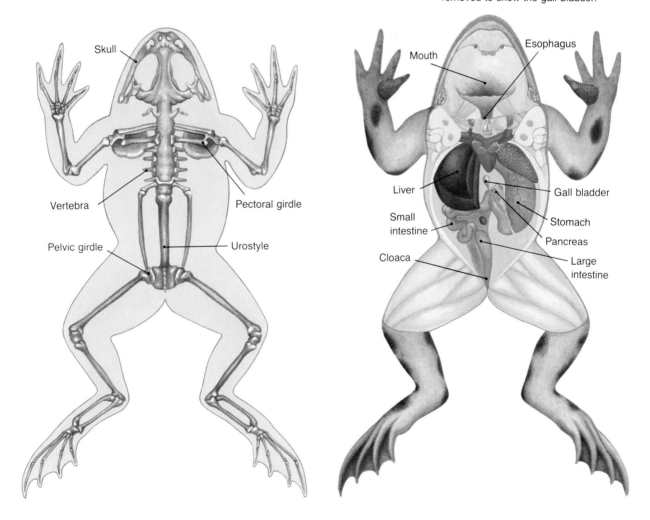

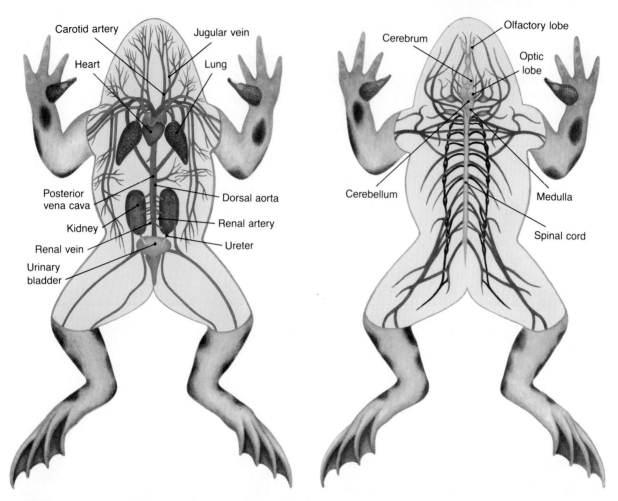

FIGURE 25·15 The circulatory and excretory systems of a frog are shown on the left. Note the three-chambered heart and the spongy lungs that help deliver oxygen to the blood and remove carbon dioxide. The nervous system of a frog is shown on the right. Note how small the cerebrum is compared to other areas of the brain.

Figure 25·15 illustrates some of the other internal organs of the frog. The reddish pear-shaped heart is found in the center of the chest. In amphibians the heart has three chambers—two atria above and a ventricle below. Deoxygenated blood returning from most of the body through the veins enters a sac called the **sinus venosus.** From the sinus venosus, blood enters the right atrium. Oxygenated blood returning from the lungs enters the left atrium. Both atria then contract, forcing the blood into the ventricle.

When blood enters the single ventricle, the deoxygenated blood from the right atrium does not entirely mix with oxygenated blood from the left atrium. The ventricle contains structures that prevent much mixing of blood from the right and left sides. When the ventricle contracts, blood is pumped into a large blood vessel called the **conus arteriosus,** which has two channels. The oxygenated blood is pumped out through blood vessels to the body systems. The deoxygenated blood is pumped out through blood vessels to the lungs and skin, where gas exchange occurs. This freshly oxygenated blood then returns to the left atrium.

The kidneys are two slim dark organs located against the dorsal wall of the frog's body cavity. The kidneys filter wastes, excess salt, and water from the blood. The liquid wastes pass down tubes called ureters and enter the cloaca. The wastes can either be released or stored for a time in the urinary bladder.

The frog's spinal cord is protected by a backbone. The spinal cord is connected to the frog's brain, which is shown in Figure 25·15. Two large olfactory lobes at the front of the brain function in the sense of smell. Two larger optic lobes in the middle of the brain function in vision. Between these lobes is the cerebrum. The cerebrum is involved in sensory interpretations and muscle responses. Behind the optic lobes are the cerebellum and the medulla. These parts of the brain control the frog's balance, coordination, and other important functions.

The reproductive organs of the male and female frogs are ventral to the kidneys. The reproductive organs of the female are two ovaries, each containing numerous eggs. Two oviducts transport the eggs to the cloaca. The reproductive organs of the male are two bean-shaped testes lying on the ventral side of the kidneys. The testes produce millions of sperm. Sperm produced in the testes pass into the sperm ducts, which lead to the kidneys. Then they pass through the ureter, which leads to the cloaca, and finally out of the body. Fertilization in the frog is external. During the mating season, male frogs court the females. Amplexus is the holding of the female frog by the male frog. By stroking the abdomen of the female, the male aids in releasing the jelly-coated eggs from the cloaca of the female. The male deposits sperm over the eggs in the water. In most species, development of the embryos takes place in the water, and there is little or no parental care.

25·14 METAMORPHOSIS IN AMPHIBIANS

Most frogs and toads mate in bodies of fresh water. During mating, the male frog or toad climbs onto the female's back and holds on firmly with his front legs. He discharges sperm to fertilize the eggs as the female lays them.

Each egg is in a jellylike substance that swells in the water. This covering helps to protect the eggs. Most females lay many eggs, either in strings or clumps. The larvae develop in the water.

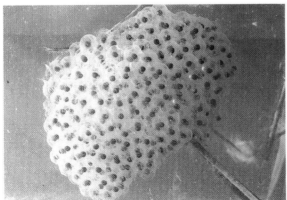

FIGURE 25·16 Frogs lay eggs in masses (left). Frog larvae, called tadpoles, hatch from these eggs and develop in the water (right).

FIGURE 25·17 Frog eggs are fertilized externally. Young tadpoles have gills, long tails, and eat plants. As the tadpole matures, it develops legs and lungs and loses its tail, gills, and most of the length of its intestine. Adult frogs eat insects and live on land.

Salamander larvae have four legs and look much like little adults. They have external gills, however, and most have a soft fin over the tail. The larvae of frogs and toads, called tadpoles or polliwogs, do not look like the adults. They do not have legs. They have a roundish body and a long tail. A tadpole's tiny mouth ends in "lips," which are used to scrape algae from rocks and stems.

The change of the larval amphibian to the adult form is called *metamorphosis*. The time required for metamorphosis to occur varies with the species, ranging from two weeks to two years. The changes that occur are major. A tadpole has a long, coiled intestine, which is a trait of a plant eater. By the time it is a flesh-eating adult, its intestine is short. The external gills have disappeared, and internal lungs have developed. The tail has disappeared and legs have developed, first the hind legs and then the front legs.

REVIEW QUESTIONS

11. What are the three orders of living amphibians, and how do they differ from each other in external appearance?

12. Discuss some adaptations of frogs related to food getting and locomotion.

13. Describe the heart of a frog and the passage of blood through the heart.

14. Describe the life cycle of a frog.

15. Metamorphosis in amphibians is brought on by a chemical called thyroxine. What effect would a lack of thyroxine have on a larval amphibian?

Investigation

What Are the Similarities and Differences Between the Frog and the Toad?

Goals
After completing this activity, you will be able to
- Compare the external features of a frog and a toad.
- Compare some behavioral traits of a frog and a toad.

Materials (for groups of 4)
1 preserved frog
1 preserved toad
2 living frogs in glass tank (for class)
2 living toads in glass tank (for class)
probe   2 dissecting pans
hand lens   toothpicks
live mealworms   metric ruler

Procedure

A. On a separate sheet of paper, set up a chart on the comparison of the frog and toad. As you work through each step, enter your observations in this chart.

B. Place the preserved frog in a dissecting pan. Run your fingers along the length of the body. How does the skin feel? Record your findings. Carefully observe the coloration of the frog. Record the color or colors and any type of pattern you note.

C. Carefully observe the hind foot of the frog. Spread the toes open. Note the webbing (**1**). Describe your findings in the chart. Rub your fingers from the eyes down the back of the frog. You will notice two ridges, called the dorsolateral folds (**2**), running from the eyes to the end of the trunk.

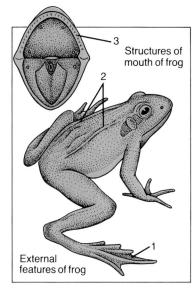

Structures of mouth of frog

External features of frog

D. Open the frog's mouth. Examine the tongue and note where it is attached. Run your fingers along the edge of the upper jaw. You should be able to feel small maxillary teeth (**3**). These teeth are used for gripping only.

E. Place the preserved toad in the dissecting pan. Record your observations on the texture of the skin, the structure of the hind foot, and the coloration.

F. Look at the toad's back just behind the eyes. Note the two oblong or round bumps. These bumps are the parotoid glands (**4**), which secrete a sticky white substance. The substance irritates and is sometimes harmful to the toad's enemies. Look again at the frog. Does it have parotoid glands? Record your observations.

G. Look for ridges framing the inside of the eyes. These ridges are called the cranial crests (**5**).

H. Open the toad's mouth. Note the position of the tongue. Find out whether it has maxillary teeth.

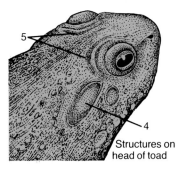

Structures on head of toad

I. Observe the live frogs and toads on display. Pick up a mealworm on the end of a toothpick. Wiggle it in front of the frog and the toad. Observe how the amphibians use their tongue to take in the food. Record your observations on how the frog and the toad detect food, eat, and move.

J. Follow your teacher's instructions for the proper cleanup and disposal of all materials.

Questions and Conclusions

1. How does the frog's skin compare with the toad's skin?
2. What are the dorsolateral folds? Are they found in both the frog and the toad?
3. What are the parotoid glands? How do they look? Do both frogs and toads have them?
4. How do the tanks containing the frogs and toads differ?
5. How does a frog use its tongue to obtain food?

Chapter Review

SUMMARY

25·1 All chordates have a notochord, a dorsal nerve cord, and paired gill slits at some stage in the life cycle.

25·2 The two subphyla of nonvertebrate chordates are made up of small marine organisms.

25·3 Fish are aquatic vertebrates with gills. Most fish have fins and scales.

25·4 Hagfish and lampreys are jawless fish, making up class Agnatha.

25·5 Sharks and rays, class Chondrichthyes, both have skeletons of cartilage. However, a typical shark has a streamlined body shape and lives in open water; a ray has a flat body and lives on or near the sea floor.

25·6 Bony fish, class Osteichthyes, are the most abundant and most varied of all the animals with backbones.

25·7 The perch, a typical bony fish, has fins, scales, and a lateral line system.

25·8 The perch has a two-chambered heart, breathes with gills, and has well-developed muscles, nerves, and senses.

25·9 All fish hatch from eggs that, varying with the species, are laid in the water, or carried inside the female's body until they hatch.

25·10 Marine fish tend to excrete salt. Freshwater fish excrete much water. In both cases a proper balance of water and solutes is maintained.

25·11 The three orders of living amphibians are Anura, made up of frogs and toads; Caudata, made up of salamanders; and Apoda, made up of caecilians.

25·12 The external structures of the frog illustrate the characteristics of order Anura.

25·13 The frog has lungs, a three-chambered heart, and well-developed skeletal, muscular, nervous, and digestive systems.

25·14 Eggs are fertilized externally in most amphibians. The eggs of frogs and toads hatch into gilled tadpoles that do not resemble the adults. Salamander larvae have gills and legs, and look like little adults.

BIOLOGICAL TERMS

amphibians
appendicular skeleton
atrium
axial skeleton
chordates
cloaca
conus arteriosis
cyclostomes
ectotherm
external nares
fins
fish
gills
gill slits
hibernation
internal nares
lancelets
lateral line system
lungs
maxillary teeth
nerve cord
notochord
operculum
oviparous
ovoviviparous
scales
sinus venosus
swim bladder
tunicates
tympanic membrane
ventricle
vertebrates
vomerine teeth

USING BIOLOGICAL TERMS

1. Which structures are found at some time in all members of the phylum Chordata?
2. Which term refers to the gill covering of bony fish?
3. What is the lateral line system?
4. Distinguish between oviparous and ovoviviparous.
5. What does the term *ectotherm* mean?
6. Distinguish between an atrium and a ventricle.
7. Into which large blood vessel does blood flow after leaving the ventricle of a frog's heart?
8. What is the eardrum of a frog called?
9. Distinguish between the axial skeleton and the appendicular skeleton.
10. What is the function of the swim bladder?

UNDERSTANDING CONCEPTS

1. Describe the ways lower chordates resemble vertebrates. (25·1)
2. What are two characteristics of vertebrates? (25·1)
3. Identify the typical chordate characteristics of a tunicate. (25·2)
4. Describe how *Amphioxus* obtains food. (25·2)
5. How do the skeletons of members of class Chondrichthyes and class Osteichthyes differ? (25·3)
6. Describe how cyclostomes feed. (25·4)
7. In what ways do sharks differ from rays? (25·5)
8. Compare the two major groups of bony fish. (25·6)
9. State the names of a perch's fins and discuss their functions. (25·7)
10. Describe the circulation of blood through the perch. (25·8)
11. What is the function of the swim bladder in the bony fish? (25·8)
12. Identify a type of parental-care behavior exhibited by nest-building fish. (25·9)
13. How do fish compensate for differences in solute concentration between their bodies and the water in which they live? (25·10)
14. What are the characteristics of class Amphibia? (25·11)
15. Name and describe the three orders of amphibians. (25·11)
16. How is a frog's tongue adapted to catching prey such as small insects? (25·12)
17. How does a frog's heart differ from a perch's heart? (25·8, 25·13)
18. Identify the parts of a frog's brain. (25·13)
19. Describe the major changes that occur when a tadpole undergoes metamorphosis. (25·14)

APPLYING CONCEPTS

1. Describe the advantages of the swim bladder in terms of energy efficiency in water-dwelling organisms.
2. Why was the discovery of *Latimeria*, the coelacanth, an exciting discovery for biologists?
3. Describe the advantages and disadvantages of both internal fertilization and external fertilization. Also describe advantages and disadvantages of both oviparous and ovoviviparous development. Give examples in both descriptions.
4. In what ways is the heart of a frog more efficient than the heart of a perch? In your answer, consider whether a two-chambered heart would be efficient enough for a land-dwelling organism.
5. Describe three adaptations that were necessary for the evolution of amphibians.

EXTENDING CONCEPTS

1. Find out when and to what places the parasitic sea lamprey spread in the Great Lakes. Show the estimated numbers of fish caught in commercial and sport fishing during the lamprey's uncontrolled spread and the numbers caught in the years since the lamprey has been under control.
2. Give a report on the most dangerous sharks and the most dangerous shark waters in the world. Tell what can be done to avoid a shark attack and what to do if an attack occurs.
3. In the library, find a description of the four-eyed fish and tell how its eyes differ from those of other fish. Why is having four eyes a useful adaptation?
4. Give a report on how and where freshwater eels spawn and about the journey of the young back from the sea. How long does the trip take young American eels as compared with the time it takes young European eels?
5. Give a report on the spawning runs of the Pacific grunion.

READINGS

Ford, John I., and Robert A. Kinzie III. "Life Crawls Upstream." *Natural History,* December 1982, p. 61.

Halliday, Tim, and Kraig Adler (Eds.). *The Encyclopedia of Reptiles and Amphibians.* Facts on File, 1986.

Partridge, B.L. "The Structure and Function of Fish Schools." *Scientific American,* June 1982, p. 114.

Randall, Robert A. "Fishery on a Fuse." *Natural History,* December 1983, p. 10.

26. Reptiles and Birds

Great Gray owls are the largest owls in North America, with a wingspan of about 1.5 m. An owl is adapted to hunt by night. Its large eyes enable it to see in the dark. The disks of feathers around the eyes channel sound to its ears. An owl's soft, downy feathers enable it to fly silently. Birds, such as the owl, evolved from small running reptiles millions of years ago. In this chapter, you will discover more about reptiles and birds.

CHAPTER OBJECTIVES

After completing this chapter, you will be able to

- **List** the major traits of reptiles.
- **Describe** the distinguishing characteristics of birds.
- **Explain** how reptiles and birds are classified.
- **Compare** the anatomy and behavior of reptiles and birds.

The Reptiles

26·1 TRAITS OF REPTILES

Members of class *Reptilia* (rehp TIHL ee uh), called **reptiles**, are ectothermic vertebrates that are covered with dry scaly skin. They are adapted for reproduction on land. Most species are land-dwelling, but some species spend much time in water. Turtles, snakes, lizards, crocodiles, and alligators are reptiles. Reptilian skin is dry, thick, and waterproof, protecting the body from drying out even in very dry climates. The skin is covered by tough scales, or plates, that protect the animal from injury.

Reptiles are ectotherms. Recall that an ectotherm is an animal that gets its body heat from the environment. Thus its body temperature goes up and down as the environmental temperature varies. However, a land-dwelling reptile can have some control over its body temperature by the way it behaves. For example, an iguana may lie in the morning sun, which quickly raises its body temperature and thus increases its metabolic activity. Later in the day the iguana may move to the shade of rocks. The shade causes a drop in the reptile's body temperature and slows down its metabolic activity.

Reptiles are fully adapted to life on land. This adaptation is best shown by their way of reproducing. Unlike many amphibians, reptiles do not need water in which to reproduce. Reptiles have reproductive organs that are adapted for internal fertilization. The eggs are fertilized inside the body of the female. Since fertilization is internal, reptiles do not depend on an external water source for moving the sperm to the egg.

Reptiles also produce shelled eggs that do not dry out on land. An embryo passes through its early period of development within an egg. When the reptile hatches, it can breathe on land. The young reptile looks like a small adult. There is no larva-in-water stage, as occurs in amphibians.

SECTION OBJECTIVES

- Explain the evolutionary significance of the amniote egg.
- State the common names of the four living orders of reptiles.
- Describe the organ systems of the turtle.

reptilis (crawling)

Reptiles differ from amphibians in other ways. Recall that an amphibian's heart has a single ventricle, in which there is some mixing of oxygenated and deoxygenated blood. A typical reptile's heart has an incomplete muscular wall that divides the left side of the ventricle from the right side. The partial division of the ventricle separates the oxygenated blood that enters from the lungs and the deoxygenated blood that enters from the body.

26·2 THE REPTILIAN EGG

The reptilian egg is a very important adaptation to life on land. It has a shell, which protects the embryo, and a large yolk, which provides enough food to nourish the embryo through its development.

This land egg is called an amniote (AM nee oht) egg. An **amniote egg** is an egg that has a fluid-filled sac, in which the embryo develops, and a protective shell. Thus, even though the egg is laid on land, the embryo develops in the protection of a fluid environment. To better understand the structure of this type of egg, look at the drawing of the amniote egg in Figure 26·1. Note that it has a shell and four embryonic membranes. The shell and each of the four membranes has an important function during the development of the embryo.

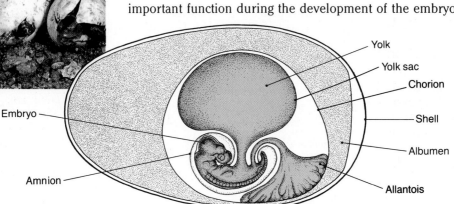

FIGURE 26·1 The amniote egg was a critical adaptation for life on land. The shell prevents water loss and protects the embryo. The albumen cushions the embryo and provides a watery environment for development. Locate the four embryonic membranes. What are the functions of these membranes?

The **shell** is a protective covering for the egg. Sperm cannot pass through the shell. Thus fertilization must occur within the reproductive tract of the female, before the shell is formed and the egg is laid. The shell is leathery and allows air, but not water, to pass through. Gas exchange takes place through the shell. Because the shell does not allow the passage of water, it protects the embryo from drying out. The shell also offers some protection against mechanical injury, such as being knocked about.

The four membranes inside the egg develop from and grow along with the developing embryo. The **yolk sac** is a membrane that grows out from the digestive tract of the embryo and surrounds the yolk. Special cells in the yolk sac give out enzymes that digest the yolk. Blood vessels form in the yolk sac and carry the nutrients to the embryo. The yolk is a rich source of proteins and lipids for the embryo.

The **amnion** (AM nee uhn) is a membrane that grows around the embryo and encloses it in a liquid called *amniotic fluid*. Amniotic fluid is the liquid environment in which the embryo floats during its development. This fluid prevents dehydration and cushions the embryo.

The **chorion** (KOR ee ahn) is a membrane that grows out from the embryo and lines the inner surface of the shell. It surrounds the embryo and the other membranes. This membrane is permeable to such gases as oxygen and carbon dioxide. The embryo gets oxygen from the environment through the chorion.

The **allantois** (uh LAN toh ihs) is a membrane that forms from the lower end of the embryo's digestive tract. It grows outward, toward the chorion. The allantois serves for deposit of wastes, such as uric acid. It also serves for gas exchange between the blood of the embryo and the outside atmosphere. The allantois is well supplied with blood vessels, which carry embryonic blood to the chorion. During development, the allantois fuses with the chorion.

Reptiles lay their eggs on land. Many of the reptiles dig holes in which they lay their eggs. Usually these eggs receive no care from the female parent. Animals that lay eggs are *oviparous*, which means that embryonic development occurs in an egg outside the body of the mother. Turtles, alligators, crocodiles, and most snakes and lizards are oviparous.

The females of some snakes and some lizards keep their eggs in a special chamber of their reproductive tract. Animals whose embryos develop within eggs inside the body of the female are said to be *ovoviviparous*. All the food the developing embryo needs is supplied by the egg. Reptiles that are ovoviviparous give birth to live young. Ovoviviparous development should not be confused with the type of development found in most mammals. Most mammals are **viviparous** (vī-VIHP uhr uhs), which means that the embryos develop within the female and are born live. Nutrients and oxygen are supplied to the embryo by diffusion from the female parent's bloodstream.

26•3 CLASSIFICATION OF REPTILES

The fossil record shows that in the past at least 16 orders of reptiles lived on the earth. Twelve of those orders are now extinct. The four surviving orders of the class Reptilia are *Rhynchocephalia* (rihng-koh suh FAY lee uh), the tuatara; *Chelonia* (kih LOHN ee uh), the turtles; *Crocodilia*, the crocodiles and alligators; and *Squamata*, the lizards and snakes.

Order Rhynchocephalia is a nearly extinct order of reptiles. It contains only one species, *Sphenodon punctatus*, commonly known as the tuatara (too uh TAHR uh). The tuatara often is called a "living fossil," because it is the only surviving species of an order of reptiles that is as old as the dinosaurs. All other species of this order became extinct 125 million years ago.

The discovery of living tuataras in 1831 was very exciting to scientists. Fossil evidence indicates that similar forms were plentiful on

---

Discovery

Were Dinosaurs Caring Parents?

Scientists have long assumed that dinosaurs gave their young very little care. The female dinosaur laid her eggs in a prepared nest, buried them, and left. When the young hatched, they emerged from the ground and began to take care of themselves.

Remains of dinosaur nests that have been found in Montana indicate that one kind of dinosaur, the hadrosaur, took care of its young. Apparently the hadrosaurs lived in colonies and laid their eggs in crowded nest clusters. They built their nests of mud. The bowl-shaped nests could hold from 20 to 25 eggs. Because they were placed about one body length apart, several nests could be guarded by one adult. Most surprisingly, food was provided for the young dinosaurs until they could find food for themselves.

The abundance of hadrosaur remains in sediments from late in the Cretaceous shows that they were among the most successful of all dinosaur groups. Perhaps the care they gave their young was one of the reasons for this success.

the earth before the dinosaurs. Today, tuataras are found on a few small islands off the coast of New Zealand. On those islands, the tuataras do not have to compete with mammals. They are slow-moving animals with a very low metabolic rate.

Order Crocodilia is an order of reptiles that contains crocodiles, alligators, caimans (KAY muhnz), and gavials (GAY vee uhlz). The members of this order, called crocodilians, live in the tropics and subtropics. In the United States there is one species of alligator that is native to the southeastern United States. There is one species of crocodile found in the southern part of Florida. Some crocodilians grow to lengths of over 6 m and weigh over 450 kg. All the animals in this order have four strong legs, the hind legs being the more powerful. The powerful tail is used for swimming.

Figure 26·2 shows some members of the family of the crocodilians. All are alike in some ways, yet each has distinctive features. For example, the American alligator can be distinguished from the American crocodile by its snout. The alligator has a much broader snout than the crocodile has. Caimans and gavials live in tropical regions also and look much like crocodiles and alligators.

Crocodilia is an old order, dating back to the time of the dinosaurs. Despite the age of the order, crocodilians have many structural features that are quite advanced. Unlike the rest of the reptiles, the crocodilians have a complete four-chambered heart. Their body cavity is divided into a chest cavity and an abdominal cavity by a muscle that aids in breathing.

FIGURE 26·2 Crocodiles (top) are more aggressive than alligators and have pointed snouts and a characteristic toothy "grin." Only a few of an alligator's teeth are visible when its mouth is closed (bottom).

Order Chelonia is an order of reptiles containing about 250 species of turtles. Because of its shell, the turtle is seldom confused with any other species of animals. You may hear both *tortoise* and *turtle* used to refer to members of order Chelonia. For convenience, in this book the term *turtle* will be used for all members of this order.

Turtles are believed to have appeared on the earth some 200 million years ago. They have probably changed very little over that period of time. Although many have become adapted to living in water, turtles are air-breathing and must return to land to lay their eggs. Turtles vary in size from very small species, weighing only a few grams, to the giant sea turtles, weighing over 450 kg at maturity. The life span of some species is believed to be very long, probably over 150 years. The anatomy of the turtle will be described later.

Order Squamata is the latest order of the living reptiles to have evolved. This order has the greatest number of reptile species today. The order can be divided into two suborders: the *lizards* (2500 species) and the *snakes* (2700 species). Members of Squamata can be found in all areas of the earth except the polar regions.

One trait that distinguishes this order is the presence of a lower jawbone that is not joined directly to the skull. This flexible attachment allows the lower jaw to drop down and forward during feeding. This trait is more pronounced in snakes than in lizards. Another distinctive trait of this order is the presence in the males of paired reproductive organs used in internal fertilization.

Lizards are adapted to a variety of habitats and show a great diversity of forms. Most lizards are terrestrial—they live on dry land. However, the marine iguanas of the Galapagos Islands are adapted to swimming in the ocean and feeding on seaweed. The most common lizards in the United States belong to the iguana family. The horned lizard, also known as the "horned toad", is an iguana that lives in the deserts of the Southwest. Geckos live in trees and have special foot pads for climbing. Some burrowing lizards have greatly reduced limbs and eyes. The Gila (HEE lah) monster and Mexican beaded lizard have small round scales that look like glass beads. The only two species of poisonous lizards, they have grooved teeth which draw venom into the wound as the lizard chews its prey.

Some lizards are herbivores, feeding on plants and algae. Others are carnivores and eat insects, spiders, crustaceans, and other small animals. The largest lizard, the Komodo dragon, preys on pigs and small deer, and is even able to kill a water buffalo.

Through millions of years of evolution, snakes have become highly specialized. They lack outer ears, tympanic membranes, movable eyelids, and limbs. All snakes have long tubelike bodies that are covered with tough scales.

All snakes are carnivores. They have several well-developed sense organs that aid them in finding prey. Snakes have inner ears, through which they can detect vibrations in the ground. The eyes of snakes are especially adapted to detect quick movement.

FIGURE 26·3 This glass lizard closely resembles a snake. However, snakes lack eyelids and cannot blink. One way to tell the difference between a legless lizard and a snake is to wait for the animal to blink. If it blinks, it's a lizard.

FIGURE 26·4 The horned lizard (top) is adapted for life in the desert. Its spiny skin conserves water and helps the lizard blend in with its surroundings. The gecko (bottom) has suction-cuplike toes adapted for climbing. It can even climb up vertical panes of glass. Geckos can shed their tails when in danger. The shed tail wriggles, distracting predators. This gives the gecko a chance to escape. Later, it grows a new tail.

FIGURE 26·5 A rattlesnake uses its fangs to inject a toxin into its prey.

Nerve endings in the nasal cavity of the snake are important for smell. The perception of smell is aided by the long forked tongue. The tongue picks up molecules in the air and then transfers these molecules to two cavities called Jacobson's organ. **Jacobson's organ,** which is at the front of the roof of the snake's mouth, is a sense organ specialized for smell.

Snakes feed on animals that are fairly large compared with the width of their bodies. Since all snakes swallow their prey whole, the mouth is well adapted for taking in a fairly large animal. The lower jaw is not firmly attached to the skull. It can thus drop down and forward during ingestion. Also, the lower jaw has two halves. They are joined by an elastic ligament and thus can separate. These adaptations allow the mouth to be opened wide to take in large prey.

A few snakes are poisonous. Such snakes kill their prey by injecting them with a poisonous substance called venom. Venom is produced by two special glands, located on either side of the head.

Some snakes produce a venom that is primarily a hemotoxin. A **hemotoxin** (hee muh TAHK suhn) is a poison that damages blood cells and bursts blood vessels. Rattlesnakes, water moccasins, and copperheads are examples of North American snakes that produce hemotoxins. These snakes inject the toxin into their victims through hollow fangs. The fangs can be seen in Figure 26·5.

Other snakes produce a venom that is a neurotoxin. A **neurotoxin** (nur oh TAHK sihn) is a poison that affects nerve centers, causing paralysis of the muscles, including those involved in breathing. Members of the cobra family produce neurotoxins. The coral snake is the only snake of this type in the United States.

26·4 THE TURTLE: EXTERNAL ANATOMY

Turtles belong to one of the oldest orders of reptiles. They have survived for 200 million years with little change in body structure. The turtle is characterized by a bony shell, a jaw that lacks teeth, and four legs.

The turtle is unique in having a protective shell around its body. The shell consists of two bony shields. The **carapace** (KAR uh pays) is the turtle's dorsal shield, or shell. It is covered with tough scales called *scutes.* The **plastron,** or ventral shield, is flatter than the carapace and covers the under surface of the turtle. The plastron is also covered with scutes. The dorsal and ventral parts of the shell are joined on each side by a bridge of bone.

The head of the turtle is covered with scales. A clear membrane covers and protects each eye. At the tip of the snout are the *external nares,* the pair of openings for breathing. A tympanic membrane is located on either side of the head. The turtle lacks teeth. It cuts and crushes its food with the tough horny beak that covers its jaws.

The limbs of the turtle are short and stubby and extend out laterally from the body. Land turtles have legs with claws. Many water turtles have clawless legs that are paddles for swimming.

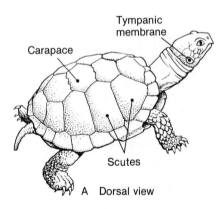
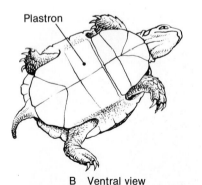

FIGURE 26·6 The dorsal shell of a turtle is the carapace. The ventral shell is the plastron. Both parts of the shell are made of bone covered with tough scales called scutes.

26•5 THE TURTLE: INTERNAL ANATOMY

The endoskeleton is the internal skeleton of the turtle. The axial skeleton consists of the skull, the vertebrae, and the ribs. The appendicular skeleton consists of the limbs and the shoulder and hip bones.

The digestive system of the turtle is similar to that of other vertebrates. Locate each of the organs of the digestive system in Figure 26•7. Also find the pancreas and the large two-lobed liver.

The turtle's respiratory system is also shown in Figure 26•7. Leading into the mouth are the external nares. Inside the mouth the internal nares can be seen in the upper jaw. Air passes from the mouth into the glottis. The glottis leads into the long trachea, which branches to form the two bronchi. Each bronchus leads to a lung.

In the turtle the heart has two atria and a partially divided ventricle. There is very little mixing of blood in the heart. The mixing is further reduced because a separate artery extends from each side of the ventricle. Deoxygenated blood is carried to the lungs, and oxygenated blood is carried to all parts of the body.

The excretory system of both the male and female turtles consists of two long narrow kidneys that filter the excretory wastes from the blood. These filtered wastes form the urine. From the kidneys, urine passes through two tubes called ureters. The urine then empties into the cloaca. The cloaca also serves as a passage for solid wastes and sperms or eggs. Urine may be stored in the large urinary bladder from which water is reabsorbed. Reabsorption of water helps to prevent excess loss of water in terrestrial turtles.

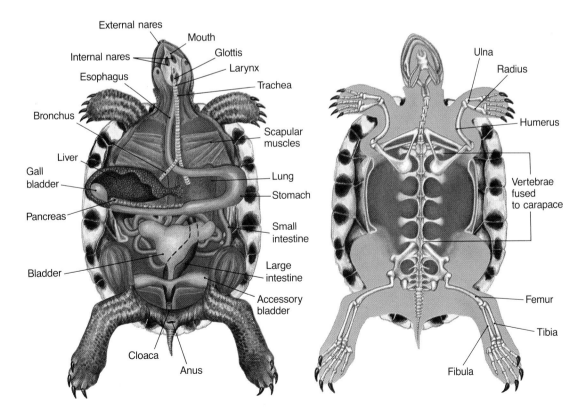

FIGURE 26•7 The respiratory and digestive systems of a turtle can be seen on the left. Part of the liver has been removed to show the gall bladder. The bones inside a turtle's shell are shown on the right.

The nervous system of the turtle is similar to that of other vertebrates. Nerves branch off the brain and spinal cord. The spinal cord is encased in the vertebrae, some of which are joined to the carapace. The skull encloses and protects the brain. The functions of the parts of the brain are similar to the functions of those parts of the brain in other vertebrates.

Turtles are adapted to reproducing on land. Eggs develop in the paired ovaries; sperm are produced in the testes. Fertilization occurs in the oviducts, which are coiled tubes leading from the ovaries. The lower end of the oviduct contains shell glands, which deposit the materials for the shell. The round shell-encased eggs enter the cloaca and leave the female's body. In general, the female deposits the eggs in a hole dug in loose soil and then covers the eggs.

REVIEW QUESTIONS

1. Identify three traits that are characteristic of reptiles.
2. Describe the amniote egg and explain how it is an adaptation of land animals.
3. List the four orders of reptiles and give an example of each order.
4. In what way is the skeleton of a turtle different from the skeleton of other vertebrates?
5. A friend brings you a small four-legged vertebrate. How can you determine whether the creature is a salamander or a lizard?

The Birds

SECTION OBJECTIVES

- Describe a bird's modifications for flight.
- Discuss the organ systems of the pigeon.

26•6 TRAITS OF BIRDS

Members of class *Aves* (AY veez), called **birds,** are warm-blooded vertebrates that are covered with feathers and have wings. Birds have certain traits that are similar to those of reptiles. These traits are internal fertilization, amniote eggs, scales on the feet, and clawed toes. Birds are different from reptiles in many ways. Birds lack teeth, and their mouths are modified into bills. Their bodies are covered with feathers, and they can keep a constant body temperature. Their forelimbs are modified as wings. Their hind limbs support their body weight when standing. Birds have a four-chambered heart, well-developed lungs, and thin hollow bones. Most birds can fly.

An important adaptation of birds is the maintenance of a high and constant body temperature in spite of changes in the external environment. An animal that maintains a constant body temperature as a result of heat produced by metabolism is called an **endotherm** (EHN-doh therm). An endothermic animal is often called a warm-blooded animal. Mammals and birds are the only endothermic organisms.

A bird's body is insulated by feathers. Scientists believe that feathers evolved from scales as an adaptation for conserving body heat. Birds have three types of feathers: contour feathers, down feathers, and filoplumes. **Contour feathers** are the large feathers that give a bird's body its contour, or outline. Locate the parts of the contour feather in Figure 26·8. The contour feather is made up of a central *shaft* and a flat *vane*. The hollow part of the shaft that extends under the skin is called the **quill.** The solid exposed part of the shaft is called the **rachis** (RAY kihs). Note that the vane is composed of several hundred **barbs,** which are thin extensions of the shaft. Each barb locks into the next barb by tiny hook-bearing branches called **barbules.** If the barbs separate, the bird will draw the feather between its beak, hooking the barbules together again. This rehooking helps to maintain the sleek, strong structure of the feather.

Down feathers are short fluffy feathers. In adult birds they are usually located under the contour feathers. In young chicks they are the chief body covering. The main function of these feathers is insulation. Note in the drawing of the down feather in Figure 26·8 that there is no rachis, and the barbules lack hooks.

Filoplumes are sparsely scattered feathers, each composed of a long shaft with a few barbs at the tip. They usually grow around the base of contour feathers and probably give added insulation. Filoplumes are also known as pinfeathers.

Birds that swim have a coating of oil on their feathers. The oil makes the feathers waterproof. The oily feathers also help a bird stay afloat. Most birds have a large **tail gland** located behind the tail feathers on the rump. The tail gland secretes an oily fluid that a bird transfers to its body feathers to keep them smooth.

26·7 ADAPTATIONS FOR FLIGHT

Flight is one of the most outstanding characteristics of birds. A bird's body is adapted to meet the major requirements for flight, which are low weight and high power.

Unlike other vertebrates, the bird has a skeleton that is very light in relation to body size. For example, a pigeon's skeleton accounts for 4.4 percent of its body weight, but a rat's skeleton is 5.6 percent of its body weight. You can see in Figure 26·9 that the bird's wing bone is thin and hollow, with criss-cross reinforcements on the inside. These reinforcements provide support but have little weight.

The presence of feathers distinguishes birds from other vertebrates. In addition to keeping a bird's body warm, feathers function for flight. The long, overlapping contour feathers streamline the body. These feathers form the wings and much of the tail. Wings made of contour feathers have much surface area and little weight. They are well adapted for fanning the air and for gliding on air currents.

Several of the bird's body systems are adapted to contribute to low weight. This low weight, in turn, is part of the bird's adaptation for flight. The bird has a series of air sacs throughout its body. These sacs

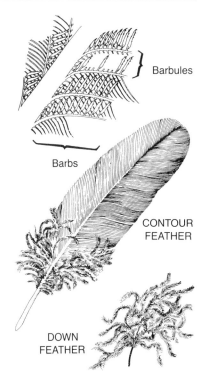

FIGURE 26·8 Contour feathers streamline a bird's body and are used for flight. Locate the rachis and vane on the large contour feather. Note the tiny hooks on the barbules which enable them to hook together. Down feathers are used for insulation and have no hooks on the barbules.

FIGURE 26·9 Notice the spaces and the reinforcing tissue in this bird bone.

cause much of the body cavity to be filled with air. The bird lacks a urinary bladder. Thus it does not store body wastes. The female has only one functional ovary and oviduct for egg production. Eggs are not stored in the body but are laid soon after they are formed. During the breeding season, the reproductive organs of both the male and female are large and functional. After the breeding season these organs shrink.

Birds also have several adaptations that add to their power. They have a high body temperature (as high as 43.5°C) and so have a high metabolic rate. The high metabolic rate allows for the release of the energy needed for the muscular contractions of flight. Birds have powerful flight muscles that may account for 50 percent of their body weight. Oxygenated and deoxygenated blood are kept separate in the heart. Only oxygenated blood is pumped throughout the body. A bird's heart is large and powerful, and it beats rapidly.

A bird's respiratory system is well adapted for providing the greater amount of oxygen needed for flight. The respiratory system has not only highly developed lungs but also, as mentioned, air sacs. You will learn more about the lungs and air sacs later in this chapter when you read about the anatomy of the pigeon.

26•8 CLASSIFICATION OF BIRDS

Birds are classified as members of phylum Chordata, subphylum Vertebrata, class Aves. They are endothermic feathered vertebrates. Most biologists recognize 8500 species grouped into 27 orders. In Figure 26•10 you can see birds that represent four of the major orders. Most birds have similar body structures. Taxonomists use variations in bone structure, muscles, and internal organs to distinguish one order from another. Variations in the shape and structure of the bill, wings, tail, and feet are useful for classification and identification.

FIGURE 26•10 Four of the 27 orders of birds are represented here. Although these birds are adapted to different habitats, their basic structure is the same. Why do birds show less variation in structure than reptiles or amphibians?

Investigation

What Is the Structure of a Chicken Wing?

Goals

After completing this activity, you will be able to
- Describe how the various tissues in a chicken wing are arranged.
- Assemble the bones of a chicken wing.

Materials (for groups of 2)

uncooked chicken wing
wing bones of cooked chicken
scalpel   probe   forceps
dissecting pan   hand lens
paper towel   glue

Procedure

A. Place the chicken wing in a dissecting pan. Feel the wing to locate the bones. Bend the wing to see how many sections it has. Note the texture of the skin.

B. Examine the covering of skin, which is a tough tissue that covers the wing. Locate openings from which feathers have been plucked.

C. Use your forceps to gently pull back the skin. Note the fat, or yellowish tissue found in small masses under the skin. Remove a small clump of fat and look at it with your hand lens.

CAUTION: Use great care in handling the scalpel. Direct the edge of the blade *away* from your body. Work on a firm surface.

D. Identify the muscles, which are bundles of pink tissue that surround the bones. Note where the muscles are attached to the bone. The muscles are covered with a thin white tissue, the connective tissue. Use your probe to pierce through the connective tissue so that the muscles can be seen.

E. Locate a tendon, which is the tough white tissue found at both ends of a muscle. Note where the tendon is joined to the muscle and bone.

F. With your probe carefully separate the bundles of muscle fibers. Look for a nerve, which is white and about the diameter of a pencil lead. You are likely to find a nerve between the muscles near a bone.

G. Look for a blood vessel, a structure that is about the same size as a nerve. The blood vessel is probably located beside a nerve. Note that it is a hollow tube and may still have a small amount of blood in it.

H. Use your scalpel to remove completely all of the skin. You will then be able to see the ligaments, which are a tough white tissue that holds two bones together. Look for the firm but flexible cartilage covering the ends of the bones.

I. Pull back the muscles with your forceps. Examine the bones, which are the hard supporting structures of the wing.

J. On a separate sheet of paper, draw the chicken wing. Label the skin, fat, connective tissue, tendon, nerve, blood vessel, ligament, cartilage, and bone.

K. Place the dry bones of the wing on a paper towel. Assemble them as shown in Figures A and B. Carefully using drops of glue, join the wing bones together.

L. Follow your teacher's instructions for the proper cleanup of all materials.

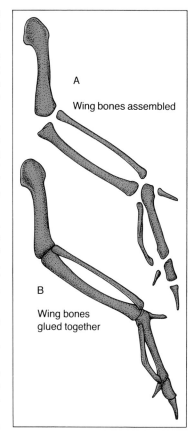

A Wing bones assembled

B Wing bones glued together

Questions and Conclusions

1. When you speak of chicken meat, which tissue are you referring to?
2. What function do you think is served by cartilage?
3. What tissues of the body are included in gristle?
4. Which of the tissues you examined would be directly involved in the movement of a bone?
5. In what ways is a wing adapted for flight?
6. What does the wing tip in the chicken correspond to in the human skeleton?

26·9 THE PIGEON: SKELETAL SYSTEM

The skeletal system of the pigeon, and other flying birds, is adapted for flight. Many of the bones are hollow, some of the bones found in other vertebrates are lacking, and many bones are fused. In spite of its light weight, a bird's skeleton is extremely strong. Refer to Figure 26·11 as you read the description of the skeleton.

A pigeon's skeleton has two major parts: the axial skeleton and the appendicular skeleton. The axial skeleton includes the skull, vertebral column, ribs, and breastbone. The appendicular skeleton includes the legs, wings, shoulders, and hip area.

Bones of the skull are lightweight, thin, and fused. The lower and upper jaws are modified as a bill. Teeth are lacking. The absence of teeth may be an adaptation for less mass.

The skull connects to the body by the neck vertebrae. Neck vertebrae are individual bones that allow for great flexibility of head movement. In fact the pigeon can extend its head to all parts of its body when smoothing and oiling its feathers. Some of the small vertebrae of the tail are also individual bones. The rest of the backbone is fused, providing a rigid supporting structure for flight.

The sternum, or breastbone, is formed into a keel. The **keel** is that part of the breastbone to which large breast muscles for flight are attached. The ribs protect the heart, lungs, and other internal organs. Locate the ribs and sternum in the drawing.

Three bones form each of the pigeon's shoulders: the *scapula*, the *coracoid*, and the *clavicle*. The clavicles come together with a third bone, forming the V-shaped "wish bone." Each *humerus* is attached to

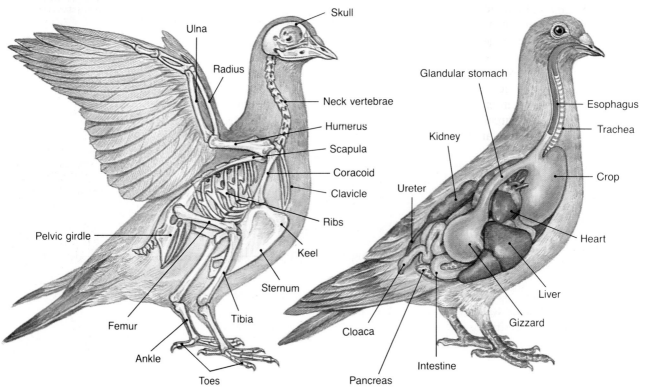

FIGURE 26·11 The skeleton of a pigeon is shown on the left. Note the large number of vertebrae in the flexible neck and the large keel bone. Some of the pigeon's major internal organs are shown on the right.

a shoulder. Below the humerus are the *radius* and *ulna*, as well as several small bones in the wing tip. The humerus, radius, ulna, and digits, along with skin and feathers, form the wing.

The fused bones of the hip area are called the pelvic girdle. The *femur*, or thigh bone, is relatively short and meets the pelvic girdle in a ball-and-socket joint. The lower leg contains the longest bone—the *tibia tarsus*. The pigeon's ankle and foot bones are fused.

26·10 THE PIGEON: OTHER BODY SYSTEMS

The digestive system of the pigeon is adapted for the digestion of seeds, the pigeon's main food. Refer to Figure 26·11 as you read about the parts of the digestive system. Food passes from the mouth through the tubular esophagus. The base of the esophagus is enlarged to form a crop. The *crop* is the region in which food is stored and hard seeds are softened with mucus. The crop allows the bird to feed rapidly and to store large amounts of food. The crop also secretes "pigeon's milk," which the bird uses to feed its young.

From the crop, food passes into the first part of the stomach, where some chemical digestion takes place. Next the food passes into the second part of the stomach, called the gizzard, where it is ground up. Pigeons swallow small stones that pass into the gizzard and aid in grinding up food. From the gizzard, food passes into the long, coiled intestine, which is the main organ of digestion and absorption. From the intestine, wastes are emptied into the cloaca, a common chamber into which excretory and reproductive materials also empty. From the cloaca, wastes are discharged from the body.

The pigeon has several other digestive organs. A large two-lobed liver with a gall bladder is located on the right side of the body. The pancreas is located along the wall of the intestine. These organs produce and store digestive enzymes.

The excretory system of the pigeon consists of a pair of large three-lobed kidneys. The kidneys are joined to the cloaca by the long tubular ureters. Since there is no urinary bladder, there is no storage of wastes. The main excretory waste of the pigeon is uric acid.

The pigeon's circulatory system, like the circulatory system of other birds, can provide large amounts of oxygenated blood to body muscles during flight. The heart is a strong four-chambered pump in which oxygenated and deoxygenated blood are kept separated.

The respiratory system of the pigeon, shown in Figure 26·12, is composed of several pairs of air sacs connected to the lungs. In the lungs, gas exchange with the blood occurs. Air is inhaled through the nares on the upper bill. It passes through the mouth, to the glottis, and then to the trachea. The enlarged area at the end of the trachea is a voice box called the **syrinx** (SIHR ihnks). The syrinx is used to produce various sounds.

The trachea branches to form two bronchi. The bronchi further divide, directing air to the lungs and to the air sacs. Air ducts connect the air sacs to the lungs. When the pigeon inhales, some of the air

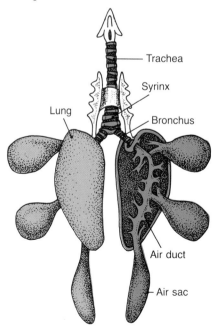

FIGURE 26·12 The respiratory system of birds includes air sacs as well as lungs. Gas exchange occurs both in the lungs and in the air sacs.

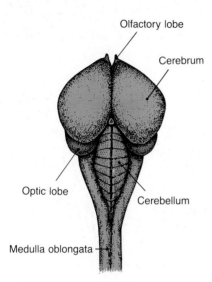

FIGURE 26·13 The birds' brain has a fairly large cerebrum and a well-developed cerebellum. The optic lobes receive information from the eyes, while the olfactory lobes control the sense of smell. What sense do you think is keenest in a bird?

enters the lungs and some enters the air sacs. When the pigeon exhales, air from the lungs is forced out, and air from the air sacs is forced into the lungs. The lungs of the pigeon thus fill with air both when the bird inhales and when it exhales.

The nervous system of the pigeon is made up of a brain, spinal cord, and nerves branching off the brain and spinal cord. Locate the parts of the pigeon brain, shown in Figure 26·13. The large cerebrum interprets sensory information and directs instinctive behavior. The optic lobes, which control the pigeon's keen sense of vision, are fairly large. The small olfactory lobes control the sense of smell. The highly developed cerebellum is the center for control of muscle movement, coordination, and balance. The medulla oblongata, which controls the internal organs, leads into the spinal cord.

The male reproductive system consists of a pair of testes, long tubelike sperm ducts, and the cloaca. Fertilization is internal in birds. Sperm pass through the sperm ducts, into the male's cloaca, and then are transferred into the female's cloaca. Then sperm swim from the cloaca into the oviduct, where fertilization occurs. The female reproductive system contains only one ovary that produces eggs, one long twisting oviduct, and a cloaca. The oviduct is lined with various glands that add materials to the egg, including the shell. From the oviduct the egg enters the cloaca and passes out of the body.

26·11 DEVELOPMENT OF THE BIRD EGG

The structure of the bird egg is similar to that of a reptile egg. The egg consists of the embryo, the yolk, and accessory structures produced in the oviduct.

The **albumen** (al BYOO muhn), often called the white of the egg, is a watery solution of protein secreted in the upper part of the oviduct. It provides a moist, fluid area around the yolk and the developing embryo. The albumen contains water for the early embryo and food for the embryo in its later phases of development.

The **chalazae** (kuh LAY zee) are a pair of dense twisted cords of albumen that are attached to opposite ends of the yolk. These cords help to keep the yolk and the developing embryo suspended in the albumen.

The egg has two *shell membranes.* The inner membrane encloses the albumen, embryo, chalazae, and yolk. The outer membrane lines the shell. Both are tough, permeable to gases, and relatively impermeable to water. They separate at the blunt end of the egg to form an air chamber. The shell is formed in the lower end of the oviduct. It contains protein and minerals, mainly calcium carbonate.

All birds are oviparous—that is, they develop from eggs laid outside the female's body. Development begins soon after fertilization and continues only if the egg is incubated. During incubation one or both of the parent birds sit on the eggs, providing them with a source of heat. Most bird eggs are incubated at a temperature of about 38°C. The period of incubation varies, depending on the species of bird.

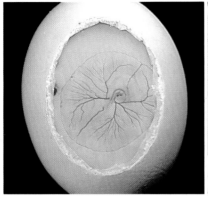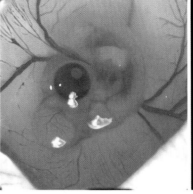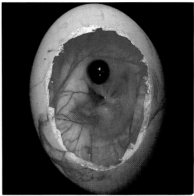

Look at Figure 26•14, which shows a chicken embryo at different stages in its development. Notice the development of organs as the embryo grows. At 21 days the embryo is fully formed and ready to hatch.

FIGURE 26•14 At 72 hours after fertilization, a chick embryo has a visibly beating heart (left). After 5 days, the limb buds and eyes are well-developed (center). A 10 day-old embryo (right) can be recognized as a chick.

REVIEW QUESTIONS

6. Describe three distinguishing traits of birds.
7. Identify three types of feathers and describe how they differ from one another.
8. Briefly describe two adaptations for low weight in birds.
9. Briefly describe two adaptations for increased power in birds.
10. Look at the foot and toes of the pigeon in Figure 26•11. Of what adaptive value to the bird is the position of the four toes?

Bird Behavior

26•12 REPRODUCTIVE BEHAVIOR

Birds lay few eggs and each egg receives care and protection. After hatching the young birds are cared for. Complex behavioral adaptations, called instincts, have evolved that help ensure survival of the species. *Instincts* are complex, inborn behavior patterns. Behavior in birds is mainly instinctive.

Territoriality is the instinctive behavior in which an organism stakes out and defends a territory. This behavior is often related to breeding. The establishment of a territory is linked with the fierce competition between birds of the same species at breeding time. Members of the same species compete for food and compete for mates. A territory lessens interference when the birds are involved with activities such as courtship, mating, nest building, and rearing of young.

SECTION OBJECTIVES

- Describe reproductive behavior in birds.
- Relate the structure of a bird's beak and feet to its habitat and preferred foods.
- Discuss the advantages of migration in birds.

Courtship involves many behaviors by which animals attract each other for mating. The birds behave in this way by instinct. The courtship acts in birds are usually, but not always, performed by males. The male may strut, showing his fanciest feathers. Songbirds may sing loudly and continuously to announce their territory and call for a mate.

Courtship displays are mainly for choosing a mate. In many species the female chooses the male. The display by the male must be specific enough to attract only females of the same species. Courtship displays also serve to ensure sexual readiness. These displays stimulate the release of sex hormones in both the male and the female. This ensures that both birds will be fertile when mating occurs.

Birds build nests that provide protection for the birds, their eggs, and their developing young. Nest building is an instinctive activity. Often the species of bird can be determined by the type and placement of a nest.

The care that birds give their young depends on whether the species is precocial or altricial. **Precocial** (prih KOH shuhl) birds are those that are well developed at the time of hatching and that maintain a constant body temperature even when young. Chickens, ducks, and other types of fowl are precocial. When these birds hatch, the mother warms them to dry their down feathers. They will then follow the mother from the nest to a nearby place to feed. The learning process involving following and imitating the first moving object seen after hatching is a form of behavior called **imprinting**. It is a behavior found in many precocial birds.

Altricial (al TRIHSH uhl) birds are birds that are very dependent on their parents at hatching because they are sightless, lack feathers, and are not able to feed on their own. When just hatched, all they can do is hold open their unsteady gaping mouths for food. Parents of such birds take turns *brooding,* or warming, the young and searching for food. Food is stored in the parents' crops, spit up, and then poked into the gaping mouths of the young. Robins, sparrows, hawks, and pigeons are examples of altricial birds.

FIGURE 26·15 A newly hatched Canada goose (top) is better developed than baby robins (bottom). Soon, the baby goose will be able to swim after its parent and look for food. The baby robins will be dependent on their parents until their feathers grow in.

26·13 ADAPTATIONS FOR FEEDING

Birds are found in most land areas of the earth. They inhabit areas as different from each other as the icy arctic and the tropics, the deserts and the forests. Some birds roost in trees; others spend much time on the ground. Some birds are waders; others are swimmers.

The foods that birds eat are as varied as the environments they inhabit. Some birds are herbivores and eat such foods as algae, seeds, nuts, fruits, or nectar. Others are carnivores and eat worms, insects, mollusks, or small vertebrates such as frogs, lizards, or mice. Differences in the structures of birds' feet and bills are related to the habitats and food of the birds. For example, a bird that swims may have paddle-shaped feet. Birds that roost in trees have feet that curve to hold the branch. Birds of prey have clawed feet.

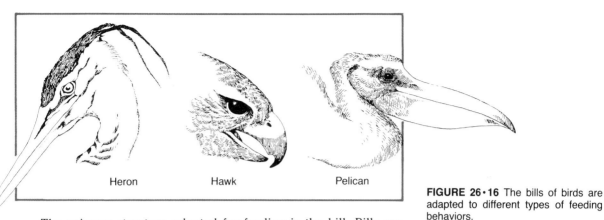

FIGURE 26·16 The bills of birds are adapted to different types of feeding behaviors.

The primary structure adapted for feeding is the bill. Bills are somewhat like tools. Look at Figure 26·16. Which is the bill of a bird that spears its food? Which bill strains food from the water? Which is used to hook and tear the food? For each of these bills, think of a tool that has a function similar to the function of that bill.

26·14 MIGRATION

One of the most interesting types of behavior in birds is migration. **Migration** is the instinctive movement of animals, usually between their wintering grounds and their breeding grounds. Look at the map in Figure 26·17, showing the migration route of the golden plover. This bird migrates between its wintering grounds in Hawaii and South America and its breeding grounds in North America.

In the Northern Hemisphere, many birds breed and raise their young in the spring and summer. As winter approaches in the breeding grounds, food is less available. Birds that migrate have a better chance of surviving. By instinct, migrating birds make their long journeys each year to reach more favorable habitats. For many species, the wintering grounds are in areas of mild climate where there is enough food. In the spring the birds return to their breeding grounds. There are usually fewer predators in the breeding grounds.

Migration is probably set in motion by several external stimuli, such as temperature and length of daylight. It is also triggered by the secretion of hormones. For many species the direction of migration is probably based on such factors as the birds' observation of the sun and stars and by the earth's magnetic field.

FIGURE 26·17 The Atlantic golden plover migrates between breeding grounds in Canada and the northern United States and wintering grounds in South America.

REVIEW QUESTIONS

11. Define courtship in birds; give examples of two forms of courtship behavior.
12. Explain the differences between precocial birds and altricial birds.
13. Of what adaptive value is territoriality to birds?
14. List two advantages of migration in birds.
15. What types of structural adaptations are likely to be found in birds of prey?

Chapter Review

SUMMARY

26·1 Class Reptilia includes those scaly ectothermic vertebrates adapted to life on land. Members of this class include turtles, snakes, lizards, and crocodiles.

26·2 The shell-covered, fluid-filled amniote egg is an important adaptation to life on land.

26·3 There are four surviving orders of reptiles. The order with the most species is Squamata, which includes the lizards and snakes.

26·4 The turtle is a member of an ancient group of reptiles and in many ways is representative of the class Reptilia. A distinctive external trait of the turtle is the presence of a two-part bony shell.

26·5 A study of the internal anatomy of the turtle reveals much about the internal anatomy of all reptiles.

26·6 The main traits of birds are wings, feathers, hollow bones, bills, and the condition of warm-bloodedness.

26·7 A relatively low body weight and a respiratory system allowing for high oxygen levels are two adaptations of birds for flight.

26·8 Birds belong to the class Aves, which contains 27 orders and at least 8500 species. The various species are distinguished by such characteristics as the shape of the bill and the structure of the wings, tail, and legs.

26·9 A study of the pigeon's skeletal system shows how it is adapted for flight.

26·10 A pigeon's internal anatomy is typical of the anatomy of most birds and shows a number of adaptations for flight.

26·11 Birds are oviparous. They produce amniote eggs that must be incubated in order for development to occur.

26·12 In birds, territoriality, courtship, and care of the young are instinctive behaviors concerned with reproduction.

26·13 A bird's adaptations for feeding are closely related to its diet and habitat.

26·14 Migration in birds is the seasonal movement between the breeding grounds and the wintering grounds.

BIOLOGICAL TERMS

albumen
allantois
altricial
amnion
amniote
  egg
barbs
barbules
birds
carapace
chalazae
chorion
contour
  feathers
down
  feathers
endotherm
filoplumes
hemotoxin
imprinting
Jacobson's organ
keel
migration
neurotoxin
plastron
precocial
quill
rachis
reptiles
shell
syrinx
tail gland
territoriality
viviparous
yolk sac

USING BIOLOGICAL TERMS

1. Which terms are names of the four embryonic membranes formed in an amniote egg?
2. What is the keel?
3. What is imprinting?
4. What is Jacobson's organ?
5. Distinguish between a hemotoxin and a neurotoxin.
6. Distinguish between the three types of feathers.
7. What is an endotherm?
8. What do the terms *carapace* and *plastron* have in common?
9. Distinguish between the terms *altricial* and *precocial*.

UNDERSTANDING CONCEPTS

1. Describe the major traits of reptiles. (26·1)
2. List the four embryonic membranes of the amniote egg and briefly state the function of each membrane. (26·2)
3. What is the evolutionary significance of the amniote egg? (26·2)
4. Name the four living orders of reptiles and give an example of one organism that belongs to each order. (26·3)
5. How is a snake's mouth adapted for feeding on animals larger than itself? (26·3)
6. Distinguish between the carapace and the plastron. (26·4)
7. Trace the flow of urine through the excretory system of the turtle. Name the structures through which the urine passes. (26·5)
8. How does a turtle's heart differ from the heart of an amphibian? (26·5)
9. List the traits of birds. (26·6)
10. Describe the types of feathers found on birds and explain the function of each. (26·6)
11. Describe two adaptations for low weight and two adaptations for increased power in birds. (26·7)
12. What traits do taxonimists use to distinguish different orders of birds? (26·8)
13. Name the major bones of a bird's wing. (26·9)
14. What traits of a bird's respiratory system make it highly efficient for providing oxygen to the body? (26·10)
15. How do the reproductive systems of the male and female bird differ from those of other vertebrates? (26·10)
16. Name and state the functions of the structures of a bird's egg. (26·11)
17. Distinguish between precocial birds and altricial birds. (26·12)
18. Write a description of the bill and feet of each of the following: (a) an aquatic bird that strains plankton from the water for food and (b) a bird that eats insects out of the cracks in trees. (26·13)
19. Name several stimuli that trigger migration in birds. (26·14)

APPLYING CONCEPTS

1. Turtles have changed little over the 200 million years of their existence. Suggest a reason to account for such little change.
2. Give a reason why some birds are found in arctic areas while no reptiles are found there.
3. Would you expect to find a bird that is ovoviviparous? Explain why or why not.
4. Poisonous snakes, like other snakes, swallow their prey whole. Why, do you think, is the snake not affected by the toxin?
5. Reptiles probably evolved from amphibians. Explain why the evolution of more efficient heart and lungs coincided with the evolution of a tough, dry skin.
6. The skeleton and feathers of a bird are found. The sternum is flat and the bones are solid and heavy. From this information, hypothesize as to whether the bird was able to fly well when alive. Explain.

EXTENDING CONCEPTS

1. Obtain a preserved turtle. Examine the external and internal anatomy of this organism. Identify the structures described in the text. Use sections 26·4–26·5 as a guide for this investigation.
2. Snake venoms are being researched for a number of different medical uses. Using reference materials, report on some of the experimental uses of snake venoms.
3. Write a report on one of the poisonous snakes that can be found in North America. Include pictures or descriptions of the snake.

READINGS

Colbert, Edwin H. *The Great Dinosaur Hunters and Their Discoveries,* Revised ed., Dover, 1984.

Diamond, Jared. "The Bower Builders." *Discover,* June 1984, p. 52.

Ferguson, Mark, and Ted Joanin. "Alligators: Clue to Dinosaurs' Demise." *Science News,* May 15, 1982, p. 326.

Newman, E.A., and P.H. Hartline. "The Infrared 'Vision' of Snakes." *Scientific American,* March 1982, p. 116.

27. The Mammals

These Japanese macaques usually live and feed in the forest. However, they can adapt to new conditions. Biologists began feeding the monkeys sweet potatoes on a sandy beach. One day, a female monkey began to wash the sand off her potatoes in the ocean. Soon, the rest of the monkeys washed their potatoes. How are mammals capable of learning? In this chapter, you will learn more about mammals.

CHAPTER OBJECTIVES

After completing this chapter, you will be able to
- **Identify** the characteristics of mammals.
- **Compare** reproduction in the different types of mammals.
- **Describe** specific mammalian adaptations to different environments.
- **Give examples** of mammalian behavior.
- **Explain** how the physiology and behavior of mammals has contributed to their success.

About Mammals

27•1 TRAITS OF MAMMALS

Mammals make up a class in the subphylum Vertebrata. A mammal is a vertebrate that has hair, nourishes its young with milk, and maintains a relatively constant body temperature. Mammals are adapted to many different environments and show a diversity of size, form and behavior. However, mammals share traits which distinguish them from other types of vertebrates.

Mammals are endotherms. Because they are able to maintain a constant body temperature, they can live in places that are too cold or too hot for reptiles and amphibians.

The four-chambered heart of a mammal is an efficient double pump that keeps deoxygenated blood separate from oxygenated blood. In many species the rate at which the heart pumps blood varies a great deal, depending on the animal's activity. For example, the resting heart rate of the big brown bat is 400 beats per minute. When the bat flies, the rate increases to 1000 beats per minute. How is this increase in rate of heartbeat important to the bat?

The lungs of a mammal are large and almost fill the chest cavity. A specialized muscle called the **diaphragm** (DĪ uh fram) helps to move large volumes of gases into and out of the lungs. What adaptive advantage is there to having a large supply of oxygen available to cells?

In general, mammals are more active and move more quickly than ectotherms. Mammals can sustain high levels of activity because they are endotherms and have a high rate of cellular respiration. In addition, many mammals have skeletons that are adapted for easy and rapid movement on land. Their limbs are directly beneath the body and perpendicular to the ground. The position of the limbs holds the body off the ground and improves the mammal's mobility.

The skin contains glands that are unique to mammals. **Mammary glands,** which are found in female mammals, secrete milk. The milk

SECTION OBJECTIVES
- Describe the adaptive advantages of mammalian traits.
- Explain why mammals are important to humans.

FIGURE 27·1 The pelage of the snowshoe hare is white in winter and brownish in summer, providing protective coloration at both times. In the mountain goat, notice how much thicker the pelage is in winter than in summer.

pilus (hair)

uterus (womb)

serves as food for the young during the early stages of growth. The watery secretions of structures in the skin called **sweat glands** help cool the body and also eliminate wastes. Secretions from *scent glands* and *musk glands* serve to attract mates and may also be used for protection and communication.

In most mammals the skin is covered with hair. Hair is made of dead epidermal cells that are strengthened by a tough protein material called *keratin*. A mammal's **pelage** (PEHL ihj), or coat of hair, gives insulation against both heat loss and overheating. Notice in Figure 27·1 that the pelage is thicker in winter than it is in summer. The color of the pelage also can change. For example, the coat of the snowshoe hare is white in winter and brownish in summer. The pelage of many small mammals is the same color as the soil in which they live. How do these adaptations aid survival?

In a mammal's brain, the cerebrum, which is the center for sensory interpretation and control of muscle responses, is greatly enlarged. With such an enlarged cerebrum, a mammal is better at retaining information and learning from past behavior than are other vertebrates. In general, mammals are more intelligent than other vertebrates.

Most mammals give birth to live young. The embryo develops in the mother's body within a muscular organ called a **uterus**. In most mammals the embryo receives nutrients from the mother's blood and transfers wastes to the mother's blood through a specialized organ called the **placenta** (pluh SEHN tuh).

Every mammal has teeth of different types. For example, narrow, sharp teeth are specialized for piercing the flesh of prey. Flatter, wider teeth are specialized for grinding. The types, number, and arrangement of teeth is called **dentition.** An animal's dentition is a good indicator of the animal's diet. Meat-eating mammals, called **carnivores,** have a type of dentition that is adapted for tearing meat. Plant-eating mammals, called **herbivores,** have a type of dentition that is adapted for clipping plants and grinding vegetable matter. Figure 27·2 shows the teeth of a wolf (a carnivore) and a deer (a herbivore). How do the teeth of these mammals differ? A mammal's dentition tells a great deal about what it eats.

The digestive system of a mammal is adapted to the animal's diet. Although all mammals have similar digestive tracts, the relative sizes of the digestive organs may vary.

Several waste products of cell metabolism are filtered in the kidneys. Nitrogenous wastes are eliminated as urea. In most mammals, urine leaves the body through a duct called the urethra.

27·2 IMPORTANCE OF MAMMALS

Mammals can have a great effect on their environments. For example, beavers change woodlands and streams by cutting down whole trees and building dams, as shown in Figure 27·3. Mammals that dig burrows and tunnels change the soil structure in their environments. Mammals also disperse plant seeds. Some mammals, such as the Australian honey possum, pollinate flowers. Herbivores such as elephants remove young leaves from plants. Light pruning encourages plant growth, but heavy pruning can harm plants. Humans cause major changes in their environment by farming, hunting, building, and other activities.

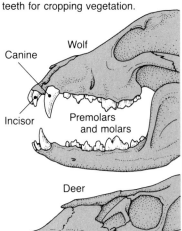

FIGURE 27·2 Compare the dentition of the wolf and the deer. The wolf has teeth used to tear meat. The deer has teeth for cropping vegetation.

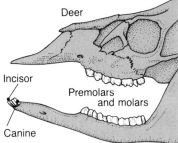

FIGURE 27·3 Beavers build dams across streams to form beaver ponds. These ponds create new types of habitats for wildlife.

546 THE MAMMALS

FIGURE 27·4 Because they feed on small animals, foxes help control the populations of such creatures. When predators such as foxes are heavily hunted or poisoned, the numbers of mice, rats, and other small animals increases.

In many environments, mammals are the most varied and numerous vertebrates. They are often the most active organisms in an environment. For these reasons mammals are said to be the dominant animals in many environments. The feeding behavior of mammals can have a great effect on the environment. Mammals depend on the consumption of large amounts of food because they are highly active endotherms. Plants and other animals in the environment are sources of food for mammals. Because they eat large amounts of food, mammals have a great effect on the sizes of the populations of other species around them.

Mammals are an important part of the food chain in their environments. In addition to preying on other species, mammals are preyed upon by birds, other mammals, and invertebrate parasites. How do mammals influence the populations of other species by being a source of food?

Various mammals are also of economic value. Since prehistoric times, humans have domesticated mammals for food and work. Humans have taught such mammals as dogs, horses, camels, and elephants to work for them.

REVIEW QUESTIONS

1. What traits are characteristic of mammals?
2. What traits make mammals of special value to humans?
3. What adaptations permit mammals to survive in cold climates?
4. How might the color of the pelage be of adaptive value?
5. The skull of a fossil mammal is found to have small front teeth and large cheek teeth. What type of diet did this animal probably have?

Classification of Mammals

SECTION OBJECTIVES

- List some orders of mammals and give examples of each.
- Compare reproduction in mammals.

27·3 MONOTREMES

Mammals belong to class Mammalia, subphylum Vertebrata. Recall that the vertebrates are a subphylum of phylum Chordata, the chordates. The most primitive of mammals, called monotremes (MAHN uh treemz), are placed in order *Monotremata*. A **monotreme** is a mammal that lays eggs. The skeleton of a monotreme is similar in many ways to that of a reptile. Monotremes are so different from all other mammals that scientists think they probably are not ancestral to the other mammalian orders. Monotremes share with the higher mammals the characteristics of hair, mammary glands, and endothermy.

Monotremes are the only mammals to lay eggs. Like birds, they incubate the eggs until hatched. The mother's mammary glands, which secrete milk for the young, lack nipples. The young suck the milk from two areas on the mother's skin called *lobules.*

The only living monotremes are the duckbilled platypus and the echidna (ih KIHD nuh), or spiny anteater. Both live in Australia and Tasmania; the echidna also lives in New Guinea. Both animals have some features that are very different from those of other mammals.

FIGURE 27·5 The echidna (left) and the duckbilled platypus (right) are monotremes. If threatened, an echidna rolls into a spiky ball or digs straight down until only its spines are visible. Both males defend themselves with poison spurs on their hind limbs.

The echidna is an insect eater with a short tail and a spine-covered back. Like most insect eaters, it has a long narrow snout. The echidna is adapted for burrowing and for fairly quick movement.

The duckbilled platypus, also shown in Figure 27·5, lives in burrows along stream banks and feeds in the water. Its long snout is covered with a bill like a duck's, and its toes are joined by webbing.

27·4 MARSUPIALS

marsipos (pouch)

The **marsupials** (mahr SOO pee uhlz), which are members of order *Marsupialia,* are the most primitive mammals to give birth to live young. Familiar examples of marsupials are the kangaroo and the opossum. Many types of marsupials are herbivorous animals that feed at night.

Although marsupials give birth to live young, the young are tiny and not fully developed at birth. The embryos spend only a short amount of time inside the mother's uterus. After they leave the uterus, the young complete their development inside an abdominal pouch called a **marsupium.** Immediately after birth, the tiny young marsupials crawl into this pouch. There they attach to nipples and are fed from the mother's mammary glands. A newly born marsupial, such as a kangaroo, will remain in the marsupium and continue to grow. Many marsupials carry their young in their pouches for several months. The marsupial pattern of reproduction provides more protection and greater chance of success for the young than egg-laying provides.

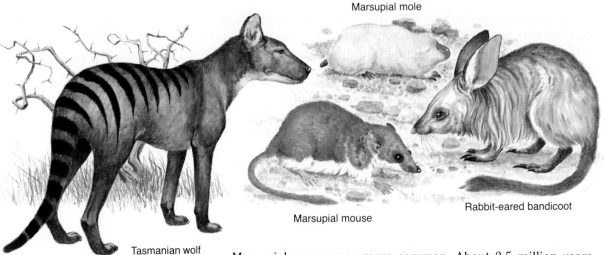

FIGURE 27·6 As a result of convergent evolution, many marsupials resemble familiar placental mammals. Today, the Tasmanian wolf is probably extinct due to hunting, and many other Australian marsupials are endangered.

Marsupials were once more common. About 2.5 million years ago, there were a large number of marsupial species in South America and Australia. These marsupials were isolated from other types of mammals. They were adapted to a wide range of environments and feeding habits. One herbivore was the same size and shape as a rhinoceros, and there was a carnivore very similar to the placental saber-toothed cat. Most of the prehistoric South American marsupials became extinct because they could not compete with the better-adapted placental mammals that migrated into their area.

However, many marsupial species are still found in Australia. These animals look much like certain placental mammals living elsewhere. These resemblances are the results of convergent evolution. Recall that convergent evolution is an increase in similarities among species derived from different ancestors as a result of similar adaptations to a certain type of environment (Section 10·8). Marsupials are adapted to habitats much like those of placental mammals on other continents. Many marsupials are named for the placental mammals they look like. For example, each of the animals in Figure 27·6 is a marsupial named for its placental counterpart. Both a marsupial mole and a placental mole live underground and are burrowers.

Three of the most familiar marsupials are the koala, kangaroo, and opossum. The koala lives in wooded areas of southeastern Australia, where it feeds only on the leaves of eucalyptus trees. The young koala is carried in the mother's pouch for six months after birth and then rides on the mother's back for several more months before it is independent.

Kangaroos are very similar in certain ways to antelopes, which are placentals. Both are herbivorous and have similar digestive tracts, skulls, and dentition. Kangaroos and antelopes are both fast runners with highly specialized limbs. However, the kangaroo usually moves on two legs. Its small forelimbs are used only for slow, four-legged movement or for handling food. The elongated hindlimbs are adapted for rapid jumping and running.

Opossums are the only type of marsupial found outside of Australia and New Guinea. They are native to North and South America. Most of them are good climbers, with feet and tails adapted for grasping. The cat-sized Virginia opossum is common in the eastern half of the United States and continues to extend its range.

27•5 PLACENTALS

The most diverse mammals are the placentals. A **placental mammal** is one that develops inside the mother's uterus while attached to a placenta. The majority of mammals are placental mammals.

In a placental mammal, all organ systems develop while the animal is still inside the mother's uterus. The offspring is therefore more fully developed and more self-sufficient when it is born than a marsupial is. The length of time the animal spends developing in the uterus is called the **gestation** (jehs TAY shuhn) **period.** During the very early stages, the developing animal is called an embryo. In later stages it is called a fetus. Gestation periods vary from species to species. Figure 27•7 lists the gestation periods for a few placentals.

Recall that the placenta is an organ that connects the developing animal and the mother. Through the placenta, the growing animal receives food and oxygen from the mother's bloodstream and gets rid of waste materials. The placenta provides a constant source of food for the developing fetus.

A long gestation period, which is made possible by the placenta, provides certain advantages. A prolonged period inside the uterus gives the fetus protection and nourishment throughout its entire development before birth. Also, a long gestation period allows more complete development of the fetus inside the uterus. At birth the placental infant is more complete and thus is stronger and more self-sufficient than the marsupial infant at birth.

FIGURE 27•7 This table lists the gestation periods for some placental mammals. Which animal has the shortest gestation time? How does the gestation period of an elephant compare to that of a human?

MAMMAL	GESTATION IN DAYS (average)	MAMMAL	GESTATION IN DAYS (average)
Bear	208	Hamster	16
Camel	406	Hippopotamus	230
Chimpanzee	243	Horse	336
Chipmunk	31	Human	267
Cow	281	Lion	108
Coyote	63	Mouse	20
Deer	215	Rabbit	31
Elephant	624	Raccoon	63
Fox	52	Sheep	151
Giraffe	442	Whale	450
Guinea pig	68	Zebra	357

Computers in Biology

Computers in Animal Research

Each year in the United States many animals die in experiments. Most such experiments are concerned with ways to improve the health of humans. Dogs, cats, monkeys, sheep, and pigs are among the many animals that are the subjects of biological experiments. The animals are used in testing cleaning products, cosmetics, drugs, and thousands of other chemicals that ultimately may be used by humans. They also are used in medical schools.

As public concern for animal welfare grows, scientists are finding new ways to experiment. One of these ways involves computers. The use of computers instead of animals saves not only millions of animal lives but also much research time and money.

Experiments can be done by using equations to represent an animal's body processes. Those equations are entered into a computer. Information also is entered about the experiment being done, such as the effects of a new drug. The computer will calculate and print out the effects of the drug on body processes.

In the United States the Food and Drug Administration (FDA) carefully tests all chemicals before they can be made available for humans to use. One test that is required on all chemicals produced is called an LD_{50}. LD_{50} means "lethal dose 50 percent," or the dose of the chemical that will kill 50 percent of the animals in the study. LD_{50} tests are being replaced by tests done on a computer. A program has been written that contains information about past LD_{50} experiments and information on the properties of about 6000 chemicals. Without killing any animals, the computer can use the data to find the toxicity (poison level) of new chemicals.

At Johns Hopkins University a more advanced program is being used. The computer gives the poison level of a chemical and the effects of the chemical on several kinds of animals. For example, it can give the effects of a drug on both a monkey and a gorilla.

Computers are being used in place of animals not only for chemical and product safety tests but also in medical schools. In the past, many animals were used by medical students as they learned about diseases. The procedures caused much animal suffering and death. Now a computer can calculate how an animal's systems will respond to certain conditions. Changes in heart rate and brain waves are among the responses the computer can determine. For example, studies of the effects of changes in body temperature once required dogs to be packed in ice. Now the computer can predict the response to cold temperatures without the use (and death) of any dogs.

Surgical procedures can be practiced using a computer. Such procedures can be repeated so that students can correct their errors. Also, this type of practice on the computer can be interrupted and can be resumed whenever the computer is available.

Scientists are sure that the use of computers will reduce the number of animals used in biological experiments in the future. This use of computers represents a new development in scientific research.

550 THE MAMMALS

27·6 ORDERS OF PLACENTALS

The variety in adaptations among the placental mammals makes the placentals a diverse group. Most scientists recognize 16 orders of placental mammals. Within each order, members are specialized in similar ways for particular environments and ways of life.

Order *Insectivora* (ihn sehk TIHV uhr uh) includes the most primitive placentals. **Insectivores** (ihn SEHK tuh vohrz) are small mammals belonging to order Insectivora and are adapted to eat small invertebrates, especially insects. Many insectivores, such as the shrew and mole, are adapted for life underground. They have small cylindrical bodies, small ears and eyes, and velvety fur, which aids movement underground. Notice in Figure 27·8 that the mole has spadelike forelimbs that are specialized for burrowing. The eyes of some kinds of moles are covered with a thin skin and are functionless. Why is sight not an important sense to such an animal?

Both moles and shrews have extremely high rates of metabolism and require large amounts of food. Moles and shrews are active both day and night, although some other insectivores are nocturnal, or active at night.

Order *Chiroptera* (kī RAHP tuhr uh), the bats, is the second largest order of mammals. Bats are the only flying mammals. Their wings are formed by elongated fingers that are joined by webbed skin. Most bats are nocturnal. Different kinds of bats feed on insects, fruit, pollen and nectar, or fish. One group of bats, called vampire bats, feeds on the blood of various vertebrates, such as cattle. Fruit-eating bats rely on sight and do not fly as well as do other bats.

The smaller, more agile bats sense their environment by a process called echolocation. **Echolocation** is a means of sensing the environment by using sound. The bat emits ultrasonic pulses. Echoes of these pulses are received by the bat, enabling the bat to detect objects in its environment. Echolocation allows the bat to catch insects in its wings even in total darkness. Fish-eating bats use echolocation to locate fish that break the surface of water.

Order *Edentata* (ee dehn TAH tuh) is a diverse group of mammals. Members of this order are called edentates (ee DEHN tayts). **Edentates** are mammals that have either very reduced teeth or no teeth at all. This group includes anteaters, tree sloths, and armadillos. A tree sloth is shown in Figure 27·8.

A much less diverse group than the edentates is order *Lagomorpha*. Members of order Lagomorpha, called lagomorphs, include rabbits, hares and pikas. **Lagomorphs** are all small herbivores with chisel-shaped front teeth. Most lagomorphs are specialized for running, jumping, and, often, burrowing. A pika is shown in Figure 27·8. These animals are smaller than rabbits and are not as well adapted for running. They are found in mountainous areas and in deserts. Rabbits and hares have long hindlimbs that are specialized for running and jumping. They are found in a wide variety of environments, including arctic tundra, forest, and desert.

FIGURE 27·8 A mole (top) is an insectivore that is specialized for burrowing. A sloth (middle) is a slow-moving edentate. It lives in trees in South American rain forests and rarely comes down to the ground. A pika (bottom) is a lagomorph. During the summer and fall, pikas collect plants and store them for winter food.

rodere (to gnaw)

The largest group of mammals is order *Rodentia*. Members of order Rodentia, called rodents, are a highly diverse and successful group. **Rodents** are small, mainly herbivorous mammals with sharp chisel-shaped front teeth. Species of rodents differ from each other in diet, in movement, and in habitat. The rodents native to North America include mice, rats, pocket gophers, beavers, squirrels, woodchucks, woodrats, porcupines, and muskrats. Although rodents resemble lagomorphs in appearance, the ancestries of the two groups are separate.

Members of order *Cetacea* (suh TAY shuh), called the **cetaceans** (suh TAY shuhnz), are mammals that are adapted to live in the ocean. Cetaceans include whales, porpoises, and dolphins. They are among the fastest swimmers in the sea. In whales the forelimbs are specialized as paddle-shaped flippers. The hindlimbs are greatly reduced and do not extend from the body. The whale has a streamlined shape. The tail of a cetacean is very powerful. The respiratory system of a cetacean is adapted for long periods of submersion without oxygen. Echolocation is used for underwater guidance. Many cetaceans are highly social animals. Cetaceans also appear to be highly intelligent.

Most members of order *Carnivora,* the carnivores, are meat-eating animals. (Note that the term *carnivore* can be used to mean any animal that eats meat. It is used here to mean any member of order Carnivora.) Carnivores have an extremely good sense of smell. Their teeth are large and sharp. Some carnivores, such as dogs, are highly specialized runners. Seals and sea lions are adapted for aquatic life and for eating fish. Bears, raccoons, coyotes, and foxes are not strictly carnivorous, feeding on fruit and vegetation as well as meat. Cats are familiar members of this order.

Order *Proboscidea* (proh buh SIHD ee uh) includes the elephants—large herbivorous animals with long agile trunks. Elephants are the largest living land mammals. They were once more numerous and diverse than they are now. Today they inhabit southern Africa and parts of southeastern Asia. The high daily need for water restricts elephants to areas near water.

FIGURE 27·9 A muskrat (left) is a rodent. Like beavers, muskrats build a house of branches and twigs near ponds and streams. However, they do not construct dams. A beluga whale (right) is a cetacean. Beluga whales are sometimes called "sea canaries" because they whistle and chirp as they hunt or communicate with each other.

FIGURE 27·10 Elephants (left) belong to the order Proboscidea. This order includes modern-day elephants and the extinct mammoths and mastodons. A pronghorn antelope (right) is an ungulate native to North America. It belongs to the order Artiodactyla.

The **ungulates** (UHNG gyuh lihts), or hoofed mammals, are in orders *Perissodactyla* (puh rihs uh DAK tuhl uh), odd-toed ungulates, and *Artiodactyla* (ahr tee uh DAK tuhl uh), even-toed ungulates. Ungulates are herbivores with very good running ability. They use their speed to escape predators. Long limbs and special muscle patterns are some of the adaptations they have for running. The digestive system in many types of ungulates is specialized for digestion of plant material. Most ungulates are well adapted for life on open grasslands. Included in order Perissodactyla are horses, zebras, and rhinoceroses. Order Artiodactyla, the larger group, includes pigs, hippopotamuses, camels, deer, giraffes, sheep, cows, goats, and antelopes.

Animals such as lemurs, tarsiers, monkeys, apes, and human beings belong to order *Primates*. **Primates** are mammals that have many adaptations that developed in arboreal, or tree-dwelling, ancestors. These adaptations include depth perception, grasping hands, and agility. Such adaptations are helpful to animals that move rapidly between trees, grabbing branches, and quickly avoiding obstacles. Some primates are **omnivores,** or animals that eat both plant and animal material. The sense of smell is reduced in primates, but the brain and eyes are large.

primas (first)

REVIEW QUESTIONS

6. What traits indicate that monotremes are more primitive than other mammals?
7. What explains the fact that so many marsupials resemble placentals?
8. What are the functions of a placenta?
9. What are the benefits of a long gestation period?
10. Cetaceans have very little hair. What is this lack of hair probably an adaptation for? Since they lack the insulation provided by hair, how are some cetaceans adapted to cold arctic waters?

THE MAMMALS

Investigation

How Can Teeth Be Used to Identify Mammals?

Goals

After completing this activity, you will be able to
- Use a dental formula to help identify the skull of a mammal.
- Compare the dental pattern of animals belonging to several different orders of mammals.

Materials (for groups of 2)

skull or jawbones of an animal

Procedure

A. Mammals generally have four main types of teeth: (1) incisors, (2) canines, (3) premolars, and (4) molars. Locate these teeth in the drawing of the jaw of the wolf shown below. Count the total number of each type of teeth. (Only half the jaw is shown.)

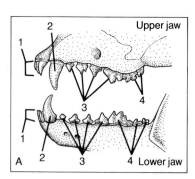

B. The four kinds of teeth are found in mammals in certain patterns. Scientists have studied these patterns and have developed dental formulas. A dental formula may be used for identifying different orders of mammals. Study the dental formula shown below:

I $\frac{3-3}{3-3}$, C $\frac{1-1}{1-1}$, P $\frac{4-4}{4-4}$, M $\frac{2-2}{3-3}$ = $\frac{20}{22}$ = 42

I stands for "incisor"; *C* for "canine"; *P* for "premolar"; and *M* for "molar."

C. Using the dental formula in step B, find the total number of premolars in the upper jaw. Find the total number of canines in both jaws. Compare the dental pattern of the animal given in the dental formula with that of the wolf, which is shown in the drawing.

D. The table below gives the dental patterns for many different mammals. (Note that *U* stands for "upper jaw"; *L* stands for "lower jaw.") Find the total number of teeth in a mouse. Compare the total number of incisors in a deer with that in a chipmunk.

E. Study the drawing in Figure B. Count the number of each type of tooth. Write a dental formula for the animal. Using the table of dental formulas, identify the animal shown in Figure B.

F. You will be looking at the jawbone of a mammal. Identify the animal by using the table of dental formulas.

Questions and Conclusions

1. Using the table write the dental formula for a cat.

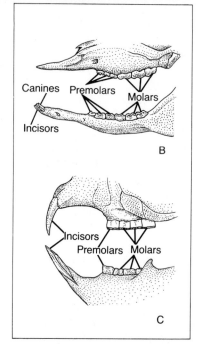

2. Based on the table, how many canines does a chipmunk have? How might this number be an advantage or disadvantage?

3. How many incisors does the animal in Figure C have? How many canines does it have? Using the table of dental formulas, what animal is shown in Figure C? Give reasons for your answer.

	INCISORS	CANINES	PREMOLARS	MOLARS	U AND L		TOTAL	LAND MAMMALS
U	3–3	1–1	4–4	2–2	20			
L	3–3	1–1	4–4	3–3	22	=	42	Dog, wolf
U	0–0	0–0	3–3	3–3	12			
L	3–3	1–1	3–3	3–3	20	=	32	Bison, deer, goat
U	3–3	1–1	3–3	1–1	16			
L	3–3	1–1	2–2	1–1	14	=	30	Cat
U	2–2	0–0	3–3	3–3	16			
L	1–1	0–0	2–2	3–3	12	=	28	Rabbit
U	1–1	0–0	1–1	3–3	10			
L	1–1	0–0	1–1	3–3	10	=	20	Chipmunk, porcupine
U	1–1	0–0	0–0	3–3	8			
L	1–1	0–0	0–0	3–3	8	=	16	Mouse, rat

Behavior of Mammals

27·7 MIGRATION AND HIBERNATION

Mammals require a great deal of food to supply the energy needed to maintain their high body temperatures and high levels of activity. During the winter, energy requirements are greatest and food, including plants and animals, is scarce. There are various ways in which some mammals find food in the winter. For many carnivores, prey is still available. Moles and shrews still can find food underground or under snow. Some mammals change their eating habits in the winter. Hares, for example, switch their diet from the leaves of green plants to twigs.

Among those mammals whose food is not available in winter, three adaptations exist: storage, migration, and hibernation. In the first of these adaptations, animals such as tree squirrels store food that will last through the winter.

Some mammals survive the winter by migrating. **Migration** is the instinctive seasonal movement of animals from one place to another. Migration occurs regularly, probably in response to environmental changes. Mammals migrate from their summer feeding and breeding grounds and spend winter in areas with higher temperatures and/or more food. For example, many whales spend the summer in the Arctic or Antarctic ocean and migrate to the tropics in the fall.

Many species of mammals migrate. Seals, sea lions, and walruses migrate to islands where they breed and raise their young. Many of the large herbivores of the north migrate in herds to new feeding grounds. Elk, moose, and deer migrate from high elevations, where they spend the summer, to lowlands, where they live in winter.

SECTION OBJECTIVES

- Describe how mammals have adapted to different environments.
- Give examples of mammalian behaviors.

FIGURE 27·11 Caribou migrate to new feeding grounds as winter approaches.

The third adaptation to a lack of food in winter is hibernation. **Hibernation** is a state of deep sleep in which the body temperature drops so that it is almost equal to that of the surroundings. The metabolic rate in the body slows down. For example, the heart rate for a ground squirrel drops from 150 beats per minute to 5 beats per minute. The need for food decreases. The mammal lives on food stored as fat in its body. Fat cells, full of mitochondria, supply the mammal with energy during its period of hibernation.

The duration and degree of hibernation vary among species. The brown bear, for example, does not truly hibernate but remains in its shelter in a state of drowsy inactivity. Its metabolic rate does not fall as much as in true hibernation. Some mammals remain in a deep sleep for over half a year. Others, such as the common hamster, wake up every few days. Some hibernating mammals, such as ground squirrels, store food in their shelters to eat during periods when they are awake.

FIGURE 27·12 Ground squirrels hibernate for the winter. What adaptations are necessary for hibernation?

27·8 TERRITORIALITY AND SOCIAL HIERARCHY

Food, shelter, and living space are essential for an animal's survival. Competition among members of a species for these essential resources can be strong. However, many mammals are adapted to live in small organized groups within well-defined areas. Such adaptation minimizes competition among members of a species. Specific patterns of behavior keep each group united within its area and prevent different groups from competing for the same resources.

The specific area in which an animal or group of animals of a species lives is called a **territory.** The marking and defending of a territory by the animals that live there is called **territoriality.** Territorial behavior differs among species. Many mammals, such as rabbits, mark their territories with scent from gland secretions or with urine or feces. These odors are recognized as boundary markers by other members of the species. Sea lions use songs or calls to claim a territory as their own. A mammal may show threatening behavior if another member of the species enters its territory. All these signs serve to keep away intruders. By defending its territory, a mammal or a group of mammals helps ensure that it will have food, shelter, and often, breeding area.

A group that lives within a territory often has a well-organized social structure. Social organization is an adaptation that causes the group to cooperate against enemies and helps maintain peace within the group. Territory is defended easily, and food, shelter, and space are shared among members with a minimum of dispute.

In most social groups a **social hierarchy,** or ranking of individuals, exists. Social hierarchy usually is determined by a series of repeated threats and fights among individuals. The largest and most powerful males usually become dominant over the other members of the group. Once ranking order has been determined, fighting within the group is minimized.

FIGURE 27·13 When on the move, a baboon troop exhibits social organization. Females and young are in the center of the troop, with the adult males forming a protective ring.

One group of mammals with a complex social hierarchy is the baboon troop. A number of such troops live in the Ethiopian desert. Each troop of baboons consists of between 10 and 100 individuals. The troop has several dominant males that are larger and stronger than the others. These dominant males lead and defend the troop. Social hierarchy of the baboon troop allows the members to live peacefully, with a minimum of competition for food, shelter, and mates.

27·9 REPRODUCTIVE BEHAVIOR

Reproduction in mammals is associated with specific types of behavior. The specific type of behavior associated with the selection of mates within a species is called **courtship behavior.**

Courtship behavior usually is performed by a member of one sex to attract a mate. In most mammal species it is the male who carries out the patterns and signals of courtship behavior. In some species the male shows courtship behavior in his territory, which then becomes a nesting area. Courtship behavior also creates a bond between the two mates.

Forms of courtship behavior differ among species. In many species the female becomes less aggressive if the male shows behavior that is normally associated with infancy. For example, male hamsters imitate the calls of nesting young while courting. In some species the males fight to attract mates.

In most mammalian species, the female will mate only when she is fertile. The fertile period of a female is known as **estrus.** Estrus occurs regularly in each species. For example, estrus occurs once a year in deer, once a month in horses, and once every few days in mice. During estrus the female shows changes in behavior and appearance that indicate to the male that she is ready to mate.

FIGURE 27·14 An extended period of parental care allows young mammals to develop and to learn. An opossum's young (left) develop in her pouch. After young leave the pouch, they follow their mother until they are weaned. Orangutans (middle) are dependent on their mothers until they are about eight years old. Lion cubs (right) learn to hunt by following their mother and other females on hunts.

The offspring that result from mating are usually given a great deal of care by one or both of their parents. Mammalian young depend on their mothers for milk. Because of nursing, the close relationship between offspring and mother is unique to mammals. The social group of mother and young, which results partly from nursing, is the primary social unit of mammalian species.

Instinct is part of most animal behavior. In addition, young mammals learn certain forms of behavior from their parents and from other adults in their social group. Lion cubs learn to hunt by following the females on hunting sessions. Young chimpanzees learn to use simple tools by imitating other adults. The behavior patterns of courtship and territoriality are partly instinctive and partly learned. The long period of bonding between parent and offspring enables young mammals to learn some of the behavior patterns of their species.

27·10 COMMUNICATION IN MAMMALS

In order for wolves to work cooperatively on a hunt, certain information must be conveyed among them. What information would this be? How is information that is necessary for social interactions conveyed among mammals? Information is conveyed among organisms through sets of behavior signals called **communication.** Communication is used to indicate the borders of the territory, the dominance of one individual over another, and the readiness of one individual to mate with another.

Mammals communicate in a variety of ways, using a combination of signals that includes odors, sounds, touching, and visual displays. Among mammals the use of scent is a common means of communication about territory and dominance. Chemical signals from urine, feces, and scent glands convey extremely specific messages that remain in place for long periods of time. Members of wolf packs use urine to mark the borders of their territory. This form of communication is understood by other wolves as well as by other species.

Visual signals used by mammals include facial expressions, the postures of body, tail, and ears, and the raising and lowering of hairs. Common examples of visual signals are the wagging of a dog's tail, which conveys friendliness, and the arching of a cat's back and raising of its fur, which indicate defensiveness. In many mammals, facial expressions are used in combination with different ear and head positions. Facial expressions are very important visual signals for primates. Notice in Figure 27·15 the kinds of information a chimpanzee communicates by using facial expressions. Visual signals are also important means of communication for herd animals. The gazelle, for example, runs in a stiff-legged gait and flares the hair on its white rump, signaling the approach of a predator to other herd members. What behavior might such signals trigger in the other members of the herd?

FIGURE 27·15 Chimpanzees use facial expressions to communicate.

Attention

Excitement

Anger

Sounds often are used as signals to maintain contact among members of a group. For example, the constant sounds of teeth grinding and stomach rumbling in a herd of grazing animals let the members of the herd locate each other. Many species use vocal signals to maintain contact. The howling of wolves establishes contact between individuals. The songs of whales may serve to maintain contact or to coordinate behavior among members of a group.

The vocal signals of humans are a complex form of communication. Humans use *language*—a system of vocal signals that allows complex and new ideas to be expressed and understood. Though other primates can make up to 25 vocal sounds, none have been shown to have the capacity for vocal communication.

REVIEW QUESTIONS

11. Describe three adaptations that allow mammals to survive the winter.
12. What are the benefits to a species that shows territorial behavior?
13. What information is conveyed through courtship behavior?
14. What is the significance of the long period of bonding between mammalian young and their parents?
15. In some species that have a wide range from north to south, the higher the latitude, the earlier the individual organism enters hibernation. Identify as many factors as you can that might trigger hibernation, based on this information.

Chapter Review

SUMMARY

27·1 Mammals are endothermic vertebrates that have hair and that bear live young and nurse them with milk.

27·2 Mammals are the dominant organisms in many environments. Mammals are of important economic value to humans as work animals and for food.

27·3 The monotremes are primitive mammals that lay eggs.

27·4 Marsupials are mammals whose young are born immature and complete their development outside the uterus in a pouch called a marsupium.

27·5 Before birth, placental mammals complete development in the mother's uterus, receiving nourishment through the placenta.

27·6 The placental mammals are a large and diverse group, composed of animals adapted to a variety of environments.

27·7 Many mammals survive winter by migrating to other locations or by hibernating.

27·8 Many mammals live in well-marked and defended territories in organized groups generally having a social hierarchy.

27·9 In mammals the selection of mates and care of the young are associated with specific kinds of behavior.

27·10 Mammals communicate by means of a variety of signals, including odors, sounds, touching, and visual displays.

BIOLOGICAL TERMS

carnivores
cetaceans
communication
courtship behavior
dentition
diaphragm
echolocation
edentates
estrus
gestation period
herbivores
hibernation
insectivores
lagomorphs
mammal
mammary glands
marsupials
marsupium
migration
monotreme
omnivores
pelage
placenta
placental mammal
primates
rodents
social hierarchy
sweat glands
territoriality
territory
ungulates
uterus

USING BIOLOGICAL TERMS

1. What is echolocation?
2. Distinguish between a uterus and a marsupium.
3. In what way are migration and hibernation similar?
4. What are cetaceans?
5. Distinguish between territoriality and territory.
6. Distinguish between sweat glands and mammary glands.
7. Distinguish between a carnivore, a herbivore, and an omnivore.
8. Distinguish between monotremes, marsupials, and placental mammals.
9. What is dentition?
10. What are edentates?
11. Distinguish between lagomorphs and rodents.
12. Distinguish between ungulates and insectivores.
13. Distinguish between estrus and gestation period.
14. What is courtship behavior?
15. To which group of placental mammals do humans belong?

UNDERSTANDING CONCEPTS

1. Identify the traits of mammals. (27·1)
2. What does dentition indicate about a mammal? (27·1)
3. Why do mammals tend to have such a great effect on other populations of organisms? (27·2)
4. List some ways in which mammals are of economic value to people. (27·2)
5. How does reproduction in monotremes differ from that in other mammals? (27·3)
6. Describe reproduction in marsupials. (27·4)
7. Identify a possible cause for the decline in the number of species of marsupials formerly living in South America. (27·4)
8. What adaptive advantage do placental mammals have over other mammals? (27·5)
9. How are primates adapted to an arboreal lifestyle? (27·6)
10. How are ungulates adapted for life on open grasslands? (27·6)
11. Identify some adaptations of cetaceans. (27·6)
12. What is the adaptive value of hibernation? (27·7)
13. How does migration aid survival? (27·7)
14. What functions do territoriality and social hierarchy serve in mammalian societies? (27·8)
15. What is the function of courtship behavior? (27·9)
16. What is the value of parental care? (27·9)
17. How does communication in humans differ from communication in other mammals? (27·10)
18. Identify some ways in which mammals communicate. (27·10)
19. List some functions of communication among mammals. (27·10)

APPLYING CONCEPTS

1. Why does endothermy demand advanced respiratory and circulatory systems?
2. Many herbivores, such as caribou and antelope, can run faster than their carnivorous predators. How, then, do the predators ever catch prey?
3. Consider the dentition of humans. Which teeth are used for initially tearing off pieces of food? Which are used for grinding? How are their shapes and locations in the mouth adapted for the various functions of the teeth?
4. Internal fertilization occurs in many plants and animals. What are some adaptive advantages of internal fertilization?

EXTENDING CONCEPTS

1. Monitor the behavior of a pet cat or dog for a week. In what ways does the animal communicate with humans? With other animals? Illustrate your findings with sketches of the animal, showing facial expressions and body positions.
2. Research the migration patterns of one of the following: the gray whale, the caribou, the arctic wolf, or a species of bat. Include in your report a map of the animal's migration paths, the reasons for its migration, a description of its yearly feeding and mating habits, and any information available on the control of navigation and/or the urge to migrate.
3. Build or draw a model of a herbivore's digestive tract. Be sure to include dentition as well as internal organs. The model should explain how the digestive tract is adapted to the herbivore's diet.
4. Research the history of marsupials. You will have to understand the geologic concept of plate tectonics, or continental drift. Include maps in your report. Focus on the question of why present-day marsupials are limited mostly to Australia.
5. Many studies have been done on the possibilities of communication between humans and mammals, especially higher primates and cetaceans. What have they found, in terms of mammals' capacities for language? After learning about these studies, how would you define *language*?

READINGS

Ferry, G. (ed.). *The Understanding of Animals*. Basil Blackwell, 1984.

Goodall, Jane. *In the Shadow of Man*. Houghton Mifflin, 1983.

MacDonald, D. (ed.). *The Encyclopedia of Mammals*. Facts on File, 1984.

Savage, R.J.G. *Mammal Evolution: An Illustrated Guide*. Facts on File, 1986.

28. Comparing Vertebrates

Each spring, adult salmon leave the open ocean and swim up freshwater rivers and streams. The tireless salmon swim until they reach the very place they hatched years before. Once they reach their birthplace, the female salmon lay their eggs. The males then fertilize the eggs. During their swim upstream, the spawning salmon provide food for many animals. In this picture, a bear has caught a salmon. Both the bear and the salmon are vertebrates. In this chapter, you will examine how vertebrates compare to one another.

CHAPTER OBJECTIVES

After completing this chapter, you will be able to

- **Compare** fish, amphibians, reptiles, birds, and mammals.
- **Discuss** the major organ systems in vertebrates.
- **Contrast** reproductive adaptations in the various vertebrate groups.
- **Describe** patterns of development in vertebrates.

Body Plans of Vertebrates

28·1 UNITY IN DIVERSITY

Bears, salmon, eagles, frogs, and snakes are very different from one another. Despite their diversity of form, size, habitat and behavior, all vertebrates share certain characteristics. In other words, vertebrates show "unity in diversity." One unifying vertebrate trait is the backbone that gives the subphylum Vertebrata its name. Vertebrates are unified by shared biological needs as well as structural similarities. Every type of vertebrate needs to obtain energy, remove wastes, coordinate body systems, and reproduce.

In this chapter you will study the structures and functions of vertebrates. You will see how each body system compares in different vertebrates and how, though these systems vary, they carry out the same basic functions. Remember as you go through the sections that certain functions are common to these animals. Note differences in how different animals carry out the same basic functions.

SECTION OBJECTIVES

- Identify some vertebrate characteristics.
- Compare the skeletal, muscular and integumentary systems of vertebrates.

28·2 SKELETAL SYSTEMS

The body plans of all vertebrates are similar in some ways. All vertebrates are bilaterally symmetrical. All have a **coelom,** or internal body cavity. And all have a tube-within-a-tube body plan. The outer tube is the body wall, including the external body covering, and the inside tube is the digestive system.

A major trait of vertebrates is the presence of a notochord during some stage of life. The **notochord** is a rodlike structure running lengthwise along the dorsal side of the body. The notochord is a supporting structure in all vertebrate embryos. In jawless fish the notochord remains during the whole lifetime of the animal. In the adult stage of other vertebrates, including bony fish, amphibians, reptiles, birds, and mammals, the notochord is replaced by the bony backbone.

koiloma (cavity)

noton (back)
chorde (string)

FIGURE 28·1 This zebra-tailed lizard's reptilian stance allows it to hold its body away from the hot stone as it looks around.

The skeleton of a vertebrate is an **endoskeleton,** or internal skeleton. It is made up, in part, of living cells and thus is able to grow. It is not shed. The endoskeleton protects the internal organs. It is also a site for the attachment of muscles. In jawless fish—the lampreys and hagfish—and in sharks and rays, the endoskeleton is made of cartilage. Bony fish, amphibians, reptiles, birds, and mammals have endoskeletons of bone, with small amounts of cartilage present.

The endoskeleton can be divided into two parts. The axial skeleton consists of the skull and vertebral column. The appendicular skeleton consists of the pectoral girdle, the pelvic girdle, and any limbs. A vertebra, a series of which make up the backbone, is made up of a *neural arch* and a *centrum.* Within each vertebra there is a canal through which the spinal cord passes. The vertebrae of a land vertebrate interlock, but those of a fish do not interlock. Such interlocking lessens flexibility but gives more support.

The pectoral and pelvic girdles are much larger in land vertebrates than they are in fish. The positions of the pectoral and pelvic girdles and of the limbs differ among the classes of land vertebrates. The sprawling stance of the early amphibian resulted from the positions of the pectoral and pelvic girdles and front and hind limbs. The early amphibian tended to drag its body when walking on land. In modern reptiles the girdle and limb patterns allow the body to be lifted off the ground more. In many mammals the pectoral and pelvic structures are further modified. This improves movement because the limbs are directly below the trunk of the body.

Vertebrate limbs are adapted to a variety of uses. The jawless fish lack limbs. The cartilaginous fish and bony fish have two pairs of ventral fins used in steering. In land vertebrates the limbs are used for moving.

28·3 MUSCULAR SYSTEMS

Movement in vertebrates is due to the interaction of bones and muscles. Three types of muscles are found in vertebrates. **Skeletal muscles,** also called voluntary muscles, aid in movement. **Smooth muscles,** also called involuntary muscles, make up part of the walls of the internal organs. **Cardiac muscle** makes up the heart.

The main skeletal muscles of fish consist of blocks of muscles called **myomeres** (MĪ oh meerz). In bony fish the myomeres alternate with the vertebrae. Such muscles make up most of the muscle mass of a fish. Have you ever watched a fish swim? The back-and-forth movement of the body and tail, which moves the fish, is caused by the alternate contraction and relaxation of the myomeres.

In most land vertebrates there are more limb muscles than in fish. Many of these muscles work in pairs. One muscle's action opposes that of the other. Muscles that work in pairs are called **antagonistic muscles.** Usually a pair consists of a flexor and an extensor. When the flexor in a pair contracts, it brings the two bones on either side of a joint closer. When the extensor contracts, it causes the bones to move apart. The two muscles usually contract alternately.

FIGURE 28·2 A horse's limbs are adapted for running and jumping.

28·4 INTEGUMENTARY SYSTEMS

The **integument** is the skin, or outer covering, of an organism. The skin of vertebrates has many functions. One function is protection against injuries. In some vertebrates, glands within the skin produce secretions that give protection against predators. Others have glands whose secretions attract mates.

The skin of vertebrates is usually divided into the epidermis, or outer layer, and the dermis, or inner layer. Lampreys and hagfish have smooth skin covered with mucus. Many vertebrates have skin cells that become hardened with a protein called keratin. Sharks and rays have tough outer skin because of the large amounts of keratin present. In addition, sharks have scales bearing toothlike spines in their skin.

Among bony fish the skin varies. Most often, the skin is covered with many scales. The scales form from the dermis. Some bony fish lack scales but have very thick tough skin. In most fish, the skin produces a layer of mucus.

In most amphibians, the skin is thin and contains many mucous glands. Recall that gas exchange occurs through the skin of many amphibians. How are the mucous glands an adaptation for gas exchange through the skin?

A reptile has a body covering of dry epidermal scales. In addition to providing protection, the dry scaly skin prevents loss of water, an adaptation for living on land. Turtles have scales on the legs and even on the shell. Snakes usually shed the upper surface of their scales all at once, whereas turtles do not.

Birds have scales on their legs. The beaks of birds are formed from the integument. Feathers, which provide insulation, cover large areas of the body. Feathers that grow on the wings and tail are used in flight.

Mammals show great diversity in adaptations of the integument. Mammals have hair. The skin also contains a layer of fat that serves to insulate the body. Many types of glands, such as scent glands and sweat glands, are found in the skin. Sweat glands help to maintain a constant body temperature. In addition, in some mammals, skin produces claws, nails, and horns. Claws and nails protect the ends of limbs. Horns are used for defense and for attracting mates.

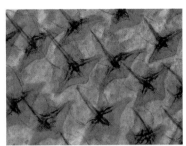

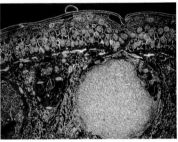

FIGURE 28·3 In sharks (top) the skin bears spines. Most bony fish have scales (middle). The skin of amphibians is thin and contains many mucous glands (bottom).

REVIEW QUESTIONS

1. Identify some traits of vertebrates.
2. Distinguish between the skeletons of a shark and a perch.
3. Compare the integuments of a shark, a perch, a frog, a snake, a pigeon, and a rabbit.
4. Compare the muscle arrangements in a fish and in a land vertebrate.
5. Amphibians have thin, moist skins, while reptiles have tough, waterproof skins. Why, then, does a reptile require better developed lungs than an amphibian?

Systems for Body Maintenance

SECTION OBJECTIVES

- Compare the digestive, circulatory, gas exchange, and excretory systems in vertebrates.
- Discuss the evolution of the vertebrate heart and lungs.
- Explain how a vertebrate's nitrogenous wastes are an adaptation to its environment.

28•5 DIGESTIVE SYSTEMS

All vertebrates have a tube-within-a-tube body plan, the internal tube being the digestive system. Vertebrates have one-way digestive tracts. Digestion is usually extracellular, occurring within the cavity of the digestive tube. The digestive systems of vertebrates vary. The digestive system of a herbivore, a plant eater, differs from that of a carnivore, a meat eater, and that of an omnivore, which eats both plant and animal matter.

Plant-eating mammals have adaptations that help them digest tough plant materials, such as cellulose. Plant eaters have large flat molars. These teeth grind the tough plant material, breaking it apart. Also, plant eaters have long digestive tracts. Food takes a long time to pass through, and thus there is more time for enzymes to act. Plant eaters also have many microbes in their intestines. These organisms break down cellulose. Vertebrates do not have enzymes to break down cellulose. The digestive system contains *ceca* (SEE kuh), or blind sacs. Food remains trapped in these sacs for long periods of time, thus giving the microbes more time to act.

Tadpoles of frogs are usually plant eaters and have long digestive tracts. As adults, frogs are meat eaters and have short digestive tracts. Food remains in the digestive tract for only a short time. However, meat is more easily digested than plant foods.

Note in Figure 28•4 that the digestive tract of a bird contains a crop and gizzard. Food is stored in the crop. The muscular gizzard has tiny stones that help to break the food apart. This action increases the surface area of the food, aiding enzyme action.

Despite the differences that are found in these digestive systems, the basic structures are similar. In addition, all vertebrates have similar digestive enzymes, which break down food before absorption.

FIGURE 28•4 A cow's digestive system allows it to break down tough plant materials (A). Adult frogs have short intestines (B). A bird's gizzard contains tiny stones that help grind food (C).

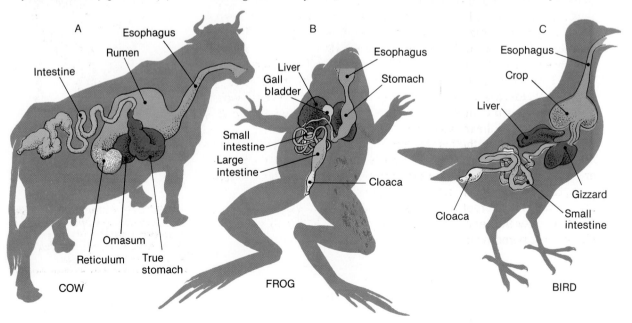

28·6 CIRCULATION

Circulatory systems in vertebrates consist of blood, blood vessels, and heart. All vertebrates have closed circulatory systems. The blood serves as the transport medium and consists of a liquid part and a cellular part. The liquid portion of the blood, called plasma, carries dissolved nutrients, wastes, and other substances throughout the organism. Red blood cells transport oxygen to the cells of the body. White blood cells protect the organism against disease. Platelets, which are found only in mammals, are involved in clotting the blood.

The major differences in circulation in vertebrates involve the path of the blood through the heart and body and the number of heart chambers. The heart is two-chambered in fish, three-chambered in amphibians and most reptiles, and four-chambered in birds and mammals.

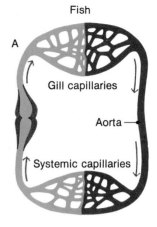

A fish has a single-circuit pathway. A **single-circuit pathway** is one in which the blood moves through the heart, then passes on to the rest of the body, and finally returns to the heart. A fish's heart has two chambers—an **atrium** and a **ventricle.** As shown in Figure 28·5, oxygen-poor blood from all parts of the body enters the atrium. It then passes into the ventricle. When the ventricle contracts, blood is pumped to the gills. The exchange of oxygen and carbon dioxide takes place between the gill capillaries and the water. The blood then moves throughout the body. Oxygen, carbon dioxide, nutrients, and wastes diffuse between the cells and the blood as the blood passes through capillaries. Blood returns through veins to the atrium of the heart. Thus blood passes through the heart once during each trip around the body.

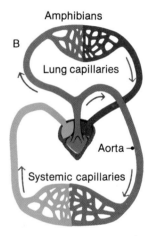

An amphibian has a heart with three chambers: two atria and one ventricle. Note in Figure 28·5 that veins carry oxygen-poor blood from the body to the right atrium. Oxygen-rich blood from the lungs and skin enters the left atrium. Blood from both atria enters the ventricle. Some mixing of blood occurs in the ventricle. As the ventricle contracts, blood is forced into a large artery. Most of the oxygen-poor blood is carried to the lungs and the skin, where oxygen and carbon dioxide exchange occurs. This blood returns to the left atrium. Most of the oxygen-rich blood moves through blood vessels to all other parts of the body. This blood returns to the right atrium. Thus in some amphibians there is a working double-circuit pathway. A **double-circuit pathway** is one in which blood moves through the heart, to the lungs, back to the heart, and then throughout the body.

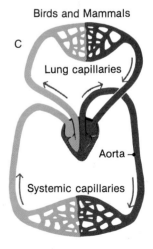

In most reptiles the heart is three-chambered. The single ventricle, however, is partly divided. Structurally, a reptile's heart is like an amphibian's heart. Because of the position of the wall within the ventricle, however, there is very little mixing of oxygen-rich and oxygen-poor blood. Thus a reptile has a double-circuit pathway. In alligators and crocodiles the ventricle is completely divided. These animals have a four-chambered heart.

FIGURE 28·5 Compare these vertebrate circulatory systems. Fish have a single-circuit pathway (A). Amphibians (B), reptiles, birds, and mammals (C) have double-circuit pathways.

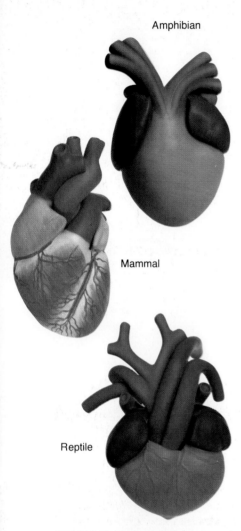

FIGURE 28·6 Amphibians and most reptiles have three-chambered hearts. Birds and mammals have four-chambered hearts. Some reptiles, such as alligators, also have four-chambered hearts.

ecto (outside)
therm (heat)
endo (inner)

Birds and mammals have a four-chambered heart. A wall divides the lower part of the heart into a right and left ventricle. Oxygen-poor blood enters the right atrium. Contraction of the right atrium causes the blood to fill the right ventricle. As the right ventricle contracts, blood flows into the pulmonary arteries. The blood travels through the pulmonary arteries to the lungs. In the capillaries of the lungs, exchange of oxygen and carbon dioxide takes place. Oxygen-rich blood is carried by pulmonary veins to the left atrium. Contraction of the left atrium forces blood into the left ventricle. From the left ventricle, the blood is pumped to all parts of the body. Thus in birds and mammals, blood makes two circuits through the heart. The circuit from the right ventricle to the lungs and then to the left atrium is called the *pulmonary circuit*. The larger circuit, from the left ventricle to all parts of the body and then to the right atrium, is called the *systemic circuit*. Thus birds and mammals have a double-circuit pathway.

28·7 GAS EXCHANGE

The cells of vertebrates need oxygen to carry on cellular respiration. The structures used by an animal for gas exchange are related to the type of environment in which the animal lives. Gas exchange in aquatic vertebrates occurs between the water and the blood in gills. Gas exchange in most land vertebrates occurs between the air and the blood in capillaries surrounding the air sacs in the lungs.

In bony fish, water enters the mouth, is forced over the gills, and then flows back into the environment. The gills have many capillaries. As water flows over the gills, oxygen diffuses into the capillaries, and carbon dioxide diffuses from the capillaries into the water.

Gas exchange varies among amphibians. Frogs, toads, and salamanders have gills in the larval stage. Some salamanders have gills in the adult stage, but most adult amphibians have lungs. In addition, many amphibians have thin, moist skin through which gas exchange occurs. Gas exchange in amphibians may also occur through the lining of the mouth and pharynx.

Gills dry out quickly in air. The lung is an internal gas-exchange surface that does not dry out, because it is not directly exposed to the dry environment. Also, lungs secrete substances called surfactants, which help maintain moist surfaces for rapid gas exchange.

Reptiles, birds, and mammals have body coverings that prevent gas exchange through the skin. Lungs are the chief means of gas exchange in these animals. In the lungs, tiny air sacs called *alveoli* (al-VEE uh lī) are surrounded by capillaries. The exchange of oxygen and carbon dioxide takes place by diffusion.

The amount of oxygen needed by an animal is related to the animal's metabolic rate, activity level, and whether the animal is an ectotherm or an endotherm. An **ectotherm** is an animal that derives its body heat from the environment. Fish, amphibians, and reptiles are ectotherms. An **endotherm** is an animal that maintains a constant internal body temperature as a result of heat produced by body metab-

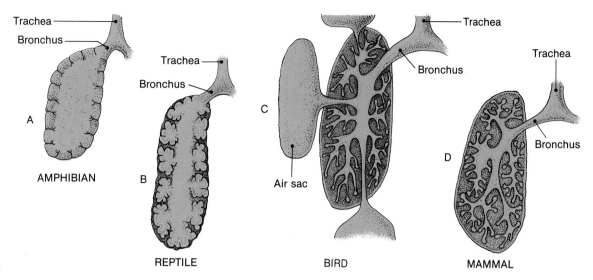

FIGURE 28·7 Mammals and birds require more oxygen than amphibians and reptiles. The large lung surface in birds and mammals allows them to maintain their high metabolic rate.

olism. Birds and mammals are endotherms. The larger and more active the animal, the greater the surface area required for gas exchange.

An increase in surface area of the lungs is an adaptation by which land vertebrates obtain enough oxygen to support their more active lifestyle. Figure 28·7 is a comparison of the lung surfaces of some terrestrial vertebrates. There is less surface area in the amphibian lung than in the reptilian lung. Amphibians may use the skin and the mouth cavity for gas exchange. Most reptiles rely solely on their lungs for gas exchange.

Notice the increased surface area in the bird's lung. Birds are endotherms with high metabolic rates. Thus they need much oxygen. Note that the bird's lung has additional air sacs attached to it. The air sacs are also gas exchange surfaces. They keep air moving across the lung surface and aid in releasing heat.

A mammal's lung, like that of a bird, has a greater amount of relative surface area than the lung of a reptile. It has many alveoli surrounded by capillary networks. The greatly increased surface area can supply enough oxygen for these highly active organisms.

28·8 EXCRETION

The excretory systems of most vertebrates are similar. Each vertebrate has two kidneys near the dorsal surface of the body. The kidneys filter wastes from the blood. Connected to each kidney is a ureter, a tube that carries the wastes from the kidney to the cloaca in fish, amphibians, reptiles, and birds. In Figure 28·8 identify those animals that have cloacas. Notice that in the mammal the ureters lead directly into the urinary bladder, where wastes are stored. Fish, amphibians, and reptiles also have a urinary bladder attached to the side of the cloaca. In these animals the cloaca opens directly to the exterior. In mammals there is a urethra. The urethra carries wastes from the urinary bladder to the exterior.

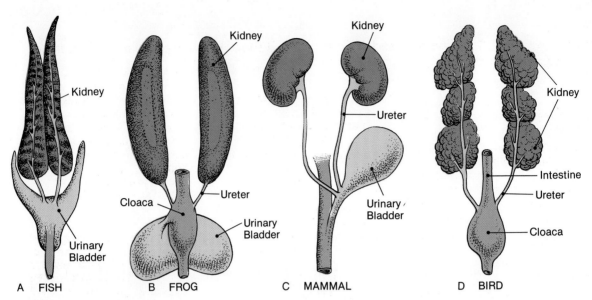

FIGURE 28·8 Compare the excretory systems of these four vertebrates: a fish (A), a frog (B), a mammal (C) and a bird (D). Why doesn't the bird have a urinary bladder?

The three major types of nitrogenous wastes are ammonia, urea, and uric acid. Ammonia is a poison. Many fish and many amphibian larvae excrete ammonia by diffusion through the gills. Since these organisms live in water, ammonia excreted is quickly diluted.

Land vertebrates—such as adult amphibians, reptiles, birds, and mammals—do not excrete large amounts of ammonia. To do so would require large amounts of water, which is limited on land. In amphibians and mammals, ammonia is changed to the less toxic urea in the liver. The urea, along with some water and salts, is then filtered from the blood by the kidneys. Ureters carry the urine to the cloaca in amphibians and to the urinary bladder in mammals.

Reptiles and birds excrete uric acid. It is only slightly soluble in water and it is not very toxic. The kidneys filter out dilute solutions of uric acid, which is then passed on to the cloaca. In the cloaca, water is reabsorbed, and the uric acid crystals that remain combine with the feces and are eliminated.

REVIEW QUESTIONS

6. Distinguish between a single-circuit pathway and a double-circuit pathway. Name an organism that has each type of pathway.
7. In what type of blood vessel does exchange of gases, nutrients, and wastes occur?
8. Compare the gas exchange surfaces of a fish, an amphibian, a reptile, a bird, and a mammal.
9. Compare the nitrogenous wastes and the structures for their removal in a bony fish, a reptile, and a mammal.
10. A newly discovered organism is found to excrete uric acid. It has a three-chambered heart and a relatively short digestive tract. Identify the kind of vertebrate and indicate its probable diet.

Investigation

What Are the Similarities and Differences Between Two Vertebrate Hearts?

Goals
After completing this activity, you will be able to
- Identify the major structures of the heart of a fish and frog.
- Describe the pathway of the blood through the heart of the fish and frog.

Materials (for groups of 2)
models of a fish heart and frog heart

Procedure

A. Examine the model of the fish heart. You may also refer to the drawing of the fish heart on this page. Locate the two chambers. The upper chamber, or atrium **(1)**, receives blood that is entering the heart. Find the lower chamber, or ventricle **(2)**. Identify the saclike structure called the sinus venosus **(3)**. This structure holds blood coming from the body before it enters the atrium.

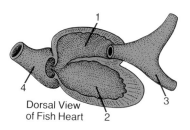
Dorsal View of Fish Heart

B. Find the large vessel called the ventral aorta **(4)** which leads from the ventricle. Blood passing through the ventral aorta passes into smaller branching arteries that lead to the gills. From the gills, blood passes to the dorsal aorta and then to the body. Blood returns to the heart through a system of veins and collects in the sinus venosus before reentering the heart.

C. On a separate sheet of paper, draw the fish heart and label the atrium, ventricle, sinus venosus, and the ventral aorta.

D. Examine the model of the frog heart. Count the number of chambers. On the dorsal surface of the heart find a triangular shaped chamber, which is the sinus venosus **(5)**. The sinus venosus receives blood from the entire body.

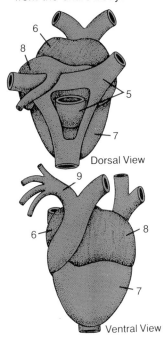

E. Locate the right atrium **(6)**, which is the upper right-hand chamber of the heart. Blood enters the right atrium from the sinus venosus. From the right atrium, blood passes to the single lower chamber, or ventricle **(7)**. Find the ventricle on the model and in the drawing. If your model comes apart, open it so that you can see the inside of the ventricle. Find out whether there is a wall dividing the ventricle into two parts.

F. Identify the left atrium **(8)**, which is the upper left-hand chamber of the heart. Blood from the lungs reenters the heart through the left atrium. Blood again flows into the ventricle and is then pumped out to the body.

G. Find the large branching blood vessel that leaves the ventricle. This branching blood vessel is called the conus arteriosus **(9)**. It directs the flow of blood toward the head region of the frog.

H. On a separate sheet of paper, draw the frog heart and label all numbered terms in steps D, E, F, and G. Use arrows to show the flow of blood into and out of the heart.

I. Follow your teacher's instructions for the proper cleanup of all materials.

Questions and Conclusions

1. Of the two hearts you examined, which is the simpler? Give reasons for your answer.
2. In the frog and the fish heart, what is the structure in which blood collects before it enters the heart?
3. In the frog heart, is there a separation between the right and left sides of the ventricle? What happens to blood in the ventricle?
4. In which of the hearts studied is there more separation of oxygen-poor and oxygen-rich blood? How does this separation take place?
5. Which of the hearts studied is more similar to a human heart? Give reasons for your answer.

Control Systems

SECTION OBJECTIVES

- Contrast the brains of different vertebrates.
- Compare the control systems in vertebrates.

28•9 NERVOUS SYSTEMS

Each vertebrate has a dorsal nerve cord connected to a brain. The brain can be divided into three areas—the *hindbrain,* the *midbrain,* and the *forebrain*—as shown in Figure 28•9.

In the hindbrain, the medulla oblongata is the control center for involuntary activities, such as breathing and beating of the heart. The cerebellum is the part of the hindbrain that is the center for coordination of muscle activities and balance. Compare the sizes and shapes of the cerebellum in the five vertebrates. Note the increase in size of the cerebellum from fish to mammal.

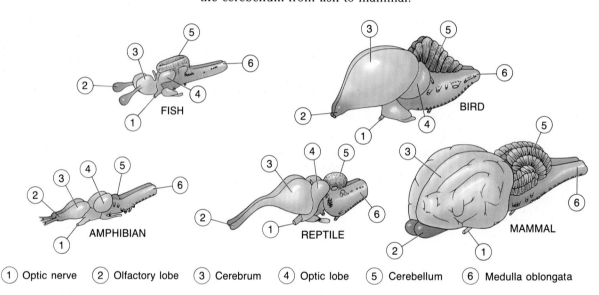

FIGURE 28•9 Compare the size of the cerebrum in the fish, amphibian, reptile, bird, and mammal. Why are mammals more capable of learning than fish?

The midbrain contains the optic lobes. In fish, amphibians, reptiles, and birds the optic lobe receives information from the eyes. In mammals the optic lobe is greatly reduced, and much of its function has been taken over by other areas of the cerebrum. The optic lobe in mammals controls the eyelids and the irises of the eyes.

Carefully compare the forebrains in Figure 28•9. Note the increase in size of the cerebrum from fish to mammal. The cerebrum is involved in the interpretation of the senses of smell, taste, hearing, touch, and pain. Depending on the type of vertebrate, the cerebrum may also include centers for learning, memory, and thinking. More complex behavior becomes increasingly evident in the reptiles, birds, and mammals. In these organisms the cerebrum is increasingly large and complex.

Vertebrates show a high degree of **cephalization**—the development of a head with sensory organs. Sensory receptors vary among vertebrates. The **lateral line system** is a receptor that helps a fish maintain balance and detect the presence of objects in the water. The system allows a fish to "sense" the size and location of its surroundings by the bouncing of pressure waves off objects. This system is an adaptation to life in water, where pressure waves carry well.

In most land vertebrates the sense of hearing is well developed. Ears are an adaptation for detecting prey and predators and also for finding mates. For example, many animals have calls that are specific to their own species. With their keen sense of hearing, these animals can detect the presence of other members of their own species.

The ears of bats are used in echolocation. Echolocation is a method of sensing the environment by using sound waves that reflect off objects. In this way, the animal that emits the sound receives information on the location of the object. Echolocation helps bats to sense prey, such as insects, as they fly at night.

Eyes are receptors specialized to form images using light. The vision of most kinds of birds is far superior to that of mammals. The large size of a bird's eyes and the high concentration of vision cells account for this superiority. Of what advantage is well-developed vision to birds?

FIGURE 28·10 Notice the large eye of this bird of prey. Good vision is vital to the animal's survival.

28·10 HORMONES IN VERTEBRATES

Regulation of body functions in vertebrates involves the nervous system. In addition, regulation also involves a chemical control system, the endocrine system. The endocrine system is made up of a number of glands that secrete hormones. **Hormones** are chemicals that are produced in one part of the body, travel in the blood, and control activities of other parts of the body. In vertebrates, hormones regulate growth, reproduction, metamorphosis, and molting, as well as many other activities. Often the interaction of hormones from many glands affects an organ of the body.

The thyroid gland in vertebrates produces the hormone thyroxine, which affects the rate of metabolism in these animals. In birds and mammals, this hormone also helps to control the amount of heat produced. In amphibians, thyroxine influences metamorphosis. The molting of amphibians, reptiles, and birds is regulated by hormones secreted by the thyroid and other glands.

The adrenal glands are endocrine glands found above the kidneys. They are very important in land vertebrates. A hormone secreted by the adrenal glands helps to prevent dehydration of land vertebrates. Another hormone produced by the adrenals is adrenaline, which increases the heartbeat, blood pressure, and blood sugar level. How might this response be of adaptive value to an animal?

The pituitary gland is a small gland found at the base of the brain. The pituitary gland produces many different hormones. Some of these hormones regulate other endocrine glands. Growth hormone controls the normal growth of the vertebrate. The pituitary gland also produces a hormone that causes skin color to darken in fish, amphibians, and some reptiles when these animals are in dark areas.

In addition, the pituitary gland secretes prolactin, which has many functions in vertebrates. In some amphibians this hormone functions in metamorphosis, and in some reptiles it influences growth. In birds prolactin stimulates behavior involved with the care of offspring. In mammals it stimulates milk production.

The gonads are the reproductive organs. In addition to producing gametes, they also serve as endocrine glands. The testes of the male secrete the hormone testosterone (tehs TAHS tuh rohn). Testosterone stimulates the production of sperm during the breeding season. This hormone also regulates the development of the male secondary sex characteristics.

The ovaries of the female produce estrogen and progesterone. These hormones are produced in response to secretions from the pituitary gland. These hormones influence the production of eggs and the maturation of reproductive structures.

REVIEW QUESTIONS

11. Compare the fish, amphibian, reptile, bird, and mammal in terms of the size and function of the three divisions of the brain.
12. Identify two sensory receptors common to land vertebrates.
13. Compare the role of the pituitary hormone prolactin in amphibians, birds, and mammals.
14. What hormones help to regulate reproduction in vertebrates?
15. A student with a pet lizard notices that the reptile is sluggish even when the air temperature is warm. Someone suggests that the animal's metabolic rate may be below normal. List a possible hormonal deficiency that could be causing the problem.

Reproduction and Development

SECTION OBJECTIVES

- Compare the reproductive structures in vertebrates.
- Contrast amphibian, bird and mammalian development.

28•11 REPRODUCTIVE SYSTEMS

Vertebrates reproduce sexually. Recall that sex cells are called gametes. Each gamete contains half the number of chromosomes of each parent and thus half the genetic information of each parent. Fertilization unites the genetic information from the two parents. The offspring will show traits that are similar to, but not identical to, the parents' traits.

The male reproductive organs are the testes, of which there are usually two. Testes produce the male gametes, sperm. Connected to each testis is a tube that carries the sperm to the cloaca in fish, amphibians, reptiles, and birds. In some fish and amphibians, these tubes are also part of the urinary system and carry wastes. In reptiles and birds, these ducts function only in reproduction.

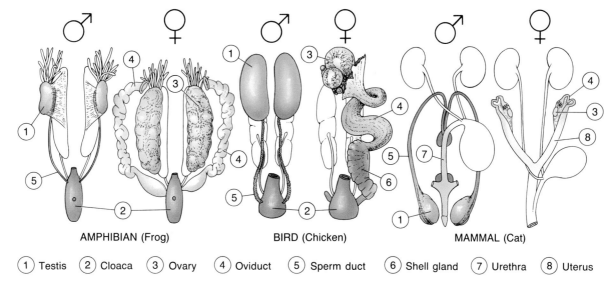

① Testis ② Cloaca ③ Ovary ④ Oviduct ⑤ Sperm duct ⑥ Shell gland ⑦ Urethra ⑧ Uterus

AMPHIBIAN (Frog)   BIRD (Chicken)   MAMMAL (Cat)

FIGURE 28·11 Compare the reproductive systems of a cat (A), a chicken (B), and a frog (C). Why do birds have only one ovary?

The female reproductive organs are the ovaries. Most vertebrates have two ovaries. Birds have only one ovary. Ovaries produce the female gametes, called ova, or eggs. Locate the ovaries in each female vertebrate in Figure 28·11. Eggs are released from the ovaries and then enter the oviducts. The oviducts are tubes that lead away from ovaries. Eggs pass through these tubes to the cloaca in fish, amphibians, reptiles, and birds. In some fish and amphibians and in reptiles, birds, and mammals, fertilization occurs in the oviducts.

In birds and reptiles, the end of the oviduct is enlarged. Within this structure, shells are deposited on eggs. In mammals the oviducts lead to the uterus, where development takes place.

28·12 REPRODUCTIVE ADAPTATIONS

In most bony fish and many amphibians, gametes are deposited in water, and fertilization occurs in water. The union of egg and sperm outside the body of the female is called **external fertilization.** Most species that have external fertilization produce eggs and sperm in large numbers. This helps to ensure that fertilization will take place, since many of the eggs and sperm can be destroyed by predators.

Spawning is the release of eggs and sperm into the water. In some species of fish, either or both parents make a shallow hollow in the bottom of a lake or stream. The eggs and sperm are then deposited in the depression. Fertilization is more likely to occur because of the nearness of the sperm and the eggs.

Internal fertilization is the union of egg and sperm inside the body of the female. In many species in which internal fertilization occurs, the male develops a special organ that transfers sperm into the reproductive tract of the female. Internal fertilization occurs in most land vertebrates.

28·13 MOVEMENT TO LAND

ovum (egg)
parere (give birth)

Many vertebrates are oviparous. The fertilized eggs of an **oviparous** species develop outside the female's body. One of the major adaptations of the vertebrates was the development of an egg that could survive on land. The eggs of fish, frogs, and toads remain moist because of the water in which they are laid. These eggs may be surrounded by a jellylike coating. However, the function of this coating is to prevent bacteria from destroying the egg.

Since it is laid on land, the egg of a reptile must maintain a moist environment for the embryo. The **amniote egg** of a reptile or a bird contains a fluid-filled space in which the embryo develops. The egg is surrounded by a shell. In a reptile the shell is a leathery covering; in a bird the shell is hard.

In some fish and reptiles and in most mammals, development of the embryo occurs inside the body of the female. In some fish and reptiles, the young develop within eggs, but the eggs remain within the female's body. Development of an embryo within an egg that is retained inside the mother's body occurs in **ovoviviparous** organisms. The developing embryo does not receive nourishment directly from the mother's body. Rather, the developing embryo receives nourishment from the yolk.

vivus (alive)

Most mammals are viviparous. In **viviparous** organisms the embryo develops inside the body of the female, receiving nourishment and excreting wastes through a structure called a **placenta**. Oxygen, nutrients, and wastes diffuse between the circulatory system of the mother and the circulatory system of the embryo through the placenta. Since food is constantly available and wastes are removed, development of the embryo can take place over a long period of time. This period of time is called the gestation period.

FIGURE 28·12 This female snake wraps her body around her young to keep them warm and protect them from predators. Many snakes are ovoviviparous. The embryos develop in eggs that are retained inside the mother's body. The young are born alive.

Monotremes, such as the platypus, lay amniote eggs. They lack a placenta. Marsupials, such as the opossum, begin embryonic development within the uterus. There is no placenta. They complete development in an external pouch.

28·14 PATTERNS OF DEVELOPMENT

In all vertebrates, the zygotes undergo a series of rapid cell divisions called **cleavage.** Because of the presence of yolk, the cell divisions in amphibians are unequal. As the mitotic divisions occur, a hollow ball of cells forms, called the **blastula.** The internal hollow cavity of a blastula is called the *blastocoel*.

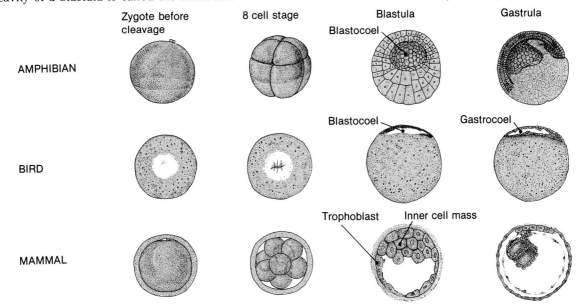

FIGURE 28·13 The cell divisions in an amphibian egg are unequal because the yolk is unevenly distributed. There is so much yolk in a reptile or bird egg that cleavage can only occur in a small region. In mammals, the inner cell mass becomes the embryo, while the other cells help form the placenta.

In an amphibian zygote there are two distinct regions: the animal pole, which is darker in color, and the lighter vegetal pole, where the yolk is concentrated. As cleavage continues, the cells of the animal pole divide more rapidly than the yolk-filled cells at the vegetal pole, resulting in an unequal division.

A crescent-shaped opening, the **blastopore,** forms. As mitosis continues, some cells from the outer surface enter the blastopore. The structure that results is called a gastrula. The **gastrula** consists of the **ectoderm,** the outer layer of cells; the **endoderm,** the inner layer of cells; and the **mesoderm,** a layer of cells that forms between the ectoderm and the endoderm. The endoderm surrounds a cavity called the *archenteron* (ahr KEHN tuh rahn). The ectoderm, endoderm, and the mesoderm, the three primary germ layers, are found in all vertebrates.

In birds and reptiles, cleavage takes place only in the cells on top of the yolk. This group of cells is called the *blastodisc*. The blastodisc is made up of two layers of cells: ectoderm and endoderm. The cavity between the ectoderm and the endoderm is the blastocoel. Along the surface of the ectoderm, a line called the *primitive streak* will form. Cells migrate from the surface ectoderm, through the primitive streak, into the blastocoel. These cells become the mesoderm.

In mammals, cleavage begins after fertilization, as the zygote passes through the oviduct. Cleavage results in the formation of a blastocyst. The blastocyst consists of an outer mass of cells—the *trophoblast* (TRAHF uh blast)—and an inner mass of cells. The embryo develops from the inner mass of cells. As cell divisions occur, some cells near the surface of the inner cell mass move inward. Three primary germ layers form. These primary germ layers will result in the formation of the embryo.

The primary germ layers give rise to all the organs and tissues of the developing vertebrate. The ectoderm gives rise to the nervous system, including the sense organs. In addition the ectoderm forms the epidermis of the skin. The mesoderm gives rise to the dermis of the skin as well as the muscles and bones. The mesoderm also forms circulatory, reproductive, and excretory structures. The endoderm gives rise to the linings of the digestive and respiratory tracts as well as the liver, pancreas, and many endocrine glands.

The embryos of different kinds of vertebrates may go through different types of development. Development is often classified as indirect and direct. **Indirect development** is the production of a larval form that is able to feed itself. Such a larval form does not look like the adult form and usually feeds on foods different from those required by the adult. Thus, competition for food between parents and offspring is eliminated. What is the biological advantage of this? Production of a free-moving larva also aids species dispersal.

Indirect development requires a metamorphosis to an adult form. Amphibian larvae, the tadpoles, slowly change their form, reabsorb their tails, and become sexually mature adults.

Direct development is the production of miniature offspring that are similar to the adults. The young of reptiles, birds, and mammals undergo direct development. Direct development occurs when the egg contains sufficient yolk for the developmental period or when development is internal and nourishment of the embryo takes place through a placenta.

REVIEW QUESTIONS

16. Compare the reproductive structures of a bird and a mammal.
17. What is the difference between external fertilization and internal fertilization?
18. What biological adaptation in reptiles allows reproduction on land?
19. Compare the formation of the three primary germ layers in an amphibian and a bird.
20. The animal pole of an amphibian zygote is dark, and when viewed from above it often resembles the murky bottom of a pond. The light-colored vegetal pole when viewed from below often blends in with the light sky. Why is this coloration advantageous to the amphibian?

Careers in Biology

Veterinarian

A veterinarian diagnoses and treats diseases and other medical problems of animals. He or she may perform surgery, carry out laboratory tests, and administer medications. Veterinarians also give advice on the care and breeding of healthy animals.

Urban veterinarians usually specialize in the care of small animals, such as pet cats, dogs, and birds. Rural veterinarians usually specialize in the care of large farm animals.

A veterinarian must have a Doctor of Veterinary Medicine degree and a state license. To become a veterinarian, a person must complete a college undergraduate program and a four-year veterinary medicine program. For additional information write to the American Veterinary Medical Association, 930 North Meacham Road, Schaumburg, IL 60196.

Nature Photographer

Nature photographers play an important role in education and research. They photograph various subjects to illustrate magazines, books, research articles, and teaching materials. The job of a nature photographer may range from tracking and photographing herds of migrating caribou on the arctic tundra to the microphotography of protozoa or bacteria in a laboratory culture dish.

No formal education is required for the career of nature photographer. A strong background in biology and ecology, however, is very helpful in understanding the subjects to be photographed. Knowledge of photographic equipment and techniques is essential. Many high schools and community colleges offer courses in basic photography and art. For additional information write to the Professional Photographers of America, 1090 Express Way, Des Plaines, IL 60018.

Fisheries Biologist

Fisheries biologists are involved in the study and management of fish populations. Some fisheries biologists conduct research in such areas as the breeding habits and life cycles of fish. Fisheries biologists who manage fish populations work for government agencies to maintain the population balance in lakes and streams used for sport and commercial fishing. Some fisheries biologists work at fish hatcheries.

Most jobs as a fisheries biologist require a bachelor's degree in aquatic science, ichthyology, or fisheries biology. Other jobs require advanced degrees. A Ph.D. is usually needed for jobs in research or teaching. For additional information write to the American Fisheries Society, 5410 Grosvenor Lane, Bethesda, MD 20014.

How to Prepare for a Job Interview

A job interview gives an employer the chance to meet you and evaluate your qualifications. The interview is a deciding factor in whether you will get the job, so it is important to be prepared. First, you should find out as much information as you can about the company you would like to work for. Your local library has information on large national companies and successful local companies. Your local newspaper may also have a file of information on companies in your area.

Second, be prepared to answer questions about yourself and why you are qualified to fill the job you are applying for. Do not be afraid to speak positively about your skills and personal background.

Third, practice role-playing an interview. Your school guidance counselor should be able to help you develop your interview skills.

Chapter Review

SUMMARY

28·1 Although the structure of a particular body system will vary from vertebrate to vertebrate, its basic function will be the same in all vertebrates.

28·2 All vertebrates have a notochord during embryonic development. In most adult vertebrates the notochord is replaced with vertebrae consisting of cartilage or bone.

28·3 In fish, the skeletal muscles form myomeres. Amphibians, reptiles, birds, and mammals have skeletal muscles that usually function as antagonistic pairs, consisting of a flexor and an extensor.

28·4 The integument of vertebrates is composed of the same tissue layers, the epidermis and dermis. Vertebrates have many adaptations of the integument, such as scales on fish and reptiles, feathers on birds, and hair on mammals.

28·5 The digestive systems of vertebrates are similar in some ways. Various modifications, such as the length of the digestive tract, improve the effectiveness of digestion and absorption in a species.

28·6 Fish have two-chambered hearts; amphibians and most reptiles have three-chambered hearts; and birds and mammals have four-chambered hearts.

28·7 Vertebrates carry out gas exchange through skin, gills, and/or lungs, depending on their environments and their metabolic rates.

28·8 The types of nitrogenous wastes produced by vertebrates are related to the environments in which the organisms live.

28·9 The nervous systems of vertebrates differ in the size of the forebrains, midbrains, and hindbrains. The more complex vertebrates have larger forebrains.

28·10 Endocrine secretions, or hormones, are similar in all vertebrates, although the effects may vary depending on the species.

28·11 All vertebrates reproduce sexually and have similar reproductive systems.

28·12 In aquatic vertebrates, such as most fish and some amphibians, fertilization usually is external; in most terrestrial vertebrates, fertilization is internal.

28·13 The amniote egg prevents dehydration of the developing embryo and is a major adaptation of reptiles and birds.

28·14 Although the zygotes of all vertebrates undergo cleavage, early embryological development varies.

BIOLOGICAL TERMS

amniote egg
antagonistic muscles
atrium
blastopore
blastula
cardiac muscle
cephalization
cleavage
coelom
direct development
double-circuit pathway
ectoderm
ectotherm
endoderm
endoskeleton
endotherm
external fertilization
gastrula
hormones
indirect development
integument
internal fertilization
lateral line system
mesoderm
myomeres
notochord
oviparous
ovoviviparous
placenta
single-circuit pathway
skeletal muscles
smooth muscles
ventricle
viviparous

USING BIOLOGICAL TERMS

1. Distinguish between *oviparous* and *ovoviviparous*.
2. Name the three primary germ layers.
3. What are myomeres?
4. What is cleavage?
5. Distinguish between an atrium and a ventricle.
6. What is the integument?
7. How are the blastula and gastrula related?
8. What is the lateral line system?
9. Distinguish between external fertilization and internal fertilization.
10. What is a placenta?

UNDERSTANDING CONCEPTS

1. Why can it be said that vertebrates, as a group, show unity in diversity? (28·1)
2. Describe the major differences between the appendicular skeleton of an aquatic vertebrate and that of a land vertebrate. (28·2)
3. Compare the muscular system of a fish with that of a land vertebrate. Comment on the arrangement of the muscles and the differences in movement found in these two groups of vertebrates. (28·3)
4. Compare fish scales and reptile scales with regard to structure and function. (28·4)
5. Describe the relationship between the structure of a vertebrate's digestive system and the diet of the animal. (28·5)
6. Compare the number of chambers in the hearts of fish, amphibians, reptiles, birds, and mammals. (28·6)
7. Compare a single-circuit pathway in circulation with a double-circuit pathway. How is a double-circuit pathway an adaptation to a more active life style? (28·6)
8. Describe how the type of environment in which a vertebrate lives is related to the adaptations it has for gas exchange. (28·7)
9. Compare the main excretory waste of fish with the wastes of adult amphibians, reptiles, birds, and mammals. How is this waste product of fish related to the environment of fish? (28·8)
10. Name the three sections of the vertebrate brain and give one function of each. (28·9)
11. Describe the roles of the pituitary gland in vertebrates. (28·10)
12. How are the oviducts of reptiles and birds adapted to oviparous development? (28·11)
13. What relationship is there between the number of gametes produced and whether fertilization is external or internal? (28·12)
14. Of what adaptive advantage is the amniote egg to land vertebrates? (28·13)
15. What reproductive adaptation is characteristic of most mammals? (28·13)
16. Compare early embryonic development in an amphibian and a mammal. (28·14)
17. What are some adaptive advantages of indirect development? (28·14)

APPLYING CONCEPTS

1. Name five criteria that could be used to distinguish the different vertebrate classes.
2. The red spotted newt, a type of salamander, begins life as an aquatic larva with external gills. It then leaves the water and lives for one or two years on land in an immature form. Thereafter it returns to the water and develops into the adult form. Upon observing adult newts, you notice that periodically they swim to the surface, quickly turn, and swim back down again. Account for this behavior.
3. While walking along a muddy stream bank in a tropical country, you notice the unusual tracks of an animal that lowered its body to rest every five or six meters. You observe that the groove from the trunk of the body is between the footprints that are to each side. To what group of vertebrates might this animal belong? Give reasons for your answer.
4. Both birds and reptiles lay shelled eggs. Birds sit on them. Reptiles often bury them. Identify two reasons why such behavior is adaptive.

EXTENDING CONCEPTS

1. Using the library, prepare a report on *Archaeopteryx,* explaining why this vertebrate is called an intermediate form.
2. Construct a display of the hearts of vertebrates. Using modeling clay or another medium, present each heart in cross section, showing the chambers and valves. Identify all structures in the display.
3. Design a poster that visually compares the origin and shows the development of integument coverings of the bony fish, amphibians, reptiles, birds, and mammals. Label all parts and include brief explanations where necessary.

READINGS

Fincher, Jack. "A Home where the Equids Can Roam." *Smithsonian,* May 1987, p. 138.

Gwinner, Eberhard. "Internal Rhythms in Bird Migration." *Scientific American,* April 1986, p. 84.

Lawrence, R.D. *In Praise of Wolves.* Henry Holt, 1986.

McMenamin, Mark A.S. "The Emergence of Animals." *Scientific American,* April 1987, p. 94.

Schoen, Miller, and Reynolds. "Last Stronghold of the Grizzly." *Natural History,* January 1987, p. 50.

Unit VI Skills

RESEARCHING A SCIENCE TERM PAPER

At some time in your educational career, you will probably be asked to write a term paper. Once your topic has been selected, you need to research and gather information on the topic. The main tool you will use in your research is a library—your school, public, or nearby college or university library. Most college and university libraries allow the general public to use the library but not to check out books.

Once at the library, the problem arises of where and how to begin your research. A library is well organized, so your task will not be as difficult as it may seem. Most libraries have a separate reference section that contains general directories and guides, such as encyclopedias, dictionaries, collections of biographies, almanacs, maps, and atlases. As a starting point in your research, look up your topic in an encyclopedia such as *World Book Encyclopedia* or *Encyclopaedia Britannica*. Encyclopedias provide general background information and usually include a list of related topics and a bibliography on the subject. In most cases an encyclopedia does not provide enough detail and depth to satisfy the requirements of a term paper.

Once you have gathered general information, you can begin to look for books, magazines, newspapers, pamphlets, and special publications that contain specific information on your topic. All books in a library are cataloged in a card or microfilm catalog. The catalog listings are arranged alphabetically according to author, title, and subject. You can see an illustration of a subject catalog card on this page.

Many magazine articles or special publications that are issued on a regular basis are indexed in the *Readers' Guide to Periodical Literature*. Arranged by subject and author, this guide is updated and published monthly. It provides an ongoing index of topics of general interest to the public. The *Readers' Guide* also lists additional related topics to investigate. Some large libraries also have indexes of certain American newspapers. Often libraries store back issues of magazines and newspapers on microfilm. For articles relating to biological topics, look for *Biological Abstracts*.

Most libraries also have pamphlet or vertical files organized by topic. These files contain booklets, pamphlets, and magazine or newspaper clippings on a wide variety of subjects. The information in these files may range from very general to highly technical.

One of the most useful resources in a library is the librarian. If you are unfamiliar with the organization of the library or the use of any of the tools already discussed, the librarian will be able to provide assistance. Many large libraries have reference librarians who are familiar with the numerous reference books and guides that are available.

To practice locating information in a library, pretend you are writing a term paper on methods of asexual reproduction in plants. Begin your research. Then answer the questions below on a separate sheet of paper; do not write in this book.

1. What are five of the related topics you found in the encyclopedia?
2. What are five specific references, either book or magazine or both, that you found in the card or microfilm catalog and the *Readers' Guide to Periodical Literature*?
3. What reference sources, besides those given in the question above, did you use?

```
                BOTANY

581.02    Van Dersal, William Richard, 1907--
            Why does your garden grow?  The facts
          of plant life.  Quadrangle c 1977

211 p     illus


1.  BOTANY   2.  GARDENING   I.  Title
```

UNIT VII

Human Biology

As he goes through his routine, all parts of this gymnast's body work together in perfect harmony. Nerves transmit messages, the heart pumps blood, muscles and bones perform complicated movements. In this unit, you will learn about the structure and function of the systems of the body and discover how the body's activities are coordinated.

Chapters

29 *Body Structure*

30 *Nutrition and Digestion*

31 *Circulation*

32 *Respiration and Excretion*

33 *The Nervous System*

34 *The Endocrine System*

35 *Reproduction and Growth*

29. Body Structure

The interaction of the skeletal and muscular systems makes it possible for you to move in a variety of ways. Motion results when bones are pulled into position by the muscles that are attached to them. By reading this chapter, you will learn how the skeletal and muscular systems work together.

CHAPTER OBJECTIVES

After completing this chapter, you will be able to

- **Classify** the four basic kinds of tissue.
- **Describe** the structure and function of the human skeleton and the development and structure of bone.
- **Compare** the types of joints found in the human skeletal system.
- **Identify** the three types of muscle tissue.
- **Describe** the structure and function of the human muscular system.
- **Relate** the structures of the skin to the functions of the skin.

The Body Framework

29•1 SPECIALIZATION IN THE BODY

The human body can be thought of as a living machine. It is made up of millions of parts that work together. Over 50 trillion cells are produced as a result of repeated divisions and specialization of cells starting from a single fertilized egg. As human development takes place, tissues are formed. A **tissue** is a group of cells, similar in structure, and specialized for a particular function.

Not all biologists classify the tissues of the human body in the same way. One widely used system recognizes four basic kinds of tissues: epithelial, connective, muscle, and nerve.

Epithelial tissue consists of sheets of cells that cover a surface or line a cavity. The major functions of epithelial tissue are protection, absorption, and secretion. Epithelial cells vary in structure, depending on where they are found. The epithelial cells of the skin are flattened near the surface and tightly packed and box-shaped beneath the surface. The surface cells are rather hard and dry. The epithelial cells lining the inside of the mouth are soft and flexible. Some epithelial cells produce secretions. For example, the epithelial tissue that lines the windpipe, lungs, stomach, and small intestine secretes mucus that lubricates the internal surfaces of these organs. Other epithelial cells in the stomach and small intestine secrete digestive enzymes.

Connective tissue is a type of tissue in which material called the *matrix* is found between the cells. The two basic kinds of connective tissues are solid and fluid, depending on the type of matrix. The solid connective tissues have a solid matrix. These tissues have many different forms and functions. In general they support, bind, connect, and hold different parts of the body in position. Examples of solid connective tissues are bone and cartilage. In fluid connective tissues the cells are suspended in a liquid matrix. Blood and lymph are examples of fluid connective tissues.

> **SECTION OBJECTIVES**
> - Identify the parts of the human skeletal system.
> - Describe the characteristics of bone.

586 BODY STRUCTURE

Muscle tissue is made up of cells that are able to contract, or shorten. The movement of some muscle tissue changes the shape or position of parts of the body.

Nerve tissue is made up of highly specialized cells that conduct messages. The brain, spinal cord, nerves, and sense organs contain this type of tissue. Nerve cells in the sense organs respond to changes in the environment. Messages are then sent by nerve cells to the brain or spinal cord, where they are interpreted. Messages also are carried from the brain and spinal cord to muscles and glands.

An **organ** is a group of different tissues specialized to carry out a particular function. For example, the stomach is an organ of digestion made up of muscle tissue, which churns food, and epithelial tissue, which lines the organ. The stomach also contains connective tissue and nerve tissue. **Organ systems** are groups of organs having related functions. Organ systems, in turn, make up the complete human organism. Some of the major systems of the human body are the skeletal system, the muscular system, and the nervous system.

The human body has an organization similar to that of other vertebrates. The *trunk* is the region of the body that contains most major internal organs. The upper portion of the trunk is known as the *thorax*, and the lower region is called the *abdomen*. In Figure 29·1, note that there are two major cavities, or spaces, in the human body. The **dorsal cavity** contains the brain and the spinal cord. The **ventral cavity** contains the heart and lungs as well as organs of the digestive, excretory, and reproductive systems. This cavity is divided by a muscular partition called the *diaphragm* (DĪ uh fram). The *thoracic cavity,* lying above the diaphragm, contains the heart and lungs. The *abdominal cavity,* below the diaphragm, contains most of the digestive organs, the kidneys, and the urinary bladder.

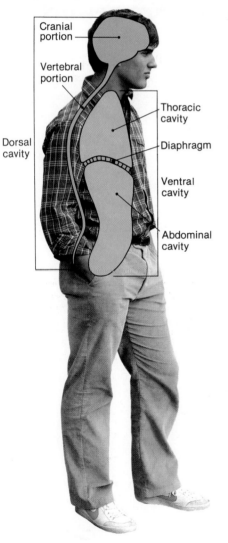

FIGURE 29·1 There are two major cavities, or spaces, in the human body. The dorsal cavity contains the brain and spinal cord. The ventral cavity contains the heart, lungs, and digestive, excretory, and reproductive organs.

29·2 THE SKELETAL SYSTEM

The human skeletal system has several important functions including those listed below.

1. The skeleton forms a living framework that supports other organs and maintains the shape of the body.
2. The skeleton provides an attachment for the muscles, which enable movement to occur.
3. Vital organs, such as the brain, heart, and lungs, are protected by the skeleton.
4. Minerals, such as calcium and phosphorus, are stored in the bones. At times these mineral reserves are used by tissues elsewhere in the body.
5. Blood cells are formed within the marrow in bones.

As in other vertebrates, the human skeleton lies within the body. Thus it is an endoskeleton. There are usually 206 bones in the human skeleton, although there may be a few more or a few less. Some of the important bones of the human skeleton are identified in Figure 29·2.

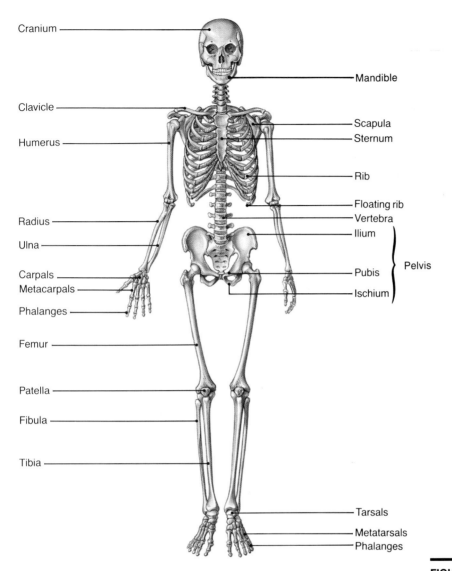

FIGURE 29·2 The human skeleton is divided into two major parts. The axial skeleton is made up of the bones of the skull, the vertebral column, and the thorax. The appendicular skeleton is made up of the bones of the arms, shoulders, legs, and pelvis.

The bones in the human skeleton are usually grouped into two divisions. The **axial skeleton** is made up of the bones of the skull, the vertebral column, and the thorax. These bones form the axis, or framework, of the trunk and head. The **appendicular skeleton** is made up of the bones of the arms and shoulders and the bones of the legs and pelvis.

One part of the axial skeleton is the *skull*. The skull is made up of two parts: the *cranium* and the facial bones. The cranium is the bony case enclosing the brain and sense organs of the head. The facial bones include the bones of the cheeks, nose, and upper and lower jaws. Of the 28 bones in the skull, only the *mandible,* which is the bone of the lower jaw, is movable.

BODY STRUCTURE 587

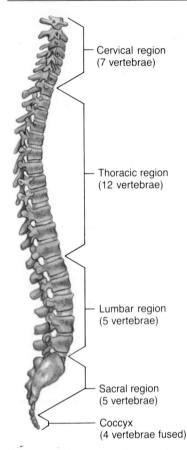

FIGURE 29·3 The 33 vertebrae, or bones of the vertebral column, are stacked one on top of another. Notice the five regions into which the vertebral column is divided.

The **vertebral column**, or backbone, is the series of bones enclosing the spinal cord. Thirty-three bones called *vertebrae* are stacked one on top of another to make up the vertebral column. Figure 29·3 shows the arrangement of the bones in the vertebral column. The neck is made up of seven *cervical vertebrae* that support the head. Twelve *thoracic vertebrae* form the bases to which the ribs are attached. The lower back consists of five *lumbar vertebrae*. Notice in Figure 29·3 that the vertebrae in the cervical, thoracic, and lumbar regions are separated from one another by disks of cartilage. The lower end of the vertebral column consists of the *sacrum* and the *coccyx* (KAHK sihks). These two regions have no cartilage disks. The sacrum is made up of five vertebrae fused together, and the coccyx is made up of four fused vertebrae.

The thoracic section of the axial skeleton consists of *ribs* and the *sternum*. There are usually twelve pairs of ribs. These bones form the cage that protects the heart and lungs. The sternum is the breastbone to which the upper seven pairs of ribs are attached. Below these ribs, and attached to them, are another three pairs of ribs. The two lowest pairs of ribs, called floating ribs, have no attachment in front. However, all 12 pairs of ribs are attached to the vertebral column, one pair of ribs to each thoracic vertebra.

The appendicular skeleton contains two sets of bones called girdles that provide the links to the axial skeleton. The **pectoral girdle** consists of two *clavicles* (collar bones), and two *scapulas* (shoulder blades). It is the structure to which the arms are attached. Locate the bones of the pectoral girdle in Figure 29·2. The **pelvic girdle** consists of three fused bones: the *ilium* (IHL ee uhm), the *ischium* (IHS kee uhm), and the *pubis* (PYOO bihs). The pelvic girdle is the structure to which the legs are attached.

29·3 DEVELOPMENT OF BONE

When they first form, most of the bones in the body are made of cartilage. **Cartilage** is a type of solid connective tissue in which the cells are embedded in a gellike matrix containing protein fibers. Because of the protein fibers that lie between the cells, cartilage is strong and flexible. Recall that in some vertebrates, such as sharks, all of the skeleton is made of cartilage. In humans and other higher vertebrates, the skeleton of the embryo is first formed of cartilage. But most of the cartilage in the skeleton is replaced by bone early in life. **Bone** is a type of solid connective tissue in which the cells are embedded in a matrix of protein fibers and minerals. Bone is formed by *osteoblasts,* or bone-producing cells. These cells move into the cartilage and deposit calcium compounds. Slowly the flexible cartilage becomes strong, less flexible bone. The process by which cartilage is changed into bone is called **ossification.**

By the time a child is born, many of its bones have ossified. However, ossification is not completed until the early twenties. A child cannot walk until about 1 year of age. At this time the bones of the legs have been largely ossified. However, cartilage in some areas is never replaced by bone. The tip of the nose, the external ear, and the lower end of the breastbone contain cartilage that does not change into bone.

Not all bones are formed from cartilage. Some bones, such as the flat bones of the skull, are formed from fibrous tissue membranes in the embryo. You will read about fibrous tissues in Section 29·6. At birth, an infant's skull has a "soft spot" on top because the cranial bones have not fused. This area of fibrous tissue membrane, seen in Figure 29·4, is called a *fontanelle*. As the child grows, the membrane becomes ossified to form a plate that joins the cranial bones. Notice in Figure 29·4 that the skull of an adult has irregular seams called *sutures* where the bones interlock.

Many people think that bone formation stops when growth is completed. However, bone is constantly reshaped and reformed throughout life. As osteoblasts deposit new bone, other cells absorb calcium and phosphorus from bone for use in other parts of the body. Throughout most of life, the body maintains a balance between the depositing of bone and the reabsorption of bone minerals. In old age the depositing of bony material slows. There is also a loss of protein fibers. The bones become brittle and are more easily broken.

29·4 STRUCTURE OF BONE

In humans most bones are made up of both living tissue and nonliving substances. The structure of a typical bone is shown in Figure 29·5, parts A through D. The bone shown is the *femur*, one of the long bones of the skeleton. Where is the femur located in the body? (If you are not sure, find the femur in Figure 29·2.)

The bone is covered by a tough membrane called the **periosteum** (pehr ee AHS tee uhm). The drawing also shows that bone is made up of two kinds of tissues. **Compact bone** is the hard outer part of the bone. Notice that it is thickest in the long shaft of the bone. Compact bone is dense, and it is formed from layers of calcium compounds. **Spongy bone** is softer and contains many small spaces. It is located toward the inside of the bone. Where is spongy bone thickest?

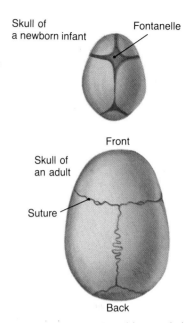

FIGURE 29·4 At birth and for a period of time after birth, the bones of the skull are not totally fused. Fontanelles, or soft spots, result. In the adult, the fontanelles ossify, or change into bone.

FIGURE 29·5 A typical long bone is made up of soft spongy bone and hard compact bone covered by a tough membrane called the periosteum. The center of the bone is filled with marrow. The Haversian canals, small channels that run through the compact bone, contain nerve and blood vessels.

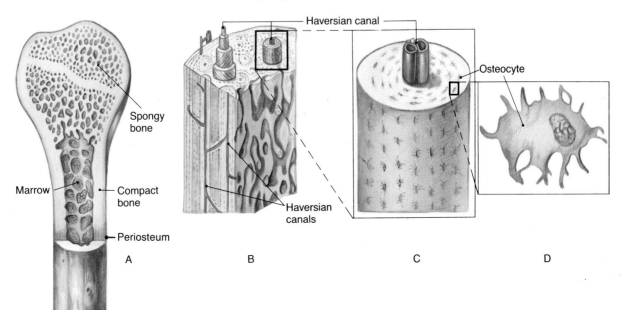

Discovery

Bone Grafts

For many years, doctors have been able to repair damaged bone by grafting, or implanting bone in small patches. Large grafts were not successful because they did not receive enough blood flow to support the growing cells.

A new type of graft, called the *vascularized fibular graft,* has made large bone grafts possible. For small grafts, bone tissue is taken from one bone in the body and applied to the damaged bone. For the vascularized bone graft, blood vessels are taken as well as bone so that the graft will have an adequate blood supply.

The fibula, the smaller of the two large bones in the lower leg, is used as a source of the tissue used in all bone grafts. After the section of the fibula has been removed, it must be attached in its new location. Using microsurgical techniques, doctors connect the blood vessels of the graft to the blood vessels of the damaged host bone. Because the graft tissue comes from the patient's own body, there is no risk of rejection. Since the circulatory pattern is complete, the graft and host tissues grow together firmly.

The center of the bone is filled with either red bone marrow or yellow bone marrow. Red bone marrow is a network of fibers, blood vessels, and the cells that produce red blood cells. This type of marrow is red in color because of the numerous red blood cells being formed there. Yellow bone marrow consists mostly of fat cells. It does not usually produce blood cells, but it can do so under certain circumstances.

Now locate the magnified wedge of bone, in part B. Note that there are many small channels called Haversian (huh VER shuhn) canals running through the compact bone. **Haversian canals** are interconnecting channels containing the nerves and blood vessels that supply the bone cells.

If you viewed a piece of bone in cross section under a microscope, you would see the structures shown in part C. Notice that the Haversian canals are surrounded by rings of nonliving matrix. Bone cells, called osteocytes (AHS tee uh sīts), are branched living cells found in this hard bony material. **Osteocytes** are mature bone cells that secrete the hard bony matrix. Part D shows an osteocyte in high magnification. Notice the many long branches extending out from the cell body. These branches extend into tiny canals within the bony matrix. The branches provide indirect contact with blood vessels branching off the Haversian canals. Food and oxygen reach the osteocytes by diffusing through this system of tiny canals.

29•5 JOINTS AND MOVEMENT

An examination of the human skeleton shows that there are many places where bones come together. The point where two or more bones meet is a **joint.** Joints permit different degrees of movement, ranging from no movement at all to extreme mobility. Figure 29•6 shows the several types of joints discussed here.

The *fixed joint* permits no movement at all. The sutures of the cranium are fixed joints. Though the bones of the adult cranium lie directly against each other, the thin layer of connective tissue in each suture can absorb some shock. This helps to prevent the bones of the skull from breaking. Most *slightly movable joints* have a pad of cartilage between the bones, against which they can move slightly. This kind of joint is found between the bones of the vertebral column. Cartilage disks between vertebrae allow only a slight bending or twisting at these joints.

There are four main kinds of freely movable joints, all of which allow considerable movement. The simplest of these joints is the *gliding joint.* In a gliding joint, the bones slide against one another. Gliding joints are found in the wrist. In the hip, great freedom of movement is permitted by the *ball-and-socket joint.* In this joint, the rounded end of one bone fits into the hollow space of the adjacent bone. A ball-and-socket joint provides for swinging and rotating movements. The *pivot joint* is a joint where two bones twist against each other in a rotating motion. A pivot joint can be found where the first two cervical verte-

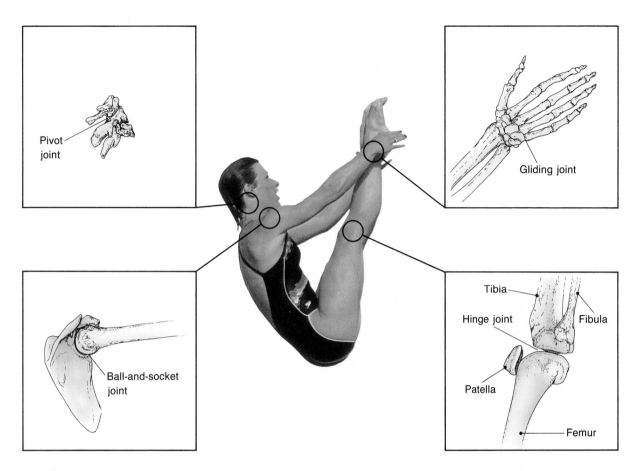

FIGURE 29·6 The actions involved in diving require the use of many types of joints. Describe the kinds of movement possible with each type of joint.

brae meet. Locate this joint in Figure 29·3. You can rotate your head, or turn it from side to side, because of the pivot joint. Finally, the *hinge joint* permits a back-and-forward movement in only one plane. The knee is an example of a hinge joint.

Although the freely movable joints provide for great freedom of movement, problems can arise because of this movement. Friction between the moving surfaces plus the weight of the body causes considerable stress and strain at joints. Joints are protected against wear in several ways. The ends of the bones that move against one another are covered by caps of cartilage. This cartilage absorbs shock and permits smoother movement of the bones. The joints are also lubricated by synovial (sih NOH vee uhl) fluid. **Synovial fluid** is a lubricant secreted by a connective tissue membrane surrounding a joint. Synovial fluid is a liquid resembling egg white. In large joints, such as the knee, bursae give added protection. **Bursae** are sacs of fluid that cushion a joint against shock.

592 BODY STRUCTURE

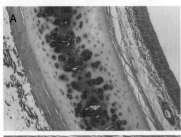
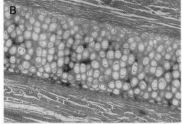
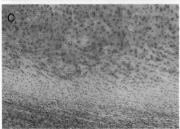

FIGURE 29·7 The three types of cartilage are grouped by the kinds and amounts of fibers found in the matrix, or gellike substance. Hyaline cartilage (A) does not have any fibers; fibrous cartilage (B) contains the most collagenous fibers; and elastic cartilage (C) is made up of many branching elastic fibers.

29·6 CARTILAGE AND FIBROUS TISSUES

In addition to bone, other types of connective tissues are part of the body framework. Cartilage cushions the bones where they meet in a joint. It is also found in such places as the nose, external ear, and windpipe. It is made of cells scattered through an intercellular matrix. In cartilage, the matrix is made of a gellike substance that does not have the minerals that make bone hard. The matrix may have many fibers scattered throughout. There are three types of cartilage. They are grouped by the different kinds and amounts of fibers in the matrix.

You can see the three types of cartilage in Figure 29·7. *Hyaline* (HĪ uh lihn) *cartilage* is clear because it does not contain fibers. Hyaline cartilage is found in many joints of the body, between the nostrils, and in the windpipe. *Fibrous cartilage* has many protein fibers. It makes up the cartilage disks found between vertebrae. *Elastic cartilage* is made up of many elastic fibers. Elastic cartilage is the type found in the external ear.

Fibrous tissues are connective tissues containing a high proportion of fibers made up of a protein called collagen. Ligaments and tendons are made of this kind of tissue. **Ligaments** are strong bands of connective tissue that hold bones together at joints. **Tendons** are bands of inelastic connective tissue that join muscles to bones.

29·7 DISORDERS OF JOINTS AND BONES

When some condition interferes with the smooth movement of bones in the joints, *arthritis* may be the result. Arthritis is inflammation of the joints. Arthritis may be relieved by pain-killing drugs or by injection of cortisone drugs into the affected joints. Some damaged joints may be replaced with artificial joints.

Sometimes bones and joints are subjected to more stress than they can withstand. This excess stress may then result in a *sprain*, a *dislocation*, or a *fracture*. A sprain is an injury in which the ligaments around a joint have been torn or stretched. A sprain usually heals after a period of rest. A dislocation is an injury in which bones move out of their normal positions in a joint. A fracture is a break in a bone. When a dislocation or fracture occurs, the bones must be returned to their normal positions and immobilized while healing takes place.

REVIEW QUESTIONS

1. Explain the relationship between cells, tissues, and organs.
2. Distinguish between the axial and appendicular skeletons.
3. Explain how bone replaces cartilage as the skeleton forms.
4. How are nutrients and oxygen supplied to bone cells?
5. The knee is the joint that probably most often suffers athletic injuries. Examine the structure of the knee in Figure 29·6 and explain why the knee is so frequently injured.

Investigation

What Is the Internal Structure of Bone?

Goals
After completing this activity, you will be able to
- Identify what percentage of bone is water.
- Explain how the removal of calcium affects the flexibility of bone.
- Identify the parts of a cross section of bone.

Materials (for groups of 2)
2 cooked chicken thigh bones (cleaned of meat and gristle)
100 mL HCl, 20% solution
forceps   250-mL beaker   scale
incubator   microscope
prepared slide of cross section of human bone

Procedure

A. Find the mass of one of the thigh bones. Record this mass on a separate sheet of paper.

B. Place one of the bones in an open dish. Heat the bone overnight in an incubator at 65°C. Allow the bone to cool.

C. After the bone has cooled, find its mass again. Record the mass. Calculate the percentage of mass that was lost or gained.

D. Test the flexibility of the other chicken bone by trying to bend and twist it. Record your observations.

E. Place the bone used in Step D in a 250-mL beaker. Add 100 mL of a 20% solution of HCl. Soak the bone for several days.

CAUTION: If any chemicals spill on your skin or clothing, flush the affected area with running water.

F. Using forceps, carefully remove the bone from the acid. Rinse it under running tap water to remove all the acid.

G. Again test the flexibility of the bone that was soaked in acid. Record any changes you note.

H. Examine a prepared slide of a cross section of bone under low power. Adjust the focus until you see several concentric circles.

I. Switch to high power. Focus on the groups of concentric circles. Look for the light central core of these circles. The core is called a Haversian canal **(1)**. The concentric circles around each canal are called lamellae (luh MEHL ee) **(2)**. Find the long dark spots between the lamellae. These are openings called lacunae (luh KYOO nee) **(3)** and contain the bone cells, or osteocytes.

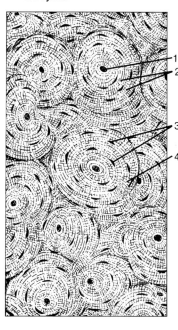

J. Extending out from the central canals are thin dark lines that look like the spokes of a wheel. Locate these lines, which are called canalicules **(4)**. Fluids pass from one part of the bone to another through these canalicules.

K. Move your slide around on the stage so that you can see the edge of the bone. Look for a thin layer of tissue covering the bone. This tissue is called the periosteum.

L. On a separate sheet of paper, draw the cross section of the bone. Label the periosteum, Haversian canal, lacunae, lamellae, and canalicules.

M. Follow your teacher's instructions for proper disposal of all materials.

Questions and Conclusions

1. Did the bone lose or gain mass from heating? What was the percentage lost or gained? What material was lost or gained?

2. How did soaking the bone in HCl affect its flexibility?

3. How did soaking the bone in HCl affect its chemical make-up?

4. Many people think of bone as a nonliving tissue. Using the terms Haversian canal, periosteum, and osteocyte, explain how bone is like other living tissues.

5. Describe any similarities you may have noticed between bone tissue and vascular tissue in the stems of plants. Base your answer on your previous studies of plants.

The Muscular System

SECTION OBJECTIVES

- Classify the three types of muscle tissue.
- Describe how muscles contract.

29·8 TYPES OF MUSCLES

The bony framework of the body is covered by hundreds of muscles. While both the skeletal system and the muscular system are involved in body movements, the bones of the skeleton are not able to move by themselves. Only muscle tissue is able to contract, or shorten. As contractions occur, the bones attached to the muscles are pulled into new positions, causing body parts to move.

Muscle is a tissue made up of long slender cells that are capable of contracting. Muscle tissue makes up the body's muscles. Muscle cells are called **muscle fibers.** As these muscle fibers contract, chemical energy stored in the cells is changed to energy of motion. There are three types of muscles, and each of these is shown in Figure 29·9.

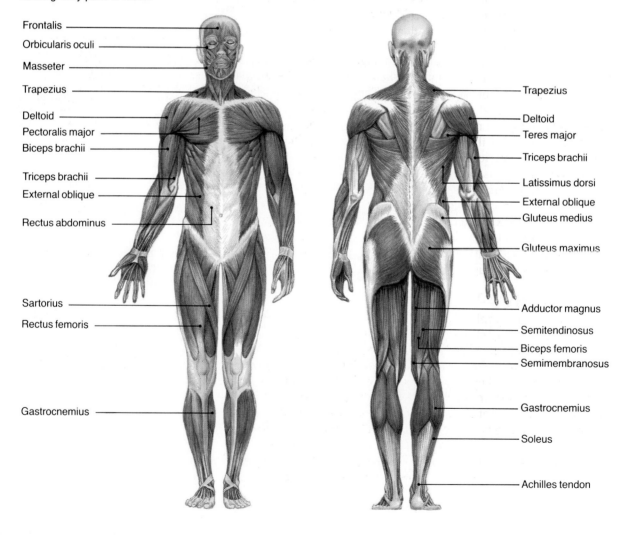

FIGURE 29·8 The skeleton is covered by the muscles of the muscular system. As these muscles contract, they pull on the bones that are attached to them, causing body parts to move.

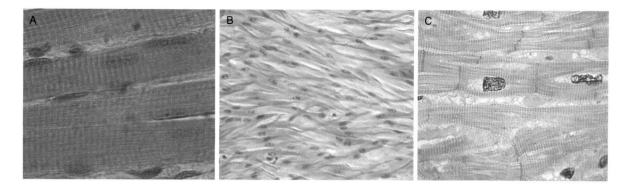

Skeletal muscle is the type of muscle that is attached to the skeleton, and it is under voluntary control. Skeletal muscles act during voluntary body movements. In Figure 29·9, you can see the type of fiber that makes up skeletal muscle. Each fiber is a single cell. Each fiber has several nuclei. Dark stripes, or striations (strī AY shuhnz), or light and dark cross bands mark the fibers. Skeletal muscle is also known as *striated muscle* because of these cross stripes.

Smooth muscle is the type of muscle found in internal organs and blood vessels, and its action is involuntary. Compare the smooth muscle and skeletal muscle shown in Figure 29·9. The fibers of smooth muscle are long and spindle shaped. They lack the striations and numerous nuclei of skeletal muscle. There is only one nucleus in each smooth muscle cell. Smooth muscle lines the stomach and intestines. As it contracts it moves food along the digestive tract. It is also found in the walls of the blood vessels and in the urinary bladder. Smooth muscle contracts much more slowly than skeletal muscle.

Cardiac muscle is the type of muscle found only in the heart, and its action is involuntary. In Figure 29·9, you can see that cardiac muscle consists of short cylindrical fibers that connect with each other to form a branching network. You can also see that cardiac muscle is striated like skeletal muscle, but the striations are not as clear. Also notice that each cell contains one oval nucleus. Cardiac muscle continues to contract throughout life.

29·9 MUSCLES AND MOVEMENT

Both ends of a skeletal muscle are usually attached to bones by tendons. Figure 29·10 shows the attachment of the biceps and triceps muscles of the upper arm. These muscles permit you to raise and lower your forearm. When the biceps contracts to pull your hand toward the shoulder, the bones of the lower arm move, and the bones of the upper arm remain still. The end of the muscle attached to a bone that moves little or not at all is called the **origin**. The origin of the biceps is the end attached to the scapula and the humerus. The end of the muscle attached to a bone that moves is called the **insertion.** Notice that the insertion end of the biceps is on the radius, in the lower arm. The body of the muscle, between the origin and insertion, is called the *belly*.

FIGURE 29·9 Skeletal muscle (A) smooth muscle (B), and cardiac muscle (C) are the three types of muscle tissue in the human body. Notice the difference in the structure of each type of muscle tissue.

FIGURE 29·10 Like most skeletal muscles, the biceps and triceps are attached to bones. The fixed end of the muscle is called the origin, and the movable end is called the insertion. The body of the muscle, between the origin and insertion, is called the belly.

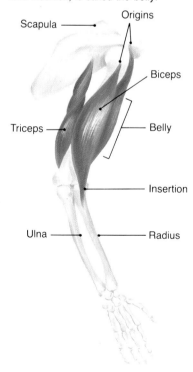

BODY STRUCTURE   595

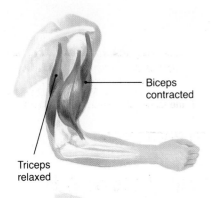

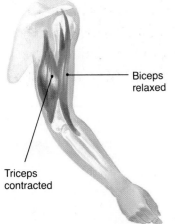

FIGURE 29·11 The biceps and triceps work in opposition to each other. When the biceps contracts, the triceps relaxes and the forearm is raised (top). When the triceps contracts, the biceps relaxes and the forearm is lowered (bottom).

When the triceps contracts, the biceps relaxes, and your forearm moves down. You may have thought that raising and lowering the arm was accomplished by the same muscle. However, most body movements are brought about by a pair of muscles that work in opposition to each other. As one muscle contracts, the other relaxes. In this way one muscle of a pair bends a joint and the other straightens it. The muscle that flexes, or bends, a joint is called the *flexor*. The muscle that extends, or straightens, a joint is called the *extensor*. Refer to Figure 29·11. When the forearm moves back and forth at the elbow, which muscle is the flexor? Which one is the extensor?

29·10 MUSCLE CONTRACTION

Recall that when viewed under the compound light microscope, skeletal muscle appears to be made up of alternating light and dark bands called striations. When viewed with an electron microscope, skeletal muscle can be seen to be made up of units smaller than the striations. Many experiments as well as studies with the electron microscope have shown how these units function when skeletal muscle contracts.

Figure 29·12 shows the structure of skeletal muscle in increasing detail. If you examine a section of skeletal muscle under a microscope, you will see that it is composed of long thin fibers. Recall that each of these fibers is a cell with many nuclei. Each fiber may be up to 30 cm long. A large muscle may have several hundred thousand fibers. Within the cytoplasm of the fibers are bundles of smaller units called myofibrils. **Myofibrils** are very fine threadlike structures running lengthwise through muscle fibers. Note the arrangement of the myofibrils within the fiber in Figure 29·12. The electron microscope has shown that each myofibril is made up of thick and thin filaments that form repeating units within the myofibril. Where these filaments overlap, they form the dark parts of the striations of skeletal muscle.

FIGURE 29·12 Muscles are made up of bundles of muscle fibers. Each muscle fiber contains bundles of smaller units called myofibrils that are made up of repeating units of thick and thin filaments.

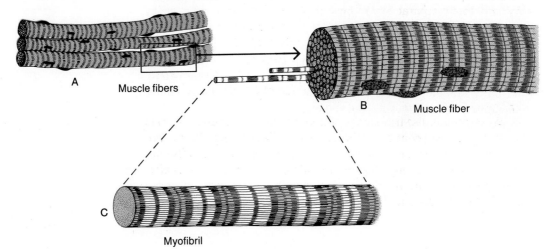

The filaments of myofibrils are made of two proteins: myosin and actin. **Myosin** is the protein that makes up the thick filaments. **Actin** is the protein that makes up the thin filaments. In Figure 29·13, notice that the myosin filaments have many projections, known as *cross bridges*. These cross bridges are believed to be links to the actin filaments. Also note that the ends of the actin filaments are attached to cross bands known as *Z lines*.

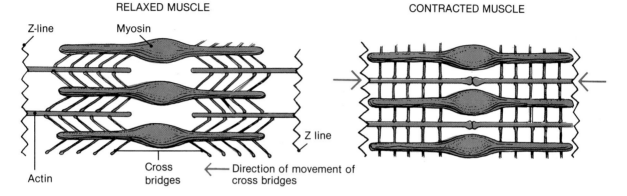

FIGURE 29·13 When a muscle contracts, actin filaments slide past the myosin filaments. This action causes the length of the muscle to shorten.

The arrangement of the two kinds of filaments within myofibrils is believed to be partly responsible for the ability of muscle to contract. One widely accepted hypothesis explaining muscular contraction suggests that the thin filaments slide past the thick filaments. This action is shown in Figure 29·13. It is believed that the cross bridges between the molecules of myosin and actin produce the sliding action of the filaments. Each cross bridge is thought to swivel around a fixed position on the myosin filament. The myosin filaments do not move. The swiveling cross bridges pull on the thin actin filaments, causing them to slide in the direction shown in Figure 29·13. The actin filaments move toward one another until they meet.

29·11 ENERGY AND MUSCLE CONTRACTION

Most activities of the human body are controlled by the nervous system. Muscle contraction is begun by nerve impulses carried by nerve cells to the muscle fibers. At the place where a nerve cell contacts a muscle, the nerve cell separates into a number of branches. Each branch makes contact with a muscle fiber by means of a small buttonlike plate. A nerve cell and the muscle fibers to which it carries impulses make up a *motor unit*. A photograph of a motor unit is shown in Figure 29·14.

When an impulse travels over a nerve cell, all the muscle fibers of that motor unit will contract at the same time. Each muscle fiber either contracts completely or not at all. A muscle fiber cannot partially contract. The type of response that occurs with full strength or not at all is called an *all-or-none response*. Note that the all-or-none response applies only to individual motor units and not to an entire muscle. The

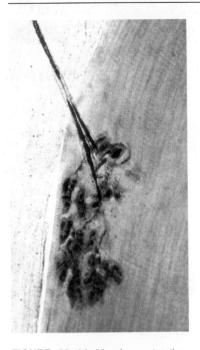

FIGURE 29·14 Muscle contraction, like most activities of the human body, is controlled by the nervous system. This photograph shows a motor unit, which consists of a branched nerve cell and the muscle fibers to which it carries nerve impulses.

strength of a muscle contraction depends on the number of motor units that are stimulated at one time. Lifting a book may activate only a few motor units. Lifting weights may stimulate so many motor units that nearly every fiber of a muscle will contract at once.

When a nerve impulse reaches the nerve endings at the muscle fibers, small amounts of a chemical called *acetylcholine* (as uh tuhl-KOH leen) are released from the nerve endings. Acetylcholine stimulates skeletal muscle fibers. The chemical also appears to increase the permeability of muscle cell membranes to calcium ions. Calcium ions, along with an enzyme, catalyze the breakdown of ATP to ADP and P. Recall that energy is released when the terminal phosphate group is removed from ATP during its conversion to ADP (Section 5·5). Apparently the energy released by the breakdown of ATP into ADP activates the cross bridges between the myosin and actin filaments. When the cross bridges are activated, muscle contraction takes place.

29·12 MUSCLE DISORDERS

When a muscle becomes tired, it is subject to a number of injuries. Muscle injuries are especially common during exercise and participation in sports or hard work. Soreness in muscles often occurs as a result of strenuous exercise, especially when the muscles have not been strengthened properly by conditioning. Muscle *cramps* are also common in muscles that are overused. Cramps are painful muscular contractions that may affect single muscles or groups of muscles.

Another common muscle disorder is a strained muscle. A *strain* is an injury in which muscles or tendons are stretched or torn. Medical attention should be obtained to learn the extent of a strain. The extent and location of a strain will determine the specific treatment needed.

Muscular dystrophy is a disease in which the skeletal muscles atrophy, or waste away. Although the disease is known to be caused by a genetic defect, the exact reason for the wasting away of muscle tissue is not known. It is believed that the disease is due to one or more enzymes that may be absent or defective. There is no effective treatment for muscular dystrophy. Physical therapy and braces are used to lessen some of the disabilities caused by the disease.

REVIEW QUESTIONS

6. List the three types of muscle and state the function of each.
7. Briefly describe the events involved in the "sliding filament" explanation for muscle contraction.
8. What is the source of energy for the contraction of skeletal muscle?
9. What is a strain?
10. Certain poisons are known to cause muscle paralysis, including paralysis of the respiratory muscles. However, evidence indicates that the poisons do not directly affect muscle fibers. How might such poisons cause paralysis?

The Integumentary System

29•13 STRUCTURE OF THE SKIN

One of the largest organs in the body is the skin. An average adult's skin has a total area of more than 1.5 m². Most people do not think of the skin as being an organ, but you should recall that an organ is a group of different tissues that are specialized for a particular function. By this definition, the skin is an organ. Biologists commonly refer to the skin as the integument (ihn TEHG yoo muhnt). **Integument** is a body covering, made up of one or more tissues, that protects the body from the environment. The skin, hair, nails, and sweat glands make up the *integumentary system*.

The main layers of the skin are shown in Figure 29•15. The two main layers are the thin outer epidermis and the thicker dermis lying beneath. Although these layers fit tightly together, they can be stretched slightly. For example, you lift both layers when you pinch the skin on the back of your hand.

Epidermis is the tough outer layer of skin. In Figure 29•15 you can see that it is composed of two layers of cells. The cells of the upper epidermis are dead or dying. The cells in the lower layer are alive and constantly dividing. Note that there are no blood vessels in the epidermis. The lower epidermal cells are supplied with nutrients that diffuse short distances from the blood vessels in the dermis below. But since nutrients can diffuse only short distances, the cells of the upper epidermis die of starvation. These dead cells become filled with a tough protein material called *keratin*. Keratin provides a barrier to keep water from entering or leaving the body. It is also the material from which hair and nails are formed. Dead skin cells are continuously worn away at the surface. They are replaced by newly formed cells from the lower epidermis that move up toward the surface.

SECTION OBJECTIVES

- Describe the structure and function of the tissues making up the skin.
- List the functions of the skin.

epi (on)
derma (skin)

FIGURE 29•15 The skin, or integument, is composed of two layers. The epidermis is the tough outer layer. The dermis is the thick inner layer.

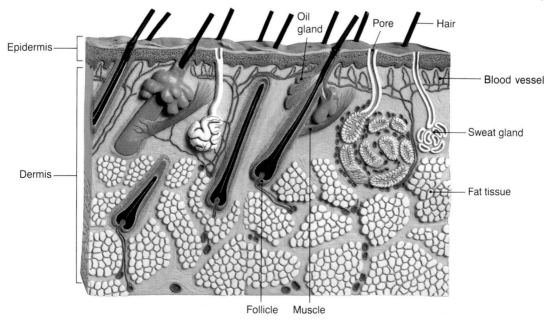

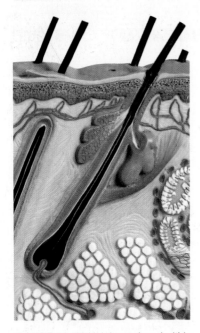

FIGURE 29·16 Hair is produced within a tubelike structure called a follicle. Notice the saclike oil glands on either side of the upper part of the hair's root.

Within the lower epidermis are cells that produce *melanin*. Melanin is a dark pigment that helps give skin its color. The amount of melanin determines the color of the skin. People with dark skin have a great amount of melanin, while light-skinned people have very little. Exposure to sunlight causes an increase in melanin production, which darkens the skin. Melanin gives some protection against harmful ultraviolet rays.

The **dermis** is the inner skin, beneath the epidermis. Figure 29·15 shows that the dermis is supplied with blood vessels and nerves. Oil glands, sweat glands, and hair follicles also are embedded in this layer. Notice that the upper surface of the dermis forms a series of ridges projecting up into the epidermis. These projections, called *papillae,* help to hold the dermis and epidermis together. They also form the ridges that produce fingerprints.

Hair grows from structures called follicles. **Hair follicles** are tubelike pockets of epidermal cells that extend downward into the dermis and produce hair. The hair, which is mostly a tube of keratin, is produced by cells at the base of the follicle. The part of the hair within the follicle is called the root. The part above the surface of the skin is called the shaft. In Figure 29·16, you can see that an oil gland opens into the follicle. Oil passes from this gland, into the hair follicle, and then to the skin surface. The oil keeps the hair and skin from becoming dry. Oil glands are especially numerous on the face and scalp.

The dermis also contains many sweat glands. A sweat gland is a tubular coiled gland with a pore opening on the surface of the skin. Water, salt, and some body wastes are eliminated from these pores as sweat. As sweat evaporates, the skin is cooled. Sweat glands are found all over the body, but they are most numerous in the armpits, on the forehead, on the palms of the hands, and on the soles of the feet. Finally, note that there is a layer below the dermis made up mainly of fat tissue. This layer serves as a cushion and an insulator.

29·14 FUNCTIONS OF THE SKIN

Though the skin has several important functions, its main function is protection. Recall that the cells of the upper epidermis are filled with keratin. Since keratin is waterproof, it prevents the loss of water through the skin. As long as the skin remains unbroken, it keeps bacteria, fungi, parasites, and most chemicals from entering the body. Further, the secretions from the sweat glands and oil glands are poisonous to certain bacteria and fungi living on the skin. Thus these organisms are killed before they can enter the body.

The skin plays an important role in the regulation of body temperature. Heat is continuously produced as the cells break down glucose. During hard exercise, the muscle cells release large amounts of heat, causing the temperature of the blood to rise. Some of this heat is carried to the skin. During exercise the skin releases an increased amount of sweat. Evaporation is a physical change that requires heat energy. As sweat evaporates from the skin, heat is lost from the body, causing it to be cooled. If a person becomes chilled, the blood vessels in the skin become narrow, or constrict. This narrowing reduces the amount of blood flowing near the skin's surface, which in turn reduces the amount of heat lost from the body.

The skin also contains sense organs. Several kinds of nerve endings, or receptors, are distributed through the skin and supply information about the environment. These nerve endings are sensitive to touch, pressure, pain, warmth, and cold (Section 33·14).

The skin functions as an excretory organ. Perspiration is about 99 percent water. The remaining 1 percent is made up of salt and body wastes. There are over 2 million sweat glands in the skin, and they excrete about 1 L of liquid daily.

The skin also contains a substance from which it synthesizes vitamin D in the presence of ultraviolet light. Vitamin D synthesized in the skin aids in the absorption of calcium. Calcium is needed for the normal growth and development of bones.

29·15 SKIN DISORDERS

Many skin disorders have been identified. A common skin ailment is *dermatitis,* an inflammation of the skin that causes itching or burning. Some skin problems are associated with hair follicles or oil glands. Boils and acne are examples of this type of disorder.

Sometimes an injury results in the skin being broken. Open breaks in the skin are called *wounds.* All wounds, even small cuts or scrapes, should be cleansed carefully. Any break in the skin could lead to an infection.

Burns are injuries resulting from the exposure of the skin to intense heat, certain chemicals, electrical shock, or radiation. Severe burns are extremely serious because the body's protective barrier against germs and fluid loss is damaged. Treatment of such burns requires intravenous replacement of lost fluids. Antibiotics are also used to combat infections.

If an injury such as a wound or a severe burn damages a large area of skin, a skin graft may be used to repair the damage. Researchers have developed a wide variety of artificial skins. One kind of artificial skin is made of a type of plastic. This artificial skin prevents fluid loss and infection while the patient's own skin grows back.

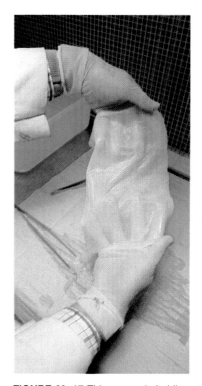

FIGURE 29·17 This person is holding a piece of artificial skin, which is made from cowhide, shark cartilage, and plastic. Artificial skin is used to help prevent infection and fluid loss when a large area of a person's own skin has been damaged.

REVIEW QUESTIONS

11. Name the two main layers of the skin and describe their appearance.
12. List five functions of the skin.
13. Why are the sweat glands considered to be organs of both excretion and temperature regulation?
14. Name four skin disorders.
15. After running a 10-km race on a brisk day, a runner's body is covered with sweat. What could happen if this runner fails to put on a jacket after the race?

Chapter Review

SUMMARY

29·1 The tissues making up the human body may be classified as epithelial, connective, muscle, and nerve tissues.

29·2 The skeleton is the basic framework of the human body, and its two divisions are the axial skeleton and the appendicular skeleton.

29·3 The cartilage and fibrous tissue membranes in the skeleton of the embryo are gradually replaced by bone through a process called ossification.

29·4 A bone is made up of compact bone tissue, spongy bone tissue, and marrow, which fills the center.

29·5 Joints are the places where two or more bones come together. Different kinds of joints permit the bones to move in various ways.

29·6 Cartilage and fibrous tissues, such as ligaments and tendons, are connective tissues found in the framework of the body.

29·7 Disorders and injuries involving bones and joints often occur because of wear or because excess stress is placed on them.

29·8 Skeletal muscle, smooth muscle, and cardiac muscle are the three types of muscles found in the body.

29·9 Movement of the bones at the joints is accomplished by paired muscles working in opposition to each other.

29·10 Muscle fibers are composed of myofibrils, which, in turn, are composed of myosin and actin filaments. During muscle contraction the actin filaments are believed to slide past the myosin filaments.

29·11 The energy for muscle contraction is derived from the conversion of ATP to ADP+P in the muscle fibers.

29·12 Muscles are sometimes injured during physical activity.

29·13 The two main layers of skin are the epidermis at the surface and the underlying dermis.

29·14 The skin has several important functions, including protecting the body, regulating body temperature, serving as a sense organ, excreting wastes, and synthesizing vitamin D.

29·15 Many skin disorders have been identified, and these have a wide variety of causes. An injury that breaks the skin results in a wound.

BIOLOGICAL TERMS

actin
appendicular skeleton
axial skeleton
bone
bursae
cardiac muscle
cartilage
compact bone
connective tissue
dermis
dorsal cavity
epidermis
epithelial tissue
hair follicles
Haversian canals
insertion
integument
joint
ligaments
muscle fibers
muscle tissue
myofibrils
myosin
nerve tissue
organ
organ system
origin
ossification
osteocytes
pectoral girdle
pelvic girdle
periosteum
skeletal muscle
smooth muscle
spongy bone
synovial fluid
tendons
tissue
ventral cavity
vertebral column

USING BIOLOGICAL TERMS

Explain the difference between each pair of related terms.
1. pectoral girdle, pelvic girdle
2. axial skeleton, appendicular skeleton
3. origin, insertion
4. skeletal muscle, smooth muscle
5. epidermis, dermis

Choose the term that does not belong and explain why it does not belong. Define each term.

6. myofibrils, actin, bursa, myosin
7. periosteum, compact bone, spongy bone, epidermis
8. muscle fibers, cartilage, ligaments, compact bone
9. compact bone, Haversian canals, synovial fluid, osteocytes
10. ligaments, joint, insertion, synovial fluid

UNDERSTANDING CONCEPTS

1. Identify the two major cavities of the human body. (29·1)
2. Describe four types of tissue found in the human body. (29·1)
3. List three functions of the skeleton. (29·2)
4. Identify the parts of the axial skeleton and the appendicular skeleton. (29·2)
5. What is the function of osteoblasts? (29·3)
6. Distinguish between compact bone and spongy bone. (29·4)
7. Describe the structure of compact bone. (29·4)
8. Identify four types of joints and give an example of each. (29·5)
9. How does cartilage differ from bone? (29·6)
10. Distinguish between a sprain, a dislocation, and a fracture. (29·7)
11. In what parts of the body are these muscles found: skeletal, smooth, cardiac? (29·8)
12. How do muscles work in opposition to each other? (29·9)
13. Describe the structure of the myofibrils and filaments that make up muscle fibers. (29·10)
14. Explain what is meant by the all-or-none-response of muscle fibers. (29·11)
15. What is muscular dystrophy? (29·12)
16. Distinguish between a sprain and a strain. (29·7, 29·12)
17. Describe the structure of the skin. (29·13)
18. Describe four functions of the skin. (29·14)
19. Why do wounds and burns cause a risk of infection? (29·15)

APPLYING CONCEPTS

1. What do you think might be the adaptive value of the fontanelles in the skull of an infant?
2. Mitochondria are more numerous in muscle cells than in some other types of cells. Give a reason why mitochondria are so numerous in muscles.
3. How is synovial fluid similar in function to motor oil?
4. Examine the human skeleton shown in Figure 29·2. Describe how the joint of the shoulder permits a person to throw a baseball.
5. Suppose smooth muscles rather than skeletal muscles could be attached to the skeleton and given voluntary control. How might the movement of the skeleton be affected?

EXTENDING CONCEPTS

1. Using appropriate resources prepare a report on the contributions of one of the following scientists to our knowledge of muscle metabolism.
   Luigi Galvani (1737–1798)
   Albert Szent-Gyorgyi (1893–1986)
2. Locate drawings or photographs showing the skeletons of the gorilla and chimpanzee. Compare the skeletons of these primates with the skeleton of a human. Note the similarities and differences that you see between these skeletons.
3. Remove the organic material from the leg bone of a chicken or turkey. To do this, place the bone in an oven or pottery kiln for several hours at high temperature. Allow the bone to cool and then examine it. Describe what happens when you touch the bone.
4. If your school has a human skeleton or a chart of a human skeleton, locate examples of the different kinds of joints in the skeleton.

READINGS

Harrington, W.F. *Muscle Contraction,* Carolina Biological Reader. Carolina Biological Supply Co., 1981.

Podolsky, Doug M. *Skin: The Human Fabric.* U.S. News Books, 1982.

Tiger, Steven. *Arthritis.* Messner, 1986.

30. Nutrition and Digestion

The villi are fingerlike projections that line the interior of the small intestine. As digested foods move through the small intestine, nutrients are absorbed into the villi. Each villus contains a network of tiny blood vessels through which nutrients enter the blood stream. In this chapter, you will learn how food is changed by the digestive system into nutrients that the villi can absorb.

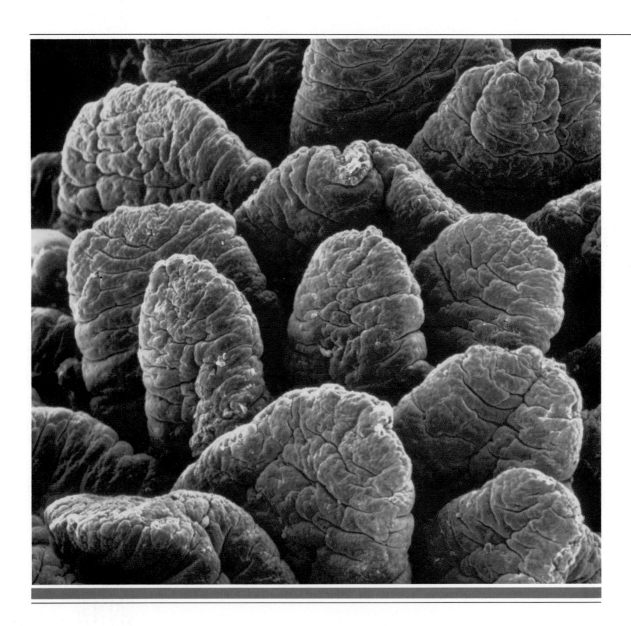

CHAPTER OBJECTIVES

After completing this chapter, you will be able to

- **Describe** the five classes of nutrients.
- **List** the four basic food groups and explain how they are important in the diet.
- **Describe** how food is digested in the organs of the digestive system.
- **Explain** the role of the liver, pancreas, and gall bladder in digestion.
- **Compare** the digestion of carbohydrates, fats, and proteins.
- **Explain** how absorption occurs in the digestive system.
- **Identify** some disorders of the digestive system.

Nutrition and Diet

30·1 NUTRIENTS IN FOOD

Nutrition (noo TRIHSH uhn) is the process by which living things obtain, digest, and assimilate food. Like all heterotrophs, humans need food to carry on their life functions. Food is a complex substance made up of organic compounds, inorganic compounds, and ions. Most of the compounds in food are nutrients. A **nutrient** (NOO-tree uhnt) is a substance that supplies the body with energy or raw materials, or that is needed for chemical reactions. Raw materials are used for growth and tissue repair.

The five classes of nutrients are carbohydrates, lipids, proteins, vitamins, and minerals. Carbohydrates, lipids, and proteins are large organic molecules that cannot be absorbed into the blood until they have been digested. **Vitamins** are organic molecules that can be absorbed without being digested. **Minerals** are inorganic materials that also are absorbed without being digested. Carbohydrates, lipids, and proteins can be broken down to provide energy for the body. Vitamins and minerals cannot be used as sources of energy.

When those nutrients that are sources of energy are broken down in the cells, the energy is released. This energy can be used by the cells. The amount of energy that a nutrient can provide is measured in units called *Calories,* or food Calories (always capitalized). A calorie is the amount of heat needed to raise the temperature of 1 g of water by 1° C. A food *Calorie* is equal to 1 kilocalorie (1000 calories).

Carbohydrates are the body's chief source of energy. Recall that the monosaccharide glucose is broken down during cellular respiration (Section 5·6). The three main types of carbohydrates found in food are monosaccharides, disaccharides, and polysaccharides. Monosaccharides and disaccharides are sugars. Polysaccharides are complex carbohydrates, such as starches and cellulose. Foods from plant sources are often rich in sugars and complex carbohydrates.

SECTION OBJECTIVES

- List the functions of the five essential nutrients necessary for normal body metabolism.
- Evaluate a diet in terms of recommended servings from the four basic food groups.

Lipids are a highly concentrated source of energy. They provide about twice as many Calories per gram as either carbohydrates or proteins. Lipids include fatty acids, which are used by the body to make cell membranes and some hormones. Both fats and oils are lipids. Fats, composed mostly of saturated fatty acids, are found in meats, eggs, whole milk, and other animal sources. Oils, composed of unsaturated fatty acids, are found in seeds, nuts, grains, and other foods from plant sources.

Proteins in food provide the amino acids that the body cells use for making proteins. These synthesized proteins are important in many ways. Some proteins form parts of the structures of cell organelles, such as cell membranes and chromosomes. Other proteins are enzymes. Some proteins, such as hormones and antibodies, may be released by cells in one area and function throughout the body. Proteins are needed for normal growth. Recall that 20 different amino acids are used to make the body's proteins.

The human body can make 12 of the 20 amino acids it needs from other amino acids. However, there are eight amino acids that the body cannot make from other amino acids. These amino acids are called the *essential amino acids*. The body can get them only from food. Foods that have all eight essential amino acids are called *complete protein foods*. Many foods from animal sources are complete protein foods. Soybean products can be used as meat substitutes since soybeans have these eight amino acids. Most other plant products are called *incomplete protein foods*, since they are missing one or more of the essential amino acids. Some incomplete protein foods when eaten together supply all eight essential amino acids. These foods are called *complementary protein foods*. Whole grains, such as rice, and legumes, such as peas and beans, are complementary protein foods.

Vitamins are another class of nutrients found in foods. Many vitamins are coenzymes and are essential for normal functioning of cells. Figure 30·2 lists many vitamins, their functions in the body, the foods in which they are found, and the diseases resulting from not having enough of these vitamins.

You can see that each vitamin in the table is either water soluble or fat soluble. Water-soluble vitamins can be lost from foods when the foods are cooked in water. Vitamins A, D, E, and K are fat-soluble. These vitamins can only be absorbed along with lipids, so foods with these vitamins should be eaten with foods that have either fats or oils. Excess water-soluble vitamins are released from the body. Excess fat-soluble vitamins are stored in the body, however, high levels of these vitamins can be harmful.

Minerals are elements, usually found in ionic form, that are necessary for the normal functioning of the body. Look at Figure 30·3. It lists some of these minerals, their functions, and the foods that are good sources of these minerals. Only small amounts of minerals are needed in the diet.

FIGURE 30·1 Although soybeans are plants, they are used as a meat substitute because they contain the same eight essential amino acids that meats do.

FIGURE 30·2 Some of the vitamins that your body needs are listed in this table. Which foods are a source of vitamin B_{12}?

VITAMIN	NEEDED FOR	FOOD SOURCES	DEFICIENCY DISEASES AND SYMPTOMS
B_1 Thiamine (water soluble)	Carbohydrate oxidation Proper function of nerves and muscles	Whole grain foods, wheat germ, eggs, potatoes, meats, yeast, milk	Poor appetite, nerve disorders *Beriberi*—nerve disorders, pain and stiffness of lower limbs, wasting of muscle
B_2 Riboflavin (water soluble)	Cellular respiration Healthy skin and eyes	Milk, eggs, meats, wheat germ, green vegetables	Scaly skin, skin sores on lips and mouth, sensitivity to bright light
B_3 Niacin (water soluble)	Cellular respiration	Meats, fish, liver, whole grain, peanuts, yeast	*Pellagra*—red, dry skin easily irritated by sunlight and heat; diarrhea, nerve disorders
B_{12} (water soluble)	Proper development of red blood cells	Meats, milk, liver	Poor appetite, weakness *Pernicious anemia*—reduced number of red blood cells
C Ascorbic acid (water soluble)	Healthy bones and teeth Maintaining strength of capillaries Aiding healing	Citrus and other fruits, tomatoes, peppers, leafy green vegetables	*Scurvy*—Swollen, bleeding gums, loose teeth, bruise easily, slow healing
A (fat soluble)	Maintaining healthy linings of respiratory, digestive, excretory, and reproductive systems Healthy eyes	Dairy products, egg yolk, fish-liver oil, green and yellow vegetables	Dry, brittle skin, *Night blindness*—poor vision in dim light
D (fat soluble)	Growth and development of bones and teeth Metabolism of calcium and phosphorus	Fortified dairy products, fish-liver oil, saltwater fish	In adults: weakened and brittle bones In children: *Rickets*—weakened and bowed leg bones, deformed rib cage
E Tocopherol (fat soluble)	Maintaining fat component of cell membranes	Wheat germ oil, leafy green vegetables, meats, egg yolk	—
K (fat soluble)	Normal blood clotting Normal liver function	Leafy green vegetables, cauliflower Produced by intestinal bacteria	Slow blood clotting (may lead to severe blood loss)

FIGURE 30·3 According to this table of essential minerals, which minerals can be found in eggs?

MINERAL	NEEDED FOR	FOOD SOURCES
Calcium	Bone and teeth formation, muscle contraction, transmission of nerve impulses, blood clotting	Dairy products, eggs, fish, legumes
Phosphorus	Bone and teeth formation, formation of lipids in cell membranes, ATP, nucleic acids	Dairy products, meats, legumes, whole grains
Potassium	Nerve and muscle function, regulation of heartbeat, maintenance of fluid balance	Beans, bran, potatoes, bananas, apricots
Iron	Formation of hemoglobin in the red blood cells, cellular respiration	Liver, eggs, nuts, legumes, raisins
Iodine	Formation of thyroxin, which regulates metabolism	Iodized salt, saltwater fish, oysters, shrimp, broccoli
Magnesium	Cofactor for enzymes that regulate nerve and muscle function	Green vegetables, beans, bran, corn, peanuts, meat, milk, prunes
Sodium	Regulation of fluid balance, maintenance of ionic balance, transmission of nerve impulses	Table salt, seafood, most foods
Chlorine	Fluid balance and acid-base balance, synthesis of hydrochloric acid in stomach	Table salt, most foods
Copper	Formation of hemoglobin	Liver and meats, oysters, shrimp, peas, pecans
Zinc	Synthesis of insulin, part of some enzymes	Beans, liver, lentils, spinach, many foods

Water is not usually considered to be a nutrient, but it is an essential part of the diet. The average person loses about 2500 mL of water each day. This amount should be replaced in order to maintain the normal functioning of the body. Although most foods are 65 to 90 percent water, the average person should still drink several glasses of water a day. The body also produces some water as a result of cellular respiration. Water takes part in many chemical reactions in the cells. It is the solvent in which many of these reactions occur. Water makes up much of the volume of the blood and the lymph. It dilutes the wastes that the body excretes. Water also helps in maintaining a constant body temperature.

30·2 THE BALANCED DIET

The amount of various nutrients needed in a balanced diet may vary from person to person. To remain healthy and active, most people need enough food to supply them with 2000 to 4000 Calories per day. The ideal number of Calories for a person per day depends on age, sex, metabolic rate, and daily physical activity. If the amount of food one consumes contains more Calories than are required for the body's needs, the body changes the excess nutrients to fat for storage. If the amount of food one consumes contains less than the required number of Calories, the body breaks down some fat to obtain energy.

To be balanced, a diet should supply more than just the energy needed each day. It should also contain an adequate supply of nutrients for such activities as growth and tissue repair. These nutrients can be provided by a diet that contains the proper amounts of food from the four basic food groups.

Look at the foods that are contained in the dairy group in Figure 30·4. The milk group provides proteins, fats, vitamin A, and some of the B vitamins. It also provides the minerals calcium, phosphorus, potassium, magnesium, and fluorine. Vitamin D is added to foods in the milk group.

The meat group provides proteins and fats. It also provides many of the B vitamins and the minerals phosphorus, sodium, potassium, and iron. Look at the foods in the meat group. In addition to meat and eggs, this group contains beans, also called legumes. Legumes are the richest plant source of protein.

FIGURE 30·4 A balanced diet contains food from the four food groups. Foods from the dairy group (left) and the meat group (right) provide you with proteins, fats, and some vitamins and minerals.

FIGURE 30·5 Foods in the third food group, the fruit and vegetable group, provide you with carbohydrates, vitamins, and minerals. Foods in the fourth food group, the bread and cereal group, provide you with carbohydrates, proteins, and vitamins.

Foods in the fruit and vegetable group supply complex carbohydrates and sugars. They supply the vitamins A, E, K, C, and many of the B vitamins. They are also an important source of the minerals phosphorus, sodium, and potassium. If vegetables and fruits are eaten raw, they provide more vitamins than if they are cooked and then eaten.

Foods in the bread and cereal group contribute starch and protein to the diet. If these foods are made from whole grains, they provide complex carbohydrates and many B vitamins. Indigestible material in these foods, called roughage, helps stimulate the muscles of the digestive tract. Roughage consists mostly of cellulose.

You should be aware of the nutrients you need for good health. You should also know what nutrients are contained in the foods you eat. Most packaged foods contain labels listing the nutrients in the food. It is wise to read these labels to know what you are eating.

REVIEW QUESTIONS

1. What is meant by the term *essential amino acids*?
2. What are minerals?
3. Describe the nutritional needs met by each of the four basic food groups.
4. Explain why vegetables that are cooked in water are lower in vitamin C and the B vitamins than those that are raw.
5. A person has brittle skin and difficulty seeing at night. What vitamin may be lacking in the diet? What foods might cure the disorders?

The Digestive System

30·3 MOUTH, PHARYNX, AND ESOPHAGUS

The digestive system is made up of a highly specialized digestive tube and several organs. As food passes through the digestive system, it is chemically broken down into small soluble molecules. These molecules are absorbed through the lining of the digestive tube into the circulatory system.

The digestive system is made up of several organs. Each organ is adapted to carry out certain functions. Figure 30·6 shows the organs of the digestive system.

SECTION OBJECTIVES

- Describe the structure and function of the digestive system.
- Compare mechanical and chemical digestion.

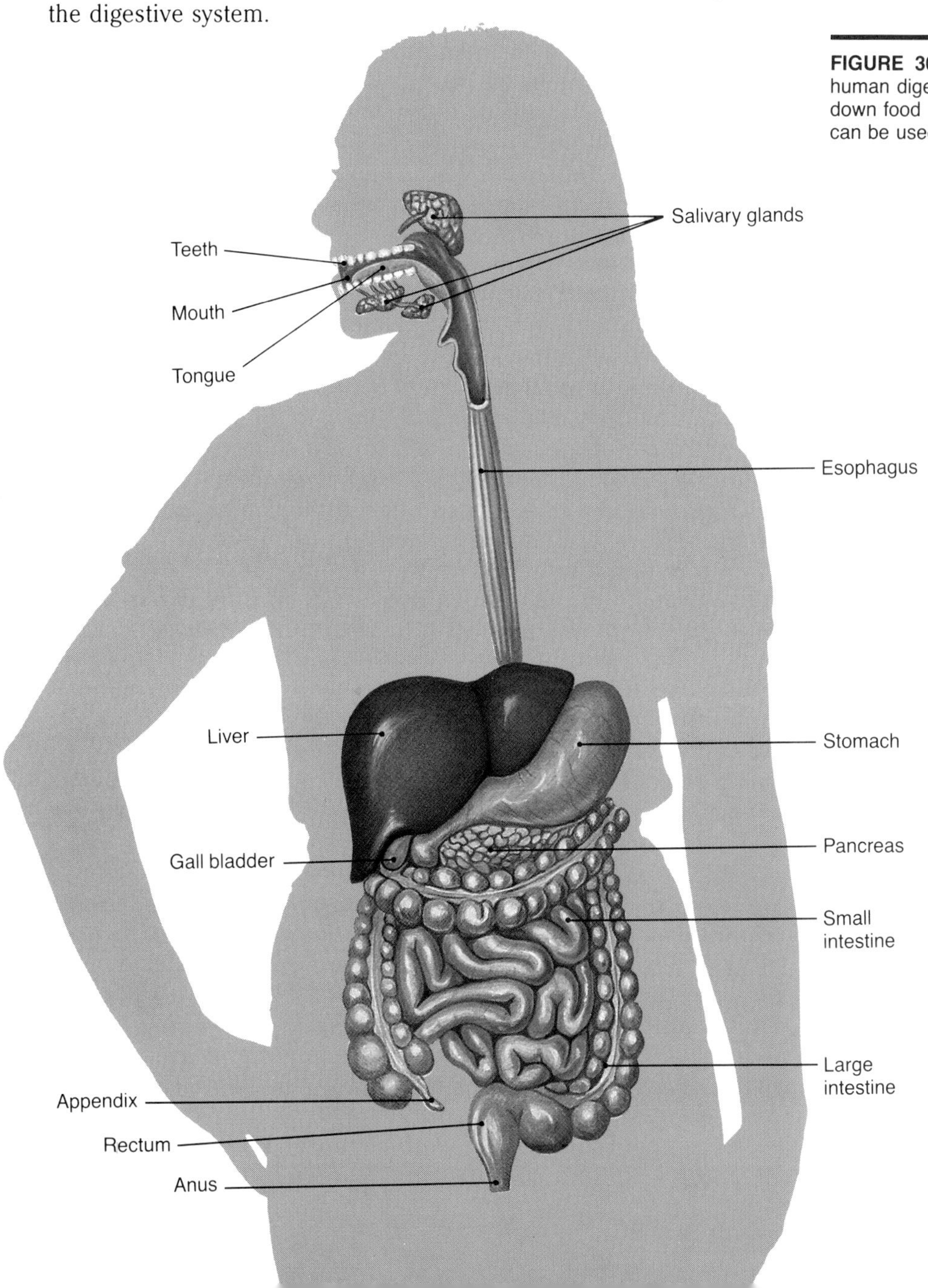

FIGURE 30·6 The structures of the human digestive system help to break down food into simple substances that can be used by the body.

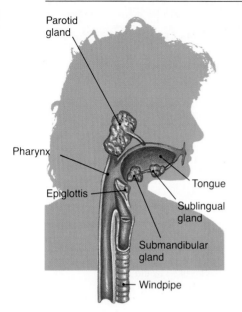

FIGURE 30·7 The salivary glands are located in the lining of the mouth. They secrete saliva which contains an enzyme that begins the chemical digestion of food.

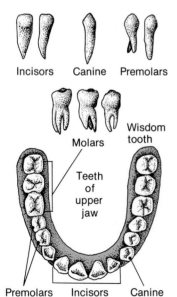

FIGURE 30·8 Different teeth perform various functions. The incisors and canines cut and tear food, respectively. The premolars and molars crush and grind the food.

Inside the mouth are several structures that aid in the early stages of digestion. The tongue is a muscular organ that has sensory structures called *taste buds* on its surface. These structures aid in the detection of chemicals that produce a sour, bitter, sweet, or salty taste. The tongue also moves food around in the mouth so that it can be chewed and swallowed.

Located in the lining of the mouth are three pairs of salivary glands. **Salivary** (SAL uh vehr ee) **glands** are glands that secrete saliva into the mouth. **Saliva** (suh LĪ vuh) is a digestive juice containing an enzyme that begins chemical digestion. Saliva also contains mucus, which lubricates the food. Such lubrication makes it easier for the food to pass through the digestive tract. Look at Figure 30·7 and find the three kinds of salivary glands. The **parotid** (puh RAHT ihd) **glands** are located under the ears. The **submandibular** (suhb man DIHB yuh luhr) **glands** are located along the lower jaw. The **sublingual** (suhb LIHNG-gwuhl) **glands** are located under the tongue.

Teeth are adapted for mechanical digestion of food. Mechanical digestion is the physical breaking apart of food, increasing the surface area of the food. An increased surface area speeds up the rate of digestion by enzymes.

Like all mammals, humans have two sets of teeth. The teeth in the first set are the *deciduous* (dih SIJ oo uhs) *teeth*. In humans this set contains 20 teeth, 10 in each jaw. The teeth in the second set are the *permanent teeth*, shown in Figure 30·8 for the upper jaw. The different types of teeth are specialized for particular purposes. **Incisors** (ihn SĪ-zuhrz) are teeth used for biting and cutting food. **Canines** (KAY nīnz) are teeth used for tearing food. **Premolars** and **molars**, with their flat surfaces, are well adapted for crushing and grinding food. The third molar on each side of both jaws is the last to appear, usually in the later teen years. These molars are called wisdom teeth.

Figure 30·9 shows the structure of a permanent tooth. The part of the tooth above the gum is called the *crown*. The crown has a thick outer layer of enamel. **Enamel** is a substance that contains calcium and protects the tooth against wear. The part of the tooth within the jaw is called the *root*. **Cementum** is a bonelike substance that forms the cover of the root of the tooth. The region between the crown and the root is called the neck of the tooth. The neck is in direct contact with the gum. Most of the inside of the tooth is made of dentin. **Dentin** is a bonelike substance that surrounds the pulp cavity. Within the pulp cavity is the pulp, which contains nerves and blood vessels.

As food is chewed, it mixes with the mucus in saliva and forms a soft mass called a *bolus* (BOH luhs). The tongue moves the bolus to the rear of the mouth. The region behind the mouth is the **pharynx** (FAR-ihngks). When the bolus touches the rear of the tongue, a reflex response produces swallowing. When food is swallowed, it passes through the pharynx and enters the esophagus (ee SAHF uh guhs). The **esophagus** is the muscular tube that connects the pharynx with the stomach. A flap of tissue called the **epiglottis** (ehp uh GLAHT ihs) closes off the windpipe during swallowing. This closing keeps food from entering the windpipe. Peristalsis (pehr uh STAL sihs) begins in the esophagus. **Peristalsis** is the series of rhythmic muscular contractions and relaxations that moves the food through the digestive system.

30·4 THE STOMACH

The **stomach** is a pouchlike enlargement of the digestive tube. It is a J-shaped structure found in the upper abdominal cavity below the diaphragm. When empty, the stomach is about the size of a large sausage. When filled with food, it is enlarged and may push against the diaphragm.

Figure 30·10 shows the structure of the stomach. Between the esophagus and the stomach is a circular muscle called the *cardiac sphincter* (SFIHNGK tuhr). **Sphincters** are muscles formed of circular fibers that close a tube when they contract. Between the stomach and the small intestine is another such muscle, the *pyloric* (pī LOR ihk) *sphincter*. When the two sphincters contract, the stomach becomes a closed sac in which food can be stored and partially digested.

Look at the three layers of muscles that form the wall of the stomach. Contractions of these muscles cause the churning movements that aid in mixing the food with digestive enzymes. In the lining of the stomach there are many gastric glands. **Gastric glands** are structures that produce a very acidic secretion called *gastric juice*. Gastric juice is a mixture of acids and enzymes that cause chemical digestion.

Because of the physical action of churning and the chemical action of the gastric juice, solid food is changed into a fluid in the stomach. This fluid mixture of food and gastric juice is called *chyme* (kīm). In the stomach, chyme is acidic. As peristaltic waves are produced in the stomach, the pyloric sphincter alternately relaxes and contracts. Small amounts of chyme are released into the small intestine each time the pyloric sphincter relaxes.

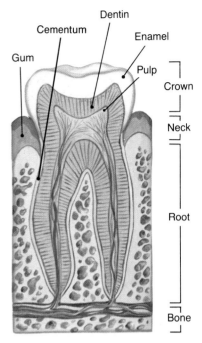

FIGURE 30·9 This drawing shows the structure of a permanent tooth. Notice that the tooth has three parts: the crown, the neck, and the root.

gastros (stomach)

FIGURE 30·10 The stomach is made up of three layers of muscle. The inner lining of the stomach has many folds that allow the stomach to expand as it fills with food.

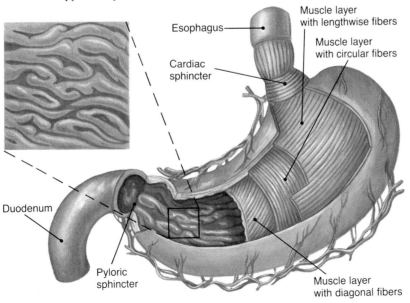

30·5 SMALL AND LARGE INTESTINES

The **small intestine** is the portion of the digestive tube in which most digestion and absorption occur. The small intestine is the longest part of the digestive tube. In an adult it is 6–7 m long and 2.5 cm wide. In Figure 30·11 note the highly coiled small intestine. The small intestine has three major sections. The first part of the small intestine, called the **duodenum** (doo uh DEE nuhm), is 25 cm long. The middle part, called the **jejunum** (jih JOO nuhm), is about 2.5 m long. The last part of the small intestine, called the **ileum** (IHL ee uhm), is over 4 m long.

Within the lining of the small intestine are *intestinal glands*. These glands produce a fluid that dilutes the chyme. This fluid also provides a watery medium for the chemical reactions of digestion. The membranes of the cells of the intestinal lining also contain proteins that function as enzymes for digestion. Digestive secretions from the gall bladder and the pancreas empty into the beginning of the duodenum and aid in the digestion of food. The muscle layers of the intestinal wall cause the churning action that mixes the chyme with the digestive juices. As digestion proceeds, peristalsis pushes the mixture along. All digestion is completed in the small intestine.

The **large intestine,** which is the lower end of the digestive tract, is about 1.5 m long and about 6 cm in diameter. You can see the large intestine in Figure 30·11. A sphincter muscle controls the passage of material from the small intestine into the large intestine. This sphincter acts like a valve. It prevents the movement of wastes back into the small intestine.

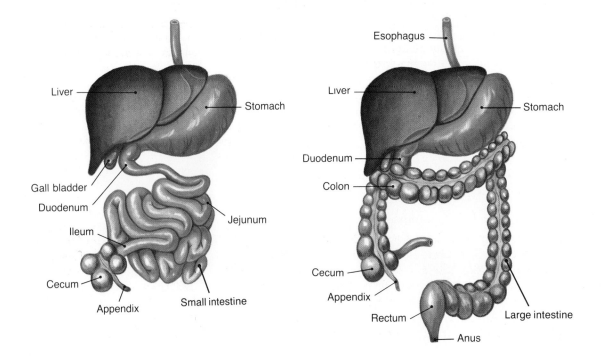

FIGURE 30·11 The small intestine, the organ where most digestion and absorption occur, has three major sections: the duodenum, the jejunum, and the ileum. The large intestine, which is the lower part of the digestive tract, is divided into three main regions: the cecum, the colon, and the rectum.

The large intestine is divided into three main regions: the cecum (SEE kuhm), the colon, and the rectum. The region below the place where the small and large intestine meet is the **cecum.** Connected to the cecum is a small tube called the **appendix.** The appendix has no digestive function in humans. Above the cecum is the **colon,** which is the longest part of the large intestine. The colon contains many bacteria that feed on the undigested waste. Some of these bacteria are helpful, since they produce certain vitamins and release them into the large intestine. The lining of the colon is smooth. It contains many glands that secrete mucus. This mucus helps to ease the movement of undigested materials along the colon. As the undigested bulk moves through the colon, water, some minerals, and some vitamins are absorbed into the blood.

The final part of the large intestine is the **rectum,** which opens to the outside of the body through the anus. Two sphincter muscles, one controlled involuntarily and the other controlled voluntarily, regulate the opening of the anus. Mass peristaltic movements force the wastes into the rectum and out of the anus.

30·6 ROLES OF LIVER AND PANCREAS

Two important glands that aid in digestion are the liver and the pancreas. Each of these glands produces digestive chemicals. These chemicals flow through ducts into the duodenal region of the small intestine.

The **liver** is the largest gland in the body. An adult's liver weighs about 1.4 kg. The liver has two main lobes. Attached to the right lobe is the gall bladder. Ducts, or tubelike vessels, connect the liver, gall bladder, and duodenum. The liver carries out many different chemical functions in the body. The liver functions with the digestive system in the following ways.

1. Digested nutrients, after being absorbed from the small intestine into the bloodstream, are carried to the liver. Glucose is stored in the liver as glycogen. As glucose is needed by the body cells, glycogen is changed to glucose and released into the blood.
2. After being absorbed from the small intestine, amino acids also are carried to the liver, where they are stored for a time. As specific amino acids are needed by body cells for protein synthesis, they are released from the liver into the bloodstream. Excess amino acids in the liver are broken down to obtain energy for metabolism, or they are converted to fats.
3. The liver stores the fat-soluble vitamins A, D, E, and K. It also stores some of the water-soluble vitamins, such as B_{12}. Carotene, the yellow pigment in vegetables, is converted to vitamin A by the liver.
4. The liver produces bile. **Bile** is a solution of water, bile salts, and bile pigments. Bile salts are produced from cholesterol, and they aid in the digestion of fats in the small intestine. Bile pigments are the products of the breakdown of old, damaged red blood cells. Bile pigments are excreted from the body through the digestive system.

Issues

Diet and Cancer

Can the foods you eat either cause or prevent cancer? According to some biomedical researchers, 30 to 70 percent of the cancer in the United States could be prevented if Americans would change their eating habits. Other researchers say there is no sure way of finding out if there is a relationship between diet and cancer in humans.

Some evidence seems to show that certain foods may prevent particular cancers. For example, a diet high in fiber may prevent colon cancer, and daily intake of vitamin A in leafy green and yellow vegetables may prevent lung cancer.

Other evidence seems to link certain foods with cancer. For example, excessive fat in the diet may cause breast cancer. Many studies by researchers in different places show similar results—that women who eat large amounts of fat and fried foods have a greater risk of developing breast cancer than women who eat smaller amounts of these foods.

On the basis of this type of evidence, the National Academy of Science has recommended that Americans reduce their intake of fat. This not only may reduce the incidence of cancer, but also may provide other health benefits.

FIGURE 30·12 The liver, pancreas, and gall bladder produce and store chemicals that are released into the small intestine. These chemicals help in the digestion of food.

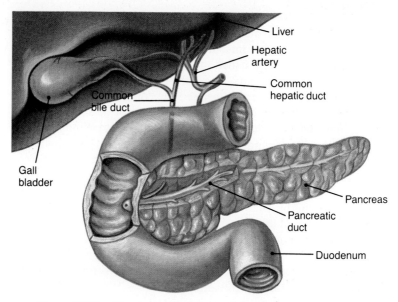

The **gall bladder** is a pear-shaped muscular organ that is connected to the liver. Bile, produced in the liver, passes into the gall bladder, where it is stored for a time. While in the gall bladder, much of the water in bile is reabsorbed, and the bile becomes concentrated. When food enters the duodenum, the gall bladder contracts, releasing bile into the duodenum. Look at Figure 30·12 and trace the flow of bile through ducts from the liver to the gall bladder to the duodenum. Also, the liver can release bile directly into the duodenum.

The **pancreas** is a narrow gland about 12 to 15 cm long. It is located next to the duodenum and beneath the stomach. Find the pancreas in Figure 30·12. You can see that there are ducts that connect the pancreas with the duodenum. The pancreas is made up of two kinds of tissues. One kind of tissue produces chemicals called *hormones,* which you will study in Chapter 34. The other kind of tissue produces *pancreatic juice,* which contains enzymes for the digestion of starch, proteins, and lipids. Pancreatic juice also contains sodium bicarbonate, which helps neutralize the acidic chyme that enters the duodenum. When food enters the duodenum, pancreatic juice is released through the pancreatic duct into the small intestine.

REVIEW QUESTIONS

6. Describe the role of each of the following in digestion: (a) teeth, (b) tongue, and (c) salivary glands.
7. Describe the functions of the liver.
8. What processes occur in the small intestine?
9. Explain the role of the cardiac and pyloric sphincters in digestion in the stomach.
10. Sometimes a person's gall bladder has to be removed surgically. How would this surgery affect the process of digestion?

Digestion and Absorption of Food

30·7 DIGESTION OF NUTRIENTS

Complex carbohydrates must be broken down in order to be absorbed from the digestive system into the bloodstream. Polysaccharides, such as starch, and disaccharides, such as sucrose, are examples of carbohydrates that must be digested (Section 3·12). The polysaccharide cellulose cannot be digested by humans since they lack the enzyme *cellulase.* Starch, which is a repeating chain of glucose, is digested to maltose by the enzyme *amylase.* Maltose, a disaccharide, is broken down to glucose, a monosaccharide. The enzyme that brings about maltose digestion is *maltase.* The enzyme that causes the breakdown of sucrose to glucose and fructose is *sucrase.* The enzyme *lactase* brings about the digestion of lactose to glucose and galactose.

Starch digestion begins in the mouth. As saliva mixes with food, salivary amylase starts the digestion of starch. When the food arrives in the stomach, the high acid content stops the action of the salivary amylase. Starch digestion begins again in the small intestine. Pancreatic juice, released into the small intestine, contains the enzyme *pancreatic amylase,* which completes the digestion of starch to maltose.

The digestion of disaccharides is completed in the small intestine by maltase, sucrase, and lactase. These enzymes are located in the membranes of the cells lining the intestine. Maltose, sucrose, and lactose bind to these enzymes. The resulting monosaccharides are moved by active transport into the cells lining the intestine.

The digestion of most lipids occurs in the small intestine. The end products of lipid digestion are fatty acids and glycerol (Section 3·13). The enzyme responsible for lipid digestion is *lipase,* which is found in pancreatic juice.

Bile salts break large fat droplets into smaller droplets. When bile salts bind to large fat droplets, the droplets break apart, a process called emulsification (ih muhl suh fuh KAY shuhn). **Emulsification** of fats makes fat digestion more efficient by increasing surface area. Bile salts also aid in the absorption of fatty acids and glycerol.

Protein digestion begins in the stomach. The hydrochloric acid and pepsinogen (pehp SIHN uh juhn) in gastric juice take part in the digestion of proteins. When *pepsinogen,* an inactive form of an enzyme, enters the stomach, the acid helps change it into *pepsin,* an active enzyme. Pepsin begins to break down proteins into polypeptides. Pepsin is most active when it is in an acidic environment. The stomach juices have a pH between 2 and 3.

In the small intestine, several enzymes complete the digestion of proteins. Three such enzymes secreted in the pancreatic juice are trypsin (TRIHP sihn), chymotrypsin (kī moh TRIHP sihn), and carboxypeptidase (kahr bahk see PEHP tih days). These three enzymes break down polypeptides to smaller polypeptides, dipeptides, and amino

---

SECTION OBJECTIVES

- Explain what occurs during the digestion of food.
- Describe how food and water are absorbed in the digestive system.

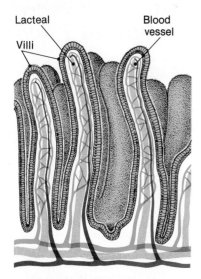

FIGURE 30·13 Most nutrients are absorbed through the fingerlike villi. Monosaccharides, amino acids, vitamins, and minerals enter the tiny blood vessels, while fatty acids and glycerol enter the lacteal.

acids. Digestion is completed by several enzymes present in the membranes of the cells lining the intestine. The amino acids produced as a result of protein digestion are absorbed through these cells.

30·8 ABSORPTION OF FOOD AND WATER

Most nutrients are absorbed through threadlike structures called **villi** (VIHL ī), which project from the lining of the small intestine. Lipid-soluble nutrients, such as fatty acids and glycerol, are thought to diffuse through the membrane of the cells of the villi. Other nutrients, such as monosaccharides, amino acids, vitamins, and minerals, are absorbed by active transport and facilitated diffusion. Specific carrier molecules in the membranes of the cells of the villi carry these nutrients into the villi.

Each villus contains microscopic blood vessels and a microscopic lymph vessel called a **lacteal** (LAK tee uhl). See Figure 30·13. In the villi, the cell membranes fold inward to form tiny *microvilli*. These microvilli produce an increased surface area, which adapts the villus for the absorption of nutrients.

Monosaccharides, amino acids, vitamins, and minerals enter the blood vessels, while fatty acids and glycerol enter the lacteal. The blood vessels from the villi unite to form the hepatic portal vein. This vein carries blood rich in nutrients directly to the liver. As nutrients are needed by the body cells, they are released from the liver through the hepatic vein. In this way nutrients enter the general circulation. Lacteals merge with larger lymphatic vessels to form a branching network. Lymph, carrying nutrients, eventually enters the blood.

Chyme entering the large intestine contains much water as well as material that cannot be digested. Some of the water in chyme has been ingested in food and beverages. Much of the water has come from the body and has been released into the digestive system in digestive juices and mucus. To maintain the body's fluid balance, some of the water must be reabsorbed. As chyme is moved through the large intestine, water is absorbed into the blood vessels in the lining of the large intestine. By the time the nondigestible material has reached the end of the colon, much of the water has been absorbed. The remaining mass is called *feces* (FEE ceez).

REVIEW QUESTIONS

11. For each of the following carbohydrates, name the digestive enzyme and the products of digestion: (a) starch, (b) maltose, (c) sucrose, and (d) lactose.
12. Describe the breakdown of lipids.
13. Where in the digestive system would enzymes for digesting proteins be found?
14. Describe the path of each of the following from inside the small intestine to the body cells: (a) glucose and (b) fatty acids.
15. Identify some advantages and disadvantages to a human who has a mutant gene for the production of cellulase.

Investigation

What Is the Effect of Pepsin on Protein?

Goals

After completing this activity, you will be able to

- Describe the effect of pepsin on the protein in albumen.
- Compare the action of pepsin in an acidic medium with its action in a basic medium.
- Predict the effect of an antacid on protein digestion.

Materials (for groups of 2)

albumen (egg white from half of a hard-boiled egg)
scalpel   water   test-tube rack
8 test tubes
glass-marking pencil
3 graduates
0.5% pepsin solution
0.2% hydrochloric acid solution
0.5% sodium bicarbonate solution
red litmus paper
blue litmus paper

Procedure

A. Place the test tubes in the test-tube rack. With the glass-marking pencil label the tubes 1 to 8.

B. With the scalpel chop the egg white into fine pieces. Place an equal amount of the chopped egg white into each test tube.

CAUTION: Use great care in handling the scalpel. Direct the edge of the blade away from your body. Work on a firm surface.

C. Treat the egg white in the test tubes by adding the substances indicated in the table below. Use one graduate for pepsin, one for hydrochloric acid, and one for sodium bicarbonate.

TEST TUBE	TREATMENT
1	10 mL water
2	10 mL pepsin
3	10 mL hydrochloric acid
4	10 mL sodium bicarbonate
5	5 mL hydrochloric acid and 5 mL sodium bicarbonate
6	10 mL pepsin and 5 mL hydrochloric acid
7	10 mL pepsin and 5 mL sodium bicarbonate
8	10 mL pepsin, 3 mL sodium bicarbonate, and 3 mL hydrochloric acid

D. Gently shake each test tube so that the liquids will mix well with the egg white.

E. On a separate sheet of paper, make a chart like the one shown below.

TEST TUBE	CONTENTS	pH	OBSERVATIONS
1			
2			
3			

F. Dip a separate strip of red litmus paper and blue litmus paper into each test tube. Find out whether the contents of each tube is acidic, basic, or neutral. Record your results in the chart.

G. Place the test-tube rack in a place where it can remain undisturbed for 3 days. Keep it at room temperature.

H. After 3 days observe the contents of each test tube. Record your observations in the chart.

Questions and Conclusions

1. In which test tube or tubes were there signs of protein breakdown?
2. In which test tube was the protein breakdown most complete?
3. Which substance was most effective in breaking down protein?
4. Which test tubes served as controls? Why are these controls needed?
5. Which solutions were acidic? Basic? Neutral?
6. Based on your results, does pepsin work better in an acidic or in a basic medium?
7. How does adding sodium bicarbonate to hydrochloric acid affect the pH of the solution?
8. Suppose an antacid such as sodium bicarbonate was taken before eating. How might it affect protein digestion?

Problems of Nutrition

SECTION OBJECTIVES
- Identify some dietary problems and their effects on health.
- List the causes and effects of some disorders of the digestive system.

30·9 OBESITY AND WEIGHT CONTROL

Obesity is a major health problem for over 20 percent of adult Americans. Obesity is a condition in which a person's body weight is above the normal weight by 10 percent or more. Overeating is a factor in causing obesity. A person may overeat to lower tension. Improper functioning of the appetite control center of the brain as well as a slow metabolic rate may be linked with obesity. An overweight person may have larger than normal fat storage cells. He or she may not be able to break down stored fat quickly. Obesity is a factor in a number of diseases. These diseases include certain types of diabetes, high blood pressure, heart and artery disease, and varicose veins.

The health hazards linked with obesity have been given wide publicity in recent years. This knowledge has sparked great interest in fad diets, most of which lack many of the nutrients the body needs for proper functioning. To maintain a healthful weight, a person should maintain the correct level of nutrients while reducing the level of Calories. Regular exercise is an important part of a weight control program. A doctor should always be consulted before a person begins a special diet or an exercise program.

For some people, dieting can become an obsession. Some people take diet pills that decrease the appetite and stimulate body metabolism. Such pills may produce harmful side effects, such as increased blood pressure, rapid heartbeat, restlessness, dizziness, and sleeplessness. An extreme obsession to lose weight is the condition called *anorexia nervosa* (an uh REHK see uh ner VOH suh). Although the exact cause of this condition is unknown, the basis of it is probably both psychological and physiological.

30·10 MALNUTRITION AND DIETARY DEFICIENCIES

Malnutrition (mal noo TRISH uhn) is the condition that occurs when the body cells do not get enough of the nutrients needed for growth and health. Some people suffer from malnutrition even though they eat a balanced diet. In such cases the persons may be unable to absorb certain nutrients from the intestine. For many people in the world, however, malnutrition is caused by not having enough food. For other people, malnutrition results from choosing foods that do not give a balanced diet.

Severe malnutrition results when a person does not have enough food. Proteins and carbohydrates are lacking in the diet. In such cases there is a breakdown of body tissues, loss of weight, lowered rate of metabolism, and lowered resistance to infection.

Choosing a diet of junk foods can also lead to malnutrition. Foods high in sugar but low in other nutrients do not make up a balanced diet. Such a diet can lead to a variety of nutritional problems.

FIGURE 30·14 If done regularly, jogging helps maintain good health by controlling weight.

Figure 30·2 lists some of the diseases that result when vitamins are lacking in the diet. For example, lack of vitamin C results in scurvy. Lack of sufficient minerals in the diet can also cause malnutrition. As shown in Figure 30·3, lack of iron can affect production of hemoglobin in the red blood cells.

30·11 DISORDERS OF THE DIGESTIVE ORGANS

Disorders may affect various digestive organs and can interfere with digestion of food. The stomach and duodenum can be affected by a condition called *peptic ulcer,* which causes mild to great pain. A peptic ulcer is a wearing away of the lining of a digestive organ. A peptic ulcer occurs when hydrochloric acid and pepsin break down an area of the lining of one of these organs. If the mucus that normally coats and protects the lining is lacking, a peptic ulcer may result. Treatment includes lowering the intake of alcohol, caffeine, and aspirin. All of these substances can irritate the stomach. Another approach is to use medicines that decrease stomach secretions.

A disorder of the digestive system that happens more often in children than in adults is *appendicitis* (uh pehn duh SĪ tuhs). This disorder is an inflammation of the lining of the appendix. In children the opening between the cecum and the appendix is quite large. Waste substances may enter the appendix and irritate the lining, causing abdominal pain and vomiting. Appendicitis may cause death if the appendix ruptures and empties its contents into the body cavity. As a person ages, the opening to the appendix becomes smaller. Thus there is less chance of developing this condition.

Another problem of the digestive system can take place in the gall bladder. The gall bladder stores bile, which is produced in the liver. Bile contains cholesterol in solution. If the cholesterol in bile solidifies, *gallstones* form. A gallstone is a stonelike mass formed in the gall bladder. Gallstones may block the duct leading to the small intestine, thus blocking the flow of bile out of the liver. If bile pigments, which have a yellow color, enter the bloodstream, they produce a yellow color in the skin and in the whites of the eyes. This condition is called *jaundice* (JAWN dihs).

Gallstones may be treated by a diet that restricts the foods that stimulate the gall bladder. For example, foods high in fats are restricted. Treatment may also involve surgery.

REVIEW QUESTIONS

16. List two causes of obesity.
17. What is anorexia nervosa?
18. What is malnutrition?
19. Describe the cause and treatment of a peptic ulcer.
20. Suppose a person has a section of small intestine surgically removed. The person loses weight. Explain why weight loss might follow such surgery.

Chapter Review

SUMMARY

30·1 The five nutrients required for good health are proteins, carbohydrates, fats, vitamins, and minerals.

30·2 Healthy people can fill their daily nutritional needs by consuming the recommended servings from the four basic food groups.

30·3 The mouth is the organ adapted for mechanical digestion. Some chemical digestion also occurs in the mouth.

30·4 The stomach is a sac in which gastric glands secrete gastric juice. Gastric juice contains pepsin and hydrochloric acid.

30·5 The small intestine consists of three sections: the duodenum, the jejunum, and the ileum. The large intestine is composed of three parts: the cecum (containing the appendix), the colon, and the rectum.

30·6 The pancreas and the liver are digestive glands that produce secretions which aid in digestion. The pancreas secretes pancreatic juice, and the liver secretes bile.

30·7 Carbohydrate digestion takes place in the mouth and in the small intestine. Lipid digestion takes place in the small intestine. The enzyme lipase and the emulsifier bile are responsible for fat digestion and absorption. Protein digestion takes place in the stomach and in the small intestine.

30·8 Most nutrient absorption occurs through the villi of the small intestine. Absorption of water occurs in the large intestine.

30·9 Obesity is a dietary problem that can lead to poor health.

30·10 When sufficient quantities of nutrients are lacking in the diet, malnutrition results.

30·11 Disorders of the digestive organs cause pain and discomfort. Ulcers, appendicitis, and gallstones are examples of such disorders.

BIOLOGICAL TERMS

appendix
bile
canines
cecum
cementum
colon
dentin
duodenum
emulsification
enamel
epiglottis
esophagus
gall bladder
gastric glands
ileum
incisors
jejunum
lacteal
large intestine
liver
minerals
molars
nutrient
nutrition
pancreas
parotid glands
peristalsis
pharynx
premolars
rectum
saliva
salivary glands
small intestine
sphincters
stomach
sublingual glands
submandibular glands
villi
vitamins

USING BIOLOGICAL TERMS

1. List all terms that name parts of the large intestine. Define the terms.
2. Distinguish between premolars, molars, incisors, and canines.
3. What is the function of saliva?
4. Distinguish between enamel, cementum, and dentin.
5. How are the duodenum, jejunum, and ileum related?
6. Distinguish between the epiglottis and the esophagus.
7. What is the connection between bile, the liver, and the gall bladder?
8. Distinguish between a villus and a lacteal.
9. What is peristalsis?
10. Salivary glands are to mouth as _____ are to stomach

UNDERSTANDING CONCEPTS

1. List the five types of nutrients and describe one function of each. (30·1)
2. What is the function of water in the body? (30·1)
3. List three foods in each of the four basic food groups and identify two nutrients provided by each group. (30·2)
4. How do the salivary glands aid digestion? (30·3)
5. In what ways do the teeth aid in the digestive process? (30·3)
6. Describe the process of peristalsis and explain its function. (30·3)
7. What are the functions of the muscles in the wall of the stomach? (30·4)
8. In what ways is the small intestine adapted for the process of digestion? (30·5)
9. Describe the structure of the large intestine. (30·5)
10. List four functions of the liver. (30·6)
11. What is the function of the gall bladder? (30·6)
12. Compare starch and cellulose and what happens to them in the digestive system. (30·7)
13. How is bile involved in the digestion of fats? (30·7)
14. Explain how proteins are digested, including digestion in the stomach and in the small intestine. (30·7)
15. In what ways is the small intestine adapted for the process of absorption? (30·8)
16. Compare the absorption of glucose and fatty acids from the small intestine. (30·8)
17. Explain the function of the large intestine in the absorption of water. (30·8)
18. Why is obesity considered a health problem? (30·9)
19. What are some of the side effects associated with the use of diet pills? (30·9)
20. What factors can contribute to malnutrition? (30·10)
21. Using Figure 30·2 identify two diseases that result from vitamin deficiencies. (30·10)
22. How do ulcers develop? (30·11)
23. Why is appendicitis more common in young people than in adults? (30·11)

APPLYING CONCEPTS

1. What special problems related to vitamins and proteins may result from a vegetarian diet?
2. Describe the processes of diffusion and active transport in the absorption of nutrients.
3. How does each of the following aid in the treatment of ulcers: (a) reducing stressful situations and (b) reducing the amount of highly acidic food?
4. In what ways does a diet high in fiber aid the digestive process?
5. What problems might arise if a person had an abnormally short small intestine?

EXTENDING CONCEPTS

1. Describe the best ways to prepare foods so as to maintain the maximum amounts of vitamins in the foods.
2. Using references, determine the role of the intrinsic factor in preventing pernicious anemia.
3. Prepare a report on anorexia nervosa, including information about who are most prone to develop this condition, what its effects are on the body, and how it is treated.
4. Using references, prepare a report on the roles of the nervous system and endocrine system in controlling the rate of digestion.
5. Using medical reference materials and interviews with physicians and hospital personnel, prepare a study correlating age and incidence of appendicitis.

READINGS

Carpenter, Kenneth J. *The History of Scurvy and Vitamin C.* Cambridge University Press, 1986.

Langone, John. "Cholesterol: The Villain Revealed" and Part I, "Heart Attack and Cholesterol." *Discover,* March 1984, p. 19.

Moog, Florence. "The Lining of the Small Intestine." *Scientific American,* November 1981, p. 154.

31. Circulation

If you were to trace the path of blood through the human body, you would find that blood always moves through a system of tubes that are called blood vessels. Arteries, like the one below, and veins are the main blood vessels. The tiny red circular objects inside the artery are red blood cells. By reading this chapter, you will discover more about these blood vessels and the other parts of the circulatory system.

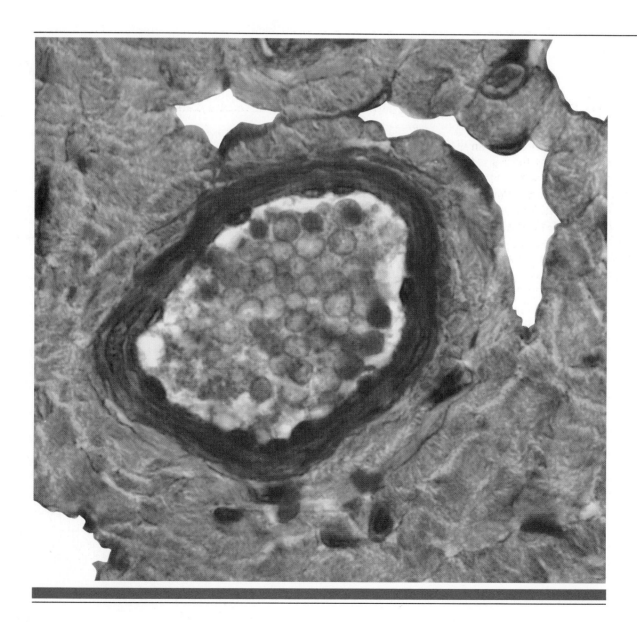

CHAPTER OBJECTIVES

After completing this chapter, you will be able to
- **List** the functions of blood and describe the composition of blood.
- **Explain** the role of blood types in transfusion.
- **Trace** the path of blood through the body.
- **Describe** the structure and function of the heart and blood vessels.
- **Explain** how circulation is controlled.

The Functions of Blood

31•1 FUNCTION AND COMPOSITION OF BLOOD

Blood, a mixture of cells and fluid, has many functions in maintaining homeostasis in an organism. Some of the blood's functions are performed by the cells. Other functions are carried out by the fluid portion of blood.

Blood travels within tubes, or blood vessels. Blood is pumped through the blood vessels by the heart. Much of the blood's work is the carrying of materials to and from parts of the body. The blood, heart, and blood vessels together make up the **circulatory system.**

What are the main functions of blood? Blood carries oxygen to the cells of the body. It carries waste products to the lungs and kidneys, where some of the waste products are removed. It moves through the digestive system, picking up dissolved nutrients and carrying them to other parts of the body. Blood also transports materials from storage areas to the areas where they are used. For example, calcium, stored in the bones, is carried to the muscles and other body parts for use.

Blood also transfers body heat. Heat is produced in the muscles. This heat is carried to the cooler surface of the body by the blood. Blood is the medium through which hormones, or chemicals with special roles, are passed from organ to organ. Finally, blood is involved in the body's defense against infectious disease (Section 15•10).

Figure 31•1 shows the proportion of solid parts to the liquid part of the blood. More than half of the blood is a fluid known as plasma. **Plasma,** the fluid part of the blood, is mostly water but also contains many dissolved substances, such as proteins and enzymes, simple food materials, waste products, and other chemicals. Some of the substances dissolved in plasma act as *buffers.* Buffers are chemicals that prevent changes in pH. The constant pH of the blood, around 7.4, helps to maintain a stable environment for the body's cells.

SECTION OBJECTIVES

- Describe the composition of blood and explain the functions of blood cells.
- Classify the four human blood types.

FIGURE 31•1 Whole blood is made up of a solid portion and a liquid portion. The solid portion is composed of cells and the liquid portion, called plasma, is made up mostly of water.

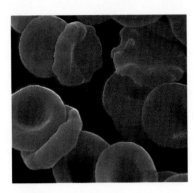

FIGURE 31·2 Red blood cells carry oxygen throughout the body.

hemato (blood)

Somewhat less than half of the blood consists of cells and cell fragments. **Red blood cells** transport oxygen and some carbon dioxide. **White blood cells** help to defend the body against infection. Also found in the blood are cell fragments called **platelets,** which are involved in blood clotting.

31·2 RED BLOOD CELLS

Red blood cells, also called *erythrocytes* (ih RIHTH roh sīts), are the most numerous cells in the blood. Each liter of blood contains 4.5 to 6 trillion red blood cells. Oxygen is supplied to cells all the time by red blood cells. Figure 31·2 shows red blood cells. Mature red blood cells are very small and have a simpler structure than most cells. Red blood cells are approximately 7.3 μm in diameter and have no nucleus or other organelles. The main component of red blood cells is the red pigment known as **hemoglobin** (HEE muh gloh buhn). Hemoglobin in the red blood cells binds and carries oxygen.

Red blood cells only live for a short time, usually about 120 days. Red blood cells that have died are replaced by new ones. Red blood cells are produced in the *bone marrow,* the soft tissue contained in cavities of certain bones.

Red blood cells passing through blood vessels in the lungs are exposed to oxygen that has been inhaled. Only the thin lining of the lungs separates the blood vessels from air. Oxygen diffuses into the blood. Hemoglobin in the red blood cells binds some of this oxygen. The hemoglobin becomes bright red and is known as *oxyhemoglobin* when oxygen is bound to it. As the oxygenated blood continues its circulation through the blood vessels, it reaches tissues that have used up their supply of oxygen. At such places the oxyhemoglobin gives up its oxygen. The oxygen leaves the red blood cells and diffuses into the tissues, where it will be used in cellular respiration. The hemoglobin that has given up its oxygen loses its bright red color and becomes a dull purplish-red. Purplish-red is the color of deoxygenated blood. In illustrations in this book, the deoxygenated blood is colored blue so that you will be able to tell it easily from oxygenated blood.

Carbon dioxide is produced as a waste product of cellular respiration. This gas is absorbed by the blood. Some of the carbon dioxide binds to the hemoglobin that has given up oxygen to the tissues. However, most of the carbon dioxide is carried away, dissolved in the plasma of deoxygenated blood. Much of this carbon dioxide is in the form of an ion. When the blood passes through the lungs, the carbon dioxide diffuses out of the blood and into the air in the lungs.

Red blood cells travel around the body an average of 15 000 times in their four-month life span. As they age, they begin to break apart. The *spleen,* an organ found in the abdomen near the stomach, removes old red blood cells from circulation. The spleen also may store reserve supplies of healthy red blood cells. During an emergency situation, such as severe blood loss, these reserve red blood cells may be used.

31·3 WHITE BLOOD CELLS

White blood cells are larger and less numerous than red blood cells. White blood cells vary from 10 to 20 μm in diameter. The blood contains only about one white cell for every 700 red cells. White blood cells, also called *leukocytes* (LOO kuh sīts), contain nuclei and other types of organelles. White blood cells may circulate in the blood for weeks before leaving the blood and entering other tissues. Some white blood cells may live for many years in tissues.

There are three basic types of white blood cells: *granulocytes* (GRAN yuh luh sīts), *monocytes,* and *lymphocytes* (LIHM fuh sīts). A granulocyte is a type of white blood cell that is named for the many granules that appear in its cytoplasm. This type of white blood cell may be identified by the irregular shape of its nucleus. Granulocytes are the most abundant type of white blood cell. They are active phagocytes, or cells that are attracted to and that destroy bacteria and other foreign matter by phagocytosis.

The monocyte, the second type of white blood cell, may be identified by the large nucleus and few granules in its cytoplasm. Monocytes are also phagocytes. They are slower to respond to the presence of bacteria than are granulocytes. Monocytes, however, can destroy many times more bacteria than can granulocytes. Monocytes can leave the blood vessels and live in other tissues. A monocyte found outside the blood may change into a type of cell called a *macrophage.*

The third type of white blood cell is the lymphocyte. A lymphocyte may be identified by its relatively small amount of clear cytoplasm. Lymphocytes are involved in the immune response (Section 15·11).

When bacteria or other foreign matter enter the body, white blood cells are attracted to the site. Granulocytes and macrophages take in the foreign matter, and lymphocytes begin to make antibodies. After a time, *pus* may form at the place of infection. Pus is a mixture of white cells, dead bacteria, and the remains of dead cells.

Granulocytes and monocytes are produced in the bone marrow. Lymphocytes, however, are made in several places in the body. Though white blood cells travel through the circulatory system, they also function in other tissues. Many white blood cells live under the skin and around the digestive system and lungs.

FIGURE 31·3 Unlike red blood cells, white blood cells have nuclei and organelles. There are three basic types of white blood cells: granulocytes, monocytes, and lymphocytes.

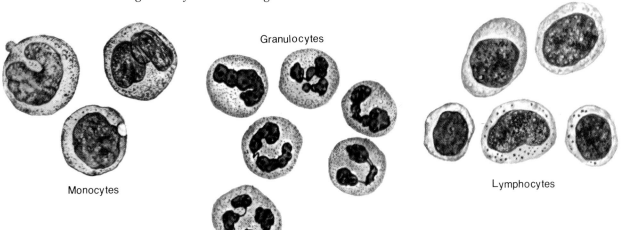

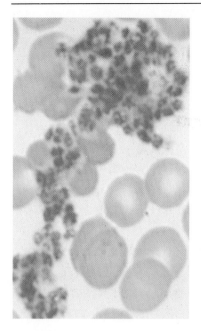

FIGURE 31·4 The tiny cell fragments that are found among the red blood cells in this photograph are the platelets. Platelets play an important role in blood clotting.

FIGURE 31·5 When a blood vessel is damaged, platelets form a plug over the opening of the blood vessel. At the same time, threads of fibrin begin to form, trapping the red blood cells and stopping the bleeding.

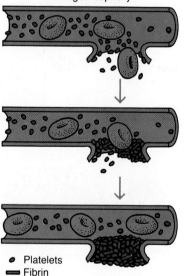

31·4 PLATELETS

When the skin is cut, blood flows for a while and then it stops. What causes the blood flow to stop? The opening is sealed by a blood clot. A **blood clot** is a tangled mass of protein fibers with trapped platelets and red blood cells. The formation of a blood clot involves *coagulation* (koh ag yuh LAY shuhn), a process in which part of the blood becomes a semisolid.

Many substances in the blood help to form a clot. Among the necessary things are platelets, also called *thrombocytes* (THRAHM buh-sīts). Figure 31·4 shows some platelets among red blood cells. Platelets are cell fragments released into the blood by cells in the bone marrow. In the center of each platelet are some granules. Several substances, including ADP, are stored in the granules.

Fibrin, a threadlike protein, has an essential role in coagulation. The clot that seals a broken blood vessel is formed of platelets, blood cells, and fibrin. Where does fibrin come from when a clot forms? Enzymes released from the walls of broken blood vessels control the reactions that produce fibrin. These reactions are as follows.

```
                THROMBOPLASTIN (enzyme from damaged cells)
                              │
                              ▼
PROTHROMBIN ─────────────▶ THROMBIN (activated enzyme)
(inactive enzyme
in plasma)                    │
                              ▼
                FIBRINOGEN ─────────▶ FIBRIN
                (plasma protein)
```

When a blood vessel is damaged, nearby platelets are stimulated to release their stored ADP. Chemicals released by broken cells probably stimulate this response. The ADP causes platelets to become sticky. They stick to each other and to the broken edges of the blood vessel. Figure 31·5 illustrates the formation of a platelet plug. The flow of blood brings platelets to the break. The platelets continue to stick to each other, causing the break in the blood vessel to be filled. At the same time, fibrin begins to form. The tough threads of fibrin hold the clot together. Red blood cells are caught in the clot, trapped in the mesh of fibrin. Figure 31·5 shows cells caught in fibrin threads. The break is sealed by a plug formed of fibrin, platelets, and trapped red blood cells. As time passes, the fibrin threads shrink, pulling the edges of the break together.

31·5 BLOOD TYPES AND TRANSFUSION

Just after the turn of this century, scientists discovered that there are four major blood types. These blood types were named A, B, AB, and O. As a group they are known as the **ABO blood types.** The name of the blood type is based on the name of the *antigen* on the surface of the red blood cell. Recall that an antigen is a substance that causes

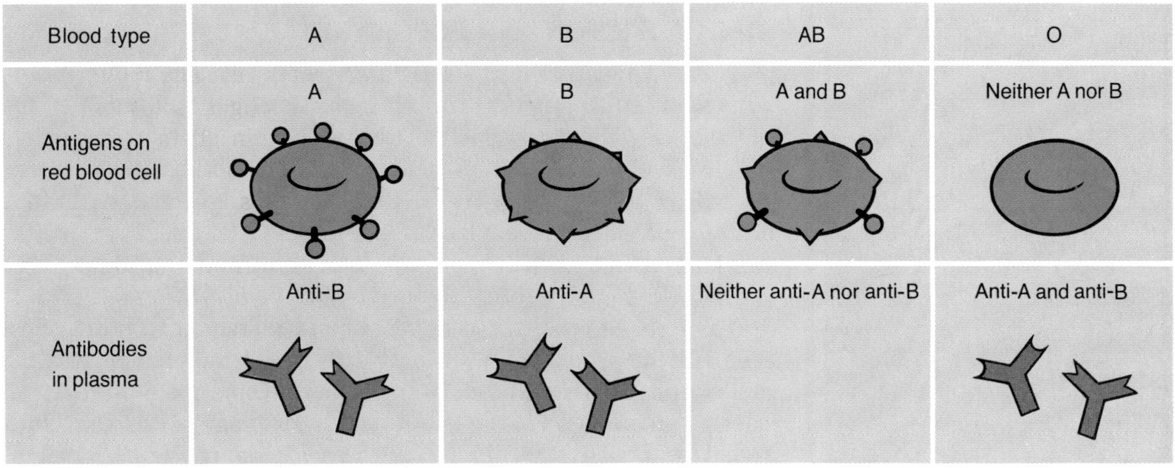

Blood type	A	B	AB	O
Antigens on red blood cell	A	B	A and B	Neither A nor B
Antibodies in plasma	Anti-B	Anti-A	Neither anti-A nor anti-B	Anti-A and anti-B

FIGURE 31·6 Each blood type is characterized by antigens on the red blood cells and antibodies in the plasma.

antibodies to be made. The antigens on red blood cells result in differences in surface features of the red blood cells. The top row of diagrams in Figure 31·6 is a representation of differences in shape.

There are two major antigens in the ABO blood type system. Blood containing red blood cells with antigen A on their surfaces is type A blood. Blood containing red blood cells with antigen B on their surfaces is type B blood. If neither antigen A nor antigen B is on the red blood cells' surfaces, the blood is type O blood. A small number of people have both antigen A and antigen B on the surfaces of their red blood cells. Blood of this type is called type AB blood.

Most people have antibodies in their plasma against the blood type antigens they *do not* have. The bottom row of diagrams in Figure 31·6 represents the antibodies. For example, persons with type A blood *do not* have antigen B. They do have an antibody against the B antigen. This antibody is called anti-B. Persons with type B blood do not have antigen A. They do produce an antibody called anti-A, which acts against antigen A. Since persons with type AB blood carry both antigens, they form neither antibody A nor antibody B. In persons with type O blood, since neither antigen is present, both antibody A and antibody B are produced.

The interaction of antigens and antibodies causes the red blood cells to clump together, or agglutinate. Testing for clumping is a method for finding a person's blood type. Figure 31·7 shows the reactions when antigens are mixed with antibodies. Antibody A will clump the cells of both type A and type AB blood. Antibody B will clump the cells of type B and type AB blood. From the figure, determine the reaction when antibody A and antibody B are mixed with type O blood.

When a blood transfusion is needed, the blood type of the donor must be matched with the blood type of the recipient. For example, a person with type A blood can receive type A blood. However, if a person with type A blood is given type B blood, the antibodies in the plasma of that person will clump the type B cells. What types of blood can a person with type O blood receive?

FIGURE 31·7 This diagram shows the reactions that occur when samples of the different blood types are mixed with antibody A (anti-A) and antibody B (anti-B). Clumping of the red blood cells is indicated by tiny dots on the slides.

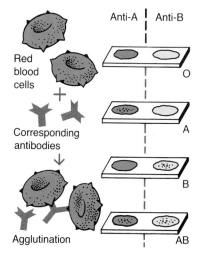

31·6 Rh FACTOR AND PREGNANCY

The A and B antigens are not the only antigens found on the surface of red blood cells. The **Rh factor** is an antigen found on the blood cells of about 85 percent of the American population. The letters *Rh* are used because this antigen was first found in rhesus monkeys. Persons with this antigen are said to be Rh-positive; those without the antigen are Rh-negative. Blood is tested and marked for both its ABO type and its Rh type. Thus a person's blood could be B+, AB−, O−, or some other combination of the two types.

Rh-related problems can occur in a pregnancy of an Rh-negative mother if the fetus is Rh-positive. Usually there is no problem with the first Rh-positive baby born to an Rh-negative mother. However, in later pregnancies involving an Rh-positive baby, the developing baby is in danger of having a disease called *erythroblastosis fetalis* (ih rihth-roh blas TOH sihs   fee TA lihs). Why does this disease occur?

Antibodies against the Rh factor do not occur in the blood of Rh-negative individuals. If Rh-positive blood cells enter the bloodstream of an Rh-negative person, antibodies will be produced by the Rh-negative person. During childbirth there may be bleeding at the point at which the baby is attached to the mother. When an Rh-positive baby is born to an Rh-negative mother, some of the baby's red blood cells may get into the mother's circulatory system during such bleeding. The mother's body will make antibodies against the Rh-positive cells and destroy them. If the mother becomes pregnant again with another Rh-positive baby, the mother's Rh antibodies may cross into the baby's bloodstream. The baby's red cells will be destroyed by these antibodies, and death may result.

Since the 1970s it has been possible to prevent this Rh disease. Artificially prepared antibodies to the Rh factor are used for this purpose. An Rh-negative mother who has been pregnant with an Rh-positive baby is given an injection of the artificial antibodies. These antibodies destroy red blood cells from the baby before the mother's immune system can make its own antibodies. The injected antibodies disappear after a while. Thus neither the artificial antibodies nor the mother's own antibodies are present in a second pregnancy.

REVIEW QUESTIONS

1. Describe the composition of blood plasma. What functions does plasma have?
2. Which blood cells carry oxygen to the tissues? What substance in those cells carries the oxygen?
3. Name the three types of white blood cells. Which types of white blood cells are phagocytic?
4. Describe the process by which platelets act to plug a break in a blood vessel.
5. Explain why an Rh− baby carried by an Rh+ mother is not subject to Rh disease.

Blood and Lymph Flow

31·7 THE HEART

The **heart** is the pump that keeps the blood in motion through the blood vessels. The heart begins beating long before birth and may beat more than 3 billion times in a person's lifetime. It is about as big as a closed fist. The heart lies between the lungs, behind the breastbone. It is enclosed in a membrane called the *pericardium* (pehr uh KAHR dee-uhm). This membrane makes a fluid that lubricates the outer surface of the heart.

Heart muscle is different from skeletal muscle in a number of ways. Skeletal muscle only contracts when it receives a signal from the nervous system. Heart muscle can contract without stimulation from the nervous system. The fibers of skeletal muscle are straight and contract in only one direction. The fibers of heart muscle branch and interlock with each other. Heart muscle thus exerts force all around the chambers it encloses. Skeletal muscle can function for short periods of time without oxygen. Heart muscle, however, must have a constant supply of oxygen.

The heart has four chambers. The two upper chambers of the heart are the **atria** (AY tree uh) (sing., *atrium*). The two lower chambers are the **ventricles.** Locate the atria and ventricles in Figure 31·8. You can see that the muscular walls of the ventricles are much thicker than those of the atria. The ventricles do more pumping work than the atria.

SECTION OBJECTIVES

- Relate structures of the heart to their functions.
- Describe the pattern of blood flow to the various organs of the body.

atria (rooms)

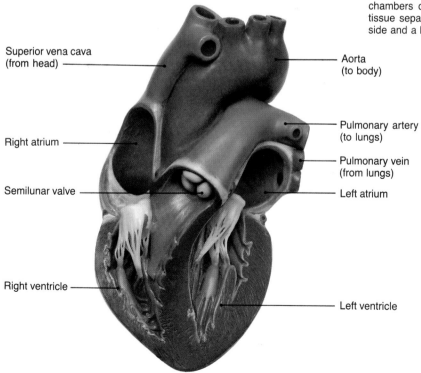

FIGURE 31·8 The heart has two upper chambers called atria and two lower chambers called ventricles. A wall of tissue separates the heart into a right side and a left side.

631

632 CIRCULATION

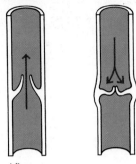

Blood flow in correct direction opens valve.

Closed valve

FIGURE 31·9 The heart and certain blood vessels contain valves. A valve is made up of soft flaps of tissue that control the direction of the blood flow. Notice how the valve keeps the blood from flowing backward.

FIGURE 31·10 Arteries are the strongest blood vessels in the body. The wall of an artery has three layers: a smooth inner lining, a thick middle layer of elastic tissue and muscle fibers, and an outer layer of connective tissue.

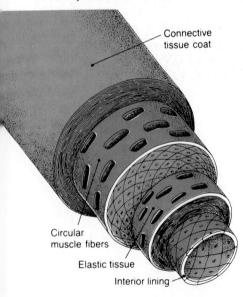

Connective tissue coat

Circular muscle fibers

Elastic tissue

Interior lining

It is sometimes useful to think of the heart as two pumps working side by side. The right atrium and right ventricle pump blood to the lungs. The left atrium and left ventricle pump blood to all other parts of the body. There is no direct connection between the right and left sides of the heart.

There are four valves in the heart. A *valve* is a structure made of soft flaps of tissue that controls the direction of blood flow. Figure 31·9 shows how the valves control the flow of blood. When blood flows in the proper direction, the flaps are pushed open. If blood begins to flow in the wrong direction, the flaps push in against each other and close off the opening. Thus blood can flow in only one direction through valves in the heart and in blood vessels.

The valves between the atria and ventricles are called **atrioventricular** (ay tree oh vehn TRIHK yuh luhr) **valves.** The valve connecting the right atrium and the right ventricle is called the **tricuspid valve.** The valve connecting the left atrium and the left ventricle is called the **bicuspid valve.** The two valves between the ventricles and the major blood vessels leaving the heart are called the **semilunar valves.**

The action of the heart can best be understood by tracing the flow of blood through the heart. Refer to Figure 31·8 to help you to understand the path of the blood through the chambers of the heart. Deoxygenated blood that has passed through the body returns to the heart at the right atrium. From the right atrium, blood is pumped through the tricuspid valve to the right ventricle. The right ventricle pumps the deoxygenated blood out through a semilunar valve to the lungs. The blood absorbs oxygen and gives up carbon dioxide as it passes through the lungs. The oxygenated blood returns from the lungs to the left atrium.

From the left atrium, blood passes through the bicuspid valve and into the left ventricle. The left ventricle is the largest and most muscular chamber of the heart. It pumps the oxygenated blood through another semilunar valve and out to the organs of the body. The blood returns to the right atrium.

31·8 BLOOD VESSELS

Arteries are large blood vessels that carry blood away from the heart and to other organs. Arteries have thick flexible walls made up of muscle cells, connective tissue, and elastic fibers. You can see the structure of an artery in Figure 31·10. Blood entering the arteries from the ventricles is under pressure. The artery walls stretch from the pressure. Each time the heart pumps, the arteries stretch because of the increased volume of blood. This stretching can be felt at points where arteries are close to the surface of the body. Place your index and middle fingers on the wrist of the opposite hand. Place the two fingertips at the part of the wrist nearest the base of the thumb. The steady beat you feel is the *pulse.* The pulse is the beat of the heart as felt through the walls of the arteries.

Arteries divide into many branches as they spread from the heart to parts of the body. The branching of arteries is similar to the branching of a tree's limbs. The smaller branches of arteries are called *arterioles*. These vessels have much thinner walls and are less flexible than large arteries.

Arterioles also branch into smaller blood vessels and finally lead into the smallest blood vessels. A **capillary** is the smallest type of blood vessel; it connects arterioles and small veins. Capillaries have very thin walls made of flattened cells. Figure 31·11 shows the structure of a capillary. The walls of a capillary are only one cell thick. Many substances can pass between the blood in the capillary and the surrounding tissues through the thin walls. Oxygen and nutrients pass out of the capillary and nourish the surrounding tissues. Carbon dioxide and other waste products enter the capillary and are carried away by the blood.

Because capillary walls are so thin, some of the liquid from the plasma leaks out. This leakage is a normal process. The liquid becomes part of the tissue fluid. *Tissue fluid* is the liquid that is found in the spaces between cells. Cells in all parts of the body are bathed by tissue fluid. It carries food and oxygen from the capillaries to the surface of cells in the surrounding tissues.

As blood flows through the capillaries, it gives up food and oxygen to the tissues. It picks up and carries away wastes. Several capillaries come together to form a *venule*. In turn, venules from several sets of capillaries come together to form a vein. A **vein** is a large blood vessel that carries blood toward the heart. Examine Figure 31·12 to see how blood vessels are connected. Locate the artery, arterioles, capillaries, venules, and vein.

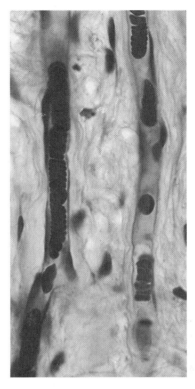

FIGURE 31·11 Because most capillaries are finer than human hairs, red blood cells are forced to pass through the capillaries in single file.

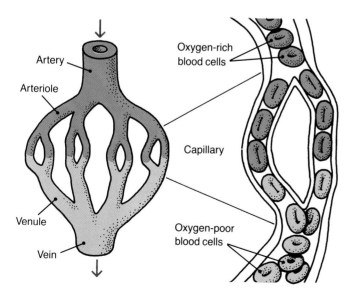

FIGURE 31·12 The oxygen-rich blood that leaves the heart is carried by arteries and then enters a dense network of capillaries. In the capillaries, oxygen is given up by the red blood cells. On its return trip to the heart, the oxygen-poor blood enters venules and then veins.

634 CIRCULATION

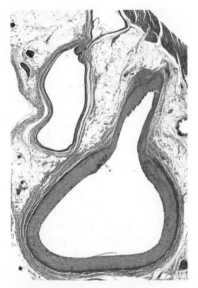

FIGURE 31·13 Although arteries and veins are both blood vessels, they each have a unique structure that enables them to perform their different functions. The photograph shows a cross section through a vein (top) and an artery (bottom).

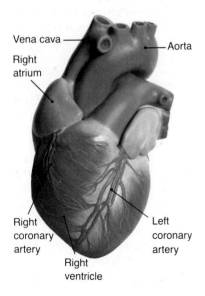

FIGURE 31·14 This model of the heart shows the structures involved in coronary circulation. Coronary circulation is the flow of blood through the blood vessels that nourish the heart.

Veins are different from arteries in several ways. Figure 31·13 shows an artery and a vein. Compare these two blood vessels. Veins have thinner walls than do arteries. The wall of a vein is mostly connective tissue. A vein has very few muscle cells or elastic fibers, which are abundant in the wall of an artery. What accounts for these differences? Remember, arteries carry blood from the heart to the tissues. Veins carry blood from the tissues back to the heart. The blood in an artery is under high pressure because it is being pumped by the heart. Arteries are strong and elastic, enabling them to withstand the pressure. By the time the blood has passed through the capillaries and reached the veins, it has lost most of its pressure.

In some veins the pressure is so low that there is a possibility the blood might flow in the wrong direction. Veins have valves in them that ensure that blood always flows toward the heart. The valves found in veins operate in the same manner as the valves in the heart.

31·9 CIRCULATION TO BODY ORGANS

Blood vessels are connected to the heart to form one continuous system of circulation. Nonetheless, it is useful to consider the circulatory system as having several parts. The **pulmonary circulation** is the flow of blood through those blood vessels that carry blood to, through, and from the lungs. The **pulmonary artery** arises from the right ventricle of the heart and gives off a branch to each of the lungs. These branch arteries lead to arterioles and then to capillaries within the lungs. In these capillaries the blood picks up oxygen and gives up carbon dioxide. Blood is returned to the heart by the **pulmonary veins.**

The **systemic circulation** is the flow of blood through those blood vessels that go to all organs and tissues *except* the lungs. The left ventricle pumps blood through the systemic circulation to all parts of the body. As blood leaves the left ventricle, it enters the **aorta,** which is the largest of all arteries. Other arteries of the systemic circulation branch off the aorta. Blood returning to the heart from systemic circulation is carried by two large veins called the **venae cavae** (VEE-nee KAY vee). The superior vena cava (VEE nuh KAY vuh) carries blood returning from the head, shoulders, and arms. The inferior vena cava, carries blood returning from the lower parts of the body. Figure 31·15 on the next page shows the major patterns of the systemic circulation.

The systemic circulation has a number of subdivisions. The **coronary circulation** is the flow of blood through the blood vessels that nourish the heart itself. **Renal circulation** is the flow of blood through the vessels that pass through the kidneys. As blood flows through the kidneys, wastes and some water are removed. The flow of blood through the veins extending from the digestive organs to the liver makes up the **hepatic portal circulation.** Blood flowing through these veins is rich in nutrients. The liver may remove food substances from the blood and store them.

CIRCULATION

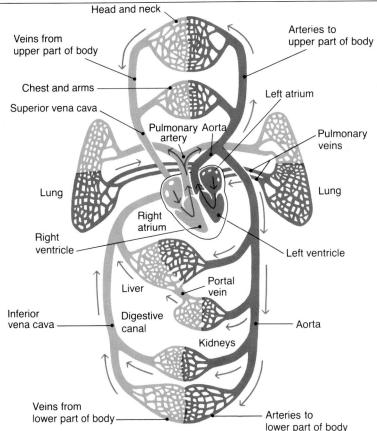

FIGURE 31·15 The path of blood through the circulatory system is illustrated in this drawing.

31·10 THE LYMPHATIC SYSTEM

Water, oxygen, and dissolved food materials pass out of the capillaries and become part of the tissue fluid. The food and oxygen are used in the tissues. Some of the water returns to the capillaries and carries wastes and carbon dioxide. But not all of the water goes back into the capillaries. For this reason, tissue fluid tends to accumulate. The **lymphatic system** is a system of "circulation" that collects excess tissue fluid. This system consists of veinlike vessels, capillaries, ducts, and glands. The fluid that flows through the lymphatic system is called **lymph.** Blood plasma, tissue fluid, and lymph have many similarities. Lymph and tissue fluid have very similar composition; when tissue fluid is inside lymphatic vessels, it is called lymph. Blood plasma contains much more protein than either tissue fluid or lymph. Most protein molecules are too large to pass out of the capillaries. These large proteins do not become part of the tissue fluid or the lymph.

The lymphatic system is a one-way system. Lymph flows only from the tissues toward the heart. The lymphatic system also lacks a pump. Lymph is kept moving by the action of skeletal muscles pressing against the lymphatic vessels. These vessels contain valves similar

---

Discovery

Laser Surgery

The surgeon holds a pencil-shaped tube and carefully aims the narrow beam of colored light at the cells to be killed. The patient feels no pain as the laser light strikes the brain tumor. Yet as the laser's tremendous energy causes water within the tumor to boil, the cells are destroyed.

Laser means "*l*ight *a*mplification by *s*timulated *e*mission of *r*adiation." A laser is made of visible light. Instead of being made of the rainbow of colors like sunlight, laser light consists of only one wavelength, or specific color of light. Laser light can travel a long distance and has very high energy. This intense beam of light can be focused on a 0.2 mm spot and kill a tumor cell while doing little harm to surrounding cells.

With a laser, an eye specialist can reattach the retina, a light-sensitive nerve tissue at the back of the eye. A laser can be used to stop the bleeding in small blood vessels in the eye and thus prevent blindness. A neurosurgeon can operate through a microscope and use the laser to cut out damaged or abnormal tissue from the brain or spinal cord.

Because a laser is finely focused, it cuts more cleanly than a scalpel, and the incision bleeds less and heals faster. For delicate operations, or those done deep in the body, laser surgery offers a very special ray of hope.

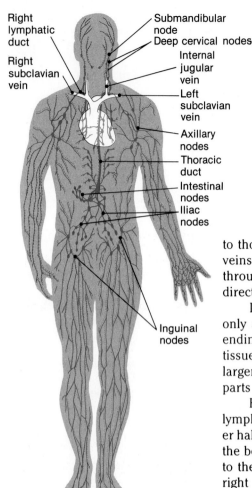
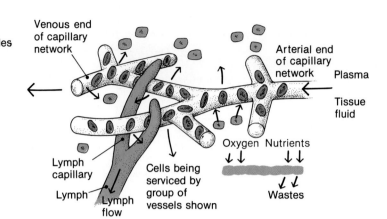

FIGURE 31·16 The lymphatic system is composed of vessels, capillaries, ducts, and glands that contain lymph. Lymph is a plasmalike fluid that leaks from the bloodstream and bathes the cells.

to those found in veins. Lymphatic vessels contain more valves than veins contain. When skeletal muscles contract, the lymph is squeezed through the lymphatic vessels. The lymph is kept flowing in the proper direction by the valves.

Lymph capillaries are similar to blood capillaries. Their walls are only a single cell thick. But lymph capillaries have blind, or closed endings. Figure 31·16 (right) shows how lymph capillaries arise in the tissues near blood capillaries. Lymph capillaries join together to form larger lymphatic vessels. The lymphatic vessels carry lymph from all parts of the body toward the upper chest region.

Figure 31·16 shows the lymphatic vessels in the body. The largest lymphatic vessel, the *thoracic duct,* collects lymph from the entire lower half of the body and the upper left side. The remaining one fourth of the body is drained by a smaller lymphatic vessel. Lymph is returned to the blood circulation by the thoracic duct on the left side and the right lymphatic duct on the right side. These two ducts release the lymph into veins just above the heart. Thus fluids lost from the plasma in the capillaries are returned to the blood near the heart.

The only type of cell commonly found in the lymph fluid is the lymphocyte. Recall that lymphocytes are one of the varieties of white blood cells. They move freely back and forth between the blood circulation and the lymphatic system. If lymphocytes come in contact with foreign matter, they may produce antibodies to destroy it.

Foreign particles that get into the tissues are often drained off with the lymph. **Lymph nodes** are small bean-shaped organs that act as filters in the lymphatic system. Most lymph nodes are found in the neck, the armpits, the groin, and around the intestines and lungs. Lymph passes through at least one lymph node before returning to the blood.

In addition to acting as filters, lymph nodes produce lymphocytes. Macrophages also are present in the lymph nodes. Foreign matter that arrives at the lymph nodes is attacked by the lymphocytes and macrophages. During some infections the lymph nodes in the neck or armpits may swell. Some people refer to this condition as swollen glands. Swollen glands are the result of lymph and cells collecting in the glands.

REVIEW QUESTIONS

6. Which chamber of the heart pumps blood to the lungs?
7. What valve controls blood flow from the left atrium to the left ventricle?
8. Which type of blood vessel has a wall that is only one cell thick?
9. What part of the circulation carries blood from the digestive organs to the liver?
10. During a physical examination a doctor sometimes presses the lymph nodes in your neck. Explain why.

Control of Circulation

31•11 HEARTBEAT

Heart muscle is different from other types of muscle. It can contract even without signals from the brain. The heart has a pacemaker that causes it to contract and controls the rate of the heartbeat. The **pacemaker,** located in the wall of the right atrium, is a bundle of tissue that produces electrical impulses. The name for this tissue is **sinoatrial** (sī noh AY tree uhl) **node,** or *S-A node* for short. The S-A node sends a regular series of electrical impulses to other parts of the heart. It is these impulses that start each heartbeat.

The **atrioventricular node,** or *A-V node,* is a second bundle of electrically active tissue. It is located near where the atria and ventricles meet in the middle of the heart. You can see in Figure 31•17 that a network of fibers from the A-V node extends throughout the right and left ventricles. When the S-A node sends out an electrical impulse, it travels throughout the atria and reaches the A-V node. The A-V node in turn passes on the impulse to all parts of the ventricles. As the impulse from the S-A node passes through the atria, the atria contract. The A-V node then passes the impulse through the ventricles and the ventricles contract. Each heartbeat has two parts. First the atria contract. A fraction of a second later, the ventricles contract.

The S-A node starts each heartbeat and can keep the heart beating regularly. However, the heart also receives control signals from the brain. These electrical signals travel through nerves from the brain to the heart. These signals from the brain can either speed up or slow down the action of the heart. In the base of the brain is the *cardiac acceleration center.* When the body requires more oxygen, this part of the brain sends signals to the heart. These signals cause the S-A and A-V nodes to produce impulses at a faster rate. The nerves from the cardiac acceleration center release a chemical called adrenaline (uh-DREHN uh lihn) onto the heart. *Adrenaline* causes the heartbeat to speed up. As the pumping action speeds up, more oxygen is delivered to the tissues of the body.

SECTION OBJECTIVES

- Describe how the heartbeat is controlled.
- Relate blood pressure to blood flow.

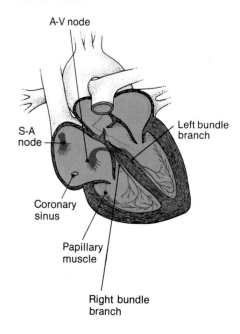

FIGURE 31•17 The heartbeat is controlled by the S-A node, which is a bundle of tissue located in the wall of the right atrium. Impulses from the S-A node cause the atria to contract. When the impulses reach the A-V node, it sends out impulses, causing the ventricles to contract.

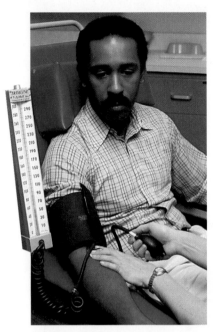

FIGURE 31·18 This person's blood pressure is being checked with a sphygmomanometer.

Also in the base of the brain is the *cardiac inhibition center*. When the body's need for oxygen decreases, the inhibition center sends signals to slow the heart. Nerves from the inhibition center release the chemical acetylcholine (as uh tuhl KOH leen) onto the heart. *Acetylcholine* inhibits the action of the S-A and A-V nodes. The heartbeat slows down, and less oxygen is carried to the tissues.

31·12 BLOOD PRESSURE AND BLOOD FLOW

Each time the heart contracts, the force of the contraction puts pressure on the blood. This pressure is exerted against the walls of the arteries and is called **blood pressure.** The period during which the ventricles are contracted is called **systole** (SIHS tuh lee). Pressure in the arteries is highest during systole. The period during which the ventricles are relaxed and are being filled with blood is called **diastole** (dī AS tuh lee). Blood pressure in the arteries is much lower during diastole than during systole. The alternation of systolic and diastolic pressures, caused by contraction and relaxation of the ventricles, can be felt as the pulse.

The units with which blood pressure is measured are millimeters of mercury. A pressure of one millimeter can support a column of mercury one millimeter high. Because blood pressure changes from systole to diastole, it is usually recorded as two numbers. The first number is the pressure during systole, or *systolic pressure*. The second number is the pressure during diastole, or *diastolic pressure*. The blood pressure of a normal, healthy adult at rest is 120/80.

Blood pressure is measured with a device called a *sphygmomanometer* (sfihg moh muh NAHM uh tur). The cuff of the sphygmomanometer usually is placed around the upper arm to measure the blood pressure. The value of 120/80 for normal blood pressure applies to large arteries. As blood flows through the arteries, it spreads out through more and more branches and loses pressure. Blood pressure in the capillaries is only about 20 millimeters of mercury.

There is not enough blood in circulation to completely fill all of the capillaries at any one time. The pattern of blood flow changes as the body's activity changes. The organs that are most active at a given time will receive the most blood supply. For example, during exercise the skeletal muscles will receive more blood. Blood flow to the digestive organs will increase after a meal. These changes in blood flow are controlled by muscle cells that surround the arterioles of all organs.

REVIEW QUESTIONS

11. What part of the heart is called the pacemaker?
12. What chemicals speed up and slow down the heartbeat?
13. In what units is blood pressure measured?
14. Explain what is meant by a blood pressure of 125/80.
15. In some people the heart does not beat with a steady rhythm. How might this condition be related to disorders of the S-A or A-V nodes?

Investigation

How Do Human Pulse Rates Differ?

Goals

After completing this activity, you will be able to

- Measure and record your pulse rate.
- Compare your pulse rate to those of other students in your class.
- State the effect of exercise on pulse rate.
- Draw a graph of the pulse rate for the entire class.

Materials (for groups of 2)

stopwatch (or clock with second hand)
2 sheets of graph paper

Procedure

A. Find your pulse. To do this, place the forefinger and middle finger of your right hand across the inside of the wrist of your left hand. Use gentle pressure until you feel a throbbing sensation. (You are pressing against the radial artery.) See the drawing for the correct way to take the pulse.

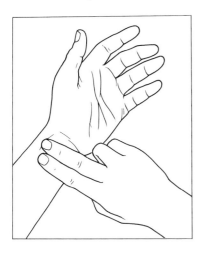

B. Take your pulse for 15 seconds. Your partner can use a stopwatch or clock to indicate a 15-second interval. Multiply the number of pulses by 4 to find the pulse rate for 1 minute. Repeat this procedure four times. On a separate sheet of paper, copy the chart below. Record your data. Find the average pulse rate for the 4 trials. (Find the total and divide by 4.)

C. Assist your partner in finding his or her pulse rate at rest. Find his or her average pulse rate.

D. At the chalkboard one student will record the average pulse rates for all class members. Record this data by listing the number of students with the same pulse rate. Record, for example, "pulse rate 72—6 students"; "pulse rate 80—4 students," and so on. Prepare a graph called "Distribution of Pulse Rates (at Rest)." Pulse rates should be indicated on the x-axis; the number of students, on the y-axis.

E. Find your pulse rate after exercise. Have your partner time you. Jog in place for 1 minute. Immediately after you finish, find your pulse rate as you did before (steps A and B).

F. Prepare a chart similar to the chart from step B. Title this chart "Pulse Rate After Exercise." Record all data as before. Find the average pulse rate.

G. One student will record the average pulse rates for all students after exercise. Use this data to prepare a second graph called "Distribution of Pulse Rates (After Exercise)."

CHART 1. Pulse Rate at Rest		
	TIME INTERVAL (15 sec)	PULSE RATE (beats/min)
TRIAL 1		
TRIAL 2		
TRIAL 3		
TRIAL 4		
Average Pulse Rate: _____		

Questions and Conclusions

1. Your pulse rate at rest for each of the four different trials was probably different. Explain why this happened.
2. What is the highest "at rest" pulse rate in the class? What is the lowest "at rest" pulse rate? Which pulse rate occurred most often in your class?
3. Compare the "at rest" pulse rates with the "after exercise" pulse rates. What are the differences? Account for the differences that you have found.
4. Suppose you were to take the "at rest" and "after exercise" pulse rates of professional athletes. How would their pulse rates compare with those of your classmates?

CIRCULATION 639

Problems of Circulation

SECTION OBJECTIVES

- Relate heart disease to the circulatory system.
- List some disorders of the circulatory system.

31·13 HEART DISEASE

Oxygen is brought into the heart muscle by coronary arteries. Any situation that interferes with blood flow in the coronary arteries may lead to damage or death of part of the heart muscle. Damage to part of the heart muscle from lack of oxygen is referred to as a *heart attack.*

Blockage of the coronary arteries may come about in several ways. It sometimes happens that deposits of fatty material form on the inside of blood vessels. Figure 31·19 shows a cross section of an artery with fatty deposits. Such deposits contain cholesterol along with several other fatty and nonfatty materials. These fatty deposits are most likely to occur in the arteries of older people. The fatty deposits may increase in thickness with time and interfere with blood flow. The buildup of fat deposits on the inside of blood vessels is called **atherosclerosis** (ath uhr oh skluh ROH sihs). Atherosclerosis in the coronary arteries is one of the causes of heart attacks.

Blood clots seal breaks in blood vessels. Occasionally, clots form in the blood when there is no break in a blood vessel. If such a clot gets trapped in a small blood vessel, it may block the flow of blood. Formation of clots in a blood vessel or in the heart is called **thrombosis.** If a clot blocks one of the coronary arteries, it can block the flow of blood to the heart muscle.

Research has shown that poor diet, lack of exercise, and smoking are factors in heart disease. Reducing the intake of fatty meats and dairy products may help lessen the risk of heart disease. Eating fruits and vegetables that are low in fat may help to reduce the risk of heart disease as well as provide other health benefits. Exercising regularly and avoiding smoking will also reduce the risk of heart disease.

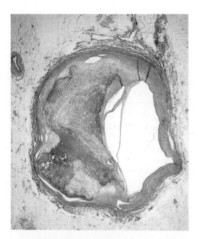

FIGURE 31·19 This photograph shows a cross section through an artery. Notice that the space within the artery has become narrowed due to the build up of fatty deposits in the inner wall of the artery.

31·14 CIRCULATORY DISORDERS

Circulatory system problems can occur in places other than the heart. Problems can result when organs of the body, such as the brain, do not receive enough blood. If a blood vessel in the brain is blocked, blood is unable to reach all the brain tissue. As a result, some of the tissue may be injured or may die. A condition called a *stroke* may result from lack of blood to the brain.

Another problem of circulation is high blood pressure. Abnormally high blood pressure is called **hypertension.** Several factors are known to contribute to hypertension. One of these factors is atherosclerosis. As fat builds up in the blood vessels, blood flow is restricted and blood pressure increases to move blood through the vessels.

High blood pressure is dangerous because it can cause other major problems such as a stroke and heart attack. Hypertension is easy to detect and can be treated.

31•15 TREATMENT OF CIRCULATORY PROBLEMS

Blood transfusions often are needed after serious injury or during surgery. Blood banks sometimes run short of blood. Artificial blood is being developed to help relieve this problem. Artificial blood is a synthetic fluid that contains no cells. Artificial blood can absorb and carry large quantities of oxygen. It can also release oxygen when needed. Thus artificial blood is really a substitute for the red blood cells of natural blood.

Several methods have been developed for treating serious heart disease. One method involves bypassing partially blocked or damaged coronary arteries. A healthy blood vessel from some other part of the body is used for this purpose. The healthy blood vessel is sewed onto the coronary artery bridging the point of blockage. Blood flows through the healthy blood vessel and the undamaged part of the coronary artery. This bypass restores oxygen supply to the heart muscle.

Bypass operations are useful if the heart muscle has not been permanently damaged. In cases where the heart is permanently damaged, it may be replaced. A *heart transplant* involves the removal of a diseased heart and replacement with a healthy one.

More recently it has become possible to replace a damaged heart with an *artificial heart*. A type of artificial heart, developed by Dr. Robert Jarvik, replaces the pumping action of both the right and left ventricles. The artificial heart is attached to the natural atria. The pumping chambers of the artificial heart, shown in Figure 31•20, contain balloonlike structures. The balloons are connected to an air pump outside the body. When the balloons are filled with air, they push the blood in the chambers out into the arteries. When the balloons are emptied by the pump, the chambers are refilled with blood from the atria.

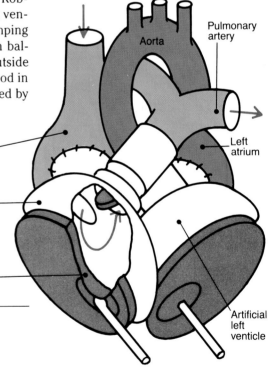

FIGURE 31•20 This drawing shows the parts of an artificial heart. The stitching at the bases of the left and right atria indicate where the artificial heart is attached to the natural heart.

REVIEW QUESTIONS

16. Describe two conditions that may cause a heart attack.
17. Identify some factors that may contribute to heart disease.
18. What problems may develop as a result of hypertension?
19. How might artificial blood ease the problem of blood shortages?
20. Artificial hearts replace only the ventricles of the natural heart. Why, do you think, is it unnecessary to replace the atria?

Chapter Review

SUMMARY

31·1 Blood consists of several types of cells suspended in a fluid called plasma.

31·2 Red blood cells carry oxygen from the lungs to the body's tissues. Carbon dioxide is carried from body tissues to the lungs via blood.

31·3 White blood cells provide defense against infectious disease. Some white blood cells are phagocytes; others produce antibodies.

31·4 Blood platelets and the protein fibrin produce blood clots.

31·5 The four major human blood types are known as the ABO blood types.

31·6 Human blood may be either Rh+ or Rh− depending on the presence or absence of an antigen on the red blood cell.

31·7 The heart contains four pumping chambers. The right side of the heart pumps blood to the lungs, and the left side of the heart pumps blood to all parts of the body except the lungs.

31·8 Blood flowing away from the heart travels through arteries. Exchange of gases, food, and wastes occurs in the capillaries. Blood is carried back to the heart by veins.

31·9 The circulatory system has several different divisions. These divisions include the pulmonary, systemic, coronary, renal, and hepatic portal circulations.

31·10 The lymphatic system carries excess tissue fluid back to the bloodstream. The lymph nodes filter the lymph.

31·11 The heartbeat is started by the S-A node, also called the pacemaker. The heartbeat is regulated by nerves from control centers in the brain.

31·12 Blood pressure is caused by the pumping of the heart and is regulated by expansion and contraction of muscle cells in the arterioles.

31·13 Atherosclerosis and thrombosis are major causes of heart disease. Proper diet, adequate exercise, and not smoking can help to prevent heart disease.

31·14 Hypertension may cause strokes, heart attacks, and damage to other organs.

31·15 Advances in treatment of circulatory disorders include the development of artificial blood, bypass operations, heart transplants, and artificial hearts.

BIOLOGICAL TERMS

ABO blood types
aorta
artery
atherosclerosis
atria
atrioventricular node
atrioventricular valves
bicuspid valve
blood clot
blood pressure
capillary
circulatory system
coronary circulation
diastole
heart
hemoglobin
hepatic-portal circulation
hypertension
lymph
lymphatic system
lymph nodes
pacemaker
plasma
platelets
pulmonary artery
pulmonary circulation
pulmonary veins
red blood cells
renal circulation
Rh factor
semilunar valve
sinoatrial node
systemic circulation
systole
thrombosis
tricuspid valve
veins
venae cavae
ventricles
white blood cells

USING BIOLOGICAL TERMS

1. Distinguish between an artery and a vein.
2. What is the relationship between blood pressure and hypertension?
3. Distinguish between atria and ventricles.
4. Distinguish between the pulmonary circulation and the coronary circulation.
5. What is a capillary?
6. Distinguish between the atrioventricular node and the sinoatrial node.
7. How are the terms blood clot and thrombosis related?
8. Select the terms that are circulatory or heart disorders and explain what they are.
9. What do diastole and systole measure?
10. What is the aorta?

UNDERSTANDING CONCEPTS

1. State three functions of the blood. (31·1)
2. Describe the main components of blood. (31·1)
3. Describe the transport function of the red blood cells. (31·2)
4. Describe the types of white blood cells. (31·3)
5. Explain the role of the white blood cells in defending the body against disease. (31·3)
6. Outline the steps in the formation of fibrin. Summarize the role of platelets in the formation of a blood clot. (31·4)
7. Compare the four main blood types in terms of the presence or absence of antigens and antibodies. Why is knowledge of blood types necessary before making a blood transfusion? (31·5)
8. What is the Rh factor? Why is it important? (31·6)
9. Describe a method for preventing erythroblastosis fetalis. (31·6)
10. Trace the path of the blood through the heart, beginning with the right atrium. (31·7)
11. Compare arteries and veins. (31·8)
12. Describe pulmonary circulation. (31·9)
13. List three subdivisions of the systemic circulation and describe each briefly. (31·9)
14. Compare the lymphatic system and the blood circulatory system with regard to movement of fluid, types of cells, and structure of capillaries. (31·10)
15. How do the A-V node and the S-A node initiate contractions of the heart muscle? (31·11)
16. Distinguish between systole and diastole. (31·12)
17. How are changes in blood flow to different areas of the body controlled? (31·12)
18. List two major causes of heart disease. (31·13)
19. How is atherosclerosis a contributing factor in hypertension? (31·14)
20. List and briefly describe three methods for treating heart disease. (31·15)

APPLYING CONCEPTS

1. What is the difference in the way oxygen and carbon dioxide are carried by the blood?
2. The walls of the ventricles are thicker than those of the atria. The left ventricle is thicker than the right. How are these differences adaptive?
3. Explain why the pulse represents the heartbeat.
4. It is said that no cell is a permanent resident of the bloodstream. Explain why.
5. The pulse can only be felt in a few places on the body, such as the wrists and neck. Explain why.

EXTENDING CONCEPTS

1. There are three different types of granulocyte white cells. Find out what they are. What are their functions?
2. Hemophilia is an inherited disorder of the blood clotting system. Find out what is missing from the blood of persons with hemophilia. Describe treatment for this condition.
3. An electrocardiogram is a recording of the electrical activity of the heart. Find out what a normal electrocardiogram looks like and what each part of it represents.
4. Contact the local chapter of the American Heart Association. Find out what the signs of a heart attack are. Find out how to care for a heart attack victim until medical help arrives.
5. How did the sinoatrial node get its name? Hint: Investigate the hearts of lower vertebrates.
6. The development of such things as artificial hearts and heart pacemakers represents a branch of science called biomedical engineering. Find out the nature of biomedical engineering. Describe two other devices developed by biomedical engineers.

READINGS

Arehart-Treichel, Joan. "Eating Your Way Out of High Blood Pressure." *Science News,* April 9, 1983, p. 232.

Edelson, Edward. "Cleansing the Blood." *Science 82,* October 1982, p. 72.

Jarvik, Robert K. "The Total Artificial Heart." *Scientific American,* January 1981, p. 74.

Moyer, Robin. "Blood Without Donors." *Science 81,* June 1981, p. 16.

Tiger, Steven. *Heart Disease.* Messner, 1986.

32. Respiration and Excretion

As this person climbs to a higher altitude, it becomes increasingly difficult to breathe because there is a decrease in air pressure. The mountain climber has to breathe harder and deeper to supply the body with useful materials and get rid of waste products. In this chapter, you will learn about the two body systems, respiratory and excretory, that perform these functions.

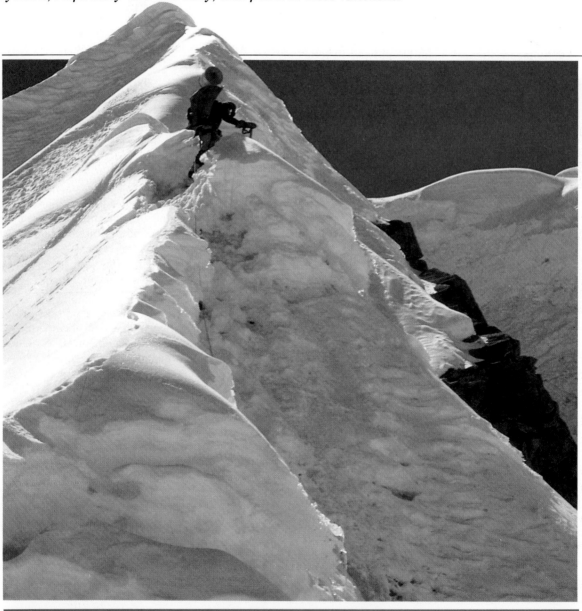

CHAPTER OBJECTIVES

After completing this chapter, you will be able to

- **Describe** the structures and functions of the respiratory organs.
- **Explain** how oxygen and carbon dioxide are exchanged in the lungs.
- **List** some common diseases and disorders of the respiratory system.
- **Describe** the structures and functions of the excretory organs.
- **List** some kidney diseases and **describe** their methods of treatment.

The Respiratory System

32·1 INTERNAL AND EXTERNAL RESPIRATION

Breathing brings oxygen into the body. You may remember from Section 5·6 that cells use oxygen to break down glucose and to make ATP. ATP is a usable source of energy for cells. The process that produces ATP is called *cellular respiration*. Glucose and oxygen are the raw materials needed for this process. Carbon dioxide and water are given off as wastes.

What is breathing? Breathing is the process by which the body exchanges gases with the environment. When you breathe, you inhale air and take the oxygen from it. You then exhale and give off carbon dioxide and water vapor. The oxygen that you inhale is used in cellular respiration. The carbon dioxide and water that you exhale are the waste products of cell respiration.

Breathing and cell respiration are the two parts of the process of *respiration*. Breathing is referred to as **external respiration** because oxygen is obtained from and carbon dioxide is released to the *external* environment. Cell respiration is referred to as **internal respiration** because oxygen is used and carbon dioxide is produced *within* cells. Both parts of this process are needed for the body to produce energy.

In this chapter you will be concerned mainly with external respiration. The organs and structures that carry out the process of external respiration make up the *respiratory system*. The respiratory system works with the circulatory system in supplying oxygen to the cells of the body and removing certain wastes.

32·2 RESPIRATORY STRUCTURES

Air can enter the respiratory system through the mouth or the nose. Air entering the nose passes into the *nasal cavity*. The bones that line the nasal cavity have folds and ridges, making the surface uneven.

SECTION OBJECTIVES

- Compare internal and external respiration.
- Trace the flow of air through the respiratory system.

646 RESPIRATION AND EXCRETION

The folds and ridges are covered with a *mucous membrane*. This membrane is made up of mucus-secreting cells that are covered with cilia. The mucus traps dust and microbes that enter with the air. The beating of the cilia moves the mucus and trapped particles back toward the throat. There they can be removed by swallowing, sneezing, or spitting. Air is warmed and moistened as it passes through the nasal cavity. Air that is warm and moist is less likely to damage other parts of the respiratory system.

The nasal cavity is connected to several spaces, called *sinuses*, within the bones of the skull. These sinuses also are lined with mucous membranes. When these membranes become infected, mucus may fill the sinuses. Clogging of the sinuses during a cold may cause pressure on either side of the nose and below the eyes.

As air passes back from the nasal cavities, it enters the part of the throat behind the mouth. This region behind the mouth is called the **pharynx.** Find the pharynx in Figure 32·1. The pharynx is the passageway for food coming from the mouth and for air coming from the nose. The pharynx leads downward toward the opening of two tubes. The tube at the front surface of the throat is the windpipe, or **trachea** (TRAY kee uh). In Figure 32·1 trace the path of air from the nose to the trachea.

pharynx (throat)

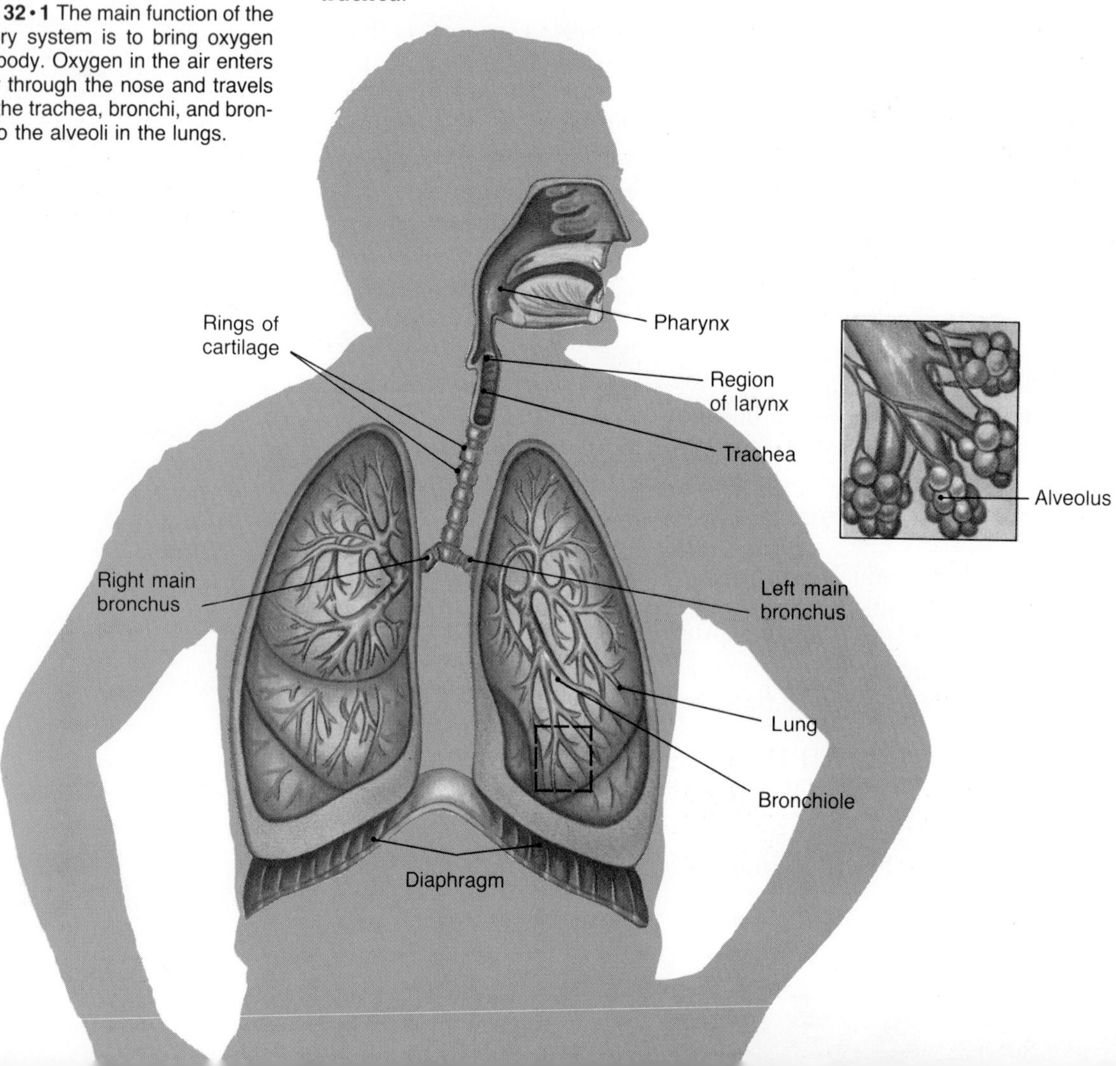

FIGURE 32·1 The main function of the respiratory system is to bring oxygen into the body. Oxygen in the air enters the body through the nose and travels through the trachea, bronchi, and bronchioles to the alveoli in the lungs.

Behind the trachea is the tube that carries food from the pharynx to the stomach. There is a fold of tissue at the bottom of the pharynx, just above the trachea. This flap of tissue is called the **epiglottis.** The epiglottis moves down over the entry to the trachea when food is swallowed. At other times the epiglottis lies up against the wall of the pharynx so that air may enter the trachea.

The voice box, or **larynx** (LAR ihngks), is at the top of the trachea. The larynx is made up of several pieces of cartilage. You can feel this cartilage as the Adam's apple. Inside the larynx are two muscular folds called *vocal cords*. When the vocal cords contract, air passing between them vibrates. In humans these air vibrations result in speech.

Below the larynx the trachea continues downward into the chest. In Figure 32·1 notice the rings of cartilage that make up the walls of the trachea. These rings of cartilage make the trachea flexible and keep it open for the movement of air.

Continue to trace the path of air through the trachea. In the chest the trachea divides into right and left branches, which are called **bronchi** (BRAHNG kī). One bronchus passes into each lung. The **lungs** are the main organs of the respiratory system.

The trachea and bronchi are lined with mucous membranes. The cilia in these air passages beat in an upward direction. Mucus and trapped particles are swept upward from the lungs to the pharynx. From the pharynx the mucus and particles can be removed.

Within the lungs the bronchi divide into smaller tubes. These smaller branches of the bronchi are called **bronchioles** (BRAHNG kee-ohlz). Figure 32·1 shows how numerous the bronchioles are. Both bronchi and bronchioles contain smooth muscle tissue in their walls. The smooth muscle controls the size of the air passages. The bronchioles continue to subdivide until they are microscopic.

The smallest bronchioles end in groups of tiny air sacs called **alveoli** (al VEE uh lī) (sing., *alveolus*). You can see in Figure 32·2 that the groups of alveoli look like bunches of grapes. It is in the alveoli that the exchange of gases between air and blood takes place. Each alveolus is surrounded by a capillary network. Gases pass between the air in the alveolus and the blood in the capillaries. This exchange is discussed in more detail in Section 32·4.

32·3 BREATHING MECHANICS

Each breath that you take consists of two major actions. During **inspiration,** air is pulled into the lungs. Air is pushed out of the lungs during **expiration.** The lungs expand and contract with each cycle of inspiration and expiration.

The space around the lungs makes up the chest cavity. As the lungs expand and contract, their outer surfaces rub against the chest wall. The lung and chest cavity wall on each side are covered by a moist membrane called the **pleura** (PLUR uh). Each pleura lubricates the outer surface of a lung and the inner surface of the chest cavity on that side. Friction between the lungs and the chest is reduced.

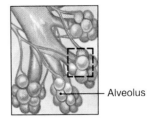

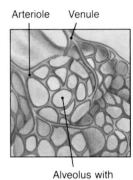

FIGURE 32·2 The exchange of gases, such as oxygen and carbon dioxide, between the air and blood occurs in the alveoli. Notice that each alveolus is surrounded by a network of capillaries.

alveus (cavity)

pleuron (rib)

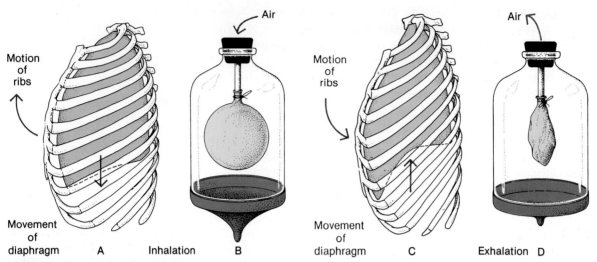

FIGURE 32·3 When the diaphragm contracts, the air pressure in the chest decreases (A). Incoming air causes the lungs to inflate. Similarly, pulling down on the elastic sheet at the bottom of the bell jar causes air to enter the balloon (B). When the diaphragm relaxes, the air pressure in the chest increases (C). This forces air out of the lungs. Similarly, when the elastic sheet on the bell jar is released, air moves out of the balloon (D).

The wall of the chest is made up of the ribs and muscles. Each rib is connected to the next by a flat layer of skeletal muscle. These muscles between the ribs are called **intercostal muscles.** The intercostal muscles provide part of the force needed for breathing.

The chest cavity is separated from the abdominal cavity by a flat sheet of muscle called the **diaphragm** (DĪ uh fram). Much of the force needed for breathing is provided by the diaphragm.

Figure 32·3 shows the movements that take place during breathing. Notice that before inspiration the diaphragm is curved upward into the chest cavity. When the muscle fibers contract, the diaphragm shortens and moves downward toward the abdomen. The downward movement of the covering on the bell jar represents this action. The result is that the volume of the chest cavity becomes larger. The intercostal muscles contract at the same time as the diaphragm. Contraction of the intercostals causes the ribs to move outward and upward. This action also causes the volume of the chest cavity to increase.

You can see that these actions of the diaphragm and the intercostal muscles enlarge the chest cavity. The pressure inside the chest cavity decreases as the volume increases. The pressure within the lungs decreases and air moves through the air passages to fill the lungs. Notice what happened to the balloon in the bell jar in Figure 32·3. Thus the contraction of the diaphragm and the intercostal muscles results in the inspiration of air.

Following inspiration, the diaphragm and the intercostal muscles relax. The diaphragm returns to its curved position. The ribs move down and inward. These actions cause the volume of the chest cavity to decrease. As the volume decreases, the pressure in the chest cavity increases. This increased pressure on the outside of the lungs causes the lungs to decrease in size. The air in the lungs is pushed out through the mouth and nose. Thus relaxation of the diaphragm and the intercostals leads to the expiration of air.

The walls of the alveoli, which contain many elastic fibers, also play a part in the process of expiration. During inspiration, the elastic fibers are stretched. When inspiration ends, the elastic fibers pull back to their original size, thus helping to force air out of the lungs.

Little muscular work is done during the expiration phase of relaxed breathing. The muscles do their work during inspiration. While the body is at rest, the diaphragm and the intercostals relax, and the chest cavity moves back to its resting position.

During exercise, breathing becomes deeper and more rapid. More oxygen is delivered to the muscles. During such breathing, the muscles of the abdominal wall assist the chest muscles. People whose work requires the intake of large amounts of air, use abdominal muscles for both restful and heavy breathing.

32·4 GAS EXCHANGE

In Section 31·7 the flow of blood from the right side of the heart to the lungs is described. The blood entering the capillaries around the alveoli contains little oxygen and much carbon dioxide. During inspiration, the alveoli fill with fresh air from the environment. This fresh air contains much oxygen and little carbon dioxide. The concentrations of the gases in the blood and in the alveoli are unequal.

Little separates the air in the alveoli from the blood in the capillaries. The wall of an alveolus and the wall of a capillary each consist of a thin layer of flattened cells. Both oxygen and carbon dioxide can pass through these layers.

Figure 32·4 shows the concentration of gases in an alveolus and a capillary. Oxygen is in high concentration in the alveolus and in low concentration in the capillary blood. In which direction does oxygen diffuse in this situation?

The concentration of carbon dioxide is high in the blood plasma and low in the alveolus. In which direction does carbon dioxide diffuse in this situation? When expiration occurs, air containing carbon dioxide and water vapor are forced out of the lungs.

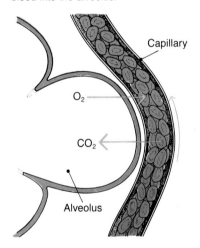

FIGURE 32·4 During the exchange of gases in the lungs, oxygen passes from the alveolus into the blood in the capillaries. Carbon dioxide moves from the blood into the alveolus.

32·5 CONTROL OF RESPIRATION

Cellular respiration occurs all the time. Some of the energy produced is in the form of heat. This heat helps to keep the body at a constant temperature. The rate of cell respiration needed to maintain life processes in a person at rest is called *basal metabolism*. Since cellular respiration uses oxygen, basal metabolism can be measured by measuring the amount of oxygen a person uses. The amount of oxygen used by a person at rest gives the basal metabolic rate. **Basal metabolic rate (BMR)** is a measure of the rate of oxygen consumption, and therefore energy production, in a person at rest. The basal metabolic rate of an average male between 14 and 16 years old is 46 kilocalories per hour per square meter of body surface. The basal metabolic rate of a female the same age is 43 kilocalories per hour per square meter of body surface.

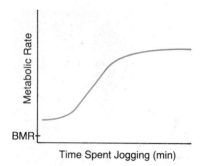

FIGURE 32·5 This graph shows the relationship between a jogger's metabolic rate and level of activity. Notice that as the rate of metabolism increases above the basal metabolism rate (BMR), more oxygen is needed by the body.

When a person exercises, the rate of metabolism increases above the basal rate. More oxygen is used, and more carbon dioxide is produced by muscles and other tissues. The concentration of gases begins to change in the tissue fluid, which bathes the cells, and in the blood. The body responds to such changes. The breathing rate goes up, and oxygen enters the blood faster. The increased oxygen in the blood, in turn, diffuses into the tissue fluid. This fresh supply of oxygen moves from the tissue fluid into the cells. Increased breathing leads to a more rapid removal of carbon dioxide from the tissue fluid.

The graph in Figure 32·5 shows the relationship between a person's metabolic rate and level of activity. What would the person be doing at the basal metabolic rate? What is happening in the jogger's respiratory system during the increase in metabolic rate? Why does jogging cause these changes inside the body?

The body's response to changing conditions maintains *homeostasis* (hoh mee uh STAY sihs), the maintenance of a relatively stable internal environment of an organism. Tissue fluid is the environment to which cells inside the body are directly exposed. The concentrations of the gases in the tissue fluid may vary from time to time. However, the levels of gases in the blood stay within a narrow range. The respiratory, circulatory, and nervous systems have important roles in maintaining proper concentrations of gases in the internal environment.

During light activity, an average adult breathes about 14 to 18 times each minute. With each breath, about half a liter of air passes in and out of the lungs. The lungs do not control the rate of breathing. Breathing is controlled by the brain. The brain responds to changes in the body's needs. When the body's need for energy increases, more oxygen is used and more carbon dioxide is produced. The brain responds by causing an increase in the rate of breathing.

The respiratory rate is controlled by a region of the brain known as the **respiratory** (REHS puhr uh tor ee) **center**. This control center

FIGURE 32·6 During inspiration, the inspiratory center in the brain sends impulses to the diaphragm causing it to contract. During expiration, the expiratory center sends impulses to the inspiratory center to stop its action. As a result, no impulses are sent to the diaphragm and it relaxes.

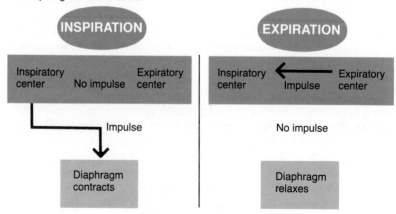

lies in the base of the brain near the centers that control the heart. The respiratory center is divided into two portions. The **inspiratory** (ihn-SPIR uh tor ee) **center** sends impulses along nerves to the intercostal muscles and to the diaphragm. The **expiratory** (ehk SPIR uh tor ee) **center** sends impulses that inhibit, or stop, the action of the inspiratory center. As shown in Figure 32·6, control shifts back and forth between these centers. This shifting of control causes the rhythmic action of breathing. When the inspiratory center sends impulses to the intercostals and to the diaphragm, these muscles contract and cause air to be drawn into the lungs. When the expiratory center inhibits the inspiratory center, the muscles relax and air is exhaled.

The amount of carbon dioxide in the blood and in the tissue fluids of the brain controls the rate at which the inspiratory center sends impulses. When metabolism increases, more carbon dioxide is produced. This increase in carbon dioxide causes the inspiratory center to send more frequent impulses to the respiratory muscles. This action results in more rapid breathing and faster elimination of carbon dioxide. Homeostasis is maintained.

Look at the athlete in Figure 32·7. Why does rapid breathing continue after physical exertion has ended? During very heavy exercise, the skeletal muscles may need more oxygen than the blood can supply. The muscles can produce energy through lactic acid fermentation, which does not use oxygen. (Section 5·7). This process produces lactic acid, some of which stays in the muscles. Most of the lactic acid, though, is carried in the blood to the liver. Carbon dioxide collects in the tissue fluid. Since oxygen is needed to change the lactic acid in the liver to glycogen, it is said that the exercising person has an *oxygen debt*.

When the exercise ends, the high level of carbon dioxide causes the respiratory center to send impulses for a continued high rate of breathing. The rapid breathing rate supplies the liver with enough oxygen to convert the lactic acid to glycogen. The oxygen debt is repaid. Think of the last time you ran fast for a long distance. You may remember that you continued to breathe heavily for some time after you stopped running. Your breathing returned to normal after the oxygen debt was eliminated.

FIGURE 32·7 When a person engages in vigorous activities, the skeletal muscles may not get enough oxygen. Under these conditions, the muscles produce energy through lactic acid fermentation. The buildup of carbon dioxide in the body causes an increase in the breathing rate. This continues until the lactic acid is removed.

REVIEW QUESTIONS

1. What gases are exchanged during breathing? Where does the exchange of gases take place?
2. What happens to the air as it passes through the nasal cavities?
3. Describe how the intercostal muscles and the diaphragm work during inspiration and expiration.
4. Explain how carbon dioxide concentration in the tissue fluid affects the rate of breathing.
5. In a crowded room it is noted that people's respiratory rates are elevated. Explain what is happening in terms of concentrations of gases.

Investigation

What Is the Effect of Temperature on the Rate of Breathing in Fish?

Goals

After completing this activity, you will be able to

- Describe the effect of temperature on the rate of breathing in fish.
- Relate gill movement to carbon dioxide production.

Materials (for groups of 4)

3 live fish (about the same size)
6 flasks with 6 stoppers
bromthymol blue solution
dropper   3 thermometers
2 ice baths
fish net
hot plate (for class)

Procedure

A. On a separate sheet of paper, set up a data chart on the breathing rate of fish. Head one column *Time for Color Change* and the other column *Gill Count*. You will be recording data for fish at 10°, 20°, and 30°C.

B. You will need 6 flasks of water. Three will serve as controls. Put the same amount of water in each flask.

C. Adjust the temperature in two flasks so that the temperature in each is 30°C. (Your teacher will have water heating on a hot plate.) Check the temperature with a thermometer.

D. Place two of the other flasks in an ice bath to cool to 10°C. Do not allow the temperature to drop below 8°C.

E. Keep the other two flasks at room temperature—about 20°C.

F. To one of the flasks with water at room temperature, add bromthymol blue solution, drop by drop. Count the number of drops you add until you produce a light blue color. (Later you will be looking for a color change. A change from blue to yellow indicates that carbon dioxide has been added to the water.)

G. Add the same number of drops of bromthymol blue to each of the other flasks. Stopper one flask at each of the temperature levels and set each flask aside.

H. Add one fish to each of the other flasks. Stopper the flasks. Allow 3 minutes for the fish to adjust to the water. While one student keeps track of the time, the other student in the group should begin counting the gill movements of the fish in each flask.

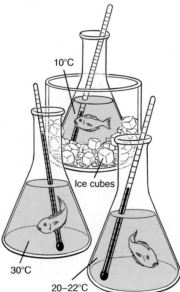

I. Observe the color of the water in each flask. Note when the blue color disappears and a slightly yellow color appears. In the data chart, record the amount of time needed and the gill movement count at that point.

J. After a color change has been noted and recorded for each flask, use a net to remove the fish. Return it to its original container.

K. Look at the three flasks that had no fish. Note any color change in these flasks. Record your observations.

Questions and Conclusions

1. In which flask or flasks did the bromthymol blue change most quickly? In which flask or flasks did it change least quickly? Were there any flasks in which there were no color changes?

2. What is the effect of temperature on the time required for bromthymol blue solution to become yellow?

3. What substance caused the color change in bromthymol blue? What can you conclude about the effect of temperature on the production of this substance?

4. What is the effect of temperature on the rate of gill movement? Relate the rate of gill movement to the time required for the color change.

5. What can you conclude about the effect of temperature on the rate of breathing in fish?

6. Why did you need three flasks without any fish? What happened in these flasks?

Respiratory Problems

32·6 DISEASES OF THE RESPIRATORY SYSTEM

Even clean air contains dust particles. Pollen grains, the spores of fungi, and many microbes and viruses also are found in "clean" air. With every breath that is taken, the dust and other particles enter the respiratory system. Most of these particles are filtered out in the nasal cavities, trachea, and bronchi. But some of the particles reach the lungs. Large numbers of macrophages are found in the alveoli of the lungs. These cells eat most of the microbes and viruses that enter with the air. Sometimes a few microbes are not killed. These organisms can start an infection of the respiratory system.

Infectious diseases of the respiratory system are common. The common cold is a viral infection of the upper portion of the respiratory system. Cold viruses reproduce in the cells of the mucous membrane of the nasal cavities and pharynx. As a result, the mucous membrane secretes more mucus than usual. Coughing and sneezing result from the extra mucus. Coughs and sneezes clear the air passages of extra mucus.

Other infections of the respiratory system may occur in the lungs themselves. *Pneumonia* is an infection of the lungs that can be caused by viruses, bacteria, or fungi. As a result of the infection, fluid collects in the alveoli. This fluid reduces the amount of air that can enter the alveoli. The fluid in the alveoli blocks the diffusion of oxygen from the alveoli into the blood. Since less air is in the alveoli, less oxygen passes to the blood and less carbon dioxide is removed from the blood. The extra carbon dioxide in the blood stimulates the respiratory center of the brain. The brain sends signals that cause breathing to become heavier and more rapid. Pneumonia is a serious disease. Some types of pneumonia can be treated with medicines that kill the infecting organism.

32·7 INTERRUPTION OF BREATHING

Sometimes the air passages may become blocked by an object, such as a piece of food. Food may get past the epiglottis and get stuck in the larynx. As a result, the flow of air through the larynx to the lungs will be blocked. Such a blockage will cause death in a few minutes unless the blockage is removed.

The *Heimlich maneuver* is a way of forcing objects out of the air passages. It involves a sharp upward pressure on the diaphragm. Air pushed out then dislodges the object.

Some types of poison and some types of accidents cause the respiratory system to fail. If such an event occurs, artificial respiration is required. An effective method of artificial respiration is *mouth-to-mouth resuscitation*. This method involves blowing into the victim's mouth to force air into the lungs.

SECTION OBJECTIVES

- Identify some respiratory problems.
- Relate smoking to problems of the respiratory system.

Discovery

Influenza

Every year, influenza, or flu, afflicts countless people with fever and muscular aches. The misery lasts only a few days and the infection is not usually dangerous. But periodically an outbreak is widespread and deadly. In 1918, a worldwide flu epidemic infected two billion people and killed 20 million of them. A serious outbreak of the Asian flu occurred in 1957, and in 1968 the Hong Kong flu struck.

In 1980, at the Institute of Experimental Pathology in Iceland, scientists were studying a flu virus that had caused an epidemic among seals along the New England coast. While a researcher was examining a seal that had been infected with the virus, the animal sneezed into the face of a lab technician who was holding it. Within 48 hours, the technician's eyes were inflamed and showed high levels of the virus found in the seals. When researchers examined the virus, they discovered that its eight genes were identical to genes occurring in viruses that were known to infect ducks and other birds. It was the first time that scientists were able to document the transmission of flu virus from birds to mammals. The discovery helped scientists to better understand how influenza viruses beat the body's immune system and cause infection.

Influenza viruses are constantly changing into new forms. In one laboratory alone, scientists have collected nearly 7000 kinds of flu virus. Two viral proteins are the key to the nature of influenza. Mutations cause changes in the amino acid segments in these proteins. As a result, the viruses change. When a new virus appears, some people may not be able to resist the infection.

Sometimes a major change occurs in a virus, and a new strain develops to which no one is immune. Scientists believe this happens when two different flu viruses infect the same cell and the genes of the two viruses become scrambled. For example, Hong Kong flu probably developed when a person with Asian flu came in contact with a flu-infected duck. The new flu virus that developed caused an epidemic.

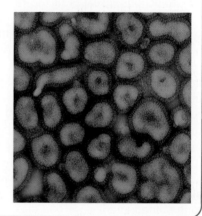

32•8 SMOKING AND RESPIRATORY DISORDERS

Cigarette smoking is a factor in hypertension and in heart disease. Smoking also is thought to be a cause of several serious respiratory problems. Smoking damages the cilia that line the bronchi and bronchioles. When these cilia are damaged, mucus, dust, and smoke particles are not carried away from the lungs as they normally would be. The many chemicals contained in cigarette smoke remain in the lungs and cause irritation. If this irritation continues for a long time, damage may result.

Study the graph in Figure 32•8. From the early part of the century through the 1960s, the average number of cigarettes smoked by Americans increased. Since then, this number has gone down. What trend do you see in the incidence of respiratory cancers? In the past, these disorders have mostly affected men. In recent years, however, more women have begun to smoke. More women are now developing lung diseases. While it is now known that lung cancer is associated with smoking, the relationship between smoking and emphysema (ehm-fuh SEE muh) is less well known.

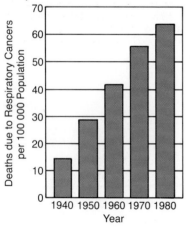

FIGURE 32•8 This graph shows the incidence of respiratory cancers. How many people per 100,000 died of respiratory cancers in 1980?

Emphysema is a disease of the respiratory system that mostly affects the alveoli. In the early stages of this disease, the elastic fibers in the walls of the alveoli are damaged. You will recall that these elastic fibers help to push air out of the lungs during expiration. If these fibers become damaged, the alveoli do not completely empty during expiration. Extra effort is required to push the air out of the lungs.

In the later stages of emphysema, parts of the alveolar walls are destroyed. Less alveolar surface means that less oxygen can be absorbed into the blood. A person with emphysema breathes hard and rapidly to make up for the damaged alveoli.

32·9 ENVIRONMENTAL FACTORS AND RESPIRATION

At high altitudes the respiratory and circulatory systems work harder because there is less oxygen present. Breathing becomes deeper and more rapid. The heart pumps more blood, sending more oxygen to the tissues. Since there is less oxygen in the air, enough will not reach the tissues despite the increased heart and breathing rates. A shortage of oxygen in the tissues is referred to as **hypoxia** (hī PAHK-see uh). Hypoxia causes altitude sickness. Symptoms include dizziness, blurred vision, and nausea. Hypoxia affects all parts of the body.

A person who stays at a high altitude for several weeks usually will recover from the altitude sickness. Two kinds of changes take place in the body. First, as a result of hypoxia, the bone marrow produces greater numbers of red blood cells than usual. As the number of red cells increases, the ability of the blood to transport oxygen also increases. Hypoxia also stimulates the respiratory center of the brain. The brain, in turn, causes deeper and more rapid breathing. If hypoxia continues over a long time, the size of the alveoli may increase. Larger alveoli can take in more air. Taking in more air helps to make up for the reduced amount of oxygen in the air.

Inhaling carbon monoxide gas results in a kind of hypoxia called *carbon monoxide poisoning*. Carbon monoxide combines more readily with the hemoglobin in blood than does oxygen. It keeps hemoglobin from carrying oxygen. Enough oxygen does not reach the tissues, so hypoxia occurs. If too much carbon monoxide is inhaled, few red cells will be carrying oxygen and the tissues will not get enough oxygen. As a result, the person may die. Carbon monoxide results from the incomplete burning of fossil fuels and other materials.

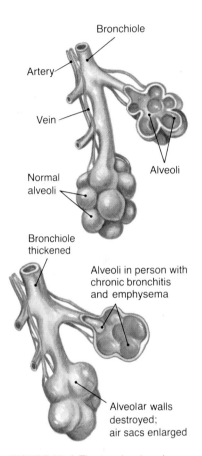

FIGURE 32·9 The top drawing shows the alveoli of a healthy person. The bottom drawing shows the alveoli of a person who has emphysema. Notice the difference between the alveolar walls in the healthy lung and those in the lung affected by emphysema.

REVIEW QUESTIONS

6. How does pneumonia affect the lungs?
7. What respiratory diseases are associated with smoking? What damage to the respiratory system does smoking do?
8. What damage occurs in the later stages of emphysema?
9. What is hypoxia? What conditions lead to hypoxia?
10. If an accident victim had stopped breathing and had a severe cut, which would you do first: stop the bleeding or give mouth-to-mouth resuscitation? Explain your choice.

The Excretory System

SECTION OBJECTIVES

- List the excretory organs of the human body and explain their functions.
- Describe how nephrons work.

32·10 ORGANS OF EXCRETION

Excretion is the process of ridding the body of chemical wastes and any materials that are in excess supply. A number of organs take part in excretion. These organs include the kidneys, the lungs, the skin, and the digestive organs. Excretion is an important part of maintaining homeostasis. To maintain a stable internal environment the body must remove wastes and surplus substances.

Urea is a nitrogen-containing waste produced as a by-product of protein breakdown. Urea is a poison. Removal of urea is a major function of the excretory system.

Balancing the amount of water in the body is another of the excretory system's functions. Water enters the body in foods and drinks. Many foods, such as fruits, contain a great deal of water. Water is produced in the body by cellular respiration. Some of the extra water is released from the lungs. The rest is released by other excretory organs.

The amount of salt and other minerals in the body also is balanced by the organs of excretion. Salt and calcium are taken in with many foods. Small amounts of salt are needed in the blood and in many organs. Salt that is above the amount needed is excreted. Extra calcium builds up in the blood and is excreted.

The skin is one of the organs of excretion. Perspiration, or sweating, is a process that is familiar to all of us. Perspiration is accomplished by *sweat glands* that lie within the skin. Locate the sweat glands in the skin in Figure 32·10. You can see that the sweat glands open to the surface of the skin through tiny *pores*. The liquid produced by the sweat glands contains water along with salt and very small amounts of urea.

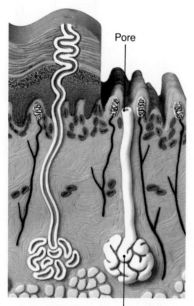

FIGURE 32·10 The skin contains sweat glands which connect to tiny pores on the skin's surface. Sweat glands excrete perspiration, a liquid that contains water, salts, and small amounts of urea.

32·11 KIDNEY STRUCTURE

The **kidneys** are the major organs of excretion. One kidney is found on each side of the backbone in the lower back. The kidneys filter the blood and produce **urine,** a liquid waste. Urine is mostly water, with some urea, salts, and other substances in it. Each kidney receives blood through a *renal artery* and returns blood to circulation by a *renal vein,* as shown in Figure 32·11. Urine formed in the kidneys is carried away through tubes called **ureters** (yu REE tuhrs). Notice in Figure 32·11 that the ureters connect the kidneys to the **urinary bladder.** The urinary bladder is a saclike organ that stores urine. The walls of the bladder are muscular and help to push urine out of the body. Urine is carried from the bladder to the outside of the body through a tube called the **urethra** (yu REE thruh).

The job of filtering the blood and forming the urine is done by microscopic units in the kidneys called **nephrons.** There are about a million nephrons in each human kidney. The structure of a single nephron and its blood vessels is shown in Figure 32·11.

Part of a nephron consists of an arteriole that forms a ball of capillaries. This ball of capillaries is called a **glomerulus** (gluh MEHR-u luhs). Surrounding the glomerulus is a double-walled structure called **Bowman's capsule.** The space inside Bowman's capsule connects with a small tube, or *tubule*. The tubule is divided into three sections. The section nearest Bowman's capsule is the **proximal convoluted tubule.** The next section is called **Henle's loop.** You can see in Figure 32·11 that this section takes the form of a U-shaped loop. The third section is the **distal convoluted tubule.** The tubule then connects with a larger tube called a **collecting tubule.** Several tubules lead into a single collecting tubule. Urine passes from the collecting tubules into a cavity and then into a ureter.

Figure 32·11 also shows a kidney that has been sliced in half. The outer portion of the kidney is called the *cortex*. In the cortex are the convoluted tubules, glomeruli, and Bowman's capsules of the nephrons. The center part of the kidney is the *medulla.* In this region are the Henle's loops and collecting tubules of the nephrons.

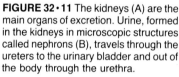

glomus (ball)

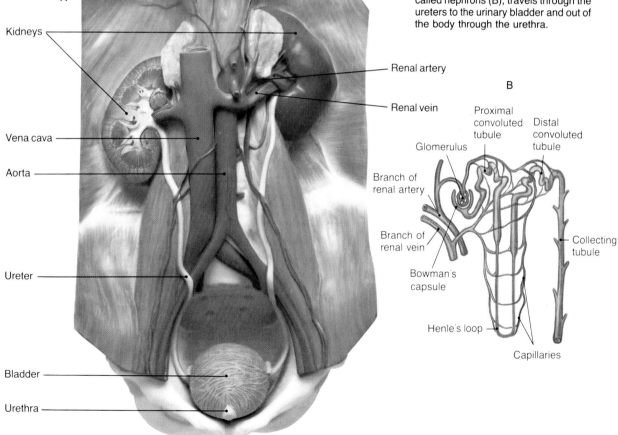

FIGURE 32·11 The kidneys (A) are the main organs of excretion. Urine, formed in the kidneys in microscopic structures called nephrons (B), travels through the ureters to the urinary bladder and out of the body through the urethra.

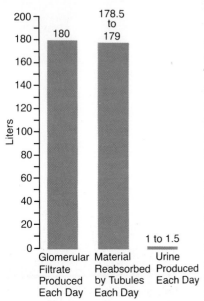

FIGURE 32·12 This bar graph shows the daily production of glomerular filtrate, material reabsorbed by tubules, and production of urine.

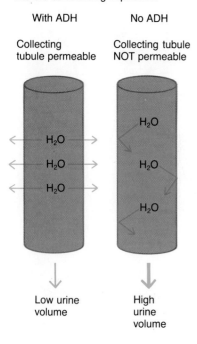

FIGURE 32·13 Antidiuretic hormone, or ADH, controls the reabsorption of water in the collecting tubules. In the presence of ADH, the collecting tubules become permeable to water, thereby increasing the reabsorption of water into the surrounding capillaries.

32·12 HOW NEPHRONS WORK

Blood enters the capillaries of the glomerulus under high pressure. In most tissues the pressure of capillary blood is about 20 mm of mercury. In the glomerulus, however, the pressure is about 60 mm. This high pressure forces much plasma out through the porous walls of the glomerular capillaries and into Bowman's capsule. Thus the glomerulus acts as a filter. Much of the plasma and most of the substances dissolved in the plasma pass through this filter. The blood cells and the plasma proteins remain behind in the capillaries. The fluid that accumulates in Bowman's capsule is the **glomerular filtrate.**

Much blood flows through the kidneys and is filtered. Glomerular filtrate is produced in large volumes. The kidneys of an average human produce about 180 L of glomerular filtrate each day. Glomerular filtrate is different from urine. It is similar to the blood plasma from which it is filtered. Glomerular filtrate contains many substances that the body needs. Sugars, amino acids, vitamins, and many other useful materials are found in this liquid. If these substances were not recovered by the kidneys, they would soon be in short supply. Study Figure 32·12. About how much of the daily production of filtrate is reabsorbed? About how much urine is produced each day?

Reabsorption of useful substances is the job of the tubules. As glomerular filtrate passes through the tubules of the kidney, many of the needed substances are reabsorbed. The cells that form the walls of the tubules carry on the reabsorption. The proximal and distal convoluted tubules and Henle's loop are surrounded by capillaries. The substances reabsorbed by the tubules pass into the blood in these capillaries. Some of the substances in the glomerular filtrate are reabsorbed by diffusion.

32·13 CONTROL OF WATER BALANCE

The last step in the formation of urine by the kidneys is the recovery of water. The reabsorption of water occurs mainly in the collecting tubules. This process is under the control of a *hormone*. (Hormones are discussed in detail in Chapter 34.) A hormone is a chemical substance produced in one organ that controls one or more activities in other organs. The hormone that regulates the reabsorption of water is called **antidiuretic** (an tee dī yu REHT ihk) **hormone** (abbreviated as ADH). ADH is released into the blood by the pituitary gland, which is at the base of the brain.

Diuresis (dī yu REE sihs) is the production of large volumes of urine. Diuresis would occur if one were to drink several glasses of water in a short time. Such action would create an oversupply of water in the body. The collecting tubules would not reabsorb as much water. The extra water would pass out of the body as dilute urine.

ADH causes the walls of the collecting tubules to become *permeable* to water. When the walls are permeable, water diffuses out of the collecting duct and into surrounding capillaries. Less water leaves with the urine and more water returns to the blood.

32·14 KIDNEY DISEASE

Kidney failure is a condition in which the kidneys do not rid the blood of urea or excess water or salt. This condition is serious because urea is poisonous. Excess water or salt may interfere with the functioning of many organs. Causes of kidney failure may be divided into two categories, those that produce low blood pressure in the kidneys and those that produce damaged tubules.

Recall that high blood pressure in the capillaries of the glomerulus forces the glomerular filtrate out into Bowman's capsule. If the pressure in these capillaries is not high enough, filtration will not occur. As a result, wastes cannot be removed from the blood.

Damage to the nephrons, the second kind of kidney failure, may result from many different factors. Certain drugs and heavy metals such as lead or mercury can damage the kidney tubules. Infections or allergic reactions in the kidneys also may cause such damage. Blockage of the ureters may cause urine to back up into the tubules of the kidney. The backed-up urine causes pressure in the wrong direction inside the nephrons. This reverse pressure can damage the tubules. The kidneys have some ability to repair damaged tubules. If the cause of damage lasts too long, the damage may be permanent.

A human can live with only one kidney. But if both kidneys fail, wastes rapidly build up in the blood. Unless the wastes are removed, death will result. One method of treating kidney failure is by a *kidney transplant*. A kidney transplant involves replacing the defective kidney with a healthy kidney obtained from a donor. The donor may be someone who died recently or a live person willing to donate a kidney. Kidney transplants are the most successful type of transplant surgery. Thousands of such operations have been performed in North America. The fact that a human can survive with only one kidney has led people to donate a kidney to a relative whose kidneys have failed.

Artificial filtering of the blood, or **hemodialysis** (hee moh dī AL-uh sihs), can be used to treat kidney failure. Hemodialysis is performed by machine. Such devices often are referred to as "kidney machines" or "artificial kidneys." The machine is connected to an artery and a vein, usually in the patient's arm or leg. The blood flows through a great length of cellophane tubing. A liquid called *dialysis fluid* flows on the other side of the cellophane. As blood and dialysis fluid flow past each other on opposite sides of the cellophane, urea diffuses from the blood to the dialysis fluid.

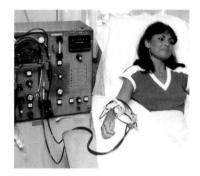

FIGURE 32·14 When a person's kidneys are not able to remove urea or water or salt from the body, a condition called kidney failure results. To treat kidney failure, hemodialysis, or artificial filtering of blood, is sometimes used.

REVIEW QUESTIONS

11. What waste substances are excreted by the body?
12. Describe the structure of a nephron.
13. Explain how ADH regulates water absorption.
14. What is hemodialysis?
15. A certain medicine a patient is taking inhibits ADH. What effect will this medicine have on the patient's volume of urine?

Chapter Review

SUMMARY

32·1 Breathing, or external respiration, provides the exchange of gases necessary for internal, or cellular, respiration.

32·2 Air passes through the nose, pharynx, trachea, and bronchi to reach the lungs.

32·3 The diaphragm and the intercostal muscles provide the force necessary for breathing movements.

32·4 Oxygen diffuses across the walls of the alveoli into the capillaries of the lungs. Carbon dioxide and water vapor diffuse in the opposite direction.

32·5 Breathing is controlled by the respiratory center. Excess carbon dioxide in the tissue fluids causes the brain to increase the rate of respiration.

32·6 Bacteria, viruses, pollen grains, and other foreign matter inhaled with the air cause several diseases of the respiratory system.

32·7 Blockage of the air passages may be relieved by the Heimlich maneuver. Mouth-to-mouth resuscitation can revive some persons whose breathing has stopped.

32·8 Smoking is thought to cause several problems in the respiratory system, including emphysema, lung cancer, and chronic bronchitis.

32·9 Exposure to high altitude creates hypoxia in the tissues. Both the respiratory and circulatory systems are involved in the response to hypoxia. Too much oxygen also creates problems of respiration and circulation.

32·10 The organs of excretion rid the body of urea and other wastes.

32·11 Each kidney is composed of about 1 million functional units called nephrons.

32·12 Formation of urine by the nephrons occurs in two stages: filtration of the blood by the glomeruli and reabsorption of useful materials by the tubules.

32·13 Reabsorption of water occurs in the collecting tubules. This process is controlled by the hormone ADH.

32·14 Kidney failure is a significant health problem that can be corrected by transplantation or by hemodialysis.

BIOLOGICAL TERMS

alveoli
antidiuretic hormone (ADH)
basal metabolic rate (BMR)
Bowman's capsule
bronchi
bronchioles
collecting tubule
diaphragm
distal convoluted tubule
diuresis
epiglottis
excretion
expiration
expiratory center
external respiration
glomerular filtrate
glomerulus
hemodialysis
Henle's loop
hypoxia
inspiration
inspiratory center
intercostal muscles
internal respiration
kidneys
larynx
lungs
nephrons
pharynx
pleura
proximal convoluted tubule
respiratory center
trachea
urea
ureters
urethra
urinary bladder
urine

USING BIOLOGICAL TERMS

1. What are nephrons?
2. Define *basal metabolic rate*.
3. How are diuresis and ADH related? Define both terms in your answer.
4. Distinguish between the proximal convoluted tubule and the distal convoluted tubule.
5. Identify and define a term that is a treatment for kidney failure.
6. Cellular respiration is to carbon dioxide as protein metabolism is to _____.
7. Distinguish between expiration and inspiration.

8. What is hypoxia?
9. Glomerulus is to kidney as _____ is to lung.
10. Distinguish between bronchi and bronchioles.

UNDERSTANDING CONCEPTS

1. How are external respiration and internal respiration linked? How do they differ? (32·1)
2. List all the structures that oxygen must pass through to get from the nose to the capillaries in the lungs. (32·2)
3. Describe what happens during inspiration. (32·3)
4. The muscles do less work during expiration than they do during inspiration. Explain why. (32·3)
5. What causes gases to be exchanged between the alveoli and the capillaries? (32·4)
6. How is the respiratory rate regulated? (32·5)
7. How is a person's basal metabolic rate affected by activity? (32·5)
8. Describe two diseases of the respiratory system. (32·6)
9. What is mouth-to-mouth resuscitation? (32·7)
10. How does the Heimlich maneuver force foreign objects out of the air passages? (32·7)
11. What structures in the respiratory system are likely to be the first to be damaged by cigarette smoke? (32·8)
12. What condition arises from travel to high altitudes? Explain how it occurs. (32·9)
13. How does the body respond to high altitude living over a period of time and to mild poisoning with carbon monoxide? (32·9)
14. What organs participate in the process of excretion? (32·10)
15. What parts of the nephron are found in the cortex of the kidney? What parts are in the medulla? (32·11)
16. The work of the kidneys is done in three stages. What are they, and in what parts of the nephron does each occur? (32·12, 13)
17. Explain the action of the antidiuretic hormone. (32·13)
18. Explain how hemodialysis works. (32·14)
19. Describe two causes of kidney failure. (32·14)

APPLYING CONCEPTS

1. How do respiration and excretion relate to homeostasis?
2. Why can metabolism be measured in terms of oxygen consumed?
3. List all the structures that a molecule of urea must pass through to get from the blood in the kidney to the outside of the body.
4. A large proportion of the body's daily use of energy occurs in the kidneys. Why do the kidneys require so much energy? What process in the kidneys do you think uses most of this energy?
5. What are some differences between the way a nephron works and the way a hemodialysis device works?

EXTENDING CONCEPTS

1. Some of the air in the lungs is never exchanged during respiration. Find out why and how much air remains in the lungs.
2. Some world-class runners train at high altitudes. What are some of the physiological effects of such a training routine?
3. Urea is formed as a result of protein breakdown. What is the chemical formula of urea? Why is urea not produced from the breakdown of fats or carbohydrates?
4. Contact the local chapter of the American Lung Association or of the Red Cross. Gather more information about lung disease and about cardiopulmonary resuscitation.
5. Prepare a report about the ill effects of smoking. Some have been mentioned in this chapter and some in the preceding chapter. Be as complete as possible.

READINGS

Arehart-Treichel, Joan. "Advances in Treating Kidney Disease." *Science News,* March 5, 1983, p. 150.

Dixon, Bernard (ed.). *Health, Medicine, and the Human Body.* Macmillan, 1986.

Gorman, L. "Chemical Key to Emphysema Found." *Science News,* March 26, 1983, p. 199.

Perlman, Eric. "For a Breath of Thin Air." *Science 83,* January/February 1983, p. 52.

33. The Nervous System

Nuclear magnetic resonance scanning (NMR) is a diagnostic technique that is used to study the fine details of the internal organs of the body. This colored picture of the brain was produced as a result of an NMR scan. As part of the human nervous system, the brain controls most of the activities that occur in the body. By reading this chapter, you will learn more about the brain and the other parts of the human nervous system.

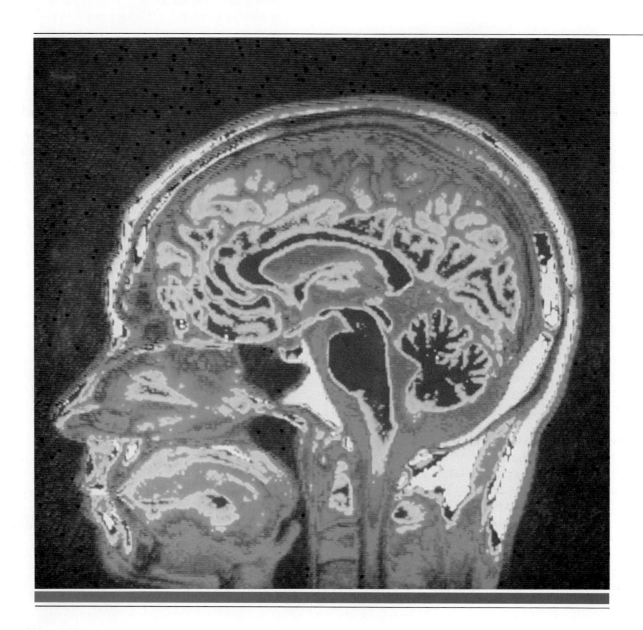

CHAPTER OBJECTIVES

After completing this chapter, you will be able to
- **Describe** the structures and functions of the nervous system.
- **Identify** the parts of the central and peripheral nervous systems.
- **Explain** the processes by which nerves conduct impulses.
- **List** the effects of certain drugs on the nervous system.
- **Describe** the structures and functions of the major sense organs.

Structure of the Nervous System

33•1 DIVISIONS OF THE NERVOUS SYSTEM

The **nervous system** is made up of the brain, the spinal cord, and nerves that extend throughout the body. Coordination of the different actions of the body is one of the two major functions of the nervous system. The second major function of the nervous system is coordination of the body's response to the outside world. The nervous system gets information about the outside world. It can send messages to parts of the body to respond to that information. The information and messages are in the form of electrical signals. The structures that receive information from the environment are called **receptors.** For example, the eyes have receptors that gather visual information about the outside world. In addition to receiving, relaying, and processing information from the environment, signals from the nervous system may cause some action or response to the information. The structures or organs that produce the response are referred to as **effectors.** Muscles and glands are examples of effectors. Information from the outside world may result in movement of the body through the action of effectors. Other information may result in a change within the body through secretions from glands.

The brain and spinal cord make up the **central nervous system.** This system receives and processes information from receptors. The processing of information includes determining what action to take in response to the received information. The central nervous system coordinates the body's organ systems.

Some actions controlled by the central nervous system are *voluntary*. Voluntary acts, such as movement of the skeletal muscles, are under conscious control. Other actions controlled by the central nervous system are *involuntary*. They occur automatically without conscious control over them. Maintaining the body at a steady temperature is an involuntary function involving the central nervous system.

> **SECTION OBJECTIVES**
> - Identify the structures of the nervous system.
> - Describe how the nervous system controls the internal functions of the body.

664 THE NERVOUS SYSTEM

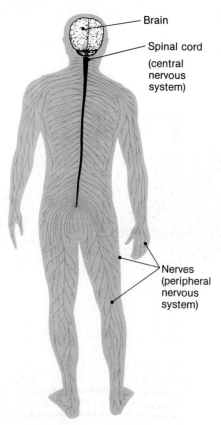

FIGURE 33·1 The human nervous system is divided into two parts. The central nervous system consists of the brain and spinal cord. The peripheral nervous system consists of the nerves that connect the central nervous system to all body parts.

FIGURE 33·2 A neuron, or nerve cell, carries nerve impulses from the dendrites, through the axon to the axon-endings.

The central nervous system is enclosed by protective membranes known as **meninges** (muh NIHN jeez). There are three layers of meninges around the brain and spinal cord. The space between the second and third layers is filled with *cerebrospinal* (sehr uh broh SPI-nuhl) *fluid*. Cerebrospinal fluid is a liquid cushion that protects the central nervous system from injury. This fluid flows around the brain and spinal cord. It also flows through a series of chambers and canals inside the brain and spinal cord.

The **peripheral** (puh RIHF uhr uhl) **nervous system** consists of the nerves that extend from the brain and spinal cord to all parts of the body. The function of this system is to relay messages between the central nervous system and the receptors and effectors that are found throughout the body. Figure 33·1 shows the central and peripheral nervous systems.

33·2 NERVE CELLS AND NERVES

The specialized cells of the nervous system are called **neurons** (NUR ahnz). Neurons carry electrical signals, called *nerve impulses*, from place to place within the nervous system. Look at the neuron shown in Figure 33·2. Notice that the nucleus is found within a region of the neuron called the *cell body*. Note also the many projections from the cell body. Most of these projections are dendrites. **Dendrites** are extensions of the neuron that carry nerve impulses toward the cell body. The **axon** is the extension of the neuron that carries nerve impulses away from the cell body. Some axons are more than a meter in length. The axon ends in several small knoblike structures called **axon-endings**. Section 33·9 will explain how impulses are transferred from one neuron to the next.

The axons of some neurons are coated with material called a *myelin sheath*. A type of cell known as a *Schwann cell* wraps around the axon many times. Numerous Schwann cells coat the axon along its length, forming the myelin sheath. Long dendrites of some neurons also may have myelin sheaths. The myelin sheath serves to insulate axons and to prevent impulses in one axon from interfering with impulses in other axons.

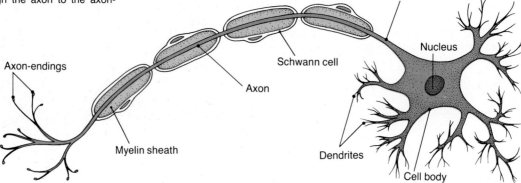

Axons and dendrites with myelin sheaths conduct impulses much more rapidly than do uncoated axons. Some myelinated axons conduct impulses at 100 m/s. Unmyelinated axons may conduct impulses as slowly as 1 m/s.

In the peripheral nervous system, axons and dendrites are grouped in bundles called *nerves*. Look at the cross section of the nerve in Figure 33·3. The bundle is wrapped in a sheath of connective tissue fibers that provides protection.

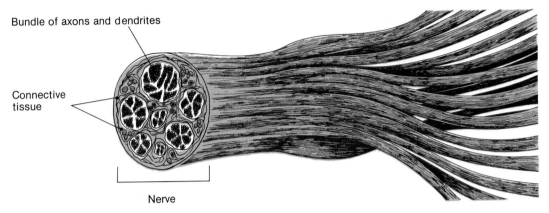

FIGURE 33·3 This cross section of a nerve shows that it is made up of bundles of axons and dendrites held together by connective tissue.

33·3 THE BRAIN

The **brain** is the major controlling structure of the nervous system. The human brain is an extremely complex organ. It is thought to contain over 10 billion neurons. Much progress has been made in recent decades in learning the details of brain function. But much more remains to be done. For example, it is known that memory is stored in the brain. But there is as yet no clear explanation of how memory storage occurs.

Figure 33·4 shows the brain. Notice that three major regions of the brain can be seen in this diagram. The largest part of the brain is the cerebrum (SEHR uh bruhm). Conscious sensation and voluntary control of the skeletal muscles are among the many activities controlled by the **cerebrum.** Below the cerebrum and to the rear of the brain is the cerebellum (sehr uh BEHL uhm). The **cerebellum** is concerned with coordination of muscular activities, thus producing smooth movements. Beneath the brain is the brainstem, which is continuous with the spinal cord. The **brainstem** regulates a number of internal functions, including the heartbeat.

The outer layer of the cerebrum is the *cerebral cortex*. This layer is the distinctive part of the human brain. Human thought and ability to reason are believed to be functions of the cerebral cortex. Lower vertebrates have no cerebral cortex at all. Mammals other than humans have a cortex, but it is not as developed as it is in humans. The cerebral cortex is made up of a tissue called *gray matter*. Gray matter contains the cell bodies of neurons along with many dendrites and axons that do not have myelin sheaths. Gray matter is a tissue in which neurons

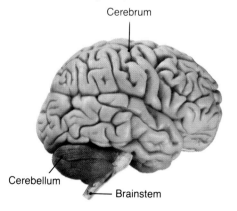

FIGURE 33·4 The human brain is divided into three regions. The cerebrum is the largest part of the brain and controls voluntary movement and conscious sensation. The cerebellum controls coordination and balance. The brain stem controls involuntary actions, such as breathing and peristalsis.

666 THE NERVOUS SYSTEM

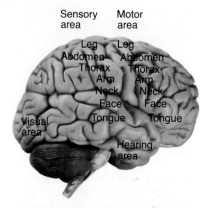

FIGURE 33·5 The cerebral cortex, the outer layer of the cerebrum, has many convolutions or folds. Certain areas of the cerebral cortex control specific functions.

interact with one another. The inner parts of the cerebrum, along with other parts of the brain, are made up of *white matter*. This tissue consists mostly of myelinated axons, which are whitish. White matter is a tissue through which impulses pass on the way to other areas.

Notice the twisted grooves and ridges in the surface of the cerebrum. The ridges are called *convolutions*. Because of this folding, a large amount of brain tissue can fit into the skull. The grooves serve as "landmarks" in mapping the functions of the different parts of the cerebrum. The cerebrum is divided into right and left halves by a large groove. These halves are the *cerebral hemispheres*. In general, the right cerebral hemisphere controls activities on the left side of the body. The left hemisphere controls activities on the right side of the body. Some functions, however, seem to be centered in only one hemisphere. The control of speech and the use of language are thought to take place in the left cerebral hemisphere of most people. Artistic and musical abilities are believed to be centered in the right hemisphere.

Activities equally divided between the two cerebral hemispheres include controlling the skeletal muscles and interpreting sensations from the surface of the body. These activities are controlled from the cerebral cortex. Figure 33·5 shows a map of the functions of the cerebral cortex.

Figure 33·6 shows an inner view of the brain. Notice the hypothalamus (hī puh THAL uh muhs) near the base of the brain. The **hypothalamus** is concerned with involuntary control of many basic body functions. It is a feature of the brains of all vertebrates. In endothermic animals, including humans, the hypothalamus regulates body temperature. It is also thought to regulate food intake and sexual activity. In the next chapter you will learn how the hypothalamus controls many of the body's glands.

FIGURE 33·6 Located within the brain is a structure called the hypothalamus. The hypothalamus controls body temperature, blood pressure, sleep, and emotions. Notice that the brain stem is divided into two parts: the pons and the medulla oblongata.

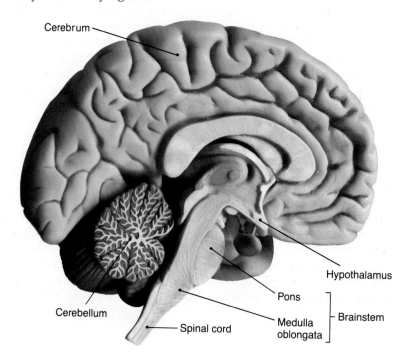

The cerebellum works with the cerebrum to control the skeletal muscles to produce smooth, useful movement. Impulses that cause the skeletal muscles to contract begin in the cerebrum. These impulses pass through the cerebellum, where action is taken to achieve coordination.

The brainstem contains several structures. The **medulla oblongata** (mih DUHL uh ahb lahng GAHD uh) contains centers that control breathing and heartbeat. The medulla connects the other structures of the brain to the spinal cord. Many of the nerves pass through a region called the *pons*.

33·4 THE SPINAL CORD

The **spinal cord** is a cord of nerve tissue located within the vertebral column. A cross section of the spinal cord is shown in Figure 33·7. The white matter of the spinal cord consists mainly of myelinated axons that carry impulses to or from the brain. The gray matter contains cell bodies. It is a region in which many contacts between neurons occur.

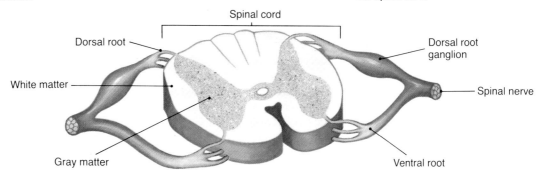

FIGURE 33·7 The cross section of the spinal cord shows that it is composed of gray matter and white matter. The gray matter consists of nerve cell bodies. The white matter consists mainly of myelinated axons. Thirty-one pairs of spinal nerves exit through openings in the bony vertebral column that encases the spinal cord.

The spinal cord gives rise to 31 pairs of spinal nerves. The spinal nerves exit between the bones of the vertebral column and connect to all parts of the body. Spinal nerves are connected to the spinal cord by two *roots*. The *dorsal root* of each spinal nerve contains sensory neurons. **Sensory neurons** are neurons that carry impulses from receptors to the central nervous system. The *ventral root* of each spinal nerve contains motor neurons. **Motor neurons** are neurons that carry impulses from the central nervous system to the effectors.

Within the gray matter of the spinal cord are found association neurons. **Association neurons,** also called *interneurons,* lie totally within the central nervous system and serve as connections between other neurons.

The dorsal root of the spinal nerve contains a thickening called the *dorsal root ganglion* (GANG glee uhn). A **ganglion** is a mass of cell bodies of neurons found outside the brain or spinal cord. The dorsal root ganglion contains the cell bodies of sensory neurons. The dendrites of these sensory neurons may reach such distant locations as the skin of the fingers. The axons of these sensory neurons pass from the dorsal root ganglion into the gray matter of the spinal cord.

ganglion (a swelling)

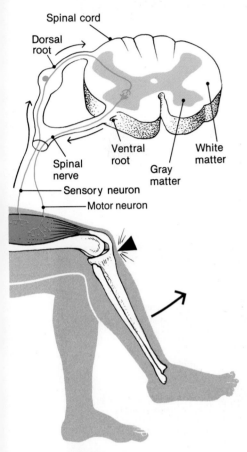

FIGURE 33·8 The knee jerk is an example of a reflex arc. A reflex arc is the pathway over which nerve impulses travel in a reflex.

auto- (self)
nomos (law)

33·5 PERIPHERAL NERVOUS SYSTEM

The peripheral nervous system consists of the nerves that connect the central nervous system to other parts of the body. It includes many ganglia, the 31 pairs of spinal nerves, and the 12 pairs of *cranial nerves.* Each cranial nerve enters the brain directly. The peripheral nervous system consists entirely of motor and sensory neurons.

Some activities are handled automatically by the peripheral nervous system and spinal cord without involving the brain. Such activities are called *reflex actions.* Lifting the foot in response to having stepped on a sharp object is an example of a reflex action. The brain becomes conscious of this action, but it neither causes nor controls the raising of the foot.

Figure 33·8 diagrams the pathway of the impulses that cause the leg to be raised in response to being tapped. The peripheral nervous system pathway for the impulses in an automatic response is called a **reflex arc.** The receptors for touch are the dendrites of sensory neurons under the surface of the skin. Tapping the knee acts as a stimulus to the sensory neuron.

Use Figure 33·8 to trace the steps in the knee jerk, which is an example of a reflex arc. The stimulus causes an impulse to be carried toward the spinal cord. A sensory neuron carries the impulse. The cell body of the sensory neuron is in the dorsal root ganglion. Its axon ends in the gray matter of the spinal cord. In most reflex arcs, but not in the knee jerk, the impulse next travels along an association neuron within the gray matter. Notice that a gap separates the axon-ending of the sensory neuron from the dendrites of the association neuron. The gap that an impulse must cross to get from one neuron to the next is called a **synapse** (sih NAPS).

The association neuron carries the impulse to a synapse with a motor neuron. The axon of this motor neuron exits the spinal cord through the ventral root and reaches a muscle in the leg. The reflex arc is completed as the impulse causes the muscle to contract and the leg to be raised.

33·6 AUTONOMIC NERVOUS SYSTEM

The **autonomic nervous system** is the portion of the peripheral nervous system that provides involuntary control of the internal organs. Control of such organs as the heart and lungs occurs independently of conscious thought. The autonomic nervous system contains two subdivisions. One subdivision, the **parasympathetic nervous system,** controls the internal organs during routine conditions. The second subdivision, the **sympathetic nervous system,** controls the internal organs during stressful situations and increased activity.

Most internal organs are contacted by axons from both the parasympathetic and sympathetic nervous systems. The two systems generally have opposite effects on a given organ. For example, during routine conditions, impulses carried by motor neurons of the parasympathetic nervous system act to slow the heartbeat. During heavy

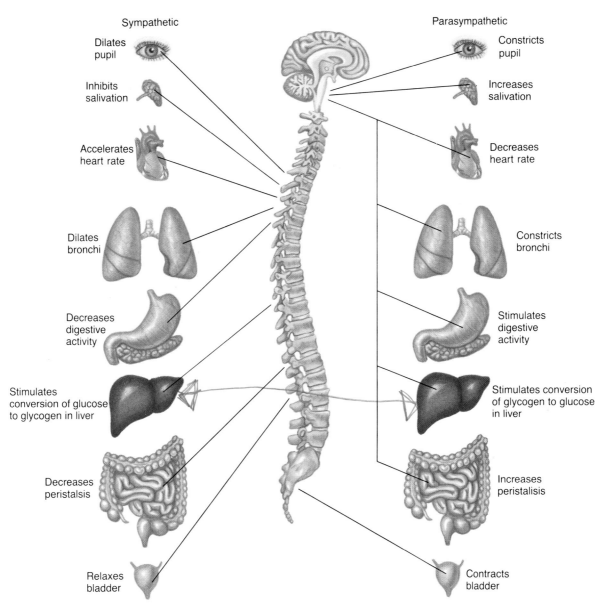

exercise or during a stressful situation, such as an injury, the sympathetic nervous system takes over. Impulses carried by motor neurons of this subdivision cause the heartbeat to speed up.

FIGURE 33·9 The autonomic nervous system has two subdivisions: the sympathetic nervous system and the parasympathetic nervous system. The sympathetic nervous system helps the body deal with emergency situations, whereas the parasympathetic nervous system controls normal, relaxed functioning of the body.

REVIEW QUESTIONS

1. What structures are part of the central nervous system? What structures are part of the peripheral nervous system?
2. What are the three main parts of a neuron?
3. What part of the brain is more highly developed in humans than in other mammals? What are its functions?
4. What is a reflex arc?
5. What effects do you think the two subdivisions of the autonomic nervous system would have on the rate of breathing?

Conduction of Impulses

SECTION OBJECTIVES

- Identify the structures of a neuron.
- Describe a nerve impulse.

33·7 RESTING STATE OF NEURONS

A nerve impulse is an electrical signal that travels along the surface of a neuron. A neuron not conducting such a signal is said to be in the *resting state*. A neuron in the resting state is not completely at rest. Although it is not producing impulses, it is active in other ways. For example, ions are moved through the cell membrane.

Ions are moved through the cell membrane of a neuron by a mechanism called the sodium-potassium pump. The **sodium-potassium pump** is a process that moves sodium ions from the inside of the neuron to the tissue fluid outside. Potassium is pumped in the opposite direction, from the tissue fluid into the neuron. The sodium-potassium pump uses energy in the form of ATP. Experiments have suggested that three sodium ions are pumped out of the neuron for every two potassium ions pumped inward.

The action of the sodium-potassium pump produces concentration gradients for sodium and potassium ions. As a result of pumping, there is a high concentration of sodium on the outside of the neuron and a high concentration of potassium on the inside of the neuron. These concentration gradients are shown in Figure 33·10. Both sodium ions and potassium ions carry a positive electrical charge.

In addition to being pumped, sodium and potassium ions can move through the cell membrane of a neuron by diffusion. Diffusion of sodium and potassium is in the opposite direction from pumping. Sodium diffuses inward and potassium diffuses outward because of their concentration gradients. Because the neuron's cell membrane is many times more permeable to potassium than it is to sodium, more potassium diffuses outward than sodium diffuses inward. For this reason, it is said that the membrane is "leaky" to potassium.

FIGURE 33·10 As a result of the action of the sodium-potassium pump, the outside of the membrane of a resting neuron has a high concentration of sodium ions and the inside of the membrane has a high concentration of potassium ions.

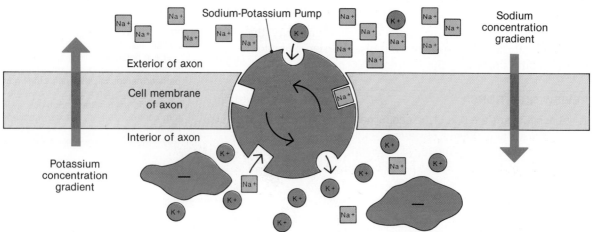

The combined result of the sodium-potassium pump and the leaking of potassium is that there are more positive ions on the outside of the neuron than on the inside. Compared to the outside, the inside of the membrane is negatively charged. The distribution of charges on either side of the membrane of a resting neuron is shown in Figure 33·11A. This condition, in which the two sides of the membrane are oppositely charged, is referred to as *polarization*. The cell membrane of a neuron is always polarized during the resting state. The condition in which the outside of the membrane is positively charged with respect to the inside is called the **resting membrane potential**.

33·8 THE NERVE IMPULSE

When a neuron is stimulated, an impulse is conducted from the point of stimulus to the opposite end of the neuron. Normally, neurons are stimulated by receptors or by other neurons. The stimulation of a neuron causes the properties of the cell membrane to change suddenly. The membrane becomes highly permeable to sodium ions. Remember that during the resting state, there is a high concentration of sodium on the outside of the membrane. When the membrane suddenly becomes permeable, many sodium ions rapidly diffuse inward. The result is that there are more positive charges on the inside of the neuron than on the outside. The neuron, at the point of stimulation, has lost polarization. It is said to be *depolarized*. See Figure 33·11.

The cell membrane of the neuron first becomes depolarized at the point of stimulus. The depolarization disturbs the next portions of the membrane, causing them to become permeable to sodium and thus depolarized. A wave of depolarization spreads away from the point of stimulation.

The depolarization of the membrane travels over the neuron as a wave. This wave of depolarization is the **action potential**. The action potential is the nerve impulse. As the action potential moves along, the area of the membrane that was first depolarized begins to recover. The sodium-potassium pump moves the sodium ions back to the outside of the membrane. Figure 33·11C shows the repolarization of the membrane. The polarized condition of the resting state is restored in a wave that follows behind the action potential.

The production of an action potential by a neuron is an all-or-none response. *All-or-none response* means that a neuron either experiences an action potential or it does not. There is no intermediate response. A stimulus is either strong enough to produce an action potential or it is not. The smallest stimulus that can produce an action potential in a neuron is called a threshold stimulus.

Neurons are capable of responding to many stimuli in succession. However, if the stimuli occur too close together in time, a neuron may not be able to respond. After producing an action potential there is a brief period during which a neuron cannot respond to further stimuli. This **refractory period** is the time during which the sodium-potassium pumps are restoring the polarization of the cell membrane.

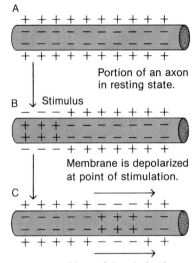

Portion of an axon in resting state.

Membrane is depolarized at point of stimulation.

Wave of depolarization travels along axon. Membrane is re-polarized at point where stimulus was applied.

FIGURE 33·11 A resting neuron has a positive electric charge on the outside of the cell membrane and a negative charge on the inside (A). When the neuron is stimulated, a section of the neuron becomes depolarized (B). Depolarization occurs when the outside of the cell membrane becomes negatively charged and the inside becomes positively charged. This change moves down the axon in a wave (C).

672 THE NERVOUS SYSTEM

33·9 CONDUCTION AT THE SYNAPSE

In the discussion of the reflex arc (Section 33·5), it was noted that impulses traveling from neuron to neuron must cross a gap called a synapse. Although action potentials can move in any direction over a neuron, they can cross a synapse in only one direction. Impulses crossing a synapse *always* begin at an axon-ending. Synapses usually occur between an axon-ending of one neuron and the dendrite of another neuron or between an axon-ending and a cell body, as shown in Figure 33·12. Less frequently, synapses may occur between an axon-ending and an axon. Neurons of the peripheral nervous system may form one, two, or several synapses. Neurons of the central nervous system form several thousand synapses each.

While conduction of an impulse along a neuron is electrical, conduction across synapses is achieved by chemicals. **Neurotransmitters** are the chemicals that carry impulses across synapses. The axon-ending contains structures called *synaptic vesicles,* which store the neurotransmitter.

The events that occur as an impulse crosses a synapse are shown in Figure 33·13. Follow the events in the diagram as you read this paragraph. When an action potential arrives at an axon-ending of one neuron, several of the synaptic vesicles move toward the cell membrane. The membranes of the vesicles fuse with the cell membrane. The neurotransmitter is released into the space between neurons. The neurotransmitter diffuses across the space and attaches to a *receptor protein* on the membrane of the second neuron. (The receptor protein of the synapse is not to be confused with receptors that receive information from the environment.) Attachment of the neurotransmitter to the receptor protein causes an action potential to be produced in the second neuron. An enzyme then breaks the neurotransmitter molecule into pieces. The pieces diffuse back into the axon-ending from which they were released. They are reassembled into neurotransmitter molecules that can be used again.

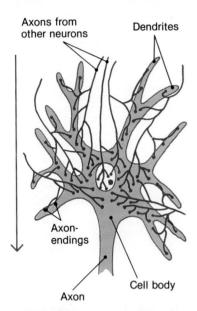

FIGURE 33·12 A synapse occurs between the axon-ending of one neuron and the dendrite or cell body of another neuron.

FIGURE 33·13 When a nerve impulse reaches an axon-ending, some synaptic vesicles release neurotransmitters. The neurotransmitters diffuse across the synapse and initiate the nerve impulse in the adjacent neuron.

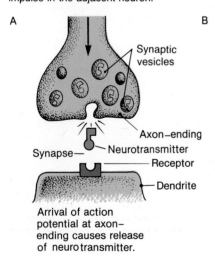

A

Arrival of action potential at axon-ending causes release of neurotransmitter.

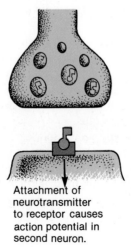

B

Attachment of neurotransmitter to receptor causes action potential in second neuron.

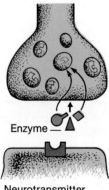

C

Neurotransmitter is split by enzyme. Pieces are reused in axon-ending.

There are two neurotransmitters in the peripheral nervous system. *Acetylcholine* (uh see tuhl KOH leen) is the neurotransmitter of synapses of the motor neurons that control skeletal muscles. Acetylcholine is also a neurotransmitter of some neurons of the autonomic nervous system. The neurotransmitter *noradrenaline* (nor uh DREHN-uh lihn) is found in some of the synapses of the sympathetic nervous system. It appears that a great variety of substances can function as neurotransmitters within the brain (Section 33·10).

The type of synapse described in this section is a *stimulatory synapse*. The central nervous system also contains *inhibitory synapses*. These synapses are similar to stimulatory synapses except that they contain inhibitory neurotransmitters. In an inhibitory synapse the neurotransmitter does not cause an action potential to be set up in the second neuron. Instead this neurotransmitter causes the second neuron to be *resistant* to forming action potentials. Thus an inhibitory synapse "turns off" the second neuron.

33·10 BRAIN CHEMISTRY

In recent years much research has been done on the chemistry of the brain. As a result a number of substances have been found that act as neurotransmitters only within the central nervous system. These substances include amino acids and peptides.

It was once thought that amino acids function only in the building of proteins. Aspartic acid and glutamic acid are amino acids that are now believed to serve as stimulatory neurotransmitters in the brain. Glycine appears to function as an inhibitory neurotransmitter in the spinal cord.

Another neurotransmitter found in the central nervous system is dopamine, which is similar to noradrenaline. Noradrenaline is produced from the amino acid tyrosine by the following reactions.

tyrosine → DOPA → dopamine → noradrenaline

These reactions represent the addition and removal of small groups of atoms to and from tyrosine to form the neurotransmitter molecules. Some neurons within the cerebrum have dopamine rather than noradrenaline as their neurotransmitter.

Perhaps the most exciting discovery of recent years has been the identification of a number of *peptides* that may act as neurotransmitters in the brain. Recall that peptides are short chains of amino acids. Peptides active in the nervous system are called *neuropeptides*.

Substance P is a neuropeptide that is found in neurons of the spinal cord that conduct pain sensations to the brain. It also is found in neurons of the cerebrum involved with control of movement.

The *endorphins* (ehn DOR fihns) and *enkephalins* (ehn KEHF uh-lihns) are two main groups of neuropeptides that have been identified in many parts of the brain. The endorphins contain 16 to 31 amino acids in their peptide chains. Enkephalins contain 5 amino acids and are produced by the splitting of the larger endorphin molecules.

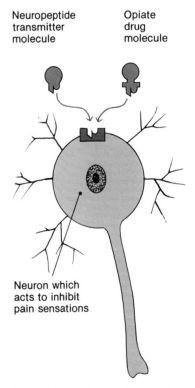

FIGURE 33·14 Certain drugs, such as opiates, may affect brain cells because their molecules are similar in shape to the molecules of neurotransmitters.

It appears that the endorphins and enkephalins serve as neurotransmitters for neurons that reduce pain and anxiety. Remember that neurotransmitters attach to receptor proteins after crossing a synapse. Research suggests that opiate drugs (opium, heroin, and morphine) affect the nervous system by attaching to receptors for endorphins and enkephalins. Further research may lead to a better understanding of drug addiction and possibly better methods for treating addiction.

33•11 DRUGS AND THE NERVOUS SYSTEM

Drugs are often used by people for medical reasons. Sometimes drugs are abused because of the effects that they have on the body. Many drugs have effects on the central nervous system.

Stimulants are drugs that increase the number of impulses conducted along neurons and at synapses. In other words, they increase the activity of the nervous system. Caffeine is a stimulant found in coffee, tea, and some colas. Caffeine stimulates activity in the cerebral cortex of the brain. Nicotine, a stimulant in cigarette smoke, increases the activity at synapses in which acetylcholine is the neurotransmitter. The shape of the nicotine molecule is similar to acetylcholine. Amphetamines (am FEHT uh meenz) are a group of drugs that are prescribed as diet pills to help obese people lose weight. Amphetamines increase activity at synapses where noradrenaline is found.

Depressants are drugs that decrease the activity of the nervous system. Ethyl alcohol, the depressant found in alcoholic beverages and liquors, interferes with transmission of nerve impulses at synapses. Barbiturates (bahr BIHCH uh rayts), often prescribed as sleeping pills, slow transmission in the center of the brain and then in the cerebral cortex. Opiates are a group of drugs made from the poppy plant. They include opium, heroin, and morphine. These drugs dull pain and slow impulse movement at various places in the nervous system.

Hallucinogenic (huh loo suh nuh JEHN ihk) **drugs** affect the user's perception of reality. Marijuana is a drug that affects electrical activity, nervous system enzymes, and the chemistry of the brain. The exact nature of the effects is not fully understood. LSD is a hallucinogenic drug that is similar to a certain brain transmitter. However, it does not transmit signals in the same way the brain transmitter does. The result is mixed and distorted signals.

Each of these types of drugs affects more than just the nervous system. Stimulants increase heartbeat, blood pressure, and respiration rate. Depressants slow these body processes. Other drugs have a variety of effects on different body systems. Because of the effects that drugs have on the nervous system and other body systems, they are very dangerous when abused. Also, there is much that is not yet known about drug effects. Using a drug without knowing the effects is dangerous. Many drugs that are used legitimately as medicines also have side effects. Side effects are effects other than the intended ones. Side effects may be harmful.

Discovery

Alzheimer's Disease

Nearly two million Americans past the age of 50 have a progressive and incurable brain disorder called *Alzheimer's* (ALTS hī muhrz) *disease*. Alzheimer's disease has three psychological phases: forgetfulness, confusion, and dementia. During the forgetfulness phase, there are annoying problems with learning and memory.

During the confusion phase, personality disturbances may be accompanied by hallucinations or by delusions. The person has trouble concentrating on purposeful activity.

In the dementia (mental deterioration) phase, victims lose all ability to carry on normal social or occupational activity. They cannot hold onto a thought long enough to understand or communicate.

The symptoms of Alzheimer's disease are related to two changes in the brain. First, twisted strands of protein fibers clog some of the nerve networks. Second, many nerve endings break down, forming spots called *senile plaques*. As these plaques increase, a reduction occurs in the ability to reason and communicate. The causes of these abnormal changes in the brain are as yet unknown.

33·12 NERVOUS DISORDERS

With its billions of neurons, the brain is in a state of constant electrical activity. This activity results in electrical waves that can be detected on the outside of the head. A device called an *electroencephalograph* (ih lehk troh ehn SEHF uh luh graf), or EEG, is used to record these brain waves. Several types of brain waves are recorded by an EEG. One type of wave is most apparent during restful activity. Another sort of wave appears during sleep. A third type appears when an individual is under emotional stress. Recording changes in brain waves by an EEG is an important method for identifying disorders of the brain. A variety of disorders can affect the brain and other parts of the nervous system.

Epilepsy (EHP uh lehp see) is a serious and often permanent disorder. Abnormal electrical impulses throughout the brain are a feature of this disorder. The electrical activity can be detected as abnormal waves on an EEG recording. In severe cases this abnormal electrical activity may cause a seizure, or violent contractions of the skeletal muscles, and unconsciousness. Seizures occur with periods of normal activity in between. Many cases of epilepsy can be controlled with medication, allowing the victims to live normal lives.

Parkinson's disease is a disorder most often found in the elderly. It involves contraction of the skeletal muscles, which results in rigid joints and uncontrollable trembling. Parkinson's disease appears to affect the region of the cerebral cortex that controls skeletal muscles. Some of the neurons in this region have dopamine as their neurotransmitter. It appears that not enough dopamine is produced in individuals with Parkinson's disease. The disease can often be controlled by medication.

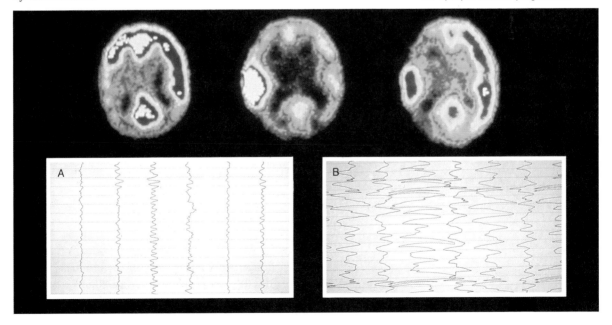

FIGURE 33·15 Compare the electroencephalograph, or EEG, reading of the brain (bottom) under normal conditions (A) and during an epileptic seizure (B). The changing colors in the brain scans (top), from left to right, indicate the change in activity of the neurons as the epileptic seizure progresses.

Multiple sclerosis (sklih ROH sihs) is a fairly common disorder of the nervous system. It results from destruction of the myelin sheaths of neurons in the central nervous system. When this insulation is lost, electrical impulses may jump from one neuron to another without crossing at a synapse. This incorrect routing of impulses generally leads to confused sensations and loss of control of the muscles. The cause of multiple sclerosis is unknown.

REVIEW QUESTIONS

6. How do neurons become polarized?
7. How do neurons become depolarized?
8. What structures are found in axon-endings but not in other parts of neurons?
9. How do stimulatory and inhibitory neurotransmitters differ?
10. Action potentials travel across neurons from dendrites or the cell body to the axon and axon-endings. Explain why action potentials ordinarily do not travel in the opposite direction.

The Senses

SECTION OBJECTIVES

- Identify the types of receptors that gather information about the environment.
- Describe the sense organs that contain the different types of receptors.

33·13 TASTE AND SMELL

Recall that receptors obtain information from the environment. Receptors are structures specialized for receiving particular stimuli. There are several types of receptors in the body. **Chemoreceptors** are receptors that are stimulated by chemicals in air, water, or food. The nose and mouth contain chemoreceptors. **Mechanoreceptors** are receptors that respond to vibrations, pressure, or other mechanical stimuli. The skin and the ears contain mechanoreceptors. **Photoreceptors** are receptors that are stimulated by light. The eyes contain these receptors. Receptors may exist individually or they may be gathered together in special structures known as sense organs.

The tongue contains many *taste buds,* which are the sense organs of taste. There are many small projections on the tongue's upper surface. These projections are called *papillae* (puh PIHL ee). The taste buds, which are microscopic, are contained within the papillae. Examine the taste bud shown in Figure 33·16C. Observe that the receptor cells in the taste bud are open to the surface of the tongue through a pore. These cells are chemoreceptors. Molecules in food stimulate the microvilli, which are tiny fingerlike projections, on the exposed surface of the receptor cells, causing impulses to be sent to the brain. The impulses are interpreted in the brain to produce the sensation of taste.

THE NERVOUS SYSTEM 677

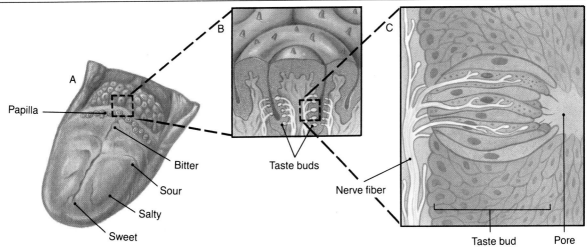

FIGURE 33·16 The taste buds are examples of chemoreceptors, or receptors stimulated by chemicals, that are located in the tongue.

Not all taste buds are sensitive to the same molecules. Figure 33·16A shows that taste buds in different areas of the tongue are sensitive to different substances. Where on the tongue is the greatest sensitivity to sweet flavors?

The *olfactory epithelium* is the sense organ for detecting odors. Like taste buds, olfactory epithelium contains chemoreceptor cells. Figure 33·17A shows the location of the olfactory epithelium. Notice the *olfactory nerve,* which carries impulses to the brain. The olfactory nerves make up one of the 12 pairs of cranial nerves.

Figure 33·17B gives a detailed view of the olfactory epithelium. This epithelium is coated with mucus, as are all of the nasal linings. Notice the olfactory hairs in the diagram. It is believed that odor molecules are detected when they attach to receptor proteins on the surface of the olfactory hairs.

The senses of smell and taste are closely related. Both involve stimulation of chemoreceptor cells. Much of the taste of food is actually a combination of taste and odor. For this reason, food often does not taste right to a person who has a head cold.

33·14 TOUCH AND PAIN

Mechanoreceptors that detect touch, pressure, and pain are distributed throughout the skin. As noted in the discussion of reflex arcs, the receptors for pain are the dendrites of sensory neurons. Such dendrites are usually found at the base of the epidermis of the skin.

Several types of receptors detect touch. One such receptor is a network of dendrites surrounding the base of a hair follicle. Hairs are found on almost all parts of the skin, although many are too small to be seen easily. When contact with an object causes a hair to move, the dendrites around the base of the hair are stimulated. The dendrites then send an impulse to the central nervous system.

FIGURE 33·17 When the olfactory hairs in the nose are stimulated by odor molecules, impulses travel from the olfactory cells, to the olfactory nerve and then to the brain.

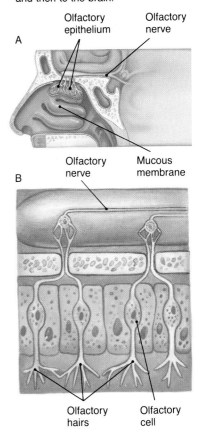

678 THE NERVOUS SYSTEM

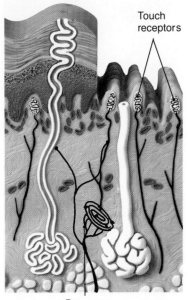

FIGURE 33·18 In addition to touch and pressure receptors, the skin also contains receptors for heat, cold, and pain.

FIGURE 33·19 Muscles attached to the eyeball (left) enable it to move in a circular motion. Notice the structures that make up the eye (right).

In addition to touch receptors, the skin contains receptors for pressure. Observe in Figure 33·18 that pressure receptors are located much deeper in the skin than are touch receptors. It is partly for this reason that a person can distinguish between a light touch on the skin and strong pressure. The skin also contains separate receptors for detecting heat and cold. Receptors for heat, cold, and pressure are less abundant in the skin than are touch receptors.

33·15 VISION

The eye is the sense organ of vision, and it contains a large number of photoreceptor cells. Examine the diagram of the eye in Figure 33·19. You will see that most of the eye is surrounded by a layer of connective tissue fibers called the *sclera*. The sclera is tough and protective, yet it is flexible and allows the eye to maintain its shape. Toward the front of the eye, the sclera merges with the *cornea*. The cornea also serves as a protective outer covering, but it is transparent and allows light to enter the eye. The sclera and cornea together form the outermost of the three layers of tissue that form the eye.

The second layer of tissue in the eye is the *choroid*. This layer contains the many small blood vessels that nourish the eye. The choroid is very darkly colored on its inner surface. The dark pigment absorbs stray light and prevents it from being reflected within the eye. At the front of the eye the choroid layer forms the *iris*. The pigment of the iris is the brown, green, or blue color seen from the outside. Notice in Figure 33·19 that there is a hole in the middle of the iris. This hole, the *pupil,* allows light to pass to the interior of the eye. The iris can enlarge or constrict the pupil, which regulates the amount of light entering the eye. In dim light the pupil opens wide to let in more light.

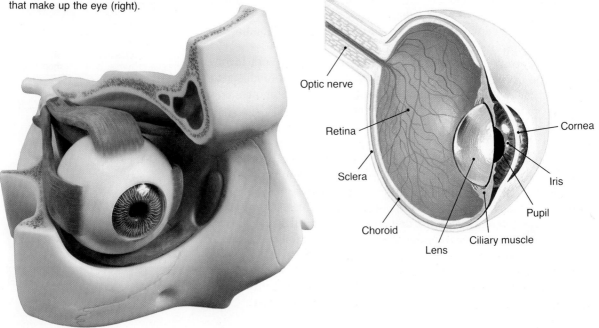

The ciliary muscle is circular and serves to control the shape of the lens. The *lens* is the transparent structure that focuses an image at the back of the eye just as the lens of a camera focuses an image on film. When the ciliary muscle contracts, the lens becomes wider and flatter. In this shape the lens is focused on distant objects. When the ciliary muscle relaxes, the lens becomes thicker and more curved. In this shape the lens is focused on nearby objects.

The innermost of the three layers of the eye is the retina. The *retina* is the layer of the eye that contains the photoreceptor cells. Images focused onto the retina by the lens are received by two types of photoreceptors. The *rods* are photoreceptors adapted to function in dim light. They cannot detect color. You may have noticed that it is difficult to distinguish colors at night or in a dimly lit room. *Cones* are the second type of photoreceptor in the retina. These receptors work best in bright light. They are responsible for color vision. Both rods and cones form synapses with sensory neurons in the retina. The axons of the sensory neurons exit each eye as the *optic nerve*.

33•16 HEARING AND BALANCE

The ear is a sense organ with two functions. It contains the mechanoreceptors that detect the vibrations that the brain interprets as sound. It also contains other mechanoreceptors that are responsible for the sense of balance, or *equilibrium*. The ear consists of three regions. The *external ear* is the portion that can be seen from the outside. The *middle ear* and the *inner ear* are within the skull.

Sound consists of vibrations of the air. The function of the external ear is to gather sound waves and pass them to the middle ear. The external ear consists of three parts. The *auricle* is the external structure commonly referred to as the ear. The *auditory canal* carries sound waves inward from the auricle. The auditory canal ends at the eardrum, or *tympanic membrane*. Sound waves in the air cause the tympanic membrane to vibrate.

The *Eustachian* (yoo STAY kee uhn) *tube* connects the cavity of the middle ear to the pharynx. Air enters the middle ear through the Eustachian tube. Because of this, air pressure is the same on both sides of the tympanic membrane. There are three small bones within the middle ear. These bones are known as the *hammer, anvil,* and *stirrup*. The hammer is attached to the tympanic membrane. All three bones are attached in a row by small ligaments. Sound vibrations pass from the tympanic membrane through each of the three bones.

The inner ear contains a structure that resembles a snail shell. This structure is the **cochlea** (KAHK lee uh), a bony chamber containing fluid and the mechanoreceptors for hearing. The end of the cochlea adjoining the middle ear contains a membrane-covered window. The stirrup rests against this window. When the bones of the middle ear are set to vibrating by sound waves, the vibrations are passed from the stirrup to the window of the cochlea. The vibration of the window, in turn, causes the fluid within the cochlea to vibrate.

kochlias (snail shell)

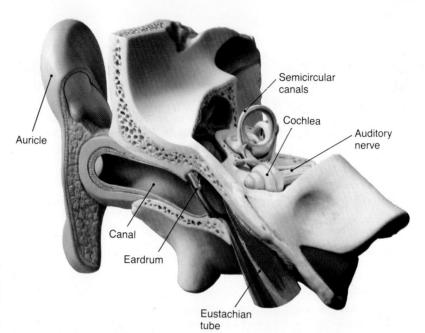

FIGURE 33·20 Sound vibrations entering the ear travel through the outer, middle, and inner parts of the ear to the auditory nerve. The auditory nerve carries the impulses to the brain.

Vibration of the fluid within the cochlea stimulates the mechanoreceptor cells. Receptors at the end of the cochlea nearest the stirrup respond to high-pitched sounds. Receptors at the far end of the cochlea respond to low-pitched sounds. When the mechanoreceptors are stimulated, they send impulses through the *auditory nerve* to the brain. The auditory nerves make up another of the 12 pairs of cranial nerves.

Notice in Figure 33·20 that the inner ear includes three semicircular canals alongside the cochlea. The **semicircular canals** are the structures associated with equilibrium. Like the cochlea, the semicircular canals are filled with fluid and contain mechanoreceptor cells.

The three semicircular canals are positioned at right angles to one another. Changes in position of the head or body cause the fluid in the canals to move. Mechanoreceptor cells within the canals are stimulated by the moving fluid. Movement in a particular direction causes a particular amount of stimulation to each of the three canals. The cerebrum receives and interprets the three sets of impulses from the three canals. Based on the interpretation, the brain sends motor impulses to the skeletal muscles to keep the body in balance. Information from the semicircular canals also is used by the cerebellum in its coordination of muscle action.

REVIEW QUESTIONS

11. What are the three categories of receptor cells? What is detected by each of these cells?
12. What receptors are found in the skin? What are their functions?
13. Which two senses are based on chemoreception?
14. What is the function of the ciliary muscle of the eye?
15. Space is said to have three dimensions. How does this relate to the arrangement of the three semicircular canals?

Investigation

How Does the Eye Perceive Images?

Goals

After completing this activity, you will be able to
- Give several examples of how the eye is "tricked" by optical illusions.
- Demonstrate a simple way to test peripheral vision.

Materials (for groups of 2)

metric ruler
2 large sheets of white paper
pencil

Procedure

A. Look closely at Figure A. Do you have a sensation of movement? On a separate sheet of paper, describe the image you are seeing.

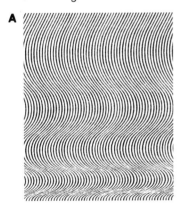

A

B. Study the pairs of vertical lines in Figure B. Measure the disance between each pair of lines. Determine whether the lines are parallel to each other. Record your observations.

C. Work with a partner. Stand with your back against the wall. Raise your arms to shoulder height and stretch them out so that they touch the wall.

B

D. Look straight ahead. Do not move your head from side to side. Wiggle your fingers. If you cannot see your fingers, move your arms slightly forward. Again wiggle your fingers. Have your partner measure how far your arms are from the wall when you can first see your fingers move. On a separate sheet of paper, record this distance.

E. Repeat step D with your partner standing aginst the wall. Record the results.

F. On a large sheet of white paper, draw an X about 4 cm in height. Place the X a little to the left of the center of the paper. Have your partner hold the paper about 30 cm from your face. Close your left eye and stare at the X with your right eye.

G. Have your partner bring the point of a pencil about 10 cm to the right of the X. Ask him or her to move the pencil point outward from the X until you can no longer see the point. Have your partner mark this spot on the paper.

H. Beginning at a point about 10 cm from the X, have your partner move the pencil point out in a circle around the X. Mark the places at which the pencil point disappears. Measure the distance from the X. The area around the X indicates how large your blind spot is. The blind spot is the region in which you have no vision. It corresponds to a region on the retina where there are no cells to pick up light signals. There are no cells in that region because at that point the optic nerve leaves the retina.

Questions and Conclusions

1. How did the image in Figure A appear to you?
2. How did the measurement of the lines in Figure B compare with what you believed you saw?
3. How far were your arms from the wall when you could see your fingers wiggle? Was this distance the same as that of your partner?
4. In step D you were measuring your peripheral vision, or the ability to see on the outside boundary of your body. Describe a situation in which good peripheral vision is needed.
5. What was the boundary of your blind spot? How did your results compare with that of your partner?
6. Figures A and B are examples of optical illusions. Describe a situation in which optical illusions have a practical application.

THE NERVOUS SYSTEM

Chapter Review

SUMMARY

33·1 The nervous system coordinates the body's activities. The nervous system receives and responds to information from the outside environment.

33·2 Neurons carry information as electrical impulses.

33·3 The brain is the central controlling structure in the nervous system. The cerebral cortex processes sensory information and controls the skeletal muscles.

33·4 The spinal cord contains sensory neurons that carry impulses to the brain, motor neurons that carry impulses from the brain, and association neurons that connect the other two types.

33·5 The peripheral nervous system includes the cranial nerves, spinal nerves, and ganglia. Together with the spinal cord, the peripheral nervous system performs some involuntary responses.

33·6 The autonomic nervous system consists of two subdivisions. The parasympathetic subdivision controls the internal organs during routine conditions, and the sympathetic subdivision takes over during emergencies and increased activity.

33·7 Neurons in the resting state are polarized, with the outer surface being positively charged and the inner surface being negatively charged.

33·8 When stimulated, neurons produce action potentials that travel along the membrane.

33·9 Impulses are carried across synapses by neurotransmitters.

33·10 A large number of neurotransmitters of the brain have been identified, including neuropeptides such as the endorphins and enkephalins.

33·11 Depressant drugs reduce conduction of impulses in the nervous system, and stimulant drugs increase conduction of impulses.

33·12 Disorders of the nervous system include epilepsy, multiple sclerosis, Parkinson's disease, and Alzheimer's disease.

33·13 Most sensations are detected by mechanoreceptors, chemoreceptors, and photoreceptors. Tastes and odors are detected by chemoreceptors.

33·14 The skin contains mechanoreceptors that detect touch, pressure, and pain. Other receptors in the skin detect heat and cold.

33·15 The retina of the eye contains photoreceptor cells, which are stimulated by light.

33·16 The inner ear contains two sets of mechanoreceptors. Receptors in the cochlea are responsible for hearing. Receptors in the semicircular canals are concerned with maintaining equilibrium.

BIOLOGICAL TERMS

action potential
association neuron
autonomic nervous system
axon
axon-endings
brain
brainstem
central nervous system
cerebellum
cerebrum
chemoreceptors
cochlea
dendrites
depressants
effectors
ganglion
hallucinogenic drugs
hypothalamus
mechanoreceptors
medulla oblongata
meninges
motor neuron
nervous system
neurons
neurotransmitters
parasympathetic nervous system
peripheral nervous system
photoreceptors
receptors
reflex arc
refractory period
resting membrane potential
semicircular canals
sensory neurons
sodium-potassium pump
spinal cord
stimulants
sympathetic nervous system
synapse

USING BIOLOGICAL TERMS

1. Distinguish between a motor neuron and a sensory neuron.
2. What is a ganglion?

3. Distinguish between receptors and effectors.
4. How are dendrites and axons related?
5. What is the resting membrane potential?
6. What is the relationship between a neurotransmitter and a synapse?
7. Distinguish between the sympathetic nervous system and the parasympathetic system.
8. Which term means "nerve impulse"?
9. What is the peripheral nervous system?
10. Distinguish between stimulants and depressants.

UNDERSTANDING CONCEPTS

1. What are the functions of the nervous system? (33·1)
2. How do the central nervous system and peripheral nervous system operate together in carrying out the functions of the nervous system? (33·1)
3. Describe the structure of a neuron. (33·2)
4. What is a nerve? (33·2)
5. What types of activities are controlled by the cerebral cortex? (33·3)
6. What are the functions of the cerebellum and the brainstem? (33·3)
7. Describe how the spinal nerves connect to the spinal cord. (33·4)
8. Describe how a reflex arc functions. (33·5)
9. What is the function of the autonomic nervous system? How do the two subdivisions of this system differ? (33·6)
10. Describe how the sodium-potassium pump produces a resting membrane potential. (33·7)
11. Describe a nerve impulse in terms of depolarization of the cell membrane. (33·8)
12. What is the relationship between the size of a stimulus and the action potential it produces? (33·8)
13. Explain how an impulse crosses a synapse. (33·9)
14. What kinds of neurotransmitters have been discovered in the brain? (33·10)
15. Explain how stimulants act on the nervous system, and give examples of some stimulants. (33·11)
16. Explain how depressants act on the nervous system and identify some depressants. (33·11)
17. What kind of neural damage occurs in multiple sclerosis? How do the symptoms of the disease relate to this damage? (33·12)
18. Describe a hypothesis for the detection of odors through the olfactory epithelium. (33·13)
19. How do receptors for touch and pressure differ? (33·14)
20. What are the three layers of the eye? Which layer contains the photoreceptors? (33·15)
21. Explain how sound waves are converted to nerve impulses. (33·16)

APPLYING CONCEPTS

1. Curare is a poison that acts by binding to the receptor proteins on dendrites. Explain how this binding results in poisoning.
2. If a neuron is stimulated at the cell body, the action potential will travel outward in all directions. Contrast what happens at the ends of the axons and the ends of the dendrites when the action potential arrives.
3. The "receiving" side of synapses always contains an enzyme that destroys the neurotransmitter after it has caused an action potential to be produced. What do you suppose would happen if the enzymes were missing?

EXTENDING CONCEPTS

1. Only some of the functions of the hypothalamus have been mentioned in this chapter. Investigate and find additional functions.
2. There is no clear understanding about how memories are stored in the brain, but there are several theories that attempt to explain this process. Try to find what these theories are.
3. Many differences in function between the left and right cerebral hemispheres have been discovered. Prepare a report on these differences.

READINGS

Maranto, Gina. "The Mind Within the Brain." *Discover.* May 1984, p. 34.

Stafford, Patricia. *Your Two Brains.* Atheneum, 1986.

34. The Endocrine System

Hormones secreted by the endocrine glands control every aspect of your life. The growth hormone is the main controller of height. Not only does this hormone stimulate bone and muscle growth, it also regulates the release of hormones from other glands. By reading this chapter you will find out more about the actions of the pituitary gland and the other glands that make up the endocrine system.

CHAPTER OBJECTIVES

After completing this chapter, you will be able to
- **List** the endocrine glands and **describe** the function of each gland.
- **Identify** some major disorders of the endocrine system.
- **Explain** the feedback mechanism and how it maintains homeostasis.
- **Describe** how hormones change the activity of their target organs.

Endocrine Glands and Hormones

34•1 ENDOCRINE GLANDS

Humans have two systems of coordination and control. You learned that the nervous system is one of these systems. The second system is the endocrine system. The **endocrine** (EHN doh krihn) **system** is made up of a number of ductless glands that secrete chemicals that control certain body activities.

Endocrine glands secrete chemicals into the capillaries that pass through the glands. These chemicals then are carried in the blood to other parts of the body. **Exocrine** (EHK suh krihn) **glands** are glands that release chemicals into ducts. Salivary glands and sweat glands are examples of such glands. Exocrine glands are not part of the endocrine system.

The chemicals released by endocrine glands are called hormones. A **hormone** is a chemical that is produced in one part of an organism and is transported to other parts where it controls certain activities of tissues and organs. Each of the endocrine glands produces different hormones. Each hormone, in turn, controls the activities of certain organs or tissues. The organ that is acted on by a hormone often is called the **target organ.**

Both the endocrine system and the nervous system help to maintain homeostasis in the body. Both systems help to coordinate the activities of the various organs. How does the action of the nervous system differ from that of the endocrine system? The impulses of the nervous system travel quickly and produce responses almost instantly. However, the response of an organ to an impulse from the nervous system may not last very long. A series of impulses is needed to continue the response. The endocrine system, on the other hand, is slower in causing responses. It takes time for a hormone to travel through the blood and to stimulate the target organ. But the response to the hormone may last much longer than the response to a nerve impulse.

> **SECTION OBJECTIVES**
> - Identify the endocrine glands.
> - Describe the function of each endocrine gland.

685

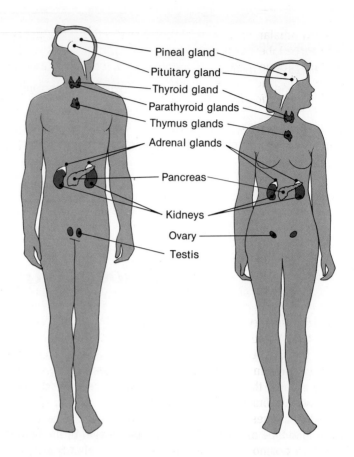

FIGURE 34·1 This diagram shows the endocrine glands and their location in the body.

The endocrine system does not usually react directly to changes in the outside environment. Endocrine organs do, however, respond to stimuli from the internal chemical environment. For example, changes in the chemical makeup of the blood may act as stimuli to endocrine organs to release hormones. Other changes in the blood may stop the action of endocrine organs.

Figure 34·1 shows where endocrine glands are found in the human body. Because hormones move in the blood and affect distant organs, the location of an endocrine gland is often not a clue to its function. For example, the hormones released by the endocrine glands in the head and neck affect all parts of the body.

34·2 PITUITARY GLAND

The **pituitary gland** is a gland found near the base of the brain. It is a small gland that releases many hormones. Many of these hormones control the activities of other endocrine glands. Because its hormones control the activities of other glands, the pituitary has been called the master gland.

From Figure 34·2 you can see that the pituitary is attached to the hypothalamus. Recall that the hypothalamus is involved in the involuntary control of many body functions. The brain controls the pituitary through the hypothalamus. Figure 34·2 also shows that the pituitary is divided into three lobes, or parts. The *anterior lobe,* or foremost lobe, of the pituitary produces the most hormones, including all those that control other endocrine glands. The *intermediate lobe* secretes a single hormone. The *posterior lobe,* or hindmost lobe, secretes two hormones.

Human growth hormone, or **HGH,** is a hormone that is produced by the anterior lobe of the pituitary. The pituitary secretes HGH throughout life. HGH is a protein that is necessary for normal growth. Almost all vertebrates make protein hormones much like it. HGH stimulates the growth of bones. It also stimulates growth of muscle and many other tissues. Scientists once thought that growth hormone was needed only during periods of growth. They now know that HGH has important functions in adults as well. In adults HGH stimulates the building of proteins, especially in the muscles.

Figure 34·3 is a graph of the results of an experiment with growth hormone. The experiment involved a number of rats whose pituitary glands were removed soon after birth. Rats survive without their pituitary glands. Half of the rats were then treated with regular doses of growth hormone. The other half of the rats were left untreated. The length of the rats' bodies was measured regularly as an indication of growth. You can see in the graph that the untreated rats grew little during the course of the experiment. The rats that received growth hormone, however, showed large increases in body length. What can be concluded from these results?

The anterior lobe of the pituitary also secretes prolactin. **Prolactin** is a protein hormone involved in the growth of the mammary glands. The mammary glands are the exocrine glands that produce milk. During pregnancy, prolactin results in the development of milk-producing tissue in these glands. At the time of childbirth, prolactin causes the production of milk.

Four other hormones produced by the anterior pituitary gland control the activities of other endocrine tissues. **Thyroid-stimulating hormone,** or **TSH,** controls secretion by the thyroid gland. The relationship between this hormone and the thyroid gland will be discussed in the next section. **Follicle-stimulating hormone (FSH)** and **luteinizing** (LOO tee uh nīz ing) **hormone (LH)** control secretion of hormones by the reproductive organs. The endocrine activities of the reproductive organs will be discussed in Section 34·5 and in the next chapter. **Adrenocorticotropic** (uh dree noh kor tuh koh TROHP ihk) **hormone,** or **ACTH,** controls some activities of the adrenal glands, which will be discussed in Section 34·4. ACTH is a protein. TSH, FSH, and LH are proteins with certain sugars attached.

The hormone that is produced by the intermediate lobe of the pituitary is *melanophore-stimulating hormone,* or *MSH.* This hormone acts on pigment cells, or *melanophores* (MEHL uh nuh forz), which are found in the skin. MSH causes the skin to become darker in color. No additional functions of this hormone have been discovered. This hor-

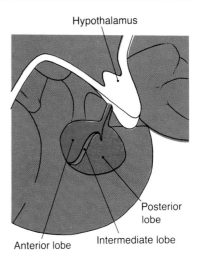

FIGURE 34·2 The pituitary gland, located near the base of the brain, is connected to the hypothalamus by means of a stalklike structure. Notice the three lobes of the pituitary gland.

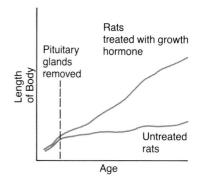

FIGURE 34·3 This graph shows the effect of growth hormone on rats that have had their pituitary glands removed. What happened to the untreated rats?

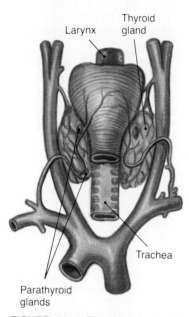

FIGURE 34·4 The thyroid gland is found straddling the trachea. Located on, or sometimes in, the thyroid gland are the four pea-shaped parathyroid glands.

para (near)

mone seems to be more important in lower vertebrates than it is in mammals. In lower vertebrates the skin may lighten or darken, blending with the environment.

The two hormones of the posterior lobe of the pituitary are peptide hormones. These hormones are short chains of amino acids. **Antidiuretic hormone,** or **ADH** (also called vasopressin), is a hormone that controls reabsorption of water in the tubules of the kidney (Section 32·13). **Oxytocin** (ahk sih TOH sihn), the other hormone of the posterior lobe, stimulates contractions of the muscles of the uterus during labor. Thus, it has an important role in childbirth. It also stimulates the release of milk from the mammary glands when an infant is being nursed.

34·3 THYROID AND PARATHYROID GLANDS

The **thyroid gland** is an endocrine gland that lies across the trachea. Notice the shape and location of the thyroid gland in Figure 34·4. The main hormone of the thyroid gland is the iodine-containing hormone called **thyroxine** (thī RAHK seen). Thyroxine is the only hormone to contain iodine. The thyroid gland contains hollow, round structures called *follicles*. The cells that line the follicles produce thyroxine, which is then stored within the follicles. Like growth hormone, thyroxine acts on almost all of the body's tissues.

Thyroxine has two kinds of effects. One of the actions is the control of energy production in the body. Thyroxine is needed to maintain the basal metabolic rate (BMR) at a normal level. Recall that the BMR is a measure of the rate of energy production in a person at rest (Section 32·5). The other action of thyroxine occurs during the growing years. While growth hormone stimulates growth in size, thyroxine causes the tissues to take the proper *form* as they grow. That is, thyroxine causes tissues to develop the correct shapes and proportions.

The activities of the thyroid follicles are under the control of TSH from the anterior pituitary gland. TSH stimulates the follicles to "trap," or remove, iodine from the blood. The iodine is then used in making thyroxine. TSH also stimulates the release of thyroxine from the thyroid gland.

Calcitonin (kal suh TOH nuhn), a second hormone of the thyroid gland, helps control the amount of calcium and phosphorus in the blood. Calcium and phosphorus form the mineral matter of bones and teeth. They are also needed for many processes in other organs. Phosphorus is used in building RNA and DNA.

Identify the four parathyroid glands in Figure 34·4. The **parathyroid glands,** found on the dorsal side of the thyroid gland, secrete parathyroid hormone. **Parathyroid hormone,** or **PTH,** acts with calcitonin to control the distribution of calcium and phosphorus in the body. PTH causes calcium to be released from the bones into the blood. Calcitonin has the opposite effect. The two hormones working together maintain homeostasis of calcium levels in the blood.

34·4 ADRENAL GLANDS

A human has two adrenal (uh DREE nuhl) glands. The **adrenal glands** are endocrine glands found above the kidneys. In Figure 34·5 one of the glands is shown in cross section. You can see that it has an outer layer, or *cortex,* and an inner layer, or *medulla* (mih DUHL uh). The adrenal cortex and the adrenal medulla function as two separate endocrine organs.

Adrenaline (uh DREHN uh lihn), also called epinephrine (ehp uh-NEHF rihn), is a hormone secreted by the adrenal medulla. The effects of adrenaline are similar to the effects of the sympathetic nervous system. Adrenaline is released during times of stress and heavy exercise. It stimulates increases in heart rate, breathing rate, and in the amount of blood flowing to the brain and skeletal muscles.

The adrenal cortex secretes a number of hormones that are chemically alike. These hormones belong to a class of fatty chemicals called *steroids*. Steroids secreted by the adrenal cortex are **corticosteroids** (kor tuh koh STEHR oidz). These are compounds of carbon, hydrogen, and oxygen and are chemically different from the protein type of hormone, such as HGH.

There are two groups of corticosteroids. The hormones of the first group are called *glucocorticoids* (gloo koh KOR tuh koidz). These hormones are named for their effects on blood glucose levels. **Cortisol** (KOR tuh sohl), the most important of this group of hormones, stimulates the production of glucose from fats and amino acids. As a result there is an increase in the level of glucose in the blood. This production of glucose seems to be an adaptation that makes more energy available during times of stress. Cortisol and other glucocorticoids also reduce inflammation. Inflammation is a reddened, swollen condition

ad- (near)
renes (kidney)

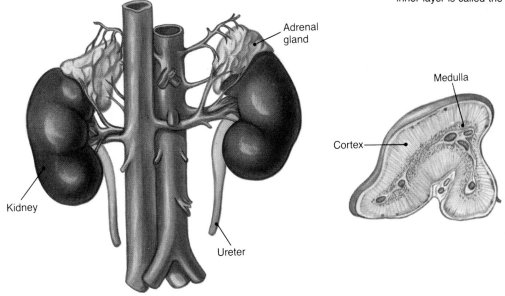

FIGURE 34·5 The drawing on the left shows the location of the two adrenal glands, one atop each kidney. In the drawing on the right, you can see a cross section of one adrenal gland. The outer layer is called the cortex and the inner layer is called the medulla.

in tissues resulting from infection or other disease processes. Cortisone (KOR tuh zohn), which is much like cortisol, sometimes is used to treat arthritis, which is inflammation of the joints.

Secretion of the glucocorticoid hormones is controlled by ACTH. ACTH influences the secretion of the glucocorticoid hormones in several ways: (1) ACTH helps to maintain the cells of the adrenal cortex at a normal size. Without ACTH, the cells of the cortex shrink and do not function. (2) ACTH stimulates these cells to make the glucocorticoids. (3) ACTH stimulates the release of hormones from the adrenal cortex into the blood.

Mineralocorticoids (mihn uhr uh luh KOR tuh koidz) are a second group of corticosteroids. They help to regulate the body's mineral balance. **Aldosterone** (al DAHS tuh rohn), the most important of these hormones, acts on the tubules of the kidney to stimulate absorption of sodium into the blood. Scientists believe that mineralocorticoid secretion is *not* controlled by ACTH.

34·5 OVARIES AND TESTES

The *ovaries,* which are the egg-producing organs of females, and the *testes,* which are the sperm-producing organs of males, are both endocrine glands. Both the ovaries and the testes produce hormones involved in the development of sexual characteristics. These organs are controlled by hormones from the pituitary gland.

The ovaries produce two hormones: estrogen (EHS truh juhn) and progesterone (proh JEHS tuh rohn). **Estrogen,** often called the female sex hormone, stimulates the development of female reproductive organs and female secondary sexual characteristics. **Progesterone** maintains the lining of the uterus during pregnancy.

The egg cell is contained within a structure called a *follicle.* It is the follicle that secretes estrogen. The growth of the follicle, and its secretion of estrogen, is regulated by FSH from the pituitary gland. Notice the *corpus luteum* (LOO tee uhm) in Figure 34·6. Progesterone is secreted by the corpus luteum. This activity is stimulated by LH from the pituitary.

Testosterone (tehs TAHS tuh rohn) is the male sex hormone. This hormone is produced by cells within the testes. Testosterone stimulates development of the reproductive organs and of male sexual characteristics.

FSH and LH from the pituitary are named for their functions in females. They also control the activities of the testes in males. In males, FSH stimulates growth of the sperm-producing cells. LH stimulates the endocrine cells that secrete testosterone.

Estrogen, progesterone, and testosterone are all steroids. They all are chemically similar to the corticosteroids of the adrenal gland. It is believed that the adrenal cortex produces small amounts of estrogen and testosterone. It is probably because of this secretion that small amounts of testosterone are present in the blood of females. Small amounts of estrogen normally are found in the blood of males.

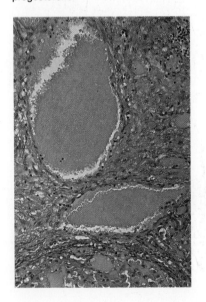

FIGURE 34·6 The two round structures in the photograph are the corpus luteum. These structures are found in the ovaries and secrete the hormone progesterone.

Investigation

What Is the Effect of Hormone Treatment on the Heartbeat of *Daphnia*?

Goals

After completing this activity, you will be able to

● Describe the effect of adrenaline on the heartbeat of *Daphnia*.

Materials (for groups of 2)

culture of *Daphnia*, a water flea
microscope   2 depression slides
2 coverslips   dropper
0.01% solution of adrenaline
stopwatch   two 100-mL beakers
glass-marking pencil

Procedure

A. Study the drawing of *Daphnia*. Note the position of the heart, which is in the dorsal part of the body.

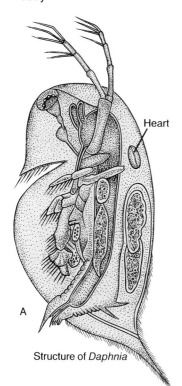

Structure of *Daphnia*

B. With the dropper, transfer a small amount (25 mL) of the *Daphnia* culture to each of two beakers. Label one beaker "Untreated *Daphnia*" and the other beaker "Treated *Daphnia*."

C. Make a hanging drop slide of *Daphnia* from the culture in the beaker marked "Untreated." (Figure B shows a hanging drop preparation.)

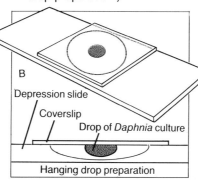

Hanging drop preparation

D. Look at the slide under low power. Refer to the drawing of *Daphnia*. Then locate the beating heart in one of the organisms you can see.

E. On a separate sheet of paper copy the chart shown at the bottom of this activity.

F. Using a stopwatch, find the number of heartbeats per minute. Find the heart rate three times. Record the results in the chart. Compute the average (add the results of the three trials and divide by 3).

G. Using a dropper, add 5 drops of 0.01% adrenaline solution to the beaker labeled "Treated *Daphnia*." Swirl the contents of the solution around so that the solution is thoroughly mixed.

H. Make a hanging drop slide of *Daphnia* from the beaker of treated organisms. Repeat steps D and F.

I. Compare your results with those of other groups in the class.

J. Follow your teacher's instructions for the proper disposal of all materials.

Questions and Conclusions

1. What was the average heart rate of *Daphnia* in the untreated beaker? What was the average heart rate after the *Daphnia* was treated with adrenaline?

2. How do the heart rates before and after treatment compare?

3. Why were the rates taken three times for treated and untreated *Daphnia*?

4. Why do you think *Daphnia* was chosen as the organism to be studied in this activity?

5. Adrenaline is a hormone secreted under certain conditions. In humans, under what conditions is it secreted? What changes take place in the body?

HEART RATE IN *DAPHNIA* (beats/min)				
	TRIAL 1	TRIAL 2	TRIAL 3	AVERAGE
UNTREATED *DAPHNIA*				
TREATED *DAPHNIA*				

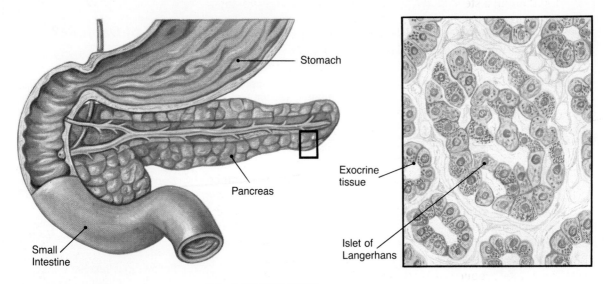

FIGURE 34·7 Located near the beginning of the small intestine is the pancreas. It contains cells organized into groups called islets of Langerhans, which secrete insulin and glucagon.

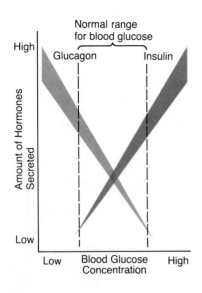

FIGURE 34·8 This graph illustrates the effects of insulin and glucagon on the level of glucose in the blood.

34·6 PANCREAS

The **pancreas** is a large gland found below the stomach. This gland contains both exocrine and endocrine tissues. The exocrine tissues of the pancreas produce many enzymes that are used in digestion. Scattered among the exocrine tissues are small groups of endocrine cells. You can see both types of tissues in Figure 34·7. The patches of endocrine cells within the pancreas are the **islets of Langerhans.** These endocrine cells secrete two hormones that control the levels of glucose in the body. The islets of Langerhans do not appear to be controlled by the pituitary gland.

The first of the pancreatic hormones to be discovered was insulin. **Insulin** is a protein produced by cells of the islets of Langerhans called *beta cells.* A high level of glucose in the blood, such as that which might occur after a meal is digested, stimulates the beta cells to secrete insulin. Insulin, in turn, causes glucose to be removed from the blood and stored in the liver and skeletal muscles. The glucose is stored as glycogen. When the glucose level in the blood falls to a normal or below-normal level, the beta cells decrease their secretion of insulin.

The second pancreatic hormone to be discovered was **glucagon.** Glucagon is produced by cells called *alpha cells.* It is released when the blood glucose level is low. Glucagon causes the removal of glycogen from the liver and the skeletal muscles and the release of glucose into the blood. When glucose in the blood reaches a normal or above-normal level, the alpha cells decrease their glucagon secretion.

Insulin and glucagon are a pair of hormones with opposite effects. Figure 34·8 shows the relationship between blood glucose concentration and insulin and glucagon secretion. You can see that much glucagon is secreted when the blood glucose level is low. Large amounts of insulin are secreted when blood glucose levels are high. When is the secretion of both insulin and glucagon very low?

Maintaining a steady supply of glucose in the blood is an important part of homeostasis. Glucose is the fuel used in many organs and tissues. While some tissues can use other sources of energy, the brain is limited to using glucose. Thus the control of the blood glucose level by insulin and glucagon ensures a steady supply of energy to the brain.

34·7 OTHER ENDOCRINE HORMONES

Several organs that had not previously been thought of as endocrine organs have been found to produce hormones. For example, the **thymus,** an organ found in the chest above the heart, is believed to produce a hormone that is needed for normal production of lymphocytes. Recall that lymphocytes are a type of white blood cell found in both blood and lymph. Lymphocytes are involved in the body's defense against disease. The thymic hormone is thought to stimulate production of lymphocytes only during childhood and adolescence. The thymus decreases in size after adulthood is reached.

The kidney produces two hormones. *Erythropoietin* (ih rihth roh-POI uh tihn) is produced when there is a lack of oxygen in the body tissues. Such a deficiency occurs, for example, after a large loss of blood resulting from an injury. This hormone stimulates the bone marrow to produce more red blood cells. The increased number of red blood cells results in the transport of more oxygen to the tissues. The increase in red blood cells eases the oxygen deficiency.

The second hormone produced by the kidney is *angiotensin* (an-jee oh TEHN sihn). Angiotensin is a hormone produced when blood pressure in the kidney is below normal. This hormone causes contraction of smooth muscle in the walls of blood vessels. As the same amount of blood flows into vessels that have become narrower, the blood pressure increases. Angiotensin also stimulates the adrenal cortex to secrete aldosterone. Aldosterone causes the sodium concentration in the blood to increase. The increased sodium leads to increased blood pressure. Thus both of angiotensin's effects lead to increases in blood pressure.

The **pineal gland** is found in the head, between the two cerebral hemispheres. It produces the hormone melatonin. Melatonin seems to have different effects in different animals. In some cases it acts on the melanophores to cause lightening of the skin. This effect is opposite to that of MSH. In other animals, perhaps including humans, melatonin appears to inhibit the secretion of FSH and LH. Some researchers think that melatonin may inhibit the release of these hormones until the teen years. It is at that time that secretion of FSH and LH causes the reproductive organs to become mature.

Hormonelike substances called *prostaglandins* (prahs tuh GLAN-duhnz) are produced by almost all tissues. Prostaglandins are fatty acids that have hormonelike effects. But they seem to act on tissues near to their site of production. The effects of prostaglandins vary widely, depending on what tissue they are produced in. The effects

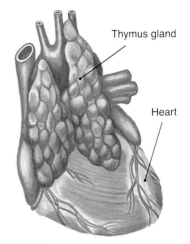

FIGURE 34·9 The thymus gland is located in the upper chest cavity near the heart. After the start of adolescence, the thymus gland begins to decrease in size. In adulthood, the thymus gland may disappear entirely.

include contraction of the uterus during childbirth and opening of the passages of the respiratory tract. They also include regulation of blood clotting, inhibition of stomach secretions, and many other effects. The relation between prostaglandins and other hormones remains a topic for further research.

REVIEW QUESTIONS

1. Distinguish between exocrine and endocrine glands.
2. List the hormones of the pituitary gland and indicate from which pituitary lobe each hormone is secreted.
3. Identify and give the function of each hormone produced by the ovaries and testes.
4. Identify two pairs of hormones that act as opposing pairs. What functions do these hormones control?
5. If a person moves from sea level to a place at a high altitude, his or her cells may not be receiving sufficient oxygen at first. The body later adjusts to the high altitude. What hormone response is responsible for the adaptation?

Endocrine Disorders

SECTION OBJECTIVE

- List some major disorders of the endocrine system.

34·8 PITUITARY AND THYROID DISORDERS

Disorders of the pituitary gland are not very common among humans. When such disorders do occur, however, their effects can be quite dramatic. A low level of HGH during the childhood years leads to *dwarfism*. In this condition there is usually a shortage of all of the anterior pituitary hormones. The result is that the body grows to less than normal size but with all its parts in proportion. A high level of HGH during childhood leads to *giantism*. As with dwarfism, all of the body parts tend to be in proportion, but they are large.

Disorders of the thyroid gland are more common than those of the pituitary. The condition in which excess thyroxine is released is called *hyperthyroidism*. Increased BMR results from hyperthyroidism. Persons with this disorder tend to be nervous and sensitive to heat. Undersecretion of thyroxine is called *hypothyroidism*. The BMR is low in those with this condition. Appetite decreases while body weight tends to increase. Persons with this disorder are sensitive to cold.

The effects of hypothyroidism vary according to the stage of life at which the condition occurs. Hypothyroidism during infancy leads to the condition known as *cretinism* (KREE tuh nihz uhm). Distorted growth and mental retardation often result from this condition. Severe hypothyroidism in the adult is known as *myxedema* (mihk suh DEE-muh). This condition affects almost every organ in the body. Most body functions slow down in persons with myxedema. Both myxedema and cretinism can be treated by giving thyroxine. In cretinism,

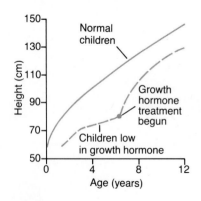

FIGURE 34·10 This graph compares the growth rates of normal children and children low in growth hormone. Notice the effect of growth hormone treatment on children low in growth hormone.

however, treatment must start right after birth in order to prevent brain damage.

Goiter is a condition in which the thyroid gland is enlarged. Goiter is often visible as a swelling of the lower neck. Goiter may occur in persons with either hypothyroidism or hyperthyroidism. In the past the most common cause of goiter was lack of iodine in the diet. The thyroid gland in a person whose diet lacks iodine grows larger and larger but fails to produce enough thyroxine. This type of goiter has become less common because most table salt has iodine added to it. The small amount of iodine found in the salt is enough to supply the thyroid gland.

34·9 PANCREATIC DISORDERS

The most common of the endocrine disorders affects the pancreas. *Diabetes mellitus* (MEHL uh tuhs) is a disorder in which the beta cells of the islets of Langerhans do not secrete enough insulin. Without the insulin, glucose from the blood cannot be absorbed into liver and muscle cells. Glucose collects in the blood of those with diabetes.

Diabetes has been known since the time of ancient Rome. Diabetics excrete large amounts of sugar in their urine. Recall from Chapter 32 that the kidneys normally reabsorb nearly all glucose. Normal persons excrete no sugar in their urine. In diabetes, however, so much glucose is in the blood that the kidneys cannot reabsorb all of it. The extra glucose is excreted in the urine. Since the extra glucose is in solution the kidneys excrete more water than is normal. The result is that many diabetics suffer from constant thirst.

Little glucose is able to enter the body's muscle and liver cells when the insulin level is low. These cells do not have their normal source of energy. Because the usual source of energy (glucose) is not available, proteins and fats are broken down to obtain energy. Waste products from the breakdown of proteins and fats collect in the blood. In extreme cases these wastes may cause a form of poisoning.

FIGURE 34·11 With proper treatment and diet control, people who have diabetes can lead normal, active lives.

Diabetes mellitus occurs in two forms. *Juvenile-onset diabetes* begins during childhood and is thought to be caused by hereditary factors. Diabetes that begins in childhood is the more severe form of the disease. Diabetes of this type generally requires lifelong treatment with insulin.

Adult-onset diabetes begins during the adult years and is a less severe form of the disease. Adult-onset diabetes very often occurs in overweight persons. Heredity may be a factor. Some adult-onset diabetics require treatment with insulin. Others can keep the disease in check by controlling their intake of fats and carbohydrates.

Diabetics have long been treated with insulin taken from the pancreases of cattle or hogs. Although these hormones are effective in treating diabetes, they are slightly different from human insulin. Human insulin is somewhat more effective in treating diabetes than are animal insulins. Recently, human insulin produced by genetically engineered bacteria has become available.

Hypoglycemia (hī poh glih SEE mee uh), or a low level of glucose in the blood, is the opposite of diabetes. This condition is caused by oversecretion of insulin. Excess insulin causes most of the glucose to move from the blood into muscle and liver cells. Insulin does not appear to affect absorption of glucose by the brain cells. The danger of a deficiency of blood glucose is that not enough glucose may be available for the brain. Without a good supply of energy, the brain cannot function properly.

REVIEW QUESTIONS

6. List several endocrine disorders that are caused by undersecretion, or deficiency, of hormones.
7. List several endocrine disorders that are caused by oversecretion or excess of a hormone.
8. Why is most table salt iodized?
9. Describe what causes diabetes mellitus.
10. If a diabetic mistakenly takes too much insulin, he or she may go into a coma. In such a case, sugar may be given to bring the person out of the coma. Explain what happens in this situation.

Endocrine Control Mechanisms

SECTION OBJECTIVES

- Describe the feedback mechanism.
- Explain the roles hormones play in maintaining homeostasis.

34·10 FEEDBACK MECHANISMS AND HOMEOSTASIS

Some of the activities of the endocrine system control growth and development. Most endocrine actions, however, help to maintain homeostasis. Insulin, glucagon, and the glucocorticoids are involved in homeostasis of the body's sugar levels. Calcitonin and PTH maintain calcium and phosphorus homeostasis. ADH, mineralocorticoids, and angiotensin are involved in maintenance of homeostasis of water and sodium levels in the body.

Many of the processes of homeostasis work through a particular type of mechanism. This mechanism is called a feedback mechanism. In a **feedback mechanism** the last step or end product of a series of steps controls the first step in the series. The concept of the feedback mechanism is represented in Figure 34·12A. The diagram shows the last step of the process inhibiting, or stopping, the first step. When the last step inhibits the first step, the feedback mechanism is *negatively* controlled. The mechanism is called **negative feedback.**

A common example of negative feedback control is a thermostat. In most houses and other buildings, the furnaces or heaters are controlled by thermostats. The thermostat detects changes in temperature and turns on the furnace when the temperature falls. The furnace burns some type of fuel. The heat produced warms the air in the

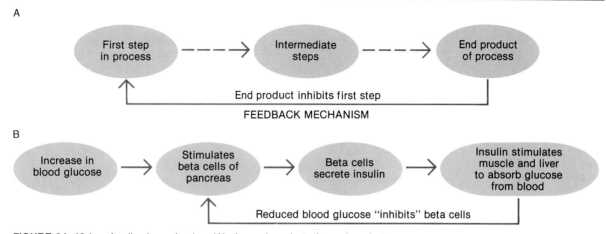

FIGURE 34·12 In a feedback mechanism (A), the end product of a series of steps controls the first step in the series. For example, as an increase in blood sugar stimulates the beta cells in the pancreas to secrete insulin, the level of sugar in the blood decreases, causing the beta cells to stop secreting insulin (B).

house. Increased air temperature, the last step in this series of events, shuts off the thermostat. The heating process is stopped.

The secretion of most hormones is controlled by negative feedback processes. Figure 34·12B reviews the action of insulin secreted by the pancreas. Increases in blood glucose stimulate the beta cells to secrete insulin. Insulin causes the liver and muscles to absorb glucose and store it as glycogen. The resulting decrease in blood glucose acts as a negative feedback signal. It causes the beta cells to stop secreting insulin. The last step in the process inhibits the first.

The control of insulin secretion is a simple example of negative feedback in the endocrine system. The secretion of glucagon is controlled by a similar mechanism. Both of these control mechanisms involve a single endocrine gland. Some more complex examples of negative feedback, involving more than one gland, will be discussed in the next section.

In positive feedback control the last step of a process *stimulates* the first step. This type of control is less common than negative feedback in living things. Positive feedback does not maintain homeostasis. Negative feedback, which prevents excessive stimulation, is effective in maintaining homeostasis.

34·11 PITUITARY AND HYPOTHALAMUS INTERACTION

The pituitary has long been thought of as the master gland because of its many secretions and its control over other glands. It is now known, however, that the pituitary is regulated by another center of control. If you refer to Figure 34·2, you will notice that the pituitary is attached to the portion of the brain known as the *hypothalamus*. Some of the functions of the hypothalamus have been described (Section 33·3). The hypothalamus controls all the activities of the pituitary gland through chemical and electrical signals.

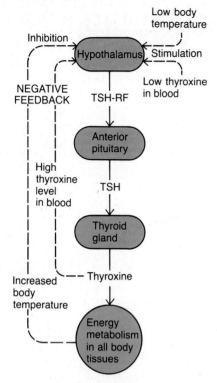

FIGURE 34·13 The level of various hormones in the body are controlled by a feedback mechanism. This diagram shows the negative feedback mechanism among hormones of the hypothalamus, and pituitary and thyroid glands.

The hypothalamus is an unusual organ in that it functions as three different tissues. First, the hypothalamus includes nervous tissue. It contains neurons that conduct impulses and have synapses with other neurons. Second, the hypothalamus is an endocrine tissue. It secretes releasing factors into the capillaries. **Releasing factors** are hormonelike substances that control the release of hormones by the anterior and intermediate lobes of the pituitary gland. Third, the hypothalamus is a type of sensory organ. It appears to detect the concentration of many chemicals in the blood. The hormones secreted by other glands are among the chemicals that the hypothalamus detects.

The control of the thyroid gland is an example of how the hypothalamus functions. The hypothalamus secretes a releasing factor that stimulates the anterior pituitary to secrete TSH. This releasing factor is *TSH-RF* (thyroid-stimulating hormone-releasing factor). Observe in Figure 34·13 that there are two conditions that stimulate the hypothalamus to secrete TSH-RF. These are low body temperature and/or low concentration of thyroxine in the blood. TSH-RF travels through blood vessels from the hypothalamus to the anterior lobe of the pituitary. TSH-RF then stimulates the pituitary to release TSH into the blood. TSH in turn travels through the bloodstream to the thyroid gland. There it stimulates the production and release of thyroxine.

Thyroxine circulates from the thyroid gland to all of the body's tissues. Thyroxine stimulates the tissues to increase their metabolic rate. Heat is produced as a by-product of such metabolism. Notice in Figure 34·13 that the secretion of TSH-RF by the hypothalamus is inhibited by increased body temperature and increased thyroxine levels in the blood. This negative feedback mechanism acts to prevent excessive increases in body temperature or thyroxine secretion.

The hypothalamus secretes releasing factors that control secretion of each of the hormones of the anterior and intermediate lobes of the pituitary. Each of these releasing factors functions in a manner similar to TSH-RF. Secretion of each of the hypothalamic releasing factors is regulated by negative feedback mechanisms. For example, growth hormone in the blood produces negative feedback on secretion of growth hormone-releasing factor by the hypothalamus.

Secretion of oxytocin and ADH by the posterior pituitary is controlled in a different manner. The posterior pituitary differs from other endocrine glands in that it is made of neurons rather than more typical endocrine cells. Oxytocin and ADH are made by neurons that extend between the hypothalamus and the posterior pituitary. The hormones are stored in the axon-endings of these neurons. The release of ADH and oxytocin into capillaries is caused by electrical impulses rather than by releasing factors. Impulses that start in the hypothalamus control the release of hormones from the posterior pituitary.

Although the control mechanisms differ, all three lobes of the pituitary gland are regulated by the hypothalamus. The anterior and intermediate lobes are controlled by the hormonelike releasing factors. The posterior lobe is controlled by electrical signals.

Computers in Biology

Computers and Medical Care

Efficient information management is vitally important in medical science. It is estimated that a doctor must be familiar with about 2 million facts! What doctor can keep track of so much information? SUPERDOC can. SUPERDOC is computer software that helps doctors diagnose diseases. When a computer is fed a patient's medical history, results of a physical exam, and laboratory data, SUPERDOC sifts through its knowledge of about 600 diseases and 4000 symptoms to give a diagnosis.

Some computer programs analyze actual clinical data. PUFF is a program that evaluates about 250 lung disorders and gives results only 90 seconds after a patient exhales into an instrument that measures the functioning of the lungs. A cancer center system prints out guidelines for tests and medication based on clinical data entered into the computer.

Computer use in hospitals is no longer limited to handling forms and bills. Today, computers are being used in almost every phase of medicine. They assist in medical diagnoses and decisions. They also are used to monitor critically ill patients, to aid persons with handicaps, and to carry out medical research and instruction.

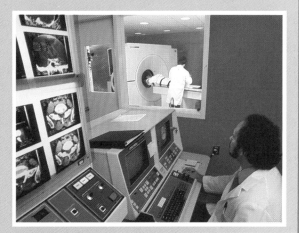

In monitoring the critically ill, data about the patients are analyzed by computer, and the results assist doctors in planning treatment for the patients. Bedside terminals are linked to computers in labs and staff areas so that medical personnel need not be at bedside all the time.

People with handicaps are being aided by computers. Computer descriptions of human movement help doctors assist persons who have artificial limbs or are paralyzed. Wires that are attached to a person with an artificial limb can carry data to a computer. The computer then directs the movement of the artificial limb. Computer signals can help people with paralyzed hands to move them. To aid the blind, a "seeing-eye" computer verbally reports objects that are "seen" by a tiny TV camera carried on the person. Another device to aid the blind contains a computer that uses impulses to stimulate the brain. The sightless person then "sees" light patterns that resemble the images.

Computers are being used in medical research. An "imaging system" called Digital Subtraction Angiography (DSA), now used to diagnose heart disease, may someday improve our understanding of how blood vessels supply body tissues. In research on the brain, a computer equipped with a camera was used to study different kinds of brain cells and their connections. The computer was then used to construct a model of the brain. The computer model led to a hypothesis about the electrical connections in the brain.

Computers are being used for instruction in medical schools. Lab animals no longer need to be used to study some physiological responses. An animal's responses can be programmed into a computer; then when data enter the computer, the computer responds as the animal would. Software is also being used to study human responses.

The uses of computers in medicine are already astonishing. The field of medicine also represents one of the most promising areas for improving the quality of life through computer use.

THE ENDOCRINE SYSTEM

34·12 MOLECULAR AND CELLULAR CONTROLS

Hormones secreted by endocrine organs travel through the bloodstream to their target organs. When a hormone arrives at its target organ, what does it do? How do hormones cause changes in the activities of their target organs? The answers to these questions are being discovered through research into the interaction of hormones with other molecules.

Recall from the previous chapter that neurotransmitter chemicals must attach to *receptor proteins* to complete the action at a synapse. Hormones work in a fashion similar to neurotransmitters. Hormones attach to receptor proteins in their target cells. This attachment stimulates those cells. Like enzymes and substrates, hormones and the receptor proteins on target cells may fit together in the manner of a lock and key (Section 5·3).

In some cases a hormone attaches to a receptor protein on the surface of the cell membrane of the target cell. This type of attachment is often true of protein hormones, such as glucagon and the hormones of the anterior pituitary. Such hormones need not enter their target cells to take effect. In other cases a hormone attaches to a receptor contained within the cytoplasm or nucleus of the target cells. Steroid hormones, such as the corticosteroids and the hormones of the gonads, work in this manner. Such hormones must enter their target cells before they can take effect.

When a hormone attaches to a receptor on the surface of a cell, how are the hormone's effects produced on the inside of the cell? Think of a hormone as a messenger that carries a signal from one organ to another. The hormone is the *first messenger*. A *second messenger* is sometimes needed to carry the signal from the surface to the interior of a target cell. Refer to Figure 34·14A, which illustrates the general principle behind the second-messenger concept.

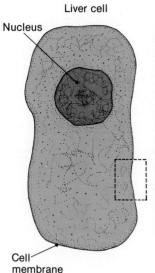
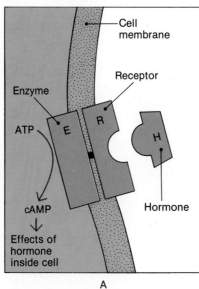
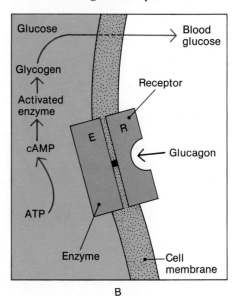

FIGURE 34·14 According to the second-messenger concept of hormone action, protein hormones produced by endocrine glands cannot pass through cell membranes of target cells. Instead, the protein hormones react with receptors on target cell membranes, which then produce a second messenger. The second messenger, cAMP, enters the cell and produces the hormonal effect.

A substance known as **cyclic adenosine monophosphate** (or cAMP) acts as the second messenger for many of the protein hormones. Adenosine triphosphate (ATP) is converted to cAMP by an enzyme. Notice in Figure 34·14A that this enzyme is located on the inner surface of the cell membrane. It is believed that the enzyme is connected to the hormone receptor on the outside of the membrane. When the appropriate hormone attaches to the receptor, the enzyme is stimulated to convert ATP to cAMP. Acting as the second messenger, cAMP causes the effect of the hormone on the inside of the cell.

Consider glucagon as an example of a hormone involving the use of cAMP as a second messenger. The liver is one of glucagon's target organs. Look at Figure 34·14B. As you can see, glucagon attaches to a receptor protein found on the surface of liver cells. The receptor protein stimulates the enzyme on the inner surface of the cell membrane. The enzyme converts ATP to cAMP, and then cAMP activates a glycogen-splitting enzyme in the cytoplasm of the liver cell. Glucose is released when glycogen is split. Glucose diffuses out of the liver cell into the blood.

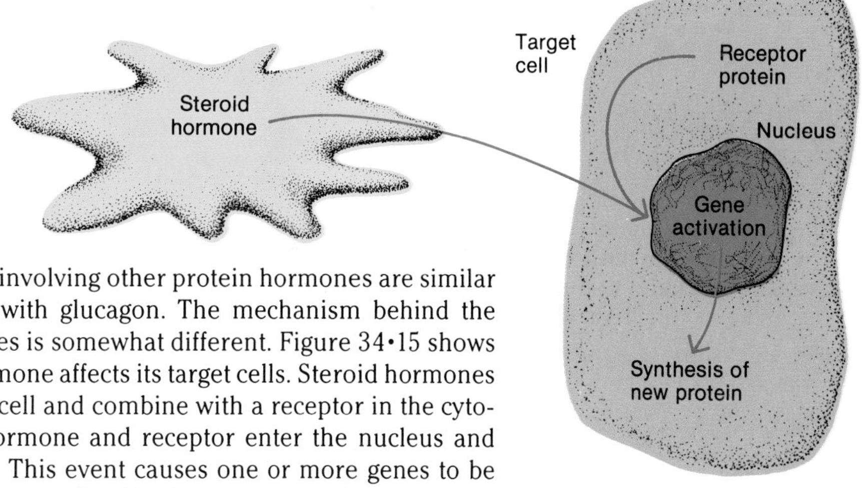

The series of events involving other protein hormones are similar to the events occurring with glucagon. The mechanism behind the action of steroid hormones is somewhat different. Figure 34·15 shows how a typical steroid hormone affects its target cells. Steroid hormones penetrate into the target cell and combine with a receptor in the cytoplasm. The combined hormone and receptor enter the nucleus and attach to a chromosome. This event causes one or more genes to be activated. The information contained in the genes is used to make new proteins in the cell. The production of new proteins is the effect of the steroid hormone.

FIGURE 34·15 Unlike other protein hormones, steroid hormones enter the target cells and combine with receptor proteins in the cytoplasm. The combined steroid hormones and receptors enter the nucleus and join to a chromosome, which activates one or more genes. This causes the genes to make new proteins.

REVIEW QUESTIONS

11. Explain the concept of feedback control.
12. How does the hypothalamus control the activity of the anterior pituitary?
13. How do receptors for protein hormones differ from receptors for steroid hormones?
14. Why is cAMP called a second messenger?
15. Suppose that a person is found to be unable to move glucose, stored as glycogen, from the liver into the blood. Tests show that glucagon levels are normal and the hormone itself is normal. Suggest a reason for the inability of glucagon to carry out its function.

Chapter Review

SUMMARY

34·1 Endocrine glands are ductless glands that secrete hormones into the bloodstream. The endocrine system, along with the nervous system, coordinates the activities of internal organs and maintains homeostasis.

34·2 The anterior lobe of the pituitary gland secretes growth hormone and prolactin, along with several hormones that regulate other endocrine glands. The posterior lobe of the pituitary gland secretes oxytocin and antidiuretic hormone.

34·3 Thyroxine from the thyroid gland regulates the body's metabolic rate. Calcitonin and parathyroid hormone maintain homeostasis of blood calcium and phosphorus levels.

34·4 The adrenal medulla secretes adrenaline, which stimulates the body's responses to stress and exertion. The adrenal cortex secretes glucocorticoids, which reduce inflammation and increase blood glucose.

34·5 Estrogen, which is produced in the ovaries, is the sex hormone of females. Testosterone, which is produced in the testes, is the sex hormone of males. The ovaries also produce progesterone, which prepares the uterus for pregnancy.

34·6 The islets of Langerhans secrete insulin and glucagon. These hormones oppose each other in maintaining steady concentrations of blood glucose.

34·7 The thymus, kidneys, and pineal gland produce hormones.

34·8 The pituitary gland and the thyroid gland may both be subject to disorders involving oversecretion and undersecretion.

34·9 Diabetes mellitus is caused by deficiency of insulin. In this disease, glucose cannot be absorbed by liver and muscle cells and accumulates in high concentration in the blood.

34·10 Many hormones act through negative feedback mechanisms to maintain homeostasis.

34·11 The hypothalamus regulates the anterior lobe of the pituitary through releasing factors that control the secretion of each of the anterior pituitary hormones. The hypothalamus controls secretion of posterior pituitary hormones by electrical impulses.

34·12 Hormones exert their effects by combining with receptor proteins in their target cells. Protein hormones often work through a second messenger, cAMP.

BIOLOGICAL TERMS

adrenal glands
adrenaline
adrenocorticotropic hormone (ACTH)
aldosterone
antidiuretic hormone (ADH)
calcitonin
corticosteroids
cortisol
cyclic adenosine monophosphate (cAMP)
endocrine glands
endocrine system
estrogen
exocrine glands
feedback mechanism
follicle-stimulating hormone (FSH)
glucagon
hormone
human growth hormone (HGH)
insulin
islets of Langerhans
luteinizing hormone (LH)
negative feedback
oxytocin
pancreas
parathyroid glands
parathyroid hormone (PTH)
pineal gland
pituitary gland
progesterone
prolactin
releasing factor
target organ
testosterone
thymus
thyroid gland
thyroid-stimulating hormone (TSH)
thyroxine

USING BIOLOGICAL TERMS

1. What is the endocrine system?
2. Distinguish between exocrine glands and endocrine glands.
3. Identify the gland that has a medulla and a cortex.
4. Identify the gland that contains both endocrine and exocrine tissue.

5. Distinguish between thyroxine and thyroid-stimulating hormone.
6. Insulin is to glucagon as calcitonin is to _____.
7. What are releasing factors?
8. What do the islets of Langerhans have to do with insulin and glucagon?
9. Define the term *negative feedback*.
10. Classify the hormones on the list of biological terms as steroids or proteins.

UNDERSTANDING CONCEPTS

1. How do endocrine glands distribute their secretions? (34·1)
2. How does the action of the endocrine system differ from that of the nervous system? (34·1)
3. Where is the pituitary gland located? Describe the structure of the pituitary gland. (34·2)
4. Which of the anterior pituitary hormones regulate other endocrine glands? Which do not? (34·2)
5. How are the thyroid and parathyroid glands related, aside from being next to each other? (34·3)
6. Describe the effects of two hormones secreted by the thyroid gland. (34·3)
7. Describe the three major types of hormones that come from the adrenal gland. Which are secreted by the cortex and which by the medulla? (34·4)
8. What hormones are produced by the gonads? How is the secretion of these hormones regulated? (34·5, 34·2)
9. What hormones are secreted by the pancreas? What are their functions? (34·6)
10. What hormones are produced by organs that are not primarily endocrine organs? (34·7)
11. What are the functions of prostaglandins? (34·7)
12. What disorder of the thyroid has become less common in recent years? Explain why. (34·8)
13. What is the effect of undersecretion of HGH? What is the effect of oversecretion? (34·8)
14. Describe the two types of diabetes, and explain how diabetes is treated? (34·9)
15. What condition is the opposite of diabetes mellitus? (34·9)
16. Explain what a feedback mechanism is. (34·10)
17. How does the hypothalamus control secretion by the anterior and intermediate lobes of the pituitary? (34·11)
18. How does the hypothalamus regulate secretion of posterior pituitary hormones? (34·11)
19. Explain the second-messenger concept. (34·12)
20. How do protein hormones differ from steroid hormones in their effects at the molecular level? (34·12)

APPLYING CONCEPTS

1. Would you expect glucagon secretion to be increased or decreased in an individual with diabetes mellitus? Explain why.
2. Both HGH and thyroxine are needed for normal growth. Explain the different effects of these hormones on growth.
3. Homeostasis of blood glucose levels is regulated by at least three hormones. Explain.
4. How do the endocrine system and nervous system work together in maintaining homeostasis?

EXTENDING CONCEPTS

1. Several of the Nobel Prizes in Physiology or Medicine have been awarded for work with endocrine glands and hormones. Prepare a report summarizing the work of one of the winners.
2. Individuals with defective adrenal glands have difficulty in tolerating stress. Investigate this fact and write a report about the role of the adrenal gland in coping with stress.
3. Vitamin D is converted into a hormone by the kidney. This hormone is involved with calcium homeostasis along with PTH and calcitonin. Find out what this hormone is and what its effects are.

READINGS

Bloom, A. *Diabetes Explained.* Kluwer-Boston, Inc., 1982.

Elting, Mary. *The Macmillan Book of the Human Body.* Aladdin (Macmillan), 1986.

Herbert, W. "Hormone Aberration—Anorexia Link Found." *Science News,* May 28, 1983, p. 340.

35. Reproduction and Growth

In every family it is always possible to identify similarities and differences between parents and their children. This is because humans reproduce sexually and the offspring receive genetic material from each parent. In this chapter, you will discover how each parent transmits genetic material and how the genetic information is combined into one new individual.

CHAPTER OBJECTIVES

After completing this chapter, you will be able to

- **Compare** the structures and functions of the male and female reproductive systems.
- **Describe** the hormonal interactions that occur in the male and female reproductive systems.
- **Identify** the stages involved in embryonic development.
- **List** the stages of development after birth.

Reproductive Systems

35·1 SIGNIFICANCE OF REPRODUCTION

Reproduction is the process by which offspring are produced. The most important function of reproduction is to continue the species. Reproduction may also serve to increase the number of individuals in a species.

Asexual reproduction involves a single parent. This type of reproduction continues the species and may quickly increase its number. Mitosis maintains the continuity of the genetic material from parent to offspring. For this reason, offspring are usually exact copies of the parent. Asexual reproduction produces little variation among offspring. Variations can arise because of mutations.

Sexual reproduction involves meiosis followed by the union of two sex cells, or gametes, to produce an offspring. This offspring receives genetic material from both parents. Although this genetic blueprint is similar to that of other members of the species, there are differences. These differences account for the variations found among members of a species that reproduces sexually. Such variations may allow the species to adapt to changes in the environment.

35·2 MALE REPRODUCTIVE SYSTEM

Humans reproduce sexually. Gametes are produced in specialized sex organs, or gonads. The gonad of the human male is the **testis** (pl., **testes**). The two functions of this organ are the production of **sperm cells,** which are the male gametes, and the production of the male hormone *testosterone* (tehs TAHS tuh rohn).

There are two testes, located outside the body cavity in a sac of skin called the **scrotum** (SKROH tuhm). The testes develop inside the pelvic cavity before birth. They descend into the scrotum just prior to birth or shortly thereafter. In the scrotum the temperature is 1–2°C lower than normal body temperature.

> **SECTION OBJECTIVES**
>
> - List the structures of the male and female reproductive systems and describe the function of each.
> - Identify the stages of the menstrual cycle.

Sperm cell production, or spermatogenesis, begins in human males at puberty. This event marks the onset of reproductive capability. Although the age varies, puberty most often occurs in boys between the ages of 9 and 16. Sperm cells are produced and mature within the testes in structures called **seminiferous** (sehm uh NIHF-uhr uhs) **tubules,** shown in Figure 35·1. Sperm cells are produced by meiosis. Thus they have half the number of chromosomes found in body cells. Mature sperm cells are made up of three parts. The head is made up of the chromosomes and other nuclear materials. The midpiece is packed with mitochondria that provide energy. The flagellum is a long whiplike tail that allows the sperm cell to swim. See Figure 35·2. Sperm cells have little cytoplasm and no stored food. Therefore, their life span is short once they are released.

The testes also serve as endocrine glands. Within each testis, scattered around the seminiferous tubules, are *interstitial* (ihn tuhr-STIHSH uhl) *cells.* These cells produce the male hormone testosterone. It is testosterone that causes the secondary sex characteristics of the male to develop at puberty. These characteristics include the deepening of the voice, increased muscle growth, and the growth of body hair, including a beard. Throughout adulthood, the level of testosterone affects spermatogenesis.

FIGURE 35·1 In the male reproductive system, the sperm cells are produced within the testes in structures called seminiferous tubules. The photograph is a cross section of a seminiferous tubule. Notice the tadpole-shaped sperm cells inside the tubule.

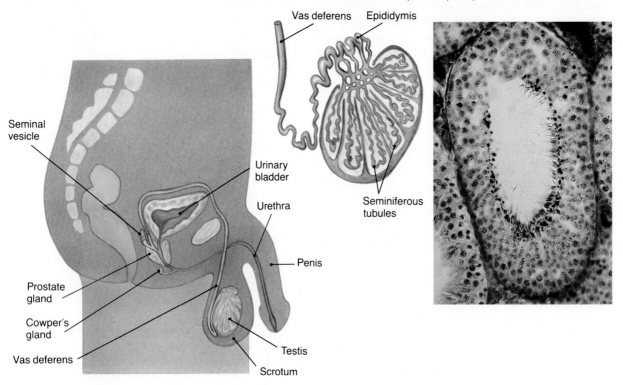

Mature sperm cells pass from the testis to the **epididymis** (ehp-uh DIHD uh mihs), the series of coiled tubules where sperm are stored. Before sperm cells leave the epididymis, they become motile, or gain the ability to move.

A long tube called the **vas deferens** (vas DEHF uh rehnz) connects each epididymis with the urethra, just below the urinary bladder. The *urethra* is a tube contained within the penis. The **penis** is the external reproductive organ of the male. Besides carrying urine from the urinary bladder, the urethra of the male also carries sperm.

The **seminal vesicles,** the **prostate gland,** and the **Cowper's glands** are glands that secrete seminal fluid. Find these glands in Figure 35·1. **Seminal fluid** is the substance that serves as a transport medium for sperm. In addition, the seminal fluid contains basic substances that help to neutralize the acidic conditions of the female reproductive tract. The seminal fluid also contains fructose, an energy source for the sperm cells. Together, the seminal fluid and the sperm cells are referred to as *semen* (SEE muhn).

During sexual intercourse, semen is released through the urethra of the male and deposited in the female reproductive tract. Approximately 300 million sperm cells are released at this time. Only one sperm cell will fertilize an egg cell.

35·3 FEMALE REPRODUCTIVE SYSTEM

The gonad of the human female is the **ovary.** There are two ovaries. They are found within the pelvic cavity. Normal body temperature does not affect the production of gametes in the female as it does in the male. The ovaries produce **egg cells,** which are the female gametes, and also secrete female sex hormones. Notice the location of the ovaries in Figure 35·3.

Each ovary has many follicles. A **follicle** (FAHL uh kuhl) is a structure that encloses a single egg cell, or ovum. Oogenesis, or egg cell production, begins before birth. At birth a female already has all of her egg cells. However, they are not fully mature. Their maturation and release begin at puberty. This event usually occurs between the ages of 9 and 16.

Besides producing gametes, the ovaries also produce the two female hormones: *estrogen* (EHS truh juhn) and *progesterone* (proh-JEHS tuh rohn). It is estrogen that influences the development of the secondary sex characteristics of the female at puberty. These traits include breast development, widening of the pelvis, and the redistribution of body fat. Estrogen and progesterone both function in the menstrual cycle of the female. This cycle will be further discussed in Section 35·4.

Beginning at puberty, the human female releases an egg cell from one of the ovaries approximately once each month. The release of an egg cell is called **ovulation** (oh vyuh LAY shuhn). The development and release of an egg cell are shown in Figure 35·3. The human

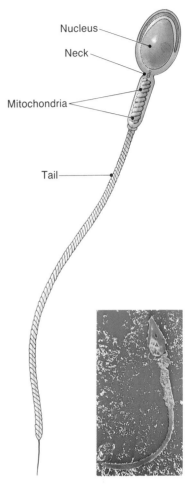

FIGURE 35·2 A mature sperm cell has three parts. The head contains the nucleus. The midpiece is packed with mitochondria. The flagellum is a long whiplike tail.

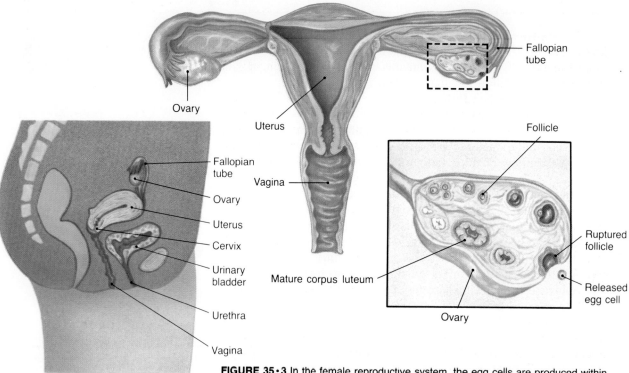

FIGURE 35·3 In the female reproductive system, the egg cells are produced within the ovaries in structures called follicles. In the cross section of the ovary, an egg cell is released from the follicle during ovulation. Then the empty follicle becomes a mass of yellow tissue called the corpus luteum.

female does not release gametes over her entire life span. **Menopause** is the time at which a female stops releasing eggs. Usually, menopause occurs between the ages of 40 and 50.

After ovulation, the egg cell is swept into the *oviduct,* or **Fallopian** (fuh LOH pee uhn) **tube,** by the action of ciliated cells that line the funnellike opening of the oviduct. Although not attached directly to the ovary, each Fallopian tube provides a passageway from an ovary to the uterus. The **uterus,** or womb (woom), is a strong muscular organ in which the embryo will develop if fertilization takes place. The **cervix** is the lower end, or neck, of the uterus and connects to the vagina (vuh-JĪ nuh). The **vagina** is a muscular tube that receives the semen during intercourse and serves as the birth canal.

35·4 MENSTRUAL CYCLE

The **menstrual** (MEHN stroo uhl) **cycle** is the hormone-controlled reproductive cycle of human females. It involves the periodic development and release of an egg and the periodic shedding of the uterine lining. The menstrual cycle involves interactions of the hypothalamus, pituitary, ovary, and uterus. The cycles usually occur regularly between puberty and menopause. The cycle may be interrupted by pregnancy, illness, or other factors. This cycle is about 28 days long but varies from one female to another.

The menstrual cycle has four stages, as shown in Figure 35·4. The first stage is *menstruation,* during which the uterine lining is shed. This stage lasts about 4 to 6 days. During this time, some blood along with some cells of the uterine lining are shed through the vagina.

As menstruation ends, the *follicle stage* begins. *Follicle-stimulating hormone (FSH)* is the pituitary secretion that causes development of egg cells within a number of follicles of an ovary. Normally, only a single follicle fully matures. As the follicles enlarge, they secrete the hormone estrogen. Estrogen causes the uterine wall to thicken. This thickening prepares the uterus to receive a fertilized egg. The follicle stage continues for about 9 to 10 days.

Ovulation, the release of the egg cell from the follicle, is a very short stage in the cycle. It takes place about in the middle of the cycle. The only time in the cycle when fertilization is possible is during the 15 to 24 hours after ovulation. *Luteinizing* (LOO tee uh nīz ihng) *hormone (LH)* and FSH are the pituitary secretions that stimulate ovulation.

After ovulation takes place, the *corpus luteum* (KOR puhs LOO-tee uhm) *stage* begins. Luteinizing hormone causes the follicle to become a yellow tissue called the corpus luteum. The **corpus luteum** is an ovarian tissue that secretes the hormone progesterone. Progesterone causes further thickening of the lining of the uterus. This stage lasts for about 13 to 15 days.

As the corpus luteum disintegrates, the thickened lining of the uterus is shed during menstruation. This stage occurs next only if fertilization has not taken place. In Figure 35·4, menstruation is shown twice. Remember, the end of one cycle is the beginning of the next.

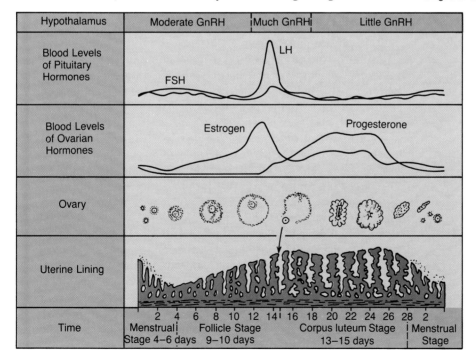

FIGURE 35·4 The menstrual cycle is the monthly cycle of change that occurs in the female reproductive system. Notice the amount of GnRH, or gonadotropic releasing hormone, produced by the hypothalamus during the four stages of the menstrual cycle.

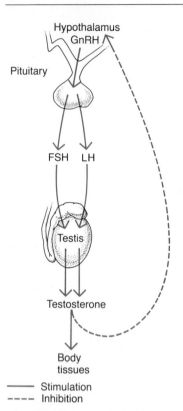

FIGURE 35·5 In males, GnRH, produced in the hypothalamus, stimulates the anterior pituitary gland to secrete FSH and LH. These hormones stimulate the testes to produce testosterone. The increased level of testosterone is a negative feedback signal. It inhibits GnRH production.

35·5 HORMONE CONTROL

The hypothalamus of the brain is involved in the control of many endocrine functions (Section 34·11). In both males and females, sexual maturation occurs through the interaction of several hormones. It is thought that beginning at puberty the hypothalamus secretes *gonadotropic releasing hormone* (GnRH). The GnRH stimulates the pituitary gland to secrete FSH and LH. In males, LH stimulates the testes to produce testosterone. Increased testosterone levels, along with FSH, causes sperm production to begin and continue throughout the life of the male.

Recall that biological systems often are regulated by *feedback mechanisms* (Section 34·10). In Figure 35·5 you can see that in males, FSH and LH from the pituitary gland stimulate the testes. As a result, testosterone is produced. The increased level of testosterone acts as a negative feedback signal. This hormone inhibits GnRH production by the hypothalamus. Thus the amount of FSH and LH produced by the pituitary gland is reduced.

In males, the feedback system usually keeps the level of testosterone within a narrow range. In females, however, hormone levels normally vary. If you refer to Figure 35·4, you will see sharp changes in some of the hormone levels as the menstrual cycle proceeds. What causes these changes? Using Figure 35·6, follow the steps in the feedback control of the female reproductive system.

The hypothalamus of the female releases GnRH, which causes the pituitary to secrete FSH. FSH causes the development of eggs within the follicles of an ovary. The follicles secrete estrogen, which causes the development of the uterine lining. The high levels of estrogen also cause the pituitary gland to secrete more LH. The interaction of FSH and LH causes ovulation.

After ovulation occurs, the corpus luteum begins to secrete progesterone, which causes the uterine lining to continue to thicken. Progesterone also inhibits GnRH production in the hypothalamus. This removal of the stimulus to the pituitary gland stops FSH and LH production. Progesterone thus acts as a negative feedback signal.

At this point in the cycle, the uterus is prepared if fertilization should take place. The corpus luteum continues to secrete progesterone. This hormone maintains the uterine lining for about 11 days. After this time the corpus luteum will break down if there has been no fertilization. As the progesterone level declines, the uterine lining is lost during menstruation. The hypothalamus begins to secrete GnRH again, thus beginning the next cycle.

If an egg has been fertilized, a special hormone is produced by the *chorion* (KOR ee ahn), an embryonic membrane. *Chorionic gonadotropic hormone* prevents the breakdown of the corpus luteum and the resulting decrease in the progesterone level. With progesterone secretion maintained, menstruation does not occur, and a new egg does not mature. Chorionic gonadotropic hormone is produced for about 3 months. It is the hormone detected in some pregnancy tests. By the

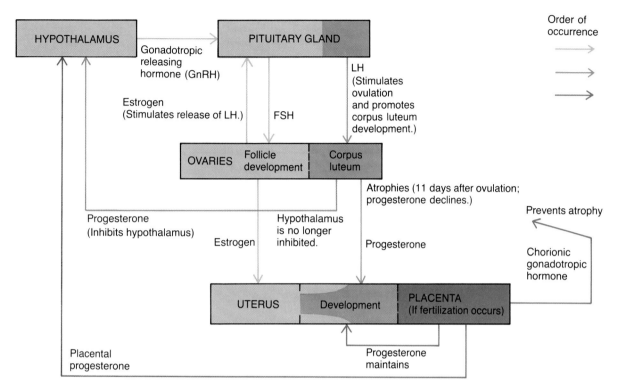

FIGURE 35·6 The female reproductive system is controlled by feedback mechanisms. The different hormones interact to produce the events of the menstrual cycle and pregnancy.

time chorionic gonadotropic hormone production ends, special tissue in the uterus, called the placenta (Section 35·10), can produce enough progesterone to maintain the pregnancy.

35·6 ADAPTATIONS FOR FERTILIZATION

In living things, some adaptations help to ensure that reproductive efforts will succeed. Among these adaptations are structures that allow fertilization to occur within the body of the female rather than outside of the body. Not only are the gametes brought closer together, but they are less likely to be destroyed by environmental factors.

Drying out of the sperm is a possible problem in animals that live on land. The seminal fluid produced by the male and the moisture of the lining of the female's reproductive tract protect against the sperm's drying out.

The production and release of millions of sperm is an adaptation that helps to ensure that some sperm will reach the egg and one might fuse with it. Also, recall that sperm have very little cytoplasm (Section 35·2). The small mass of the sperm and the presence of a long tail are adaptations that increase motility. Increased motility increases the likelihood that sperm will be able to reach the egg.

REPRODUCTION AND GROWTH

Adaptations of the female include muscular contractions of the vagina and uterus. These movements help to move sperm upward into the oviducts. The ciliated cells of the oviduct help to move the egg toward the uterus. These adaptations help to bring egg and sperm together.

REVIEW QUESTIONS

1. What is sexual reproduction?
2. What are the functions of seminal fluid?
3. What is the function of the corpus luteum?
4. During which stage of the menstrual cycle can fertilization occur?
5. If a female did not produce sufficient quantities of FSH and LH, how would her ability to have children be affected?

Development Before Birth

SECTION OBJECTIVE

- Describe the changes that occur between fertilization and birth.

35·7 FERTILIZATION

Although only one sperm will fertilize an egg, millions of sperm are released. Why are so many sperm needed? A closer look at the process of fertilization may provide some answers.

The egg cell releases a chemical that attracts sperm cells. The sperm cells near the egg then release an enzyme that helps to loosen the layer of extra cells. If these extra cells remained, the sperm cells would not be able to reach the egg cell's surface. The membrane of the egg cell is covered with a jellylike coat. The sperm cells release chemicals that act on the jelly coat. As can be seen in Figure 35·7, a tube forms from a sperm cell toward the egg cell's membrane. After the tube connects with the cell membrane, the nuclear material of the sperm enters the egg cell. Fertilization occurs in the oviduct.

After the nucleus of one sperm enters the cytoplasm, the egg cell secretes a complex layer of carbohydrates and proteins around the cell membrane. This new layer is called the *fertilization membrane*. This layer forms a chemical barrier that prevents more sperm from entering the fertilized egg.

Gametes are produced by meiosis. In humans, the completion of meiosis in the female takes place at the time of penetration by the sperm. The large, immature egg cell divides, forming the mature egg cell and one small polar body. The mature egg cell is much larger than the functionless polar body.

Recall that the mature egg nucleus contains the monoploid (n) number of chromosomes, and the sperm nucleus contains another monoploid (n) number of chromosomes (Section 6·8). Fertilization is not complete until the contents of these two nuclei fuse and reestab-

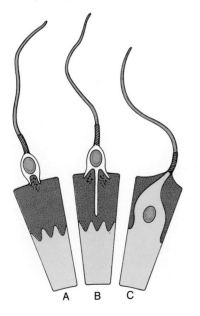

FIGURE 35·7 This drawing shows how a sperm cell attaches to an egg cell (A) and penetrates it (B and C).

lish the species ($2n$) number of chromosomes. This fertilized egg cell is referred to as a **zygote** (ZĪ goht).

It has recently become possible to fertilize eggs of human females outside their bodies. This type of fertilization in a laboratory culture dish is known as *in vitro* (ihn VĪ troh). It is used when the oviducts of the female are blocked due to malformation, disease, or another cause that prevents the sperm from reaching the egg. The fertilized egg is returned to the uterus of the mother, where it continues to develop. This technique has helped many couples who otherwise would not have been able to have children.

35·8 CLEAVAGE AND IMPLANTATION

Soon after fertilization, the zygote goes through a series of rapid mitotic divisions known as **cleavage** (KLEE vihj). These mitotic divisions result in the formation of a solid ball of cells called the **morula** (MOR yuh luh). The morula is no larger than the original zygote. Thus it is clear that cell growth does not follow each mitotic division. Look at Figure 35·9. Note the changes that occur during cleavage. As the cells continue their rapid mitotic divisions, they begin to secrete a fluid into the center of this cell mass. This fluid pushes the cells to the outside. The ball of cells that results is a **blastocyst**. The blastocyst is made up of an outer layer of cells called the **trophoblast** and an **inner cell mass** that will later become the embryo. The **embryo** is the organism in the early stages of development. The cells surround an inner fluid-filled cavity, the **blastocoel** (BLAS tuh seel).

The trophoblast has two main functions. One function is the production of chorionic gonadotropic hormone. This hormone prevents the corpus luteum from stopping its production of progesterone. As a

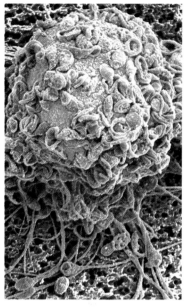

FIGURE 35·8 Although the enzymes from hundreds of sperm cells are needed to remove the protective coat on an egg cell, only one of the sperm cells enters and fertilizes the egg cell.

FIGURE 35·9 During the early stages of cleavage, the fertilized cell (1) undergoes mitotic divisions (2-4) and forms a solid ball of cells called the morula (5). As the cells continue to divide, a hollow ball of cells called the blastocyst is formed (6).

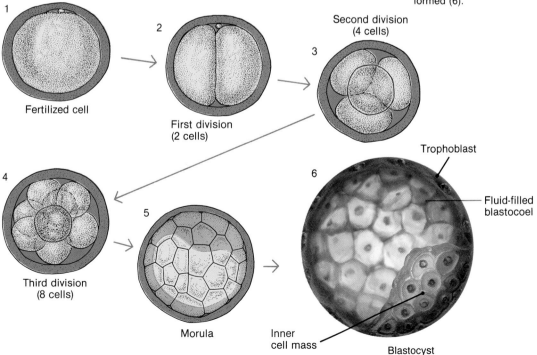

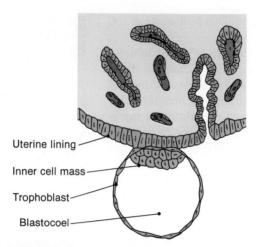
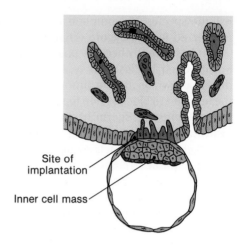

FIGURE 35·10 Implantation takes place about 8 to 10 days after fertilization. The blastocyst secretes an enzyme that digests some of the cells in the uterine wall. This permits the blastocyst to embed itself in the uterine wall.

FIGURE 35·11 As the cells of the blastocyst continue to undergo mitotic division, the cells of the inner cell mass change to form a double layer of cells called the embryonic disk. In a process called gastrulation, the cells move and form three primary germ layers called the ectoderm, the mesoderm, and the endoderm.

result, the uterine lining is maintained (Section 35·5). The second function is the production of enzymes. At the point where the embryo touches the uterus, these enzymes digest some of the cells of the uterine lining. The embryo then embeds itself in this location. The process by which the embryo attaches to the uterus is called **implantation.**

During implantation, the embryo becomes attached to the uterine wall. Two membranes, the *amnion* and the *chorion*, develop from the trophoblast. These membranes surround the embryo. They are similar in structure to those of a bird embryo. Implantation and membrane formation take place about 8 to 10 days after fertilization.

35·9 GASTRULATION

The cells of the blastocyst continue to multiply by mitotic division. After a few hundred cells have formed, the inner cell mass begins to change. Two cavities form in the inner cell mass, each surrounded by a layer of cells. The region where these two layers of cells are in contact is the *embryonic disk*. Find the embryonic disk within the blastocyst shown in Figure 35·11. The entire embryo forms from the embryonic disk. All of the other structures in the blastocyst form membranes that enclose and protect the embryo.

The embryonic disk is made up of two layers of cells. In a process called *gastrulation* (gas truh LAY shuhn), some cells from the upper layer sink in through the center of the embryonic disk. As a result of these cell movements, the embryonic disk becomes a three-layered structure. The upper layer is the **ectoderm.** The middle layer is the **mesoderm.** The lower layer is the **endoderm.** These three layers are referred to as the *three primary germ layers* because they give rise to all of the structures in an embryo. The ectoderm gives rise to the nervous system, the epidermis, and the sweat glands. Muscles, bones, blood, blood vessels, the reproductive system, kidneys, and dermis develop from the mesoderm. The endoderm forms the liver, some endocrine glands, and the linings of the digestive organs, lungs, trachea, and bronchi.

The process by which the cells in the primary germ layers become specialized tissues is called *differentiation*. At the end of gastrulation, certain ectodermal cells are the first to differentiate. These cells begin to thicken and form a *neural plate*. The edges of this neural plate then begin to rise, as shown in Figure 35·12. These raised outer edges are called *neural folds*. The neural folds gradually come together and form a hollow *neural tube*. The anterior part of the neural tube will become the brain. The posterior part will become the spinal cord.

Scientists do not know just what causes certain cells to develop into the specialized tissues of the embryo. Differentiation may be due to mechanisms that "turn on" certain genes in these cells (Section 8·10). It has been observed that some cells can influence the differentiation of nearby cells. Due to this ability, such cells have been named *organizers*. The process in which one cell influences the differentiation of another cell is called **embryonic induction**. How this induction of nearby cells takes place is not completely known. Relative positions of certain embryonic tissues seem to be important, as are certain chemicals that pass between the cells. These chemicals, called *inducers*, can cause a cell to express certain genes. Thus these chemicals induce differentiation in the embryo.

35·10 EMBRYONIC DEVELOPMENT

The normal gestation period in humans is about nine months, or 266 days. For the sake of reference, pregnancy is usually divided into three trimesters of three months each.

During the first trimester, the fertilized egg develops into a blastocyst and then becomes implanted. Gastrulation is followed by a great deal of differentiation. By the end of the first trimester, the embryo has become a **fetus** (FEE tuhs), an unborn organism that has developed all of its major organs. The fetus is surrounded by a membrane called the amnion. The amnion contains amniotic fluid which protects the developing fetus from dehydration and infection, and absorbs shocks.

The fetus is attached to the uterine wall by means of the placenta and the umbilical (uhm BIHL uh kuhl) cord. The **placenta** is a structure through which nutrients and oxygen pass from the mother to the developing young. Wastes pass in the opposite direction. The placenta is made up of maternal and embryonic tissue. It is first formed from small *chorionic villi*, fingerlike projections of the chorion, another membrane surrounding the embryo. Another membrane, the *allantois* (uh LAN toh ihs) provides the blood vessels in the embryo's portion of the placenta. Unlike the chorion and amnion, the allantois does not surround the human embryo.

The **umbilical cord** is the structure that connects the embryo to the placenta. It contains two arteries and one vein formed from the allantois. The blood systems of the mother and the embryo are separated by a thin layer of cells called the *placental barrier*. Here, nutrients and oxygen diffuse from the mother to the embryo. Carbon dioxide and other wastes diffuse from the embryo to the mother.

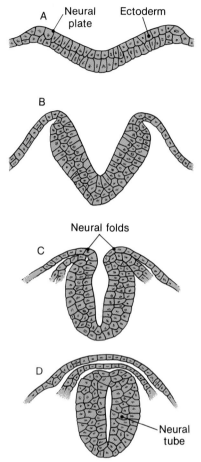

FIGURE 35·12 At the end of gastrulation, ectodermal cells divide and form a neural plate (A) with two raised edges called neural folds (B). The neural folds come together and form the neural tube (C and D), which will later form the brain and spinal cord.

umbilicus (navel)

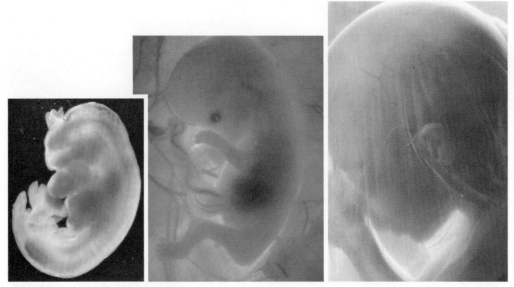

FIGURE 35·13 During the fourth week, the embryo is about the size of a pea (left). At ten weeks (center) and eighteen weeks (right), the fetus, as it is now called, looks more like a human.

Some substances and diseases can harm the developing embryo, particularly during the period of organ formation. Rubella (German measles), when contracted in the fourth through twelfth weeks of pregnancy, may cause damage to the heart, eyes, ears, or brain of the fetus or embryo. Recent studies suggest that caffeine and alcohol, as well as drugs such as heroin and morphine, may cause harm to developing fetuses. Drinking alcohol during pregnancy can cause slowed growth of the fetus.

In Figure 35·13, compare the first trimester fetus with later stages. In the second trimester a bony skeleton forms and the fetus becomes covered with soft, downy hair called *lanugo* (luh NOO goh). In the fourth month, the fetus is about 14 cm long and has a mass of about 115 g. If it is born at the end of the sixth month, the fetus has a small chance of surviving, but only if it receives special care. At this time, the fetus is about 35 cm long and has a mass of about 680 g.

The final trimester is a period of rapid growth. The fetus greatly increases in size and mass, more so than in the previous trimesters. Antibodies (Section 15·11) that the fetus acquires from its mother allow it to resist infection during the first few months after birth. It has been observed that as the fetus floats in its amniotic environment, it sucks its thumb and kicks its limbs.

35·11 BIRTH

After approximately 266 days of gestation, the fetus is pushed out of the uterus by a process known as *labor*. Labor is made up of three stages. In the first stage, the cervix thins and opens. This change is accomplished through rhythmic contractions of the uterus. During this stage of labor, the cervix dilates, or enlarges, from 1 cm to 10 cm.

The second stage of labor is the expulsion phase. During this stage the baby is delivered as a result of more uterine contractions. After birth, the baby is still attached by the umbilical cord to the placenta, which remains attached to the uterine wall. The umbilical cord

FIGURE 35·14 When a fetus is about to be born, the muscles of the uterus begin slow, rhythmic contractions. This process is known as labor and is illustrated in the two models.

is clamped and then cut. The piece of the umbilical cord that is left attached to the newborn will dry and fall off. The lifelong scar that remains is the navel, or belly button.

In the third stage of labor the afterbirth is expelled by more uterine contractions. The placenta and attached umbilical cord plus the empty amnionic sac make up the *afterbirth*.

Sometimes multiple births occur. When two babies are born at the same time, they are referred to as *twins*. There are two possible types of twins. *Fraternal twins* are the result of a double ovulation and two separate fertilizations. Two zygotes develop and each embryo forms its own membranes and placenta. The twins produced may be of the same sex, or the twins may be a boy and a girl. These twins each receive separate genetic material. Therefore they may be as alike or as different as any two children of the same parents.

Identical twins result from a single fertilized egg. Early in development the cells of the developing embryo become separated and develop into separate individuals. Identical twins may share a common placenta, and both may grow within one set of membranes. Since these twins share the same chromosomal makeup, they must be of the same sex and will show similar physical traits.

Multiple births include the births of more than two babies. In humans, such naturally occurring multiple births are rare. However, the use of fertility drugs can cause the release of several eggs at a time. These multiple births may be combinations of both types of twinning.

REVIEW QUESTIONS

6. What is fertilization? Where does it occur in humans?
7. Describe the differences between a morula and a blastocyst.
8. What are the three primary germ layers, and what body systems does each become?
9. Describe the function of each membrane of the human embryo.
10. Suggest a possible combination of fraternal and identical twinning that could produce a set of quintuplets with three girls and two boys.

Investigation

How Is the Chicken Egg Adapted to Support an Embryo?

Goals
After completing this activity, you will be able to
- Locate the nuclear material in a chicken egg from which an embryo develops.
- Describe the nutrient and water source present in the chicken egg.
- Identify the parts of a chicken egg that protect the embryo.

Materials (for groups of 2)
unfertilized chicken egg
hand lens
petri dish
metric ruler

Procedure
A. Refer to the figure as you locate the parts of the egg.

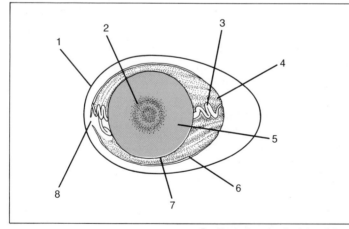

B. Gently crack an egg crosswise. Separate the two halves of the shell as you would to put an egg into a frying pan. Allow the contents of the egg to drop into a petri dish. Be careful not to break the yolk. Examine the shell **(1)** with your hand lens. Do not discard the shell.

C. On the surface of the yolk locate a small, white round dot. This dot is the blastodisc **(2)**, from which an embryo would form after fertilization. Use the metric ruler to find the diameter of the blastodisc.

D. Find two dense, white, cordlike structures. These structures are called the chalazae (kuh-LAY zee) **(3)**. Using forceps, gently pull on one of these.

E. The white of the egg contains water and a protein called albumen (al BYOO muhn) **(4)**. Notice that the white surrounds the yolk and is fairly thick.

F. Observe the yolk **(5)**. The yolk of the egg contains fats, minerals, and vitamins. These materials, which come from the liver of the hen, pass to the ovary and become part of the egg.

G. Examine the inside of the shell and the membranes **(6, 7)** attached to it. Try to locate an air space **(8)** between the membranes and shell. The chick uses this air only when it is ready to hatch.

H. On a separate sheet of paper, draw the egg you have examined. Label the albumen, yolk, chalazae, blastodisc, air space, membranes, and shell.

I. Follow your teacher's instructions for the proper disposal of all materials.

Questions and Conclusions
1. What is the approximate diameter of the blastodisc?
2. To what are the chalazae attached? What is the likely function of the chalazae?
3. What might the egg white supply for the developing embryo?
4. At which end of an egg is the air space—the blunt or the pointed end?
5. If the shell membranes are removed from the shell and water is added, the water will seep out of the shell. What feature of the shell would allow this to occur?
6. If the membranes remain in a shell, very little water will seep out. What function do the membranes probably perform for the developing embryo?
7. What function does the shell perform for the embryo?
8. In the developing human, there is no albumen serving as a source of nutrients. How does the human embryo obtain nutrients for growth and development?

Development After Birth

35·12 INFANCY AND CHILDHOOD

In humans, the fastest rate of growth occurs before birth. During infancy and childhood, the growth of various organs continues at different rates. The brain of a newborn baby is only 23 percent of its full adult size. During the first years of infancy and childhood, the rapid rate of growth of nervous tissue continues. Brain development proceeds at this rate until the age of 3. Then the rate becomes slower until, by the age of 7, the brain reaches its full adult size.

Similarly, the bones of a baby are not fully grown at birth. The growth rates of different parts of the body vary. A newborn baby has a large head size in proportion to total body length. The long bones of the legs continue their rapid growth rate for about 10 years after birth.

Motor development, or coordinated movements, begins in infancy. At the end of the first month, an infant begins to smile and can usually turn its head and follow moving objects. By the age of 3 months, the baby can control its own head movements and reach for objects. At 10 months, the baby sits unassisted, crawls, and attempts to stand. At 1 year, the child walks with one hand held and attempts a few words. By 2 years, the child runs, walks up and down stairs, feeds itself, and uses simple sentences. At 3 years, the child can play simple games. By age 4, he or she can hop on one foot and throw a ball. At 5, the child can skip.

35·13 ADOLESCENCE AND ADULTHOOD

Most of the growth of body systems is completed by the end of childhood. The reproductive system is the exception. It is not until puberty that the rapid development and maturation of this system occurs.

Adolescence is a period of life that begins with puberty and ends when full body size is reached. It usually includes the teenage years and sometimes extends into the early twenties. Muscle development increases during early adolescence. It is completed in late adolescence. At this time muscle performance and athletic ability peak. Although the brain was fully formed in childhood, further development of neurological pathways continues through adolescence.

Adulthood is the period when organisms have reached their full physical size. There is no additional growth, only replacement of damaged cells. A gradual decrease in physical abilities occurs. Proper nutrition and adequate exercise may help to slow this decrease in physical abilities. Between the ages of 40 and 50, most human females experience menopause, the cessation of the menstrual cycle. At this time the ovaries stop producing estrogen and progesterone. Some women experience *osteoporosis*, a thinning and weakening of the bones. This disorder has been connected with a low level of estrogen.

SECTION OBJECTIVE

- Describe the changes that occur in the development of humans after birth.

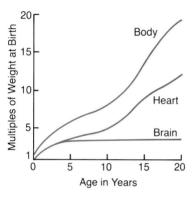

FIGURE 35·15 The graph shows the growth rates of the body, heart, and brain after birth.

35·14 AGING

Aging is a biological process that is usually described by the characteristics seen in older persons. Those characteristics include features such as gray hair, wrinkled skin, and lowered resistance to disease. Although signs of aging are easy to see at the organismic level, the process of aging is thought to occur at the cellular level. Several hypotheses have been proposed to explain aging in cells.

One hypothesis, the *error-catastrophe hypothesis of aging,* suggests that over a period of time errors occur in the steps of protein synthesis. Faulty proteins are produced as a result. These enzymes no longer can control the cell's metabolic activities. It is known that cells can repair errors in faulty proteins. However, young cells appear better able to correct errors than old cells can. Thus, as a cell ages, it accumulates errors until the metabolic activities become disrupted. The catastrophe part of the hypothesis is cell death.

Research suggests that certain reactive chemicals, called *free radicals,* cause DNA mutations. Free radicals may be made within the cell. They also can be taken into the body in food, in cigarette smoke, and in other substances that reach cells. Free radicals may also form in cells as a result of exposure to radiation. By changing the DNA, free radicals cause the production of proteins containing errors.

Transposons are segments of DNA that move around on a chromosome or from one chromosome to another. DNA altered as a result of transposons could also be the source of some errors in proteins. These errors may be some of those that cause the eventual disruption of metabolic processes in the cell.

A second hypothesis suggests that aging is a normal cellular event programmed into the genetic message. According to the hypothesis, genes controlling aging direct the cell to shut down certain metabolic pathways. These decreases in metabolic function then result in both decreased resistance to disease and decreased ability to repair injuries. The cell eventually dies.

Although many other ideas about how aging occurs have been proposed, the two hypotheses outlined here are given the most support. Research on aging continues in an effort to increase the years of quality life of humans.

REVIEW QUESTIONS

11. How does the growth rate before birth compare with that of an infant and a child?
12. Compare the rate of growth and development of the reproductive system in a child and an adolescent.
13. What is menopause?
14. How are the two hypotheses of aging similar? How do they differ from one another?
15. A growth spurt occurs at the beginning of puberty. What might trigger a growth spurt at this time?

Careers in Biology

Professional Medical Services

Professionals in the medical services are responsible for the diagnosis, treatment, and prevention of illnesses and injuries.

Physicians diagnose health problems and administer the proper treatment. They also perform routine examinations and advise patients on ways to stay healthy. To practice medicine a person must graduate from medical school and have advanced training in a specialized area, such as surgery.

Nurses assist physicians in the treatment and prevention of illness. They monitor and record patients' conditions and administer medication as directed by a physician. To become a nurse, a person must complete at least a two-year nursing degree and pass a state licensing examination.

Diet-Related Staff

Dietitians are experts on the role of nutrition in health and in disease. They may specialize in one of several areas, such as education, research, or nutrition care. Clinical dietitians, for example, are responsible for the nutritional needs of hospital patients. They consult with doctors to arrange proper diets, and they educate patients and the families of patients about proper nutrition at home. Other dietitians manage hospital kitchens, conduct research at medical centers, or work as educators for public health services. A registered dietitian must have a bachelor's degree, with a major in nutrition or institutional management. He or she must also pass a national certification exam. Dietetic assistants provide day-to-day food service or nutrition care under the direction of a dietitian.

Medical Support Staff

The medical support staff is made up of many people doing jobs that aid physicians and nurses in the diagnosis and treatment of illness. For example, electrocardiograph technicians operate machines that provide information for the diagnosis of heart ailments. These medical technicians and others must have a high school diploma and at least one year of training in their specialty at a hospital or technical institute.

Physical therapists plan and administer treatment programs for patients with physical disabilities. They use exercise, massage, and special electrical equipment to help relieve pain and restore function after injury or disease. A physical therapist must have at least a bachelor's degree, with courses in anatomy, physiology, and physics.

How to Find Out About Health Careers

There are over 200 different health professions. Health professionals work in hospitals, nursing homes, psychiatric institutions, neighborhood clinics, research centers, and public health agencies.

There are several ways to find out about health careers. First, look in your school or public library for the *Occupational Outlook Handbook,* published by the U.S. Department of Labor. Also look for the *Health Careers Guidebook,* published by the U.S. Department of Labor and the U.S. Department of Health and Human Services. Second, contact government health agencies, professional organizations, and the personnel departments of local health facilities for information about employment opportunities. Finally, the National Health Council, 1740 Broadway, New York, NY 10019, can provide a copy of *200 Ways to a Health Career.*

Chapter Review

SUMMARY

35·1 Reproduction continues the species and may increase the number of individuals in a species.

35·2 The male gonad, the testis, produces sperm cells and releases hormones that control the male secondary sex characteristics.

35·3 The female gonad, the ovary, produces egg cells and releases hormones that control both the secondary sex characteristics and the menstrual cycle.

35·4 The menstrual cycle is a hormone-controlled series of events that involves the release of a mature egg and preparation of the uterus for possible pregnancy.

35·5 The production and release of gametes in males and females is controlled by the interaction of hormones from numerous endocrine glands.

35·6 Special adaptations of the reproductive systems of males and females help to ensure that fertilization will take place.

35·7 Fertilization is the fusion of the nuclear contents of a sperm with the nuclear contents of an egg.

35·8 Mitotic cell divisions, called cleavage, allow the embryo to increase the number of cells and produce structures that aid implantation.

35·9 The development of the three primary germ layers provides the cells that eventually differentiate into the body systems.

35·10 The nine-month gestation period is divided into three trimesters, which include the differentiation of cells into organs, formation of the bony skeleton, and a final period of rapid growth.

35·11 Birth follows the labor process, consisting of three main stages: opening of the cervix, delivery of the baby, and expulsion of the afterbirth.

35·12 During infancy and childhood, 0 to 12 years, growth of the brain, bones, and other organs continues.

35·13 During adolescence, the reproductive systems of males and females mature and growth continues.

35·14 Aging may involve errors in protein synthesis or may be a genetically determined event.

BIOLOGICAL TERMS

blastocoel
blastocyst
cervix
cleavage
corpus luteum
Cowper's glands
ectoderm
egg cells
embryo
embryonic induction
endoderm
epididymis
Fallopian tube
fetus
follicle
implantation
inner cell mass
menopause
menstrual cycle
mesoderm
morula
ovary
ovulation
penis
placenta
prostate gland
scrotum
seminal fluid
seminal vesicles
seminiferous tubules
sperm cells
testis
trophoblast
umbilical cord
uterus
vagina
vas deferens
zygote

USING BIOLOGICAL TERMS

1. What is the relationship between the follicle and the corpus luteum?
2. Distinguish between an embryo and a fetus.
3. How are the trophoblast and the inner cell mass related?
4. What is implantation?
5. What is the relationship between the placenta and the umbilical cord?

Choose the term that does not belong and explain why it does not belong. Define each term in your explanation.

6. seminal vesicle, testis, Cowper's gland, prostate gland
7. mesoderm, ectoderm, blastocyst, endoderm
8. penis, cervix, epididymis, prostate gland

9. uterus, vagina, vas deferens, Fallopian tube
10. morula, trophoblast, follicle, blastocyst

UNDERSTANDING CONCEPTS

1. Why does sexual reproduction result in more variation in a species than does asexual reproduction? (35•1)
2. Describe the pathway taken by sperm from the site of production to the outside of the body. (35•2)
3. What are the functions of the seminal vesicles, prostate gland, and Cowper's glands? (35•2)
4. Identify the structures of the female reproductive system. (35•3)
5. Describe the path of the egg from the ovary to the uterus. (35•3)
6. List the stages of the menstrual cycle and briefly describe the events of each stage. (35•4)
7. What is the effect of gonadotropic releasing hormone on males? (35•5)
8. Describe hormonal regulation of the menstrual cycle. (35•5)
9. List three adaptations that help to ensure fertilization. (35•6)
10. Describe how the actions of the sperm cells lead to one sperm cell fertilizing the egg cell. (35•7)
11. Describe the events that follow fertilization, from the beginning of cleavage through the time of implantation. (35•8)
12. Describe what occurs during gastrulation. (35•9)
13. What is the importance of the primary germ layers to the developing embryo? (35•9)
14. Explain the roles of organizers and inducers in embryonic induction. (35•9)
15. What is the function of the placenta? (35•10)
16. Identify the three membranes of the human embryo and name their function(s). (35•10)
17. Describe the three stages of labor. (35•11)
18. Compare the origins of fraternal and identical twins. (35•11)
19. Compare the growth of the head with the growth of the leg bones in children. (35•12)
20. What changes occur during adolescence? (35•13)
21. Explain the error-catastrophe hypothesis of aging. (35•14)

APPLYING CONCEPTS

1. Mothers who are nursing their babies do not usually resume ovulation for some time compared with mothers who do not nurse their babies. What biological advantage would the absence of ovulation during nursing be to the species?
2. What would the effects be on the female reproductive system if a woman did not have a corpus luteum stage following ovulation?
3. After a multiple birth the doctor may examine the afterbirth in detail. What would the doctor be looking for in this special situation?
4. Compare the function of the membranes of a bird embryo with the function of the membranes of a human embryo.

EXTENDING CONCEPTS

1. Prepare a report detailing some examples of Siamese twins. Determine how such twins are formed and what can be done to correct this situation.
2. Infertility is a problem faced by some couples who desire to have children. Identify possible causes of this problem. Is it simply the inability to produce sperm or eggs? What corrective measures can be taken for males and females?
3. Prepare a report on human chorionic gonadotropic hormone. Be sure to include where and when it is produced and what its function is.
4. Do library research on *in vitro* fertilization. Be sure to include why the technique is used, what procedures are involved, and how successful the technique is.
5. Prepare a report on the effect of caffeine intake or smoking during pregnancy.

READINGS

Lenard, Lane. "High-Tech Babies." *Science Digest,* August 1981, p. 86.

Rodgers, Joann. "Me, Myself, and Us: Twins." *Science Digest,* November/December 1980, p. 6.

Wyatt, Richard Jed. *After Middle Age: A Physician's Guide to Growing Old and Staying Healthy.* McGraw-Hill, 1985.

Unit VII Skills

INTERPRETING SCIENTIFIC DRAWINGS

The illustrations in this textbook are of two major types—photographs and drawings. Both kinds of illustrations are used to show clearly how something looks or how it works. Drawings are not as realistic as photographs, but the scientific drawings in this unit represent the human body and its parts. Learning to interpret scientific drawings can help you to understand biological structures and ideas. Drawings are used in many ways: to simplify, to enlarge and show details, to show various views, and to show relationships of one part to another.

Through drawings an artist can simplify structures that in reality are very complicated. Figure 31·12 shows how various blood vessels are connected. It also uses color to show the relative oxygen content of the blood as it passes through the vessels. Study the figure and answer the questions on a separate sheet of paper.

1. What is the color of the blood as it passes from an artery to an arteriole? Why is the blood shown in this color?
2. What color does the blood appear as it passes from a capillary into a venule? Why is the blood shown in this color?
3. Why is a drawing of the blood vessels more useful than a photograph?

Sometimes a drawing is used next to a photograph to clarify the details of the photo. Look at Figure 29·6, which shows a diver in motion. Around her are several drawings of types of joints. The drawings show how the diver's joints look when she is in the position shown. Answer the following questions.

4. What details have been removed from the joints in the drawings? How does this simplification help?
5. How do these drawings differ from drawings of joints at rest?

Art can be used to enlarge something so that you can see details of a structure. For example, a series of drawings may show an organ system, a section of one organ, the cells making up that organ, and the molecules of a cell. Look at Figure 34·5, which shows a small section of an adrenal gland enlarged to show details of internal structure. Answer the questions.

6. What details of the adrenal gland do you see?
7. How does the drawing help you understand how the gland works?

Drawings can show different views of the same structure. You learned about both longitudinal and cross sections in Unit I Skills. In Figure 30·9 you can see a drawing of the inside of a tooth. Answer these questions.

8. Does this drawing show a longitudinal section or a cross section?
9. If the tooth were cut in the opposite direction, what structures might not be visible?

It is helpful to see the organ systems in relation to the rest of the body. In several places in this unit you find drawings of an organ system over the silhouette of a human body. This technique lets you see the organs in relation to each other and to the rest of the body. Study Figure 30·6 and answer the following questions.

10. On which side of the body is the appendix? To what is it attached?
11. Where is the liver in relation to the stomach? How does the liver compare in size to the stomach?

Making your own drawings can help you understand the relationships of body organs to each other. Using Figures 30·6 and 32·1, draw the digestive and respiratory systems in the following way. On a sheet of white paper, draw a body outline. Draw in the organs of digestion in the correct positions. On a separate sheet of tracing paper, draw a body outline with the organs of respiration in the correct positions. Be sure that both outlines are the same size. Why is this important?

On both drawings, label each organ or part of the body system. Draw straight lines called tag lines from each label to the corresponding body part. Now tape the second drawing over the first drawing to produce a finished overlay drawing. Then answer these questions.

12. Which parts of the two body systems share the same areas of the body?
13. How do drawing overlays help clarify details of these body systems?

UNIT VIII

Ecology

No organism is completely self-sufficient. Organisms depend upon other organisms and upon the environment for survival. This porcupine depends on plants for food and can live only in an environment that has the right conditions for porcupines. Unit VIII identifies and explains the relationships that exist among organisms and the interaction that occurs between these organisms and the environment.

Chapters
36 *The Environment*
37 *Ecological Changes*
38 *Biomes*
39 *Your Environment*

36. The Environment

The beauty of this delicate coral is greatly enhanced by the green alga living on it. But this relationship between the coral and the alga has a more practical value. The alga produces food for the coral through photosynthesis. The alga also aids the coral in building its enduring limestone skeleton. How do other organisms interact with one another? What relationships exist between the living and nonliving world? These questions will be explored in this chapter.

CHAPTER OBJECTIVES

After completing this chapter, you will be able to

- **Describe** how organisms affect their environment and how the environment affects them.
- **Give examples** of the ways in which nutrient elements cycle in the environment.
- **Discuss** the roles of producers, consumers, and decomposers.
- **Explain** how energy flows through ecosystems.
- **Identify** three kinds of symbiotic relationships.

Organisms and the Environment

36·1 INTRODUCTION TO ECOLOGY

Ecology is the study of the relationships between living things and their environment. An organism's environment is made up of all the living and nonliving factors that surround the organism.

The **abiotic factors,** or nonliving factors, include physical factors such as temperature, water, light, and minerals. The **biotic factors** are living things, such as plants, fungi, and animals.

Interaction is a key idea in ecology. All the living and nonliving things in an environment affect each other. For example, water is needed by plants to grow. The plants give shade. Shaded areas may remain cooler and more moist. When plants die and decay, their organic material is added to the soil. This material helps the soil hold more water.

36·2 ECOSYSTEMS

Living things can be found only within a limited region on the earth, called the **biosphere.** The biosphere extends from the deepest part of the ocean to a few thousand meters into the atmosphere. The biosphere is made up of units called ecosystems. An **ecosystem** is any area in which energy is transferred as organisms interact with each other and with nonliving things. Ecosystems are the basic units of ecology. Ecosystems include oceans, ponds, swamps, forests, farms, and cities. Each of these ecosystems consists of organisms interacting with each other and with abiotic factors. An ecosystem can be small, like the surface of a leaf, or huge, like an ocean.

A salt marsh is an example of an ecosystem. A salt marsh has abiotic factors such as water, sunlight, soil, and minerals. Organisms such as marsh plants, small fish, and birds such as snowy egrets are found in this ecosystem. These living things interact with each other and with the abiotic factors of the ecosystem. The plants grow in soil,

SECTION OBJECTIVES

- Give examples of biotic and abiotic factors in the environment.
- Distinguish between a biosphere, an ecosystem, a habitat, a community, and a population.

FIGURE 36·1 An ecosystem can be as large as a planet. An ecosystem can also be very small. These tiny aphids and this leaf are also an ecosystem.

which gives them minerals. The sun provides the plants with energy, and water provides them with moisture and minerals such as nitrogen, calcium, and phosphorus. Certain fish eat the plants. These fish serve as food for birds such as egrets. As the organisms in the marsh die, bacteria living in the soil break down the dead organic matter. The bacteria obtain energy and nutrients during this process. In addition, the bacteria return minerals stored in organic matter to the soil, where they are available to support new plant growth.

A forest ecosystem has abiotic factors such as soil, moisture, and sunlight. Organisms such as various species of shrews, owls, woodpeckers, spruce trees, and ants can be found in a particular forest ecosystem. These organisms interact with each other. They depend on proper amounts of minerals, moisture, and sunlight in much the same way as organisms in the salt marsh ecosystem depend on these abiotic factors.

Even the surface of a leaf can be considered an ecosystem. Insects called aphids, which live on the leaf, depend on the leaf for food. Abiotic factors such as moisture and temperature are also part of this ecosystem.

An ecosystem can be further broken down into habitats. A **habitat** is the place in which an organism lives. In the forest ecosystem the habitat of a spruce tree includes the soil and the space above the ground occupied by the tree. The habitat of a certain kind of ant is in the tree. The habitat of a shrew is the forest floor and the soil. How is the habitat of an egret different from the habitat of a fish?

36·3 POPULATIONS AND COMMUNITIES

Ecosystems include a variety of organisms. A group of organisms of the same species living in an ecosystem at a specific time is called a **population.** For example, a salt marsh ecosystem might include a population of snowy egrets. A forest ecosystem might include populations of white spruce trees and great horned owls.

A population is influenced by many factors. Nearly any factor that influences a population can also limit the size of that population. A population is influenced by abiotic factors such as climate and nutrients. How crowded a population is can also affect the population. This factor affects a population because members of the population compete with each other for energy, nutrients, and space. As a population becomes more crowded, some organisms receive less of these things. Biotic factors also affect a given population. Other populations living in the ecosystem might compete with the first population for food. Or members of a second population might feed on members of the first population. Either of these conditions could affect the growth of the first population.

All the populations in a given area form a **community.** A community consists only of living things. Abiotic factors are not part of the makeup of a community. A forest community might consist of white spruce trees, downy woodpeckers, great horned owls, short-tailed

FIGURE 36·2 A forest community would include various species of trees. It might also include populations of short-tailed shrews, long-eared owls, and downy woodpeckers.

shrews, and all the other populations of organisms living in the forest. A salt marsh community might consist of various species of marsh plants, fish, egrets, bacteria, and all the other populations of organisms living in the salt marsh.

Each organism in a community has a specific biological role. The role of an organism in a community is called a **niche.** The niche of a certain species of bacteria in a salt marsh includes decomposing organic matter. Green plants fill niches that involve food making.

Two different species can occupy the same habitat, but they usually cannot occupy the same niche. If two different species attempt to fill the same niche, they will compete with each other. The species that is better adapted to the environment may eventually occupy the niche.

REVIEW QUESTIONS

1. Distinguish between a population, a community, and an ecosystem.
2. Describe a salt marsh ecosystem.
3. Distinguish between a habitat and a niche.
4. How do members of a population affect each other?
5. Suppose overhunting results in the extinction of a particular species of bird. How would the theory of natural selection explain how the unoccupied niche would be filled?

Abiotic Factors in the Environment

SECTION OBJECTIVES

- Explain how climate and soil affect the living things in an ecosystem.
- Describe the water, carbon, nitrogen, and oxygen cycles.

36•4 CLIMATE

One major influence on ecosystems is the interaction of temperature, moisture, and radiant energy. Together these three abiotic factors largely determine the *climate* in an ecosystem. The climate in an area is the long-term weather conditions in the area.

Radiant energy, mainly sunlight, is an abiotic factor that determines climate and affects ecosystems. Sunlight heats air and the surface of the earth. It is the source of energy for much of the movement of air on the earth. It is also the source of energy for the process of evaporation, through which liquid water is changed to water vapor.

The amount of radiant energy in an ecosystem is affected by many factors. The latitude of the ecosystem determines the strength of the sunlight. Ecosystems that are nearer the equator receive more intense sunlight than do those that are nearer the poles. Other factors, such as whether something is preventing the sunlight from reaching the ecosystem, can also affect the amount of energy in the ecosystem. For example, the ecosystem in a cave will be much different from the ecosystem in a desert found at the entrance to the cave.

Radiant energy affects organisms in an ecosystem in several ways. Green plants, on which all living things in an ecosystem depend, need sunlight to make food. Sunlight also affects the activities of both plants and animals. Many plants will not flower until a certain day length is reached. Day length also triggers such things as spawning in certain fish, mating in certain mammals, and migration.

Temperature is another factor that directly influences climate. Temperatures vary with latitude. In the Northern Hemisphere, in general, the farther north an ecosystem is located, the lower the temperatures will be. Temperature also varies with altitude. Temperatures decrease about 10°C with every 1-km increase in altitude.

To understand the importance of temperature as a factor that determines climate, consider the difference between an arctic tundra and a hot desert. Both places receive about the same amount of precipitation each year. But the average temperature in the tundra is no more than 6°C in any given month. The temperature in a desert can be more than 43°C on many days during the year.

All organisms must live within a certain range of temperatures. In general, warm-blooded animals are active in a wide range of temperatures. Cold-blooded animals are active in a narrow range of temperatures. Animals generally cannot survive in temperatures that exceed 52°C. Some types of algae survive in hot springs, where temperatures may be 73°C or higher. Some types of algae live in arctic ecosystems.

Moisture is another factor that determines climate. Moisture includes both water that falls from clouds, or precipitation, and water vapor in the air, or humidity.

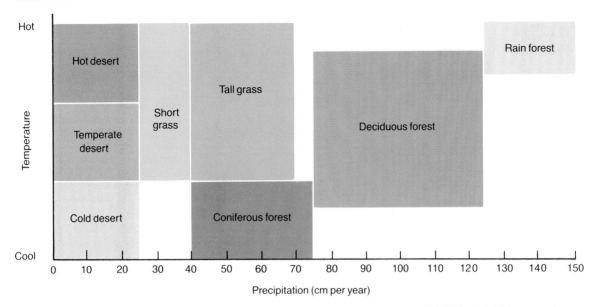

FIGURE 36·3 Moisture and temperature are important factors in determining the type of ecosystem that will be found.

Precipitation comes from clouds, which are formed when water vapor in the air cools and condenses. Precipitation in an area is affected by many factors, including wind patterns and landforms. Wind moves clouds and water vapor. Areas that are downwind from a large body of water generally receive more precipitation than areas that are upwind from the same body of water. Landforms can block the movement of clouds and thus affect patterns of precipitation. For example, ecosystems located on the western slopes of the Sierra Nevada receive more precipitation than do those found on the eastern slopes. The availability of water in an environment often is linked to the amount of precipitation received.

Humidity in an area is affected by the amount of precipitation that falls in the area. Two extreme examples are deserts and rain forests. Deserts receive little rain and have low humidity. Rain forests receive much more rain and have high humidity. Humidity can also be affected by plant life. In forests, trees form a living canopy, or roof, above the ground. This canopy tends to trap water vapor. The humidity under such a canopy is much higher than that in an open area just a few meters outside the canopy.

All organisms need moisture to survive. However, different organisms need different amounts of moisture. For example, many invertebrates and amphibians require moist air to prevent drying out. Many reptiles, on the other hand, can live in dry environments. Many plants have developed adaptations that enable them to live and reproduce in areas where there are long periods of dry weather. Some have developed short life cycles for growth and reproduction. Others have developed deep root systems or may become dormant during the dry season. Other organisms, such as some types of algae, must live in water.

Investigation

How Does the Water-Holding Capacity of Three Different Materials Compare?

Goals
After completing this activity, you will be able to
- Compare the water-holding capacity of sand, topsoil, and sphagnum.

Materials (for groups of 2)
water   graduate
3 glass funnels   funnel support
3 beakers (250-mL)
3 glass slides
microscope or stereomicroscope
3 sheets of filter paper
clock   glass-marking pencil
dry topsoil   dry sand
dry sphagnum moss

Procedure

A. Place the three funnels into the funnel support as shown in Figure A. Place a beaker under the lower end of each funnel. With the marking pencil label the first beaker *Topsoil;* the second beaker, *Sand;* and the third beaker, *Moss.*

Set-up of funnel and funnel support

B. Fold one sheet of filter paper to form a cone as shown in Figure B. Open one section of the folded filter paper and insert the cone into the funnel. Fold the other pieces of filter paper and insert them into the other funnels.

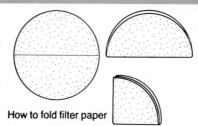

How to fold filter paper

C. About one fourth of the way up from the bottom of each funnel, make a mark with the glass-marking pencil. To the first funnel, add topsoil to this mark. (Remember: The topsoil is being poured into the *filter paper,* which is inside the funnel.)

D. In the second funnel add sand to the mark. In the third funnel add sphagnum moss to the mark.

E. On a separate sheet of paper set up a chart with these headings: *Time Water Added, Time Beaker Removed,* and *Amount of Water Collected.* Provide space for recording data on *Topsoil, Sand,* and *Sphagnum Moss.*

F. Using a graduate, measure 150 mL of water. Pour this water into the funnel with the topsoil. Record the time in the chart. Note what the time will be in 10 minutes. After exactly 10 minutes, remove the beaker in which water has collected.

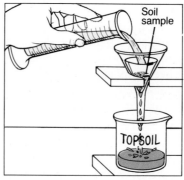

Soil sample

G. Measure out 150 mL of water and add this water to the second funnel. Add 150 mL of water to the third funnel. In each case, record the time. Wait exactly 10 minutes and remove the beaker into which the water has drained.

H. Measure the quantity of water collected in each beaker. Record this amount.

I. On each of three slides, place a small amount of topsoil, sand, and sphagnum. Look at each of these samples under the microscope.

J. On a separate sheet of paper, draw each sample as you see it under the microscope.

K. Follow your teacher's instructions for the proper disposal of all materials.

Questions and Conclusions

1. In which beaker was the most water collected? In which was the least amount collected?
2. Which substance tested has the greatest water-holding capacity? Explain the reason for your answer.
3. Describe how sand looked under the microscope. How does its structure affect the quantity of water it holds?
4. Describe how sphagnum moss looked under the microscope.
5. Sometimes gardeners are told to add sphagnum moss to their potting soil. How does this addition affect the soil?
6. Suppose a plant grows in a soil with poor water-holding capacity. What kind of a root system would be most adaptive—a deep tap root or a shallow wide-spreading system? Explain.

36·5 SOIL

Soil is another abiotic factor of ecosystems. Soil is important to plants as a source of minerals and as a material in which to anchor their roots. Many animals also depend on soil for a place to live and for food.

Soil is a combination of weathered rock materials, organic matter, water, and gases. Soil forms when the actions of wind, water, and organisms break apart rocks and minerals. These actions, along with slow chemical reactions that take place, change the rocks and minerals into tiny particles. These tiny particles combine with particles of organic matter, called humus, which forms from decomposing organisms. The result is soil. Soil known as sand is made of large particles and little humus. Loam is soil made of smaller particles and much humus. Clay is made of tiny particles and little humus.

The exact type of soil found in an area depends on several factors. One of these factors is the parent material, or the material from which the soil was formed. For example, the mineral quartz forms sandy soil, while feldspar helps form clay. A second factor that affects the type of soil formed is climate. For example, places that are cool and moist generally contain more vegetation than do hot, dry places. Soil formed in a cool and moist ecosystem, therefore, will contain much more humus than will soil formed in a hot and dry ecosystem. Biotic factors also help determine soil type. Plants often break down rocks and minerals into smaller particles, thus helping to form soil. Animals such as earthworms disturb soil as they move through it. These factors change the makeup of the soil. Bacteria living on roots of certain plants add nitrogen compounds to the soil. In these ways, biotic factors both determine and depend on soil, an abiotic factor.

FIGURE 36·4 Clay (top), loam (middle), and sand (bottom) are types of soil. Different plant populations are adapted to different types of soil.

The type and chemical makeup of soil in an ecosystem determine the plant populations living in the ecosystem. For example, plants like willows and alders grow in soil from which water drains poorly. Cactuses grow in sandy soil from which water drains quickly. Beets and cotton grow best in soil that contains high amounts of salts. Beans, oats, and peaches grow best in soil that contains low amounts of salts. Cranberries grow in soil that is very acidic. Alfalfa grows in soil that has a neutral pH.

36·6 CYCLES IN THE ENVIRONMENT

In addition to climate and soil, there is another link between the living and nonliving parts of an ecosystem. This link is the need that organisms have for certain chemicals, such as oxygen, carbon, hydrogen, nitrogen, and water. These chemicals, along with about 30 others, are essential for life. For example, nitrogen is found in chlorophyll and proteins. All organisms need water to carry on basic life processes. Water also provides an environment in which many organisms live.

There is a limited amount of any given element on the earth. However, elements can be found in different forms and compounds. For example, nitrogen can be found in the air, in ammonia, in com-

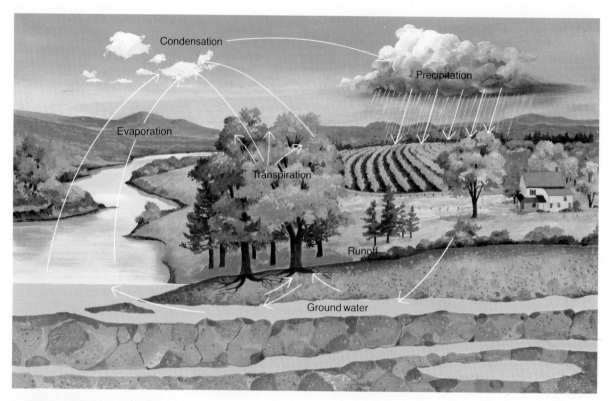

FIGURE 36·5 The water cycle moves water from the earth to the air and back again. Water enters the air through evaporation and transpiration. After undergoing condensation in the atmosphere, water returns to the earth through precipitation.

pounds called nitrates, and in proteins making up the tissues of living things. Sometimes nitrogen is found in living things and sometimes it is found in nonliving things. Nitrogen moves back and forth between living and nonliving things. This movement forms a cycle. All the chemicals essential for life move in cycles. This movement recycles the chemicals and helps ensure a constant supply.

36·7 THE WATER CYCLE

There is nearly 1.5 billion km³ of water on the earth. Less than 1 percent of this water is fresh water that is available to land organisms. This fresh water is constantly being replaced. Fresh water is replaced through the water cycle. The **water cycle** is the movement of water from the air to the earth and back again.

e- (out)
vapor (steam)

The water cycle involves a number of processes. One of these is evaporation. **Evaporation** is a process in which liquid water changes to water vapor. In the water cycle, most evaporation occurs from bodies of water such as oceans and lakes. A second process in the water cycle is condensation. **Condensation** is a process in which water vapor changes to liquid water or ice. Clouds form due to condensation. A third process is precipitation. **Precipitation** includes all the forms of moisture that fall to the earth from clouds.

Since a cycle such as the water cycle has no beginning or end, one can begin studying it at any point. Follow the water cycle shown in Figure 36·5, beginning with evaporation. As you can see, evaporation of water due to solar energy leads to the condensation of the water and the formation of clouds. Precipitation falls from these clouds. The precipitation might fall on the ground. There it might flow, either on the ground or underground, to lakes or oceans. Water on or in the ground might also be taken in by organisms. It might reenter the air by evaporating. Evaporation of water from the leaves of plants is called *transpiration.* Water might also be released or excreted by an organism and, eventually, evaporate. In these ways, water moves from living to nonliving things.

36·8 THE CARBON CYCLE

Carbon is found in all organic compounds and in carbon dioxide. In the **carbon cycle,** carbon moves between organic compounds that make up living tissues and carbon dioxide in the air. Refer to Figure 36·6 as you read about the carbon cycle.

In photosynthesis, autotrophs take in carbon dioxide from the air. Sugars and a number of other organic compounds made from sugars result from this process. Some of these compounds become stored in the plant. Heterotrophs eat autotrophs and thus obtain these organic compounds.

Carbon can return to the air, and thereby complete the carbon cycle. One way in which carbon can return is during respiration. For example, it returns to the air after organisms die, during decomposition. In such cases the carbon is returned as carbon dioxide.

FIGURE 36·6 Carbon moves between organic compounds in living things and carbon dioxide in the air. Carbon dioxide is released during respiration and decomposition. Carbon dioxide from the air is made into organic compounds during photosynthesis.

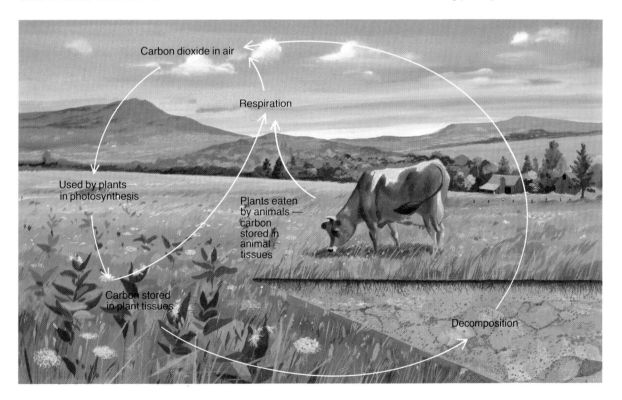

36·9 THE NITROGEN CYCLE

Nitrogen is a basic element found in the tissues of every organism. Although nitrogen gas makes up 78 percent of the atmosphere, it is not usable by most organisms. Most organisms must obtain the nitrogen they need in the form of nitrates or proteins. Nitrogen is changed into usable forms such as these during the nitrogen cycle. The **nitrogen cycle** is a process in which nitrogen moves between organisms and the atmosphere.

The nitrogen cycle involves four main processes. Bacteria are needed for all four processes. In one of these processes, known as **nitrogen fixation,** nitrogen gas, N_2, is changed into ammonia, NH_3. This process is carried out mainly by bacteria living with certain plants, especially legumes. Nitrogen fixation is also carried out by blue-green bacteria, by some types of free-living soil bacteria, and by certain kinds of lichens. Lightning also causes nitrogen fixation.

A second process in the nitrogen cycle is **ammonification,** a process in which amino acids are changed to ammonia. This process is carried out by bacteria that break down animal wastes and the bodies of dead organisms.

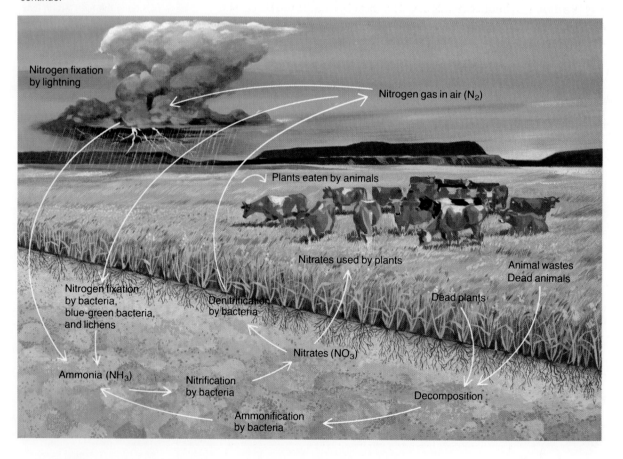

FIGURE 36·7 In the nitrogen cycle, nitrogen gas from the air and amino acids from living things are changed into ammonia by bacteria. The ammonia from these two sources is changed to nitrates, which are used by plants. Eventually, anaerobic bacteria change nitrates to nitrogen gas, which returns nitrogen to the air so that the cycle can continue.

The ammonia formed by these first two processes is used during a third process, called nitrification. During **nitrification,** ammonia is changed by bacteria into compounds containing nitrates, NO_3^-. Nitrification actually involves two kinds of bacteria. The first kind changes ammonia into nitrites, NO_2^-. The second kind changes nitrites into nitrates. The nitrogen in nitrates is in a form that can be used by autotrophs to make proteins.

In the fourth process in the nitrogen cycle, **denitrification,** nitrates are changed into nitrogen gas, N_2. This process is carried out by certain species of anaerobic bacteria. Denitrification returns nitrogen gas to the atmosphere.

The nitrogen cycle is shown in Figure 36·7. Refer to this figure, beginning with nitrogen gas in the air. As you can see, nitrogen in the air is fixed by some bacteria, blue-green bacteria, certain types of lichens, and lightning. Ammonia is formed. Through nitrification, ammonia is changed by certain bacteria into nitrates. These nitrates are absorbed by autotrophs, and the nitrogen in them becomes part of the protein in that organism's tissue. Heterotrophs that eat the autotrophs ingest this nitrogen. Decomposition of animal wastes and of plant and animal tissues produces ammonia through ammonification. The nitrogen can be returned to autotrophs following nitrification. The nitrogen cycle is completed when anaerobic bacteria carry on denitrification of nitrates. This process produces nitrogen gas, which enters the atmosphere.

36·10 THE OXYGEN CYCLE

Oxygen gas makes up 21 percent of the atmosphere. There is also some oxygen dissolved in water. Oxygen is required by most living things for cellular respiration. ATP is produced during cellular respiration. The movement of oxygen from living things to the air and water and back again is called the **oxygen cycle.**

As with the other cycles that have been described, oxygen cycles between the nonliving environment and living things. Oxygen from the air and water is taken in by organisms and used for cellular respiration. A product of this process is water. This water is released as water vapor by land organisms. It enters the water cycle. Water is taken in by autotrophs, which use it to carry out photosynthesis. As a result of this process, oxygen is released. This oxygen returns to the air or water, completing the oxygen cycle.

REVIEW QUESTIONS

6. List three factors that help determine climate.
7. How does climate influence organisms in an ecosystem?
8. How does soil influence a community?
9. Why are chemical cycles important?
10. Predict the long-term effects resulting from the extinction of those bacteria and blue-green bacteria capable of nitrogen fixation.

Biotic Factors in the Environment

SECTION OBJECTIVES

- Describe the roles of living things in the environment.
- Explain how food webs, food chains, and ecological pyramids are used to describe ecosystems.
- Give examples of mutualism, commensalism and parasitism.

herba (plant)
vorare (devour)

carnis (flesh)

omni (all)

36·11 THE ROLES OF LIVING THINGS

The living things in an ecosystem can be divided into several groups. These groups are determined by the role of a given organism in the transfer of energy in an ecosystem. All organisms need energy to survive. The sun is the main source of energy for all organisms. Autotrophs change the sun's energy into chemical energy in organic molecules during photosynthesis. Organisms that produce their own food in this way are called **producers.** Green plants and algal protists are examples of producers.

Organisms that cannot manufacture their own food must meet their energy needs by feeding on other organisms. Organisms that eat other organisms are called **consumers.** All animals are consumers, getting the food they need in a variety of ways.

Consumers can be divided into three groups—herbivores, carnivores, and omnivores. Each group of organisms has a different way of obtaining its energy and nutrients. **Herbivores** are plant eaters. Herbivores include organisms as small as zooplankton and as large as elephants. Most insects are herbivores. Animals that eat other animals are called **carnivores.** Some carnivores, called predators, hunt and kill their food. Still other carnivores eat dead animals. Animals that get food in this way are called scavengers. Lions, wolves, spiders, frogs, and snakes are predators. Vultures are scavengers. Some organisms, like humans, are called omnivores. **Omnivores** eat both plants and animals. The diets of omnivores are varied. Raccoons, red foxes, and some types of fish are omnivores.

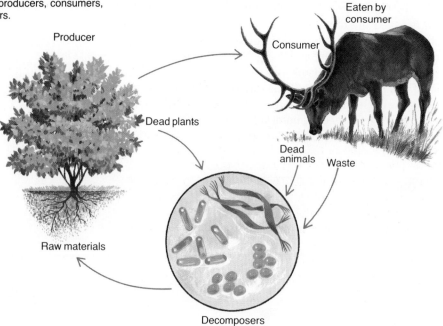

FIGURE 36·8 Organisms in an ecosystem include producers, consumers, and decomposers.

FIGURE 36·9 The beaver (A) is a herbivore. The snapping turtle (D), wood ibis (B), and spider (E) are carnivores. The grizzly bear (C) is an omnivore.

In addition to producers and consumers, there is a third group of organisms in an ecosystem. These organisms are called decomposers. **Decomposers** are organisms that break down dead organic material into inorganic matter. They obtain energy and nutrients in the process. Fungi and many types of bacteria are decomposers. Decomposers play an important role in the chemical cycles you have read about.

36·12 ENERGY FLOW IN THE ENVIRONMENT

Organisms in almost every ecosystem depend on the sun for energy. Some organisms directly use the sun's energy to make food, while other organisms use this energy indirectly. The series of steps through which energy is transferred from the sun to organisms in an ecosystem is called a **food chain.**

Almost all food chains begin with the sun. Energy from the sun is stored in tissues of producers, such as green plants. Consumers make up the next steps in all food chains. Herbivores, which obtain energy by eating plants, are often called first-order consumers. They feed directly on plants. Herbivores are followed by second- and third-order consumers. These organisms may be either carnivores or omnivores. Decomposers feed on producers and consumers. This group of organisms obtains energy by breaking down the tissues of dead organisms. They return simple materials to the environment.

An ocean food chain is shown in Figure 36·10. Phytoplankton, tiny autotrophs, fill the role of producer. Zooplankton, tiny consumers, and herring, gulls, and bacteria make up the rest of this food chain. Which of these organisms is a herbivore? Which is a second-order consumer?

FIGURE 36·10 In a food chain, energy is transferred from the sun to a series of organisms.

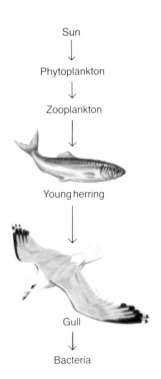

Sun
↓
Phytoplankton
↓
Zooplankton
↓
Young herring
↓
Gull
↓
Bacteria

THE ENVIRONMENT

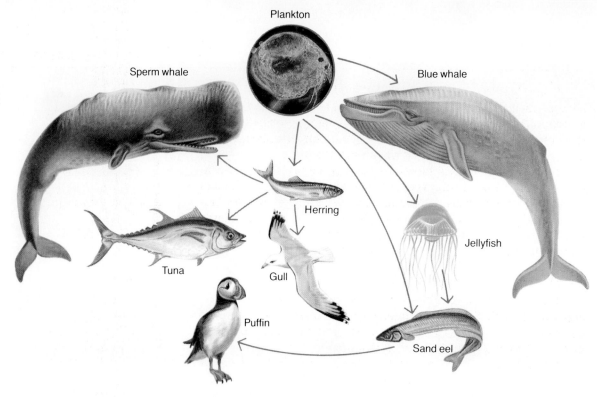

FIGURE 36·11 This ocean food web is made up of interrelated ocean food chains.

FIGURE 36·12 Energy is lost with each successive step of a food chain. The size of each level in this energy pyramid represents the available energy at that level. Which level has the most energy? Which level has the least?

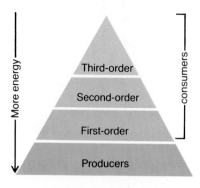

An ecosystem usually contains more than one food chain. Most animals have more than one source of energy. Because of this, food chains in an ecosystem usually interconnect. This results in a food web. A **food web** is made up of interrelated food chains in an ecosystem.

An example of an ocean food web is shown in Figure 36·11. This food web includes the ocean food chain consisting of phytoplankton, zooplankton, herrings, gulls, and bacteria. It also shows other organisms that feed on plankton, such as blue whales. It includes other food chains, such as one made of plankton, jellyfish, sand eels, and puffins. Which organisms in this food web feed on plankton?

36·13 ECOLOGICAL PYRAMIDS

Energy is lost with each step of a food chain. Plants do not store all the food they make. Some of this food is broken down by the plant to get energy to carry out life processes. Herbivores do not use all the energy stored by plants. All consumers use some of the energy they take in to carry out their life processes. Some energy is given off as heat. Thus this energy is not available to the next level of consumers. These losses from one level in a food chain to the next can be shown in a *pyramid of energy*. A pyramid of energy is shown in Figure 36·12. The base of the pyramid consists of producers. Consumers make up each of the other levels. The size of each level represents the amount of energy available at each level. The pyramid shown here shows that much less energy is found in the top level than is found in the bottom level.

740 THE ENVIRONMENT

Other types of pyramids can be used to show relationships between members of a food chain. For example, the *pyramid of numbers* shows the relationships between the numbers of organisms in the steps of a food chain. In Figure 36·13A the first-order consumers are most numerous, although this is not always the case.

A *pyramid of biomass* shows the mass of the organic material at each level. Since less energy is available at each higher level, biomass at each higher level is usually limited. Figure 36·13B shows how biomass usually becomes less with each level. As you can see, a large mass of organic matter at the base is needed to support a small amount of matter at the top of the pyramid.

36·14 SYMBIOTIC RELATIONSHIPS

Different organisms in the same ecosystem often interact in several ways. Sometimes this interaction involves one organism depending on another as a source of food. Any long-term interaction involving two organisms is called **symbiosis.**

Mutualism is a symbiotic relationship in which both organisms benefit. For example, ants live in the thorns of bull's-horn acacia trees and are fed by special structures on the plant. In return, the ants eat herbivorous insects and drive away large herbivores with painful stings. Mutualism is also seen in large herbivores and tickbirds. A tickbird lives on the parasites it removes from its host's skin.

Commensalism is a symbiotic relationship in which one organism benefits and the other is neither benefited nor harmed. Orchids and the trees on which they grow are an example of this kind of relationship. The orchids use the trees for support. The trees are not harmed, nor do they receive any benefits from the orchids. Remoras are fish that attach themselves to sharks. They get a free ride from the shark and feed on fragments of the shark's food.

Parasitism is a symbiotic relationship in which one organism benefits from and harms another organism. Parasites usually do not kill their host, but a weakened host may die from the relationship. There are many types of parasites. Some, like fleas and ticks, live on the surface of their host. Others, like tapeworms and roundworms, live inside their host. Some organisms are parasites during part of their life cycle, while others are parasites during their entire life cycle.

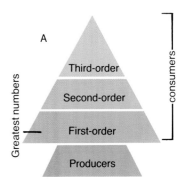

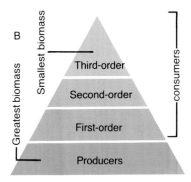

FIGURE 36·13 Usually, producers are the most numerous organisms in a food chain. However, in this pyramid of numbers, herbivores, first-order consumers, are the most numerous (A). In a pyramid of biomass, a large mass of organic matter at the base is necessary to support the pyramid (B).

REVIEW QUESTIONS

11. Distinguish between herbivores, carnivores, and omnivores. Give an example of each.
12. What is the role of decomposers in an ecosystem?
13. How does energy get into food chains?
14. Distinguish between a pyramid of biomass and a pyramid of energy.
15. Both second-order and third-order consumers depend indirectly on the same producers. Why is less energy available to third-order consumers than to second-order consumers?

Chapter Review

SUMMARY

36·1 The environment consists of biotic, or living, and abiotic, or nonliving, factors.

36·2 Ecosystems are areas in which organisms interact with each other and with nonliving things.

36·3 A group of organisms of the same species living in an ecosystem is called a population. All of the populations in a given area form a community.

36·4 Climate is an abiotic factor that influences ecosystems. Climate is determined by temperature, moisture, and radiant energy.

36·5 Soil, which provides organisms with minerals and places to live, is another abiotic factor that influences ecosystems.

36·6 A number of chemicals that are essential for life are recycled through the environment.

36·7 The water cycle involves evaporation, condensation, and precipitation.

36·8 Carbon moves from organic compounds in living tissues to carbon dioxide in the air, as described by the carbon cycle.

36·9 In the nitrogen cycle, nitrogen in the air is changed into forms that are usable by organisms and then returned to the air.

36·10 The oxygen cycle involves oxygen moving from air and water to living things and back again.

36·11 In an ecosystem, all organisms are producers, consumers, or decomposers. Consumers may be herbivores, carnivores, or omnivores, depending on their diet.

36·12 Energy is transferred through an ecosystem by a series of steps known as a food chain. Overlapping food chains form food webs.

36·13 Pyramid-shaped diagrams can be used to illustrate the relationships between the numbers of organisms in an ecosystem, the distribution of energy in an ecosystem, and the biomass in an ecosystem.

36·14 The long-term interaction between two organisms is called symbiosis. Mutualism, commensalism, and parasitism are types of symbiosis.

BIOLOGICAL TERMS

abiotic factor
ammonification
biosphere
biotic factor
carbon cycle
carnivores
commensalism
community
condensation
consumers
decomposers
denitrification
ecology
ecosystem
evaporation
food chain
food web
habitat
herbivores
mutualism
niche
nitrification
nitrogen cycle
nitrogen fixation
omnivores
oxygen cycle
parasitism
population
precipitation
producers
symbiosis
water cycle

USING BIOLOGICAL TERMS

1. What is the difference between nitrification and denitrification?
2. What is a niche?
3. Explain the difference between a population and a community.
4. What is the difference between evaporation and condensation?
5. Distinguish between herbivores and carnivores.

Choose the term that does not belong and explain why it does not belong. Define each term.

6. evaporation, nitrogen fixation, condensation, precipitation
7. herbivore, omnivore, producer, consumer
8. nitrification, mutualism, parasitism, commensalism
9. consumer, decomposer, habitat, producer
10. precipitation, ammonification, nitrification, nitrogen fixation

UNDERSTANDING CONCEPTS

1. How is a biotic factor different from an abiotic factor? Give examples of each. (36·1)
2. What is an ecosystem? Give three examples of ecosystems. (36·2)
3. Distinguish between an ecosystem and a habitat. (36·2)
4. Describe two factors that can influence a population. (36·3)
5. Describe the niches of three organisms that might be part of a forest community. (36·3)
6. Describe how temperature helps determine climate. How does temperature affect living things in an ecosystem? (36·4)
7. How does radiant energy help determine climate? Describe how sunlight affects living things. (36·4)
8. What is soil? What factors determine the exact type of soil in an area? (36·5)
9. Describe the importance of soil to living things in an ecosystem. (36·5)
10. Why is it important that chemicals essential to life are recycled through the environment? How are the chemicals recycled? (36·6)
11. Describe the water cycle. (36·7)
12. How is carbon recycled in nature? (36·8)
13. What is ammonification? (36·9)
14. Describe the nitrogen cycle. (36·9)
15. How is oxygen recycled in nature? (36·10)
16. Describe the roles of producers, consumers, and decomposers. Give an example of each. (36·11)
17. How does energy flow in an ecosystem? (36·12)
18. Give an example of a food chain. (36·12)
19. Distinguish between a food chain and a food web. (36·12)
20. Explain how a pyramid can be used to explain the relationships between the numbers of organisms in a food chain. (36·13)
21. Describe three types of symbiotic relationships. Give an example of each. (36·14)

APPLYING CONCEPTS

1. As energy moves through a food chain, only 10% of the energy is available to the next level in the chain. Take a hypothetical food chain that has 10 000 kilocalories of energy available to plants. Determine the amount of energy available to first-order consumers, second-order consumers, and third-order consumers in this food chain.
2. Feeding the world's human population is becoming a major concern. Consider two types of human diets—one that consists almost completely of plants and one that consists mainly of meat. Which type of diet would better meet the food/energy needs of the world's population? Explain your answer.
3. How do terrestrial ecosystems differ from aquatic ecosystems? How do freshwater ecosystems differ from marine ecosystems?
4. Parasites usually do not kill their hosts. Why is this advantageous to the parasite?
5. Design a spacecraft that is a self-contained ecosystem. Include all the items needed for astronauts to stay for an extended time in space.

EXTENDING CONCEPTS

1. Describe your habitat. What is your niche? What will your niche be in 10 years?
2. Establish an ecosystem in an aquarium or large glass jar. Carefully choose the organisms that you include. What abiotic factors can you control? What biotic factors can you control? Observe changes in the ecosystem for 8 to 10 weeks and prepare a written report.
3. Contact the local extension service and obtain information about the energy budgets for agriculture in your area. Develop an energy flow diagram that includes energy inputs (electricity, fossil fuels, solar radiation, fertilizers) and energy outputs (crops, meat, milk, eggs). Try to determine how much energy is lost in the process.

READINGS

Adams, John. *Dirt*. Texas A&M University Press, 1986.

Childress, James J., Horst Feldbeck and George N. Somero. "Symbiosis in the Deep Sea." *Scientific American,* May 1987, p. 114.

Reisner, Marc. *Cadillac Desert: The American West and its Disappearing Water*. Viking, 1986.

37. Ecological Changes

Once every hundred years, century bamboo plants produce seeds and then die. The death of the bamboo results in the death of many giant pandas, for these rare animals feed almost entirely on bamboo. How do ecological changes affect wildlife? How does the availability of food affect a population? These questions will be answered in this chapter.

CHAPTER OBJECTIVES

After completing this chapter, you will be able to

- **Identify** the factors affecting population size and population growth.
- **Discuss** the types of behavior in populations, stressing the adaptive functions of each.
- **Describe** how communities change over time.
- **Appraise** the effects that humans can have on communities and on the earth's capacity to support populations.

Changes in Populations

37·1 THE STRUCTURE OF POPULATIONS

Recall that a population is a group of organisms of the same species living in a certain place. Examples of populations are the people in the United States, the mallard ducks in North America, and the bacteria of a particular species in a test tube. The total number of members of a population is called the **population size.**

Two populations of equal size may use different amounts of space. The number of individuals of a population per unit of area or volume is called **population density.** Population densities are usually stated as the number of organisms per square meter or per hectare (ha). Small organisms often are found in greater numbers per unit area than are larger organisms. For example, the population density of whitetail deer in a forest might be 39/ha. The population density of a kind of soil arthropod in a field might be 100 000/m .

A population can be divided into three classes based on age. The juvenile class consists of the youngest individuals, which cannot yet reproduce. The mature class consists of those individuals capable of reproduction. The post-reproductive class is made up of older individuals, those no longer able to reproduce.

The ratio of these classes is important to the future of the population. The juvenile and post-reproductive classes tend to have a higher percent of deaths than the mature class. The mature class produces offspring, adding to the population. A population that has a large proportion of individuals in the post-reproductive class may be producing few offspring. In extreme cases, extinction of the population could occur. A population with a very large proportion of juveniles often indicates a healthy mature population. The age structure of a population can be shown as an age pyramid. Figure 37·1 is an example of an age pyramid of a healthy population of voles. What is the ratio of the classes in this population?

SECTION OBJECTIVES

- Explain how biotic potential and environmental resistance influence population growth.
- Distinguish between density-dependent and density-independent factors.

FIGURE 37·1 A healthy population of voles has a very large proportion of juveniles.

Post-reproductive — 7%

Mature 28%

Juvenile 65%

The ratio of males to females is another part of a population's structure. In many vertebrate species more males than females are born. In many species, however, more males than females die as juveniles. Thus in the mature class there are often more females than males. In many species a high ratio of females to males does not lead to a decrease in the number of offspring produced. What effect might there be on the number of offspring produced if the ratio of females to males were reversed?

37·2 POPULATION GROWTH

Several factors cause changes in population size. Births increase the size of a population. Immigrants also lead to an increase in the size of a population. **Immigrants** are organisms that move into a population. Deaths decrease the size of a population. Emigrants also cause a decrease in the size of a population. **Emigrants** are organisms that move out of a population. If the number of organisms that are born plus the number of immigrants is larger than the number of organisms that die plus the number of emigrants, the population will increase in size. Such an increase in size is called **population growth.**

immigrare (to move into)

emigrare (to move out)

Living things have a very large potential for population growth. The unrestricted growth of bacteria would cover the earth in a short time. A single pair of houseflies might have 25 million descendants in two months. The common bluegill, or sunfish, produces 60 000 young in a single nest. Oysters produce millions of eggs each time they reproduce. The maximum population size that a species could reach in a certain time, if all offspring survived and reproduced, is called its **biotic potential.**

The earth is not covered with bacteria or flies or fish or oysters. Certain factors keep species from reaching their biotic potential. Factors that keep species from reaching their biotic potential are called **limiting factors.** The limiting factors on populations vary. For example, predation is an important limiting factor for plant eaters but less important for many meat eaters. Water is an important limiting factor. It often determines both the numbers and kinds of living things in an area. The net effect or sum of the limiting factors on a population's growth is the **environmental resistance.**

A line graph of a population's growth through time is called a **population growth curve.** Figure 37·2 is a population growth curve. Notice that the population grows slowly at first but then grows more rapidly. The population size finally becomes stable at a certain level. The population size may then change slightly, but it remains around the same level. How would the number of births plus immigrants compare with the number of deaths plus emigrants at this level?

When population size remains nearly steady, it has reached the carrying capacity of the environment. **Carrying capacity** is the population size of a particular species that a particular environment can support. Carrying capacity varies for different environments and for different populations.

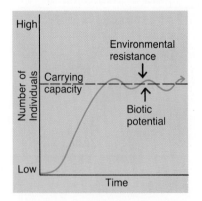

FIGURE 37·2 A typical population growth curve shows that a population grows slowly at first, then grows very rapidly. At a certain level, the population size becomes relatively stable.

Improvements in farming methods, industry, and sanitation have enabled some human populations to increase the carrying capacity of parts of the environment. However, the carrying capacity remains lower than the biotic potential. Figure 37·3 compares a normal population growth curve (S-curve) with a growth curve reaching the biotic potential of the population (J-curve).

37·3 HUMAN POPULATIONS

The same kinds of conditions that affect other populations also affect human populations. Limiting factors, such as supplies of food and water, prevent the human population from reaching its biotic potential. As mentioned, some human populations have increased the natural carrying capacity of the land through the use of technology. Cities are good examples. Cities are supported by bringing in energy and natural resources from outside areas. The carrying capacity of some regions has been reduced by human activities.

The ratio between the number of births and the number of individuals in a population within a specified time is called **birth rate.** On a world basis, human birth rates are declining. For example, India's birth rate has gone down from over 40 births per thousand around 1960 to about 34 per thousand in 1976. As shown in Figure 37·4, birth rates vary considerably from country to country. Countries with high birth rates tend to have large populations of young people. These young people are ready to enter the reproductive, or mature, period of life. The high birth rate and large juvenile population contribute to continued growth of the world's population.

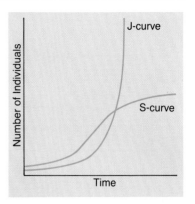

FIGURE 37·3 A normal growth curve (S-curve) compared with a growth curve that reaches the biotic potential of the population (J-curve).

FIGURE 37·4 The birth rate is the ratio between the number of births and the number of individuals in a population within a specific time. The birth rates of a number of countries are shown here.

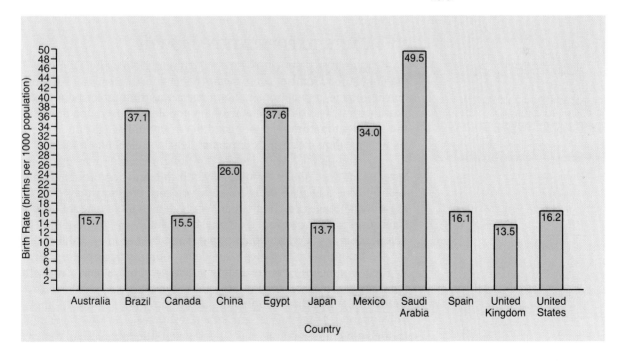

748 ECOLOGICAL CHANGES

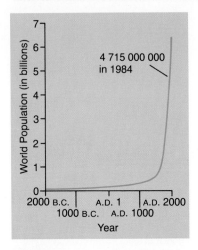

FIGURE 37·5 A growth curve for the human species shows that the population is still increasing. The earth's carrying capacity is not known.

FIGURE 37·6 If the human population continues to grow at the current rate, the population will double to 10 billion in 37 years.

When your grandparents were born, the world population was about 1.6 billion. The human population had taken perhaps a million years to reach that size. Now, that many people will probably be added to the world population within 20 years. There are over 5 billion people on the earth today, and there may be 6 billion as the twenty-first century begins.

Population growth rates are often described using doubling time. The time it takes a particular population to double in size can be determined by using the following formula.

$$\text{Doubling Time} = \frac{0.7}{\text{percent growth}}$$

The figure 0.7 in the equation is a constant. In one doubling time the population would double in size, if the growth rate remains constant. The population of the United States is growing at a rate of about 1 percent per year. At that rate the doubling time is 0.7/1 percent (0.01) = 70 years. The world's population of 5 billion is growing at a rate of 1.9 percent per year. At that rate the population would double to 10 billion in 37 years.

Figure 37·5 is a growth curve for the human population. It is thought that the earth's carrying capacity, increased by technology, has not yet been reached. What is the carrying capacity of the earth? No one really knows. Scientists are working to increase the carrying capacity by using methods such as improved crop production. However, the food supply is only one of many limiting factors working against unlimited population growth. The problem may not be how many people the earth can support but what kind of life those people will enjoy.

37·4 LIMITS TO POPULATION GROWTH

Populations do not grow forever. Limiting factors affect the size of the population. Some of these factors affect growth no matter what the density of the population is. Limiting factors that affect population growth regardless of the density of the population are called **density-independent factors.** Weather conditions, amount of light, and events such as earthquakes are examples of density-independent factors.

Density-independent factors are often abiotic. For example, severe winter weather may kill many bobwhite quail. A drought might destroy breeding places for frogs. A volcanic eruption may destroy most living things in a large region. While density-independent factors can cause changes in population size, they usually are not important in determining the level at which population size becomes stable.

As a population becomes more dense, other limiting factors begin to affect population growth. Limiting factors whose effects vary with the density of the population are called **density-dependent factors.** Density-dependent factors include predation, disease, and competition for food, living space, and other resources. Density-dependent factors affect populations the most when the population density approaches or exceeds the carrying capacity. As the population density falls, density-dependent factors become less important in limiting population growth.

The availability of food can be an important density-dependent limiting factor. For instance, as the population density of a herd of antelopes increases, there is competition for available grass for food. At some point the amount of grass may be insufficient to feed the entire herd. In extreme cases some animals may starve. Others may become weak from lack of food. Weak individuals are more susceptible to disease. In addition, disease spreads more rapidly in a dense population. Some individuals will die from disease. Thus population growth is limited.

Predation is a factor in controlling the population growth of many kinds of plant eaters. In the example of the antelope population that outgrew its food supply, the weak animals become easier prey. Also, a dense population makes it easier for predators to find their prey. So the growth of the herbivore population becomes limited. At the same time, the large number of prey may lead to an increase in the number of predators. Such an increase further limits the growth of the herbivore population. Figure 37·7 shows changes in population size for the snowshoe hare and its chief predator, the Canada lynx. How does the population size of the lynx change as the population size of the hare changes?

Competition among members of a population for limited resources, such as food and water, is called **intraspecific competition.** As population density increases, so does intraspecific competition for available resources. Intraspecific competition can limit population density, since some individuals may be unable to obtain sufficient food or adequate shelter, and do not survive.

intra- (within)

FIGURE 37·7 This graph shows the cyclical variation in the population sizes of the snowshoe hare and the Canada lynx over time.

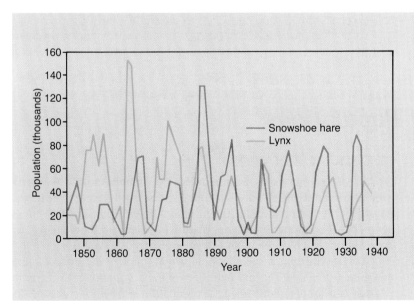

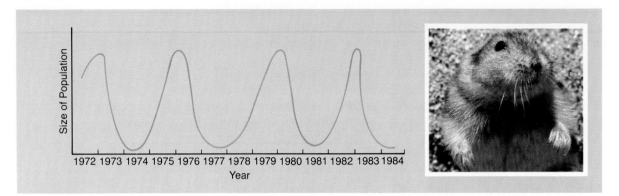

FIGURE 37·8 The population density of lemmings changes in predictable cycles.

In some species, population density changes in predictable cycles. Populations of the snowshoe hare and the Canada lynx are known to change in 9- to 10-year cycles. Figure 37·8 shows changes in lemming populations. How long are these population cycles?

Reasons for these cycles are not well understood. Some scientists think that environmental factors, such as food availability, cause the cycles. Others think predator-prey interactions are responsible. Another hypothesis is that high population density causes physiological changes within organisms that result in population decrease.

REVIEW QUESTIONS

1. What is the difference between population size and population density?
2. In what sense are biotic potential and environmental resistance opposing forces?
3. What is carrying capacity?
4. What is the difference between density-dependent and density-independent factors?
5. How could the carrying capacity of a desert be increased?

The Behavior of Populations

SECTION OBJECTIVES

- Discuss the adaptive advantages of social behaviors in populations.
- Give examples of rhythmic behaviors in populations.

37·5 SOCIAL BEHAVIOR

All of an organism's actions make up its **behavior.** If you have ever observed a cat stalking a mouse or a bird building a nest, you have observed animal behavior. Animal behavior that involves interaction between members of a population is called social behavior. Two dogs fighting each other is an example of social behavior.

Territoriality is a social behavior that involves defense of a limited area, often against other members of the population. Territoriality is most commonly seen during a population's breeding season. For instance, a pair of robins will drive all other robins from the area near the nest they have built. Such behavior restricts reproductive activity, since only males that have established territories attract females. So territoriality limits population growth. Territorial behavior is observed in many groups of animals.

The size and shape of territories vary with species and with population density. Individual territories tend to be large when population density is low. As density goes up, the territories shrink, but only to a certain point. There seems to be a minimum individual territorial size for many species. Once this size is reached, animals tend to become more aggressive. As a result, weaker members of the population do not establish territories. For example, the weaker males in a rabbit population are forced to live on the edges of the dominant males' territories. These weaker males do not mate with females. The idea that territoriality is a factor in human behavior is being studied.

Home range is the area over which an animal travels to obtain food and, in some species, to reproduce. Home range is usually not defended, except in the special case where home range and breeding territory occupy the same area. Blue jays will vigorously defend the territory around their nests, but they share feeding areas with other members of the population. Mockingbirds tend to show territorial behavior throughout their home range, especially during the nesting season.

FIGURE 37·9 Different species of birds occupy different heights, or territories, in a tree. Such behavior reduces competition between species.

Olive-sided Flycatcher
Territory: 12 m–21 m

Golden-crowned Kinglet
Territory: 3 m–12 m

Red-breasted Nuthatch
Territory: 1 m–9 m

White-throated Sparrow
Territory: 0 m–3 m

FIGURE 37·10 Meerkats live together in packs. These meerkats are on the lookout for predators.

In many vertebrate species, animals live together in groups, called flocks, schools, herds, or packs. These groups often have a social structure. The structure may be based on aggressiveness, which leads to dominance of some individuals over others. A *hierarchy*, or order of dominance, is established. In populations that have such a social hierarchy, each member has a position based on the amount of aggressiveness the member exhibits. Social rank may change because of death, fighting, and, occasionally, alliances between members. The leader, or dominant member of the group, is usually also the male that fathers the most offspring.

In a baboon troop, the largest, most aggressive males are the dominant members. These males defend the troop against other troops and against predators. They keep peace within their troop by stopping fights among other males. Dominant baboons usually are the only males to mate with the females. Because the role of these male baboons is accepted by the rest of the baboons, unity and order exist within the troop. How might such organization aid the troop in fighting off predators or in finding food?

Communication among members of a population is part of social behavior. Animals have many ways of communicating. These may involve the senses of sound, smell, and sight.

Many animals use sound to communicate. Mosquitos use their buzzing sounds to attract mates. Male crickets rub specialized parts of their wings together to attract females and establish social hierarchy. Male frogs attract females to their territories by calling.

Bird songs have been studied by biologists for some time. The main function of bird songs seems to be territorial. A male bird establishes a territory and announces his presence. The song is a warning to other males to stay out of the territory. It may also attract a mate.

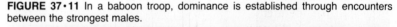

FIGURE 37·11 In a baboon troop, dominance is established through encounters between the strongest males.

Sound communication among marine mammals is being studied. Evidence indicates that dolphins can communicate with each other. Records have been made from whale songs. Scientists are not sure what the sounds mean. They may help schools of marine mammals travel together, or they may be warning messages or mating calls.

Some animals and plants use chemicals to communicate. They secrete substances that can influence other members of the population or other populations in the community. The chemicals used to send these messages are **pheromones** (FEHR uh mohnz). There are two kinds of pheromones. *Primer pheromones* cause long-term and often major changes in animals that receive them. Primer pheromones are important in the population structure of social insects. The social hierarchy of insects like ants, bees, and termites is largely determined by primer pheromones. They are secreted by the dominant members of the population and help determine the roles played by other members.

Releaser pheromones produce immediate responses, but the responses are not permanent. Releaser pheromones are often used by mammals to mark territory. You may know that dogs use urine to mark "scent posts." The urine contains a releaser pheromone. This pheromone signals the dog's presence to other dogs. Ants make use of releaser pheromones by leaving chemical trails to and from food supplies. Other ants are able to follow the trail. Pheromones also are involved in mating behavior. For example, female silk moths release pheromones that attract males.

Visual displays are another form of communication. Pigeons fluff their necks, spread their tails, and strut. The tropical birds of paradise have bright feathers that are useful in visual communication. In all these cases the behaviors and the color patterns serve to attract mates, and this leads to breeding.

Not all visual displays are related to breeding. In populations in which a social hierarchy exists, visual displays communicate other signals. Figure 37·12 shows a visual signal used by baboons. This signal is a threatening gesture.

37·6 RHYTHMIC BEHAVIOR

Most organisms show periodic behavior that seems to be related to periodic changes in the environment. A behavior that occurs periodically is called a **rhythmic behavior.** Rhythmic behaviors often are related to length of daylight versus darkness and to the change of seasons.

The rhythmic behavior of organisms seems to be regulated by an internal biological clock. This biological clock is probably a series of chemical responses to environmental stimuli. The flowering of plants is an example of a periodic behavior regulated by a biological clock (Section 20·2).

Many short-term rhythms involve a period of about one day. Rhythms that show a cycle of about 24 hours are called **circadian**

FIGURE 37·12 Animals communicate through visual displays. This baboon is making a threatening signal, warning other baboons to stay away.

FIGURE 37·13 Both animals and plants show periodic behavior. The flowers of this morning glory open in the morning.

(ser KAY dee uhn) **rhythms.** Circadian rhythms mark the active and rest phases of the behavior of most animals. For example, chipmunks and gray squirrels are active by day and sleep at night. Animals that are active by day are said to be diurnal (dī ER nuhl). Flying squirrels and bats are active at night. Animals that are active by night are said to be nocturnal.

Competition between different species is called **interspecific competition.** Differences in patterns of activity result in less competition between members of different species for available resources. For example, both hawks and owls are birds of prey. However, hawks are diurnal, and most owls are nocturnal, so interspecific competition is limited.

An example of a long-term rhythmic behavior is hibernation. **Hibernation** is a period of dormancy accompanied by a low metabolic rate and a much lowered body temperature. Animals that hibernate do so during the coldest part of the year. Both cold-blooded and warm-blooded animals hibernate. Hibernation appears to be an adaptation to surviving harsh environmental conditions, such as cold temperatures and the scarcity of food.

Another example of rhythmic behavior in response to harsh conditions is estivation (ehs tuh VAY shuhn). **Estivation** is a period of dormancy, like hibernation, during which metabolic rate and body temperature are lowered. Unlike hibernation, which occurs partly in response to cold conditions, estivation occurs in response to hot, arid conditions. Several desert mammals, such as ground squirrels and pocket mice, estivate during the hottest, driest season. Such behavior helps the organisms to avoid overheating and prevents excessive loss of water.

FIGURE 37·14 The arctic tern has the longest migration of any bird. It breeds in the Arctic and flies to the Antarctic during the northern winter. A year's journey is more than 30,000 km.

Migration is another example of rhythmic behavior. **Migration** is the movement of a population to new feeding grounds or to breeding grounds, with the population later returning to the original area. As with hibernation, migratory behavior seems to be triggered by seasonal changes. Thus many birds in the Northern Hemisphere fly south in the fall. In the spring the birds return north, where they breed and raise their young. The migratory route of the arctic tern is shown in Figure 37·14.

Migration also occurs in some species of fish, mammals, and insects. It seems to be an adaptation that enhances reproductive success and the survival of the young.

REVIEW QUESTIONS

6. How does territorial behavior limit population growth?
7. What are some adaptive advantages of a social hierarchy?
8. What functions does communication serve within a species?
9. Describe two kinds of rhythmic behavior that occur in animals.
10. How might circadian rhythms in humans explain the phenomenon called jet lag?

Investigation

How Can a Serial Dilution Be Used for the Study of a Microscopic Population?

Goals
After completing this activity, you will be able to
- Prepare a serial dilution of yeast cells for immediate examination.
- Compare the turbidity of tubes of serially diluted yeast cells.

Materials (for groups of 4)
three 1-mL micropipettes + bulbs
stoppered tube of yeast culture
1 stoppered test tube with 10 mL glucose solution
3 stoppered test tubes with 9 mL glucose solution
microscope   4 slides
4 coverslips   test tube rack
glass-marking pencil

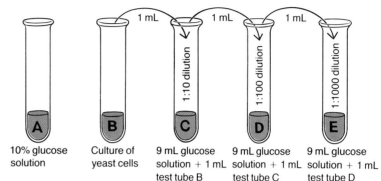

Procedure

A. Obtain a stoppered test tube with 10-mL glucose solution. With your glass-marking pencil, label this tube A. Obtain a stoppered test tube with a culture of yeast cells. Label this tube B. Place the tubes in a rack.

B. Obtain three test tubes, each containing 9 mL glucose solution. Label these tubes C, D, and E.

C. Populations of microorganisms cannot be studied easily if the concentration of organisms is too dense. In tubes C, D, and E, you are going to prepare a serial dilution of the yeast culture for immediate examination. In this dilution, you will have several tubes, each containing fewer and fewer yeast cells. You will then be able to observe organisms in each culture. Refer to the drawing, which shows the correct way to prepare the dilutions.

D. Swirl the contents of test tube B until the yeast cells and solution are thoroughly mixed. Using a clean micropipette, withdraw 1 mL of the contents of test tube B and transfer it to test tube C. Gently draw the fluid into the pipette again and release it into test tube C to rinse the pipette.

CAUTION: Place a bulb at the end of the micropipette and squeeze the bulb to draw up material. Do not place the micropipette in your mouth.

E. Swirl the contents of test tube C. With a second clean pipette, withdraw 1 mL of the contents of test tube C and transfer it to test tube D. Swirl the contents of test tube D. Using a third clean pipette, withdraw 1 mL from test tube D and transfer it to test tube E. Place the tubes in a rack. From the drawing, note the dilution of each tube.

F. Note any differences you observe in the *turbidity* (cloudiness) of each tube. On a separate sheet of paper, record the contents, dilutions, and appearances of test tubes A–E.

G. Using a clean micropipette, withdraw a small quantity of the contents of test tube B. Make a slide of this material. Place a coverslip over the material. Examine the slide under the microscope. Prepare and examine slides of material from tubes C, D, and E. Compare the number of cells visible in each of the slides.

H. Follow your teacher's instructions for the proper cleanup and disposal of all materials.

Questions and Conclusions

1. Describe the appearance of test tubes A–E after you prepared the serial dilution. Which tube was the most turbid? Which was the least turbid?
2. Compare the numbers of yeast cells in slides made from the contents of test tubes B–E.
3. What was the purpose of having one test tube (A) with no yeast cells?
4. The glucose solutions you have been using have not been sterile. Suppose you incubated the yeast cultures overnight. What do you think might be growing in the cultures after incubation?
5. What was the purpose of diluting the yeast culture?

Changes in Communities

SECTION OBJECTIVES

- Describe primary succession in lakes, ponds, and on land.
- Relate human activity to succession.

37·7 PRIMARY SUCCESSION

Recall that all the populations within an ecosystem make up a community. The organisms within a community influence one another and also the environment that supports them. While many communities, such as forests, appear stable, communities can undergo changes. Such changes can involve both the populations that make up the community and the environment itself. Scientists have found that the changes follow a regular pattern. The orderly sequence of changes in communities is called **ecological succession.**

There are two kinds of succession, primary succession and secondary succession. **Primary succession** is the establishment and development of communities in newly formed areas. Cooled lava, rocky surfaces, and sandy areas are examples of regions in which primary succession occurs. **Secondary succession** is succession that occurs on disturbed areas, such as abandoned farmland and vacant lots. Changes in populations and changes in the abiotic environment all occur during succession.

Primary succession often begins on bare rock. As the rocks erode, organisms begin to grow in the cracks and crevices. The first organisms to enter a new area are called *pioneers*. Pioneer organisms like lichens may release mild acids that help erode the rock. Dust particles and bits of dead vegetation begin to build up, and a thin layer of soil forms. Small invertebrates colonize the soil. Mosses may begin to grow in the soil. Gradually other plants invade the area and the community changes. Ferns and grasses may become more abundant than the lichens and mosses. They dominate until other plants replace them. Eventually the conditions may change sufficiently for trees and shrubs to grow where there was only bare rock earlier.

At some point the community usually becomes stable. The dominant plants, perhaps oak and hickory trees, remain from generation to generation. A community that is no longer undergoing succession is called a **climax community.** However, should the climax community be destroyed, succession usually begins again.

The nature of the climax community varies. In the eastern United States the climax community is characterized by hardwood forests throughout much of the region. In many areas of Canada, coniferous forests predominate. In the central regions of North America the natural climax community is made up of various grasses.

37·8 SUCCESSION IN LAKES AND PONDS

Primary succession also occurs in newly formed lakes and ponds. At the earliest stage a pond is without life. Figure 37·15 shows one possible sequence of succession in a pond. Microscopic algal protists and plants invade the water and begin to multiply. Soon protozoans and small animals are found and simple food chains develop. As the organisms die and sink, organic material begins to build up on the pond bottom. When the pond becomes shallow enough, plants such as pondweed become rooted in the bottom. Cattails and water lilies may develop later. In addition, insects, amphibians, reptiles, fishes, and birds become part of the pond community. As material continues to build up and the pond becomes shallower, reeds develop. The pond has become a marsh. Eventually the pond may fill in completely.

The addition of nutrients to lakes is called **eutrophication** (yoo-truh fuh KAY shuhn). Eutrophication is usually a slow natural process. It is a part of the succession process. In a natural aquatic ecosystem, nutrient production and use are normally in balance.

Human activity can change the rate of eutrophication in a lake. Dumping organic materials into water speeds up the process of eutrophication. The nutrients stimulate growth of aquatic autotrophs. Dense growth of algae, called *algal blooms,* occurs. As the algae die, they are decomposed by rapidly growing populations of bacteria. This process can deplete the water of dissolved oxygen, resulting in the death of organisms in the lake. The speeding up of the eutrophication by human activities is called cultural eutrophication.

FIGURE 37·15 Lakes and ponds pass through stages of succession. The body of water gradually fills in as fallen leaves and other debris collect on its bottom. Cattails and other plants rooted along the shore gradually spread further out into the water as the pond or lake becomes shallow. Eventually, the pond becomes a marsh. Later, it may fill in completely and become indistinguishable from the surrounding forest.

758 ECOLOGICAL CHANGES

FIGURE 37·16 Stages in secondary succession: grasses in a field (left), a maturing pine forest (middle), and a hardwood forest (right).

37·9 SECONDARY SUCCESSION

Secondary succession begins on land where the previous community has been destroyed. Abandoned farmland, forests destroyed by fire, and new construction sites are areas where secondary succession occurs.

Figure 37·16 shows some stages in secondary succession on abandoned farmland. Grasses such as crabgrass are the first plants to colonize the soil. Within a few years, asters and then broom sedge, a type of grass, are found. Pine seedlings grow within the field of grasses, and within 10 years they make the area less suitable for the growth of the grasses. The pine forest matures in about 100 years. Beneath it, a hardwood forest containing maple, dogwood, and oak is growing. As the pines die, the forest changes to a hardwood forest. The climax community may have mostly oak and hickory trees, or other hardwoods may be abundant, depending on the region.

REVIEW QUESTIONS

11. Describe primary succession on land.
12. Describe primary succession in a lake.
13. What is eutrophication?
14. Under what conditions does secondary succession occur?
15. What kinds of factors might determine the climax community that develops in an area?

Computers in Biology

Computers in Ecology

All Americans should share the responsibility of protecting our natural heritage—the prairies and rivers, deserts and wetlands, mountains and shorelines—for tomorrow. One Washington, D.C., organization is leading the way by preserving valuable lands today. The Nature Conservancy is locating and providing long-term protection for unique and rare natural areas throughout the United States and Canada. The Conservancy owns and manages about 700 sanctuaries, the largest system of privately owned refuges in the world. This network protects everything from ohia (oh HEE uh) trees in Hawaii to manatees in Florida.

To help scientists determine which lands were most in need of protection, the Conservancy designed the Natural Heritage Database. This computerized inventory gives scientists, policymakers, and landowners information about threatened areas and the animals and plants in those areas. Thus natural resource managers can identify important natural areas and set priorities for protecting them.

The Natural Heritage Database contains information about the status and distribution of what the computer calls "elements." Elements include rare plant and animal species, natural communities, and even unusual geological formations. The computer serves as a clearinghouse, providing data about the existence, numbers, conditions, status, and locations of the elements. Clearly, no one scientist could have such a wealth of information available in seconds without the use of the computer.

Another advantage of this database is that the stored information is continually being updated. Old data is checked in the field. New unique natural areas are being surveyed, and facts about them are being entered into the computer. These processes allow individuals to change policies when necessary to protect threatened species and places of beauty.

The Natural Heritage Database is also of importance in preventing accidental destruction of land. In addition, by identifying ecologically important land before costly investments have been made to build on the land, the computer plays a key role in preventing disagreements when determining land use.

After identifying a valuable area, the Conservancy gets ownership of the land, through gift or purchase, and looks after the long-term condition of the area. Often the land is then given to a nature organization, a government agency, or a university to continue the supervision of the area.

The Natural Heritage program is one of the most successful applications of computer use in the field of ecology. With the use of the database, the Nature Conservancy has provided living space for many of the endangered "elements" of North America.

ECOLOGICAL CHANGES

Chapter Review

SUMMARY

37·1 Populations differ in size and density. The ratios between numbers of individuals in age classes and the ratio of males to females affect the size of a population.

37·2 Population growth is determined by birth and death rates, immigration and emigration, and limiting factors in the environment.

37·3 Various factors influence populations. Humans can change the carrying capacity of land, but the earth's carrying capacity is not known.

37·4 Population size is limited by both density-dependent and density-independent factors.

37·5 Members of a species interact with one another through social behavior, such as territoriality and mating, using various means of communication.

37·6 Many organisms exhibit periodic behavior in response to daily or seasonal changes in the environment.

37·7 Newly formed habitats are colonized by organisms through a series of changes called primary succession.

37·8 Lakes and ponds undergo a series of changes in which more and more organisms become established as part of the aquatic community.

37·9 Secondary succession accounts for the gradual redevelopment of areas where previous communities had been destroyed.

BIOLOGICAL TERMS

behavior
biotic
  potential
birth rate
carrying
  capacity
circadian
  rhythm
climax
  community
density-
  dependent
  factors
density-
  independent
  factors
ecological
  succession
emigrants
environmental
  resistance
estivation
hibernation
immigrants
interspecific
  competition
intraspecific
  competition
limiting
  factors
migration
pheromone
population
  density
population
  growth
population
  growth
  curve
population
  size
primary
  succession
rhythmic
  behavior
secondary
  succession
territoriality

USING BIOLOGICAL TERMS

1. What is carrying capacity?
2. Which two terms together make up environmental resistance?
3. How do immigrants and emigrants affect population growth?
4. What is the difference between population size and population density?
5. What is biotic potential?
6. What is a population growth curve?
7. Which terms describe long-term rhythmic behaviors?
8. Define the term *pheromone*.
9. Which term describes a type of behavior involving the defense of a limited area, often against other members of the population?
10. Does interspecific competition involve members of only one species or members of different species?

UNDERSTANDING CONCEPTS

1. What elements describe a population's structure? (37·1)
2. What conclusions can be drawn about a population in which there is a low proportion of juveniles? (37·1)
3. Explain the interaction of biotic potential and environmental resistance. (37·2)

4. Write an equation showing how birth rate, death rate, immigrants, and emigrants affect population growth. (37·2)
5. What happens to a population once it has reached the carrying capacity of its environment? (37·2)
6. How have humans possibly affected carrying capacity with respect to their own population? (37·3)
7. If a population is growing at a rate of 10 percent per year, how long will it take that population to double? (37·3)
8. Give an example of how a density-independent factor might limit population growth. (37·4)
9. Explain why density-dependent factors become more important as population density increases. (37·4)
10. In what ways is communication an adaptive behavior? (37·5)
11. Describe the social hierarchy of a baboon troop. How is this behavior advantageous to the troop? (37·5)
12. How might differences in circadian rhythms reduce interspecific competition? (37·6)
13. What are the adaptive advantages of hibernation and migration? (37·6)
14. How is estivation different from hibernation? (37·6)
15. What is a climax community? (37·7)
16. Explain the role pioneer plants play in primary succession. (37·7)
17. Describe the natural process of succession in a pond. (37·8)
18. How can humans change the rate of eutrophication in a lake? (37·8)
19. Under what conditions does secondary succession occur? (37·9)
20. Describe the process of secondary succession. (37·9)

APPLYING CONCEPTS

1. A stable population suddenly experiences a tremendous growth in size. Speculate on some factors that may be involved.
2. Your local newspaper runs an editorial calling for your town to grow at a rate of 5 percent per year. Write a letter to the editor discussing this growth rate.
3. A new volcanic island has just emerged from the Atlantic Ocean. Use your knowledge of ecology to discuss what will most likely happen to the island.
4. A person who flies from America to Asia may have trouble adjusting to the living schedule. Discuss what is happening in terms of circadian rhythms.
5. How does cultural eutrophication interfere with natural succession?
6. Why is the size of the moose population of Canada probably not regulated by density-independent factors?

EXTENDING CONCEPTS

1. Study historical records and observe nearby areas to determine what type of community dominated your town's area 50, 100, and 300 years ago. What will it be like in 100 more years?
2. Assume the population in your town or city has doubled. Write a science-fiction story describing the effects of an expanding human population on the natural community. Design an animal or plant that would be adapted to these changes.
3. Prepare a report on the island of Surtsey. How long has the island been in existence? How was it formed? What population and community changes have occurred on the island? What changes will most likely occur in the next 100 years?

READINGS

Brown, et al. "The Children's Fate." *Natural History,* April, 1986, p. 41.

Ehrlich, P. "Human Carrying Capacity, Extinctions And Nature Reserves." *BioScience,* May 1982.

Fisher, Ron. *Our Threatened Inheritance: Natural Treasures of the United States.* National Geographic Society, 1984.

Omang, Joanne. "The Hands-on Level of Deforestation." *Smithsonian,* March 1987, p. 56.

Van Lawick, Hugo. *Among Predators and Prey.* Sierra Club Books, 1986.

Zaslowsky, Dyan. *These American Lands: Parks, Wildernesses and the Public Lands.* Henry Holt, 1986.

38. Biomes

The leopard's spots make it difficult to see in dappled sunlight. This camouflage and the leopard's quiet movements prevent prey from sensing the leopard until it is too late. Leopards are one of the many animals that live in a grasslands biome. In this chapter, you will learn about the major biomes.

CHAPTER OBJECTIVES

After completing this chapter, you will be able to
- **Identify** the earth's major biomes.
- **Explain** how climate affects the distribution of organisms.
- **List** the characteristics of the earth's biomes.
- **Discuss** the differences between biomes.
- **Describe** changes in biomes resulting from human influence.

The Earth's Biomes

38·1 WHAT IS A BIOME?

The biosphere can be divided into regions called biomes. A **biome** (BĪ ohm) is a large region that has a distinct combination of plants and animals. Climate is a factor in determining the type of biome that occurs. A terrestrial biome is usually identified by the types of plants that make up a climax community within it. The dominant types of plants are called the climax vegetation. However, a biome includes all stages of succession leading up to the climax community. In the deciduous forest biome, for example, deciduous trees are the climax vegetation. However, earlier stages such as grasslands and shrubs will also be found in the biome. A biome also includes all the animals in the community, which often are largely determined by the kinds of plants.

Ecologists have identified several biomes in the world. The major terrestrial biomes are tundra, coniferous forest, temperate deciduous forest, tropical rain forest, temperate rain forest, grasslands, and desert. Bodies of water such as oceans, lakes, and rivers make up the aquatic biomes. Aquatic biomes can be identified by their dominant animal populations.

The world's biomes are shown on the next page in Figure 38·1. Notice that the same kind of biome is found at the same *latitude,* or distance from the equator, in different parts of the world. For example, there are grassland biomes in South America, Africa, and Australia. The specific plants and animals found in these different grassland biomes are not identical, however. Populations in different biomes have similar characteristics and are sometimes related, but they are different species. Plants and animals from the same type of biome resemble each other because they are adapted to nearly identical physical and climatic conditions. Such similarities are an example of convergent evolution.

SECTION OBJECTIVES
- List the major terrestrial biomes.
- Discuss the factors that determine the kind of biome in an area.

bio- (life)
oma (group)

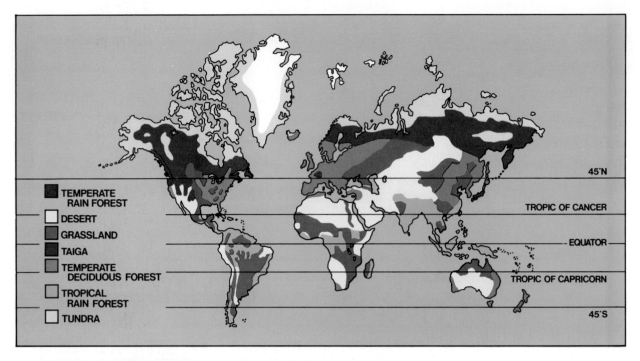

FIGURE 38·1 A biome is a large region that has a certain climate and a distinct combination of plants and animals. The world's biomes are shown here.

FIGURE 38·2 A climatogram shows precipitation data and temperature data on a monthly basis.

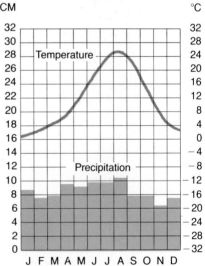

38·2 BIOMES AND CLIMATE

The main factor that determines the kind of biome in a certain area is climate. Recall that climate is determined mainly by temperature and precipitation (Section 36·4). These factors vary with latitude and altitude. Average temperature decreases from the equator to the poles. A decrease also can be seen from sea level to the mountains. Features such as mountain ranges and the nearness of large bodies of water affect precipitation. Average monthly temperature and precipitation can be plotted on a graph called a **climatogram,** such as shown in Figure 38·2.

Temperature and precipitation largely determine a biome. Figure 38·3 lists the average yearly precipitation range and the average annual temperature range for the major terrestrial biomes. These figures are averages. The same type of biome in different places will not have exactly the same amount of precipitation or the same temperature range.

FIGURE 38·3 Precipitation and temperature are important in determining the kind of biome in an area. Which biome has the most precipitation?

BIOME	AVERAGE PRECIPITATION (yearly)	AVERAGE TEMPERATURE RANGE (yearly)
Tundra	<25 cm	−26°C to 4°C
Coniferous forest	35–75 cm	−10°C to 14°C
Deciduous forest	75–125 cm	6°C to 28°C
Tropical rain forest	>200 cm	25°C to 27°C
Grassland	25–75 cm	0°C to 25°C
Desert	<25 cm	24°C to 32°C

REVIEW QUESTIONS

1. What is a biome?
2. List the major terrestrial biomes.
3. What factors determine the type of biome that will be found in a particular area?
4. Why do plants and animals from separate biomes of the same type resemble each other?
5. How might the climatogram of a biome in the middle-latitude region in the Northern Hemisphere compare to that of a similar biome in the Southern Hemisphere?

Terrestrial Biomes

38•3 TUNDRA

Throughout much of the year, the northern polar region is a sea of ice that is several meters to more than 30 meters thick. In summer there are open stretches of water in which large pieces of ice float. Just south of the ice-covered waters lies the **tundra,** the northernmost terrestrial biome. The tundra encompasses areas of North America, Asia, Europe, Greenland, and many small islands. Both precipitation and average temperature are low in the tundra.

For six to nine months of the year, tundra lands are covered with snow. During this long winter the temperature is below freezing most of the time and there is little daylight. Spring and summer last only three months, with a brief growing period of just 60 days. However, there is almost constant daylight during this season, which allows a few centimeters of the topsoil to thaw. The melting snow cannot soak into the permanently frozen layer of soil—called **permafrost**—just below the thawed surface. The water collects on the plains, forming small lakes, ponds, and marshes throughout the tundra.

During the short arctic summer, organisms take advantage of the warmth, sunlight, and food. Hundreds of species of lichens, mosses, and flowering herbaceous plants grow in the soggy soil. Some woody shrubs that grow close to the ground, such as dwarf birch and willow, also begin growing. The tundra has no trees. Deep-rooted plants cannot penetrate the permafrost. Most tundra plants are dormant in winter. Therefore during the brief summer they grow rapidly, store food for the whole year, and quickly produce flowers and seeds.

Few animals can be found year-round in the tundra. Most move into the tundra during the summer and then go south again before winter sets in. Among these animals are the caribou, which travel some 1000 km annually. Their principal food is reindeer moss, a lichen. Wolves, musk oxen, arctic hares, arctic foxes, and lemmings are among the other mammals of the tundra.

SECTION OBJECTIVES

- Describe the major terrestrial biomes.
- Compare the plants and animals in the different terrestrial biomes.

FIGURE 38·4 Musk oxen are found year-round in the tundra. There are no tall trees in the tundra because their roots cannot penetrate the permafrost.

Many waterfowl, such as ducks and geese, migrate to the tundra in the summer. Great flocks with their young then fly south before the long winter begins. Insects are also abundant during the warm days. Mosquitoes, black flies, and other biting insects infest the area and are the most numerous animals on the tundra.

The arctic tundra does not have a corresponding terrestrial biome in the Southern Hemisphere. Oceans cover the southern latitudes where a tundra climate could develop. The nearest landmass is Antarctica, some 14 million square kilometers in size. Most of this continent is always frozen, windswept, and uninhabitable. Warm water currents near the continent do make it possible for a few mosses, lichens, and several species of flowering plants to grow in limited coastal areas. There are also a few kinds of insects, mites, and other small animals. For most of the year—and sometimes for several years—they are dormant. Most life is confined to the seas surrounding Antarctica. Some animals are abundant in these seas during the year. Several species of penguins are coastal residents, and during the short summers, sea-going birds such as shearwaters appear. Several species of seals, whales, and dolphins also visit the coastal waters.

37·4 CONIFEROUS FOREST

South of the tundra is the coniferous forest biome, which stretches across the northern sections of North America, Europe, and Asia. In the **coniferous forest** biome the vegetation is almost totally made up of conifers, or cone-bearing trees, such as pines, spruces, hemlocks, and firs. The northern part of this biome is also known as the **taiga** (TĪ guh), a Russian word meaning "swamp forest." Precipitation and average temperatures are higher than in the tundra. A few broad-leaved trees grow where the land has been disturbed by fire or logging, but these are only an intermediate stage in the forest succession. Eventually they are replaced by conifers.

Needles, which break down very slowly into the poor soil, cover the forest floor. Conifers cause a dense shade year-round, which prevents the growth of shrubs and herbs beneath the trees. Winter in this biome is long and cold, with heavy snowfall. Needle-leaved evergreens, the predominant vegetation, are adapted to these harsh conditions. Evergreen needles have a waxy coating and a small surface area which prevents water loss and freezing. The branches of the trees are able to bend and shed heavy snow without breaking.

The summer growing season in the coniferous forest is longer and warmer than that of the tundra. The soil partially thaws and becomes swampy, often forming mats of moist, spongy ground. Animal life is abundant. Many small insects feed on the needles and bark of trees, and small birds consume the insects. Waterfowl inhabit the numerous lakes and ponds of this biome, while predatory birds such as hawks, eagles, and owls live within the forests. Many kinds of mammals live in the coniferous forest biome, including lynxes, wolverines, black bears, moose, elks, snowshoe hares, porcupines, and red squirrels. Some of these animals are residents of the forest all year. These species build up layers of body fat before winter. Some store food and others go into hibernation. Other animals migrate to warmer climates where food is more plentiful.

Coniferous forests are among the world's great sources of lumber. Extensive harvesting has destroyed large areas, but areas that are mostly untouched still remain, particularly in North America. These wilderness tracts have become a last stronghold for some species, such as the timber wolf.

FIGURE 38·5 This bull moose is eating water plants. Because spruce trees and moose are common, the coniferous forest is sometimes called the "spruce-moose" biome.

FIGURE 38·6 The coniferous forest is characterized by conifers such as the red pine (inset) and by long, cold, snowy winters.

38·5 TEMPERATE DECIDUOUS FOREST

The **temperate deciduous forest** biome occurs in the Northern Hemisphere south of the coniferous forest and in the Southern Hemisphere in New Zealand, Tasmania, and the eastern coast of Australia. The typical vegetation is broad-leaved deciduous trees, which shed their leaves seasonally when the climate becomes unfavorable. The chief northern trees include maples, birches, oaks, cherries, and beeches. Although they are not as extensive as coniferous forests, deciduous forests once covered large areas of Europe and China, eastern North America, parts of Japan, and the tip of South America. Large areas have been eliminated by logging and clearing for agriculture.

The climate of the temperate deciduous forest has a distinct pattern of four well-defined seasons: cold winter, hot summer, moderate spring, and moderate fall. Precipitation, mostly in the form of rain, is distributed evenly throughout the year and ranges between 75 and 150 cm annually. The growing season in this biome is long, lasting up to six months, and there is a rich diversity of plant and animal species.

FIGURE 38·7 In a deciduous forest, the vegetation grows in layers. This is called vertical stratification (top). In autumn, the leaves of deciduous plants change color and are later shed (bottom).

The vegetation in a deciduous forest grows in a series of layers, or *strata*. This distribution of vegetation in layers is known as **vertical stratification.** The tallest trees make up the **canopy,** the top layer of vegetation. These trees receive the greatest amount of sunlight and produce a lush layer of broad, flat leaves over the forest. The leaves are thin, however, and enough light filters through to support a layer of shorter trees called the **understory.** Beneath the understory is the **shrub layer,** which is made up of short, woody plants. Close to the ground are grass and ferns, which make up the **herb layer.** Finally there is the **forest floor,** which is a layer of decaying plant and animal matter. Patches of moss grow on the forest floor.

Animal life also tends to be stratified. Mice, shrews, salamanders, snails, some reptiles, and other small animals inhabit the forest floor, sometimes burrowing into the soil for food or shelter. Larger mammals such as white-tailed deer also live on the ground but feed in several layers on herbs, shrubs, and short understory trees.

Other species are found in the upper layers. Flying squirrels and many birds live in the canopy, while wood thrushes inhabit the understory. Insects are numerous throughout the different layers of the forest and serve as food for many birds. Birds can be found moving between several layers or concentrated in just one. Hooded warblers and ruffed grouse are found near or on the ground but sometimes move into the trees to nest or feed. Red-eyed vireos live in the canopy, and woodpeckers live on the trunks of trees between the shrub and canopy layers. Large predators such as black bears and mountain lions are no longer common in this biome, but smaller predators, like foxes, hawks, and owls, are still found.

Human civilization has developed more in this biome than in any other. The temperate deciduous forest, therefore, has been changed greatly in many regions. In some places, forests have been replaced entirely by other types of ecosystems, such as farmland. As a result, certain animal species have disappeared. A famous example is the once-numerous passenger pigeon. At one time the passenger pigeon was the most abundant bird in the world; it was forced into extinction by human settlers, who destroyed the bird's forest habitat, and by hunters. However, as some cut forestlands have grown into woods again, other nearly extinct species, such as the wild turkey, have managed to recover.

38·6 TEMPERATE RAIN FOREST

The temperate rain forest is found along the west coast of North America from central California to Canada and southern Alaska. This biome has features in common with both the tropical rain forest and the northern coniferous forest. In fact, it is sometimes called the *moist coniferous forest*.

The **temperate rain forest** biome is a region in which temperatures are moderate, humidity is high, and rainfall is often more than 300 cm each year. At the more southern end of the biome, fog rolling off the ocean provides more water each year than precipitation does. Most of the trees are conifers. These trees reach much greater size than conifers in the taiga because of warmer temperatures and greater amounts of moisture. Redwoods grow to more than 100 m, and Douglas firs may be 80 m tall and 3 m wide. Ferns and mosses grow on the forest floor wherever sunlight filters through the leaves overhead. There are far fewer plant species in this biome than in the tropical rain forest.

There are few tree-dwelling mammals, although there are many birds and insects. Most animals are found on the ground. Deer, elk, rodents, and other species found in the northern coniferous forest are common in this biome.

38·7 GRASSLANDS

A **grassland** is a treeless biome marked by a wide range of temperatures and an uneven distribution of precipitation. Grasslands occur in the middle of continents at about the same latitude as temperate deciduous forests. Grasslands make up the largest biome in North America, where they are also known as the *prairie*. They stretch from central Canada south to Mexico City and from Indiana west to the Rocky Mountains. Similar biomes are found in South America (the *pampas*), Asia (the *steppes*), and Africa (the *veldt*).

Several factors cause the development of grasslands instead of forests in these areas. The main reason is the difference in the amount of precipitation. Grasslands receive between 25 and 75 cm of rainfall each year. Sometimes grasslands are subject to droughts. Trees are unable to tolerate these dry conditions. High rates of evaporation and strong winds increase the dryness. Also, periodic fires, often caused by lightning, are very damaging to trees. Grasses, however, have deep roots that will easily sprout new growth the following season, even though the tops of the plant have been burned away by fires.

FIGURE 38·8 This fox squirrel is one of the many animals that live in the upper layers of a deciduous forest.

FIGURE 38·9 Conifers are found in the temperate rain forest. Compare the height of these giant sequoias with the height of the person in the lower left-hand corner.

Grassland vegetation can be placed into two groups based on growth patterns: sod formers and bunch grasses. *Sod formers,* such as Kentucky bluegrass, grow in a solid mat that completely covers the ground. They have an extensive network of underground stems by which they reproduce and spread. *Bunch grasses,* such as big bluestem, grow in clumps.

Three different layers are evident in the grassland biome. Beneath the surface is the **root layer,** which is more developed in grasslands than in any other biome. In some species the roots are so deep that at least half of the plant can be considered to lie under the soil. Buffalo grass, for example, has roots that reach up to 1 m deep. The root layer stores food for the winter and is the base from which new shoots grow.

FIGURE 38·10 Grasslands are the largest biome in North America. Prairie dogs and snakes are examples of grassland animals.

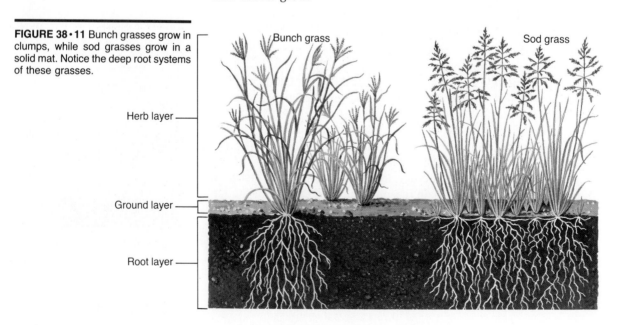

FIGURE 38·11 Bunch grasses grow in clumps, while sod grasses grow in a solid mat. Notice the deep root systems of these grasses.

Just at the surface of the soil is the **ground layer,** which consists basically of mulch. *Mulch* is made up of dead vegetation, much of which is in a state of decomposition. Mulch is important to the soil of the grasslands. Mulch retains water, stabilizes soil temperature, and provides the proper conditions for seed germination.

Above the ground is the herb layer, consisting of plants that vary from season to season. In the spring the herb layer is made up of plants that grow close to the ground, such as wild strawberry and dandelions. In the summer, short grasses and some herbs—for example, wild mustard—appear. Tall grasses and other herbs take over in the fall.

At one time, grasslands had more herbivorous animals than any other biome. In North America, herds of buffalo, or bison, numbered between 45 and 60 million. Herds of pronghorn antelope numbered between 50 and 100 million animals. Overhunting reduced these species to near extinction, but conservation measures have enabled them to recover. They are not as abundant, however, as they once were.

Other grassland herbivores include pocket gophers, ground squirrels, mice, jack rabbits, and prairie dogs. Prairie dogs were perhaps the most abundant of all mammals living on the North American grasslands. Their population may have exceeded 100 million, and their burrows and mounds once covered many square kilometers of land. Ranchers and farmers considered the rodent to be a pest, however, and killed the animals in many areas. Today the prairie dog can only be found in small parts of the Great Plains.

Animal species in other grassland biomes, such as in Africa, are different than those of the North American prairie. They are adapted to similar conditions, though, and have similar characteristics. Burrowing or running behavior is common. Gathering into herds, which have a well-developed warning system, also serves as protection against predators. Grassland animals also tend to have acute hearing and a good sense of smell.

North American grasslands have been altered greatly by humans. In areas with high precipitation farming has replaced native grasses with grain crops, which are also grasses, for human consumption. Cattle and sheep have replaced the bison and pronghorn antelope as the dominant grazers. In some places, overgrazing, plowing, and intensive farming have caused the land to deteriorate and valuable topsoil to be lost. When topsoil is eroded, desert conditions result—as in the great Dust Bowl of the 1930s.

In South America and Australia, the grasslands are also being cultivated. Here, too, the native species are being eliminated and replaced by domesticated plants and animals. The savannas of Africa are being put under similar pressure, but large areas have been set aside as wildlife preserves, such as Serengeti National Park in Tanzania. **Savannas** are grassland regions where there is just enough rainfall to support a scattered growth of trees. Wildebeests, zebras, giraffes, and lions are familiar animals of the African savanna.

FIGURE 38·12 The graceful black-footed ferret is one of the rarest animals in North America. It preys on prairie dogs. Scientists fear that the black-footed ferret will soon be extinct. Why is the population of ferrets getting smaller?

FIGURE 38·13 Giraffes and zebras inhabit the African savannas.

38·8 TROPICAL RAIN FOREST

Tropical rain forests are found around the world in areas along the equator. A **tropical rain forest** biome is a region that has a large amount of rain and year-round warm temperatures. More than 200 cm of rain falls in this biome each year. Typically there are two distinct seasons: wet and dry. The wet season lasts about six months with rain nearly every day. The remainder of the year is the dry season. Rainfall during this period averages 13 cm or less a month. Temperatures range between 15°C and 30°C all year. Day and night temperatures usually vary more than those of summer and winter.

The largest rain forest is the Amazon Basin in South America. Other major areas include the Congo and Niger basins of west and central Africa and parts of Asia, Malay, Borneo, New Guinea, and Madagascar.

FIGURE 38·14 The distribution of vegetation in layers is called vertical stratification. Like the deciduous forest, a tropical rain forest shows strata. The tops of tall trees make up the canopy. Enough light filters through the canopy to support the shorter trees of the understory. Short, woody plants make up the shrub layer. The grasses and ferns of the herb layer grow close to the ground. The bottommost layer is the forest floor.

The warm, wet climate of this biome is ideal for plant growth. For this reason the tropical rain forest has more types of plants and animals than any other biome. For example, in only a small area, there may be hundreds of species of trees. In a temperate deciduous forest, an area of similar size might have only a dozen species.

As in the temperate deciduous forest, plants in the tropical rain forest grow in strata. The tops of tall trees form a solid canopy more than 30 m high. Some huge trees extend 60 m or more above this layer. The trees shed their leaves a few at a time throughout the year. Although they are tall, the trees have a shallow root system. They are anchored in the soil by a spreading base.

Shorter, shade-tolerant trees are found beneath the canopy layer. Woody-stemmed vines, or *lianas* (lee AH nuhs), hang from the trees and are rooted in the soil far below. *Epiphytes* (EHP uh fīts), the so-called air plants, cover the trunks and branches of many trees. They use the trees for support and draw their water and minerals from the atmosphere. Since the broad leathery leaves of the trees block out most of the sunlight, there is no well-developed undergrowth in the tropical rain forest. The forest floor may even be bare in places because organic matter is rapidly recycled by decomposers.

Contrary to popular belief, the tropical rain forest is not a jungle. The tangled vegetation commonly known as jungle usually grows at the edge of the forest and in clearings along riverbanks. Jungle growth also develops where the forest has been disturbed.

Animals also live in different vertical layers, but most species are tree-dwelling. Bats, monkeys, sloths, and colorful birds such as toucans and parrots feed on fruits and leaves in the canopy. Insect-eating birds and bats, tree snakes, lizards, frogs, and other climbing or flying animals can be found on tree trunks and branches. These species move up and down the trees to feed on the epiphytes, insects, or other small animals. There are also many tree-dwelling insects.

On the ground, large mammals that are unable to climb feed on fallen fruit, roots, and leaves of shorter plants. Armadillos and anteaters consume insects. Small mammals and some birds feed among the bases and lower parts of trees. There are also large predators, such as the tiger of Asia and the jaguar of tropical America. Ants, worms, beetles, crickets, and termites are abundant on the forest floor. These animals play an important role in wood decomposition.

Tropical rain forests are very fragile ecosystems. The soil is poor, and nutrients are easily washed away by the heavy tropical rains. Most nutrients are stored in the vegetation. The soil is not good for farming or for grazing animals, yet expanding human populations have cleared large areas of the forest at a rapid rate. After a short period of high productivity, fertility declines and the soil is quickly eroded. Once degraded, the tropical rain forest grows back very slowly or not at all. The tropical animal species are sometimes unable to recolonize the cleared areas. As this process continues, many species of plants and animals have become threatened with extinction.

FIGURE 38·15 Squirrel monkeys, arrow frogs, and jaguars are inhabitants of the tropical rain forest.

38·9 DESERTS

Deserts are biomes that get less than 25 cm of rainfall and that have high rates of evaporation. They can be hot or cold. High daytime temperatures and cool nights are typical in *hot deserts*. In *cold deserts*, the range of day and night temperatures is also extreme, but winter temperatures often drop below freezing.

Deserts are found in two bands on the pole sides of the tropic region. The largest and best-known desert in the world is the Sahara. It stretches across northern Africa from the Atlantic Ocean to the Red Sea.

A surprising number of plants and animals live in these arid lands. In the North American desert, rodents are the most common herbivores, but deer, antelope, and rabbits occur. Lizards and insect-eating birds such as woodpeckers and warblers are found. Predators include kit foxes, badgers, coyotes, and snakes.

Plants may have adaptations that allow them to survive in deserts. *Annuals* grow only when enough water is available. The plants become mature in just a few weeks, produce seeds quickly, and then die.

Perennials, such as grass and shrubs, and *succulents* (SUHK yuh-luhnts), such as cactuses, survive dry periods with many adaptations. Some perennials have widespread shallow roots that quickly soak up water during the rainy season. Others have deep tap roots that reach into moist soil above underground water supplies. Desert shrubs also have sparse, highly modified leaves. Generally the leaves are small and have a cottony or waxy coating that reduces transpiration. In times of severe drought, many perennials shed their leaves or die back to ground level. Their roots survive and put out new shoots when conditions improve.

FIGURE 38·16 During the rainy season, the desert blooms. Most of the wildflowers here will live just long enough to produce seeds. The prickly pear cactus (inset) stores water in its stems. The cactus also has deep roots which allow it to reach water far below the surface. The jack rabbit (inset) lives in the North American desert.

Many succulents have needlelike leaves. Cactuses, for example, carry on photosynthesis in their stems, which reduces water loss from evaporation. Like the perennials, succulents have shallow, widespread roots that enable them to absorb vast amounts of water during the brief desert rainy season. They do not have tap roots, however, and instead survive long droughts by storing water in their cells.

Desert animals are also either *drought evaders* or *drought resisters*. Invertebrates such as crickets and wasps avoid dry periods. Their eggs and larvae lie dormant until the rains provide enough water to allow them to complete their life cycle. Amphibians burrow into the mud and estivate, or lie dormant, during the dry season.

Drought resisters are active year-round but have particular adaptations to enable them to cope with heat and lack of moisture. Spiders, scorpions, and other arthropods have thick, hard body coverings to protect against water loss. Reptiles have scaled or armored skin, which is highly resistant to drying. The most common adaptation to heat is to restrict activity to cooler hours—dawn or dusk or during the night. While daytime temperatures are high, animals such as rodents remain in underground burrows where the temperature is many degrees cooler than the desert surface and where the humidity is much higher. Other species seek out the shade of rocks and vegetation.

Evaporation of water from the body is another way in which heat can be lost. Water can evaporate from the lungs by panting, from the skin by sweating, or from the mouth. In birds, lizards, and some mammals such as the coyote and the fox, evaporation is mostly from the lungs. Koalas, wallabies, and other Australian marsupials dribble saliva and rub it on their bodies when their body temperature gets high. Other animals have sweat glands or evaporate water directly from large areas of exposed skin.

Water is also necessary for eliminating body waste, but because water is scarce in the desert it must be conserved. Birds and reptiles excrete highly concentrated waste from which most of the water has been reclaimed. Mammals are not as efficient, but they excrete urine that is quite concentrated.

The water lost through cooling and excretion must be replaced. But because there is little water available to drink, most desert animals get the water they need from their food. Herbivores get fluid from the moisture in the plants they consume. Carnivores derive water from the tissues of their prey.

38•10 MOUNTAINS

Moving from the equator to the poles, climate becomes cooler. With increasing height above sea level, a similar effect is observed. Thus, mountains show terrestrial biomes in a sequence similar to that seen with changes in latitude. A snow-capped mountain with its base in a tropical rain forest, for example, may have a band of coniferous forest immediately above the rain forest. The coniferous forest is followed by an alpine tundra, and finally the snowy peak.

FIGURE 38•17 A kangaroo rat is adapted for life in the desert. Like many desert animals, it is active at night. It avoids the heat of the day in its underground burrow. The kangaroo rat obtains all the water it needs from the seeds it eats.

Distribution of communities with altitude varies from one mountain to the next. A mountain may not be tall enough to have a tundra zone, for example. Distribution also varies with exposure to wind, sunlight, and precipitation. In the Northern Hemisphere, the tundra zone and each of the following zones extends farther down the slope on the north side than on the south side. Animals inhabiting these zones are similar in habits and habitats to those of the major terrestrial biomes just discussed.

REVIEW QUESTIONS

6. What type of plant life characterizes the tundra?
7. In which biomes does vertical stratification occur?
8. What kinds of animals are typical of the grasslands?
9. How do succulents survive long periods of drought in the desert?
10. Suppose that rainfall in the prairie increased to more than 75 cm a year. What changes would you expect to see in the plant and animal communities?

Aquatic Biomes

SECTION OBJECTIVES

- Contrast freshwater and saltwater biomes.
- Discuss the factors that determine the kinds of organisms in a freshwater biome.
- Describe the three major zones of the marine biome.

38·11 FRESHWATER BIOMES

Aquatic biomes can be placed in two broad categories: freshwater and saltwater, or marine. **Freshwater biomes** are those habitats in which the salinity is less than 0.005 percent. The salt content of **marine biomes** is 3 to 3.7 percent. Lakes, ponds, rivers, and streams are freshwater biomes. The marine biome is made up of the oceans.

Freshwater habitats are divided into two groups. One group is the **lentic habitat,** or standing-water habitat, such as lakes, ponds, and swamps. The second type of freshwater habitat is the **lotic habitat,** or running-water habitat. Lotic habitats, which include streams and rivers, have a significant current.

In fresh water there are five conditions, or *limiting factors,* that determine the kinds of organisms that will be present. Temperature is one such factor. Because of certain physical properties, water temperature varies less than air temperature. Aquatic organisms are therefore adapted to a much narrower range of temperature than terrestrial organisms are. A sudden change of only a few degrees can often destroy aquatic populations.

Turbidity, or the amount of particles suspended in the water, is a limiting factor. Suspended particles limit the penetration of light, which in turn determines how much photosynthesis can occur.

The third limiting factor is current. In streams and other lotic habitats, the force of the water flow determines which species will be able to exist. Each species has a range of tolerance for pressure from the current. Current also affects the distribution of dissolved nutrients and small organisms.

The amount of dissolved oxygen gas is the fourth limiting factor. The oxygen level depends on the temperature of the water, the current, and the salt content. Oxygen is more soluble in cold water than in warm water. Running water has more oxygen than standing water because there is greater mixing with the air. Oxygen is less soluble, however, in water with high salt content.

The fifth limiting factor is the amount of dissolved minerals. The amounts of nitrates and phosphates in the water may limit the growth of some organisms. Increased amounts may encourage the growth of certain organisms.

Organisms in fresh water are classified according to their distribution and habits. Floating throughout the upper layers of the water are the **plankton,** small, sometimes microscopic organisms that are carried about by the motion of the water. Plankton can be further classified as phytoplankton and zooplankton. *Phytoplankton* (fī toh-PLANGK tuhn) are the primary producers (for example, algae and diatoms). *Zooplankton* (zoh uh PLANGK tuhn) are small animals and protozoans that feed on the phytoplankton. **Nekton** are swimming animals that are able to move independently of the currents. Fish, insects, and crustaceans are examples of nekton. Attached to the bottom or living in the bottom sediments are the **benthos,** which include both plants and animals. **Periphyton** (pehr uh FĪ tahn) are organisms that cling to plants or other objects on the bottom.

In lakes, ponds, and other lentic ecosystems, three regions or zones are evident. The region closest to shore is the shallow-water or **littoral** (LIHT uhr uhl) **zone.** Light penetrates to the bottom, which enables the growth of rooted plants. There are also phytoplankton and many animals. The **limnetic** (lihm NEHT ihk) **zone** is the upper open-water region. The upper portion of the water receives enough light to allow a zone of photosynthesis. There are no rooted plants, and phytoplankton consist mostly of microscopic algae. Crustaceans are the most numerous form of zooplankton, and their populations vary with the amount of phytoplankton. Fish are the dominant nekton in the limnetic zone.

With increasing depth, light penetration decreases until it is no longer strong enough for photosynthesis. The absence of photosynthesis affects oxygen levels. This region of deep water, which reaches to the bottom, is called the **profundal** (pruh FUHND uhl) **zone.** Organisms in the profundal zone depend mainly on food that sinks into the deeper water from the zones above it. Some species of worms, small clams, and insect larvae live in this zone. Bacteria and fungi are important decomposers that live on the muddy bottom. Oxygen is often a limiting factor in the profundal zone. Many living things in this region are able to survive periods of low oxygen levels. Many of the bacteria in the sediment are anaerobic.

FIGURE 38·18 Lentic habitats, such as lakes, ponds and swamps, have still water (top). Lotic habitats, such as rivers and streams, have moving water (bottom).

Running-water, or lotic, ecosystems have several important features that distinguish them from lentic habitats. The flowing water results in a greater mixing from top to bottom. Rivers and streams, therefore, have more uniform temperature and concentration of dissolved nutrients than lakes and ponds. Also, because of the shallow depth, constant mixing, and greater contact with the air, oxygen is generally not a limiting factor. Therefore, organisms living in running water are more susceptible to reduced oxygen than organisms in the profundal zone of a lake.

Current is the most significant factor determining the makeup of lotic communities. In rapidly moving streams and rivers, the water carries away small particles and debris, which leaves mostly stones and gravel at the bottom. Benthos are limited to those organisms able to withstand the force of the current. Green algae and aquatic moss are the major plant types. They grow permanently attached to the rocks. Some animals have hooks and suckers to grasp rock surfaces.

Pools and slowly moving waters resemble lentic habitats. Plankton populations here are more diverse and numerous than in rapids in

Issues

The Kissimmee River

Before the 1950s, Florida's Kissimmee River followed a meandering route from Lake Kissimmee to Lake Okeechobee. The marshland on both sides of the river was kept moist by periodic flooding. The large expanse of water in the region released a great deal of water into the air. This, in turn, caused almost daily rains.

As the tourist industry grew, the flooding was seen as a nuisance. The United States Army Corps of Engineers and the state of Florida worked together to find a way to control the water cycle and reduce the rainfall. Between 1950 and 1954, a channel was constructed, diverting the river from its 160-km meandering route into an 84-km straight canal.

The straightened river had less surface area for evaporation, so the rainfall decreased. As the surrounding marshlands became dry, habitat loss reduced populations of waterfowl, alligators, and bald eagles. The newly established grasslands supported populations of quail.

After some years of decreased rainfall, water shortages have developed. Farmlands are becoming dry, and the water table is too low to supply water for irrigation. The Everglades are drying up, and many species that live there are threatened.

The South Florida Water Management District is working to restore the original route of the Kissimmee River. Thirteen years ago a small portion of the original wetlands was reflooded. Ecologists see the rapid return of natural populations in that area as an encouraging sign that such a recovery is possible.

streams or rivers. Rooted plants are common along the riverbanks and on the surface. The slower moving waters deposit more silt, so the number of burrowers such as worms and clams increases. Nekton species are different from those of the rapids and are adapted to moving among the plants.

38•12 MARINE BIOME

The marine biome consists of the world's oceans. It is the largest continuous biome, covering about 71 percent of the earth's surface. As in fresh water, physical and chemical factors determine the type and distribution of organisms in the sea. Light, temperature, salinity, and water pressure are the most important factors.

With increasing depth there is generally less light for photosynthesis, temperatures are colder, and there is greater pressure. Salt concentration varies in the upper layers but is stable in deeper water. The availability of nutrients is also important in determining ocean communities and depends partly on the circulation of water. In some coastal regions, strong winds create currents that move surface water away from shore. These currents draw up colder, nutrient-rich water in *upwellings*. The abundance of food in upwellings attracts many fish, which makes these areas important in commercial fishing. Deep currents caused by differences in temperature and salt concentration also help to circulate nutrients. Some materials do, however, sink into deep sediments, where they are lost for very long periods.

FIGURE 38•19 The marine biome is the largest biome on earth. The coral and fish in this picture are from a tropical marine biome. What ocean zone is shown in this photograph?

The sea can be divided into three major zones. The **intertidal zone** is the region close to shore, where high and low tides occur. With the change of the tides, the zone is alternately exposed to the air and covered with water. Living things in this habitat must be adapted to changes in temperature, salinity, and moisture. They must also be able to withstand the force of the waves and resist being washed out to sea. The multicellular algae (seaweeds) attach themselves securely to rocks with *holdfast organs*. Barnacles secrete a gluelike substance that cements them to hard surfaces. Other animals, such as clams and crabs, burrow into the sand.

Extending from the intertidal zone to the edge of the continental shelf is the **neritic** (nih RIHT ihk) **zone.** This zone is the most productive area of the marine biome. Light reaches to the bottom and supports a well-developed community of phytoplankton. Zooplankton, including copepods and protozoa feed on the phytoplankton. Nekton in this region include fish, turtles, and some mammals such as seals. Benthos consist of large numbers of sessile, or stationary, animals—sponges, sea anemones, corals—and burrowers such as clams, crabs, and worms. Starfish, snails, and sea urchins also crawl about on the bottom. Bacteria are the primary decomposers.

The **oceanic zone,** or open water, is the third major zone beginning just beyond the continental shelf. This region can be further divided into several vertical zones. From the surface to about 200 m is the **euphotic** (yoo FAHT ink) **zone.** Enough light penetrates in this zone to allow photosynthesis. It is equivalent to the freshwater limnetic zone. Plankton, fishes, and larger nekton—sharks, whales, and squids—are found in this region. Seabirds such as albatrosses, petrels, and shearwaters are also characteristic of the open ocean.

Below the euphotic zone, the ocean becomes cold and dark. From about 200 to 2000 m on the continental slope is the **bathyal** (BATH ee-uhl) **zone.** The deep ocean, reaching from 2000 m to the sea floor, is the **abyssal** (uh BIHS uhl) **zone.** There are no producers here. Animals inhabiting these depths are either scavengers that feed on dead plants and animals that sink from the upper waters, or they are predators. Prey can be scarce, however. In order to catch whatever is available, some deep-sea fish have large mouths, long and sharp teeth, and stomachs that can stretch for a meal sometimes larger than the fish itself. Benthos of the deep ocean include crustaceans, echinoderms, worms, and mollusks that are adapted to great water pressure and intense cold.

38•13 ESTUARIES

FIGURE 38•20 An estuary is a shallow area where fresh water and salt water mix.

Many streams and rivers eventually flow into the ocean. The shallow areas where fresh and salt water mix are known as **estuaries** (EHS-chu ehr eez). Bays, inlets, tidal marshes, and the mouths of rivers are examples of estuaries. Such areas have strong currents and changing salinity. Salinity varies both vertically and horizontally. Where currents are strong enough to thoroughly mix the water, salt concentration may be the same from top to bottom. Otherwise there will be a layer of fresh water on top of a layer of heavier salty water. Salt levels increase horizontally from the mouth of the river to the ocean. These variations are also affected by the season, the amount of rainfall, and the flow of the tides.

Estuaries have a higher productivity than freshwater or marine environments. The shallow water allows a high rate of photosynthesis, and the mixing of waters with different salinities helps keep important nutrients in the estuary. Marsh grass, algae, and microscopic phytoplankton are the main producers. The most common animal in an estuary is the oyster. Oysters grow in reefs attached to rocks or clustered on the shells of older, dead oysters. Many other animals are associated with the oyster community: sponges, barnacles, snails, and worms. Estuaries are very important as nurseries for many fish.

REVIEW QUESTIONS

11. What factors determine the kinds of organisms that will be found in a freshwater ecosystem?
12. Why are aquatic organisms more sensitive than land organisms to temperature changes?
13. What is the main difference between a lentic and a lotic habitat?
14. Name the three major zones of the sea.
15. What is the effect of climate on aquatic biomes?

Investigation

What Are Some Limiting Factors in a Freshwater Biome?

Goals
After completing this activity, you will be able to
- Graph the relationship between oxygen concentration and depth in a lake.
- Graph the relationship between the percent of light penetration and depth in a lake.

Materials (for groups of 2)
2 sheets of graph paper

Procedure

A. Tables 1 and 2 list the data gathered in a study of a lake (freshwater biome). Study the data, which represent some of the limiting factors in a freshwater biome.

B. A group of scientists collected specimens of floating plants (phytoplankton) in the same lake studied in Tables 1 and 2. The scientists' findings on the phytoplankton populations are shown in Table 3. Study these findings.

C. Construct a graph that will show oxygen concentration in relation to depth of the lake. Plot depth on the x-axis and oxygen concentration on the y-axis.

D. Construct a graph that will show the percent of light penetration in relation to depth of the lake. Plot depth on the x-axis and light penetration on the y-axis.

E. Graph the oxygen data from Table 1 onto your graph of oxygen concentration in relation to depth. Graph the data on light penetration from Table 2 onto your second graph. Connect the dots with a curved line.

F. Examine the table on the quantity of phytoplankton in the freshwater biome studied. Construct graphs showing the sizes of the phytoplankton populations at various depths.

TABLE 1. OXYGEN CONCENTRATION AT DIFFERENT DEPTHS OF THE LAKE

Depth (m)	Oxygen Concentration (mg/L)
0	8.5
1	8.5
2	8.5
3	8.5
4	9.0
5	17.0
6	16.0
7	14.0
8	12.0
9	10.0
10	5.0
11	2.0
12	0

TABLE 2. LIGHT PENETRATION AT DIFFERENT DEPTHS OF THE LAKE

Depth (m)	Light Penetration (%)
0	100.0
1	32.0
2	23.0
3	18.0
4	12.0
5	10.0
6	9.0
7	7.0
8	5.0
9	2.0
10	1.0
11	0.5
12	0.0

TABLE 3. AMOUNT OF PHYTOPLANKTON AT DIFFERENT DEPTHS OF THE LAKE

Depth (m)	Phytoplankton (# individuals/L)
0	800
1	3 500
2	2 800
3	4 000
4	5 700
5	183 000
6	96 000
7	30 000
8	10 000
9	2 000
10	600
11	500
12	50

Questions and Conclusions

1. What happens to oxygen concentration between 0 and 4 m; between 4 and 8 m; between 8 and 12 m? At what depth is oxygen concentration greatest?

2. At what depth is the quantity of phytoplankton greatest?

3. Compare Tables 1 and 3. Is there a relationship between the amount of phytoplankton and oxygen concentration? Explain the relationship noted.

4. Oxygen is more soluble in cold water than in warm water. In the freshwater biome studied, the deep water is coldest. Does the deep water have the most oxygen? Explain.

Chapter Review

SUMMARY

38·1 Biomes are large geographic areas that have characteristic types of plants and animals.

38·2 Climate is the main factor that determines a biome.

38·3 The tundra experiences long, cold winters and short summers. Lichens, mosses, and fast-growing herbs are the typical vegetation.

38·4 The coniferous forest biome, or taiga, has freezing temperatures for at least six months of the year. The characteristic plant life is cone-bearing trees such as spruces, hemlocks, and firs.

38·5 The temperate deciduous forest has four distinct seasons and evenly distributed annual precipitation. Broad-leaved deciduous trees, which lose their leaves seasonally, are the typical vegetation.

38·6 The temperate rain forest along the northwest coast of North America has the high yearly precipitation that is typical of tropical rain forests and the cone-bearing evergreen trees.

38·7 Grasslands develop in the middle of continents where the annual rainfall is not enough to support the growth of trees. Grasslands once had more grazing animals than any other biome.

38·8 Tropical rain forests get at least 200 cm of rain annually and have year-round warm temperatures. The greatest diversity of plants and animals in the world is found in tropical rain forests.

38·9 Deserts receive less than 25 cm of rain a year. Plants and animals are adapted to hot days, cool nights, and extreme dryness.

38·10 Mountains show a sequence of terrestrial biomes from base to peak. Altitude rather than latitude determines where one biome changes into another.

38·11 Aquatic biomes are placed in two categories based on salt concentration: freshwater biomes and marine biomes. Freshwater biomes include standing-water ecosystems, such as lakes and ponds, and running-water ecosystems, such as rivers and streams.

38·12 The marine biome is made up of the world's oceans and is the largest continuous biome. Light, temperature, salinity, and water pressure are the most important factors affecting the nature of ocean communities.

38·13 Estuaries are shallow coastal regions, such as bays and inlets, where fresh and salt water mix.

BIOLOGICAL TERMS

abyssal zone
aquatic biome
bathyal zone
benthos
biome
canopy
climatogram
coniferous forest
desert
estuaries
euphotic zone
forest floor
freshwater biome
grasslands
ground layer
herb layer
intertidal zone
lentic habitat
limnetic zone
littoral zone
lotic habitat
marine biome
nekton
neritic zone
oceanic zone
periphyton
permafrost
plankton
profundal zone
root layer
savanna
shrub layer
taiga
temperate deciduous forest
temperate rain forest
tropical rain forest
tundra
understory
vertical stratification

USING BIOLOGICAL TERMS

1. What is vertical stratification?
2. How is a marine biome different from a freshwater biome?
3. What is permafrost?
4. How is a tropical rain forest different from a temperate rain forest?
5. Distinguish between a lotic habitat and a lentic habitat.

UNDERSTANDING CONCEPTS

1. What major characteristic identifies a terrestrial biome? (38·1)
2. Describe a climatogram. (38·2)
3. Why don't trees grow on the tundra? (38·3)
4. What adaptations allow evergreen trees to survive long, cold winters in the coniferous forest? (38·4)
5. Give one reason why shrubs and herbs do not grow in the coniferous forest biome. (38·4)
6. What type of climate characterizes the temperate deciduous forest biome? (38·5)
7. Explain how plant and animal life is distributed in the temperate deciduous forest. (38·5)
8. What are the characteristics of the temperate rain forest? Identify two other biomes that are similar to the temperate rain forest. (38·6)
9. What factors cause grasslands to develop instead of forests? (38·7)
10. Running or burrowing are common behavior among grassland herbivores. Explain the adaptive value of such behaviors. (38·7)
11. What factors account for the great variety of plants in the tropical rain forest? (38·8)
12. Why is the floor of the tropical rain forest usually bare? (38·8)
13. List the three types of plants adapted to the desert biome and explain how each survives such harsh conditions. (38·9)
14. How do desert animals cope with high daytime temperatures and lack of water? (38·9)
15. Explain why mountains show a series of biomes from top to bottom. (38·10)
16. Name the four groups of organisms found in fresh water. Which of these are also found in the ocean? (38·11)
17. In which parts of the lentic habitat does photosynthesis occur? (38·11)
18. What conditions determine the kinds of organisms that will be found in the intertidal zone? (38·12)
19. What is the most productive zone in the marine biome? Explain why. (38·12)
20. What are the two major characteristics of an estuary? (38·13)

APPLYING CONCEPTS

1. Describe the changes in plant and animal life that you would expect to see if a river were dammed and the water formed a small lake.
2. Why are the majority of animals in the tropical rain forest tree-dwelling?
3. Suppose scientists developed a new kind of food crop that requires about 50 cm of rainfall a year and is able to survive short periods of drought. Name at least two places where this new crop could be grown.
4. Imagine you are hiking up a mountain in the Rockies. Describe the kinds of plants and animals you observe as you climb toward the peak. What factors account for the changes you see as you move up the slope?

EXTENDING CONCEPTS

1. Determine the status of the effort to save the black-footed ferret. References earlier than 1975 list it as an extinct animal of the Great Plains, but since then a few have been seen. Include in your report a discussion of why the animal nearly became extinct and what steps were taken to save it.
2. Identify the kind of biome in which you live. Using library sources, prepare a chart comparing the kinds of plants and animals found in the biome today with those found 50 and 100 years ago. Discuss how any changes relate to human activity and population growth.
3. Visit a local fish market and make a list of the kinds of animals sold there. Using available resources, identify which are freshwater species and which are marine. In which zone is each kind of animal found?

READINGS

Durrell, Lee. *State of the Ark: An Atlas of Conservation in Action.* Doubleday, 1986.

Maltby, Edward. *Waterlogged Wealth: Why Waste the Earth's Wet Places?* International Institute for Environment, 1986.

Perry, Donald. *Life Above the Forest Floor: A Biologist Explores a Strange and Hidden Treetop World.* Simon & Schuster, 1986.

39. Your Environment

Sea otters dive to the ocean bottom for food—and for stones. Floating at the surface, they use the stones to break open clams and oysters. Once, sea otters were hunted nearly to extinction for their beautiful fur. Laws now protect sea otters from hunters. However, sea otters are still killed indirectly. Some drown when they become trapped in fishing nets; others are poisoned by water pollutants. In this chapter, you will examine how humans affect the environment.

After completing this chapter, you will be able to
- **Distinguish** between renewable and nonrenewable resources.
- **Explain** the causes of resource losses and **describe** ways of reducing these losses.
- **Describe** the role of energy in our daily lives.
- **Identify** some of the sources of energy that are available.
- **Identify** some causes of air pollution and water pollution.
- **Discuss** methods for controlling pollution.

Material Resources

39·1 THE EARTH'S RESOURCES

Materials from the earth that are used to sustain life or that are critical to economic systems are called **natural resources.** Land, air, water, and minerals such as copper, iron, and gold are all examples of natural resources. Ecosystems depend upon resources to function properly. The earth's rapidly growing human population is making ever-increasing demands on natural resources. Another reason for the growing use of resources is a rise in the standard of living around the world. People buy and use more resources than ever before. An increase in the use of resources puts a strain on the ecosystems that provide the resources. Also, the use of resources often makes waste products, some of which may be harmful to ecosystems.

Natural resources are usually classified as renewable and nonrenewable resources. **Renewable resources** are substances that can be replaced through natural cycles, such as trees, air, soil, and water. **Nonrenewable resources** are substances that cannot be replaced, such as metals. Most energy sources are nonrenewable resources. Once the energy in these fuels is used, it is no longer available. Many nonrenewable resources can be recycled. **Recycling** is a process that uses wastes as the raw material for new products. In ecosystems the wastes from one process are the resources for another process. Humans are attempting to imitate nature's method of resource recycling.

Human activities often use both renewable and nonrenewable resources. For example, farming produces crops, which are renewable, while using many nonrenewable resources. Fertilizers contain nonrenewable minerals. Farm tools and tractors are made of aluminum, copper, and iron. Energy from fuels is used to operate machinery and transport crops to market.

Although renewable resources can be replaced, their supply is not unlimited. The amounts of renewable resources available are lim-

SECTION OBJECTIVES

- Describe ways of conserving natural resources.
- Identify ways humans affect the living world.

FIGURE 39·1 Recycling glass, paper, and metal is a method of recovering nonrenewable resources.

com- (with)
servare (to preserve)

ited by the capacity of ecosystems to replace them. The amount of available water decreases if the rate of use is faster than the rate of replacement by the water cycle. Soil is renewable if it is protected from erosion and loss of organic matter. Living species are renewable resources as long as breeding populations exist. Ecosystems must be protected from pollution and destruction. The wise management and use of natural resources is called **conservation.** Recycling, reforestation, and restrictions on hunting and fishing are common conservation practices.

39•2 SOIL

Soil is needed to grow the plants that provide oxygen and food for people and animals. Recall that some important steps in the nitrogen and water cycle occur in the soil. Although soil is a renewable resource, it forms very slowly. It takes 200 to 400 years to form one centimeter of topsoil, the fertile uppermost layer of the soil. Because soil forms so slowly, it is important to conserve it.

A major conservation problem is depletion, or loss, of organic materials and minerals from farmland. Growing the same crop continuously on the same land uses nutrients faster than they can be replaced by nature. The fertility of the soil is reduced as nutrients are depleted. Growing plants such as soybeans and clover that restore nutrients to the soil and using fertilizers wisely helps prevent depletion. The practice of growing different crops in succeeding years is called **crop rotation.**

As well as losing nutrients, soil can be lost altogether. The wearing away of soil due to the action of water or wind is called **erosion.** In nature, plants slow down erosion by holding soil together with their roots and preventing water from flowing quickly over the surface of the soil. Native plant cover is destroyed as land is used for farming, housing, mining, and industry. It may also be lost because of overgrazing by livestock, fire, crop failure, and poor farming practices. Without plants to hold loose particles of soil together, soil can be washed away by water or blown away by the wind.

Several conservation methods can reduce erosion. **Terracing** involves digging a series of level areas, or terraces, into the side of a hill. The terraces are cultivated, while native vegetation and grasses are left on the slopes. Water soaks into the terraces, rather than running off and carrying valuable topsoil with it.

Contour farming involves plowing at right angles to a slope, following the contours of the land, rather than plowing straight up and down the slope. The furrows, or shallow trenches, formed by this sort of farming hold water, while the ridges and plants act like dams. In **strip cropping,** bands of cover crops such as wheat and clover are alternated with strips of row crops such as corn and lettuce. Although the row crops leave the soil partially exposed, the cover crops catch runoff and lost topsoil. Strip cropping and contour farming are often practiced together.

FIGURE 39•2 Terracing reduces runoff and erosion and allows steep slopes to be cultivated.

FIGURE 39·3 Soil can be conserved through good farming practices. In strip cropping, cover crops and row crops are planted in alternating bands (left). In contour farming, rows are plowed across slopes, following the contour of the land (right).

Wind erosion may be reduced by planting **windbreaks,** which are rows of trees and shrubs that reduce wind speed. Windbreaks are especially important in areas where the land is very flat.

39·3 WATER

Recall that during the water cycle, surface water evaporates from bodies of water such as oceans and lakes. Water vapor condenses and falls as precipitation, usually rain or snow. This cycle is the source of most fresh water. Some of the fresh water in the water cycle is found beneath the surface in the form of ground water. Ground water is especially important in dry areas, where surface water quickly evaporates. Ground water and the water in reservoirs are the main sources of water for use by humans.

If enough water is to remain available, water use cannot be larger than water supply. Surface water may be collected and stored in reservoirs. Reducing runoff increases the amount of water that seeps into the ground, and is an important conservation practice. **Watersheds** are drainage areas that supply water to lakes, ponds, and reservoirs. If the watershed is covered with vegetation, rainwater will run down the slope slowly, and more water will seep into the ground.

39·4 FOOD

Farmers today can produce more food than ever before. All this productivity is not without costs. Modern farming techniques use many nonrenewable resources, such as gasoline, artificial fertilizers, and pesticides. **Pesticides** are chemicals used to kill insects.

Food production in many areas of the world is very different from food production in North America. In areas where fertilizers are unavailable or too expensive, soil nutrients may become used up, and food production decreases. Natural fertilizers, including animal wastes, are often used. Such fertilizers are less expensive than artificial fertilizers and do not damage the environment.

FIGURE 39·4 Mariculture involves cultivating food in the sea. At farms like this one, edible seaweeds are grown on nets.

Many people believe that food from the sea will feed an increasing world population. The open sea makes up 90 percent of the ocean and has been suggested as a new place to find food. However, this area produces few fish and has little potential to produce more. In the open ocean, nutrients settle to the bottom, but light is available only near the surface. Thus the open ocean cannot support enough producers to feed a large population of fish.

Coastal zones are more productive and contain many more fish than does the open ocean. These areas have been sources of food for a long time. Some increase in production can be expected in these areas. Yet the fishing industry may already be overfishing some areas. Coastal areas must be protected from overfishing. In addition, sewage, garbage, and other substances that pollute the water and damage the coastal ecosystem must be controlled.

Farming the sea is called **mariculture.** Mariculture can involve the cultivation of seaweed and marine animals such as oysters, clams, and fish. Mariculture may allow an increased production of food from the sea.

Careful management of food resources should also include methods to reduce the amount of food wasted, especially during transport and storage. Ten to twelve percent of India's grain is eaten by rats. In some areas of the world, insects have been known to destroy as much as 50 percent of stored food. Molds, mildews, and bacteria destroy thousands of tons of food each year. Preventing food losses may be as important a strategy against hunger throughout the world as is increasing food production.

39·5 FORESTS

North America has an abundance of forest resources that contain a mixture of tree species. There are harvestable forests, noncommercial forests, and wilderness forests in national parks. Forests are sources for firewood, lumber, and wood used in paper production.

The type of trees that grow in an area depends on climate and soil conditions. Human activities also change the makeup of a forest. Certain species of trees may be removed, and other species may be planted. Fire is sometimes used to clear an area and stimulate new tree growth. This controlled burning is a method of forest management that is carefully planned and supervised. Forest fires caused by careless human behavior also destroy forests.

Although forest fires destroy many trees, they are not the largest cause of forest loss. The major cause of forest loss around the world is clearing for farming. As many countries try to produce more food, forests are cut and the land is used to grow crops.

FIGURE 39·5 The major cause of forest destruction is clearing for farming. Unfortunately, the cleared land erodes quickly and is not easily reforested.

Secondary succession will slowly replace a forest that has been cut (Section 37·7). Unused fields will be taken over by natural vegetation. Sometimes people help this process take place by planting new trees to replace the ones that were cut down. Such reforestation programs have been very successful.

39·6 WILDLIFE

As humans change an environment, the numbers and kinds of wildlife found in that environment also change. For example, clear-cutting a forest and replacing it with corn eliminates the habitat for forest animals such as squirrels. At the same time, this change increases the habitat for field wildlife such as pheasants.

The change of areas in North America from wilderness to farms or towns has caused population changes in many species of native wildlife. Species such as the woodland caribou and the ivory-billed woodpecker, which require a wilderness habitat, decline and may become extinct. Their places are taken by other animals, such as the white-tailed deer and the robin. Habitat loss is the main cause of loss of wildlife.

Another problem for wildlife is the introduction of foreign or *exotic species* by humans. These species often lack natural population controls in the form of predators. Thus they become very numerous. Competition from the new species may force native species into poorer habitats, where they may not be able to survive. Ecosystems around the world have been harmed by the introduction of exotic species. The African clawed frog is becoming a dominant species in some southern California waterways. It eats the food of native species, such as tree frogs, toads, and bullfrogs. The African clawed frog also eats the tadpoles of these native species.

Changes in the earth's communities have caused many species to become extinct. Today many organisms are identified as being *endangered species,* or species close to extinction. Organisms are classified as endangered if the breeding population is small or restricted to a very small area. Figure 39·6 shows the wide range of taxonomic groups and the numbers of species currently endangered. The Florida manatee, shown in Figure 39·7, is an example of an endangered species. The manatee lives in shallow estuaries. Pollution and human development of rivers have destroyed available habitats. In addition, many manatees have been killed by boat propellers.

The loss of any organism represents the loss of genetic material that cannot be replaced. While there are examples of failure, there are many examples of successful conservation practices that have restored fish and wildlife populations.

FIGURE 39·6 This table shows the number of species of plants and animals that are endangered. Endangered species are close to extinction.

GROUP	NUMBER OF ENDANGERED SPECIES
Mammals	274
Birds	274
Reptiles	74
Amphibians	13
Fishes	48
Snails	3
Clams	25
Crustaceans	1
Insects	7
Plants	51
TOTAL	770

FIGURE 39·7 The Florida manatee is endangered because of habitat destruction due to development, pollution, and boating. Scientists can often identify individual manatees by the scars on their backs. The scars are caused by the propellers of speeding boats.

REVIEW QUESTIONS

1. Why do consumption and production of natural resources have to be balanced?
2. List four farming techniques that reduce soil erosion.
3. What area of the ocean produces the most food for humans?
4. List two reasons why organisms become endangered.
5. Sometimes when a forest is cut for agriculture, the soil is not fertile enough to support the crops that are planted. What might happen to the field if fertilizers are not used to improve the soil?

Pollution in the Environment

SECTION OBJECTIVES

- Explain how smog and acid rain are formed.
- Discuss how biomagnification affects a food chain.

39•7 AIR POLLUTION

The various harmful gases and solid particles that are discharged into the air by motor vehicles, power plants, homes, and industries are called **air pollution.** Figure 39•8 identifies some of the major air pollutants, their sources, and their effects.

Fossil fuels, especially coal, give off many pollutants as they burn. The sulfur that is often found in coal combines with oxygen, producing a smoke that has an unpleasant smell. As it burns, coal also releases a large amount of soot, or particles of unburned carbon. During the early 1900s, some industries changed their power source from coal to cleaner-burning oil and natural gas. The air became cleaner.

Later in the 1900s, with the widespread use of the automobile, other types of pollution became common. The automobile engine does not burn gasoline completely. As a result, carbon monoxide and hydrocarbon molecules are added to the air. Various compounds of nitrogen and sulfur are also produced. These materials are all toxic. Pollutants from automobile exhaust undergo chemical reactions in the air to form **smog.** Sunlight provides the energy for these reactions, hence the name *photochemical smog.*

Smog and other forms of pollution can be reduced. More efficient automobile engines burn gasoline more completely. The catalytic converter found on many cars reduces air pollution by changing the chemical makeup of the exhaust. Filters on smokestacks can trap soot and other particles that also cause smog.

Cities produce large amounts of air pollution. Usually the air nearest the earth is warm. Since warm air is less dense than cold air, it

FIGURE 39•8 Common air pollutants come from a variety of sources and have many different effects on the environment.

POLLUTANT	SOURCE	EFFECTS
Sulfur oxides and nitrogen oxides	Burning fossil fuels	Acid rain; plant damage; potential ecosystem damage; human health damage
Carbon monoxide	Auto exhausts	Reduces O_2 levels in blood
Organic materials	Burning; wind erosion; industrial processes	Eye and nose problems; cancer causing; ecosystem impact
Inorganic metals (lead, mercury)	Mining and refining; burning; manufacturing processes	Accumulate in food chains and in body organs; closely linked to cancer; toxic to many organisms and ecosystems

tends to rise. These rising air currents, along with winds, carry many of the pollutants away. *Temperature inversions* are caused when cool air is replaced by a layer of warm air. During an inversion, warm air on top of cooler air blocks normal circulation patterns. This inverted air pattern traps pollutants near the ground and causes serious public health problems. Lungs are particularly sensitive to air pollution. Nose and eye problems, as well as coughing and asthma, are associated with air pollutants.

One of the most far-reaching effects of air pollution is called the *greenhouse effect.* The earth is heated by energy from the sun. The atmosphere acts like the glass in a greenhouse and holds in the heat. The amount of carbon dioxide in the atmosphere is increasing, mostly because of the large amounts of fuel that are burned. Carbon dioxide increases the heat-trapping effect of the atmosphere. The increased carbon dioxide could lead to a gradual increase in worldwide temperatures. There is other evidence, however, that shows that the earth's temperature may be gradually decreasing. Researchers believe that this is due to large increases in the amount of dust and other particles in the atmosphere. These particles reflect solar energy away from the earth. The potential for large changes in the earth's climate is being studied all over the world.

39·8 ACID RAIN

Sulfur dioxide (SO_2) is a gas that is toxic to both animals and plants. Most sulfur dioxide in the air is produced by power plants that burn coal to make electricity. In natural systems a small amount of sulfur dioxide is used by soil microorganisms. However, large amounts of sulfur dioxide can cause serious pollution problems.

When sulfur dioxide combines with oxygen and water vapor in the air, it changes to sulfuric acid (H_2SO_4). Acids formed in the atmosphere fall to the earth with rainwater, and this combination is called **acid rain.** Nitrogen oxides from motor vehicles and power plants form nitric acid in the atmosphere, and contribute to the problems.

The global effects of acid rain are well-known. Limestone and marble in statues and buildings are dissolved by acid rain. Metal structures, such as bridges, corrode. Evidence from forests, rivers, lakes, and agricultural areas shows that the long-term effects of acid rain include lowering soil and water pH. As a result, trees may die and lakes may become lifeless.

Acid rain was first described by a British chemist over a century ago, but the problem did not receive much attention until the 1960s. Since then more than 3000 studies have been conducted. Everyone agrees that there is a problem, but opinions differ on how to solve the problem. Acid produced in one area may be carried in the air, and it may fall far from its source. Because of the movement of pollutants by weather systems, acid rain has become an international problem. Stronger national laws and international agreements are needed to deal with the problem of acid rain.

FIGURE 39·9 During temperature inversions, a layer of warm air traps pollutants close to the ground. This polluted air causes many serious health problems.

FIGURE 39·10 Acid rain can dissolve the stone making up buildings and statues. This ornamental pillar is discolored and damaged by acid rain.

Profiles

Rachel Carson (1907–1964)

The work of Rachel Louise Carson is as important today, some 20 years after her death, as it was when originally done. In *Silent Spring*, her most famous book, Rachel Carson wrote of the potential dangers to wildlife and the environment from the uncontrolled use of pesticides, such as DDT, and other harmful chemicals.

Silent Spring was published in 1962. Rachel Carson wrote this book as the result of a friend's request. The friend was shocked by the killing of birds and harmless insects in an area that had been sprayed with DDT as part of a mosquito control program. Carson was concerned about the use of such chemicals and she researched their effects for several years before writing *Silent Spring*. Carson died before legislation went into effect banning the use of DDT.

One suggested way to reduce acid rain is to remove sulfur from coal before it is burned. This is very expensive. New methods to reduce acid rain are being tested with good results. But it will take a major commitment from people, industry, and governments to solve the acid rain problem.

39•9 WATER POLLUTION

The world's waters receive pollutants from many different sources. **Water pollution** is any substance that damages the quality of water. Toxic chemicals from agriculture, industry, and landfills are well-known causes of water pollution.

Sewage and fertilizers contain substances that serve as nutrients in ecosystems. However, natural cycles can use only a small amount of these nutrients. Large amounts of nutrients are pollutants. Considerable progress has been made in cleaning up nutrient pollution.

Heavy metals, such as lead, copper, silver, and mercury, are common water pollutants. When living things consume compounds that contain these metals, death may result. Heavy metals may cause birth defects in humans. Metals tend to interact with each other. For example, copper and zinc in combination are ten times more toxic to fish than either element alone.

It is possible to find ways to reduce heavy metal pollution. For example, silver, which is widely used in photography, can now be recovered from waste water, instead of being discharged. Such a process has both economic and ecological value because it improves the efficient use of an expensive and nonrenewable resource.

Although toxic chemicals are diluted as they are discharged into the water, they are still a serious problem. Many pollutants are measured in parts per million (ppm) or parts per billion (ppb). Two examples of one part per million would be 1 mm per 1 km, or one cent in $10,000. An example of one part per billion would be one second in 32 years. Pollutants are dangerous even in such low concentrations.

Pollutants in low concentration in the water may be found in greater concentration in the organisms that live in the water. This is especially true of secondary consumers. The increasing concentration of pollutants in a food web is called **biomagnification.** Pollutants remain in the tissues of organisms and accumulate there. When these organisms are eaten the pollutants are passed to the next consumer.

There are many examples of toxic materials being concentrated in food webs. The classic example is the insecticide DDT. DDT was widely used throughout the world to protect crops and destroy disease-carrying insects such as mosquitoes and fleas. DDT accumulated through food webs and was found to be responsible for the failure of many birds to reproduce. One of these birds, the American bald eagle, became a symbol in efforts to ban the use of DDT.

Further research developed chemicals that do not stay in the ecosystem. These **biodegradable chemicals** are substances that break down into substances that do not harm and can even enrich ecosys-

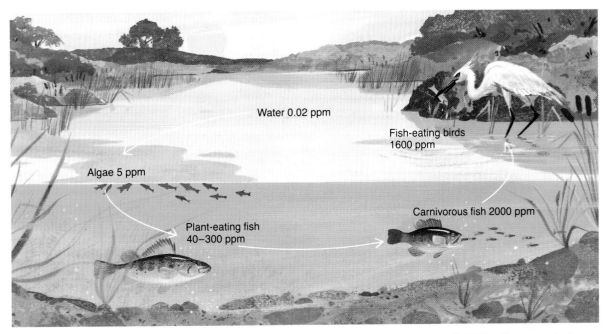

FIGURE 39·11 The concentration of this DDT-like chemical increases as it moves up the food chain. This is an example of biomagnification.

tems. Biological control of plant and insect pests is another way of preventing pollution. **Bio-controls** are techniques that use organisms to reduce pest populations. Using bacteria, fungi, and even insects to kill pests has many advantages over using chemical pesticides.

Waste heat put into the environment is called **thermal pollution.** Many industries produce thermal pollution, especially power plants. Whether the power plant uses nuclear energy or a fossil fuel, about two thirds of the heat produced cannot be used. This extra heat is released into the environment. The easiest way to give off the heat is to pump water from lakes or rivers over cooling coils. The warmed water is then returned to its source. Although the water is not contaminated by the fuel, the temperature of the river will increase by several degrees. Many aquatic communities have been destroyed or greatly changed because of thermal pollution.

As the use of electrical energy increases, people must become more aware of the problem of thermal pollution. Many laws regulate the use of water as a coolant, and lake and river temperatures are often checked. One way to reduce thermal pollution of water is to release the waste heat into the air by using cooling towers.

REVIEW QUESTIONS

6. What are some sources of air pollution?
7. Explain how smog is formed.
8. List some effects of acid rain.
9. Explain how biomagnification occurs. Give an example.
10. Why is it difficult to reduce pollution and conserve energy resources at the same time?

Energy Resources

SECTION OBJECTIVES

- Explain why alternatives to fossil fuels are needed.
- Discuss the advantages of alternative energy sources.

39•10 FOSSIL FUELS

Green plants use the energy of less than one percent of the available sunlight to produce energy-rich carbon compounds. This process has been occurring for millions of years, and some of the energy-rich compounds have accumulated in the earth. **Fossil fuels** are the oil, coal, and natural gas that formed from plants that lived long ago. Fossil fuel resources are limited in supply and cannot be recycled. Sooner or later the world will run out of fossil fuels.

Oil supplies 38 percent of the world's total energy. It has more energy per gram than any other fossil fuel. It is also an important source of chemicals, especially for the plastics industry.

Estimates of global oil reserves have changed very little recently. All of the easily found and easily obtained oil probably has been discovered. Many wells are producing less oil each year. Since it is unlikely that geologists will find many new large supplies of oil, other sources of energy will have to be used in the future.

One possible replacement for oil is natural gas. Since the 1970s, natural gas has been the fastest-growing energy source. It now supplies almost 20 percent of the world's energy. Most natural gas is found with oil. Until recently it was simply burned or allowed to escape into the air. People today know the value of natural gas as a fuel and a source of chemicals.

Natural gas, like oil, is not evenly distributed in the earth. Because of costs and safety problems, transporting natural gas is very difficult. Pipelines between suppliers and users are possible and several have been built. For example, the Trans-Canada Pipeline carries natural gas 3700 km from the Alberta-Saskatchewan border to Montreal.

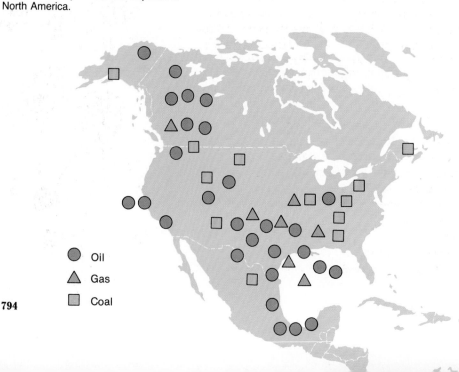

FIGURE 39•12 A map showing the locations of major fossil fuel deposits in North America.

World use of coal is also increasing each year. Many industrialized countries are planning to substitute coal for oil. The major appeal of coal is its abundance. Reserves are estimated to be at least 250 times the amount used in one year. Coal currently supplies about 25 percent of the world's energy and may be a larger source of energy in the future.

There are many problems with using large amounts of coal. Transporting coal will require major investments in ships and railroads. Power plants and coal-fired boilers are expensive to build and operate. Coal can be converted to a liquid or a gas, but this would raise energy costs two or three times.

The greatest costs of increased use of coal are health and environmental costs. More coal means more air pollution, more land destruction, and more damage to biotic communities. Burning coal increases carbon dioxide levels in the atmosphere. New methods to reduce some of these problems will increase the price of energy.

39·11 NUCLEAR ENERGY

Electric power has grown in importance. The cost of oil and gas has increased, and most suitable sites for hydroelectric plants have been developed already. As a result, coal and nuclear energy are being used to provide more electrical energy to meet the increasing demand. **Nuclear energy** is energy that is released when atoms undergo changes in their structure. Nuclear energy produces about nine percent of the world's electricity.

Most electricity generated from nuclear power plants is produced by the process of **nuclear fission,** which is the splitting of atoms to release energy. The fuel for this process is an isotope of uranium called U-235. The nucleus of an atom is held together by a tremendous amount of energy called nuclear binding energy. In some atoms, such as U-235, the binding energy can be overcome and the nucleus will split apart. The fission process occurs when a neutron enters the nucleus of the atom. As the atom splits into two smaller atoms, a small amount of mass is converted to energy. The power plant absorbs some of this energy as heat. The heat is used to produce electricity. The process of nuclear fission is shown in Figure 39·13. Note that neutrons are also given off during fission. What happens to some of these neutrons?

Breeder reactors are special nuclear plants. **Breeder reactors** produce fuel atoms during the fission process and actually produce more fuel than they use. Scientists around the world have been working on breeder technologies for many years, but the breeder reactor is not currently being used to generate electricity.

Nuclear fusion is another new technology being investigated as an energy resource. Instead of splitting large atoms, **nuclear fusion** combines small atoms to form larger atoms. Two isotopes of hydrogen are combined to form one helium atom. As in fission, small amounts of mass are converted to energy. The energy produced by the sun and

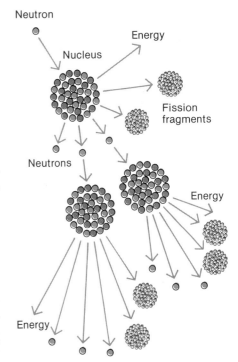

FIGURE 39·13 In nuclear fission, a neutron enters the nucleus of an atom, splitting it into two smaller atoms. As the atom splits, neutrons and energy are released. Nuclear power plants use this energy to make electricity.

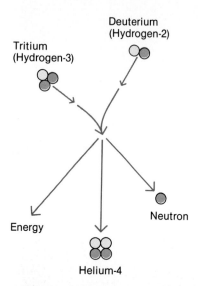

FIGURE 39·14 In nuclear fusion, two hydrogen atoms combine to form a helium atom. In addition, a neutron and energy are released. Nuclear fusion does not produce nuclear wastes.

other stars comes from the fusion process. One of the problems with fusion reactions is that they occur only at very high temperatures. Much more research needs to be completed before fusion technology can produce useful energy.

Nuclear power has had a short and troubled history. Originally it was thought to be safe and less expensive than power generated from fossil fuels. Many countries supported large nuclear research programs. As the first nuclear plants were completed, important questions about public safety and disposal of radioactive wastes came up. *Nuclear wastes* are radioactive materials left over after the fuel has been used. Some of these wastes remain dangerous for thousands of years. They must be stored in containers that do not leak.

39·12 SOLAR ENERGY

Energy from the sun is called **solar energy.** Some of the oldest and simplest uses of solar energy are still the most successful. Over 2000 years ago Socrates noted that houses could be warmed by the sun. Designing buildings to take advantage of energy from the sun without using specialized equipment is called passive solar design.

Solar energy can be absorbed by devices called solar collectors. Collectors are usually made of glass, copper, or aluminum. Some plastics can also be used. Solar energy can be used to heat water. As the hot water flows through pipes, the heat energy may be used to warm a building. The hot water itself may also be used for washing.

Another solar technology uses **photovoltaic cells,** or solar cells, which are devices that produce electricity directly from solar energy. When light energy hits a solar cell, electrons are released. This produces an electric current. Solar cells are expensive, but researchers think they will be less expensive and more common in the future. Some scientists estimate that by the year 2000, solar cells will produce more electricity than will nuclear energy.

39·13 OTHER ENERGY SOURCES

Most of the energy used in the world comes from fossil fuels, and a small amount comes from nuclear power and solar energy. Hydroelectric plants, which use the energy of falling water, are used to generate electricity. Hydroelectric plants produce only a small part of the world's energy because there are a limited number of good locations for building such plants.

geo- (earth)
thermos (heat)

Geothermal energy is energy from the earth's hot interior. As hot water and steam come to the surface from deep in the earth, the heat energy can be absorbed and used. Sources of geothermal energy are unevenly distributed over the earth, but there are some geothermal power plants in operation. The use of geothermal energy does create some environmental problems. Salts and sulfur compounds are brought to the surface in the hot water. In addition to corroding equipment, these compounds can produce both water pollution and air pollution.

Oceans contain large amounts of thermal energy. Several countries are building floating power plants to convert this thermal energy into electricity. **Tidal energy** is the kinetic energy in the movement of tides and waves. Although this energy is difficult to capture, several countries are experimenting with tidal energy.

Wind energy is the kinetic energy of moving air. Windmills have been used for many decades to produce small amounts of electrical energy. For example, many farms have small windmills to power water pumps. Windmills need open space where the wind is steady. California and Hawaii have large wind-powered electric systems.

The energy from photosynthesis long ago is used today in the form of fossil fuels. The energy in organic materials as a result of recent photosynthesis is called **biomass energy.** Plant, animal, and human wastes can be changed to methane gas (CH_4) by some kinds of bacteria. Methane from such biological processes is currently being used as an energy source in many countries. Fermentation by yeast converts plant material such as sugar cane and corn into alcohol. Alcohol can be used as a gasoline additive, or engines can be modified to operate entirely on alcohol fuels.

FIGURE 39·15 Some coastal countries are experimenting with power plants that use the energy in the movement of ocean tides and waves (left). For many years, windmills have been used to harness the energy of moving air. Modern windmills provide energy safely and efficiently in some areas (right).

REVIEW QUESTIONS

11. Where do fossil fuels come from?
12. What is the difference between nuclear fission and nuclear fusion?
13. What is the difference between a solar collector and a solar cell?
14. List some examples of biomass energy.
15. What factors should be considered when determining whether to equip a home with solar collectors?

Investigation

Which Material Stores Solar Energy the Best?

Goals

After completing this activity, you will be able to
- State which of four materials stores solar energy best.
- Suggest possible applications for the storage of solar energy.

Materials (for groups of 2)

5 thermometers
5 beakers (250-mL)
graduated cylinder
3 sheets of newspaper
large cardboard box with lid
black paint
stopwatch (or clock with second hand)
glass-marking pencil
salt sand water
paintbrush

	SAND	SALT	WATER	PAPER	AIR
ROOM TEMP.					
TEMP. AFTER 30 MIN IN SUN					
TEMP. 5 MIN AFTER REMOVAL FROM SUN					
TEMP. 10 MIN AFTER REMOVAL FROM SUN					
TEMP. 30 MIN AFTER REMOVAL FROM SUN					

Procedure

A. Paint the outside of the cardboard box black. Be sure to paint the lid. Let the paint dry for several hours.

B. Using the glass-marking pencil, label five beakers *Sand, Salt, Water, Paper,* and *Air.*

C. Make a mark on each beaker about seven eighths of the way up from the bottom. Using a graduated cylinder, pour sand up to the mark on the first beaker. Pour salt up to the mark on the second beaker. Pour water into the third beaker. In the fourth beaker, place crumpled-up sheets of newspaper. Stuff the paper into the beaker so that the beaker is seven-eighths full. Leave the fifth beaker empty.

D. On a separate sheet of paper, copy the chart above.

E. Place a thermometer in each beaker. Record the temperatures on the chart.

F. Place all beakers inside the black box. Cover the box. Place the box for 30 minutes in direct sunlight. Record the temperature in each beaker after 30 minutes in the sun.

G. Place the box in a shaded area. Remove the beakers from the box. Note the change in temperature in each beaker after 5 minutes, after 10 minutes, and after 30 minutes. Record these temperatures.

H. Follow your teacher's instructions for the proper disposal of all materials.

Questions and Conclusions

1. In which beaker did the temperature rise the most after being heated for 30 minutes? In which did the temperature rise the least?

2. Rank the materials tested from the one that increased the most in temperature to the material that increased the least in temperature.

3. In which beaker did the temperature drop most quickly? In which did the temperature drop most slowly? Which substance therefore has the best heat-storing capacity?

4. Explain why the cardboard box was painted black.

5. Why was one beaker with only air used in the activity?

6. Suggest a design for solar panels to be used in a new home. Explain why you would use certain materials in the construction.

Careers in Biology

Park Ranger

Park rangers work in national and state parks and forests. They enforce laws and regulations, collect fees, and issue permits for parking, camping, and hunting. Park rangers provide information to visitors about park use, safety requirements, and places of interest. They also direct conservation and animal management programs in the park. Sometimes, park rangers are involved in first aid and rescue activities.

To become a park ranger, a person must have a bachelor's degree in park management, forestry, biology, or a related field. On-the-job experience can sometimes count toward college credit. Positions as wildlife and forest technicians are available to people who have completed a technical school or two-year college. For additional information write to the National Park Service, United States Department of the Interior, Washington, DC 21240.

Ecologist

Ecologists study the relationships between living things and their surroundings. Some ecologists study the populations and communities that make up a given ecosystem. Some ecologists examine the effects of human activities in order to reduce or prevent damage to the environment.

The study of ecology takes researchers to many different habitats. It involves observation in the field, as well as laboratory work.

To become an ecologist, a person must have a master's degree or a Ph.D. Courses in biology, chemistry, soil science, physics, climatology, statistics, and computers are essential. Write to the Ecological Society of America, Department of Botany and Microbiology, Arizona State University, Tempe, AZ 85281.

Soil Scientist

Soil scientists study the characteristics of soils. They classify and map soils based on physical, chemical, and biological properties, and determine the best use of a particular soil. Some soil scientists specialize in soil conservation. They find ways to prevent erosion and reduce damage to soil that has been caused by incorrect use.

To become a soil scientist, a person must have a bachelor's degree in soil science, agronomy, agriculture, or a related field. Many soil scientists also study biology, chemistry, physics, and mathematics. For additional information write to the Soil Conservation Service, Information Division, United States Department of Agriculture, P.O. Box 2890, Washington, DC 20013.

How to Follow Up a Job Interview

After you have a job interview with a company, several weeks may pass before you hear the results. During this time you can do something to find out if you are still being considered for the job. You can call the person who interviewed you to find out if he or she needs any additional information on your background and skills. You should write that person a thank-you letter for taking the time for your interview.

If you do not get a job after an interview, you can still gain valuable information. Call the person who interviewed you and ask why you did not get the job. Most potential employers are willing to share this information if you explain that it will help you to learn from your experience. Any information or feedback you can gain from an unsuccessful interview will enable you to be better prepared for your next interview.

Chapter Review

SUMMARY

39·1 Natural resources are materials from the earth that are used to sustain life or that are critical to economic systems. Natural resources are either renewable or nonrenewable.

39·2 Soil is a renewable resource. Topsoil, the surface layer of soil, must be protected from the loss of too many minerals and from erosion.

39·3 The water cycle is the source of most fresh water. Water use cannot be larger than water supply.

39·4 Food is another resource. Food is produced in the ground and in the sea.

39·5 Forests are the source of firewood, lumber, and wood used in paper production.

39·6 Many types of wildlife have been harmed by humans changing their environment or introducing exotic species of organisms to their environment.

39·7 Air pollution can be caused by harmful gases and solid particles being discharged into the air by motor vehicles, power plants, industries, and homes. Air pollution affects human health and the ecosystem.

39·8 Acid rain results from the combination of sulfur dioxide, oxygen, and water vapor in the air. Acid rain corrodes metal structures, dissolves stone structures, and lowers soil and water pH.

39·9 Any substance that damages the quality of water is called water pollution. Toxic chemicals, large amounts of nutrients, and heavy metals are some examples of water pollutants.

39·10 Oil, coal, and natural gas are types of fossil fuels. Fossil fuels are formed from plants that lived long ago.

39·11 Nuclear fission and nuclear fusion are two sources of nuclear energy.

39·12 Solar collectors and photovoltaic cells are devices that use solar energy, or energy from the sun, to produce other forms of energy.

39·13 Geothermal energy, hydroelectric energy, tidal energy, wind energy, and biomass energy are other sources of energy.

BIOLOGICAL TERMS

acid rain
air pollution
bio-controls
biodegradable chemicals
biomagnification
biomass energy
breeder reactors
conservation
contour farming
crop rotation
erosion
fossil fuels
geothermal energy
mariculture
natural resources
nonrenewable resources
nuclear energy
nuclear fission
nuclear fusion
pesticides
photovoltaic cells
recycling
renewable resources
smog
solar energy
strip cropping
terracing
thermal pollution
tidal energy
water pollution
watersheds
windbreaks
wind energy

USING BIOLOGICAL TERMS

1. Distinguish between renewable resources and nonrenewable resources.
2. What is biomagnification?
3. Define the term *acid rain*.
4. Distinguish between nuclear fission and nuclear fusion.
5. What is mariculture?
6. How is recycling a form of conservation?
7. Which energy source is used with photovoltaic cells?
8. What are biodegradable chemicals?
9. How are air pollution and water pollution similar?
10. What is erosion?

UNDERSTANDING CONCEPTS

1. Give some examples of renewable resources and nonrenewable resources. (39·1)
2. Why is conservation important, even in connection with renewable resources? (39·1)
3. Why is soil a natural resource? (39·2)
4. Identify some methods for preventing soil erosion. (39·2)
5. How are watersheds important to ecosystems? (39·3)
6. In what ways do modern farming techniques, which produce more food, also use more resources? (39·4)
7. Why is the open ocean unlikely to be an important source of food? (39·4)
8. What is the major reason for the loss of forest resources? (39·5)
9. How do human activities affect wildlife resources? (39·6)
10. What are some ways of reducing air pollution? (39·7)
11. Explain how burning fuel could lead to a greenhouse effect on the earth. (39·7)
12. How is acid rain produced? (39·8)
13. What are some of the damaging effects of acid rain? (39·8)
14. Explain how biomagnification results in an increased concentration of toxic chemicals in the organisms in a food chain. (39·9)
15. Why is thermal pollution a kind of water pollution? (39·9)
16. Why are coal and natural gas likely to become more widely used in the future? (39·10)
17. How does an increased use of coal affect the environment? (39·10)
18. What problems are associated with the use of nuclear power? (39·11)
19. How are solar collectors used to heat a home? (39·12)
20. What are some of the limitations to the widespread use of geothermal energy, tidal energy, and wind energy? (39·13)
21. Explain how biomass energy is converted to useful energy sources. (39·13)

APPLYING CONCEPTS

1. Agricultural interests are trying to increase the production of fuel alcohol from grains. Discuss this in terms of world food supplies and the need for liquid fuels.
2. How can our life-styles be changed to reduce pollution?
3. Show different ways you can reduce pollution at school, at home, and in your community.
4. Why is the term *pest* an economic term and not a biological one?
5. Conduct an energy audit of your school. Prepare a report for your principal on ways the school can save energy.

EXTENDING CONCEPTS

1. Brainstorm ways to increase the wildlife habitat in your community. Contact your area forester for sources of inexpensive tree and shrub seedlings. Develop an environmental study area on your school campus.
2. Contact a local solar energy firm. Have the company send a representative to discuss solar applications with your class. Construct a model solar water- or space-heater.
3. Contact your electric utility to obtain information on the fuels used to generate electricity. Find out how they think the "fuel mix" will change by the year 2000.
4. Identify the businesses in your community that are involved with recycling. Find out what products they take and how much they pay for the products. Make a directory of recyclers and distribute it. Use recycled paper in your directory.

READINGS

Alvo, Robert. "Lost Loons of the Northern Lakes." *Natural History,* September 1986, p. 58.

Gever, et al. *Beyond Oil: The Threat to Food and Fuel in the Coming Decades.* Ballinger, 1986.

Penney, Terry and Desikan Bharathan. "Power from the Sea." *Scientific American,* January 1987, p. 86.

Moretti, Peter and Louis Divone. "Modern Windmills." *Scientific American,* June 1986, p. 110.

Tangley, L. "Groundwater Contamination." *BioScience,* March 1984, p. 142.

Unit VIII Skills

MAPPING SCIENTIFIC INFORMATION

Climate consists of long-term weather conditions. Climatologists, or scientists who study climates, keep accurate records of high, low, and average temperatures and average amounts of precipitation in an area. This information is organized and plotted on a graph called a climatogram. You saw one example of a climatogram in Section 38·2 of this book.

Major climate areas around the world can be classified into types. Climate types are distinguished by annual amounts of precipitation (rainfall, snowfall, fog, mist, dew, etc.) and average monthly temperatures. One useful system of climate classification was developed by a scientist named Köppen. The Köppen system describes the following five major climate zones of the world:

Climate Zone A—hot, rainy climates that have no month with an average temperature below 18°C

Climate Zone B—very dry climates in which total evaporation is greater than total precipitation

Climate Zone C—warm, rainy climates with moderate temperatures and enough rainfall to support tree growth

Climate Zone D—climates found only in the northern hemisphere where snow cover remains on the ground for more than 100 days per year

Climate Zone E—climates at high latitudes where the average temperature of warmest summer months is below 10°C.

These five major climate zones also can be subdivided into more-specific climate zones or regions. However, to clarify climate classification the map below shows five simplified climate zones. The map includes North America and Central America. Study the map. Answer the questions below on a separate sheet of paper.

1. What climate zones are found in the United States?
2. In which climate zone is your town or city?
3. Look for the Grand Canyon. In which climate zone is it found?
4. Which is the largest climate zone shown?
5. In what climate zone do you find Mexico and Central America?

The climate of an area affects the types of plants and animals that can survive there. Therefore, climate indirectly determines the major biomes of the earth. Turn in your book to Figure 38·3, which shows the average precipitation and temperature of the biomes. Study the table and answer the questions below.

6. In which climate zone would you be likely to find a tropical rain forest?
7. What climate zone includes the deserts?
8. In which climate zone is the tundra found?

To understand how biome maps and climate maps can be used together, do the following exercise. First, use a sheet of tracing paper to trace a rough outline of the map on this page. Then look at Figure 38·1, which is a map showing the biomes of the world. Find the area of North America that roughly corresponds to the map on this Skills page. Use crayons or pencils of seven different colors to shade in the biomes on your map. Now lay your completed map over the climate zone map. Answer these questions.

9. What climate zones are found in the taiga?
10. What climate zones are found in the temperate deciduous forest?

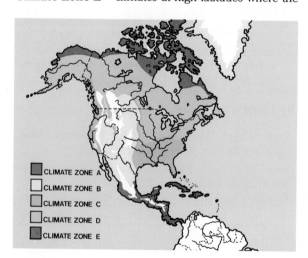

Safety in the Laboratory

APPENDIX A

Safety in the laboratory is your responsibility. Any laboratory activity has potential hazards. It is up to you to be alert to possible hazards and to prevent them from happening. By following a few basic rules, you can avoid accidents. Remember, an accident takes only a moment to happen.

GENERAL RULES

1. Be serious in the laboratory. Running, pushing, and joking are not appropriate in the laboratory. Such behavior can lead to accidents.
2. Know the location and use of all safety equipment—first aid kit, fire blanket, fire extinguisher, safety hood, and eye washes. Know the location of all exits.
3. Read all the directions for an activity BEFORE you begin work. Pay special attention to CAUTION statements in the procedure. A CAUTION statement alerts you to a potential hazard.
4. Report any accidents to your teacher immediately.
5. Do not eat or drink in the laboratory. Do not taste any substances used in the laboratory. Wash your hands before and after each activity.
6. Perform only those activities assigned by your teacher. Never work alone in the laboratory.
7. Wipe up chemical spills immediately. Report chemical spills to your teacher.
8. Keep the laboratory work area free of books, papers, and any unnecessary equipment.
9. Make sure that all hot plates, gas outlets, burners, and water faucets are turned off at the end of the period.
10. Follow your teacher's instructions for the proper cleanup and disposal of all materials.

SPECIFIC RULES

1. Wear safety goggles whenever an activity involves chemicals or other materials that could splatter and injure your eyes.
2. Wear a protective apron to cover your clothing when working with chemicals, hot materials, or preserved specimens.
3. Tie back long hair. Roll up long, flowing sleeves.
4. When using a scalpel for dissection, cut in a direction away from your body. Place specimens in a dissecting pan. Never cut a specimen while holding it in your hand.
5. Sweep up broken glass with a broom. Gather up the tiny glass fragments with wet paper towels. Dispose of the broken glass in proper containers.
6. Use only those chemicals specifically called for in the activity. To prevent contamination never pour unused chemicals back into their original containers. Dispose of chemicals according to your teacher's instructions.
7. Clearly label the contents of all containers into which you transfer materials.
8. When using acid always add acid to water. Never pour water into acid.
9. Do not use direct sunlight for microscope work. The sunlight will injure your eyes.
10. When heating materials point the test tube away from yourself and others.

SAFETY SYMBOLS

Within certain activities, safety symbols are included next to the heading PROCEDURE. These safety symbols alert you to potential hazards in the procedure and to safety measures to prevent accidents. Six safety symbols are used throughout the text. In any given activity you will see no more than three or four of these symbols. The symbols are as follows:

Body Protection

Eye Safety

Hand Safety

Electrical Safety

Poison

Fire

APPENDIX B

Biological Word Parts

WORD PART	MEANING	WORD PART	MEANING	WORD PART	MEANING
a-	without	halo-	salt	pestis	plague
ab-	away from	hemato-	blood	phage	to eat
ad-	near	hemi-	half	-phore	bearer
aeros	air	herba	plant	photo-	light
allelon	of each other	heres	heir	phyllon	leaf
alveus	cavity	hetero-	other	-phyte, phyto-	plant
amphibios	double life	histo-	tissue	pilus	hair
an-	without	homeo-	like	pino-	to drink
ana-	up	hybrida	mongrel	plankto	drifting
annulus	ring	hydro-	water	poly-	many
annus	year	hyper-	over	pseudes	false
arthron	joint	hypo-	under	primordius	original
atrium	entrance room	inter-	between	primos, proto-	first
autos	self	intra-	within	pro-	before
bacterio-	bacteria	iso-	equal	renes	kidney
bi-	two	karyo-	nucleus	reptilis	crawling
bio-	life	leuco-	white	rhiza	root
carnis	flesh	locus	place	rodere	to gnaw
chele	claw	-logy	study of	saccharum	sugar
chloros	green	lysis	to loosen	sapros	rotten
chorde	string	macro-	large	-scopy	observation
chroma	color	maris	sea	sonus	sound
-cide	killer	maxilla	jaw	soma	body
con-	with	mensis	month	sperma-	seed
cystis	pouch	mesos	middle	spirare	breathe
-cyte, cyto-	cell	meta-	between, change	stasis	position
-dermis	skin	micro-	small	stoma	mouth
di-	two	mono-	one	succedere	go after
ecto-	on the outside, outer	morphe	form	taxis	arrangement
end-	taking in	multi-	many	telo-	end
endo-	inner, inside	mykes	fungus	thallus	green shoot
epi-	on	nema	thread	-therm, thermos	heat
ergon	work	neur-	nerve	thrombos	clot
eu-	true	nomen	name	trans-	across
ex-	out	noton	back	tri-	three
exo-	outside of	-oid	like	tricho-	hair
feto-	fetus	omni-	all	trophos	feeder
filia	daughter	oo-, ovum	egg	ultra-	beyond
filius	son	oris	mouth	umbilicus	navel
ganglion	swelling	osteo-	bone	uni-	one
gastro-	stomach	paleo-	old	uterus	womb
-gen	producing	para	near	vasculum	vessel
geo-	earth	pathos	suffering	vestigium	footprint
germe	sprout	pausis	cessation	vorare	devour
glomus	ball	pedis, podus	foot	xero-	dry
gonos	producing	per-	through	-zoa, zoo-	animals
gymnos	naked	peri-	around	zygon	yoke

Five-Kingdom System for Classifying Organisms
Major Phyla

APPENDIX C

KINGDOM MONERA
(All organisms are prokaryotic.)

PHYLUM CYANOPHYTA (blue-green bacteria; includes organisms formerly called blue-green algae)
Prokaryotic; unicellular or colonial; photosynthetic; all cells contain blue pigment and green pigment; pigments not contained in chloroplasts; nucleus lacks nuclear membrane. Example: *Oscillatoria*.

PHYLUM SCHIZOMYCETES (bacteria and bacterialike organisms; includes eubacteria, mycoplasmas, rickettsias, actinomycetes, spirochetes, and others)
Prokaryotic; some flagellated, others nonflagellated; some aerobic, others anaerobic; heterotrophic, autotrophic, and chemosynthetic species; can form endospores to survive adverse conditions; reproduce by binary fission; some lack cell walls or are obligate intracellular parasites. Example: *Escherichia coli*.

KINDGOM PROTISTA
(All organisms are eukaryotic.)

Subkingdom Protophyta (algal protists)
Unicellular; plantlike organisms; mostly autotrophic; most contain chlorophyll.

PHYLUM EUGLENOPHYTA (euglenoids)
Found mostly in fresh water; move with flagella; no cell wall. Example: *Euglena*.

PHYLUM CHRYSOPHYTA (golden algae)
Found mostly in salt water; have yellow-brown pigment; silica in cell wall. Example: *Diatom*.

PHYLUM PYRROPHYTA (fire algae)
Found mostly in salt water; usually have two flagella; have cellulose in cell wall. Example: *Gonyaulax*.

Subkingdom Gymnomycota (fungal protists, or slime molds)
Show traits of both fungi and animallike protists; heterotrophic; multicellular in some stages.

PHYLUM MYXOMYCOTA (true slime molds)
Example: *Physarum*.

Subkingdom Protozoa (animallike protists)
Unicellular; heterotrophic; divided into four phyla based on means of locomotion.

PHYLUM SARCODINA (sarcodines)
Move with pseudopodia (cell extensions). Example: *Amoeba*.

PHYLUM CILIATA (ciliates)
Move with cilia. Example: *Paramecium*.

PHYLUM MASTIGOPHORA (flagellates)
Move with one or more flagella. Example: *Trypanosoma*.

PHYLUM SPOROZOA (sporozoans)
No locomotion; all parasitic. Example: *Plasmodium*.

KINGDOM FUNGI
(All organisms are eukaryotic.)

Eukaryotic; with cell walls; range in size from microscopic one-celled organisms to macroscopic multicellular organisms 25 cm in diameter, mostly saprobic; some parasitic; some symbiotic; asexual reproduction by fragmentation; sexual reproduction by spore formation; classification based on method of spore formation and kind of fruiting body.

PHYLUM OOMYCOTA (water molds, white rusts, downy mildews; also called oomycetes)
Organisms form asexual spores called zoospores; cell walls of a polysaccharide similar to cellulose. Example: *Saprolegnia* (freshwater mold).

PHYLUM ZYGOMYCOTA (conjugation fungi; also called zygomycetes)
Reproduce sexually via diploid spores called zygospores; cell walls of chitin. Example: *Rhizopus stolonifer* (bread mold).

PHYLUM ASCOMYCOTA (sac fungi; also called ascomycotes)
Spores develop in sac called ascus; cell walls of chitin. Example: *Saccharomyces cerevisiae* (baker's yeast).

PHYLUM BASIDIOMYCOTA (club fungi; also called basidiomycotes)
Spores develop in club-shaped structures called basidia; cell walls of chitin. Example: *Puccinia graminis* (wheat rust).

KINGDOM PLANTAE
(All organisms are eukaryotic.)

Eukaryotic; most species are multicellular and nonmotile; photosynthetic; most contain chlorophyll in chloroplasts.
Note: In some systems of classification, the term *Division* is used in place of *Phylum* in Kingdom Plantae.

PHYLUM CHLOROPHYTA (green algae)
Mostly freshwater species, some marine species; some live in soil or on trees; contain pigments similar to those in vascular plants. Examples: *Chlamydomonas, Ulothrix*.

PHYLUM PHAEOPHYTA (brown algae)
Mostly marine species; contain brown pigment fucoxanthin; multicellular; usually sessile. Example: *Laminaria* (kelps).

PHYLUM RHODOPHYTA (red algae)
Mostly marine species; contain red pigment phycoerythrin; mostly multicellular; generally filamentous or membranous. Example: *Porphyra.*

PHYLUM BRYOPHYTA (liverworts, hornworts, mosses; also called bryophytes)
Nonvascular plants that live on land; gametophyte is the main plant; gametophyte has no true roots, stems, or leaves. Example: *Marchantia* (liverwort).

PHYLUM TRACHEOPHYTA (vascular plants)
Contain vascular tissue of xylem and phloem; true roots specialized for water absorption; sporophyte the dominant generation; in some groups, gametophyte independent of sporophyte.

Subphylum Lycopsida (club mosses)
Primitive vascular plants; sporophyte the main plant. Example: *Lycopodium* (club moss).

Subphylum Sphenopsida (horsetails)
Primitive vascular plants; only one living genus; stem the major part of plant. Example: *Equisetum* (horsetails).

Subphylum Pteropsida (ferns)
Vascular plants more complex than club mosses or horsetails; grow abundantly in tropical rain forests; grow in moist woodlands; sporophyte is the main plant; well-adapted to life on land. Example: *Osmunda* (cinnamon fern).

Subphylum Spermopsida (seed plants)
Vascular plants that reproduce by forming seeds; found in most types of land environments and in many water environments.

- **Class Gymnospermae** (nonflowering seed plants)
  Seeds do not develop inside ovaries; mostly trees and shrubs.

  Order Ginkgoales
  Deciduous trees with fan-shaped leaves; only one living species. Example: *Ginkgo biloba* (ginkgo).

  Order Cycadales
  Small group of tropical plants that look like large ferns. Example: *Zamia.*

  Order Coniferales
  Cone-bearing trees or shrubs. Example: *Pinus strobus* (white pine).

  Order Gnetales
  Small cone-bearing plants. Example: *Gnetum* (tropical trees and vines with broad, tough leaves).

- **Class Angiospermae** (flowering seed plants)
  Seeds develop inside ovaries; comprise largest plant group on the earth; include trees, shrubs, vines, herbs, and floating plants; found in a wide range of habitats.

  - ***Subclass Monocotyledonae*** (monocots)
    Embryo with one cotyledon; leaves with parallel venation; flower parts in threes and multiples of threes; vascular bundles scattered throughout stem. Example: *Lilium canadense* (wild yellow lily).

  - ***Subclass Dicotyledonae*** (dicots)
    Embryo with two cotyledons; leaves with net venation; flower parts in fours and fives; vascular bundles form a circle. Example: *Rosa odorata* (garden rose).

KINGDOM ANIMALIA
(All organisms are eukaryotic.)

Eukaryotic, multicellular, and heterotrophic; most with tissue, organs, and organ systems of development; larval or embryonic development.

PHYLUM PORIFERA (sponges)
Sessile body consisting of system of canals and pores; lack tissues and organs; filter feeders. Example: *Grantia.*

PHYLUM CNIDARIA (cnidarians)
True tissues; two tissue layers; tentacles; threadlike stinging structures called nematocysts; food digested in gastrovascular cavity; radially symmetrical; some have alternating body forms (polyp and medusa); divided into three classes based on life cycle and body form—class Hydrozoa (hydras), class Scyphozoa (jellyfish), and class Anthozoa (sea anemones, corals). Examples: *Hydra littoralis* (hydra).

PHYLUM PLATYHELMINTHES (flatworms)
Three tissue layers—ectoderm, endoderm, and mesoderm; acoelomate; bilaterally symmetrical; distinct head end; parasitic and free-living forms. Phylum divided into three classes—class Turbellaria (planarians), class Trematoda (flukes), and class Cestoda (tapeworms). Example: *Opisthorchis sinensis* (liver fluke).

PHYLUM ASCHELMINTHES (rotifers and roundworms)
Tissues derived from ectoderm, endoderm, and mesoderm; simplest organisms to have complete digestive system from mouth to anus; pseudocoelomates; divided into two classes —class Rotifera (rotifers, or microscopic aquatic organisms) and class Nematoda (roundworms, or free-living or parasitic worms.) Example: *Ascaris lumbricoides* (ascaris).

PHYLUM ANNELIDA (segmented worms)
Body made up of segments; three body layers derived from ectoderm, endoderm, and mesoderm; coelomate; tube-within-a-tube body plan; divided into three classes—class Oligochaeta (earthworms, freshwater worms), class Polychaeta (sandworms, tubeworms), and class Hirudinea (leeches). Example: *Lumbricus terrestris* (earthworm).

PHYLUM MOLLUSCA (mollusks)
Soft-bodied invertebrates; muscular foot; mantle that often secretes a hard, calcified shell; body of most adult mollusks bilaterally symmetrical; unsegmented body. Phylum divided

into four classes—class Pelecypoda (clams, oysters, mussels), class Gastropoda (snails, slugs), class Cephalopoda (squids, octopuses, cuttlefish), and class Amphineura (chitons). Example: *Loligo pealei* (squid).

PHYLUM ECHINODERMATA (echinoderms)
Spiny-skinned marine invertebrates; bony plates form endoskeleton; radially symmetrical in adult form and bilaterally symmetrical in larval form; water-vascular system used for locomotion divided into five classes—class Crinoidea (crinoids), class Asteroidea (sea stars), class Ophiuroidea (brittle stars), class Echinoidea (sea urchins, sand dollars), class Holothuroidea (sea cucumbers). Example: *Asterias forbesi* (sea star).

PHYLUM ARTHROPODA (arthropods)
Segmented body; many species divided into head, thorax, and abdomen; exoskeleton of chitin covering body; periodic molting; paired, jointed appendages. Phylum divided into three subphyla based on specializations of segments—subphylum Trilobita (trilobites; no living species), subphylum Chelicerata (horseshoe crabs, spiders, ticks, mites), and Subphylum Mandibulata. Subphylum Mandibulata further subdivided into four classes—class Crustacea (crustaceans), class Chilopoda (centipedes), class Diplopoda (millipedes), and class Insecta (insects). Class Insecta further divided into 16 orders. Example: *Drosophila melanogaster* (fruit fly).

PHYLUM HEMICHORDATA (hemichordates)
Coelomate; bilaterally symmetrical; unsegmented body; wormlike organisms that live in V-shaped burrows; develop from eggs that may become larvae; gill slits in pharynx; nerve cord; have some traits of echinoderms and some traits of chordates. Example: *Saccoglossus* (tongue worm).

PHYLUM CHORDATA (chordates)
Notochord; dorsal nerve cord; gill slits at some time in life cycle.

Subphylum Urochordata (tunicates)
Primitive chordates; outer tough covering or tunic; show all chordate features during larval stage; some adults are free swimming. Example: *Halocynthia* (sea peach).

Subphylum Cephalochordata (lancelets)
Long, slim, flattened fishlike body; dorsal nerve cord; notochord; gill slits. Example: *Amphioxus* (lancelet).

Subphylum Vertebrate (vertebrates)
Vertebral column or backbone to enclose, support, and protect spinal cord or nerve cord; notochord generally absent in adults; endoskeleton for protection of internal organs and attachment of muscles.

- **Class Agnatha** (jawless fish)
  Most primitive class of fish; circular mouths; slim, wormlike bodies covered with slime; lack scales and paired fins; lack teeth but have toothlike structures; two-chambered heart. Example: *Petromyzon marinus* (sea lamprey).

- **Class Placodermi**
  Extinct jawed fish.
- **Class Chondrichthyes** (cartilaginous fish)
  Skeletons of cartilage, not bone; paired fins, scales, and jaws; scales with toothlike spines; slitlike gill openings on each side of body; some oviparous organisms, some ovoviviparous organisms. Examples: *Squalus* (sharks), *Raja* (skate), *Dasyatis* (ray).
- **Class Osteichthyes** (bony fish)
  Skeleton mainly on bone; breathe with gills; have scales and fins, some modified; divided into two main groups—ray-finned and lobe-finned fish. Example: *Perca* (perch).
- **Class Amphibia** (amphibians)
  First vertebrates to become adapted to life on land; spend part of life in water; have gills as larvae and lungs as adults; moist, smooth, scaleless skin. Example: *Rana pipiens* (leopard frog).
- **Class Reptilia** (reptiles)
  Ectothermic vertebrates; dry, scaly skin; most are land-dwelling, some live in water; adapted for reproduction on land; internal fertilization; amniote egg; divided into four surviving orders—order Rhynchocephalia (tuatara), order Chelonia (turtles), order Crocodilia (crocodiles, alligators), and order Squamata (lizards, snakes). Example: *Chelydra* (snapping turtle).
- **Class Aves** (birds)
  Endothermic feathered vertebrates; hollow, light-weight bones; spongy lungs with system of air sacs; scales on legs; oviparous (embryo develops in shelled egg outside of adult's body). Example: *Columba livia* (domestic pigeon).
- **Class Mammalia** (mammals)
  Vertebrate; endothermic; hair present, nourish young with milk from mammary glands; four-chambered heart; sweat glands; diaphragm in chest cavity; in most, embryo develops in uterus in female's body; in most, embryo receives nourishment from mother through placenta; some carnivorous, others herbivorous.

    Order Monotremata (monotremes)
    Most primitive mammals; egg-laying; lack nipples; young suck milk from lobules. Example: duck-billed platypus, spiny anteater.

    Order Marsupialia (marsupials)
    Most primitive mammals to give birth to live young; young not fully developed at birth; complete development in pouch called marsupium. Examples: kangaroo, opossum, koala.

    Order Insectivora (insectivores)
    Most primitive placental mammals; adapted to eat small invertebrates, especially insects; often live underground. Examples: mole, shrew.

Order Chiroptera (bats)
Flying mammals; forelimbs adapted for flight; nocturnal; sense objects in environment by echolocation. Example: brown bat.

Order Edentata (edentates)
Mammals with very reduced teeth or no teeth at all. Examples: anteaters, tree sloths, armadillos.

Order Lagomorpha (lagomorphs)
Small herbivorous mammals; adapted for running and jumping; chisel-shaped front teeth. Examples: rabbits, pikas.

Order Rodentia (rodents)
Largest group of mammals; mainly herbivorous; sharp chisel-shaped front teeth. Examples: rats, mice, squirrels.

Order Cetacea (cetaceans)
Adapted to live in the ocean; forelimbs often specialized as flippers; streamlined body form; able to undergo long periods of submersion; echolocation for underwater guidance. Examples: whales, porpoises, dolphins.

Order Carnivora (carnivores)
Mostly meat-eating mammals, some not strictly carnivorous; large sharp teeth; well-developed sense of smell; some land-dwelling, others aquatic. Examples: dogs, cats, seals, bears.

Order Proboscidea (elephants)
Large herbivorous mammals with long agile trunks; front teeth missing except for incisors. Example: elephants.

Order Perissodactyla (odd-toed ungulates)
Hoofed, herbivorous mammals; digestive system adapted for digestion of plant material. Examples: horses, zebras, rhinoceroses.

Order Artiodactyla (even-toed ungulates)
Hoofed, herbivorous mammals; digestive system adapted for digestion of plant material; more advanced than odd-toed ungulates. Examples: sheep, cows, camels, pigs.

Order Sirenia (sirenians)
Aquatic mammals; forelimbs modified as flippers; hind limbs absent; sparse hair. Examples: sea cows, manatees.

Order Primates (primates)
Posture erect or semi-erect; well-developed brain and eyes; five digits on hands and feet; well-developed social hierarchy; reduced sense of smell. Examples: monkeys, apes, lemurs, humans.

GLOSSARY

KEY TO PRONUNCIATION

Many terms and names in this book have been respelled as an aid to pronunciation. This key shows symbols used in the respellings and provides examples of words that have sounds represented by the symbols. Syllables in SMALL CAPITAL LETTERS receive primary stress.

Symbol	Sound	Symbol	Sound	Symbol	Sound
a	hat	oi	oil, boy	ks	fox, exam
ah	hot, father	oo	cool, rule, new	kw	queen
ai	air, care	or	orbit, door	n	bon (French)
aw	awful, all, off, auto	ow	cow, out	ng	bring
ay	pay, late	u	put, look	s	sit, city
ee	see, eat, funny	uh	ago, taken, pencil, lemon, circus	sh	ship, action
eh	let	ch	church	th	thin, breath
er	learn, sir, fur, her	g	game, pig	th	then, breathe
ih	fit, trip	j	just, gentle, ridge	y	young, onion
ī	cycle, life	k	key, came	z	zing, freeze, has
oh	no, flow	kh	ach (German)	zh	measure, garage, vision

abdomen In an arthropod, the posterior segments in the body.

abiogenesis (ay bī oh JEHN uh sihs) A hypothesis that states that living things can originate from nonliving matter.

abiotic factors Nonliving factors in the environment such as temperature, water, light, and minerals.

ABO blood types The four major blood types, A, B, AB, and O.

aboral surface The surface opposite the mouth.

abscisic acid A plant hormone that acts mainly to inhibit growth.

abscission (ab SIHZH uhn) **layer** A specialized layer of cells that forms where the petiole is attached to the stem.

absorptive heterotroph A heterotroph that digests its food externally before it absorbs it into its cells. Example: fungi.

abyssal (uh BIHS uhl) **zone** The deep ocean, reaching from 2000 m to the sea floor.

acid A chemical that produces hydrogen ions in a water solution.

acid rain A combination of acids formed in the atmosphere and rainwater that falls to the earth.

acoelomates (ay SEE luh mayts) Animals without a coelom.

actin The protein that makes up the thin filaments in myofibrils.

action potential The nerve impulse.

activation energy The energy that is needed to start a chemical reaction.

active immunity Immunity that develops as a result of exposure to pathogens or their toxins.

active site A special region of an enzyme that can join with a specific substrate or substrates.

active transport The movement of materials across the cell membrane that requires the use of energy by the cell.

adaptation An inherited trait that makes an organism better able to survive in a certain environment.

adaptive radiation The rapid evolution of many new species from a small number of species.

adenine (AD uh neen) One of the bases that occurs in DNA.

adenosine diphosphate (uh DEHN uh seen dī FAHS fayt) The molecule that remains when ATP loses a phosphate group.

adenosine triphosphate (uh DEHN uh seen trī FAHS-fayt) The compound that stores energy in cells.

adhesion The attraction of unlike molecules for each other.

adrenal (uh DREE nuhl) **glands** Endocrine glands found above the kidneys.

adrenaline (uh DREHN uh lihn) A hormone that stimulates increases in heart rate, breathing rate, and in the amount of blood flowing to the brain and skeletal muscles.

809

adrenocorticotropic (uh dree noh kor tuh koh TROHP ihk) **hormone (ACTH)** A pituitary hormone that controls some activities of the adrenal glands.

adsorption In viruses, the process by which a phage sticks to the surface of a bacterial cell.

aerobic (air OH bihk) **respiration** Cellular respiration that needs oxygen.

air pollution The various harmful gases and solid particles that are discharged into the air by motor vehicles, power plants, homes, and industries.

albumen (al BYOO muhn) A watery solution of protein secreted in the upper part of the oviduct, the white of an egg.

alcoholic fermentation Fermentation that forms alcohol.

aldosterone (al DAHS tuh rohn) A hormone that acts on the tubules of the kidney to stimulate absorption of sodium into the blood.

algae (AL jee) The simplest types of nonvascular plants.

algal protists The plantlike protists, grouped in subkingdom Protophyta.

allantois (uh LAN toh ihs) A membrane that forms from the lower end of the embryo's digestive tract.

alleles (uh LEELS) The different forms of the gene for a particular trait.

allergy (AL uhr jee) An extreme reaction by the immune system to the presence of antigens.

alternation of generations In many plants, the alternation between diploid and monoploid stages during one complete life cycle.

altricial (al TRIHSH uhl) Birds that are very dependent on their parents at hatching because they are sightless, they lack feathers and they are not able to feed on their own.

alveoli (al VEE uh lī) (sing., *alveolus*) Groups of tiny air sacs in the lungs.

amino (uh MEE noh) **acids** The building blocks of proteins.

ammonification A process in which amino acids are changed to ammonia.

amniocentesis (am nee oh sehn TEE sihs) A procedure used to detect genetic disorders in fetal cells.

amnion (AM nee uhn) A membrane that grows around the embryo and encloses it in a liquid called amniotic fluid.

amniote (AM nee oht) **egg** An egg that has a fluid-filled sac, in which the embryo develops, and a protective shell.

amoeba A freshwater sarcodine that moves by means of pseudopodia.

amoebocytes (uh MEE buh sīts) The amoebalike cells of a sponge.

amoeboid movement The kind of movement shown by an amoeba, in which it moves by changing the shape of its cell.

amphibians Vertebrates that have gills as larvae and, usually, lungs as adults.

anabolic (an uh BAHL ihk) **reactions** Reactions in which simpler substances join to form more complex substances.

anaerobic respiration Cellular respiration that does not need oxygen.

anal pore In a paramecium, a fixed structure that lets out wastes.

analogous (uh NAL uh guhs) **structures** Structures in different organisms that are similar in function but different in structure.

anaphase The third stage of mitosis, during which the paired chromatids separate.

angiosperms (AN jee uh spermz) Plants whose seeds develop within structures called ovaries.

Animalia (an uh MAYL yuh) The kingdom of animals.

annelids (AN uh lihds) Segmented worms.

annuals (AN yu uhlz) Plants that live and reproduce for only one season.

annulus In mushrooms, a sheetlike tissue that draws the edge of a mushroom cap toward the stipe.

antagonistic muscles Muscles that work in pairs in which one muscle's action opposes that of the other.

antennae Sensory appendages in arthropods.

anther (AN thuhr) The structure in which the male gametophytes are produced in plants.

antheridium (an thuh RIHD ee uhm) The male reproductive structure of certain organisms including mosses and liverworts.

antibiotics (an tee bī AHT ihks) Substances that are formed by microbes or fungi and that inhibit the growth of bacteria.

antibodies (AN tee bahd eez) Chemicals which protect the body by helping to destroy pathogens and neutralize their toxins.

anticodon A set of three bases on one loop of a tRNA molecule that is complementary to an mRNA codon.

antidiuretic (an tee dī yu REHT ihk) **hormone** The hormone that regulates the reabsorption of water.

antigen (AN tuh juhn) Any substance that causes the body to make antibodies.

anus The opening through which undigested food leaves an animal's digestive tract.

aorta The largest artery. In humans, blood leaves the left ventricle through the aorta.

aortic arches Muscular tubes which help to pump blood in an earthworm.

apical dominance (AP uh kuhl DAHM uh nuhns) The inhibiting effect of the terminal bud on the lateral buds.

apical meristem (AP uh kuhl MEHR uh stehm) The small dome-shaped area at the tip of the meristematic region.

appendicular skeleton A division of the skeleton consisting of the bones of the arms and shoulders and the bones of the legs and pelvis.

appendix In humans, A small functionless tube connected to the cecum.

aquatic biomes Biomes that consist of freshwater or saltwater habitats.

archegonia (ahr kuh GOH nee uh) Structures that produce the eggs in certain organisms, including mosses and liverworts.

arterial heart In a squid, the heart that pumps blood throughout the body.

arteries Large blood vessels that carry blood away from the heart and to other organs.

arthropods Animals having a segmented body, an exoskeleton, and paired, jointed appendages.

ascocarp (AS kuh kahrp) A fruiting body which can develop into fungal hyphae.

ascogonia (as kuh GOH nee uh) The female reproductive structures of ascomycotes.

ascomycotes (as kuh mī KOHTS) Fungi that are members of phylum Ascomycota.

ascospores (AS kuh sporz) Monoploid nonmoving spores of ascomycotes.

ascus (AS kuhs) In ascomycotes, a sac in which fungal spores develop.

asexual reproduction Reproduction in which a single parent gives rise to one or more individuals.

association neurons Neurons that serve as connections between other neurons.

atherosclerosis (ath uhr oh skluh ROH sihs) The buildup of fat deposits on the inside of blood vessels.

atom The smallest particle of matter that can exist and still have the properties of a particular kind of matter.

atomic number The number of protons in the nucleus of an element.

atrioventricular node A bundle of electrically active tissue in the heart.

atrioventricular (ay tree oh vehn TRIHK yuh luhr) **valves** The valves between the atria and ventricles.

atrium An upper chamber of the heart.

Australopithecines (aws truh loh PIHTH uh seenz) Human-like organisms in the genus *Australopithecus*.

Australopithecus afarensis (af ahr EHN sihs) A species of australopithecines that lived nearly 4 million years ago.

Australopithecus africanus (af rih KAN uhs) A species of australopithecines that lived from more than 3 million to 1.5 million years ago.

Australopithecus bosei (BOI zee ī) A species of australopithecines that lived about 2 million years ago.

Australopithecus robustus (roh BUHS tuhs) A species of australopithecines that lived from about 2.5 million to a little over 1 million years ago.

autonomic nervous system The portion of the peripheral nervous system that provides involuntary control of the internal organs.

autosomes (AW tuh sohms) Chromosomes other than the sex chromosomes.

autotrophs (AW tuh trohfs) Living things that make their food.

auxin (AWK sihn) A plant hormone that regulates growth.

axial skeleton A division of the skeleton consisting of the bones of the skull, the vertebral column, and the thorax.

axon The extension of the neuron that carries nerve impulses away from the cell body.

axon-endings Knoblike structures at the end of an axon.

bacilli (buh SIHL ī) Bacteria shaped like rods.

bacteria Monerans in phylum Schizomycetes.

bacteriophages (bak TIHR ee uh fayj uhz) Viruses that infect bacteria.

barbs Thin extensions of the shaft of a feather.

barbules Tiny hook-bearing branches of the barbs of a feather.

bark The outermost layer of protective tissue produced by a plant's stem.

basal metabolic rate (BMR) A measure of the rate of oxygen consumption, and therefore energy production, in a person at rest.

base A chemical that produces hydroxide ions in a water solution.

basidia (buh SIHD ee uh) Structures in which basidiomycotes produce spores.

basidiocarp (buh SIHD ee uh kahrp) The structure that bears the basidia in most basidiomycotes.

basidiomycotes (buh sihd ee uh mī KOHTS) Fungi that are members of phylum Basidiomycota.

basidiospores (buh SIHD ee uh sporz) The spores of basidiomycotes.

bathyal (BATH ee uhl) **zone** The ocean region from about 200 to 2000 m on the continental shelf.

B-cells Lymphocytes that make antibodies and give off the antibodies into the bloodstream.

behavior All of an organism's actions.

behavioral (bih HAYV yuhr uhl) **adaptations** Inherited behaviors that help an organism to survive in a given environment.

benthos Plants and animals that live in the bottom sediments of a body of water.

bias An existing idea that influences an observation.

bicuspid valve The valve connecting the left atrium and the left ventricle.

bilateral symmetry The arrangement of an organism's parts so that the parts can be separated into mirror-image halves along only one lengthwise plane.

bile A solution of water, bile salts, and bile pigments which is produced by the liver.

binary fission Asexual reproduction involving division into two cells.

binomial nomenclature (bī NOH mee uhl NOH muhn klay-chuhr) A classification system by which each species is given a two-word Latin name.
bio-controls Techniques that use organisms to reduce pest populations.
biodegradable chemicals Substances that break down into substances that do not harm and can enrich ecosystems.
biogenesis A hypothesis that living things come from other living things.
biology The study of life.
biomagnification The increasing concentration of pollutants in a food web.
biomass energy The energy in organic materials as a result of recent photosynthesis.
biome (BĪ ohm) A large region that has a distinct combination of plants and animals.
biosphere The region on the earth in which living things can be found.
biotic factors Living things in the environment, such as plants, fungi, and animals.
biotic potential The maximum population size that a species could reach in a certain time, if all offspring survived and reproduced.
birds Warm-blooded vertebrates that are covered with feathers and have wings.
birth rate The ratio between the number of births and number of individuals in a population within a specified time.
bivalves Mollusks that live within a shell made up of two sections, or valves.
blade In plants, the broad flat part of the leaf.
blastocoel (BLAS tuh seel) An inner fluid-filled cavity in a blastula.
blastocyst In humans, a hollow ball of cells that results from rapid mitotic divisions of a zygote.
blastula A hollow ball of cells formed through mitotic divisions during the development of an embryo.
blood clot A tangled mass of protein fibers with trapped platelets and red blood cells.
blood pressure The pressure which is exerted against the walls of the arteries.
blue-green bacteria Members of phylum Cyanophyta in kingdom Monera.
bone A type of solid connective tissue in which the cells are embedded in a matrix of protein fibers and minerals.
book lungs A series of air-filled plates with a large surface area in which gas exchange occurs.
botanist A person who studies plants.
Bowman's capsule A double-walled structure surrounding the glomerulus.
brain The major controlling structure of the nervous system.
brainstem The part of the brain which regulates many internal functions.
breeder reactors Special nuclear plants which produce fuel atoms during the fission process and actually produce more fuel than they use.
bronchi (BRAHNG kī) The first two branches of the trachea.
bronchioles (BRAHNG kee ohlz) The smaller branches of the bronchi.
brown algae Members of phylum Phaeophyta.
bryophytes (BRĪ uh fitz) Members of phylum Bryophyta.
budding The formation of an outgrowth on the parent's body which develops into a new individual. A method of vegetative reproduction that involves grafting buds.
buds Plant structures that produce stems, leaves, and flowers. Outgrowths on an organism's body that develop into new individuals.
bursae Sacs of fluid that cushion a joint against shock.

calcitonin (kal suh TOH nuhn) A hormone of the thyroid gland which helps control the amount of calcium and phosphorus in the blood.
calyx (KAY lihks) The ring of sepals at the base of a flower.
canines (KAY nīnz) Teeth used for tearing food.
canopy The top layer of vegetation in vertical stratification.
capillarity (kap uh LAR uh tee) The tendency of a liquid to rise inside a tube of very small diameter.
capillary The smallest type of blood vessel.
capsid The shell that encloses the nucleic acid in a virus.
capsomere (KAP suh mihr) A single protein subunit of a capsid.
capsule An outer layer found on some bacteria. A structure in some plants in which monoploid spores are formed as a result of meiosis.
carapace (KAR uh pays) The dorsal shield, or shell, of a turtle. In some arthropods, a shield of chitin which covers the dorsal surface of the cephalothorax.
carbohydrates (kahr boh HĪ drayts) Compounds, composed of carbon, hydrogen, and oxygen, that are the body's main sources of energy.
carbon cycle The movement of carbon between organic compounds that make up living tissues and carbon dioxide in the air.
cardiac muscle Muscle that makes up the heart.
carnivores Animals that eat other animals.
carrier molecules Molecules in the cell membrane that permit specific molecules on one side of the membrane to pass through to the other side.
carrying capacity The population size of a particular species that a particular environment can support.
cartilage Solid connective tissue in which the cells are embedded in a gellike matrix containing protein fibers.

catabolic (kat uh BAHL ihk) **reactions**   Reactions in which complex substances are broken down into simpler substances.

catalyst (KAT uh lihst)   A substance that controls the rate of a chemical reaction without itself being changed or used up.

cecum   The region of the large intestine below the place where the small and large intestine meet.

cell   The basic unit of structure and function of an organism.

cell membrane   The cell structure that regulates the passage of materials between the cell and its environment.

cell theory   A theory that states: All organisms are made up of one or more cells that reproduce to form new cells.

cell wall   The cell structure that gives shape and rigidity to a plant cell.

cellular respiration   The breakdown of glucose with the use of oxygen or some other inorganic substance to produce energy for cells.

cellular transport   The process by which needed materials enter the cell while waste materials and cell secretions leave.

cementum   A bonelike substance that forms the cover of the root of the tooth.

central nervous system   The brain and spinal cord.

centrioles (SEHN tree ohlz)   Structures outside the nucleus in animal cells that are involved in mitosis.

centromere (SEHN truh mihr)   A structure that holds together each pair of chromatids.

cephalization   The development of a head with sensory organs.

cephalopods (SEHF uh luh pahdz)   Marine mollusks having a ring of tentacles around the mouth.

cephalothorax (sehf uh loh THOR aks)   In arthropods, the anterior body division made up of a fused head and thorax.

cerebellum (sehr uh BEHL uhm)   The part of the brain which controls coordination of muscular activities.

cerebrum (SEHR uh bruhm)   The largest part of the brain, which controls conscious sensation, voluntary control of skeletal muscles, and many other activities.

cervix   The lower end, or neck, of the uterus.

cetaceans (suh TAY shunz)   The mammals best adapted to live in the ocean, such as dolphins and whales.

chalazae (kuh LAY zee)   A pair of dense twisted cords of albumen that are attached to opposite ends of the yolk in an amniote egg.

chelicerae (kuh LIHS uhr ee)   Appendages which are modified as claws or fangs in certain arthropods.

chelipeds (KEE luh pehds)   Large pincers adapted for capturing food and for defense in certain arthropods.

chemical equation   The symbolic language used to describe a chemical reaction.

chemical evolution   Evolution based on nonbiological chemical processes that involves changing inorganic and simple organic compounds into more complex organic compounds.

chemical reaction   A change in which one or more new substances form.

chemoreceptors   Receptors that are stimulated by chemicals in air, water, or food.

chemosynthesis (kee moh SIHN thuh sihs)   A kind of autotrophic nutrition in which chemical energy is used to make carbohydrates.

chemotherapy (kee moh THER uh pee)   The use of drugs to treat disease.

chemotropism (keh MAHT ruh pihz uhm)   The growth of a plant part toward chemicals.

chlorophyll   A green pigment in the cells of photosynthetic autotrophs that captures light energy.

chloroplast (KLOR uh plast)   The organelle in which the food-making activities of a plant cell occur.

choanocyte (KOH uh nuh sīt)   A specialized cell in a sponge that has a single flagellum surrounded by a collar.

chordates (KOR dayts)   Animals that have a notochord, a dorsal nerve cord, and gill slits at some time in their life cycle.

chorion (KOR ee ahn)   In an amniote egg, a membrane that grows out from the embryo and lines the inner surface of the shell.

chromatids (KROH muh tihdz)   Two distinct strands that make up each chromosome.

chromatin (KROH muh tihn)   A material within the nucleus that is made of protein and DNA.

chromoplasts (KROHM uh plasts)   Plastids that contain red, yellow, or orange pigments.

chromosomal alteration   A change in the structure or in the number of chromosomes.

chromosome theory of heredity   A theory stating that chromosomes are the carriers of genes.

chromosomes (KROH muh sohmz)   Threadlike structures that carry the cell's genes.

cilia (SIHL ee uh)   Hairlike structures, found in many cells, that aid in movement.

ciliates   One-celled organisms that have many cilia.

circadian (ser KAY dee uhn) **rhythms**   Rhythms that show a 24-hour cycle.

circulatory system   The body system consisting of the blood, heart, and blood vessels.

citric acid cycle   A series of reactions that completes the breakdown of glucose in aerobic respiration.

class   A group of closely related orders.

cleavage   The series of mitotic divisions that follows fertilization and results in a mass of cells.

climatogram   A graph that shows average monthly temperature and precipitation.

climax community A community that is no longer undergoing succession.

clitellum (klih TEHL uhm) The swollen band that functions in reproduction in an earthworm.

cloaca (kloh AY kuh) A chamber into which the products of the digestive, excretory, and reproductive system are emptied.

cloning The process of forming a group of genetically identical offspring from a single cell.

closed circulatory system The system in which the transport medium, blood, is always contained within vessels.

club mosses Plants of subphylum Lycopsida.

cnidarians (nī DAIR ee uhnz) Animals that belong to the phylum Cnidaria, such as jellyfish and hydras.

cnidoblasts (NI duh blasts) Cells on a cnidarian's tentacles that contain nematocysts.

coacervate (koh AS uhr vayt) A microscopic group of droplets formed by attractions between molecules.

cocci (KAHK sī) Bacteria that are shaped like spheres.

cochlea (KAHK lee uh) A structure in the inner ear containing fluid and the mechanoreceptors for hearing.

codon (KOH dahn) A sequence of three nucleotides on an mRNA molecule.

coelom (SEE luhm) An internal fluid-containing cavity that is completely surrounded by mesoderm.

coelomates (SEE luh mayts) Animals in which a true coelom forms at some stage.

cohesion The attraction of like molecules for each other.

collecting tubule A tube in a nephron that carries urine into a cavity and then into a ureter.

colloid (KAHL oid) A mixture composed of particles dispersed in a medium, the size of the particles being between the size of those in a solution and those in a suspension.

colon The longest part of the large intestine.

colony A group of similar organisms that are attached to each other.

commensalism A symbiotic relationship in which one organism benefits and the other is neither benefited nor harmed.

communicable diseases Diseases that are transmitted or transferred from one organism to another.

communication The process through which information is conveyed among organisms through sets of behavior signals.

community All the populations in a given area.

compact bone The hard outer part of the bone.

companion cells Specialized cells that are believed to provide some materials and energy used by the sieve-tube elements.

complete metamorphosis In insects, the process of change from egg to larva to pupa to adults.

compound Matter composed of two or more elements chemically bonded.

concentration gradient A measure of the difference in concentration of a substance in two regions.

condensation A process in which water vapor changes to liquid water or ice.

cones In conifers, the structures that bear the male and female gametophytes.

conidia (kuh NIHD ee uh) The asexual spores of many ascomycotes.

conidiophores (kuh NIHD ee uh forz) Specialized hyphae on which conidia are formed.

coniferous forest The biome in which the vegetation is almost totally made up of conifers.

conifers Plants that usually produce seeds in cones.

conjugation The transferring of genetic material from one organism, such as a bacterium, to another by direct cell-to-cell contact.

connective tissue A type of tissue in which material called the matrix is found between the cells.

conservation The wise management and use of natural resources.

consumers Organisms that eat other organisms.

contagious (kuhn TAY juhs) **diseases** Diseases that are transmitted from one organism to another.

contour farming A method of plowing at right angles to the slope of the land so that runoff water is held between the furrows.

contour feathers The large feathers that give a bird's body its contour, or outline.

contractile fibers Strips of protein inside the cell membrane that can stretch and contract.

contractile vacuole In some protists, a structure that collects extra water and discharges it from the cell.

conus arteriosus In a frog, a large blood vessel which has two channels.

convergence (kuhn VER juhns) An increase in similarities among species derived from different ancestors and resulting from similar adaptations to similar environments.

cork A tissue made up of thick cells that forms the outer protective tissue of woody plants.

cork cambium A meristematic layer near the surface of the vascular cylinder.

corm A round underground stem that stores food.

corolla (kuh RAHL uh) All the petals of a flower together.

coronary circulation The flow of blood through the blood vessels that nourish the heart itself.

corpus luteum An ovarian tissue that secretes the hormone progesterone.

cortex In a plant's root, a region of parenchyma tissue outside the stele.

corticosteroids (kor tuh koh STEHR oidz) Steroid hormones produced by the adrenal cortex.

cortisol (KOR tuh sohl)) A hormone that stimulates the production of glucose from fats and amino acids.

courtship behavior The specific type of behavior associated with the selection of mates within a species.
covalent (koh VAY luhnt) **bond** A bond formed when two atoms share electrons.
covalent compound A compound formed when electrons are shared.
Cowper's glands Glands that secrete seminal fluid.
Cro-Magnons (kroh MAG nahnz) A group of *Homo sapiens* that cannot be physically distinguished from modern humans.
crop A temporary food storage area in some animals.
crop rotation The practice of growing different plants in a particular field each season.
cross-pollination A process by which pollen formed by a flower of one plant is moved to the pistil in a flower of another plant of the same kind.
crossing over The exchange of pieces of chromatid material between homologs during meiosis.
cutin (KYOO tihn) A waxy material that covers the leaves of vascular plants.
cutting A plant part that is used for growing a new plant through artificial vegetative propagation.
cycads (SI kadz) A small group of tropical plants that look like large ferns or palm trees.
cyclic adenosine monophosphate (cAMP) A substance which acts as the second messenger for many of the protein hormones.
cyclostomes The living species of jawless fish.
cyst In protists, a protective shell secreted around a protozoan cell that allows the cell to survive unfavorable conditions.
cytokinins (sī toh KI nuhnz) Plant hormones that promote growth through cell division.
cytoplasm (SI tuh plaz uhm) The jellylike material found within prokaryotic and eukaryotic cells.
cytosine (SI tuh seen) One of the bases that occurs in DNA.
cytoskeleton The system of microtubules and microfilaments that maintains a cell's shape and controls certain kinds of cell movement.

dark reactions Reactions of photosynthesis that do not depend directly on light.
data Information obtained from observations and other sources.
day-neutral plants Plants in which flowering does not depend on photoperiod.
decomposers Organisms that break down dead organic material into inorganic matter.
degrees Celsius (SEHL see uhs) The basic unit of temperature in SI.
dehydration synthesis The formation of a large molecule by joining smaller molecules, removing water in the process.
deletion (dih LEE shuhn) An alteration caused by a loss of a segment of a chromosome.
dendrites Extensions of the neuron that carry nerve impulses toward the cell body.
density-dependent factors Limiting factors whose effects vary with the density of the population.
density-independent factors Limiting factors that affect population growth regardless of the density of the population.
dentin A bonelike substance that surrounds the pulp cavity of the tooth.
dentition The types, number, and arrangement of teeth.
dentrification A process in which nitrates are changed into nitrogen gas.
depressants Drugs that decrease the activity of the nervous system.
dermis The inner skin, beneath the epidermis.
desert The biome that gets less than 25 cm of rainfall and has a high rate of evaporation.
diaphragm (DI uh fram) A flat sheet of muscle which separates the chest cavity from the abdominal cavity.
diastole (di AS tuh lee) The period during which the ventricles are relaxed and are being filled with blood.
diatomaceous (di uh tuh MAY shuhs) **earth** A fine, powdery substance formed by the piled-up shells of diatoms.
diatoms Algal protists with silica shells.
dicots Angiosperms in which each seed contains two primary leaves.
diffusion (dih FYOO zhuhn) The movement of atoms, molecules, or ions from a region where they are more concentrated to a region where they are less concentrated.
digestion The metabolic processes involved in breaking down food into simpler substances.
dihybrid (di HI brihd) **cross** A cross involving two different sets of traits.
dinoflagellates (di nuh FLAJ uh layts) Protists that are members of phylum Pyrrophyta.
dipleurula (di PLUR yuh luh) A bilaterally symmetrical larva of an echinoderm.
diploid number The total number of chromosomes in a cell that contains chromosomes in homologous pairs.
direct development The production of offspring that are small copies of the adults.
disaccharides (di SAK uh rīdz) Double sugars, produced when two monosaccharides chemically combine.
disease A condition that interferes with the normal functioning of a living thing but that is not the result of an injury.
disinfectants Chemicals that kill microbes on contact.
distal convoluted tubule The third section in a nephron, located between Henle's loop and the collecting tubule.
diuresis (di yu REE sihs) Overproduction of urine.
divergence (duh VER juhns) An increase in the differences among descendants of a single ancestral species as time passes.

DNA The chemical that codes for the production of a cell's proteins and thus cellular activities.

DNA replication The process in which a DNA molecule makes copies of itself.

DNA splicing The introduction of genes into a plasmid.

dominant (DAHM uh nuhnt) The form of gene that is expressed and prevents the expression of another gene.

dormancy A condition of decreased growth and activity during unfavorable environmental conditions.

dorsal cavity The space containing the brain and the spinal cord.

double-circuit pathway A path in which blood moves through the heart to the lungs, back to the heart, and then throughout the body.

down feathers Short fluffy feathers.

Down's syndrome A genetic disorder resulting from an extra chromosome number 21.

duodenum (doo uh DEE nuhm) The first part of the small intestine.

duplication An alteration in which a segment of a chromosome is present more than once on a chromosome.

echinoderms (ih KĪ nuh dermz) Spiny-skinned marine invertebrates.

echolocation Sensing the environment via sound.

ecological succession The orderly sequence of changes in communities.

ecology The study of the relationship between living things and their environment.

ecosystem Any area in which energy is transferred as organisms interact with each other and with nonliving things.

ectoderm The outer cell layer in an embryo, which becomes the epidermis.

ectoplasm The clear outer layer of cytoplasm in an amoeba.

ectotherm An animal that obtains most of its body heat from the environment.

edentates (ee DEHN tayts) Mammals that have either very reduced teeth or no teeth at all.

effectors The structures or organs that produce the response to information.

egg cells The female gametes.

electron transport chain A series of electron carriers involved in the production of ATP.

element Matter that contains only one type of atom.

embryo The organism in the early stages of development.

embryo sac The female gametophyte within the ovule of a flower.

embryonic induction The process in which one cell influences the differentiation of another cell.

emigrants Organisms that move out of a population.

emulsification (ih muhl suh fuh KAY shuhn) A process in which large fat droplets break apart.

enamel A substance that contains calcium and protects the teeth against wear.

endemic A condition in which disease is present in a population at all times.

endergonic (ehn duhr GAHN ihk) **reaction** A chemical reaction that uses energy.

endocrine glands Glands that secrete chemicals into the capillaries that pass through the glands.

endocrine (EHN doh krihn) **system** A number of ductless glands that secrete chemicals that control certain body activities.

endocytosis (ehn doh sī TOH sihs) The process by which cells take in large materials that cannot pass through the cell membrane.

endoderm The inner cell layer in an embryo which lines the digestive cavity.

endodermis In a plant root, the inside layer of the cortex.

endoplasm In an amoeba, the granular form of cytoplasm in the interior of the cell.

endoplasmic reticulum (ehn doh PLAZ mihk reh TIHK-yuh luhm) A system of membranes that runs throughout the cytoplasm.

endoskeleton A skeleton that lies inside the body of an animal.

endosperm (EHN doh sperm) Rich food-storage tissue that nourishes the developing embryo of a plant.

endospore A spore formed inside a bacterial cell that allows the bacterium to survive unfavorable conditions.

endotherm An animal that maintains a relatively constant internal body temperature as a result of heat produced by body metabolism.

endotoxin A toxic substance made up of lipid, saccharide, and protein and usually given off when the bacterium dies.

environment All the living and nonliving factors that surround an organism.

environmental resistance The net effect, or sum, of limiting factors on a population's growth.

enzyme-substrate complex The structure formed when an enzyme joins with a substrate.

enzymes (EHN zīmz) Catalysts in cells.

epidemic A condition in which a disease is widespread in a population.

epidermal tissue In plants, a layer of cells that covers the entire plant.

epidermis The outer body layer or skin.

epididymis (eph uh DIHD uh mihs) The series of coiled tubules where sperm are stored.

epiglottis A flap of tissue which moves down over the entry to the trachea when food is swallowed.

epithelial tissue Sheets of cells that cover a surface or line a cavity.

epoch (EHP uhk) An interval of time in a geologic time chart.

era The largest interval of time in a geologic time chart.

erosion The wearing away of soil due to the action of water or wind.

esophagus (ee SAHF uh guhs) The muscular tube that connects the pharynx with the stomach.

estivation (ehs tuh VAY shuhn) A period of dormancy during which metabolic rate and body temperature are lowered in response to hot, arid conditions.

estrogen (EHS truh juhn) The hormone that stimulates the development of female reproductive organs and female secondary sexual characteristics.

estrus The fertile period of a female.

estuaries (EHS chu ehr eez) The areas where the fresh waters of rivers flow into the salt water of oceans.

ethylene A phytohormone that is involved in the falling of leaves and the ripening of fruit.

euglenoids (yoo GLEE noidz) Organisms in phylum Euglenophyta.

eukaryotes (yoo KAR ee ohts) Organisms whose cells contain membrane-bound organelles.

euphotic (yoo FAHT ihk) **zone** The region from the surface to about 200 m in a marine biome.

evaporation A process in which liquid water changes to water vapor.

evolution The process of change by which new species develop from preexisting species.

excretion The process of ridding the body of chemical wastes and any materials that are in excess supply.

excurrent siphon A structure that regulates the movement of water out of the mantle cavity of a mollusk.

exergonic (ehk suhr GAHN ihk) **reaction** A chemical reaction that releases energy.

exocrine (EHK suh krihn) **glands** Glands that release chemicals into ducts.

exocytosis (ehks uh sī TOH sihs) The release of large molecules or groups of molecules from the cell.

exoskeleton A skeleton covering the outside of the body.

exotoxin A toxic substance made of protein and given off by a living bacterium.

experimentation The scientific testing of a hypothesis.

expiration The process during which air is pushed out of the lungs.

expiratory (ehk SPIR uh tor ee) **center** The part of the brain that sends impulses that inhibit, or stop, the action of the inspiratory center.

external fertilization The union of sperm and eggs outside a female's body.

external nares (NAIR eez) Two openings that lead into the mouth, the nostrils.

external respiration The process in which oxygen is obtained from and carbon dioxide is released to the external environment.

extinction The complete elimination of a group of organisms.

extracellular digestion The process in which food is broken down in a digestive cavity or structure rather than within individual cells.

eyespot A red, light-sensitive structure in some protists.

facilitated diffusion The diffusion of materials across the cell membrane assisted by carrier molecules.

factors Mendel's term for the units of heredity that are passed on to future generations.

facultative anaerobes Bacteria that can produce ATP with or without oxygen.

Fallopian (fuh LOH pee uhn) **tube** A tube which provides a passageway from an ovary to the uterus.

family A group of closely related genera.

feedback mechanism A process in which the last step or end product of a series of steps controls the first step in the series.

fermentation The breakdown of glucose and the release of energy in which organic substances are the final electron acceptors.

ferns Plants belonging to subphylum Pteropsida.

fetoscopy (fee TAHS kuh pee) A technique that allows doctors to study the fetus and treat some problems before birth.

fetus (FEE tuhs) A human embryo after three months.

fiddlehead A tightly coiled frond.

field guide A book that contains information about a group of things found in nature.

filament (FIHL uh muhnt) In a flower, a long thin stalk that attaches the anther to the receptacle.

filoplumes Sparsely scattered feathers, each composed of a long shaft with a few barbs at the tip.

fins Membranes that extend from the body of a fish and are supported by rays or spines.

first filial (FIHL ee uhl) **generation** The offspring of a parental cross

fish Vertebrates that live in water and breathe with gills.

flagellates Animallike protists that have flagella.

flagellum (fluh JEHL uhm) (pl., **flagella**) A taillike structure, found in some cells, that aids in movement.

flatworms Worms belonging to phylum Platyhelminthes.

flower The reproductive structure of an angiosperm.

follicle (FAHL uh kuhl) In the human ovary, a structure that encloses an egg cell.

follicle-stimulating hormone (FSH) A hormone that controls secretion of hormones by the reproductive organs.

food chain The series of steps through which energy is transferred from the sun to organisms in an ecosystem.

food vacuole An organelle in which food is digested.

food web The interrelated food chains in an ecosystem.

foot In mollusks, a muscular structure for movement.

foraminiferan (fuh ram uh NIHF er ehn) A sarcodine that has a shell of calcium carbonate.

forest floor A layer of decaying plant and animal matter in vertical stratification.

fossil fuels The oil, coal, and natural gas that formed from plants that lived long ago.

fossils The remains and traces of organisms that once lived.

fragmentation The separation of an organism into parts, with each part growing into a new individual.

freshwater biomes Habitats in which the salinity is less than 0.005 percent.

frond A fern leaf.

fruit An enlarged, ripened ovary that contains one or more seeds.

fruiting bodies Reproductive structures of certain organisms which produce many spores.

fungal protists The slime molds, grouped in subkingdom Gymnomycota.

Fungi (FUHN jī) A kingdom that contains eukaryotic organisms that have cell walls and are heterotrophic.

fungus An organism made of eukaryotic cells with cell walls that gets its food by absorbing organic substances.

gall bladder A pear-shaped muscular organ, connected to the liver, that stores bile.

gametes (GAM eets) The specialized cells that join during sexual reproduction.

gametogenesis (gam uh tuh JEHN uh sihs) The formation of gametes by meiosis.

gametophyte (guh MEE tuh fīt) The monoploid plant that produces gametes.

ganglion (GANG glee uhn) A mass of cell bodies of neurons found outside the brain or spinal cord.

gastric glands Structures that produce a very acidic secretion called gastric juice.

gastrodermis The inner body layer in cnidarians.

gastropods (GAS truh pahdz) Mollusks having a well-developed head, a large flat foot, and often a one-part shell.

gastrovascular cavity A digestive cavity having only one opening, the mouth. It is found in cnidarians.

gastrula The embryonic stage in which the three primary germ layers are formed.

gene A segment of DNA that codes for a particular protein.

gene flow The movement of alleles into or out of a population.

gene frequency The percent of each allele in a gene pool.

gene mutations A change in the chemical nature of DNA.

gene pool All the genes of all the members in a population.

genetic drift The change that takes place in a gene pool, caused by chance.

genetic engineering Altering the structure of a DNA molecule by substituting genes from other DNA molecules.

genetic equilibrium A condition in which the gene frequencies remain the same from one generation to the next.

geneticist A person who studies the field of biology that deals with how characteristics are passed from an organism to its offspring.

genetics (juh NEHT ihks) The branch of biology dealing with the study of heredity.

genotype (JEEN uh tīp) The genetic makeup of an organism.

genus A group of closely related species.

geographic isolation A situation in which interbreeding between two populations of a species is prevented by a physical barrier.

geological time chart A chart summarizing the history of life.

geothermal energy Energy from the earth's hot interior.

geotropism (jee AHT ruh pihz uhm) The growth of a plant part toward or away from gravity.

germ mutations Mutations that take place in the reproductive cells of an organism.

germ theory of disease The theory that living microorganisms are the cause of many diseases.

germination (jer muh NAY shuhn) The development of an embryo into a seedling.

gestation (jehs TAY shuhn) **period** The length of time an animal spends developing in the uterus.

gibberellins (jihb uh REHL ihnz) Growth-promoting plant hormones.

gill In an aquatic animal, a structure through which the respiratory gases are exchanged between the animal's blood and the water in which the animal lives. In a mushroom, a structure where basidiospores are formed.

gill heart A heart which pumps blood through the gills in a cephalopod.

gill slits Structures located behind the mouth in the region of the pharynx which are involved in food gathering and, to some extent, gas exchange in some chordates.

ginkgoes (GING kohz) Trees with small fan-shaped leaves.

girdling A condition in which the phloem tubes that encircle a tree trunk are cut.

gizzard A structure used to grind food in some animals.

glomerular filtrate The fluid that accumulates in Bowman's capsule.

glomerulus (gluh MEHR u luhs) A ball of capillaries in a nephron.

glucagon A pancreatic hormone which causes the removal of glycogen from the liver and the skeletal muscles and the release of glucose into the blood.

glycolysis (glī KAHL uh sihs) The production of ATP by changing glucose to pyruvic acid.

gnetales (nuh TAY leez) A group of small, cone-bearing plants.

Golgi apparatus (GAHL jee ap uh RAY tuhs) An organelle that prepares materials for secretion from the cell.

gradualism An idea stating that the process of evolutionary change is slow and steady.

grafting The joining of a twig of one plant onto the body of a different plant.

grassland A treeless biome marked by a wide range of temperatures and an uneven distribution of precipitation.

green algae Members of phylum Chlorophyta.

green glands Structures which remove wastes from blood in crustaceans such as the crayfish.

ground layer The layer of vegetation found just at the surface of the soil.

growth An increase in living mass.

guanine (GWAH neen) One of the bases that occurs in DNA.

guard cells Specialized epidermal cells that control the opening and closing of a stoma by changing shape.

gullet In a paramecium, a funnellike tube that leads from the mouth opening into the interior of the organism.

guttation (guh TAY shuhn) The process by which a plant rids itself of excess water through special pores on the leaf.

gymnosperms (JIHM nuh spermz) Plants whose seeds do not develop within ovaries.

habitat The place in which an organism lives.

hair follicles Tubelike pockets of epidermal cells that extend downward into the dermis and produce hair.

half-life The time in which half the radioactive atoms of a certain kind of rock or fossils will break down into another form.

hallucinogenic (huh loo suh nuh JEHN ihk) **drugs** Drugs that affect the user's perception of reality.

halophyte (HAL uh fīt) A plant adapted to growing in salty soils.

Hardy-Weinberg Law A law stating that under certain ideal conditions genetic equilibrium is maintained.

haustoria (haw STOR ee uh) Hyphae specialized for absorbing nutrients in parasitic fungi.

Haversian (huh VER shuhn) **canals** Interconnecting channels containing the nerves and blood vessels that supply the bone cells.

head In an arthropod, the most anterior segments in its body.

heart The pump that keeps the blood in motion through the blood vessels.

hemodialysis (hee moh dī AL uh sihs) Artificial filtering of the blood.

hemoglobin (HEE muh gloh buhn) The main component of red blood cells which binds and carries oxygen.

hemophilia (hee muh FIHL ee uh) A genetic disorder that involves the failure of the blood to clot normally.

hemotoxin (hee muh TAHK suhn) A poison that damages blood cells and bursts blood vessels.

Henle's loop The U-shaped loop section in a nephron.

hepatic portal circulation The flow of blood through the veins extending from the digestive organs to the liver.

herb layer The layer of grass and ferns in vertical stratification.

herbaceous (her BAY shuhs) **plants** Plants in which the cell walls of the xylem are not thickened.

herbivores Plant eaters.

heredity (huh REHD uh tee) The passing of traits from parents to their young.

hermaphrodite (her MAF ruh dīt) An organism that produces both male and female sex cells.

heterotrophs (HEHT uhr uh trohfs) Living things that cannot make their own food and thus must take in food.

heterozygous (heht uhr uh ZĪ guhs) A condition in which the two genes for a given trait are different.

hibernation A period of dormancy accompanied by a low metabolic rate and a much lowered body temperature.

holdfast In algae, the cell modified to hold algae to stones and other objects in streams, lakes, and oceans.

homeostasis (hoh mee uh STAY sihs) The balanced regulation of metabolic processes to maintain the conditions needed for life.

hominid (HAHM uh nihd) Humanlike or human species.

Homo erectus (ih REHK tuhs) A hominid that lived from 1.5 million to 0.5 million years ago.

Homo habilis (HAB uh luhs) A hominid that lived nearly 2 million years ago.

homologous (hoh MAHL uh guhs) **chromosomes** Chromosomes in the body cells that occur in pairs.

homologous structures Parts of different organisms that are similar in structure but serve different functions.

homologs Homologous chromosomes.

Homo sapiens (HOH moh SAY pee uhnz) The species to which humans belong.

Homo sapiens neanderthalensis (nee an der thawl EHN sihs) The Neanderthals, a variety of *Homo sapiens* that became extinct about 35 000 years ago.

homozygous (hoh muh ZĪ guhs) A condition in which the two genes for a given trait are alike.

hormone A chemical that is produced in one part of an organism and is transported to other parts where it controls certain activities of tissues and organs.

horny outer layer The outermost layer of a mollusk shell.

horsetail A plant in the genus Equisetum.

host cell The cell in which a virus or other parasite lives.

human growth hormone (HGH) A hormone necessary for normal growth.

hybrid (HĪ brihd) The offspring of two parents that differ in one or more inherited traits.

hybrid vigor The superiority seen in offspring having the desirable characteristics of both parents.

hydathodes (HĪ duh thohdz) Water pores on a leaf.

hydrolysis (hī DRAHL uh sihs) A catabolic reaction that involves the addition of water.

hydrotropism (hī DROHT ruh pihz uhm) The growth response of plants to moisture.

hypertension Abnormally high blood pressure.

hypertonic (hī puhr TAHN ihk) **solution** A solution in which the concentration of dissolved substances is greater than the concentration inside the cell.

hyphae (HI fee) Branching filaments that make up most fungi.

hypothalamus (hī puh THAL uh muhs) The part of the brain which regulates involuntary control of many basic body functions.

hypothesis (hī POTH uh sihs) A possible answer to a question about nature, based on the observations, reading, and knowledge of a scientist.

hypotonic (hī puh TAHN ihk) **solution** A solution in which the concentration of dissolved materials is lower than the concentration inside the cell.

hypoxia (hī PAHK see uh) A shortage of oxygen in tissues that may result from several causes.

ileum (IHL ee uhm) The last part of the small intestine.

immigrants Organisms that move into a population.

immune system The cells and tissues involved in the body's response to infection.

immunity (ih MYOO nuh tee) The body's ability to prevent an infection from developing by having antibodies against the pathogen.

implantation The process by which the embryo attaches to the uterus.

imprinting In birds, the learning process involving following and imitating the first moving object seen after hatching.

inbreeding A type of breeding that involves crossing closely related individuals to maintain desirable characteristics.

incisors (ihn SI zuhrz) Teeth used for biting and cutting food.

incomplete dominance Inheritance in which one allele is not dominant over any other.

incomplete metamorphosis In insects, the process of change from egg to nymph to adult.

incurrent siphon A structure that regulates the movement of water into the mantle cavity of a mollusk.

indirect development A life cycle that involves the production of a larval form and metamorphosis into the adult form.

induced-fit model A model of enzyme action that states that the substrate alters the structure of the enzyme's active site.

infectious (ihn FEHK shuhs) **disease** A disease that is caused by organisms that enter the body.

inflammation (ihn fluh MAY shuhn) A tissue response to the presence of microbes, other foreign antigens, or an injury involving swelling, redness, heat, and pain.

ingestive heterotroph A type of heterotroph that takes in food and then digests it.

inner cell mass The part of a blastocyst that will become the embryo.

insectivores (ihn SEHK tuh vohrz) Small mammals belonging to order Insectivora that are adapted to eat small invertebrates, especially insects.

insertion The end of the muscle attached to a bone that moves.

inspiration The process during which air is pushed into the lungs in breathing.

inspiratory (ihn SPIR uh tor ee) **center** The part of the brain that sends impulses along nerves to the intercostal muscles and to the diaphragm.

insulin A hormone produced by the islets of Langerhans in the pancreas. It promotes the absorption of glucose.

integument The skin, or outer covering of an organism.

intercostal muscles The muscles between the ribs.

interferon A cell-produced protein that helps protect the body against viral infection.

internal fertilization The union of sperm and eggs inside a female's body.

internal nares In a frog, two openings near the vomerine teeth connected by channels to the external nares.

internal respiration The process in which oxygen is used and carbon dioxide is produced within cells.

international system of measurement (SI) The system of measurement used in science.

interphase The period in which cells carry out many activities other than mitosis.

interspecific competition Competition between different species.

intertidal zone The region close to shore, between high and low tides.

intracellular digestion The process in which enzymes break down food inside food vacuoles.

intraspecific competition. Competition among members of a population for limited resources, such as food and water.

inversion (ihn VER zhuhn) An alteration that occurs when a chromosome breaks and reattaches so that the sequence of genes is reversed.

invertebrates Animals that do not have a backbone.

ion An atom that has gained or lost one or more electrons and so has a negative or positive charge.

ionic bond A bond formed when two atoms transfer electrons.

ionic compound A compound formed when two or more ions chemically bond.

islets of Langerhans Endocrine cells in the pancreas that secrete two hormones that control the levels of glucose in the body.

isomers (ī suh muhrz) Compounds with the same molecular formula but different molecular structures.

isotonic (ī suh TAHN ihk) **solution** A solution in which the concentration of dissolved substances is equal to the concentration inside the cell.

isotopes (ī suh tohps) Atoms of the same element that have different mass numbers.

Jacobson's organ A sense organ specialized for smell found at the front of the roof of the mouth in some animals.

jejunum (jih JOO nuhm) The middle part of the small intestine.

joint The point where two or more bones meet.

karyotype (KAR ee oh tīp) A picture of chromosomes arranged as homologs.

keel In birds, that part of the breastbone to which large breast muscles for flight are attached.

kidneys The major organs of excretion in many animals, including humans.

kilogram The basic unit of mass in SI.

kingdom A group of closely related phyla.

Klinefelter's syndrome A genetic disorder in which the sex chromosome makeup is XXY.

Koch's postulates (PAHS chuh lihts) Statements used to identify an organism as a disease agent.

labium A liplike structure, such as the lower lip of an insect.

labrum The upper lip of an insect or other arthropod.

lacteal (LAK tee uhl) A microscopic lymph vessel.

lactic acid fermentation Fermentation that forms lactic acid.

lagomorphs Small mammalian herbivores with chisel-shaped front teeth such as rabbits and hares.

lancelets Animals in subphylum Cephalochordata.

large intestine The lower end of the digestive tract.

larva A pre-adult form unlike its parent in appearance, such as the wormlike stage that follows the egg stage in insects.

larynx (LAR ihngks) The voice box.

lateral line system A series of sensory pits forming a line on each side of the body from the operculum to the tail of a fish.

layering. A method of vegetative propagation in which a branch of the parent plant is bent to the ground and its tip is covered with soil.

leaf The plant organ that absorbs sunlight and carries out photosynthesis.

leaf primordium An undeveloped leaf.

lentic habitat A standing-water habitat.

leucoplasts (LOO kuh plasts) Storage plastids.

leukocytes (LOO kuh sīts) White blood cells.

lichen (LĪ kuhn) A symbiotic association between a fungus and a blue-green bacterium, an algal protist, or a green alga.

life span The length of an organism's life.

ligaments Strong banks of connective tissue that hold bones together at joints.

light reactions Reactions of photosynthesis that depend on light.

limiting factors Factors that keep species from reaching their biotic potential.

limnetic (lihm NEHT ihk) **zone** The upper open-water region in a lentic habitat.

linkage The occurrence of a number of genes on the same chromosome.

lipids Organic compounds, including fats and oils, that store energy.

liter The basic unit of volume in SI.

littoral (LIHT uhr uhl) **zone** The region of a lentic habitat closest to shore.

liver The largest gland in the body, involved in many physiological processes.

liverworts Bryophyte plants that grow flat along a surface, rather than upright.

lock-and-key model A model of enzyme action that states that enzyme action is based on a fit between enzyme and substrate.

long-day plants Plants that flower only if daylength is longer than a critical number of hours.

lotic habitat A running-water habitat.

lungs The main organs of the respiratory system.

luteinizing (LOO tee uh nīz ing) **hormone (LH)** A pituitary gland hormone that controls secretion of hormones by the reproductive organs.

lymph The fluid that flows through the lymphatic system.

lymph nodes Small bean-shaped organs that act as filters in the lymphatic system.

lymphatic system A system of "circulation" that collects excess tissue fluid.

lymphocytes (LIHM fuh sīts) White blood cells in the body that respond to the presence of antigens.

lysis (LĪ sihs) The bursting of a cell. For example, the destruction of a bacterium during viral replication.

lysogenic (lī suh JEHN ihk) **bacteria** Bacteria that carry prophages

lysogeny (lī SAHJ uh nee) The process by which viral DNA is attached to and carried by a bacterial chromosome.

lysosomes (LĪ suh sohmz) Organelles that contain digestive enzymes.

lytic (LIHT ihk) **cycle** The entire process of viral replication.

macronucleus The control center of certain protists.

macrophages (MAK ruh fayj uhz) Large phagocytic white blood cells.

Malpighian (maph PIHG ee uhn) **tubules** Excretory structures, found in insects, that remove wastes from the blood.

mammal A vertebrate that has hair, nourishes its young with milk, and maintains a relatively constant body temperature.

mammary glands Glands which secrete milk.

mandibles Mouth parts adapted for chewing and tearing food.

mantle The thin fleshy tissue surrounding most of the soft body of a mollusk.

mantle cavity A space lying between the mantle and the visceral mass in a mollusk.

mariculture Farming the sea.

marine biomes Habitats in which the salt content is 3 to 3.7 percent.

marsupials (marh SOO pee uhlz) Animals that are members of order Marsupialia, mammals whose young develop in a marsupium.

marsupium An abdominal pouch found in most female marsupials in which young marsupials complete their development.

mass The quantity, or amount, of matter in an object.

mass extinction The sudden disappearance of many species. The fossil record shows evidence of two major periods of mass extinction.

mass number The number of protons and neutrons in an element.

maxillae (mahk SIH lee) (s., **maxilla**) In arthropods, jawlike feeding organs. The upper jaw in vertebrates.

maxillary teeth In a frog, tiny teeth around the inside edge of the upper jaw.

maxillipeds (mahk SIHL uh pehds) Appendages adapted for feeding in crustaceans

mechanoreceptors Receptors that respond to vibrations, pressure, or other mechanical stimuli.

medulla oblongata (mih DUHL uh ahb lahng GAHD uh) A structure in the brainstem that contains centers that control breathing and heartbeat.

medusa A cnidarian with a bell-shaped body form.

megaspore A monoploid plant cell that produces a multicellular female gametophyte.

meiosis (mī OH sihs) Cell division in which the chromosome number is reduced by one half and gametes are formed.

meiospores Monoploid flagellated cells that result from the meiotic division of a diploid zygote in certain kinds of green algae, such as *Ulothrix*. They can develop into a new filament of green algae.

memory cells Special plasma cells that can produce a particular antibody when needed.

meninges (muh NIHN jeez) Protective membranes which enclose the central nervous system.

menopause The time at which a human female stops releasing eggs and the menstrual cycle ceases.

menstrual (MEHN stroo uhl) **cycle** The hormone-controlled reproductive cycle of human females.

meristem (MEHR uh stehm) **tissue** Unspecialized cells that are capable of frequent cell division, providing for growth of specialized plant tissues.

mesoderm The middle cell layer in an embryo from which the skeletal, muscle, circulatory, and reproductive systems develop.

metabolism (muh TAB uh lihz uhm) The sum of all the chemical activities carried out by a living thing.

metamorphosis A series of changes in body form which some animals undergo during the life cycle.

metaphase The second stage of mitosis, during which the paired chromatids move to the center of the cell.

meteorite impact hypothesis A hypothesis stating that the mass extinction in the late Cretaceous period was caused by a meteorite.

meter The standard unit of length in SI.

methanogens Very primitive organisms placed in kingdom Monera.

microbial diseases Infectious diseases caused by pathogenic microscopic organisms.

microbiologist A person who studies organisms too small to be seen with the unaided eye.

microfilaments Threadlike fibers that move materials within the cell and help in locomotion.

microfossils Fossils of small organisms.

micronucleus The small nucleus that undergoes meiosis and mitosis before conjugation, the form of sexual reproduction that takes place in ciliates. During conjugation, micronuclei are exchanged between two ciliates.

micropyle (MĪ kruh pīl) An opening at the end of the ovule through which sperm enter the embryo sac.

microscope A scientific tool that magnifies objects, using a series of lenses.

microspore A cell that produces a male gametophyte.

microtubules Hollow, pipelike structures that help give a cell its shape.

migration The instinctive movement of animals, usually between their wintering grounds and their breeding grounds.

minerals Inorganic materials that are absorbed without being digested and are necessary for the normal functioning of the body.

mitochondria (mi tuh KAHN dree uh) Organelles that carry out the chemical reactions that release energy for cellular activities.

mitosis (mī TOH sihs) Cell division in which each daughter cell receives the same number of chromosomes the parent cell had.

mixture Two or more distinct substances that are mixed together but not chemically combined.

modern theory of evolution A theory that includes the idea that evolution is a change in the genetic makeup of a population.

molars Flat teeth used for crushing and grinding food.

molecule A collection of two or more atoms covalently bonded.
mollusks (MAHL uhsks) Invertebrate animals that have a muscular foot and a mantle that often secretes a hard shell.
molting In arthropods, the periodic shedding of the exoskeleton.
Monera (muh NIHR uh) A kingdom that contains only those organisms that have prokaryotic cells.
monerans Members of kingdom Monera such as bacteria.
monocots Angiosperms in which each seed contains one primary leaf.
monohybrid (mahn uh HĪ brihd) **cross** A cross that involves one pair of contrasting traits.
monoploid number Half the diploid number of chromosomes.
monosaccharides (mahn uh SAK uh rīdz) Single sugars, the simplest carbohydrates.
monosomy (MAHN uh soh mee) The condition in which there is only one chromosome of a homologous pair.
monotreme (MAHN uh treem) A mammal that lays eggs.
morphological (mor fuh LAHJ uh kuhl) **adaptations** Structural traits that make an organism better able to live in a given environment.
morula (MOR yuh luh) A solid ball of cells that results from mitotic divisions of a zygote.
mosses Bryophytes whose gametophytes grow erect.
motor neurons Neurons that carry impulses from the central nervous system to the effectors.
mRNA (messenger RNA) The nucleic acid molecule that carries the instructions to make a particular protein from the DNA in the nucleus to the ribosomes.
multicellular Consisting of more than one cell.
multiple alleles The different forms of a gene when there are more than two alleles for a particular trait.
muscle fibers Muscle cells.
muscle tissue Tissue that is made up of cells that are able to contract or shorten.
mutagen (MYOO tuh juhn) Something that can cause mutations.
mutations (myoo TAY shuhnz) Changes that may occur in chromosomes and genes.
mutualism A symbiotic relationship in which both organisms benefit.
mycelium (mī SEE lee uhm) A network of fungal hyphae.
mycoplasmas Tiny monerans that are smaller than most bacteria and lack cell walls.
mycorrhiza (mī kuh RĪ zuh) A structure formed by the mutualistic association of a fungus and the roots of a plant.
myofibrils Very fine threadlike structures running lengthwise through muscle fibers that are made of thick myosin filaments and thin actin filaments.

myomeres (MĪ oh meerz) Blocks of skeletal muscles.
myosin The protein that makes up the thick filaments in myofibrils.

nastic movements Plant responses that do not depend on the direction of the stimulus.
natural resources Materials from the earth that are used to sustain life or that are critical to economic systems.
natural selection The process resulting in survival of those organisms best adapted to the environment.
Neanderthals A variety of *Homo sapiens* that became extinct about 35 000 years ago.
negative feedback A process in which the last step or end product of a series of steps inhibits the first step.
nekton Swimming animals that are able to move independently of the currents.
nematocyst A threadlike stinging structure in a cnidarian.
nephridium (nih FRIHD ee uhm) A structure, found in earthworms, that filters excretory wastes, ammonia, urea, salts, and excess water from the fluid in the coelom.
nephrons Microscopic units which filter blood and form urine in the kidneys.
neritic (nih RIHT ihk) **zone** The region that extends from the intertidal zone to the edge of the continental shelf.
nerve cord A tubular cord of nerves that runs along the dorsal, or upper, surface of the body.
nerve tissue Tissue that is made up of highly specialized cells that conduct messages.
nervous system The brain, the spinal cord, and nerves that extend throughout the body.
neurons (NUR ahnz) The specialized cells of the nervous system.
neurotoxin (nur oh TAHK sihn) A poison that affects nerve centers, causing paralysis of the muscles, including those involved in breathing.
neurotransmitters The chemicals that carry impulses across synapses.
niche The role of an organism in a community.
nitrification A process in which ammonia is changed by bacteria into compounds containing nitrates.
nitrogen cycle A process in which nitrogen moves between organisms and the atmosphere.
nitrogen fixation A process in which nitrogen gas is changed to ammonia.
nondisjunction (nahn dihs JUHNGK shuhn) The failure of one or more pairs of chromosomes to separate during meiosis.
nonrenewable resources Substances that cannot be replaced.
nonvascular plants Plants without vascular tissue.
notochord A stiff dorsal rod that supports the body and is found in all chordates at some point in their development.

nuclear energy Energy that is released when atoms undergo changes in the structure of their nuclei.
nuclear fission The process of splitting atoms to release energy.
nuclear fusion The process which combines small atoms to form larger atoms and energy.
nuclear membrane A membrane that surrounds the nucleus of a cell.
nucleic (noo KLEE ihk) **acids** The hereditary materials of cells.
nucleocapsid The nucleic acid and the capsid together in a virus.
nucleolus (noo KLEE uh luhs) A body within the nucleus that makes and stores RNA.
nucleotides (NOO klee uh tīdz) The units that make up a DNA molecule.
nucleus The organelle that controls a cell's activities.
nutrient (NOO tree uhnt) A substance that supplies the body with energy or raw materials, or that is needed for chemical reactions.
nutrition (noo TRIHSH uhn) The process by which living things obtain, digest, and assimilate food.
nymph The immature animal that hatches from the egg of an insect that undergoes incomplete metamorphosis.

obligate aerobes Organisms that require oxygen for respiration.
obligate anaerobes Organisms that grow only in the absence of oxygen.
observation The examination of something in nature.
oceanic zone The open water region of the ocean, beginning just beyond the continental shelf.
ocelli (oh SEHL lī) Simple eyes which detect light but do not form images.
omnivores Animals that eat both plants and animals.
oogenesis (oh uh JEHN uh sihs) The formation of egg cells.
oogonium (oh uh GOH nee uhm) The female reproductive structure of certain organisms.
oomycotes (oh eh mī KOHTS) Fungi that belong to phylum Oomycota.
oospore (OH uh spor) The spore of an oomycote.
open circulatory system The system in which the blood is not always contained within vessels and there is direct contact between the blood and the body cells.
operculum In fish, the protective bony covering over the gills.
oral groove In a paramecium, a tube through which beating cilia drive food toward the mouth opening.
oral surface The surface on which the mouth is located.
order A group of closely related families.
organ A group of different tissues specialized to carry out a particular function.

organ systems Groups of organs having related functions.
organelles Cell structures that carry out specific functions.
organic compounds Compounds that contain carbon.
origin The end of the muscle attached to a bone that moves little or not at all.
osculum (AHS kyuh luhm) A large opening at the top of a sponge, the excurrent pore.
osmosis (ahz MOH sihs) The movement of water across a selectively permeable membrane.
ossification The process by which cartilage is changed into bone.
osteocytes (AHS tee uh sīts) Mature bone cells that secrete the hard bony matrix.
ostia In arthropods, openings through which blood reenters the heart in an open circulatory system.
outbreeding The crossing of distantly related strains.
ovary In animals, the female gonad, which produces eggs. In plants, the structure that contains one or more ovules and developing gametophytes.
oviparous Producing eggs that develop outside the female's body.
ovipositor An organ used in digging a hole in the ground where eggs can be laid.
ovoviviparous Producing an embryo which develops within an egg that is retained inside the female's body.
ovulation (oh vyuh LAY shuhn) The release of an egg cell.
ovules Specialized reproductive structures in seed plants.
oxygen cycle The movement of oxygen from living things to the air and water and back again.
oxytocin (ahk sih TOH sihn) A hormone that stimulates contractions of the muscles of the uterus during labor.

pacemaker A bundle of tissue in the wall of the right atrium that controls the heartbeat.
paleoanthropology (pay lee oh an thruh PAHL uh jee) The study of past humanlike species and their societies.
palisade mesophyll A layer of rectangular parenchyma cells, elongated at a right angle to the surface of the leaf.
pancreas A large gland found below the stomach which contains both exocrine and endocrine tissues.
paramecium A ciliate often found in pond water.
parasite An organism that lives in or on a host organism and does harm to this host.
parasitism A symbiotic relationship in which one organism lives on or in another and harms it.
parasympathetic nervous system The portion of the autonomic nervous system that controls the internal organs during routine conditions.
parathyroid glands Four glands found on the dorsal side of the thyroid gland.
parathyroid hormone (PTH) A hormone which acts with calcitonin to control the distribution of calcium and phosphorus in the body.

parenchyma (puh REHNG kuh muh) **tissue** Thin-walled unspecialized cells found in roots, stems, and leaves.

parental generation The group of organisms used to make the first cross in a breeding experiment.

parotid (puh RAHT ihd) **glands** Salivary glands located under the ears.

parthenogenesis (par thuh noh JEHN uh sihs) The production of offspring from unfertilized eggs.

passive immunity Immunity caused by receiving antibodies from the blood of humans or other animals that have been exposed to the antigen.

passive transport The movement of materials across the cell membrane not requiring the use of energy by the cell.

pathogens (PATH uh jehnz) Organisms that cause disease.

pearly layer The innermost part of the shell in certain mollusks.

pectoral girdle A set of bones consisting of two clavicles and two scapulas.

pedipalps (PEHD ih pahlps) Arachnid appendages which function as feeding organs, as walking legs, as sensory organs, or as reproductive organs that transfer sperm from the male to the female.

pelage (PEHL ihj) Coat of hair.

pelecypods (puh LEHS uh pahdz) Mollusks having a compressed body within a two-part hinged shell.

pellicle The firm flexible outer covering of certain protists.

pelvic girdle A set of three fused bones; the ilium, the ischium, and the pubis.

penis The external reproductive organ of the male.

perennials (puh REHN ee uhlz) Plants that live and reproduce for an indefinite number of years.

pericycle The layer of parenchyma cells at the edge of the stele in the root of a plant.

period An interval of time in a geologic time chart.

periosteum (pehr ee AHS tee uhm) A tough membrane that covers the bone.

peripheral (puh RIHF uhr uhl) **nervous system** The nerves that extend from the brain and spinal cord to all parts of the body.

periphyton (pehr uh FI tahn) Organisms in a body of water that cling to plants or other objects on the bottom.

peristalsis (pehr uh STAL sihs) The series of rhythmic muscular contractions and relaxations that moves the food through the digestive system.

permafrost The permanently frozen layer of soil of the tundra.

pesticides Chemicals used to kill insects and other destructive or troublesome organisms.

petals (PEHT uhlz) A ring of brightly colored structures inside the sepals of a flower.

petiole (PEHT ee ohl) A stalk that connects the blade of a leaf to the stem.

phages Viruses that infect bacteria.

phagocytosis (fag uh sī TOH sihs) The process by which large, solid materials are taken into a cell.

pharynx (FAR ihngks) A tubelike muscular structure that joins the mouth with the rest of the digestive tract.

phenotype (FEEN uh tīp) The outward appearance of an organism.

phenylketonuria (fehn uhl kee tuh NYUR ee uh) A genetic disorder involving the inability to break down a chemical, resulting in mental retardation.

pheromones (FEHR uh mohnz) Chemicals released by one animal that cause a response in another animal of that species.

phloem A vascular tissue that conducts food throughout the plant.

phloem fibers Groups of very long and thick-walled cells in phloem.

photoperiod (foh toh PIHR ee uhd) The ratio of light to dark in a 24-hour time span.

photoperiodism The response of an organism to the length of light and darkness.

photoreceptors Receptors that are stimulated by light.

photosynthesis The process by which autotrophs make their own food using light, CO_2, and H_2O.

phototropism (foh toh TROH pihz uhm) Growth response of a plant to light.

photovoltaic cells Devices that produce electricity directly from solar energy.

phylum A group of closely related classes.

physiological (fihz ee uh LAHJ uh kuhl) **adaptations** Adaptations that involve the metabolic processes of an organism.

phytohormones (fī toh HAWR mohnz) Plant hormones.

phytoplankton (fī toh PLANGK tuhn) Algae that make up part of plankton.

pineal gland A gland located in the head which produces melatonin.

pinocytosis (pih noh sī TOH sihs) The process by which cells take in small particles or liquid droplets.

pistil (PIHS tuhl) The female reproductive structure of a flower.

pith Parenchyma tissue found within the vascular ring in plants.

pituitary gland A small gland that releases many hormones, sometimes called the "master gland."

placenta A structure through which the embryo of some mammals receives nourishment and excretes wastes.

placental mammal A mammal that develops inside the mother's uterus while attached to a placenta.

plankton Small, sometimes microscopic organisms that are carried about by the motion of the water.

Plantae (PLAN tee) The kingdom of plants.

planula A larval form of many cnidarians that is ciliated, free-swimming, and has a body made of two cell layers.

plasma The fluid part of the blood.

plasma cell An enlarged white blood cell that can make large amounts of an antibody.

plasmids Small ring-shaped pieces of DNA in some bacteria. In genetic engineering, genes are often spliced into a plasmid, which is later returned to a bacterial cell.

plasmodium (plaz MOH dee uhm) A mass of cytoplasm that contains many nuclei.

plasmolysis (plaz MAHL uh sihs) The shrinking of the contents of a plant cell due to loss of water.

plastids (PLAS tihdz) Organelles in plant cells that may function as chemical factories or as storehouses for food and pigments.

plastron The ventral shield that covers the under surface of the turtle.

platelets The part of the blood involved in blood clotting.

pleura (PLUR uh) A moist membrane which covers and lubricates a lung and half of the chest cavity.

pollen grain The male gametophyte produced by a microspore.

pollination (pahl uh NAY shuhn) The transfer of pollen from the anther of a stamen to the stigma of a pistil.

polygenic inheritance (pahl ee JEHN ihk ihn HER uhtuhns) Inheritance in which a single trait is determined by the interaction of two or more pairs of genes.

polyp (PAHL ihp) A tubelike form of a cnidarian.

polyploidy (PAHL ee ploi dee) The condition in which cells have extra sets of chromosomes.

polysaccharides (pahl ee SAK uh rīdz) Large molecules formed when many monosaccharides bond together.

population A group of organisms of the same species living in an ecosystem at a specific time.

population density The number of individuals of a population per unit of area or volume.

population genetics The application of genetic principles to groups of organisms rather than to single organisms.

population growth An increase in the population size.

population growth curve A line graph of a population's growth over time.

population size Total number of members of a population.

precipitation All the forms of moisture that fall to the earth from clouds.

precocial (prih KOH shuhl) Birds that are well developed at the time of hatching and that maintain a constant body temperature even when young.

premolars Flat teeth used for crushing and grinding food that are located in front of the molars.

primary succession The establishment and development of communities in newly formed habitable areas.

primates Mammals that belong to order Primates.

primordial (prī MOR dee uhl) **soup** The solution of organic molecules that made up the early oceans according to Oparin's hypothesis.

principle of dominance A principle stating that in a hybrid organism, one gene determines the expression of a particular trait and prevents the expression of the contrasting form of that trait.

principle of independent assortment A principle stating that during gamete formation, the genes for one trait are separated and distributed to the gametes independently of the genes for other traits.

principle of segregation (seg ruh GAY shuhn) A principle stating that when gametes form, the genes that control a particular trait are separated into different gametes.

prismatic (prihz MAT ihk) **layer** The thick middle layer of the shell of certain mollusks.

probability Study of the operation of the laws of chance.

producers Organisms that produce their own food and form the base of ecological pyramids. Green plants are the producers in almost all ecosystems.

product rule A rule stating that the probability of independent events occurring together is equal to the product of the probabilities of these events occurring separately.

products The substances that are produced as a result of a chemical reaction.

profundal (pruh FUHND uhl) **zone** The region of deep water in a lentic habitat.

progesterone (proh JEHS tuh rohn) The hormone produced by the ovaries that maintains the lining of the uterus during pregnancy.

proglottid (proh GLAHT ihd) A body section of a tapeworm that has male and female reproductive organs.

prokaryotes (proh KAR ee ohts) Cells that do not contain membrane-bound organelles.

prolactin A protein hormone involved in the growth of the mammary glands.

prophage Viral DNA attached to a bacterial chromosome.

prophase The first stage of mitosis during which chromosomes can be seen through a microscope.

prostate gland A gland that secretes seminal fluid.

proteins Large, complex organic compounds that contain the elements carbon, hydrogen, oxygen, and nitrogen.

prothallus (proh THAL uhs) The gametophyte that develops from a fern spore.

Protista (pruh TIHS tuh) A kingdom that contains eukaryotic, mostly unicellular organisms.

protists Organisms grouped under kingdom Protista.

protonema (proh tuh NEE muh) A simple branching filament that develops in mosses.

protoplasm (PROH tuh plaz uhm) Living material in cells.

protozoans The animallike protists, grouped in subkingdom Protozoa.

proximal convoluted tubule The section nearest Bowman's capsule in a nephron.

pseudocoelom (soo doh SEE luhm) A body cavity that is not surrounded entirely by mesoderm.

pseudocoelomates (soo doh SEE luh mayts) Animals that have a pseudocoelom.

pseudopodia (soo duh POH dee uh) Fingerlike projections of a cell that function in locomotion and feeding.

pulmonary artery The blood vessel that arises from the right ventricle of the heart and gives off a branch to each of the lungs.

pulmonary circulation The flow of blood through those blood vessels that carry blood to, through, and from the lungs.

pulmonary veins Blood vessels that carry oxygenated blood to the heart from the lungs.

Punnett (PUHN iht) **square** A chart showing the possible combinations of genes among the offspring of a cross.

pupa A stage between the larva and the adult in insects that undergo complete metamorphosis.

pure line A group of living things that produce young having only one form of a trait in each generation.

quill The hollow part of the shaft of a feather that extends under the skin.

rachis (RAY kihs) The solid exposed part of the shaft of a feather.

radial canals In a starfish, tubes that branch out from the ring canal and carry water to the many pairs of tube feet.

radial symmetry The arrangement of an organism's parts so that the parts can be separated into mirror-image halves along many lengthwise planes.

radiolarian (ray dee oh LAIR ee uhn) A protist that belongs to the protozoan phylum Sarcodina and has a lacy, glassy shell composed of silica.

radula (RAJ u luh) In snails, a strip covered with teeth which is used to scrape food.

reactants The substances that react with each other in a chemical reaction.

receptacle (rih SEHP tuh kuhl) The structure to which all flower parts are attached.

receptors The structures that receive information from the environment.

recessive (rih SEHS ihv) The form of a gene that is only expressed when paired with a gene coding for the same trait.

recombinant (rih KAHM buh nuhnt) **DNA** The DNA molecule that forms from the combining of portions of two different DNA molecules.

rectum The final part of the large intestine.

recycling A process that uses wastes as the raw material for new products.

red algae Plants in phylum Rhodophyta.

red blood cells The part of the blood that transports oxygen and some carbon dioxide.

reflex arc The peripheral nervous system pathway for the impulses in an automatic response.

refractory period The brief period following an action potential, or nerve impulse, during which a neuron cannot respond to further stimuli.

regeneration The process of regrowing missing or damaged parts.

releasing factors Hormonelike substances that control the release of hormones by the anterior and intermediate lobes of the pituitary gland.

renal circulation The flow of blood through the vessels that pass through the kidneys.

renewable resources Substances that can be replaced through natural cycles.

reproduction The process by which new individuals are produced.

reproductive isolation A situation in which two populations can no longer interbreed and produce fertile offspring.

reptiles Ectothermic vertebrates that are covered with dry scaly skin and are adapted for reproduction on land.

reservoir A flask-shaped groove at the front of the cell of certain protists.

respiration The processes by which most cells obtain energy.

respiratory (REHS puhr uh tor ee) **center** A region of the brain which controls the respiratory rate.

response A reaction of an organism to a stimulus.

resting membrane potential The condition in which the outside of a nerve cell membrane is positively charged with respect to the inside.

Rh factor An antigen found on the blood cells of about 85 percent of the American population.

rhizoids (RĪ zoidz) In fungi, small branching hyphae growing downward from the stolons. In bryophytes, rootlike structures that anchor the plant to the ground.

rhizome (RĪ zohm) A horizontal underground stem.

rhythmic behavior A behavior that occurs periodically.

ribosomes (RĪ buh somes) The organelles on which proteins are made.

rickettsias (rih KEHT see uhz) Bacterialike organisms that cannot live outside of living tissue.

ring canal In starfish, a circular tubelike structure surrounding the mouth and leading water into five radial canals.

ripeness to flower The stage at which a plant is ready to flower.

RNA A nucleic acid made of a single chain of nucleotides.

rodents Small, primarily herbivorous mammals with sharp chisel-shaped front teeth such as mice and squirrels.

root The organ that anchors a plant in the ground or other substrate.

root apex The apical meristem that gives rise to the developing root.

root cap A structure that covers the root tip.
root hairs Epidermal structures that increase the absorbing surface of the root.
root layer The layer of vegetation just beneath the surface of the ground.
root pressure The pressure in the xylem resulting from the inward movement of water.
rotifers Organisms in phylum Aschelminthes that have rotating rings of cilia.
roundworms Slender cylindrical worms in phylum Aschelminthes that are pointed at both ends.
rRNA (ribosomal RNA) A nucleic acid that makes up part of the ribosomes.
rule of independent events A rule stating that previous events do not affect the probability of later occurrences of the same event.
runners Long horizontal stems that grow above the ground.

saliva (suh LI vuh) A digestive juice containing an enzyme that begins chemical digestion.
salivary (SAL uh vehr ee) **glands** Glands that secrete saliva into the mouth.
sap The fluid inside phloem tissue.
saprobes Organisms that use dead organic matter as food.
sarcodines (SAHR kuh dīnz) Protists that are members of phylum Sarcodina.
savannas Grassland regions where there is just enough rainfall to support a scattered growth of trees.
scales In fish and reptiles, structures that cover and protect the body.
science A method of obtaining knowledge about nature.
scientific law A description of some aspect of nature.
scientific method A means of gathering information and testing ideas.
scion (SĪ uhn) A twig of a plant to be grafted.
sclerenchyma (sklih REHNG kuh muh) **tissue** Thick-walled cells specialized to strengthen plant parts throughout the plant body.
scolex A knob-shaped head of a tapeworm that has hooks and/or suckers.
scrotum (SKROH tuhm) A sac of skin that holds the testes, or male gonads.
second The basic unit of time in SI.
second filial generation The offspring resulting from the crossing of the F_1 (first filial) generation.
secondary succession Succession that occurs on disturbed areas, such as abandoned farmland and vacant lots.
sedimentary (sehd uh MEHN tuhr ee) **rock** Rock that forms from buildup of sediment, tiny grains of eroded rock, and other materials that settle to the bottom of a body of water.

seed A plant structure containing an embryo, food for the embryo, and a seed coat.
seed coat The outer layer of the ovule of a plant.
seed dispersal The transport of seeds away from the parent plant.
seed plants Vascular plants that form seeds.
selective breeding The crossing of animals or plants that have desirable characteristics to produce offspring that have a combination of the parents' desirable characteristics.
self-pollination The transfer of pollen within the same flower or between flowers of the same plant.
semicircular canals The structures associated with equilibrium in the inner ear.
semilunar valves The two valves between the ventricles and the major blood vessels leaving the heart.
seminal fluid The substance that serves as a transport medium for sperm.
seminal vesicles Glands that secrete seminal fluid.
seminiferous (sehm uh NIHF uhr uhs) **tubules** Structures that produce sperm cells within the testes.
sensory neurons Neurons that carry impulses from receptors to the central nervous system.
sepals (SEE puhlz) The small leaflike structures above the receptacle of a flower.
septa In some fungi, cross walls that divide hyphae into cells.
serum (SIHR uhm) In vaccines, an antibody-containing substance used to produce immunity.
setae (SEE tee) In an earthworm, bristles that aid in moving.
sex chromosomes The pairs of chromosomes that are different in the males and females.
sex-linked characteristic A trait determined by a gene or genes carried on the X chromosome but not on the Y chromosome.
sexual reproduction Reproduction that involves the fusion of two specialized sex cells.
sexually transmitted diseases Diseases spread by sexual contact.
shoot apex The apical meristem that gives rise to the stem, leaves, and flowers.
short-day plants Plants that develop flowers after being exposed to a light period that is shorter than a critical number of hours.
shrub layer The layer of short, woody plants in vertical stratification.
sickle cell anemia A genetic disorder resulting in the production of abnormal hemoglobin.
sieve plate In a vascular plant, a structure that connects sieve-tube elements. In a starfish, a structure through which water enters the water-vascular system.
sieve-tube elements Phloem cells that are joined end-to-end to form continuous tubes throughout the plant.

sieve tubes Cells of sieve-tube elements joined end-to-end.

single-circuit pathway A path in which the blood moves through the heart, then passes on to the rest of the body, and finally returns to the heart.

sinoatrial (SI noh AY tree uhl) **node** Tissue that makes up the pacemaker in the heart.

sinus venosus In a frog, a sac which receives deoxygenated blood returning to the heart from most of the body.

skeletal muscle Voluntary muscle that is attached to the skeleton and aids in movement.

skin gills In starfish, hollow tubes used for gas exchange that project from the lining of the body cavity.

slime molds Fungal protists.

small intestine The portion of the digestive tube in which most digestion and absorption occur.

smog A type of air pollution formed by chemical reactions of pollutants from automobile exhaust and other sources.

smooth muscle Involuntary muscle found in internal organs and blood vessels.

social hierarchy A ranking of individuals in a social group.

sodium-potassium pump A process that moves sodium ions from inside of the neuron to the outside.

solar energy Energy from the sun.

solute (SAHL yoot) The substance dissolved in a solution.

solution A homogeneous mixture in which one substance is dissolved in another.

solvent (SAHL vuhnt) The substance in which a solute is dissolved.

somatic mutations Mutations that occur in the body cells of an organism.

sori (SOR ee) Structures that are masses of spore cases on the underside of fronds.

speciation (spee shee AY shuhn) The process by which new species evolve from ancestral species.

species A group of organisms of a particular kind that can naturally interbreed and produce fertile young.

sperm cells The male gametes.

spermatogenesis (sper mat uh JEHN uh sihs) The formation of sperm cells.

sphincters (SFIHNGK tuhrz) Muscles formed of circular fibers that close a tube when they contract.

spicules (SPIHK yoolz) Hard structures which provide support for sponges.

spinal cord A cord of nerve tissue located within the vertebral column.

spindle A three-dimensional structure of microtubules that appears to guide the movements of chromosomes during mitosis.

spinnerets In spiders, specialized appendages from which silk is released.

spiracles (SPIHR uh kuhlz) Openings found on the ventral surface through which air enters an organism.

spirilla (spI RIHL uh) Bacteria shaped like spirals.

sponges Aquatic animals in the phylum Porifera.

spongin A flexible material which provides support for a sponge.

spongy bone The softer inner part of the bone which contains many small spaces.

spongy mesophyll A layer of irregularly shaped parenchyma cells around the intercellular spaces in the mesophyll.

spontaneous (spahn TAY nee uhs) **generation** A hypothesis that states that living things can originate from nonliving matter.

sporangiophores (spuh RAN jee uh forz) Stalklike hyphae involved in asexual reproduction in zygomycetes.

sporangium (spuh RAN jee uhm) A spore case.

sporophyte (SPOR uh fīt) The diploid plant that produces spores.

sporozoans Parasitic protozoans that form spores.

stamens (STAY muhnz) The male reproductive structures of a flower.

statocysts Small saclike organs that aid in balance in crayfish and other invertebrates.

stele The vascular cylinder in the root of a plant.

stem The structure that conducts water and minerals, absorbed by the roots, upward into the leaves.

sterilization The process of killing microbes by heat and pressure.

stigma (STIHG muh) The upper part of the pistil upon which pollen grains land.

stimulants Drugs that increase the number of impulses conducted along neurons and at synapses.

stimulus (STIHM yuh luhs) A change that can cause a reaction.

stipe The stem of a mushroom.

stock The rooted stem of the parent plant in grafting.

stolons (STOH lahnz) Hyphae that are parallel to the fungus' growth medium.

stomach A pouchlike enlargement of the digestive tube in which food can be stored and partially digested.

stomata (STOH muh tuh) Openings in the leaf epidermis through which water vapor and oxygen pass out of the leaf and carbon dioxide passes into it.

stone canal In a starfish, a short tube which carries water from the sieve plate to the ring canal.

strata (STRAY tuh) In geology, layers of sedimentary rock.

strip cropping A farming method that alternates cover crops such as hay with row crops such as corn.

stromatolite (stroh MAT uh līt) A mound of limestone formed by the activities of one-celled organisms.

style (stīl) The stalk between the stigma and the ovary in a flower.

sublingual (suhb LIHNG gwuhl) **glands** Salivary glands located under the tongue.
submandibular (suhb man DIHB yuh luhr) **glands** Salivary glands located along the lower jaw.
substrate The substance on which an enzyme acts.
succulent (SUHK yuh luhnt) A plant that has large fleshy leaves or stems in which water is stored.
suspension A heterogeneous mixture containing particles distributed within a liquid, gas, or solid.
sweat glands Structures in the skin which give off watery secretions to help cool the body and eliminate wastes.
swim bladder A structure that is filled with gases and helps the fish keep afloat in the water.
swimmerets Appendages that aid both in gas exchange and in reproduction in crustaceans.
symbiosis (sihm bī OH sihs) A close relationship between two species in which at least one species benefits.
symbiotic (sihm bī AHT ihk) **hypothesis** A hypothesis stating that eukaryotic cells evolved from prokaryotic cells when certain prokaryotes began to live inside other cells.
symmetry The arrangement of parts on opposite sides of a plane or around a central axis.
sympathetic nervous system The portion of the autonomic nervous system that controls the internal organs during stressful situations and increased activity.
synapse (sih NAPS) The gap that an impulse must cross to get from one neuron to the next.
synapsis (sih NAP sihs) The pairing of homologs in prophase I of meiosis.
synovial (sih NOH vee uhl) **fluid** A lubricant secreted by a connective tissue membrane surrounding a joint.
synthesis (SIHN thuh sihs) The metabolic processes by which living things combine simple substances to form more complex substances.
syrinx (SIHR ihnks) A voice box in birds.
systemic circulation The flow of blood through those blood vessels that go to all organs and tissues except the lungs.
systole (SIHS tuh lee) The period during which the ventricles are contracted.

taiga (TĪ guh) The northern part of the coniferous forest biome.
tail gland A gland which secretes an oily fluid that a bird transfers to its body feathers to keep them smooth and waterproof.
target organ The organ that is acted on by a hormone.
taxonomic key A guide to the identification of organisms based on certain distinguishing characterstics.
taxonomist (tak SAHN uh mihst) A scientist who specializes in taxonomy, the science of classification.
taxonomy (tak SAHN uh mee) The science of classification.

Tay-Sachs (tay saks) **disease** A genetic disorder involving the inability to synthesize an enzyme that prevents lipid buildup in brain cells.
T-cells Lymphocytes that join with foreign antigens and then destroy foreign cells.
technology (tehk NAHL uh jee) The use of scientific knowledge to improve the quality of human life.
telophase The last stage of mitosis, during which the chromosomes again become threadlike and the cytoplasm is divided.
telson A flat, taillike structure of certain arthropods.
temperate deciduous forest The biome in which the typical vegetation is broad-leaved deciduous trees, which shed their leaves seasonally.
temperate phages Phages that stay inside their host cells for a long time without causing lysis.
temperate rain forest The biome in which temperatures are moderate, humidity is high, and rainfall is often more than 300 cm each year.
tendons Bands of inelastic connective tissue that join muscles to bones.
terracing The grading of cultivated slopes into a series of steps to reduce runoff and erosion.
territoriality The marking and defending of a territory by the animals that live there.
territory The specific area in which an animal or group of animals of a species lives.
test cross A cross between a living thing showing a dominant trait, but of uncertain genotype, and a living thing that is homozygous recessive for that trait.
testis The gonad of the male, which produces sperm cells.
testosterone (tehs TAHS tuh rohn) The hormone that stimulates development of the reproductive organs and of male sexual characteristics.
theory A hypothesis, or explanation, of some part of nature that has repeatedly been supported by evidence.
theory of evolution by natural selection Darwin's theory explaining evolution, based on the survival of those organisms best adapted to the environment.
theory of punctuated equilibrium A theory stating that species change little or not at all for long periods of time, but sometimes give rise to new species in a relatively short time.
thermal pollution Waste heat put into the environment.
thigmotropism (thigh MAHT ruh pihz uhm) Tropic response to touch in plants.
thorax In an arthropod, the middle segments in the body.
thrombosis Formation of clots in a blood vessel or in the heart.
thymine (THĪ meen) One of the bases that occurs in DNA.

thymus An organ found in the chest above the heart, believed to produce a hormone that is needed for normal production of lymphocytes.

thyroid gland An endocrine gland that lies across the windpipe and that produces thyroxine.

thyroid-stimulating hormone (TSH) A hormone that is produced by the pituitary gland and controls secretion by the thyroid gland.

thyroxine (thī RAHK seen) A hormone of the thyroid gland that regulates the metabolic rate of the body's tissues.

tidal energy The kinetic energy in the movement of tides and waves.

tissue A group of cells, similar in structure, and specialized for a particular function.

tobacco mosaic virus A kind of virus that infects tobacco leaves.

toxin (TAHK suhn) A poison released by a pathogen.

trachea (TRAY kee uh) The windpipe, a tube at the front of the throat that carries air to and from the lungs.

tracheal tubes Tubes in terrestrial arthropods that carry oxygen and carbon dioxide and extend into the body tissue.

tracheids (TRAY kee ihdz) Elongated and thickened xylem cells that conduct water in a vascular plant.

trait A feature that a living thing can pass on to its young.

transcription (tran SKRIHP shuhn) The process of producing mRNA from the instructions in DNA.

transformation The process by which genetic material absorbed from the environment is added to or replaces part of a bacterium's DNA.

transitional form An organism intermediate in characteristics between two other kinds of organisms and showing an evolutionary link between these organisms.

translation The assembly of a protein molecule according to the code in an mRNA molecule.

translocation (trans loh KAY shuhn) In genetics, a chromosomal alteration involving the transfer of a chromosome segment from one chromosome to another nonhomologous chromosome. In botany, the movement of dissolved materials from one part of a plant to another.

transpiration The process by which water vapor escapes from plant surfaces.

transpiration-cohesion theory The theory that transpiration pulls the water up from the roots.

trichocysts (TRIHK uh sihsts) Sharp-pointed structures of certain protists for defense or for catching prey.

tricuspid valve The valve connecting the right atrium and the right ventricle.

trisomy (TRĪ soh mee) The condition in which there are three homologs of a particular chromosome.

tRNA (transfer RNA) A nucleic acid that carries amino acids to the ribosomes during protein synthesis.

trochophore (TRAHK uh for) A top-shaped, bilaterally symmetrical larva of certain mollusks and marine annelids.

trophoblast The outer layer of cells of a blastocyst.

tropical rain forest The biome that has a large amount of rain and year-round warm temperatures.

tropism (TROH pihz uhm) A movement of a plant part toward or away from a stimulus.

tube feet Hollow thin-walled cylinders found in echinoderms that are used in movement.

tundra The northernmost terrestrial biome in which precipitation and average temperature are low.

tunicates (TOO nuh kihts) Chordates in subphylum Urochordata.

turgor (TER guhr) **pressure** The pressure exerted by water and the materials dissolved in it against the cell membranes and walls of a plant.

turgor movements Movements that result from the loss and gain of turgor pressure.

Turner's syndrome A genetic disorder in which the sex chromosome makeup is XO.

tympanic membranes Sensory organs for sound.

ultrasonography (uhl truh sahn AHG ruh fee) A technique that uses sound waves to help physicians examine fetuses.

umbilical cord The structure that connects the embryo to the placenta.

umbo The oldest part of a bivalve shell.

understory The layer of shorter trees in vertical stratification.

ungulates (UHNG gyuh lihts) Hoofed mammals such as horses, antelope, deer, and rhinoceroses.

unicellular Consisting of a single cell.

univalves Mollusks that have a one-piece shell.

uracil (YUR uh suhl) A base in RNA which bonds only with adenine.

urea A nitrogen-containing waste produced as a by-product of protein breakdown.

ureters (yu REE tuhrs) Tubes which carry urine from the kidneys to the urinary bladder.

urethra (yu REE thruh) A tube which carries urine from the bladder to the outside of the body.

urinary bladder A saclike organ that stores urine.

urine A liquid waste consisting mostly of water, with some urea, salts, and other substances.

uropods Wide flat appendages in certain crustaceans that form part of a posterior paddle.

uterus A muscular organ in mammals in which the embryo develops.

vaccination The injection of weakened or dead pathogens or their toxins into a healthy person or animal to produce immunity.

vacuoles (VAK yoo ohlz) Fluid-filled structures in cells that may contain a variety of substances.
vagina A muscular tube that receives the semen during intercourse and serves as the birth canal.
variable factor The condition that distinguishes the experimental group from the control group in an experiment.
vas deferens (vas DEHF uh rehnz) A long tube which connects each epididymis with the urethra.
vascular bundles Bundles of xylem and phloem.
vascular cambium A band of meristematic cells that has phloem cells on its outer side and xylem cells on its inner side.
vascular plants Plants that have specialized tissues for carrying water and food throughout the plant body.
vascular tissue In plants, cells that conduct water and food.
vectors (VEHK tuhrz) Organisms that spread disease but are not affected by the disease.
vegetative propagation The production of new plants from a nonreproductive part of the parent plant.
vein In animals, a large blood vessel that carries blood toward the heart. In plants, a vascular bundle in the leaf.
venae cavae (VEE nee KAY vee) Two large veins in humans that carry blood to the heart from systemic circulation.
venation A pattern made by veins.
ventral cavity The space containing the heart and lungs as well as organs of the digestive, excretory, and reproductive systems.
ventricle A lower chamber of the heart.
vernalization (ver huh luh ZAY shuhn) The exposure of plants to a period of cold to induce flowering.
vertebral column The series of bones enclosing the spinal cord.
vertebrates Those animals that have a backbone.
vertical stratification The distribution of vegetation in layers.
vessel elements Open-ended xylem cells that conduct water.
vessels Vessel elements joined end-to-end to form continuous tubes.
vestigial (vehs TIHJ ee uhl) **organs** Functionless structures in an organism.
villi (VIHL ī) Threadlike structures through which most nutrients are absorbed in the small intestine.
viral replication The reproduction of viruses.
viral transduction The transfer of host DNA to another organism by a virus.
viroids (VĪ roidz) Infective strands of naked RNA.
virulent (VIHR yuh luhnt) **phages** Phages that cause lysis of their hosts.
virus A very small particle usually made of nucleic acid and protein.

visceral (VIHS uhr uhl) **mass** The region of the body that is made of most of the organs of a mollusk.
vitamins Organic molecules that can be absorbed without being digested and that are involved in many metabolic activities.
viviparous (vī VIHP uhr uhs) Producing embryos that develop within the female and are born live.
vomerine teeth In a frog, large teeth on the roof of the mouth used to hold the prey prior to swallowing it whole.

water cycle The movement of water from the air to the earth and back again.
water pollution Any substance that damages the quality of water.
water-vascular system A connected system of water-filled canals in echinoderms that serves in locomotion.
watersheds Drainage areas that supply water to lakes, ponds, and reservoirs.
weight A measure of the force of gravity between two objects.
white blood cells The part of the blood that helps to defend the body against infection.
wilting The loss of turgor in leaves, stems, and flowers of a plant due to the loss of water from its tissues.
wind energy The kinetic energy of moving air.
windbreaks Rows of trees and shrubs that reduce wind speed.
woody plants Plants in which the xylem cells have hard, thick cell walls giving strength and support to the plant.

xerophyte (ZIHR uh fīt) A plant adapted to grow where water is scarce.
xylem A vascular tissue that provides mechanical support of the plant and conducts water and minerals from the roots to the leaves.

yolk sac A membrane that grows out from the digestive tract of the embryo and surrounds the yolk.

zone of differentiation The region in plants in which newly lengthened cells differentiate.
zone of elongation The area where cell enlargement occurs in plants.
zoologist A person who studies animals.
zoospores (ZOH uh sporz) In fungi, asexual spores produced by oomycotes. In algae, asexual spores produced by certain types of green algae.
zygomycotes (zī guh mī KOHTZ) Land-dwelling fungi such as molds and blights.
zygospores (ZĪ guh sporz) In fungi, thick-walled diploid spores of zygomycotes. In green algae, diploid cells resulting from conjugation.
zygote (ZĪ goht) A fertilized egg cell.

INDEX

References to illustrations appear in italics following text references.

A

Abdomen, 586
Abdomen, of arthropods, 459
Abdominal cavity, 586
Abiogenesis, 25
Abiotic factors, 727
    climate, 730–731
    soil, 733
Aboral surface, 452
Abscisic acid, 402
Abscission layer, 401
Absorptive heterotroph, 221
Abyssal zone, 780
Acetylcholine, 598, 638, 673
Acetyl-coA, 87–88
Acid rain, 791–792
Acids, 45–46. *See also* Bases.
Acoelomates, 418, 482
Actin, 597
Action potential, 671
Activation energy, 81; *82*
Active immunity, 308
Active site, 82
Active transport, 72; *72*
Adaptation, 179–180; *179–180*
    in birds, 531–532, 538–539; *531, 539*
    in insects, 473–477; *473–477*
    in leaves, 353
    in plants and water, 368–371; *370–371*
    in reproduction, 575
    in roots, 344–345; *345*
    in stems, 349; *349*
Adaptive radiation, 198, 200; *198*
Adenine, 110, 112
Adenosine diphosphate (ADP), 86
Adenosine triphosphate (ATP), 76–81, 83, 86–91, 93; *76–77, 86–87*
Adhesion, 364
Adolescence, 719
Adrenal glands, 689–690; *689*
Adrenaline, 637, 689

Adrenocorticotropic hormone (ACTH), 687
Adsorption, 236
Adulthood, 719
Adult-onset diabetes, 695
Adventitious roots, 342, 375
Aerobic organism, 195
Aerobic respiration, 87, 246
Afterbirth, 717
Agars, 324, 325
Aging, 720
Agnatha, 505
Air Pollution, 790–791; *790*
Albinism, 153
Albumen, 536
Alcoholic fermentation, 90
Aldosterone, 690
Algae, 318–324; *318–324*. *See also* Plants.
    reproduction of, 320–324; *320–323*
Algal blooms, 757
Algal protists, 256–260; *257–258*. *See also* Protists.
Algin, 323, 325
All-or-none response, 597, 671
Allantois, 525, 715
Alleles, 136; *136*
Allergy, 310
Alligators, 526; *526*
Alpha cells, 692
Alternation of generations, 322
Altricial birds, 538
Alvarez, Luis and Walter, 201–202
Alveoli (s. alveolus), 568, 647
Amanita, 289
Amino Acids, 52–53, 606; *52*
Ammonia, 489
Ammonification, 736
Amniocentesis, 154
Amnion, 525, 714
Amniote egg, 524, 576; *524*
Amoeba, 263; *263*
    digestion in, 265
    locomotion of, 264

    reproduction in, 265; *265*
    respiration in, 265
Amoeba proteus, 264; *264*
Amoebocytes, 419
Amoeboid movement, 264
Amphibia, 513
Amphineura, 440
Amphioxus, 504; *505*
Amylase, 617
Anabolic reactions, 79; *80*
    dehydration synthesis, 80; *80*
Anaerobic organism, 195
Anaerobic respiration, 87, 89, 246
Anal fin, 509
Anal pore, 267
Analogous structures, 173
Anaphase I and II, 106; *106*. *See also* Meiosis.
Angiosperms, 339–340
    conifer reproduction, 379–381; *379–381*
    flowers, 382–383; *382–383*
    pollen grains and ovules, 383–384; *383–384*
    reproduction in, 378–379; *378–379*
    seed dispersal of, 388; *388*
    seed and fruit formation, 385–386; *386*
    seed germination, 388–390; *389–390*
    success of, 390–391; *391*
Angiotensin, 693
Animal cell, 62; *62*
Animalia, 220, 222
Annelida, 431
Annelids, 431
Annual rings, 347
Annuals, 406, 774
Annulus, 283
Anopheles mosquito, 304–305
Anorexia nervosa, 620
Antagonistic muscles, 564
Antennae, 461
Anterior adductor muscle, 442

833

Anterior end, 417
Anterior lobe, 687
Anther, 382
Antheridia (s. antheridium), 279, 328
Anthozoa, 451
Anthropology, 205
Antibiotics, 312; *312*
Antibodies, 308–309, 629; *629*
Anticodon, 114
Antidiuretic hormone (ADH), 658, 688
Antigen, 308, 628–629; *629*
Anura, 513
Anus, 429
Anvil, 679
Aorta, 634
Aortic arches, 433
Apical dominance, 400
Apical meristem, 403–404; *403–405*
Apoda, 514
Appendicitis, 621
Appendicular skeleton, 510, 587
Appendix, 615
Aquatic biomes, 776–780
Arachnida, 461
Arachnids, *463–464*
 circulation in, 464
 digestion in, 463
 reproduction in, 463
 respiration in, 464
Archaebacteria, 244
Archaeopteryx, 172; *172*
Archegonia, 328
Archenteron, 577
Aristotle, 25–26, 214
Arterial heart, 447
Arteries, 632; *633*
Arterioles, 633
Arthritis, 592
Arthropoda, 459
Arthropods
 adaptations of, 473–477; *473–477*
 arachnids, 463–464; *463–464*
 chilopods and diplopods, 469; *469*
 classification of, 460–461; *461*
 crustaceans, 465; *465*
  crayfish, 465–467; *466–467*
 importance of, 462
 insects, 469–470; *469*
  grasshoppers, 470–472; *470–471*
 success of, 462; *462*

traits of, 459–460; *460*
Artificial heart, 641; *641*
Artificial photosynthesis, 95
Artificial skin, 601; *601*
Artificial vegetative propagation, 377; *377*
Artiodactyla, 553
Ascaris lumbricoides, 430; *430*
Aschelminthes, 429
Ascocarp, 282
Ascogonia, 281
Ascomycota, 281–282; *281–282*. See also Fungi.
Ascomycotes, 281
Ascospores, 281
Ascus, 281
Asexual reproduction, 32, 104
 artificial vegetative propagation, 377; *377*
 natural vegetative propagation, 375–376; *376*
Aspergillus, 282, 288
Association neurons, 667
Astasia, 259
Atherosclerosis, 640; *640*
Athlete's foot, 304, *304*
Atmosphere, 191; *191*
Atomic mass unit (AMU), 37
Atomic number, 39
Atoms, 37–38; *38.* See also Electron; Neutron; Nucleus.
Atria, (s. atrium) 506, 567, 631
Atrioventricular node (A-V node), 637; *637*
Atrioventricular valves, 632
Auditory canal, 679
Auditory nerve, 680
Aurelia, 424; *424*
Auricle, 679
Australopithecines, 205
Australopithecus afarensis, 206; *206*
Australopithecus africanus, 206
Australopithecus boisei, 206
Australopithecus robustus, 206
Autonomic nervous system, 668–669
Autosomes, 139
Autotrophs, 79
Auxins, 400–401; *400*
Aves, 530
Axial skeleton, 510, 587
Axon-endings, 664
ABO blood types, 628–629

ACTH, 687
ADH, 688
AIDS, 307; *225*

B

Baboon troop, 557; *557*
Bacilli, 243; *243*
Bacteria (s. bacterium)
 beneficial, 251; *251*
 blue-green, 240–242; *240–242*
 conjugation and transformation in, 249–250; *249–250*
 disease caused by, 299–301; *300*
 endospore formation, 248–249; *248*
 energy production in, 246; *246*
 growth conditions for, 244–246; *245*
 locomotion of, 242
 reproduction of, 248; *248*
 respiration in, 246
 structure and function of 242–244; *242–243*
Bacteriophages, 235–236; *235–236*
Ball-and-socket joint, 590
Barbs, 531
Barbules, 531
Bark, 346, 347
Basal metabolic rate (BMR), 649
Basal metabolism, 649
Bases, 46. See also Acids.
Basidia, 283
Basidiocarp, 283–284
Basidiomycota, 283–284; *283–284.* See also Fungi.
Basidiomycotes, 283
Basidiospores, 283–284
Bathyal zone, 780
B-cells, 308–309
Beavers, 552; *545, 552*
Bees, 476–477; *476–477*
Behavior
 of birds, 537–538; *538*
 of fish, 512
 of plants, 408, 410; *408*
Behavioral adaptations, 180; *180*
Belly, 595
Beta cells, 692
Bias, 4
Bicuspid valve, 632
Bilateral symmetry, 417
Bile, 615
Binary fission, 241, 248; *248*

Binomial nomenclature, 216–217; *216*
Bio-controls, 793
Biochemical likenesses, 218
Biodegradable chemicals, 792
Biogenesis, 26–29; *27, 28*
Biology, 8
Biomagnification, 792
Biomass energy, 797
Biomes
  climate, 764; *764*
  coniferous forest, 766–767
  definition of, 763
  deserts, 774–775
  freshwater biomes, 776–779
  grasslands, 769–771
  marine biome, 779–780
  mountains, 775–776
  temperate deciduous forest, 768–769
  temperate rain forest, 769
  tropical rain forest, 772–773
  tundra, 765–766
Biosphere, 727
Biotic factors, 727
  ecological pyramids, 740–741
  energy flow in environment, 739–740
  roles of living things, 738–739
  symbiotic relationships, 741
Biotic potential, 746
Bipedal organisms, 205
Birds, 566
  classification of, 532; *532*
  egg development in, 536–537; *537*
  feeding adaptations of, 538–539; *539*
  flight adaptations of, 531–532; *531*
  migration of, 539; *539*
  pigeon, anatomy of, 534–536; *534–536*
  reproductive behavior of, 537–538; *538*
  traits of, 530–531; *531*
Birth, 716–717
Birth rate, 747; *747*
Bivalves, 442
Blade (*Laminaria*), 323
Blade (leaf), 350
Blastocoel, 577, 713; *713*
Blastocyst, 713; *713*
Blastodisc, 577; *577*
Blastula, 495, 577

Blood
  blood types and transfusion, 628–629; *629*
  function and composition of, 625–626; *625*
  platelets, 628; *628*
  red blood cells, 626; *626*
  Rh factor and pregnancy, 630. *See also* Circulation; Circulatory system; Lymphatic system.
  white blood cells, 627; *627*
Blood clot, 628
Blood pressure, 638; *637*
Blood types, *136*, 629
Blood vessels, 632–634
Blue-green bacteria, 240–242; *240–242*. *See also* Bacteria.
Body, elements found in, 39
Body cells, 104
Bolus, 612
Bone, 505; *587*
  development of, 588–589; *589*
  disorders of, 592
  joints and movement, 590–591; *591*
  structure of, 589–590; *589*
Bone marrow, 626
Bony fish, 508–509; *508–509, 565*
Book gills, 461
Book lungs, 464
Botanist, 8
Bowman's capsule, 657
Brain, 572, 665–667; *572, 666*
Brain chemistry, 673–674
Brainstem, 665
Branching of plants, 369
Breathing, mechanics of, 647–649; *648*
Breeder reactors, 795
Breeding behavior, in fish, 512
Bronchi, 647
Bronchioles, 647
Brooding, 538
Brown, Robert, 59
Brown algae, 322–324; *323*. *See also* Plants.
Bryophyta, 327
Bryophytes, 327–329; *327–329*. *See also* Plants.
Bud, 346, 399
Budding, 377, 493; *493*
Buds (hydra), 423
Buffers, 625
Bulb, 376; *376*
Bunch grasses, 770

Bundle scars, 346
Bundle sheaths, 352
Bursae, 591
Bursitis, 592

C

Calcitonin, 688
Calories, 605
Calvin cycle, 94; *94*
Calyx, 382
Cambarus, 465
Cambrian period, 198, 200
Canada lynx, changes in population of, 749–750; *749*
Cancer, 654
Canines, 612
Canis, 216–217; *216*
Capillarity, 364; *364*
Capillary, 633; *633*
Capsid, 233
Capsomere, 234
Capsule, 242, 329
Carapace, 465, 528
Carbohydrates, 49–51, 605, 617
Carbon compounds, 48–49; *48–49*
Carbon cycle, 735; *735*
Carboxyl group, 51
Carcinogens, in plants, 344
Cardiac acceleration center, 637
Cardiac inhibition center, 638
Cardiac muscle, 564, 595
Cardiac sphincter, 613
Careers
  arborist, 411
  beekeeper, 497
  biochemist, 117
  biology teacher, 117
  cytotechnologist, 117
  diet-related staff, 721
  ecologist, 799
  entomologist, 497
  finding out about, 117
  fisheries biologist, 579
  genetic counselor, 227
  health careers, finding out about, 721
  job interview, preparation for, 579
  job interview follow-up, 799
  landscape architect, 411
  medical support staff, 721
  medical technologist, 311
  microbiologist, 311
  museum technician, 227

mushroom grower, 311
nature photographer, 579
parasitologist, 497
park ranger, 799
plant breeder, 227
plant physiologist, 411
in professional medical service, 721
resume, how to write, 497
soil scientist, 799
summer job, how to find, 311
veterinarian, 579
want ads, how to use, 227
Carnivora, 552
Carnivores, 545, 552, 738
Carotenoids, 92
Carrageenans, 324, 325
Carrier molecules, 71–72; *72*
Carriers, 153
Carrying capacity, 746
Cartilage, 505, 588, 592; *591*
Cartilaginous fish, 507–508; *507*
Catabolic reactions, 79
  hydrolysis, 80–81; *80*
Catalyst, 81
Caterpillar, 473
Caudata, 514
Ceca, 471, 566
Cecropia, 474; *474*
Cecum, 615
Cell
  cell theory, 59–60, 99
  cellular reactions, 79–85
  cellular transport, 67–74; *72*
  discovery of, 59; *59*
  energy for, 86–91
  kinds of, 60–61; *61*
  methods of studying, 16–19; *16–18*
  reproduction of, 104–108; *105–107, 108*
  structure and function of, 62–66; *66*
Cell body, 664
Cell membrane, 67–68; *68*
Cell plate, 103
Cell theory, 59, 99
Cellular respiration, 87; *91*
Cellular transport, 67–74
  active transport, 72; *72*
  diffusion, 69–70; *69*
  endocytosis and exocytosis, 72–74
  facilitated diffusion, 71–72
  osmosis, 70–71; *70–71*

Cellulase, 617
Cellulose, 51
Cell wall (plant cell), 66
Cementum, 612
Cenozoic Period, 200
Centipede, 469; *469*
Central nervous system, 663
Centrioles, 101
Centromere, 101
Centrum, 564
Cephalization, 425, 490, 572
Cephalochordata, 503
Cephalopoda, 440
Cephalopods, 440, 446–448; *448*
Cephalothorax, 461
Cerebellum, 665
Cerebral cortex, 665
Cerebral hemispheres, 666
Cerebrospinal fluid, 664
Cerebrum, 665
Cervical vertebrae, 588
Cervix, 708
Cestoda, 425
Cetacea, 552
Chalazae, 536
Chelicerae, 461
Chelicerata, 460, 461
Chelipeds, 465
Chelonia, 525
Chemical change, 54; *54*
Chemical equation, 44
Chemical evolution, 192–194; *193–194*
Chemical formula, 43. *See also* Coefficient.
Chemical reaction, 43–44; *43–44*
Chemicals and plant growth, 396; *397*
Chemoreceptors, 676
Chemosynthesis, 246
Chemotherapy, 312
Chemotropism, 408
Chilopoda, 461
Chimpanzees, *559*
Chinook salmon, *512*
Chiroptera, 551
Chitin, 275, 459
Chiton, 440–441; *440*
Chlamydomonas, 319; *319*
Chlorella, 319, 320
Chlorophyll, 65, 91, 92, 240
Chlorophyta, 318
Chloroplast, 65; *65*
Choanocyte, 419
Choanoflagellate, 271; *271*

Chondrichthyes, 505
Chordata, 503
Chordates
  amphibians
    frog, anatomy of, 514–517; *514–516*
    metamorphosis in, 517–518; *517–518*
    traits of, 513–514; *514*
  fish
    bony fish, 508–509; *508–509*
    breeding behavior and development of, 512; *512*
    cartilaginous fish, 507–508; *507*
    jawless fish, 506–507
    perch, anatomy of, 509–512; *509–511*
    water balance in, 513
    traits of, 505
  lower chordates, 504; *504–505*
    traits of, 503–504
  See also Birds; Mammals; Reptiles.
Chorion, 525, 710, 714
Chorionic gonadotropic hormone, 710
Chorionic villi, 715
Choroid, 678
Chromatids, 101–102, 106
Chromatin, 62
Chromatophores, 492; *492*
Chromoplast, 65
Chromosomal alteration, 147–149
Chromosomes, 62, 99–102, 104–105; *99, 108*
  alterations in, 147–149; *148–149*
  disorders from nondisjunction, 151–152; *151–152*
Chromosome theory of heredity, 138; *138*
Chrysalis, 474
Chrysophyta, 259
Chyme, 613
Cilia, 265
Ciliary muscle, 677
Ciliata, 265
Ciliates, 265–266, *266. See also* Protozoans.
Circadian rhythms, 753–754
Circulation
  blood pressure and blood flow, 638; *637*
  circulatory disorders, 640–641

heartbeat and blood volume, 637–638
heart disease, 640; *640*
in invertebrates, 486; *486*
treatment of circulatory problems, 641
in vertebrates, 566; *566*
See also Blood; Circulatory system; Lymphatic system.

Circulatory system
blood vessels, 632–634; *633*
circulation to body organs, 634; *634*
heart, 631–632; *631*
See also Blood; Circulation; Lymphatic system.

Citric acid cycle, 88; *88*
Clams, 442; *442–443*
circulation in, 444
digestion in, 444; *485*
nervous system of, 444
reproduction in, 444; *494*
respiration of, 443
Class, 219; *218–219*
Classification, 213; *214*
binomial nomenclature, 216–217; *216*
classification groups, 218–220; *218–220*
early systems of, 214–215; *215*
identification aids, 223–224; *223, 225*
modern classification, 217–218, 220–222; *220–222*
Linnaean system, 215
Clavicles, 552, 588
Cleavage, 495, 577–578, 713; *495, 577*
Climate, 730–731
Climatogram, 764; *764*
Climax community, 756
Clitellum, 432
Cloaca, 515
Cloning, 160–161; *160*
Closed circulatory system, 486; *486*
Clostridium, 246; *246*
Clostridium botulinum, 302
Club mosses
reproduction of, 330; *330*
See also Plants.
Cnidarian, 420
Cnidoblast, 421
Coacervate, 194
Coagulation, 628

Cocci, 243; *243*
Coccyx, 588
Cochlea, 679
Cocoon, 474
Codon, 113
Coefficient, 43
Coelom, 418, 482, 563
Coelomates, 418, 482
Coenzymes, 82
Cohesion, 364
Cold deserts, 774
Coleoptile, 390
Collecting tubule, 657
Colloid, 45
Colon, 615
Colony, 240
Color blindness, 141; *141*
Commensalism, 741
Communicable diseases, 293
Communication, 558–559; *559*
Communities, 728
primary succession, 756; *756*
secondary succession, 757; *757*
succession in lakes and ponds, 756–757; *756*
Compact bone, 589
Companion cells, 363
Comparative anatomy, 172; *172*
Comparative biochemistry, 173–174
Comparative embryology, 173; *173*
Complementary protein foods, 606
Complete metamorphosis, 474, 496; *473*
Complete protein foods, 606
Compound microscope, 16–17; *16*
Compounds, 41–42
inorganic, 48
organic, 48–53; *49–52*
Computers
genetic applications, 159
use in animal research, 550
use in ecology, 759
use in genetics, 159
use in medical care, 699
Concentration gradient, 69
Condensation, 734
Cones (eye), 679
Cones (tree), 379
Coniferous forest, 766–767
Conifers, 338–339, 379–381; *339, 379–381*
Conjugation, 249, 268, 320; *249, 268, 320*

Conjugation tubes, 320
Connective tissue, 585
Constipation, 618
Consumers, 738
Contagious diseases, 293
Contour farming, 786
Contour feathers, 531; *531*
Contractile fibers, 258
Contractile vacuoles, 64, 258
Control group, 4–6
Conus arteriosus, 516
Convergence, 203; *203*
Convolutions, 666
Coracoid, 534
Coral, 424–425; *425*
Cork, 344
Corm, 376; *376*
Cornea, 678
Corolla, 382
Coronary circulation, 634
Corpus luteum, 690, 709
Cortex, 343, 657, 689
Corticosteroids, 689
Cortisol, 689
Courtship behavior, 538, 557
Covalent bond, 42
Covalent compound, 42
Cowper's glands, 687; *686*
Cramps, 598
Cranial nerves, 668
Cranium, 587
Crayfish, 465; *466–467*
circulation in, 466
digestion in, 466
excretion in, 467
nervous system of, 467
reproduction in, 467
respiration in, 467; *488*
Cretaceous Period, 200–202
Cretinism, 694
Crick, Francis, 109
Cristae, 63
Crocodiles, 526; *526*
Crocodilia, 525
Cro-Magnons, 207; *207*
Crop, 471, 535
Crop rotation, 786
Cross bridges, 597
Crossing over, 105, 141–143; *105, 141.* See also Heredity.
Cross pollination, 124–127; *125–127*
Crown, 612
Crown gall, 302; *302*
Crustacea, 461

837

Crustaceans, 465–467; *465–467*
Cutin, 330
Cutting, 377; *377*
Cyanophyta, 240
Cycads, 338
Cyclic adenosine monophosphate (cAMP), 701
Cyclostomes, 506
Cyst, 265, 279
Cytochromes, 89
Cytokinesis, 103
Cytokinins, 402
Cytoplasm, 61, 103, 106
Cytosine, 110, 112
Cytoskeleton, 64

D

Daphnia, 465
Dark reactions, 93–94; *93–94*
Darwin, Charles, 175; *176*
  and gradualism, 180–181; *181*
  and theory of natural selection, 176–177; *177*
Data, 4
Daughter cells, 99–100, 103–104
Day-neutral plants, 397–398
Deciduous plants, 371
Deciduous teeth, 612
Decomposers, 739
Degrees Celsius, 14; *14*
Dehydration, 80
Dehydration synthesis, 80
Deletion, 149; *149*
Democritus, 37
Denaturation, 83–84; *84*
Dendrites, 664
Denitrification, 736
Density-dependent factors, 748–749
Density-independent factors, 748
Dentin, 612
Dentition, 545; *545*
Deoxyribonucleic acid. *See* DNA.
Deoxyribose, 110
Depolarization, 671
Depressants, 674
Dermatitis, 601
Dermis, 600; *599*
Deserts, 774–775
Development
  after birth, 719–720; *719*
  before birth, 712–717
  in fish, 512
  in invertebrates, 495–496; *495*
  in vertebrates, 577–578; *577–578*
Diabetes mellitus, 695
Dialysis fluid, 659
Diaphragm, 543, 586, 648
Diarrhea, 618
Diastole, 638
Diastolic pressure, 638
Diatomaceous earth, 262
Diatoms, 259–260; *262; 259*
  locomotion of, 260
Dicots, 340–341; *340. See also* Seed plants.
Diet
  balanced diet, 609–610; *607–608*
  malnutrition and diet deficiency, 620–621
  obesity and weight control, 620
  See also Nutrition.
Differentiation, 715
Diffusion, 69–72, 358–361; *71*
Digestion, 31
  absorption of food in, 618
  absorption of water in, 618; *618*
  of carbohydrates, 617
  in invertebrates, 484–486; *485*
  of lipids, 617
  of proteins, 617–618
  in vertebrates, 566; *566*
  See also Digestive system.
Digestive gland, 453
Digestive system
  disorders of, 621
  liver and pancreas, 615–616; *616*
  mouth, pharynx, and esophagus, 611–613; *611*
  small and large intestines, 614–615; *614*
  stomach, 613; *613*
  See also Digestion.
Dihybrid cross, 135; *135*
Dinoflagellates, 260; *260, 262. See also* Protists.
  locomotion of, 260
  reproduction of, 260
Dinosaurs, 200, 525–526
Dipeptide, 53; *53*
Dipleurula, 451, 496; *496*
Diplococci, 243
Diplococcus pneumoniae, 243
Diploid number, 99; *99*
Diplopoda, 461
Direct development, 496, 578
Disaccharides, 50
Discovery
  Alzheimer's disease, 676
  dinosaurs as parents, 525
  fireflies, 474
  hay fever, 385
  laser surgery, 635
  malaria, 269
  mass extinction, 185
  new phylum, 433
  sharkskin and yacht racing, 9
Disease
  bacterial disease, 299–301; *300*
  barriers to pathogens, 304–305; *305*
  chemotherapy, 308
  definition of, 293
  fungal disease, 302; *302*
  germ theory of, 296–297
  immune response, 306–308
  interferon, 308
  pathogens and hosts, 294–295
  preventing spread of, 310; *310*
  protozoan disease, 302–304
  spreading of, 295–297; *296*
  viral disease, 298–299; *298*
  viroid and rickettsial diseases, 301; *301*
Disinfectants, 310
Dislocation, 592
Dissociation, 45
Distal convoluted tubule, 657
Diuresis, 658
Divergence, 202; *203*
DNA (deoxyribonucleic acid), 53, 62, 173–174
  and bacteria, 241, 250
  and genetic clock, 208–209; *209*
  replication of, 110–111; *111*
  role of, 109
  structure of, 109–110; *109–110*
  transcription process, 112–113; *113*
  translation process, 114–115
  and viruses, 234, 238
DNA splicing, 158, 160; *158*
Dobzhansky, Theodosius, 185
Dominance, 128; *128*
Dormancy, 388–389, 406–407; *407*
Dorsal cavity, 586
Dorsal fins, 509
Dorsal root, 667
Dorsal root ganglion, 667
Dorsal surface, 417
Double-circuit pathway, 567
Double fertilization, 385; *385*

Double helix, 110
Down feathers, 531; *531*
Down's syndrome, 151–152; *151–152*
Drosophila melanogaster (fruit fly), 138–141; *139–140*
Drought evaders, 775
Drought resisters, 775
Drugs and the nervous system, 674
Duckbilled platypus, 547; *547*
Duodenum, 614
Duplication, 149; *149*
Dwarfism, 694
Dynamic equilibrium, 69

E
Earthworm
   circulation in, 432–433; *486*
   digestion in, 432; *485*
   nervous system of, 433; *491*
   reproduction in, 433
Ecdysone, 474
Echidna, 547; *547*
Echinodermata, 451
Echinoderms
   economic importance of, 455
   other echinoderms, 454–455; *454*
   starfish, 451–453; *452–453*
   traits of, 451; *451*
Echolocation, 551
Ecological succession, 756
Ecology, 727
Ecosystem, 727–728
Ectoderm, 417, 481, 577, 714
Ectoplasm, 264
Ectotherm, 505, 568
Edentata, 551
Edentates, 551; *551*
Effectors, 663
Egg cells, 707
Einstein, Albert, 54
Elastic cartilage, 592
Eldredge, Niles, 180–181
Electroencephalograph, 675
Electron, 37
Electron cloud, 38
Electron microscope, 18–19
Electron transport chain, 89; *89*
Elements, 38–40; *39–40*
   See also Atomic number; Isotopes; Mass number.
Embryo, 173, 713

Embryo sac, 384
Embryological likenesses, 218
Embryology, 455
Embryonic development, 715–716; *716*
Embryonic disk, 714
Embryonic induction, 715
Emigrants, 746
Emphysema, 655
Emulsification, 617
Enamel, 612
Endangered species, 789; *789*
Endemic, 297
Endocrine system
   adrenal glands, 689–690
   endocrine glands, 685–686; *686*
   feedback mechanisms and homeostasis, 696–697
   interaction of pituitary and hypothalamus, 697–698
   molecular and cellular controls, 700–701; *700–701*
   other hormones, 693–694
   ovaries and testes, 690; *690*
   pancreas, 692–693; *692*
   pancreatic and adrenal disorders, 695–696
   pituitary gland, 686–688; *686*
   pituitary and thyroid disorders, 694–695
   thyroid and parathyroid glands, 688
Endocytosis, 72–74; *73–74*
Endoderm, 418, 481, 577, 714
Endodermis, 343
Endoplasm, 264
Endoplasmic reticulum (ER), 63; *63*
Endorphins, 673
Endoskeleton, 451, 484, 504, 564; *484*
Endosperm, 385
Endospore, 248–249, 301; *248, 301*
Endotherm, 530, 568
Endotoxin, 301
Energy, 54–55
Energy resources
   fossil fuels, 794–795
   nuclear energy, 795–796
   other energy sources, 796–797
   solar energy, 796
Enkephalins, 673
Entomologist, 8–9
Entomophthora fumosa, 287

Environment, 727
   carbon cycle, 735; *735*
   ecological pyramids, 740–741; *740–741*
   energy flow in, 739–740
   nitrogen cycle, 736; *736*
   oxygen cycle, 737
   roles of living things, 738–739
   soil and climate, 730–733
   symbiotic relationships in, 741
   water cycle, 734–735; *735*
Environmental resistance, 746
Enzyme(s), 81–82; *82*
   factors affecting activity, 83–84; *83–84*
   models of, 82–83; *83*
   RNA as an, 112
Enzyme-substrate complex (ES), 72, 82–83; *83*
Ephedra, 338; *338*
Ephedrine, 289
Epicotyl, 386
Epidemic, 297
Epidermis, 421, 599; *599*
Epididymis, 687
Epiglottis, 612, 647
Epilepsy, 675
Epiphytes, 773
Epithelial tissue, 585
Epithelium, 419
Epochs, 198
Equator (of cell), 102
Equilibrium, 679
Eras, 198
Ergotism, 288; *288*
Erosion, 786
Error-catastrophe hypothesis of aging, 720
Erythroblastosis fetalis, 630
Erythrocytes, 626
Erythropoietin, 693
Escherichia coli, 158, 160, 243, 251, 295
Esophagus, 432, 612
Essential amino acids, 606
Estivation, 754
Estrogen, 690, 707
Estrus, 557
Estuaries, 780
Ethylene, 401
Euglena, 220; *220*
Euglenoids, 257–259; *257–258*. *See also* Protists.
   locomotion of, 257
   reproduction of, 258

Euglenophyta, 257
Eukaryotes, 60, 66–67; *61, 66*
   control systems in, 164–165; *164–165*
   origin of, 196–197; *196–197*
Euphotic zone, 779–780
Euplotes, 266; *266*
Eustachian tube, 679
Eutrophication, 757
Evaporation, 734
Evolution
   and comparative anatomy, 172–173; *172*
   and comparative biochemistry, 173–174
   and comparative embryology, 173; *173*
   definition of, 169
   and fossils, 170–172; *170–172*
   mechanisms of, 183–187; *183–184, 186*
   observation of, 178–179; *178, 179*
   rate of, 180–181, 198; *181*
   theories of, 174–177; *175–177*
   types of adaptations, 179–180; *179–180*
   See also Origin of life.
Excretion
   in invertebrates, 489–490; *489–490*
   in vertebrates, 569–570; *569*
Excretory system
   and kidney disease, 659; *659*
   kidney structure, 656–657; *657*
   nephrons, how they work, 658; *657*
   organs of, 656; *657*
   and water balance control, 658
Excurrent siphon, 443
Exergonic reaction, 81; *82*
Exobiology, 9
Exocrine glands, 685
Exocytosis, 71
Exons, 164
Exoskeleton, 459, 483; *484*
Exotic species, 789
Exotoxin, 301
Experimental group, 4–6
Experimentation, 4–6
Expiration, 647
Expiratory center, 651
Extensor, 596
External ear, 679

External fertilization, 494–495, 575
External nares, 514, 528
External respiration, 645
Extinction, 31
Extracellular digestion, 485
Eyespot, 258

F
Facilitated diffusion, 71–72
Facultative anaerobes, 246
Fallopian tube, 708
Family, 219; *218–219*
Fatty acid molecule, 51; *51*
Feces, 618
Feedback mechanisms, 696–697; *696*, 710
Felis, 216–217, 220; *216, 220*
Female reproductive system, 707–708; *707–708*. *See also* Reproduction.
Femur, 535, 589
Ferguson, Lloyd N., 52
Fermentation, 89–91, 246; *90–91*
Ferns, 331–333; *331–332*
   reproduction of, 332; *332*
Fertilization, 104, 384–385, 712–713; *105, 385, 712–713. See also* Reproduction.
Fertilization membrane, 712
Fetoscopy, 155
Fetus, 715
Fibrin, 628; *628*
Fibrous cartilage, 592
Fibrous roots, 342; *343*
Fibrous tissues, 592
Fiddlehead, 331; *331*
Field guide, 223; *223*
Filament, 382
Filoplumes, 531; *531*
Filter feeders, 418
Fins, 505
First filial generation, 126
First messenger, 700
First polar body, 107
Fish, 567
   bony fish, 508–509; *508–509*
   breeding behavior and development, 512; *512*
   cartilaginous fish, 507–508; *507*
   jawless fish, 506–507
   perch, anatomy of, 509–512; *509–511*
   traits of, 505
   water balance in, 513
Fixed joint, 590
Flagella, 257, 319
Flagellates, 271; *271*
   digestion in, 271
   See also Protozoans.
Flame cell, 426, 489
Flatworms, 425
   flukes, 427; *427*
   planarians, 426–427; *426–427*
   tapeworm, 428; *428*
Fleming, Alexander, 312
Flexor, 596
Flowering plants, 382–386, 388–391; *382–386, 388–391*
Flowers, 382–383; *382–383*
Flukes, 427; *427*
Follicle, 688, 690, 707
Follicle stage, 709
Follicle stimulating hormone (FSH), 687, 709
Fontanelle, 589; *589*
Food chain, 262, 739; *739*
Food resources, 787–788
Food vacuole, 265
Food web, 740; *740*
Foramen magnum, 206
Foraminiferan, 263; *263*
Forebrain, 572; *572*
Forest floor, 708
Forest resources, 788
Fossil fuels, 794–795
Fossil record, 218
Fossils, 170–172; *170–172*
Fracture, 592
Fragmentation, 277, 493
Fraternal twins, 717; *717*
Free radicals, 720
Freshwater biomes, 776–779
Frog, 514; *514*
   circulation in, 516; *516*
   digestion in, 515; *515*
   excretion in, 516; *516*
   metamorphosis in, 517–518; *517–518*
   nervous system of, 517; *516*
   reproduction in, 517
   skeleton of, 515; *515*
Frond, 331
Fructose, 50; *50*
Fruit fly, 138–141; *139–140*
Fruiting bodies, 261, 277
Fruits, 386

Fucoxanthin, 322
Fucus, 361; *361*
Fungal protists, 256, 260–261; *260–261.* See also Protists.
Fungi (s. fungus), 221, 222
  characteristics of, 275
  classification of, 278; *278*
  digestion in, 286; *286*
  diseases caused by, 302; *302*
  importance of, 288–289; *288–289*
  nature of, 276–277; *276*
  nutrition, 286; *286*
  phylum Ascomycota, 281–282; *281–282*
  phylum Basidiomycota, 283–284; *283–284*
  phylum Oomycota, 278–279; *279*
  phylum Zygomycota, 279–280; *280*
  reproduction of, 277–284; *279–284*
  structure of, 277; *277*
  symbiotic relationships of, 287; *287*

G

Galactose, 50; *50*
Galileo, 59
Gall bladder, 616
Gallstones, 621
Gametes, 104, 107, 127–129
Gametogenesis, 107
Gametophyte, 322
Ganglia (s. ganglion), 426, 444, 491, 667
Garden pea traits, 124; *124*
Gas exchange, 568–569, 649; *569, 649*
Gastric ceca, 471
Gastric glands, 613
Gastric juice, 613
Gastrodermis, 421
Gastropoda, 440, 445
Gastropods, 440
Gastrovascular cavity, 420, 481
Gastrula, 495, 577
Gastrulation, 714–715; *714.* See also Reproduction.
Gelans, 324
Gemmules, 420
Gene flow, 187
Gene frequency, 186; *186*
Gene mutation, 147

Gene pool, 186
Genes, 109, 112, 127
  disorders from mutations, 153–154; *153–154*
  mutations in, 147, 150; *150*
  sex-linked, 138–141; *139–141*
Genetic code, 112
  DNA replication, 110–111; *111*
  DNA role in, 109
  DNA structure, 109–110; *109–110*
Genetic disorders
  detection and treatment of, 154–155; *155*
  from gene mutations, 153–154; *153–154*
  from nondisjunction, 151–152; *151–152*
  See also Genetics; Heredity.
Genetic drift, 187
Genetic engineering, 158–161; *158, 160*
Genetic equilibrium, 186
Genetic expression, 161–162
Genetic likenesses, 218
Geneticist, 9
Genetics
  environmental influences, 161–162; *161–162*
  genetic engineering, 158–161; *158, 160*
  Mendel's early experiments, 123–127; *125–126, 127*
  Mendel's hypotheses, 127–129; *128–129*
  Mendel's results, 132–134; *133–134*
  other discoveries in, 135–137; *135, 136, 137*
  and probability, 129–130; *130*
  selective breeding, 157; *157*
  See also Genetic disorders; Heredity.
Genetics and the Origin of Species (Dobzhansky), 185
Genital pore, 426
Genotype, 129
Genus, 205, 219; *218–219*
Geographic isolation, 183; *183–184*
Geologic time chart, 198; *199*
Geothermal energy, 796
Geotropism, 408; *408*
Germination, 388–390; *389–390*
Germ mutations, 147

Germ theory of disease (Koch), 296–297
Gestation period, 549; *549*
Giantism, 694
Giardia, 271
Gibberellins, 401–402; *401*
Gill bailers, 465
Gill hearts, 447
Gill pouches, 173
Gills, 283, 442, 467, 503
Gill slits, 503
Ginkgo, 338; *338*
Girdling, 366
Gizzard, 432, 471
Gliding joint, 590
Glomerular filtrate, 658
Glomerulus, 657
Glucagon, 692
Glucocorticoids, 689
Glucose, 49, 86, 88, 89; *50*
Glycercol, 51
Glycogen, 50–51
Glycolysis, 87
Gnetales, 338
Goiter, 695
Golden algae, 259–260; *259.* See also Protists.
Golgi apparatus, 63–64; *64*
Gonadotropic releasing hormone (GnRH), 710
Gould, Stephen Jay, 180–181
Gradualism (Darwin), 181; *181*
Grafting, 377
Granulocytes, 627
Granum, 65
Grasshopper, 470–471
  circulation in, 471; *486*
  digestion in, 470–471; *485*
  excretion in, 472; *490*
  nervous system of, 472; *491*
  reproduction in, 472; *494*
  respiration in, 472; *488*
Grasslands, 769–771
Gray matter, 665
Green algae 318–322; *318–322*
  locomotion of, 319
Green glands, 467
Greenhouse effect, 791
Ground layer, 771
Growth, in plants, 395–407; *396–407*
Growth lines, 442
Guanine, 110, 112
Guard cells, 352

Gullet, 267
Guttation, 371; *371*
Gymnomycota, 256
Gymnosperms, 338, 339; *338–339.* See also Seed plants.

H

Habitat, 728
Hair follicles, 600
Half-life, 170–171
Hallucinogenic drugs, 674
Halophyte, 370–371
Hammer, 679
Haploid number, 105
Hardy, G.H., 186
Hardy-Weinberg Law, 186–187
Haustoria, 286
Haversian canals, 590
Hay fever, 385
Head (arthropods), 459
Hearing, 679–680; *680*
Heart, 631–632; *631*
Heart attack, 640
Heart transplant, 641
Heartwood, 347
Heimlich maneuver, 653
Hemocoele, 464
Hemocyanin, 464
Hemoglobin, 626
Hemophilia, 154; *154*
Hemotoxin, 528
Henle's loop, 657
Hepatic portal circulation, 634
Herbaceous plants, 342
Herbivores, 545, 788
Herb layer, 768, 771
Heredity, 123
    chromosome theory of, 138
    linkage and crossing over, 141–143; *143*
    sex-linked genes, 138–141; *139–141*
    See also Genetic disorders; Genetics; Mutations
Hermaphrodites, 420, 493
Heterogeneous mixture, 44; *44*
Heterotrophs, 79
Heterozygous, 128
Hibernation, 514, 556, 754; *556*
Hierarchical system, 218–219; *218–219*
Hierarchy, 752
Hindbrain, 572; *572*
Hindmost lobe, 687

Hinge joint, 591
Hirudidea, 431
Histamines, 310
Histones, 164
HIV, 225
Holdfast, 320
Holdfast organs, 779
Homeostasis, 31, 650
Home range, 751
Hominids, 205–207; *206, 207*
Homo erectus, 206–207
Homogeneous mixture, 44
Homo habilis, 206; *207*
Homologous chromosomes, 99, 105
Homologous structures, 172, 218
Homologs, 105
Homo sapiens, 205, 207–208
Homo sapiens neanderthalensis, 207
Homozygous, 128
Honeybees, 476–477; *476–477*
Hooke, Robert, 59
Hormones, 616, 658, 685
    and invertebrates, 492–493; *492*
    and plant growth, 399–402; *399–402*
    and vertebrates, 573–574; *574*
Horny outer layer, 442
Horse, 171, 202; *171*
Horseshoe crab, *461*
Horsetails, 331; *331*
    See also Plants.
Host cell, 233
Host resistance, 294
Hosts, 294–295
Hot deserts, 774
Human growth hormone (HGH), 160, 687
Human populations, 746–747; *746–747*
Humerus, 534
Hyaline cartilage, 592
Hybrid, 126; *126*
Hybrid vigor, 157
Hydathodes, 371
Hydra, *422*
    digestion in, 422; *485*
    nervous system of, 422; *491*
    reproduction in, 423
    respiration in, 422
Hydrodictyon 319; *319*
Hydrolysis, 80–81; *80*
Hydrotropism, 368–369
Hydroxyl group, 51

Hydrozoa, 421, 422
Hyman, Libbie Henrietta, 449
Hypertension, 640
Hyperthyroidism, 694
Hypertonic solution, 71; *71*
Hyphae, 277; *277*
Hypocotyl, 386
Hypoglycemia, 696
Hypothalamus, 666, 697–698
Hypothesis, 4
Hypothesis of use and disuse (Lamarck), 174
Hypothyroidism, 694
Hypotonic solution, 70–71; *71*
Hypoxia, 655
Hyracotherium, 171, 202

I

Identical twins, 717; *717*
Ileum, 614
Ilium, 588
Immigrants, 746
Immune response, 306–308
Immune system, 308
Immunity, 306
Imperfect flowers, 383
Implantation, 714; *714*
Imprinting, 538
Imprints, 170
Inbreeding, 157
Incisors, 612
Incomplete dominance, 137; *137.* See also Genetics.
Incomplete metamorphosis, 473, 496; *473*
Incomplete protein foods, 606
Incurrent siphon, 443
Indirect development, 496, 578
Induced-fit model (enzyme), 82–83; *83*
Inducer, 162, 715
Inducible enzyme, 162
Infancy, 719; *719*
Infections, 293
Infectious disease, 293
Infestations, 293
Inflammation, 310
Ingestive heterotroph, 221
Inheritance of acquired characteristics (Lamarck), 174–175; *175*
Inhibitor, 84
Inhibitory synapses, 673
Inner cell mass, 713

Inner ear, 679
Inorganic compounds, 47. *See also* Organic compounds.
Insecta, 461
Insectivora, 551
Insectivores, 551; *551*
Insects, 469; *469*
   adaptations of, 475–476; *475*
   communication among, 477; *477*
   grasshopper, 470–472; *470–471*
   metamorphosis of, 473–474; *473–474*
   social insects, 476; *476*
Insertion, 595
Inspiration, 647
Inspiratory center, 651
Instars, 473
Instincts, 537
Insulin, 160, 692
Integument, 565, 599
Integumentary system, 565; *565*
   skin disorders, 601; *601*
   skin functions, 600–601; *601*
   skin structure, 599–600; *599*
Intercellular fluid, 308
Intercellular spaces, 347
Intercostal muscles, 648
Interferon, 160, 308
Interkinesis, 106
Intermediate lobe, 687
Internal fertilization, 495, 575
Internal nares, 514
Internal respiration, 645
International System of Measurement (SI), 10–14
   length, 11–12
   time and temperature, 14
   volume, 13
   weight and mass, 12–13
Interneurons, 667
Internode, 346
Interphase, 100–101; *101. See also* Mitosis.
Interspecific competition, 754
Interstitial cells, 706
Intertidal zone, 779
Intestinal glands, 614
Intracellular digestion, 485
Intraspecific competition, 749
Introns, 164
Inversion, 149; *149*
Invertebrates
   body forms of, 483–484; *483–484*
   body layers and coeloms, 481–483; *482*
   circulation in, 486; *486*
   cnidarians, 420–421; *421*
   development of, 495–496; *495*
   digestion in, 484–486; *485*
   excretion in, 489–490; *489–490*
   flatworms, 425–429; *426–428*
   hormones in, 492–493; *492*
   hydra, 422–423; *422*
   larval forms, 496; *496*
   nervous systems of, 490–492; *491–492*
   organization of, 417–418; *417–418*
   other cnidarians, 423–425; *423–425*
   reproduction in, 493–495; *493–494*
   respiration in, 488; *488*
   roundworms and rotifers, 429–430; *429–430*
   segmented worms, 431–433, 435; *431–432, 435*
   sponges, 418–420; *418–420*
In vitro, 713
Involuntary acts, 663
Ionic bond, 42
Ionic compound, 42; *42*
Ions, 40–41; *41*
Iris, 678
Ischium, 588
Islets of Langerhans, 692; *692*
Isomers, 49
Isotonic solution, 70; *71*
Isotopes, 40; *40*

J

Jacob, François, 162, 163
Jacobson's organ, 528
Janssen, Jans and Zacharias, 59
Jarvik, Robert, 641
Jaundice, 621
Jawless fish, 506–507
Jejunum, 614
Jellyfish, 424; *424*
Joint, 590; *591*
Juvenile-onset diabetes, 695

K

Karotype, 151; *151*
Keel, 534
Kelps, 323, 325; *323, 325*
Keratin, 544, 599
Kidney failure, 659
Kidney machine, 659; *659*
Kidneys, 656–657; *657*
Kidney transplant, 659
Kinetic energy, 55; *55*
Kingdom, 218–221; *218–219, 221*
Klebsiella, 160
Klinefelter's syndrome, 152
Koch, Robert, 296–297
Koch's postulates, 296–297

L

Labium, 470
Labor, 716–717; *716*
Labrum, 470
Lactase, 617
Lacteal, 618
Lactic acid fermentation, 90
Lactose, 50
Lagomorpha, 551
Lagomorphs, 551; *551*
Lamarck, Jean Baptiste, 174–175; *175*
Laminaria, 323; *323*
Lancelets, 504
Language, 559
Lanugo, 716
Large intestine, 614–615; *614*
Larva, 473
Larval forms, 496; *496*
Larynx, 647
Lateral buds, 346
Lateral line system, 510, 572–573
Lateral roots, 342
Latimeria, 509
Latitude, 763
Law, definition of, 7
Law of Conservation of Energy, 54
Law of Conservation of Mass, 54
Law of Conservation of Mass-Energy, 54
Layering, 377
Leaflets, 350
Leaf primordium, 404
Leaf scars, 346
Leaves, 350–353; *350–353. See also* Seed plants.
Leeches, 435; *435*
Leeuwenhoek, Anton van, 15, 255
Length, 11–12
Lens, 679
Lenticels, 346
Lentic habitat, 776

Leukocytes, 308, 627
Lianas, 773
Lichen, 287; *287*
Liem, Karel F., 506
Life, definition of, 29
Life span, 30
Ligaments, 592
Light, and plant growth, 396–398; *397–398*
Light reactions, 92–93; *93*
Limiting factors, 746, 776
Limnetic zone, 777
Linkage, 141–143; *143. See also* Heredity.
Linnaeus, Carolus, 215, 217
Lipase, 617
Lipids, 51–52; *51*, 606, 617
Liposomes, 194; *194*
Liter, 13
Littoral zone, 777
Liver, 615; *616*
Liver fluke, life cycle of, 427; *427*
Liverworts, 327; *327*
Lizards, 527; *527*
Lobe-finned fish, 508
Lobules, 547
Lock-and-key model (enzyme), 82–83; *83*
Long-day plants, 397–398
Lotic habitat, 776
Lumbar vertebrae, 588
Lumbricus terrestris, 431
Lungfish, *509*
Lungs, 503, 647
Luteinizing hormone (LH), 687, 709
Lwoff, Andre, 162, 163
Lycopodium, 330; *330*
Lycopsida, 330
Lymph, 308
Lymphatic system, 635–636; *635–636. See also* Circulatory system.
Lymph nodes, 636
Lymphocytes, 308–309, 627
Lysis, 236–237; *237*
Lysogenic bacteria, 237
Lysogeny, 237; *237*
Lysosomes, 64

M

McClintock, Barbara, 142
Macrocystis, 323
Macronucleus, 266
Macrophage, 308, 627
Magnification, 16–17
Malaria, 269
Male reproductive system, 705–707; *706–707. See also* Reproduction.
Malnutrition, 620–621
Malpighian tubules, 469, 472, 490
Maltase, 617
Malthus, Thomas, 176
Maltose, 50
Mammals
  communication among, 558–559; *559*
  evolution of, 200
  hibernation of, 556; *556*
  importance of, 545–546; *545–546*
  marsupials, 547–549; *548*
  migration of, 555; *555*
  monotremes, 546–547; *547*
  orders of, 551–553; *551–553*
  placentals, 549; *549*
  reproductive behavior of, 557–558; *558*
  territoriality and social hierarchy of, 556–557; *557*
  traits of, 543–545; *544–545*
Mammary glands, 543
Mandible, 461, 587
Mandibulata, 460, 461, 469
Mantle, 439
Mantle cavity, 442
Marchantia, 327
Mariculture, 788
Marine biomes, 776, 779–780
Marsupialia, 547
Marsupials, 547–549; *548*
Marsupium, 547
Mass, 12–13
Mass extinction, 185, 198, 200–202; *200–202*
Mass number, 39
Mastigophora, 271
Matrix, 585
Matter
  changes in, 54
  conservation of, 54
Maxillae, 465
Maxillary teeth, 514
Maxillipeds, 465
Mechanoreceptors, 676
Medulla, 657, 689
Medulla oblongata, 667
Medusa, *421*
Megaspore, 380

Megasporocytes, 380
Meiosis
  compared with mitosis, 108; *108*
  male and female gametes, 107; *107*
  meiosis I, 105–106; *105–106*
  meiosis II, 106; *106. See also* Mitosis.
Meiospores, 321
Melanin, 153, 600
Melanophore-stimulating hormone (MSH), 687
Membrane (cell), 67–68; *68*
Memory cell, 309
Mendel, Gregor
  early experiments of, 123–127; *125–126, 127*
  hypotheses of, 127–129; *128–129*
  results of, 132–134; *133–134. See also* Genetics.
Meninges, 664
Menopause, 708
Menstrual cycle, 708–709; *709*
Menstruation, 709
Meristem tissue, 342
Mesenchyme, 419
Mesoderm, 418, 482, 577, 714
Mesoglea, 421, 482
Mesophyll, 350–351
Mesozoic Era, 338
Metabolism, 30–31, 79
Metamorphosis, 473–474, 492–493, 517–518; *473–474, 517–518*
Metaphase, 102; *102. See also* Mitosis.
Metaphase I and II, 106; *106. See also* Meiosis.
Metazoans, 417
Meteorite impact hypothesis (Alvarez), 202; *202*
Meter, 11
Methanogens, 195, 222
Microbes, 293
Microbiologist, 8
Microfilaments, 64
Microfossils, 194
Micrographia (Hooke), 59
Micronucleus, 266
Microorganisms, 27, 232–271, 292–310
Micropyle, 384
Microscope
  compound, 16–17
  compound light, 16–17

electron, 18–19
Microspheres, 194; *194*
Microspore, 380
Microsporocytes, 380
Microtubules, 64; *64*
Microvilli, 618
Midbrain, 572; *572*
Middle ear, 679
Migration, 539, 555, 754; *539, 555, 754*
Miller, Stanley, 192
Mineralocorticoids, 690
Missing link, 208
Mitochondria, 62–63, 196; *63, 197*
Mitosis
    compared with meiosis, 108; *108*
    interphase, 100–101; *101*
    significance of, 103–104
    stages of, 101–103; *101–103*
    why cells divide, 100; *100*
    See also Meiosis.
Mixtures, 44–45. *See also* Solutions.
Model, of atom, 37–38; *38*
Modern theory of evolution, 185
Moist coniferous forest, 769
Molars, 612
Molecules, 42
Mollusca, 439
Mollusks
    cephalopods, 446–448; *447–448*
    clams, 442–444; *442–443*
    economic importance of, 448–449
    gastropods, 445–446; *445–446*
    other bivalves, 444; *444*
    traits of, 439–442; *439–441*
Molting, 459
Monera, 221, 222, 240
Monerans, 240
Monocots, 340–341; *340. See also* Seed plants.
Monocytes, 627
Monod, Jacques, 162, 163
Monohybrid cross, 126–127; *127*
Monoploid number, 105
Monosaccharides, 49
Monosomy, 151
Monotremata, 546
Monotremes, 546–547; *547*
Morgan, Thomas Hunt
    and linkage and crossing over, 141–143; *141*
    and sex-linked genes, 138–141; *139–141*

Morphological adaptations, 179; *179*
Morula, 713
Mosses, 327–328; *327*
Motor neurons, 667, 673; *673*
Motor unit, 597; *597*
Mountains, 776
Mouth-to-mouth resuscitation, 653
mRNA, 112–115, 163–165; *113*
Mucous membrane, 646
Mulch, 771
Multicellular organisms, 29
Multiple alleles, 136. *See also* Genetics.
Multiple sclerosis, 676
Muscle, 594; *594*
Muscle fibers, 594
Muscle tissue, 586
Muscular dystrophy, 598
Muscular system
    energy and muscle contraction, 597–598; *597*
    muscle contraction, 596–597; *596–597*
    muscle disorders, 598
    muscles and movement, 595–596; *595*
    types of muscles, 594–595; *594–595*
Mushrooms, 289; *289*
Musk glands, 544
Mutations
    chromosomal alterations, 148–149; *148–149*
    gene mutations, 150, 153–154; *150, 153–154*
    nature of, 147–148
    See also Genetic disorders; Heredity.
Mutualism, 287, 741
Mycelium, 277
Mycobacterium tuberculosis, 294
Mycoplasmas, 254
Mycorrhiza, 287
Myelin sheath, 664
Myofibrils, 596; *596*
Myomeres, 564; *564*
Myosin, 597
Myxedema, 694
Myxomycota, 260

N

Nasal cavity, 645

Nastic movements, 410; *410*
Natural resources, 785
    food resources, 787–788
    forest resources, 788
    soil resources, 786–787
    water resources, 787
    wildlife resources, 789; *789*
Natural selection, 176–177; *177*. *See also* Evolution.
Natural vegetative propagation, 375–376; *376*
Neanderthals, 207; *207*
Needham, John, 28
Negative feedback, 696
Nekton, 777
Nematocyst, 421
Nematoda, 429
Nematodes, 429
Nephridia (s. nephridium), 433, 489
Nephrons, 656–658; *657*
Neritic zone, 779
Nerve cord, 503
Nerve impulses, 664
Nerve net, 422, 490
Nerves, 665
Nerve tissue, 586
Nervous system
    autonomic, 668–669
    brain, 665–667
    brain chemistry, 673–674
    conduction at the synapse, 672–673; *672–673*
    division of the, 663–664
    drugs and the, 674
    in invertebrates, 490–492; *491–492*
    nerve cells and nerves, 664–665
    nerve impulse, 671
    nervous disorders, 675–676
    peripheral, 668
    resting state of neurons, 670–671
    spinal cord, 667
    in vertebrates, 572–573; *572*
    See also Senses.
Net venation, 340; *340*
Neural arch, 564
Neural folds, 715
Neural plate, 715
Neural tube, 715
Neurons, 664; *664*
    conduction at synapse, 672–673; *672–673*
    nerve impulse, 671; *671*
    resting state of, 670–671

Neuropeptides, 673
Neurospora crassa, 281; *281*
Neurotoxin, 528
Neurotransmitters, 672
Neutral solution, 45
Neutron, 37
Newton (N), 12
Niche, 729
Nitrification, 736
Nitrogen cycle, 736
Nitrogen fixation, 251, 736
Nitrogenous base, 110
Node, 346
Nondisjunction, 148, 151–152; *148, 151–152*
Nonhistone protein, 164
Nonrenewable resources, 785
Nonvascular plants, 317
Noradrenaline, 673
Nostoc, 287
Notochord, 173, 503
Nuclear energy, 795–796
Nuclear fission, 795; *795*
Nuclear fusion, 795; *795*
Nuclear membrane, 61, 62
Nuclear wastes, 796
Nucleic acids, 53
Nucleocapsid, 233
Nucleolus, 62
Nucleotides, 109–110; *109–110*
Nucleus
  in atom, 37
  in cells, 59, 61–62
Nutrient, 605
Nutrition, 605–606, 608–609; *607–608*
  malnutrition and diet deficiency, 620–621
  See also Diet.
Nymph, 473
NADH, 87
NADPH, 93

O

Obelia, life cycle of, 423; *423*
Obesity, 620
Obligate aerobes, 246
Obligate anaerobes, 246
Observations, 3–4
Oceanic zone, 779–780
Ocelli, 463
Oil-immersion lens, 17
Olfactory epithelium, 677; *677*
Olfactory nerve, 677

Oligochaeta, 431
Omnivores, 553, 738
On the Origin of Species by Means of Natural Selection (Darwin), 175
Oogenesis, 107; *107*
Oogonium, 279
Oomycota, 278–279; *279. See also* Fungi.
Oomycetes, 278
Ootid, 107
Oparin, A.I., 192, 194
Open circulatory system, 486; *486*
Operator, 162–163
Operculum, 510
Operon, 162–163; *163*
Opisthorchis sinensis, 427; *427*
Optic nerve, 679
Oral groove, 267
Oral surface, 452
Orchids, 287; *287*
Order, 219; *218–219*
Organ, 586
Organelles, 60
  of plant cells, 65–66
Organic compounds
  bonding of, 47–49; *49*
  carbohydrates, 49–51; *49–51*
  lipids, 51–52; *51*
  proteins, 52–53; *52–53*
Organizers, 715
Organ systems, 586
Origin, 595
Origin of life
  adaptive radiation, 198, 200; *198*
  chemical evolution, 192–194; *193–194*
  divergence and convergence, 202–203; *203*
  early earth, 191; *191*
  first organisms, 194–195; *195*
  human ancestry, 205–207; *205–207*
  mass extinctions, 200–202; *200–202*
  origin of eukaryotes, 196–197; *196–197*
  primate relationships, 208–209; *209*
  See also Evolution.
Ornithologist, 8
Oscillatoria, 242; *247*
Osculum, 418
Osmosis, 70–71; *70–71*
Ossification, 588

Osteichthyes, 505
Osteoblasts, 588
Osteocytes, 598
Osteoporosis, 592, 719
Ostia, 464
Outbreeding, 157
Ovaries, 382–383, 690; *575, 690*
Oviduct, 426, 708
Oviparous organisms, 508, 525
Ovoviviparous organisms, 508, 525, 576
Ovulation, 707, 709; *707*
Ovules, 380, 384
Ovum (ova), 104, 107
Oxygen, 48
Oxygen cycle, 737
Oxygen debt, 651
Oxyhemoglobin, 626
Oxytocin, 688

P

Pacemaker, 637
Paired jointed appendages, 460
Paleoanthropology, 205
Paleontology, 205
Palisade mesophyll, 351
Palmately compound leaves, 350
Pampas, 769
Pancreas, 616, 692–693; *692*
  disorders of, 695
Pancreatic amylase, 617
Pancreatic juice, 616
Papillae, 600, 676
Parallel venation, 340; *340*
Paramecium, 267–269; *267–268*
  See also Protozoans.
  digestion in, 267
  locomotion of, 267
  reproduction of, 268
  respiration in, 267
Paramecium caudatum, 267; *267*
Parapodia, 435
Parasite, 245, 276, 294
Parasitism, 287, 741
Parasympathetic nervous system, 668
Parathyroid glands, 688; *688*
Parathyroid hormone (PTH), 688
Parenchyma tissue, 341
Parental generation, 125; *125*
Parent cell, 99–100, 103–104
Parkinson's disease, 675
Parotid glands, 612
Parthenogenesis, 493

Passive immunity, 308
Passive transport, 68–70
Pasteur, Louis, 28–29; 297
Pathogens, 294–295, 307–308
Pearly layer, 442
Pectoral fins, 509
Pectoral girdle, 588
Pedipalps, 461
Pelage, 544; *544*
Pelecypoda, 440, 444
Pelecypods, 440
Pellicle, 267
Pelvic fins, 509
Pelvic girdle, 588
Penicillin, 282, 289
Penicillium, 282, 288, 289; *282*
Penicillium camemberti, 288
Penicillium notatum, 312
Penicillium roqueforti, 288
Penis, 707
Pentaradial, 451
Peppered moths, 178–179; *178, 179*
Pepsin, 617
Pepsinogen, 617
Peptic ulcer, 621
Peptide bond, 53, 114
Peptides, 673
Perch, 509–510
  circulation in, 511; *511*
  digestion in, 510
  excretion in, 512
  locomotion in, 509
  nervous system of, 512
  respiration in, 511
  skeleton of, 510; *511*
Perennials, 406, 774
Perfect flowers, 383
Pericardium, 631
Pericycle, 343
Periderm, 341
Periods, 198
Periosteum, 589
Peripheral nervous system, 664, 668; *663, 668*
Periphyton, 777
Perissodactyla, 553
Peristalsis, 432, 613
Permafrost, 765
Permanent teeth, 612; *611*
Permanent wilting, 368
Permian Period, 339
Pesticides, 787
Petals, 382
Petiole, 350
pH, 46, 84; *46, 84*

Phaeophyta, 318
Phages, 235–236; *235*
Phagocytosis, 74, 265; *74*
Pharynx, 426, 612, 646
Phase-contrast microscope, 17; *17*
Phenotype, 129
Phenylketonuria, 153
Pheromones, 476, 753
Phloem, 317, 363, 365–366; *363, 366*
Phosphate group, 109
Photochemical smog, 790
Photoperiod, 397
Photoperiodism, 397–398; *398*
Photoreceptors, 676
Photosynthesis
  conditions needed for, 91
  dark reactions in, 93–94; *93–94*
  factors affecting, 94; *94*
  light and pigments, 92; *92*
  light reactions, 92–93; *93*
  respiration versus, 95; *95*
Phototropism, 408; *408*
Photovoltaic cells, 796
Phycocyanin, 240
Phycoerythrin, 324
Phylum, 218–219; *218–219*
Physalia, 424
Physical change, 54; *54*
Physiological adaptations, 179
Phytohormones, 399–402; *399–401*
Phytoplankton, 318, 777
Pigeon
  circulation in, 535
  digestion in, 535
  excretion in, 535
  nervous system of, 536; *536*
  reproduction in, 536
  respiration in, 535; *535*
  skeleton of, 534–535; *534*
Pincers, 452
Pine, life cycle of, 380–381; *381*
Pineal gland, 693
Pinnately compound leaves, 350
Pinocytosis, 72–73; *73*
Pioneers, 328, 756
Pistil, 382
Pistillate flowers, 383
Pith, 347
Pituitary gland, 686–688; *686*
  disorders of, 694–695
  interaction with hypothalamus, 697–698
Pivot joint, 590
Placenta, 544, 576, 715

Placental barrier, 715
Placental mammals, 549
  carnivores, 552
  cetaceans, 552; *552*
  Chiroptera, 551
  edentates, 551; *551*
  insectivores, 551; *551*
  lagomorphs, 551; *551*
  primates, 553
  Proboscidea, 552; *553*
  rodents, 552; *552*
  ungulates 553; *552*
Planarians, 426–427; *426–427*
Plankton, 318, 777
Plantae, 220, 222
Plant cells, 65–66; *65*
Plantlet, 376; *376*
Plants
  behavior of, 408, 410; *408, 410*
  brown algae, 322–324; *323*
  bryophytes, 327–329; *327–329*
  club mosses and horsetails, 330–331; *330–331*
  evolution of, 200
  ferns, 331–333; *331–332*
  green algae, 318–322; *318–322*
  growth of, 395–407; *396–407*
  red algae, 324; *324*
  reproduction in, 378–386; *378–386*
  significance of algae, 324–325; *325*
  transport in, 357–366, 368–371; *358–366, 368, 370–371*
  vascular and nonvascular, 317
  See also Seed plants.
Planula, 424, 496; *496*
Plaques, 236; *236*
Plasma, 625
Plasma cell, 309
Plasmids, 158
Plasmodium, 260–261, 269, 295, 304; *304*
Plasmolysis, 71
Plastids, 65; *65*
Plastron, 528
Platelets, 626, 628; *628*
Platyhelminthes, 425
Pleura, 647
Pliohippus, 202; *203*
Plumule, 386
Pneumonia, 653
Point mutation, 150; *150*
Polarization, 671
Polar nuclei, 384

Pollen grains, 380, 383; *383*
Pollination, 124, 384–385; *385*
Pollution
   acid rain, 791–792
   air, 790–791; *790*
   water, 792–793
Polychaeta, 431, 435
Polychaetes, 435; *435*
Polygenic inheritance, 136
   See also Genetics.
Polyp, 421; *421*
Polypeptide, 53
Polyploidy, 149, *149*
Polysaccharides, 50–51; *51*
Polytrichum, 328–329; *329*
Population, 728–729
   behavior of, 750–754
   growth of, 746–747
   human populations, 747–748; *747–748*
   limits to population growth, 748–750; *749*
   structure of, 745–746
Population density, 745
Population genetics, 185
Population growth, 746; *746*
Population size, 745
Pore cells, 419
Porella, 327
Pores, 62, 656
Porifera, 418
Porphyra, 324, 325; *324*
Portuguese man-of-war, 424
Posterior adductor muscle, 442
Posterior end, 417
Posterior lobe, 687
Potential energy, 55; *55*
Prairie, 769
Precipitation, 734
Precocial birds, 538; *538*
Pregnancy and Rh factor, 630
Premolars, 612
Primary growth, 343
Primary oocyte, 107
Primary root, 342
Primary spermatocyte, 107
Primary succession, 756
Primates, 553
Primer pheromones, 753
Primitive streak, 578
Primordial soup, 192
Principle of dominance, 128
Principle of independent assortment, 136
Principle of segregation, 129

Prismatic layer, 442
Probability, 129
   product rule, 130; *130*
   rule of independent events, 130
   Punnet square, 132–134; *132–133*
   See also Genetics.
Proboscidea, 552; *553*
Producers, 738
Product rule, 130; *130*
Products (in chemical reaction), 54
Profundal zone, 777
Progesterone, 690, 707
Proglottid, 428
Prokaryotes, 60–61, 66–67, 220–221; *61, 66*
   control systems in, 162–163; *163*
   evolution of, 194–196; *195–196*
Prolactin, 687
Prophage, 237
Prophase, 101; *101.* See also Mitosis.
Prophase I and II, 105–106; *106.*
   See also Meiosis.
Prostaglandins, 693–694
Prostate gland, 687; *687*
Prostomium, 432
Proteins, 52–53, 606, 617–618; *53.*
   See also Amino acids.
Protein synthesis, 112–115
Prothallus, 332
Protista, 220, 222, 255
Protists
   dinoflagellates, 260; *260, 262*
   euglenoids, 257–259; *257–258*
   golden algae, 259–260; *259*
   importance of, 261–262; *262*
   nature of, 255
   slime molds, 260–261; *261*
   traits and classification of, 256–257; *256*
   See also Protozoans.
Proton, 37
Protonema, 328
Protophyta, 256
Protoplasm, 59
Protozoa, 257
Protozoans, 257
   amoeba, 264–265; *264–265*
   ciliates, 265–266; *266*
   diseases caused by, 302–304
   flagellates, 271; *271*
   importance of, 271
   paramecium, 267–269; *267–268*
   sarcodines, 263; *263*

   sporozoans, *269*
   See also Protists.
Proximal convoluted tubule, 657
Pseudocoelom, 418
Pseudocoelomates, 482
Pseudomonas aeruginosa, 245
Pseudopodia, 263
Pteropsida, 331
Pubis, 588
Puccinia graminis (wheat rust) 284; *284*
Pulmonary artery, 634
Pulmonary circuit, 568
Pulmonary circulation, 634
Pulmonary veins, 634
Pulse, 632
Punctuated equilibrium, 181; *181*
Punnett square, 132–134; *132–133*
Pupil, 678
Pure line, 125; *125*
Pus, 627
Pyloric sphincter, 613
Pyramid of biomass, 741; *741*
Pyramid of energy, 740; *740*
Pyramid of numbers, 741; *740*
Pyrrophyta, 260
Pyruvic acid, 87, 88, 90; *87*

Q

Quarantine, 312
Quill, 531

R

Radial canals, 452
Radial cleavage, 495; *495*
Radial symmetry, 417, 451
Radiant energy, 92
Radicle, 386
Radiolarian, 263; *263*
Radius, 535
Radula, 439
Rana clamitans, 514
Range, 388
Raven, Peter H., 333
Ray, John, 215
Ray-finned fish, 508
Reactants, 54
Receptacle, 382
Receptor proteins, 672, 700
Receptors, 663
Recessive genes, 128
Recombinant DNA, 158
Rectum, 615

Recycling, 785
Red algae, 324; *324*
  See also Plants.
Red blood cells, 626; *626*
Redi, Francesco, 26–27, 297
Red tide, 260; *262*
Reflex actions, 667
Reflex arc, 668; *668*
Refractory period, 671
Regeneration, 418
Regulator gene, 162, 163
Releaser pheromones, 753
Releasing factors, 698
Renal artery, 656; *656*
Renal circulation, 634
Renal vein, 656; *656*
Renewable resources, 785
Repressor, 163
Reproduction
  asexual, 32, 104, 108, 375–377; *108, 376–377*
  birth, 716–717; *717*
  cleavage and implantation, 713–714; *713*
  and courtship behavior, 557
  definition of, 31
  embryonic development, 715–716; *716*
  female reproductive system, 707–708; *707–708*
  fertilization, 711–713; *711–712*
  gastrulation, 714–715
  hormone control, 710–711; *710*
  in invertebrates, 493–495; *493–494*
  male reproductive system, 705–707; *706–707*
  menstrual cycle, 708–709; *709*
  in plants, 378–386; *378–386*
  sexual, 32, 104–108; *105–108*
  significance of, 705
  in vertebrates, 574–577; *575*
  viral replication, 236–238; *236–237*
  See also Meiosis; Mitosis.
Reproductive isolation, 184
Reptiles
  classification of, 525–528; *525–528*
  reptilian egg, 524–525; *524*
  traits of, 523
  turtle, 528–530; *528–529*
Reptilia, 523
Reservoir, 258
Resolution, 17

Respiration, 31
  in invertebrates, 488; *488*
  in vertebrates, 568–569; *569*
  See also Respiratory system.
Respiratory center, 650; *650*
Respiratory system
  blockages of breathing, 653
  breathing mechanics, 647–649; *648*
  control of respiration, 649–651; *650–651*
  diseases of, 653
  environmental factors affecting respiration, 655
  gas exchange, 649; *649*
  internal and external respiration, 645
  respiratory structures, 645–647; *646*
  smoking and respiratory disorders, 654–655; *654*
  See also Respiration.
Response, 32; *32*
Resting membrane potential, 671
Resting state, 670
Retina, 679
Rh factor, 630
Rhizoids, 280, 327, 358
Rhizomes, 331, 349, 375
Rhizopus stolonifer, 280, 288; *280*
Rhodophyta, 318
Rhynchocephalia, 525
Rhythmic behavior, 753–754
Ribonucleic acid. See RNA.
Ribose, 112
Ribosomes, 63; *63*
Ribs, 588
Rickettsiae, 244, 301
Rickettsia rickettsii, 301; *301*
Ring canal, 452
Ripeness to flower, 405
RNA (ribonucleic acid), 53, 62
  transcription process, 112–113; *113*
  translation process, 114–115; *114–115*
  and viruses, 234, 238–239
Rocky Mountain spotted fever, 303; *303*
Rodentia, 552
Rodents, 552; *552*
Rods, 679
Root apex, 403
Root cap, 343
Root hairs, 343, 358–361; *358*

Root layer, 770
Root pressure, 364; *364*
Roots, 342–345, 358–361; *343, 345, 359–360.* See also Seed Plants.
Rotifera, 429
Rotifers, 429; *429*
Rough ER, 63; *63*
Round dance, 477; *477*
Roundworms, 429; *429*
Royal jelly, 476
rRNA, 112
Rule of independent events, 130
Runners, 375

S

Saccharomyces, 282, 288
Sacrum, 588
Saliva, 612
Salivary glands, 612
Salmonella, 301; *302*
Sap, 366; *366*
Saprobes, 245, 276; *276*
Saprolegnia, 278–279; *279*
Sapwood, 347
Sarcodines, 263; *263.* See also Protozoans.
  locomotion of, 263
  reproduction of, 263
Savannas, 771
Scales, 505; *510*
Scanning electron microscope (SEM), 18–19; *18*
Scapula, 534, 588
Scent glands, 544
Schizomycetes, 242
Schleiden, Matthew, 59–60
Schwann, Theodor, 59–60
Schwann cell, 664
Scientific method,
  conclusions and theories, 6–8
  experimentation, 4–6
  hypothesis formation, 4
  observations, 3–4
Scion, 377
Sclera, 678
Sclerenchyma tissue, 341
Scolex, 428
Scrotum, 705
Scutes, 528
Scyphozoa, 421
Sea anemone, 424; *424*
Second, 14
Secondary growth, 344

Secondary oocyte, 107
Secondary spermatocyte, 107
Secondary succession, 756, 758; *758*
Second filial generation, 126
Second messenger, 700
Second polar body, 107
Secretion, 64
Sedimentary rock, 170–171; *171*
Seed coat, 386
Seed ferns, 378; *378*
Seed plants, 378–379; *378–379*
  angiosperms, 339–340
  gymnosperms, 338–339; *338–339*
  leaves of, 350–353; *350–353*
  monocots and dicots, 340–341; *340*
  roots of, 342–345; *343, 345*
  stems of, 345–347, 349; *346–347, 349*
  tissues of, 341–342; *342*
  traits of, 337
  See also Plants.
Seeds, 381
  dispersal of, 388; *388*
  formation of, 385–386; *386*
  germination of, 388–390; *389–390*
Segmented body, 459
Segmented worms
  earthworms, 431–433; *431–432*
  leeches, 435; *435*
  polychaetes, 435; *435*
Segregation, 128–129; *129*
Selective breeding, 157
Selectively permeable, 67
Self-pollination, 124
Semen, 707
Semicircular canals, 680
Semilunar valves, 632
Seminal fluid, 707
Seminal receptacle, 428
Seminal vesicles, 707; *707*
Seminiferous tubules, 706; *706*
Senses
  hearing and balance, 679–680; *679*
  taste and smell, 676–677; *676–677*
  touch and pain, 677–678; *677*
  vision, 678–679; *678*
Sensory neurons, 667
Sepals, 382
Serum, 308
Sessile, 418

Setae, 431
Sex chromosomes, 139
Sex-influenced characteristics, 162
Sex-linked characteristics, 141; *141*
Sex-linked genes, 138–141; *139–141*. *See also* Heredity.
Sexually transmitted diseases (STD), 296
Sexual reproduction, 104–108; *105–108*. *See also* Reproduction.
Shaft, 531
Sharks, 507–508; *507*
Shell, 524
Shellfish, 448
Shell membrane, 536
Shoot apex, 403
Short-day plants, 397–398
Shrub layer, 768
Sickle cell anemia, 153; *153*
Sieve pits, 363
Sieve plates, 363, 452
Sieve tubes, 363
Silica, 259
Simple microscope, 14–15; *15*
Single-circuit pathway, 567
Sinoatrial node (S-A node), 637
Sinuses, 466, 486, 646
Sinus venosus, 516
Skeletal muscle, 564, 595
Skeletal system, 586–587; *587–588*
  bone development, 588–589; *589*
  bone structure, 589–590; *589*
  cartilage and fibrous tissues, 592; *592*
  disorders of, 592; *592*
  joints and movement, 590–591; *590*
  of vertebrates, 563–564; *564*
Skin, 445
  disorders of, 601; *601*
  functions of, 600–601; *601*
  structure of, 599–600; *599*
  See also Integumentary system.
Skin gills, 452
Skull, 587
Slightly movable joints, 590
Slime molds, 260–261; *261*. *See also* Protists.
Small intestine, 614
Smell, 677; *677*
Smoking, 654–655; *654*
Smooth ER, 63
Smooth muscle, 564, 595

Snail
  locomotion in, 445; *483*
  nervous system of, *492*
  respiration in, 446
Snakes, 527
Social behavior, 750–753
Social hierarchy, 556–557; *557*
Social insects, 476; *476*
Sod formers, 770
Sodium-potassium pump, 670; *670*
Soil, 733
Soil resources, 786–787
Solar energy, 796
Solute, 45
Solutions, 45; *45*. *See also* Mixtures.
Solvent, 45
Somatic mutations, 147
Sori, 332
Spallanzani, Lazzaro, 28–29
Specialization, 585–586
Speciation, 183
Species, 214, 219; *218–219*
Sperm, 104, 107
Spermatids, 107
Spermatogenesis, 107; *107*
Sperm cells, 705
Sphagnum moss, 327–328; *328*
Sphenodon punctatus, 525
Sphenopsida, 330–331
Sphincters, 613
Sphygmomanometer, 638
Spicules, 419–420
Spinal cord, 667; *667*
Spindle, 101–102
Spinnerets, 464
Spiracles, 464, 470, 507
Spirilla, 243; *243*
Spirochetes, 243
Spirogyra, 320
Spleen, 626
Split genes, 164–165; *165*
Sponges, *418–419*
  digestion in, 418–419
  reproduction in, 420; *420*
  respiration in, 419
Spongin, 419–420
Spongocoel, 418
Spongy bone, 589
Spongy mesophyll, 351
Spontaneous generation, 25–29
Sporangium, 279
Sporazoa, 269
Spores, 277
Sporocyst, 427
Sporophyte, 322

850

Sporozoans, 269. *See also* Protozoans.
Sprain, 592
Squamata, 525
Squid, *447*
    circulation in, 447
    digestion in, 447
    locomotion in, 446
    nervous system of, 447
Stamens, 382
Staminate flowers, 383
Staphylococci, 243
Staphylococcus, 302
Staphylococcus aureus, 243
Starch, 50
Starfish, 451; *452–453, 484*
    digestion in, 453
    locomotion in, 452–453
    nervous system of, 453
    reproduction in, 453, 494
    respiration in, 452
Statocysts, 467
Stele, 343
Stems, 345–347, 349; *346–347, 349.* See also Seed plants.
Stentor, 266; *266*
Steppes, 769
Sterilization, 312
Sternum, 588
Steroids, 689
Stigma, 382
Stimulants, 674
Stimulatory synapse, 673
Stimulus, 32
Stipe, 283, 323
Stirrup, 679
Stock, 377
Stolons, 280, 349
Stomach, 613
Stomata, 350, 352–353; *352–353*
Stone canal, 452
Strain, 598
Strata, 170–171, 768; *171*
Streptobaccilli, 243
Streptococci, 243
Streptococcus thermophilus, 243
Streptomyces griseus, 312
Striated muscle, 595
Strip cropping, 786
Stroke, 640
Stromatolite, 194; *195*
Structural formula, 43
Structural gene, 162
Style, 382
Sublingual glands, 612
Submandibular glands, 612
Substrate, 82, 84
Succulents, 370, 774
Sucrase, 617
Sucrose, 50
Suspension, 45
Sutton, Walter S., 138
Sutures, 589
Swarm cells, 261
Sweat glands, 544, 656; *656*
Swim bladder, 508
Swimmerets, 466
Symbiosis, 196, 287, 741; *287*
Symbiotic hypothesis, 196–197; *196*
Symbol, for elements, 38
Symmetry, 417
Sympathetic nervous system, 668
Synapse, 672–673; *672–673*
Synapsis, 105
Synaptic vesicles, 672
Synovial fluid, 591
Synthesis, 31, 80
Syrinx, 535
Systemic circuit, 568
Systemic circulation, 634; *634*
Systole, 638
Systolic pressure, 638

T

Taiga, 766
Tail gland, 531
Tapeworms, 428; *428*
Taproot, 342; *343*
Target organ, 685
Taste, 676–677; *676–677*
Taste buds, 612, 676–677; *676–677*
Taxonomic key, 224; *225*
Taxonomist, 213
Taxonomy, 213
Tay-Sachs disease, 154
T-cells, 308–309
Technology, 9–10
Teeth, 205, 612; *205*
Telophase, 102–103; *103.* See also Mitosis.
Telophase I and II, 106; *106.* See also Meiosis.
Telson, 466
Temperate deciduous forest, 768–769
Temperate phages, 237
Temperate rain forest, 769
Temperature, 14; *14*
Temperature inversions, 791
Temperature and plant growth, 395–396
Temporary wilting, 368
Tendonitis, 592
Tendons, 592
Tentacles, 421
Terminal bud, 346
Terminal bud scars, 346
Territoriality, 537, 556, 750–751
Territory, 556
Test (sea urchin), 454
Test cross, 133; *133*
Testes, 690; *574*
Testis, 705
Testosterone, 162, 690, 705
Tetrad, 105
T-even phages, 235
Theophrastus, 214
Theory, 7
Theory of evolution by natural selection (Darwin), 177
Theory of punctuated equilibrium (Eldredge and Gould), 181; *181*
Thermal pollution, 793
Thigmotropism, 408
Thiobacillus, 246
Thoracic cavity, 586
Thoracic duct, 636; *636*
Thoracic vertebrae, 588
Thorax, 459, 586
Three primary germ layers, 714
Thrombosis, 640
Thymine, 110, 112
Thymus, 693
Thyroid gland, 688; *688*
    disorders of, 694–695
Thyroid-stimulating hormonone (TSH), 687
Thyroxine, 688
Tibia tarsus, 535
Tidal energy, 797
Time, 14
Tissue, 585
    of seed plants, 341–342; *342*
Tissue fluid, 642
Tobacco mosaic virus, 234–235; *234*
Tortoise, 527
Touch, 677–678; *677*
Toxin, 295
Trachea, 646; *646*
Tracheal tubes, 464, 488; *488*
Tracheids, 362

Tracheoles, 472
Tracheophyta, 317
Trait, 123
Transcription, 112–113; *113*
Transformation, 250; *250*
Transfusion, 628
Transitional forms, 171–172
Translation, 114–115; *114–115*
Translocation, 149; *149*
Translocation of food, 365–366; *366*
Transmission electron microscope (TEM), 18–19
Transpiration, 357–358, 365, 735; *365*
Transpiration-cohesion theory, 365; *365*
Transport
   adaptations of plants and water, 368–371; *370*
   capillarity and root pressure, 364; *364*
   diffusion in simple plants, 361; *361*
   guttation, 371; *371*
   translocation of food, 365–366; *366*
   transpiration-cohesion theory, 365; *365*
   turgor pressure and wilting, 368; *368*
   vascular plants, 361–363; *362–363*
   water enters plants, 357–361; *358–360*
   See also Plants.
Transposons, 720
Trebouxia, 287
Trematoda, 425
Trichocysts, 269
Trichophyton, 276
Tricuspid valve, 632
Trilobita, 460; *461*
Trilobites, 198; *198*
Triplets, 112
Trisomy 21, 151
tRNA, 112, 114; *114*
Trochophore, 439, 496; *439, 496*
Trophoblast, 578
Tropical rain forest, 772–773
Tropisms, 408; *408*
Trunk, 586
Trypanosoma, 271, 306
TSH-RF (thyroid-stimulating hormone-releasing factor), 698

Tube feet, 453
Tuber, 349
Tundra, 765–766
Tunicates, 504; *504*
Turbellaria, 425
Turgor movements, 410
Turgor pressure, 71, 368; *368*
Turner's syndrome, 152
Turtle, 527; *529*
   circulation in, 529
   digestion in, 529
   excretion in, 529
   external anatomy of, 528; *528*
   nervous system of, 530
   reproduction in, 530
   respiration in, 529
Twins, 717; *717*
Tympanic membrane, 470, 514, 679

U

Ulna, 535
Ulothrix, 320–321; *320–321*
Ultrasonography, 154; *155*
Ulva, 320, 322; *320, 322*
Umbo, 442
Understory, 768
Ungulates, 553; *553*
Unicellular organisms, 59
"Unity in diversity", 481
Univalves, 445–446; *446*
Upwellings, 779
Uracil, 112
Urea, 489, 656
Ureters, 656
Urethra, 656, 707
Urey, Harold, 192
Uric acid, 489
Urinary bladder, 656
Urine, 656
Urochordata, 503
Uropods, 466
Uterus, 428, 544, 708

V

Vaccination, 310
Vaccine, 310; *310*
Vacuoles, 64–65; *65*
Vagina, 430, 708
Valve, 632; *631*
Vane, 531
Variable factor, 4
Vascular bundles, 346

Vascular cambium, 344
Vascular plants, 317, 361–363; *362–363*
Vascular tissue, 342
Vas deferens, 707
Vectors, 297
Vegetative hyphae, 277
Vegetative mycelium, 277
Veins (leaf), 350
Veldt, 769
Venae cavae, 634
Venation, 340
Ventral cavity, 586
Ventral root, 667
Ventral surface, 417
Ventricle, 506, 567, 631
Venule, 633
Vernalization, 405
Vertebrae, 588
Vertebral column, 503–504, 588
Vertebrata, 503
Vertebrates, 417, 503
   circulatory stystem of, 567–568; *567*
   development of, 577–578; *577–578*
   digestive systems of, 566; *566*
   excretion in, 569–570; *569*
   gas exchange in, 568–569; *569*
   hormonal control in, 573–574; *574*
   integumentary systems of, 565; *565*
   muscular systems of, 564; *564*
   nervous systems of, 572–573; *572*
   reproduction in, 574–577; *575*
   skeletal systems of, 563–564; *564*
Vertical stratification, 768
Vessel elements, 362
Vessels, 362
Vestigial organs, 173
Viable seed, 388
Villi, 618
Viral replication, 236–238; *236–237*
Viral transduction, 238–239; *238*
Virchow, Rudolf, 60
Viroids, 239; *239*
   diseases caused by, 303
Virulence, 295
Virulent phages, 237
Viruses
   bacteriophages, 235–236; *235–236*

classification of, 234–235; *234–235*
diseases caused by, 299–300; *299*
structure of, 233–234; *233–234*
replication in, 236–238; *236–237*
transduction in, 238–239; *238*
viroids, 239; *239*
Visceral mass, 441
Visible spectrum, 92; *92*
Vision, 678–679; *678*
Vitamins, 605–606; *607*
Viviparous organisms, 525; *576*
Vocal cords, 647
Volume, 13
Voluntary acts, 663
Volvox, 319; *319*
Vomerine teeth, 514; *514*
Vorticella, 266; *266*

W

Waggle dance, 477; *477*
Walking legs, 465
Water, 48
absorprion of, 618
in diet, 608–609
and plant entry, 358–361; *358–360*
and plants, 357–358
water balance and excretion, 658
See also Transport.
Water, and plant growth, 395
Water balance (in fish), 513
Water cycle, 734–735; *734*
Water pollution, 792–793
Water resources, 787
Watersheds, 787
Water-vascular system, 451
and plant entry, 358–361; *358–360*
and plants, 357–358
See also Transport.
Watson, James, 109
Weight, 12
Weight control, 620
Weinberg, G, 186
Welwitschia, 338; *338*
Wheat rust, 284; *284*
White blood cells, 306, 627; *627*
White matter, 666
Wildlife resources, 789
Wilting, 368; *368*
Wind, and plant growth, 396; *396*
Windbreaks, 787
Wind energy, 797
Woody plants, 342
Wounds, 601

X

Xerophyte, 370; *370*
Xiphosura, 461
Xylem, 317, 362; *362*

Y

Yeast, 90
Yersinia pestis, 296, 300
Yolk sac, 524

Z

Zamia, 338; *338*
Z lines, 597
Zone of differentiation, 405
Zone of elongation, 404
Zoologist, 8
Zoology, 8
Zooplankton, 777
Zoospores, 278, 321
Zygomycota, 279–280; *280. See also* Fungi.
Zygomycotes, 279
Zygospores, 279, 280, 320
Zygote, 279

CREDITS

Contributing artists: Lee Ames & Zak, Ltd., Jean Helmer, Seward Hung, Susan Johnstone, Philip Jones, Peter Krempasky, Lartaud Design, John Lind, Davis Meltzer, Rebecca Merrilees, Taylor Oughton, Neil Paulino, Al Pucci, Dolores Santoliquido, Den Schofield, Herman Vestal, Beth Anne Willert.

Unit One 1: Wolfgang Kaehler

Chapter 1 2: M.P. Kahl/DRK Photo. 3: Jeff Foott 4: Stephen Dalton/ Animals Animals. 5: *t.l.* © Edgar Moench/Photo Researchers, Inc.; *t.r.* Mitchell Robert/Tom Stack & Associates; *b.* © Charles E. Mohr/ Photo Researchers, Inc. 7: *l.* Stephen J. Krasemann/DRK Photo; *r.* Merlin Tuttle/Photo Researchers 8: *t.* Stephen Krasemann/ Photo Researchers, Inc.; *b.* © Allan B. Cruikshank/Photo Researchers, Inc.; *b.r.* Stephen Krasemann/ DRK Photo 9: Jeff Rotman/Peter Arnold 11: *t.l.* J. Menschenfreund/ Taurus Photos; *t.m.* Grace Moore for Silver Burdett; *t.r.* © Stephen Krasemann/Photo Researchers, Inc.; *b.* E.R. Degginger. 12: *l.* Ken Sherman/Bruce Coleman; *r.* NASA 13: *t.l.* John Shaw/ Tom Stack; *t.r.* Dr. Tony Brain/ Science Photo Library Photo Researchers; *b.r.* Ken Karp for Prentice Hall 15: *t.* Warren & Genny Garsi/ Tom Stack & Assoc.; 16: Silver Burdett 17: Alfred Owczarzak/Taurus Photos. 18. *t.* J.A.L. Cooke/ Animals Animals; *m.* David Scharf/Peter Arnold; *b.* Manfred Kage/Peter Arnold. 19: *l.* Dan McCoy/Rainbow; *r.* Dan McCoy/Rainbow

Chapter 2 24: Douglas Mazonowicz/Monkmeyer Press 25: *l.* The Granger Collection; *r.* Holt Studios Inc./Earth Scenes 30: *t.r.* Werner Muller/Peter Arnold; *t.l.* Manfred Kage/Peter Arnold; *m.l.* Ken Karp for Prentice Hall; *b.l.* Eric Crichton /Bruce Coleman. 31: *t.l.* Wayne Lankinen/DRK Photo; *t.r.* Tom Bean/DRK Photo *m.t.* Zig Leszcynski/Animals Animals; *m.b.* Bill Curtsinger/Photo Researchers; *b.l.* Ed Reschke/Peter Arnold; *b.r.* Wolfgang Kaehler 32: *t.* Stephen Krasemann/DRK Photo; *m.* Stephen Krasemann/ DRK Photo; *b.* Runk/Schoenberger/Grant Heilman

Chapter 3 36: Hans Pfeletschinger/Peter Arnold, Inc. 39: E.R. Degginger. 40: Larry Mulvehill/Science Source/Photo Researchers. 43: Barry Runk/ Grant Heilman Photography. 44: Breck P. Kent. 45: Silver Burdett. 48: *t.* Silver Burdett; *b.* E.R. Degginger. 49: Silver Burdett. 52: California State University at Los Angeles. 54: *t.* Ron Moorhead/ Tom Stack & Associates; *b.* Silver Burdett. 55: Cheryl A. Traendly/ Jeroboam.

Chapter 4 58: © Michael Abbey/Photo Researchers, Inc. 59: The Science Museum, London. 60: *t.* Tom Branch/Photo Researchers; *b.* Jeff Foott/DRK Photo. 61: *l.* John J. Cardamone, BPS/Tom Stack & Associates; *r.* © Biphoto Associates/Photo Researchers, Inc. 63: © K.R. Porter/Photo Researchers, Inc.; *r.* Warren Rosenberg, BPS/Tom Stack & Associates. 64: *t.* Gary Grimes, Science Photo Library/Taurus Photos; *b.s.* Fawcett/Photo Researchers. 65: W.P. Wergin, courtesy of E.H. Newcomb, BPS/ Tom Stack & Associates. 66: *l.* Tom Branch/Photo Researchers, Inc.; *r.* Jeff Foott/DRK. 71: *l.* Phillip A. Harrington/Peter Arnold, Inc.; *r.* Alfred Owczarzak/Taurus Photos. 74: © Michael Abbey/ Photo Researchers, Inc.

Chapter 5 78: Peter David/Seaphot/Planet Earth Pictures. 84: Silver Burdett. 87: Focus On Sports. 90: *t.* Brian Parker/Tom Stack & Associates; *b.* Robert Frerck/Odyssey Productions.

Chapter 6 98: R. Langridge/Dan McCoy/ Rainbow 101: *t.* Runk-Schoenberger/Grant Heilman Photography, 117: William Patterson/Tom Stack & Associates. 120: © Dr. Don Fawcett/Photo Researchers, Inc.

Unit Two 121: H. Reinhard/Bruce Coleman.

Chapter 7 122: © Sven-Olaf Lindblad/Photo Researchers, Inc. 130: Silver Burdett. 142: David Micklos.

Chapter 8 146: Fiona Sunquist/Tom Stack & Associates, 149: E. R. Degginger. 150: © Carolina Biological Supply Co./Photo Researchers, Inc. 151: © OMIKRON/Photo Researchers, Inc. 152: N.H. Cheatham/DRK. 153: Bill Longcore/Photo Researchers. 155: Howard Sochurek/Woodfin Camp and Assoc. 158: Dr. Gopal Murti/Science Photo Library/Photo Researchers 159: Michael Salas/The Image Bank 160: *t.* Runk/Schoenberger/Grant Heilman *b.* Carolina Biological Supply Co. 164: Dugaiczyk et al./Proc. National Academy of Science. Courtesy of B.W. O'Malley.

Chapter 9 168: R. Van Nostrand/Berg & Associates. 170: *t.l.* Doug & Pat Valenti/Tom Stack & Associates; *b.l. r.* E.R. Degginger. 171: Tom Bean/Tom Stack & Associates. 172: Breck P. Kent. 177: Kevin Schafer/Tom Stack & Assoc. 178: Breck P. Kent. 180: *t.* Rod Planck/Tom Stack & Associates.; *b.* E.R. Degginger. 185: Imagery. 186. Adolf Hungry Wolf/Taurus Photos.

Chapter 10 190: E.R. Degginger. 194: Professors D. Deamer & J. Oro, University of Houston, Houston, Texas. 195: Rick Smolan/ Contact Press Images. 196: A.H. Knoll, The Biological Laboratories, Harvard University, Cambridge, Massachusetts. 197: *t.* T.J. Beveridge, BPS/Tom Stack & Associates; *b.* © K.R. Porter/Photo Researchers Inc. 198: Fred Bavendam/Peter Arnold, Inc. 206: © The Cleveland Museum of Natural History. 207 *t.* John Reader; *m., b.* © Field Museum, Chicago, Tom McHugh/Photo Researchers, Inc. 208: The Bettmann Archive. 209: *l.* John Chellman/Animals Animals *c.* Jim Tuten/Animals Animals *r.* Lewis Kramer/ DKR Photo.

Chapter 11 212: Seaphot/Planet Earth. 220: BPS/Tom Stack & Associates. 226: *l.* Grant Heilman Photography; *m.* Lynn M. Stone/Bruce Coleman; *t.r.* © S. McKeever/National Audubon Society Collection, Photo Researchers, Ind.; *b.r.* E.R. Degginger/ Bruce Coleman. 227: *l.* Pat Lanza Field/Bruce Coleman; *r.* Mark Sherman/Bruce Coleman.

Unit Three 231 © CNRI/Science Photo Library/Photo Researchers, Inc.

Chapter 12 232: © Dr. Tony Brain, SPL/Photo Researchers, Inc. 234: *l.* Grant Heilman Photography; *r.* James Somers/Taurus Photos. 235: Dennis Kunkel/Phototake. 236: L.V. Bergman & Associates, Inc. 241: *t.* Phil Degginger; *b* © Biophoto Associates/ Photo Researchers, Inc. 242: Alfred Owczarzak/Taurus Photos. 243: *l.* All CNRI/Science Photo Library & Photo Researchers. 244: Dennis Kunkel/Phototake for Silver Burdett. 245: Silver Burdett. 246: Martin M. Rotker/Taurus Photos. 251: Silver Burdett.

Chapter 13 254: Manfred Kage/Peter Arnold. 257: *l.* © Walker England/Photo Researchers, Inc.; *m., r.* © Biophoto Associates/ Photo Researchers, Inc. 259: *l.* Manfred Kage/Peter Arnold; *m., r.*

Eric V. Gravé Phototake. 261: Breck P. Kent. 262: © Carleton Ray/Photo Researchers, Inc. 263: *l.* Manfred Kage/Peter Arnold, Inc.; *r.* © Biology Media/Photo Researchers, Inc.; *b.* Eric V. Gravé/Phototake. 264: M.I. Walker/Photo Researchers. 266: *l.* Eric V. Gravé/Phototake; *m., t.r., b.r.* © Eric V. Gravé/Photo Researchers, Inc. 269: Alfred Pasieka/Taurus Photos.

Chapter 14 274: Frank E. Toman/Taurus Photos. 274: *t.l.* W.H. Hodge/Peter Arnold, Inc.; *t.m.* Stephen J. Krasemann/Peter Arnold, Inc.; *t.r.* S. Rannels/Grant Heilman Photography; *b.l.* Runk-Schoenberger/Grant Heilman Photography; *b.r.* W.H. Hodge/Peter Arnold, Inc. 279: Runk-Schoenberger/Grant Heilman Photography. 280: Breck P. Kent. 281: Sal Giordano III. 282: *t.* Manfred Kage/Peter Arnold, Inc.; *b.* W.H. Hodge/Peter Arnold, Inc. 283: *t.* E.R. Degginger; *b.* S. Rannels/Grant Heilman Photography. 284: Cereal Rust Laboratory, U.S. Department of Agriculture. 286: N. Allin & G.L. Barron, University of Guelph, Ontario. 287: *t.* Breck P. Kent; *b.* E.R. Degginger. 288: W.H. Hodge/Peter Arnold, Inc. 289: Clyde H. Smith/The Stock Shop.

Chapter 15 292: Peter J. Kaplan, Medichrome/The Stock Shop. 294: *l.* Dr. Tony Brain/Photo Researchers CNRI/Science Photo Library/Photo Researchers. 295: © OMIKRON/Photo Researchers, Inc. 296: *t.* Manfred Kage/Peter Arnold, Inc.; *b.* James C. Webb/Bruce Coleman. 298: © Lowell Georgia/Photo Researchers, Inc. 299: E.R. Degginger/Animals, Animals. 300: *t.* Tom Stack and Assoc. *b.* USDA Science Source/Photo Researchers; *b.* L.V. Bergman & Associates, Inc. 302: *l.* © Biophoto Associates/Photo Researchers, Inc.; *r.* © Science Photo Library/Photo Researchers, Inc. 303: Bob Gossington/Bruce Coleman. 304: Manfred Kage/Peter Arnold, Inc. 308: *t.* © John Durham, SPL/Photo Researchers, Inc.; *b.* James C. Webb/Bruce Coleman. 311: E.R. Degginger.

Unit Four 315: Hans Reinhard/Bruce Coleman

Chapter 16 316: Robert P. Carr/Bruce Coleman. 318: *l.* J. Robert Waaland/Tom Stack & Associates; *m.* E.R. Degginger; *r.* © Biophoto Associates/Photo Researchers, Inc. 319: *l.* Biophoto Associates/Photo Researchers, Inc.; *m.* Manfred Kage/Peter Arnold, Inc.; *r.* E.R. Degginger. 320: *t.* E.R. Degginger; *b.* © Joyce Photographics/Photo Researchers, Inc. 323: W.H. Hodge/Peter Arnold, Inc. 324: *t.* © Patrick Lynch/Photo Researchers, Inc.; *m.* J. Robert Waaland/Tom Stack & Associates; *b.* © Michael P. Gadomski/Photo Researchers, Inc. 325: Robert Evans/Peter Arnold, Inc. 327: *t.* Dwight R. Kuhn; *b.* Rod Planck/Tom Stack & Associates. 328: *t.* © Michael Gadomski/Photo Researchers, Inc.; *b.* E.R. Degginger. 329: Runk-Schoenberger/Grant Heilman Photography. 330: Rod Planck/Tom Stack & Associates. 331: *t.* Lysbeth Corsi/Tom Stack; *b.* Runk-Schoenberger/Grant Heilman Photography. 333: Missouri Botanical Garden.

Chapter 17 336: Robert P. Carr/Bruce Coleman 339: *l., m.* E.R. Degginger; *r.* Alan Pitcairn/Grant Heilman Photography. 340: *l.* Imagery; *m.* Runk-Schoenberger/Grant Heilman Photography; *r.* Runk-Schoenberger/Grant Heilman Photography. 345: *l.* Mindy E. Klarman; *r.* Runk-Shoenberger/Grant Heilman Photography. 347: *t.l.* Runk-Shoenberger/Grant Heilman Photography; *t.r.* Dwight R. Kuhn; *b.* Science Photo Library/Taurus Photos, 349: E.R. Degginger. 352: © Ray Simons/Photo Researchers, Inc.

Chapter 18 356: Ken Davis/Tom Stack & Associates. 358: © Dan Guravich/Photo Researchers, Inc. 359: E.R. Degginger. 361: Gwen Fidler/Tom Stack & Associates. 362: © Biophoto Associates/Photo Researchers, Inc. 366: Katherine S. Thomas/Taurus Photos. 368: Silver Burdett. 369: John Shaw/Bruce Coleman. 370: *l.* E.R. Degginger, *r.* L.L.T. Rhodes/Taurus Photos. 371: © P.W. Grace/Photo Researchers, Inc.

Chapter 19 374: E.R. Degginger, 376: *t.l., t.r.* © Jerome Wexler/Photo Researchers, Inc.; *b.* Breck P. Kent. 377: Silver Burdett. 378: © Townsend P. Dickinson/Photo Researchers, Inc. 379: Robert P. Carr/Bruce Coleman.380: *l.* Robert A. Ross/E.R. Degginger; *m.* Breck P. Kent; *r.* E.R. Degginger. 383: *t.* © M.E. Warren/Photo Researchers, Inc.; *b.* © Ken Brate/Photo Researchers, Inc. 388: *t.* Breck P. Kent; *b.* Robert P. Carr/Bruce Coleman.

Chapter 20 394: Imagery. 396: Brian Parker/Tom Stack & Associates. 397: *l.* Laurie Riley/Stock Boston; *r.* Breck P. Kent. 398: *l.* Lynn M. Stone; *m.* Martin M. Rotker/Taurus Photos; *r.* Bruce Coleman/Bruce Coleman. 400, 401: Robert Lyons/Color Advantage. 404: *l.* Robert Lyons/Color Advantage; *r.* E.R. Degginger/Bruce Coleman. 405: Robert Lyons/Color Advantage. 406: © Robert A. Isaacs/Photo Researchers, Inc. 408: *t.* E.S. Beckwith/Taurus Photos; *b.* Silver Burdett. 410: *t.* Breck P. Kent; *b.* D. Lyons/Bruce Coleman. 411: Silver Burdett.

Unit Five 415: Hans Pfletschinger/Peter Arnold, Inc.

Chapter 21 416: Jane Burton/Bruce Coleman. 418: Runk-Schoenberger/Grant Heilman Photography. 419: Neil G. McDaniel/Tom Stack & Associates. 422: E.R. Degginger. 424: *t.* Fred Bavendam/Peter Arnold, Inc.; *b.* Breck P. Kent. 425: Al Grotell. 429: © R. Knauft, Biology Media/Photo Researchers, Inc. 433: SEM photo by R. Kristensen, courtesy of National Museum of Natural History, Smithsonian Institution, Washington, D.C. 435: *l.* Brian Parker/Tom Stack & Associates; *r.* © J.H. Robinson/Photo Researchers, Inc.

Chapter 22 438: Eda Rogers/Sea Images. 440: Robert A. Ross/E.R. Degginger. 444: *t.* Timothy Eagan/Woodfin Camp and Assoc.; *m.* Fred Bavendam/Peter Arnold, Inc.; *b.* Breck P. Kent. 445: © Alvin E. Staffan/National Audubon Society Collection, Photo Researchers, Inc. 446: *t.* Doug Wechsler; *b.* Jeff Rotman. 447: Jeff Rotman. 448: © Douglas Faulkner/Photo Researchers, Inc.; *r.* Z. Leszczynski/Breck P. Kent. 449: University of Chicago. 453: Doug Wallin/Taurus Photos. 454: *t.l.* Ed Robinson/Tom Stack & Associates; *t.m.* Jeff Foott/Bruce Coleman; *t.r., b.l., r.* Jeff Rotman.

Chapter 23 458: Breck P. Kent. 460: *t.l.* Hans Pfletschinger/Peter Arnold, Inc.; *t.m.* Eda Rogers/Sea Images; *t.r.* John Shaw/Tom Stack & Associates; *b.l.* Dave Woodward/Taurus Photos; *b.r.* Robert Carr/Bruce Coleman. 463: *t.l.* Hans Pfletschinger/Peter Arnold, Inc.; *t.m.* Harry N. Darrow/Bruce Coleman; *r.* C. Allan Morgan, Peter Arnold, Inc.; *b.l.* Rod Planck/Tom Stack & Associates; *b.m.* E.R. Degginger. 464: Alan Blank/Bruce Coleman. 465: *t.* E.R. Degginger; *b.* William H. Amos/Bruce Coleman. 469: *t.* John MacGregor/Peter Arnold, Inc.; *b.* C. Allan Morgan/Peter Arnold, Inc. 470: Frank E. Toman/Taurus Photos. 475: Breck P. Kent. 476: Imagery.

Chapter 24 480: Al Grotell. 483: *t.* Sal Giordano III; *b.* Jeff Rotman. 485: *t.* Brian Parker/Tom Stack & Associates; *b.* E.R. Degginger. 492: Photo Researchers. 495: Photo Researchers, Inc. 497: Silver Burdett, courtesy of Mrs. Pauline Myers, Butterfly Farm, Denville, New Jersey.

Unit Six 501: Jonathan Scott/Seaphot Limited.

Chapter 25 502: Jeff Rotman 506: Harvard University/Museum of Comparative Zoology. 509: Peter Arnold/Peter Arnold, Inc. 514: *t.* Imagery; *b.* Andrew Odum/Peter Arnold, Inc. 517: *l.* © J.L. Stone/National Audubon Society Collection, Photo Researchers, Inc.; *r.* © Stephen Dalton/Photo Researchers, Inc.

Chapter 26 522: Jeff Foott. 524: Breck P. Kent. 526: *t.* Breck P. Kent; *b.* C. Allan Morgan/Peter Arnold, Inc. 527: Breck P. Kent.

855

528: C.W. Schwartz/Animals, Animals. 531: Eda Rogers/Sea Images. 537: Kim Taylor/Bruce Coleman. 538: *t.* Breck P. Kent; *b.* E.R. Degginger.

Chapter 27 542: Life Picture Service. 544: *t.l.* © Charlie Ott/Photo Researchers, Inc.; *b.l.* Stephen J. Krasemann/Peter Arnold, Inc.; *m.* Doug Wechsler, *r.* John Shaw/Tom Stack & Associates. 545: Tom Bean/Tom Stack & Associates. 546: Breck P. Kent. 547: *l.* Ken Stepnell/Taurus Photos; *r.* © Tom McHugh/Photo Researchers, Inc. 550: Richard Wood/Taurus Photos. 551: *t.* John MacGregor/Peter Arnold, Inc.; *m.* Michael Fogden/Bruce Coleman; *b.* Rod Planck/Tom Stack & Associates. 552: *l.* Leonard Lee Rue III/Bruce Coleman; *r.* E.R. Degginger. 553: *l.* Walt Anderson/Tom Stack & Associates; *r.* G.C. Kelly/Tom Stack & Associates; 555: Stephen J. Krasemann/Peter Arnold, Inc. 556: Warren Garst/Tom Stack & Associates. 557: © M.P. Kahl/Photo Researchers, Inc. 558: *l.* Sal Giordano III; *m.* © Tom McHugh/Photo Researchers, Incl; *r.* Arthus-Bertrand/Peter Arnold, Inc.

Chapter 28 562: Pro Pix/Monkmeyer Press. 564: *t.* John Gerlach/Tom Stack & Assoc.; *b.* S.L. Craig/Bruce Coleman. 565: *t.* © Chuck Brown/Photo Researchers, Inc.; *m.* © Eric Gravé/Photo Researchers, Inc.; *b.* © Gennaro & Grillone/Photo Researchers, Inc. 573: E.R. Degginger. 576: Breck P. Kent. 579: Phil Degginger.

Unit Seven 583: Diane Johnson/Focus West

Chapter 29 584: David Madison/Bruce Coleman. 586: Pam Hasegawa/Taurus Photos. 591: Kelly McCormick/Duomo. 592: E.R. Degginger. 595: *l.* L. V. Bergman & Associates; *m.* Terry Kirk/Tom Stack & Associates; *r.* Ed Reschke/Peter Arnold, Inc. 598: © Biophoto Associates/Photo Researchers, Inc. 601: Nubar Alexanian/Stock Boston.

Chapter 30 604: L. Nilsson/Boehringer Ingelheim International. 606: Robert Barclay/Grant Heilman Photography. 609: Barry L. Runk/Grant Heilman Photography. 610: *l.* Alan Pitcairn/Grant Heilman Photography; *r.* Breck P. Kent. 611, 612: Silver Burdett. 614: Beth Anne Willert for Silver Burdett. 620: Silver Burdett.

Chapter 31 624: Ed Reschke/Peter Arnold, Inc. 625: Martin M. Rotker/Taurus Photos. 626: CNRI/Science Photo Library/Photo Researchers. 628: Manfred Kage/Peter Arnold, Inc. 632: © Biophoto Associates/Photo Researchers, Inc. 633: Alfred Owczarzak/Taurus Photos. 634: Alfred Pasieka, Science Photo Library/Taurus Photos. 638: Grace Moore for Silver Burdett. 640: Martin M. Rotker, Science Photo Library/Taurus Photos.

Chapter 32 644: Peter Cummings/Tom Stack & Associates. 646: Silver Burdett. 650: D.P. Hershkowitz/Bruce Coleman. 654: © Dr. Gopal Murti, Science Photo Library/Photo Researchers, Inc. 655, 657: Beth Anne Willert for Silver Burdett. 659: Grace Moore/Taurus Photos.

Chapter 33 666: CNRI/Science Photo Library/Photo Researchers. 669: Beth Anne Willert for Silver Burdett. 675: *t.* © NCI, Science Source/Photo Researchers, Ind.; *b.l., b.r.* Grace Moore/Taurus Photos. 677: Beth Anne Willert for Silver Burdett.

Chapter 34 684: Jerry Wachter/Focus on Sports. 688: Beth Anne Willert for Silver Burdett.690: E.R. Degginger. 693: Beth Anne Willert for Silver Burdett. 695: Courtesy of The American Diabetes Association, Greater Philadelphia Affiliate, Inc., Camp Firefly. 699: *l.* Martin M. Rotker/Taurus Photos; *r.* David York, Medichrome/The Stock Shop.

Chapter 35 704: Silver Burdett. 706: Ed Reschke/Peter Arnold, Inc.; art by Beth Anne Willert for Silver Burdett. 707: CNRI/Science Photo Library/Photo Researchers, Inc. 708: Beth Anne Willert for Silver Burdett. 713: *t.* David Scharf/Peter Arnold, Inc.; *b.* © Russ Kinne/Photo Researchers, Inc. 716: © Petit Format, Nestle, Science Source/Photo Researchers, Inc. 717: Reproduced with permission of the Maternity Center Association, New York. 721: © Joseph Nettis/Photo Researchers, Inc.

Unit Eight 725: Stephen J. Krasemann/DRK Photo.

Chapter 36 726: Eda Rogers/Sea Images. 728: *t.* © J.H. Robinson/Photo Researchers, Inc.; *b.* Charles Marden Fitch/Taurus Photos. 729: *background* Alfred Owczarzak/Taurus Photos; *l. inset* Dwight R. Kuhn; *t.r. inset* Breck P. Kent; *b.r. inset* E.R. Degginger. 733: Lee Foster/Bruce Coleman. 739: *t.l.* Joy Spurr/Bruce Coleman; *t.m., t.r.* E.R. Degginger; *b.l.* Breck P. Kent; *b.m.* E.R. Degginger; *b.r.* Milton Rand/Tom Stack & Associates. 738: J. Alex Langley/DPI. 739: E.R. Degginger. 739: *inset in art,* 740: William H. Amos/Bruce Coleman.

Chapter 37 744: Robert C. Simpson/Tom Stack & Associates. 745: E.R. Degginger. 748: J. Alex Langley/DPI. 750: © Tom McHugh/Photo Researchers, Inc. 751: *l.* E.R. Degginger; *m.l.* Edgar T. Jones/Bruce Coleman; *m.r.* Stephen J. Krasemann/Peter Arnold, Inc.; *t.r.* Edgar T. Jones/Bruce Coleman; *b.r.* John S. Bunning/Bruce Coleman. 752. *l.* Sullivan and Rogers/Bruce Coleman *r.* Craig Packer/Bruce Coleman. 753: *t.b.* James H. Carmichael/Bruce Coleman. Fawcett/Tom Stack & Associates. 758: E.R. Degginger. 759: Eda Rogers/Sea Images.

Chapter 38 762: Karl Ammann/Bruce Coleman. 766: *l.* Stephen J. Krasemann/Peter Arnold, Inc.; *r.* Scott Ransom/Taurus Photos. 767: *l.* Ron Sherman/Bruce Coleman; *r.* E.R. Degginger. 768: *t.* Rod Planck/Tom Stack and Assoc. *b.* E.R. Degginger. 769: Breck P. Kent/Tom Stack & Associates. 770: *l., r.* E.R. Degginger; *m.* Jeff Foott/Bruce Coleman. 771: Jeff Foott. 773: *t.* E.R. Degginger; *m.* Alan Blank/Bruce Coleman; *b.* C. Allan Morgan/Peter Arnold, Inc. 774: *l.* Imagery; *m.* Rod Allin/Tom Stack & Associates; *r.* Stephen J. Krasemann/Peter Arnold, Inc. 775: Wardene Weisser/Bruce Coleman. 777: *t.* Ronald F. Thomas/Taurus Photos; *b.* Robert McKenzie/Tom Stack & Associates. 778: Robert Pelham/Bruce Coleman. 779: Dave Woodward/Tom Stack & Associates.780: C.C. Lockwood/Bruce Coleman.

Chapter 39 784: © F. Gohier/Photo Researchers, Inc. 785: Frances Bennett/DPI. 786: Mike Yamashita/Woodfin Camp & Associates. 787: *l.* E.R. Degginger, *r.* Phil Degginger. 788: *t.* W.H. Hodge/Peter Arnold, Inc. *b.* Jack Swenson/Tom Stack & Associates. 789: Fred Bavendam/Peter Arnold, Inc. 791: Mickey Gibson/Tom Stack & Associates. 792: Alfred Eisenstaedt/Life Picture Service. 797: *l.* W. Gontscharoff/Shostal Associates; *r.* Kevin Schaffer/Tom Stack & Associates. 799: Courtesy of U.S. Department of the Interior, National Park Service, Special Programs and Populations Divisions.